MICROTUBULES

MODERN CELL BIOLOGY

RECENT VOLUMES PUBLISHED IN THE SERIES

Volume 11 — Antisense RNA and DNA
James A.H. Murray, Editor

Volume 12 — Genes in Mammalian Reproduction
Ralph B.L. Gwatkin, Editor

Volume 13 — Microtubules
Jeremy S. Hyams and Clive W. Lloyd, Editors

MICROTUBULES

Editors

Jeremy S. Hyams

Department of Biology
University College London
London, England

Clive W. Lloyd

Department of Cell Biology
John Innes Institute
Norwich, England

WILEY-LISS

A JOHN WILEY & SONS, INC., PUBLICATION
New York • Chichester • Brisbane • Toronto • Singapore

Address All Inquiries to the Publisher
Wiley-Liss, Inc., 605 Third Avenue, New York, NY 10158-0012

Printed in United States of America

Library of Congress Cataloging-in-Publication Data

Microtubules / editors, Jeremy S. Hyams, Clive W. Lloyd.
p. cm.
Includes bibliographical references and index.
ISBN 0–471–56193–2
1. Microtubules. I. Hyams, Jeremy S. II. Lloyd, Clive W.
QH603.M44M523 1993
574.87′34—dc20 93-30114
CIP

The text of this book is printed on acid-free paper.

Contents

Contributors

Peter M. Bayley, Division of Physical Biochemistry, National Institute for Medical Research, Mill Hill, London NW7 1AA, United Kingdom **[111]**

Michel Bornens, Department of Biology of Cell Cycle and Cell Motility, Centre de Génétique Moléculaire du CNRS, 91 198 Gif-sur-Yvette, France **[303]**

Jeannette Chloë Bulinski, Department of Anatomy and Cell Biology, and Department of Pathology, Columbia University College of Physicians and Surgeons, New York, NY 10032 **[167]**

Roy G. Burns, Biophysics Section, Blackett Laboratory, Imperial College of Science, Technology and Medicine, London SW7 2BZ, United Kingdom **[3]**

Don W. Cleveland, Department of Biological Chemistry, Johns Hopkins University School of Medicine, Baltimore, MD 21205 **[47]**

Christine A. Collins, Department of Cell, Molecular, and Structural Biology, Northwestern University Medical School, Chicago, IL 60611 **[367]**

R.A. Crowther, Medical Research Council Laboratory of Molecular Biology, Cambridge CB2 2QH, United Kingdom **[183]**

W.C. Earnshaw, Department of Cell Biology and Anatomy, Johns Hopkins University School of Medicine, Baltimore, MD 21205 **[393]**

Craig C. Garner, Center for Molecular Neurobiology, University of Hamburg, D-2000 Hamburg 20, Germany; present address: Department of Cell Biology, Neurobiology Research Center, University of Alabama, Birmingham, AL **[141]**

M. Goedert, Medical Research Council Laboratory of Molecular Biology, Cambridge CB2 2QH, United Kingdom **[183]**

Lawrence S.B. Goldstein, Department of Cellular and Developmental Biology, Harvard University, Cambridge, MA 02138 **[269]**

Erika L.F. Holzbaur, Worcester Foundation for Experimental Biology, Shrewsbury, MA 01545; present address: Department of Biochemistry, University of Pennsylvania School of Veterinary Medicine, Philadelphia, PA 19104 **[251]**

R. Jakes, Structural Studies Division, Medical Research Council Laboratory of Molecular Biology, Cambridge CB2 2QH, United Kingdom **[183]**

Didier Job, Laboratoire du Cytosquelette, Unité INSERM 244, Département de Biologie Moléculaire et Structurale, Centre d'Etudes Nucléaires de Grenoble, 38041 Grenoble Cedex, France **[221]**

Mary Ann Jordan, Division of Molecular and Cellular Biology, Department of Biological Sciences, University of California, Santa Barbara, CA 93106 **[59]**

Stefan Kindler, Center for Molecular Neurobiology, University of Hamburg, D-2000 Hamburg 20, Germany; present address: California Institute of Technology, Pasadena, CA **[141]**

The numbers in brackets are the opening page numbers of the contributors' articles.

Stephen M. King, Cell Biology Group, Worcester Foundation for Experimental Biology, Shrewsbury, MA 01545; present address: Department of Biochemistry, University of Connecticut Health Center, Farmington, CT **[229]**

Anne Marie Lambert, Institut de Biologie Moléculaire des Plantes, Centre National de la Recherche Scientifique, Université Louis Pasteur, 67084 Strasbourg Cedex, France **[325]**

Richard W. Linck, Department of Cell Biology and Neuroanatomy, University of Minnesota, Minneapolis, MN 55455 **[201]**

Clive W. Lloyd, Department of Cell Biology, John Innes Institute, Norwich NR4 7UH, United Kingdom **[325]**

Robert L. Margolis, Institut de Biologie Structurale, 38027 Grenoble Cedex 1, France **[221]**

Stephen R. Martin, National Institute for Medical Research, Mill Hill, London NW7 1AA, United Kingdom **[111]**

Andrew Matus, Department of Neurobiology, Friedrich Miescher Institute, and Department of Cellular Neurobiology, Zoological Institute, University of Basel, Basel, Switzerland **[155]**

J. Richard McIntosh, Department of Molecular, Cellular, and Developmental Biology, University of Colorado, Boulder, CO 80309-0347 **[413]**

Atsushi Mikami, Worcester Foundation for Experimental Biology, Shrewsbury, MA 01545 **[251]**

Timothy J. Mitchison, Department of Pharmacology, University of California, San Francisco, CA 94143-0450 **[287]**

Reinhold Müller, Center for Molecular Neurobiology, University of Hamburg, D-2000 Hamburg 20, Germany; present address: Faculty of Biology/Chemistry, University of Bremen, Bremen, Germany **[141]**

Jan M. Norrander, Department of Cell Biology and Neuroanatomy, University of Minnesota, Minneapolis, MN 55455 **[201]**

Berl R. Oakley, Department of Molecular Genetics and Biotechnology Center, Ohio State University, Columbus, OH 43210-1292 **[33]**

Bryce M. Paschal, Worcester Foundation for Experimental Biology, Shrewsbury, MA 01545; present address: Scripps Research Institute, La Jolla, CA 92037 **[251]**

Andrea Pereira, Department of Cellular and Developmental Biology, Harvard University, Cambridge, MA 02138 **[269]**

Elizabeth C. Raff, Department of Biology and Institute for Molecular and Cellular Biology, Indiana University, Bloomington, IN 47405-3680 **[85]**

Winfield S. Sale, Department of Anatomy and Cell Biology, Emory University School of Medicine, Atlanta, GA 30322 **[381]**

Jonathan M. Scholey, Section of Molecular and Cellular Biology, Division of Biological Sciences, University of California, Davis, CA 95616 **[343]**

Kam K. Sharma, National Institute for Medical Research, Mill Hill, London NW7 1AA, United Kingdom **[111]**

Elizabeth F. Smith, Department of Anatomy and Cell Biology, Emory University School of Medicine, Atlanta, GA 30322; present address: Departments of Genetics and Cell Biology, University of Minnesota, Minneapolis, MN 55455 **[381]**

M.G. Spillantini, Medical Research Council Laboratory of Molecular Biology, Cambridge CB2 2QH, United Kingdom **[183]**

Christopher D. Surridge, Biophysics Section, Blackett Laboratory, Imperial College of Science, Technology and Medicine, London SW7 2BZ, United Kingdom **[3]**

Nicholas G. Theodorakis, Department of Biological Chemistry, Johns Hopkins University School of Medicine, Baltimore, MD 21205 **[47]**

Frédéric Tournier, Centre de Génétique Moléculaire du CNRS, 91 198 Gif-sur-Yvette, France **[303]**

Ronald D. Vale, Department of Pharmacology, University of California, San Francisco, CA 94143 **[343]**

Richard B. Vallee, Worcester Foundation for Experimental Biology, Shrewsbury, MA 01545 **[251]**

Curtis G. Wilkerson, Cell Biology Group, Worcester Foundation for Experimental Biology, Shrewsbury, MA 01545 **[229]**

Leslie Wilson, Division of Molecular and Cellular Biology, Department of Biological Sciences, University of California, Santa Barbara, CA 93106 **[59]**

George B. Witman, Cell Biology Group, Worcester Foundation for Experimental Biology, Shrewsbury, MA 01545 **[229]**

Linda Wordeman, Department of Pharmacology, University of California, San Francisco, CA 94143-0450 **[287]**

Preface

It is 30 years since Ledbetter and Porter gave us the first full description of microtubules in plant cells. At the halfway mark of the brief history of these organelles, Roberts and Hyams (*Microtubules,* 1979, Academic Press) summarized the achievements of the first, largely descriptive phase of microtubule research: the structure of the microtubule as deduced from negative staining and optical diffraction, the properties of the tubulin dimer, the *in vitro* assembly of tubulin and the recognition of the importance of associated proteins, the role of microtubule organizing centers, the beginnings of immunofluorescence microscopy to characterize microtubule arrays, the use of cilia and flagella as model systems, and the mysteries of chromosome movement at mitosis.

At that time, the vision of the coming decade included the primary structure of tubulin, the mechanism and control of microtubule assembly *in vivo,* and the function of microtubules in cell motility and mitosis. In documenting the achievements of the second, molecular phase, this present book confirms the enormous advances that have indeed been made in some of these areas, but hindsight also underlines just how difficult it is to predict the path of scientific progress. How many of us could have foreseen that advances in microtubule research through the 1980s would owe as much to developments in light microscopy as to the application of restriction enzymes? Immunofluorescence microscopy in particular continued to blossom as antibody probes specific for tubulin isotypes, MAPs, etc., became increasingly available and an increasing diversity of cell types yielded themselves to their application. Confocal microscopy added, literally, another dimension to this type of analysis. The use of AVEC DIC to visualize the gliding of microtubules across a glass slide and the bidirectional movement of particles along them provided assays for the identification and characterization of motor proteins that were undreamed of a few years earlier. Another great conceptual leap was the discovery of the dynamic instability of microtubules and the realization that growing and shrinking microtubules coexist in the cytoplasm. This phenomenon, originally detected *in vitro* by darkfield microscopy, was revealed in living cells by low-light-level cameras that could resolve the behavior of microtubules labeled by the incorporation of microinjected exogenous tubulin. The molecular dynamics of living microtubules, unsuspected during the era of *in vitro* biochemistry and electron microscopy, was now visible in real time. The catastrophic assembly and disassembly of microtubules has particularly influenced how we think about mitosis, and although it will be a long time before we can reduce mitosis to a one-line summary, microtubule dynamics and motors have stimulated enormous conceptual advances.

But overwhelmingly, this second phase has been dominated by molecular biology and the impact of genetics. The primary sequences of α- and β-tubulin are now established for animals, plants, and microorganisms, and in all of these it is clear that tubulins constitute a multigene family. Similarly, the cloning and sequencing of the major MAPs has revealed familial relationships such as the microtubule-binding domains shared by the structural MAPs such as MAP2, MAP4, and tau. MAP1c brain has been demonstrated by a combination of structural, biochemical, and molecular techniques to be the long-sought cytoplasmic homologue of axonemal dynein, the retrograde microtubule motor. The deeper molecular understanding of microtubule proteins has revealed that molecules have more in common than the cells they come from, i.e., that genetic similarities run through evolutionarily diverse organisms. Because of this, microbial genetics has had a major impact. Nowhere is this better demonstrated than in the discovery in *Aspergillus nidulans* of the third member of the tubulin super-family, γ-tubulin. γ-Tubulin appears to have diverged from an ancestral tubulin at roughly the same time that α- and β-tubulin went their separate ways. Its localization at microtubule-organizing centers may be of profound importance for the establishment of microtubule polarity. Similarly, analysis of cell cycle mutants of fission and budding yeast led to the identification of catalytic components (kinases and phosphatases) that appear to influence the microtubule nucleating capacity of centrosomes. This interface between genetics and cell biology is one of the most exciting in current biology and will undoubtedly tell us more about the way in which the cell cycle and the microtubule cycle are intergrated.

Jeremy S. Hyams
Clive W. Lloyd
June 1993

ABOUT THE EDITORS

JEREMY S. HYAMS is Professor of Cell Biology and Head of the Department of Biology at University College London, where he teaches undergraduate courses in cell biology and microbiology. Following a B.Sc. from the University of London, he received his Ph.D. from the University of East Anglia in 1972 for work on the genetics of *Chlamydomonas*. Postdoctoral work at the University of Edinburgh with Sir John Randall, at the University of Wisconsin with Gary G. Borisy, and at the University of Exeter with Howard Stebbings focused on microtubule systems as diverse as *Chlamydomonas* flagella, yeast mitotic spindles, and insect nutritive tubes. His research over the past several years has been directed to the cytoskeleton and cell division cycle of the fission yeast, *Schizosaccharomyces pombe*. He has taught the Cold Spring Harbor course *The Molecular Genetics of Fission Yeast* and is European editor of the journal *Cell Motility and the Cytoskeleton* and a member of the editorial boards of the *Journal of Cell Science* and *Trends in Cell Biology*.

CLIVE W. LLOYD is Senior Principle Scientific Officer in the Department of Cell Biology at the John Innes Institute, Norwich, England. He is engaged in full-time research into the role of the cytoskeleton in plant morphogenesis. After receiving first-class honors in zoology at the University College of Wales, Aberystwyth, he continued in that department to investigate the involvement of cell surface glycoproteins in the adhesion of animal cells. After being awarded his Ph.D. in 1971, he went, as Beit Memorial Research Fellow, to Dr. G.M.W. Cook's laboratory at the Strangeways Laboratory, Cambridge, England, to continue his research into cell surface carbohydrates in animal cell adhesion. Next he joined Dr. D.A. Rees at the Unilever Research Sharnbrook Laboratory, where ideas on the importance of the underlying cytoskeleton were beginning to emerge. In 1979 Dr. Lloyd turned his attention to plants and the part that the different cytoskeletal arrays play in regulating cell shape and, in particular, the plane of cell division. This theme was maintained when he moved, as Royal Society Senior Research Fellow in the Biological Sciences, to the John Innes Institute, where he is now a staff member. He is a member of the editorial boards of the *Journal of Cell Science, Protoplasma,* and *Cell Motility and the Cytoskeleton*.

TUBULIN BIOCHEMISTRY AND ASSEMBLY

Microtubules: 3–31

Tubulin: Conservation and Structure

Roy G. Burns and Christopher D. Surridge

I. INTRODUCTION

As all eukaryotic cells contain microtubules, the microtubule subunit proteins probably originated when the eukaryotes first evolved approximately 1.5 billion years ago. These subunit proteins comprise the α-, β-, and γ-tubulins, a distinct protein family that shares 35%–40% identity with each other [e.g., Little and Seehaus, 1988; Oakley and Oakley, 1989] but has minimal sequence homology with any other known protein. It is also probable that a variety of proteins that associate with microtubules, including members of the kinesin and dynein families (see chapters by Scholey and Vale and by Collins, this volume), also date to the origin of the eukaryotes, although other microtubule–associated proteins (MAP), such as the neuronal-specific tau and MAP2 families, are clearly less ancient.

Given this ancient ancestry, tubulins from different organisms might be expected to have highly divergent sequences. This is indeed true for the C-terminal peptides of both α- and β-tubulin, but these peptides represent the exception rather than the rule. The N-terminal peptides (α:1–436, β:1–430) are remarkably highly conserved [e.g., Little and Seehaus, 1988; Burns, 1991], and this minimal sequence divergence is almost certainly due to the structural constraints imposed by the requirements of microtubule assembly and disassembly, coupled with those imposed by the binding of associated proteins such as kinesin and dynein.

II. CONSERVATION OF α- AND β-TUBULINS

The conservation of a particular tubulin sequence has generally been quoted by comparing the percentages of residues that are identical to a specific "standard" sequence. This approach is limited in that the "standard sequence" may itself be atypical and because any sequencing errors will result in a lower calculated value. We have adopted an alternative approach and have subdivided the known α- and β-tubulin sequences into those from the metazoa, the plants and algae, the protozoa, and the fungi and have then calculated consensus sequences (omitting the highly variable C-terminal peptides) for each of these phylogenetic orders to use as the "standards." This alternative approach, which is based on a synthesis of the available sequences, has the benefit of minimizing both the effects of any sequencing errors (and the independent sequencing of several tubulins has revealed a number of such errors) and the effects of using a "standard" that may have unusual properties. For example, the sensory cells of *Caenorhabditis elegans* have microtubules with 15 rather than 11 protofilaments, and this is due to the expression of the mec-7 β-tubulin gene [Savage et al., 1989; Hamelin et al., 1992], while specific isoforms may have particular functional properties (see Raff, this volume). One disadvantage of this approach of using the consensus sequences is that they are inevitably derived from biased samples and may change as additional sequences become available. For example, the metazoan β-tubulin consensus sequence

A

```
M     MRECISIHVG  QAGVQIGNAC  WELYCLEHGI  QPDGQMPSDK  TIGGGDDSFN   50
P     ...V......  ...I.V....  ...F......  ..........  ........A..
P/A   ...V....I.  ...I.V....  ..........  ..........  ........A..
F     ...V...N..  ..C.......  ..........  ....YHLEDG  LSK..EEG.S
      ***      *  *   *  * *    *  *** *    **

M     TFFSETGAGK  HVPRAVFVDL  EPTVIDEVRT  GTYRQLFHPE  QLITGKEDAA   100
P     ..........  .......L..  ....V.....  ..........  ...S......
P/A   ..........  .....I.L..  ..........  ..........  ...S......
F     .......Q..  F....IY...  ..N.......  .P.KD.....  .........S
      ***  *       ***    **   *   *  *        *  **  *   ****

M     NNYARGHYTI  GKEIIDLVLD  RIRKLADQCT  GLQGFLVFHS  FGGGTGSGFT   150
P     ..F.......  ....V..C..  .......N..  .........N.  V.......LG
P/A   ..F.......  ....V..C..  .......N..  .........NA  V.......LG
F     .........V  ......E...  ...RM..N.S  ..........  F.......L.
      ** *** *    * *               * *    ****        *** ***

M     SLLMERLSVD  YGKKSKLEFS  IYPAPQVSTA  VVEPYNSILT  THTTLEHSDC   200
P     ...L......  ........G.T  V..S......  .......V.S  ..SL...T.V
P/A   ...L......  ........G.T  V..S......  .......V.S  ..SL...T.V
F     ...L....ME  ..........  V........S  .......V..  .......A..
       ** * *       **  * *     *         ******  *   **      *

M     AFMVDNEAIY  DICRRNLDIE  RPTYTNLNRL  IGQIVSSITA  SLRFDGALNV   250
P     .V.L......  ..........  ..........  .A.VI..L..  ..........
P/A   .SL.......  .....S....  ..........  .A.VI..L..  ..........
F     T.........  ..........  ..S.E.....  .A.VV.....  ......S...
          ***  *  * *   *     **   * * *    *  *  *    * * *  **

M     DLTEFQTNLV  PYPRIHFPLA  TYAPVISAEK  AYHEQLSVAE  ITNACFEPAN   300
P     .V........  .......M.S  S.........  ..........  ...SA....S
P/A   .I........  .......M.S  S.........  ..........  ....A....S
F     ..N.......  .........V  S.S.IV..A.  .F..SN..S.  ........G.
      *  *******  * ***** *    *    *       **         *   **

M     QMVKCDPRHG  KYMACCLLYR  GDVVPKDVNA  AIATIKTKRT  IQFVDWCPTG   350
P     M.A.......  .......M..  ..........  .V........  ..........
P/A   M.V.......  ......M...  ..........  .V........  ..........
F     ........N.  ....T.....  .....R..Q.  .VE...N...  V.........
       * * **  *   ***    **  ***   **         *       ****  ***

M     FKVGINYQPP  TVVPGGDLAK  VQRAVCMLSN  TTAIAEAWAR  LDHKFDLMYA   400
P     ..C.......  ..........  .M.....I..  S.....VFS.  I.....Y...
P/A   ..C.......  ..........  .......I..  S..V..VFS.  I.........
F     ..I..C.E..  QHI..SE...  .D........  ..S.....K.  .........S
       * **    *          *     *  ** **   *             ***

M     KRAFVHWYVG  EGMEEGEFSE  AREDMAALEK  DYEEVG                   436
P     ..........  ..........  ....L.....  ......
P/A   ..........  ..........  ....L.....  ......
F     ..........  ..........  ....L....R  ......
       ********   ******** *  ***  * **   *  ***
```

Fig. 1. *Consensus sequences of the α-tubulins* **(A)** *and β-tubulins* **(B)** *subdivided into metazoa (M), the protozoa (P), plants and algae (P/A), and fungi (F) based on the available sequences (June 1992) and the predominant residue at each position in each phylogenetic order. Asterisks indicate residues that are absolutely conserved in all sequences. The highly variable C-terminal peptides (from α:437 and β:431) and any insertions have been omitted from this analysis; deletions were treated as the equivalent of amino acid substitutions.*

is heavily weighted in terms of the vertebrates even though it includes examples from insects, nematodes, echinoderms, and a mollusc, while the plant and alga α-tubulin consensus sequence currently only includes examples of green alga and monocotyledons.

B

```
M    MREIVHIQAG  QCGNQIGAKF  WEVISDEHGI  DPTGTYHGDS  DLQLERINVY   50
P    ..........  ..........  .........V  ..........  ..........
P/A  ........G.  ..........  ...V......  ..........  ..........
F    ........T.  ........A.  .QT..G...L  .PS.V...T.  ......M...
     ***      *  ** **    *        **           *

M    YNEATGGKYV  PRAILVDLEP  GTMDSVRSGP  FGGIFRPDNF  VFGQSGAGNN  100
P    .......R..  .....M....  .......A..  ...L......  ..........
P/A  F....A.R..  ...V......  ..........  Y.Q.......  ....T.....
F    F...S.N...  ...V......  ....A..A..  F.QL......  ..........
       *      *  **   ****                     *           ***

M    WAKGHYTEGA  ELVDSVLDVV  RKEAESCDCL  QGFQLTHSLG  GGTGSGMGTL  150
P    ..........  ..I.......  .....G....  ....I.....  ..........
P/A  ..........  ..I.......  ...EA.....  ....VC....  ..........
F    ..........  ....Q.....  .R...G....  ....I.....  ....A.....
      ********        * **          * *  * **   * *  **** *****

M    LISKIREEYP  DRIMNTFSVV  PSPKVSDTVV  EPYNATLSVH  QLVENTDETY  200
P    ..........  ....E.....  ..........  ..........  .....A..VM
P/A  ..........  ..M.L....F  ..........  ..........  .....A..CM
F    ........F.  .RM.A....V  ..........  ..........  .....S...F
     *   *   *   *     *          *      ****     *   * *

M    CIDNEALYDI  CFRTLKLTTP  TYGDLNHLVS  ATMSGVTTCL  RFPGQLNADL  250
P    ..........  ..........  ..........  .A.....C..  .......S..
P/A  VL........  ..........  .F......I.  .......C..  .......S..
F    ..........  .M.....SN.  S.........  .V......S.  .......S..
       ** **       ** *          ** *          *     * *   * *

M    RKLAVNMVPF  PRLHFFMPGF  LPLTSRGSQQ  YRALTVPELT  QQMFDAKNMM  300
P    ......LI..  .......I..  A.........  ..........  ..........
P/A  ......LI..  .......V..  A.........  ..........  ...W......
F    ..........  .......V..  A......AHS  F..VS.....  .....P....
     * ****  *   ****         **                            * *

M    AACDPRHGRY  LTVAAVFRGR  MSMKEVDEQM  LNVQNKNSSY  FVEWIPNNVK  350
P    C.S.......  ..AS.L....  ..T.......  ..........  ........I.
P/A  C.A.......  ..AS.M....  ..T.......  ..........  ..........
F    ..S.F.N...  ..CS.I...K  V.....ED..  R.........  .........Q
            **   *     **          *                 *  *** *

M    TAVCDIPPRG  LKMSATFIGN  STAIQELFKR  ISEQFTAMFR  RKAFLHWYTG  400
P    SS......K.  ...AV.....  ......M...  VG........  ..........
P/A  SS......K.  ....S.....  ......M...  V.........  ..........
F    ..L.S.....  ....S.....  ..S.......  VGD.......  ..........
            * *        *  *   *  *    *      *  **   *  * ** *

M    EGMDEMEFTE  AESNMNDLVS  EYQQYQDATA                          430
P    ..........  ..........  ..........
P/A  ..........  ..........  ..........
F    ..........  ..........  ........GI
     ****  ** *         *    * **
```

Figure 1B.

The remarkable conservation of both α- and β-tubulin is clearly demonstrated by a comparison of the phylogenetic consensus sequences (Fig. 1A,B). The four α-tubulin consensus sequences are identical at 72% of all residue positions and the β-tubulin at 78% of all positions. Furthermore, inspection of the individual sequences shows that many of the nonconserved residues involve highly conservative amino acid substitutions.

This conservation is highlighted still further by considering which residues are totally conserved in *all* of the sequences from these orders (see Fig. 1A,B). Approximately 41% of the α- and 30% of the β- tubulin residues are totally conserved, with the apparent difference being due to the smaller number of available α-sequences. The fractions are probably underestimates as only single exceptions are found at many additional residue positions, some of which may be sequencing errors. The conservation is not restricted to specific domains, as the totally conserved residues are distributed throughout the two sequences. The similarity of the α- and β- tubulin sequences suggests that the two subunits must have closely related tertiary structures and that the sequence differences relate to the subunit-specific properties, including (see sections VI and VIII, below) the ligand binding and the GTP hydrolysis by β-tubulin.

The fidelity of the phylogenetic consensus sequences can be assessed by expressing as a percentage the fraction of the available sequences with the consensus residue at each position (Table I), averaged over the entire sequence after excluding the variable C-terminal peptides. These mean fractions vary between 87.6% and 96.6% for the α-tubulins and 90.5% and 96.4% for the β-tubulins. This analysis minimizes the contribution by specific sequences, such as the *C. elegans* mec-7 β-tubulin, so that individual tubulins can be significantly more heterogeneous than implied by the mean fractions, while the high values of these fractions imply that few neutral mutations have been accommodated during the prolonged evolution of the different phylogenetic orders.

TABLE I. Analysis of the Sequence Homologies of the α-Tubulins (A) and From β-Tubulins (B), Metazoa (M), Protozoa (P), Plants and Algae (P/L), and Fungi (F)

	Variable sequence			
Consensus sequence	M	P	P/A	F
α-Tubulin				
M	96.6	88.3	85.8	76.4
P	86.4	95.4	94.1	71.6
P/A	85.5	93.2	96.0	71.6
F	80.2	76.0	76.1	87.6
β-Tubulin				
M	96.4	89.2	86.7	80.6
P	89.9	92.3	88.5	79.6
P/A	88.1	88.6	92.8	77.8
F	83.1	82.1	79.2	90.5

Each residue (α:1–436, β:1–430) was compared for each tubulin within a particular phylogenetic order with the equivalent residue of the stated consensus sequences (shown in Fig. 1A,B), scoring an identity as 1.0 and an amino acid subsitution as 0. Values are means averaged over all positions and all individual sequences, expressed as a percentage for each pair of comparisons, i.e., the protozoan α-tubulins have a homology index of 88.3% when compared against the metazoan consensus sequence but only 76% when compared against the fungal consensus sequence.

This analysis can be extended by calculating the mean fraction of one phylogenetic order using the consensus sequences derived from the other orders. The mean fraction of the metazoan α-tubulin sequences decreases from 96.6% when compared with the metazoan consensus sequence to 86.4%, 85.5%, and 80.2% when compared with the protozoan, plant and algal, and fungal consensus sequences, respectively. Fairly comparable values (difference of 0.7%–3.8%) result from the reciprocal analysis: the differences are due to the fact that the phylogenetic orders are not equally heterogeneous. The fungal α- and β-tubulins are distinctively different: they are less conserved within the order than the tubulins of the other orders and are more different from tubulins from other phylogenetic orders. More surprising is the close similarity of the protozoan tubulins to those from the plants and alga. In particular, the fractional means using the homologous consensus sequences are 95.4% and 96.0% for the α-tubulins but 93.2% and 94.1% for the heterologous sequences, i.e., the protozoan α-tubulins only differ from those from the plants and alga by less than 2%, the equivalent of only about nine residue positions. Such subtle differences must contribute to the specific properties of tubulins from different phylogenetic orders, such as the drug-binding properties (see section VIII, below). The individual members of the tubulin family therefore have evolved exceptionally slowly, at a rate comparable to that for the histones and actin [Little et al., 1981; Doolittle, 1992].

Two conclusions can be drawn from this remarkable conservation of the primary sequences. First, the functional properties of both the α- and β- subunit (and presumably the γ-tubulin, although a detailed analysis will require more sequences to be available) must impose immense constraints on any sequence diversification, and mutations can only be accommodated at relatively few positions without having deleterious effects. Second, any differences that have been conserved within a phylogenetic order are likely to have been functionally advantageous and are therefore likely to relate to the specific properties that characterize the tubulins from the different orders.

III. STRUCTURE OF MICROTUBULES

The protein–protein interactions between the subunits must constitute a primary constraint on the tertiary structure of the α- and β- tubulins. Conventional thin section electron microscopy of glutaraldehyde-fixed cells, particularly in the presence of tannic acid, established that the microtubule normally consists of 13 linearly aligned protofilaments [Tilney et al., 1973]. Variants of this morphology have been reported, with in vivo protofilament numbers varying between 12 and 15 [e.g., Burton et al., 1975; Eichenlaub-Ritter and Tucker, 1984; Mogensen et al., 1989].

Greater structural heterogeneity has been observed following the *in vitro* assembly of either neuronal tubulin or neuronal microtubule protein. In particular, pH affects the mean number of protofilaments [Burton and Himes, 1978], while differing assembly conditions can result in the formation of a variety of polymorphs [e.g., Mandelkow et al., 1984; Karecla et al., 1989]. High resolution electron microscopy, coupled with image analysis and low angle x-ray diffraction, has established that these diverse structures are each formed from a single basic unit, the protofilament [Mandelkow et al., 1984; Mandelkow and Mandelkow, 1983].

Such polymorphs have been examined by both image analysis following negative staining or cryoelectron microscopy [e.g., Amos and Baker, 1979; Mandelkow et al., 1984, 1985; Crepeau et al., 1978; Downing and Jontes, 1992], by low angle x-ray diffraction [Mandelkow et al., 1977; Mandelkow and Mandelkow, 1983; Beese et al., 1987], and by scanning tunneling microscopy [Simic-Krstic et al., 1989]. The protofilament consists of alternating globular units with dimensions of approximately 50 × 40 × 40 Å, with a repeating unit along the protofilament axial length of about 80 Å. Such dimensions are consistent with each globular unit consisting of a single 50 kD tubulin monomer and with the 80 Å repeat being formed from an end to end arrangement of α/β heterodimers. Low angle x-ray diffraction of assembled microtubules, at a resolution of 18 Å, suggests that each monomeric subunit has three structural domains (see Fig. 2) [Beese et al., 1987]. This trilobed substructure is not apparent following image analysis to 4 Å of negative-stained, glucose-embedded Zn^{2+} sheets [Downing and Jontes, 1992], a discrepancy that may reflect differences in sample preparation. While the substructure of the native tubulin monomer is unknown, the three-domain model might well provide a structural basis for the evidence (see section V, below) that the tubulin subunit undergoes a conformational change following microtubule assembly that markedly affects the intersubunit interactions. This potential subdivision of the monomers into three structural domains must, however, be distinguished from the similar evidence based on the presence of two sites, positioned at approximately equal intervals along the primary sequence, that are highly sensitive to proteolytic digestion [de la Vina et al., 1988]. There is, for example, no evidence that these sensitive sites lie between structural domains equivalent to the "hinge region" of myosin. Indeed, although trypsin specifically digests α-tubulin at α-arg^{339} and α-chymotrypsin digests β-tubulin at β-tyr^{281} [Mandelkow et al., 1985], it has not been possible to purify under native conditions the individual cleavage products.

As the protofilaments form the basis of each of the tubulin assembly products, it is probable that the longitudinal interactions between the heterodimers is more stable than the lateral, interprotofilament interactions. This is strongly supported by the elegant images of microtubules "fraying apart" at their ends into individual protofilaments following the induction of microtubule disassembly [Mandelkow et al., 1991] and by the variation in the number of protofilaments along the lengths of single microtubules [Chretien et al., 1992].

The essential protein–protein interactions involved in microtubule assembly can be broken down into three classes: those involved in forming 1) the α/β-heterodimer, 2) the longitudinal interactions that result in the formation of the protofilament, and 3) the lateral, interprotofilament interactions. The assembly of tubulin into a variety of polymorphic structures, the variability in the number of protofilaments, and the antiparallel polarities of Zn^{2+}-sheet protofilaments all suggest that the lateral interaction is not only the weakest but is probably the most plastic. The fraying apart of the protofilaments on cooling [Mandelkow et al., 1991] suggests that this interaction has a strong hydrophobic contribution.

Low angle x-ray diffraction and image analysis studies (Fig. 2) indicate that the longitudinal intra- and interdimer interfaces are likely to be structurally similar. The α/β-dimer has a dissociation constant of $0.15–0.38 \times 10^{-6}$ M [Mejillano and Himes, 1989; Sackett and Lippoldt, 1991; Panda et al., 1992]. This interaction probably involves a hydrophobic contribution as dimer formation is enhanced at higher ionic strengths and at higher temperatures [Panda et al., 1992]. The primary hydrophobic contribution must, however, relate to the lateral interactions. In particular, dimers do not fully dissociate into monomers at 4°C: most microtubules are cold labile, and the protofilament-derived single and double gyre rings are common components of cold-dissociated microtubules. In summary, the protein–protein interactions involved in the lateral association of tubulin heterodimers appear to involve a significant hydrophobic contribution, while those at the α/β-intradimer and the longitudinal dimer–dimer interfaces are probably less hydrophobic.

IV. TUBULIN–TUBULIN INTERACTIONS

The ability of tubulins from different sources, despite their divergent sequences (Fig. 1A,B), to coassembly *in vitro* demonstrates the high conservation of the intersubunit interactions. For example, yeast tubulin [Kilmartin, 1981; Barnes et al., 1992], but not *Dictylostelium* tubulin [Trivinos-Lagos et al., 1993], coassembles *in vitro* with neuronal tubulin. Similarly, microinjection and transfection studies have shown the coassembly of certain of the chick–yeast β-tubulin chimeric constructs [Bond et al., 1986; Fridovich-Keil et al., 1987], *Physarum* tubulin [Prescott et al., 1989], and *Paramecium* tubulin [Geuens et al., 1989] with the microtubules of tissue culture cells and of neuronal tubulin in, for example, sea urchin [Salmon et al., 1984], sand dollar [Hamaguchi et al., 1985], and *Drosophila* cells [Kellogg et al., 1988]. Unfortunately, the association and dissociation rate constants governing such heterologous tubulin assembly have not been examined, since such studies would provide direct evidence about the conservation of the subunit–subunit interfaces.

The precise tubulin sequence can, however, markedly affect the assembly properties. Chick–yeast chimeric β-tubulins in which residues from β:282 are derived from the yeast sequence exhibited minimal assembly in vivo [Fridovich-Keil et al., 1987]. Unsurprisingly, there was no detectable *in vivo* incorporation of a Chinese hamster ovary β-tubulin containing two major deletions, between β:250–255 and β:336–345 and from about β:325 to the C terminus [Boggs et al., 1988]. Indeed, peptides α:390–436 and β:386–430 are highly conserved (Fig.

Fig. 2. *Stereopairs of three subunits of one protofilament based on x-ray fiber diffraction of microtubules assembled from unfractionated microtubule protein, showing the possible subdomain structure of the individual α- and β-subunits.* **A:** *Viewed from the outside of the microtubule.* **B:** *Viewed in a direction tangential to the microtubule wall. The bar represents 20 Å. [Reproduced from Beese et al., 1987, with permission of the publishers.]*

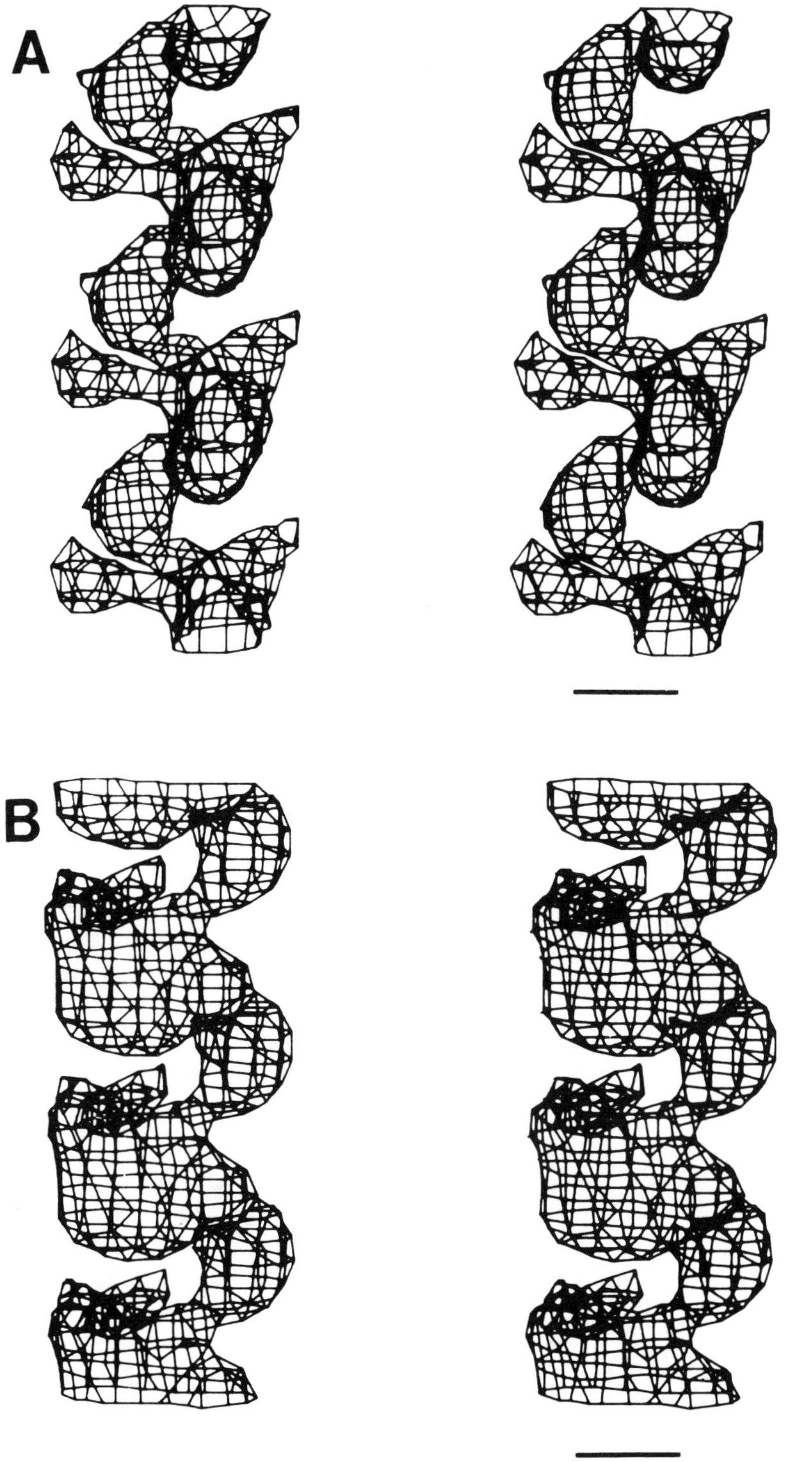
A
B

1A,B), indicating that they have considerable functional importance. This is further indicated by the inhibition of microtubule assembly following the methylation of α-lys^{394} [Sherman et al., 1983], a residue that lies within the absolutely conserved α-lys^{394}-asp^{396} peptide (Fig. 1A). As this inhibition is pH modulated [Szasz et al., 1986], α-his^{393} may also be important, particularly since microtubule assembly can be inhibited by reacting three histidines with diethylpyrocarbonate [Lee et al., 1976]. This peptide may therefore contribute to an intersubunit interaction, either within the α/β-heterodimer or between the dimers of the assembled microtubule. The equivalent β-peptide (β:387–391; see Burns [1991] for alignment) has a noticeably different sequence and charge (met–phe–arg–arg vs. lys–phe–glu–leu), is also highly conserved (Fig. 1B), and may also contribute to the interface interactions.

Immunological methods have implicated a number of other peptides in the subunits–subunit interactions. For example, an antibody raised against the α:155–168 peptide recognizes iodinated (and consequently partially denatured) tubulin dimers but not assembled microtubules [Arevalo et al., 1990]. By contrast, an antibody raised against the equivalent β-tubulin peptide (β:153–165) binds to both iodinated tubulin and assembled microtubules [Arevalo et al., 1990], while a further antibody against β:154–165 inhibits microtubule assembly [Hesse et al., 1987] and does not immunostain assembled microtubules (Ponstingl, personal communication). These α- and β-peptides therefore probably do not contribute directly to the intersubunit interaction, despite being moderately hydrophobic [de la Vina et al., 1988]. By contrast, antibodies raised against the α:214–226 and β:241–256 peptides recognize iodinated tubulin, fixed microtubules, but neither unfixed taxol-stabilized microtubules or Zn^{2+} sheets [Arevalo et al., 1990], observations that suggest that these peptides may lie close to an interdimer interface. Similarly, the β:28–38 peptide may lie close to interface between the dimers forming the longitudinal, or protofilament, interactions, since a polyclonal antibody against this peptide labeled native α/β-heterodimers but not assembled microtubules or protofilament-derived polymorphs [Chene et al., 1992]. Significantly, β:28–38 is far less highly conserved than α: 155–168, β:154–165, α-214–226, or β:241–256 (Fig. 1A,B); the essential interactions, may, however, be affected by a limited number of highly or totally conserved residues within such peptides. Residue β:288 may also contribute to the lateral interactions, since a β-glu^{288} to β-lys^{288} substitution in the testis-specific β_2-tubulin of *Drosophila melanogaster* assembles into aberrant microtubules [Rudolf et al., 1987]. This residue is conserved in 92.6% of all β-tubulin sequences, with the only exception, apart from the conservative substitution to β-asp^{288}, being the β-ala^{288} found within the divergent sequence of a *Porphyra* β-tubulin. Similarly, the neighboring β:287 has been suggested to contribute to specifying 13 protofilament microtubules on the basis of a β-pro^{287} to β-ala^{287} substitution in the *C. elegans* tub-1 and mec-1 tubulins [Driscoll et al., 1989]; other tubulins, however, contain this, or equivalent (β-pro^{287} to β-gly^{287}, β-ser^{287} or β-val^{287}), substitutions.

Finally, the C terminus of α-tubulin may also lie close to an interdimer interface. Many α-tubulin isoforms can be modified by the post translational addition of tyrosine (see section IX, below) and must lie on the exposed surface of the assembled microtubule, as microtubules are recognized by a monoclonal antibody that specifically recognizes the tyrosinylated form [Wehland et al., 1983]. This C-terminal tyrosine can be phosphorylated by insulin receptor kinase, with additional phosphorylation of both α- and β-subunit tyrosines [Wandosell et al., 1987]. There was, however, selective loss on taxol-induced assembly of tubulin phosphorylated on the C-terminal tyrosine, indicating that such subunits were assembly incompetent and that the C-terminal tyrosine lies close to an interdimer interface. Indeed, there was minimal labeling of this tyrosine when preassembled microtubules were phosphorylated with the insulin receptor kinase.

Additional information about the nature of these tubulin–tubulin interactions may be derived by studying the effects of lactoperoxidase (LPO), an inhibitor of microtubule assembly [Rousset and Wolff, 1980a,b]. In particular, while some LPO co-pellets with assembled microtubules, LPO also dissociates the α/β-heterodimer, with the formation of LPO–α and LPO–β complexes. LPO may therefore bind close to the intradimer α/β-interface, and, although the site of this interaction has not been mapped, it has been postulated to include a cysteine residue on either the α- or β-subunit or the LPO [Rousset and Wolff, 1980b].

Such studies highlight the intrinsic difficulty of identifying the inter- and intradimer interfaces. Chemical modifications that inhibit microtubule assembly, such as C-terminal tyrosine phosphorylation, reacting cysteine residues with ethacrynic acid [Xu et al., 1992] and other sulfhydryl reagents, modifying α-lys^{394} [Sherman et al., 1983], or about half of the eight tryptophan residues with 2-hydroxy-5-nitrobenzyl bromide [Maccioni and Seeds, 1982], may act by altering residues that are directly involved in these interactions. Such residues may, however, lie elsewhere within the tertiary structure, with the observed inhibition resulting from either pertubation of the tertiary structure or the prevention (see sections V and VI, below) of a conformational change that is essential for subunit addition. In contrast, the immunological and LPO-binding approach lacks the required resolution, as indicated by the differing effects of the anti-β:153/4–165 antibodies and because an antibody against an epitope lying close to an interface may be prevented by steric effects from binding to the assembled subunits. Finally, the predicted high conservation of these interface peptides is itself a handicap, since any mutations are likely to be lethal; certain of the cold-sensitive mutations of *Saccharomyces cerevisiae* and *Schizosaccharomyces pombe,* which will be considered later, may map to the subunit–subunit interfaces.

V. NUCLEOTIDE BINDING

The tubulin dimer binds 2 mol·mol^{-1} guanosine nucleotide: one is at an exchangeable site, while the other is effectively nonexchangeable. Purified neuronal tubulin contains 1 mol·mol^{-1} of both GTP and GDP, of which the GDP is bound to the exchangeable site. Photocrosslinking of 8-azido-GTP [Geahlen and Haley, 1977] or GTP [Nath and Himes, 1986; Hesse et al., 1987; Linse and Mandelkow, 1988] has demonstrated that the exchangeable site is located on the β-subunit, while the high homology of the α- and β-sequences has led to the unproven assumption that the nonexchangeable GTP is bound to the α-subunit. One consequence of this difference is that virtually all aspects of the role of nucleotides in microtubule assembly specifically relate to the β-subunit.

The association constants governing the binding of GTP and GDP to the exchangeable site on the β-tubulin subunit are approximately $6.4–9.0 \times 10^7$ and $2.5–2.7 \times 10^7$ M^{-1} in the presence of Mg^{2+} and 1.6×10^7 and 1.4×10^4 M^{-1} in its absence [Correia et al., 1987]. A variety of GTP and GDP analogues can also bind to this site, including the 2′- and 2′3′deoxy-derivatives, which bind with similar affinities as GTP and GDP [Hamel et al., 1984; Hamel and Lin 1984], and nonhydrolyzable analogues such as GMPPNP and GMPPCP, which bind with much lower affinities [Mejillano et al., 1990; Hamel and Lin, 1990]. There is also some evidence for the binding of nonguanosine nucleotides, although it is essential, in such studies, to distinguish between a direct effect of the nucleotide and the generation of GTP by nucleoside diphosphate kinase from the added nucleotide triphosphate and the residual GDP bound to the exchangeable site [Duanmu et al., 1986]. Examples of a direct interaction between tubulin and a nonguanosine nucleotide include the evidence for the weak binding of ATP [Zabrecky and Cole, 1982; Hamel and Lin, 1990; but also see O'Brien and Erickson, 1989] and for ATP-induced assembly of tubulin, which had been depleted of E-site GDP [O'Brien and Erickson, 1989; Mejillano et al., 1990; Hamel and Lin, 1990].

Microtubule assembly is, under almost all conditions, dependent on GTP or a nonhydrolyzable GTP analogue, being bound to the exchangeable site, although the direct incorporation of tubulin with bound GDP has been reported [Hamel et al., 1986]. This GTP is subsequently hydrolyzed to GDP, which remains bound to the assembled microtubule and only becomes exchangeable following the disassembly of the microtubule. This change in the exchangeability may be due to the packing of the subunits within the assembled microtubule or to an assembly-dependent conformational change [Kirchner and Mandelkow, 1985].

One consequence of the assembly-dependent GTP hydrolysis is a marked increase in the dissociation rate constant governing the rate of loss of tubulin subunits from the microtubule end (see details, this volume). This strongly suggests, but does not formally prove, that a conformational change is induced by either the GTP hydrolysis or the consequential release of the phosphate product. As the increase in the dissociation rate constant underlies the behavior known as *dynamic instability* (see details, this volume), it is highly probable that the structural requirements for this change contribute to the high conservation of β-tubulin.

VI. THE GTP-BINDING SITE OF β-TUBULIN

Two approaches have been used to identify which peptides define the GTP-binding site. The first assumes that the mechanism of GTP hydrolysis resembles that of the various members of the GTPase superfamily, such as the elongation factor EF-Tu and the G-proteins, which use GTP hydrolysis to induce a conformational change. Indeed, this is supported by the stereoselectivity of certain GTP analogues: the assembly of pure tubulin is specific for the S_p diasterisomer of GTPβS [Roychowdhury and Gaskin, 1988] and for the Δ pseudoaxial configuration of chromium GTP [Carlier et al., 1991]. Furthermore, studies with various members of the GTPase superfamily have suggested that aluminum and beryllium fluorides bind to the site that is transiently occupied by the inorganic phosphate (Pi)-product of GTP hydrolysis and thereby inhibit the hydrolysis-dependent conformational change. These salts also stabilize microtubules and inhibit dynamic instability [Carlier et al., 1988, 1989; Combeau and Carlier, 1989; Hamel et al., 1991], which not only suggests that Pi release, rather than GTP hydrolysis, is the essential prerequisite for dynamic instability, but also that the mechanism of GTP hydrolysis by β-tubulin may resemble that of the GTPase superfamily proteins.

Comparison of the β-tubulin sequences with those of members of the GTPase superfamily has suggested that the guanine moiety interacts with β:295–298 and the phosphates with a β:102–108 and β:205–208 [Sternlicht et al., 1987; Burns et al., 1993]. Further analysis has suggested that thr^{143} (or ser^{145}) and phe^{133}, each of which is totally conserved (Fig. 1B), may also be involved in coordinating the Mg^{2+}/γ-phosphate and the guanine residue [Burns et al., 1993]. One unique feature of this comparison of the β-tubulin sequences with those of other members of the GTPase superfamily is the requirement to reverse the polarity of peptides β:102–108 and β:295–298. The validity of this polarity reversal has been tested by site-directed mutagenesis of the β-tubulin gene. While changes to β-asn^{297} failed to have the predicted effect on GTP binding [Farr and Sternlicht, 1992], mutating β-thr^{106} yielded a phenotype in *S. cerevisiae* with properties consistent with a predicted change in the rate of the assembly-dependent GTP hydrolysis [Sage et al., 1991].

The alternative model of the GTP-binding site is based on the assumption that the β-tubulin site is analogous to that of other nucleotide-binding proteins [Mandelkow et al., 1985; Burns et al., 1993]. It predicts that a phosphate interacts with β:140–146, and that the guanine moiety interacts with β:67–70 (or β:241–245) and the ribose sugar with β:178–181. There is no direct evidence for the involvement of either β:178–181 or β:241–245: indeed, the latter peptide corresponds to one postulated to be involved in forming the subunit–subunit interactions. This second model is, however, supported by the ultraviolet photoaffinity labeling by GTP and 8-azido-GTP of β:63–77 [Linse and Mandelkow, 1988; Kim et al., 1987] and by the 10,000-fold reduction of GTP and GDP binding following the site-directed mutagenesis of β-asp[67] to β-asn[67] [Farr and Sternlicht, 1992].

Neither of these models, however, readily accommodates the observed ultraviolet photolabeling by GTP of β:154–165 [Hesse et al., 1987], a peptide already implicated in lying close to a dimer–dimer interface, and its inhibition by an antibody raised against this peptide. The lack of spatial resolution of immunological methods does not, however, preclude the involvement of β:154–165 in both GTP binding and an intersubunit interface. Indeed, such an interface must be altered by GTP hydrolysis and phosphate release to account for the properties of dynamic instability. Similarly, neither sequence homology model would predict the stoichiometric ultraviolet photocross-linking of GTP to the totally conserved β-cys[12] (Fig. 1B) [Shivanna et al., 1993], even though this is consistent with the demonstration that GTP removal is essential for the cross-linking of β-cys[12] to β-cys[201] or β-cys[211] by *N,N'*ethylene*bis*(iodoacetamide) [Little and Luduena, 1987] and with the effects of maytansine (see section VIII, below), which locates β:cys[12] close to the GTP-binding site.

Significantly, most of the peptides identified by the competing models and by the ultraviolet cross-linking studies are highly conserved in all known β-tubulin sequences, while there are significant, yet conserved, differences between the α- and β-tubulins that may relate to the differing properties of the nonexchangeable and exchangeable nucleotide-binding sites [Burns et al., 1993]. This dichotomy between the two models based on sequence homologies and the experimental ultraviolet cross-linking evidence has led to the suggestion that the nature of the interaction between β-tubulin and the nucleotide may change in response to the hydrolysis-dependent, and other, conformational changes. In particular, the interaction when β-tubulin is in one conformational state may possibly resemble that of the other GTPases, while the interaction in a second conformational state may be more akin to that of other nucleotide-binding proteins. Such a "dual personality" model offers an explanation for a variety of other anomalous observations, including the promotion of microtubule assembly by non-GTP nucleotides and the differential effects of 2′dGTP and 2′3′ddGTP on nucleation compared with the subsequent elongation reaction [Hamel et al., 1984] and the differences in ultraviolet photoaffinity cross-linking by 8-azido-GTP and 8-azido-GDP [Geahlen and Haley, 1979]. Finally, the conformational state of tubulin with bound GTP does appear to be different from that with bound GDP, as there is only minimal direct assembly of the latter subunits and there is evidence for a conformational change following the GTP–GDP nucleotide exchange and prior to subunit addition to the microtubule end [Carlier, 1983].

VII. SUBSTRUCTURE OF α- AND β-TUBULINS

The peptides implicated in defining the GTP-binding site are widely dispersed along approximately two-thirds of the primary sequence, while cross-linking with *N,N'*-ethylene*bis*(iodoacetamide) indicates that β-cys[12] lies within 9 Å of either β-cys[201] or β-cys[211], and β-cys[239] within 9 Å of β-cys[354] [Little and Luduena, 1985, and 1987]. The tubulin subunits may therefore consist of subdomains separated by a cleft that binds the GTP and that is flanked by the reactive cysteine residues. This is further indicated by the interactions of tubulin with phenyl-Sepharose [Hanssens et al., 1990]: tubulin without exogenous nucleotide binds, and is desorbed with either water or, at a higher ionic strength, with either GTP or GDP. The water-eluted α/β-heterodimers have only 1 $mol \cdot mol^{-1}$ bound GTP but can bind, presumably at the exchangeable site, an additional $mol \cdot mol^{-1}$ GTP. Removal of the exchangeable nucleotide therefore appears to expose a hydrophobic surface, which has been postulated [Hanssens et al., 1990] to correspond to β:141–154, a peptide implicated in binding the triphosphates in the conformational state associated with GTP hydrolysis [Sternlicht et al.,

1987; Burns et al., 1993], and which is in part highly hydrophobic. The GTP-binding site therefore appears to be located, at least in part, within a cleft between two structural domains.

The presence of such a cleft is further supported by studies of the effects of detergents and of various ligands that report hydrophobic surfaces. The binding of Nile Red to tubulin results in an enhanced fluorescence and a wavelength shift [Sackett et al., 1990]. Measurement of the fluorescence decay suggests that Nile Red binds to two sites, one that permits energy transfer from a tryptophan residue and that is partially shielded from the solvent and a second that is exposed to a polar environment. The dependence of the fluorescence derived from the former site on the tubulin dimer concentration has indicated that it is located at the intersubunit interface. Binding Nile Red may, however, alter such an interaction: the calculated dimer dissociation constant (1×10^{-6} M) is somewhat higher than most of the values obtained by other methods. As the intradimer interface is not particularly hydrophobic, the two Nile Red–binding sites may correspond to the homologous clefts in the α- and β-subunits, with the observed concentration dependence of the blue shift resulting from an enhanced dimer dissociation in response to the binding of the Nile Red. Indeed, the dimer dissociation constant is affected by colchicine (see section VIII, below), and the α-subunit affects the binding of GTP to β-tubulin synthesised *in vitro* using a reticulocyte expression system [Farr et al., 1990].

The binding of 8-anilinonaphthalene-1-sulfonate also consists of more than one component, one of which is slow and partially inhibited by GTP [Hanssens et al., 1990], consistent with the evidence from phenyl-Sepharose binding that the exchangeable nucleotide masks a hydrophobic surface. Neither this probe nor 2,6-anilonaphthalene-sulfonate affects the alkylation with iodoacetamide [Luduena et al., 1986]. In contrast, *bis*(1,8-anilinonaphthalene)-sulfonate binds to two sites and enhances the alkylation by iodoacetamide [Luduena et al., 1986; Prasad et al., 1986]. The increase in this reactivity and the fluorescence changes are slow and may reflect the time dependent conformational change underlying the loss of the assembly competency of tubulin [Prasad et al., 1986]. This ligand has proved useful in monitoring the stabilization of tubulin by various antimitotic ligands (see section VIII, below), and, while the alkylation sites have not been mapped, they presumably correspond to the highly reactive β:239 or β:354 [Little and Luduena, 1985; Bai et al., 1989]. One of the sites binding *bis* (1,8-anilinonaphthalene)-sulfonate also binds 6-propionyl-2-(*N, N'*-dimethyl)-aminonaphthalene [Mazumdar et al., 1992]. Fluorescence studies using both probes indicate that this site is protected from the solvent, consistent with the hydrophobic environment of β:239, and that the enhanced fluorescence involves energy transfer from neighboring tryptophan residue(s) [Prasad et al., 1986; Mazumdar et al., 1992].

Tubulin cooperatively, and reversibly, binds at least 60 mol·mol^{-1} octyl glucoside, and hydrodynamic studies indicate that, in the presence of this detergent, tubulin is highly flexible and there is an equilibrium between a water-soluble and an amphipathic state [Andreu and Munoz, 1986]. The critical concentration for assembly is higher in the latter state, and the tubulin is more susceptible to proteolytic digestion [Andreu et al., 1986]. Similarly, tubulin binds about 95 mol·mol^{-1} deoxycholate, and the hydrodynamic studies indicate that this detergent fails to dissociate the α/β-heterodimer but does induce a major conformational change and enhances the proteolytic sensitivity [Andreu and Munoz, 1986; Andreu et al., 1986]. These detergents therefore appear to weaken the tertiary structure by minimizing hydrophobic interactions within the α- and β-monomers by binding to apolar faces between tubulin domains [Andreu et al., 1986]. Both fluorescence and spectrophotometric methods confirm that octyl glucoside and deoxycholate perturb the tertiary structure of tubulin and suggest that aromatic residues lie close to the binding site(s) for certain of these amphiphilic molecules [Andreu, 1982]. Finally, their inhibition of colchicine binding [Andreu, 1982] strongly suggests (see section VIII, below) that they bind within a hydrophobic cleft.

Various yeast strains with mutations in α- and β-tubulin have been isolated that fail to grow at low temperatures [Umesono et al., 1983; Hiraoka et al., 1984; Thomas et al., 1985; Schatz et al., 1988]. Some of the *S. cerevisiae* mutations have been mapped [Thomas et al., 1985; Huffaker et al., 1988] to five β-tubulin peptides. Remarkably, almost all of the altered residues are highly conserved in all β-tubulins; the notable exception is the C-terminal peptide due to the presence of ser^{429} in many plant β-tubulins (Fig. 1B). Two of the mutated peptides (β:241, β:233/242/245, and β:391), which show an increase in hydrophobicity relative to the wild-type sequence (unpublished observations), correspond to peptides previously identified as being potentially involved in forming or being close to the subunit–subunit interfaces (α:393/4; β:241–256). The mutation of β:318 (which [see section VIII, below] is associated with the binding of certain antimitotic ligands) also increases the local hydrophobicity while those at β:372/374/375 each reduce it. Consequently, inhibition of growth at a low temperature can result from both an increase and a decrease in the calculated hydrophobicity of specific peptides: cold sensitivity can result from both an enhancement and a weakening of the local hydrophobicity. This indicates that cold sensitivity is not necessarily due to a change at the intersubunit interfaces, but may also result from a requirement during microtubule assembly for the displacement of one domain relative to another.

In summary, the α- and β-tubulin subunits appear to contain a cleft that includes both the nucleotide-binding site and features that are moderately hydrophobic. The "dual personality" of the nucleotide-binding site and the conformational change in response to GTP hydrolysis may simply represent the relative displacement of the domains defining this cleft. Indeed, buffer-dependent differences in the kinetics of microtubule assembly [e.g., Waxman et al., 1981; Barton et al., 1987], the assembly products [Karecla et al., 1989], and ligand binding [e.g., Sackett et al., 1990] indicate that the tubulin subunit is remarkably plastic. This conformational flexibility highlights three additional points. First, nucleotide binding may be of considerable importance in defining the tertiary structure of the protein: for example, GTP protects tubulin against denaturation at pH 8.0 [Croom et al., 1986]. Furthermore, the probable location of the nonexchangeable, and nonhydrolyzable, GTP on the α-subunit strongly suggests that it has a vital structural, rather than an enzymatic, role. Indeed, there is increasing evidence that a chaperone, possibly TCP1 [Yaffe et al., 1992], is required for the correct folding of the nascent polypeptide chain [Zabala and Cowan 1992; Frydman et al., 1992]. Second, the pre-assembly conformational change [Carlier, 1983] and dynamic instability dictate that the precise tertiary structure must depend on whether GDP or GTP is bound to the tubulin dimer, whether, following assembly, the exchangeable GTP has been hydrolyzed to GDP, and whether the phosphate product has been released. Furthermore, as the hydrolysis-dependent conformational change must affect the subunit–subunit interactions, the structure of microtubules assembled with GTP (Fig. 2) may relate to the conformation following this hydrolysis. Finally, the conformational flexibility, and particularly that associated with the assembly-dependent GTP hydrolysis, may (see section VIII, below) relate to the mechanism(s) by which the antimitotic drugs inhibit microtubule assembly.

VIII. ANTIMITOTIC LIGANDS

Antimitotic ligands that target tubulin are characterized by four striking features: they are chemically very divergent, certain of them exhibit phylogenetic specificity, many inhibit microtubule assembly at substoichiometric concentrations, and most, if not all, bind to the β-subunit. The phylogenetic differences, particularly of the benzimidazole derivatives, are of great commercial and medical importance, particularly in the control of fungal and helminthic infections, although the fungal and metazoan tubulins also differ in their sensitivities to colchicine (see Burns [1992] for examples of species-dependent differences of the K_d for colchicine binding), rhizoxin [Takahashi et al., 1990] and taxol [e.g., Barnes et al., 1992]. Such differences must relate, in part, to the phylogenetically based differences between the primary sequences of the β-tubulins (Fig. 1B). The chemical heterogeneity, illustrated by active analogues of the colchicine, podophyllotoxin, benzimidazoles, maytansine, rhizoxin, phomopsin, dolastatin, vinblastine, and taxol analogues (see Lin and Hamel [1981] and Bai et al. [1990a,b] for their structures), is particularly remarkable since almost all of these ligands inhibit microtubule assembly *in vitro*. Taxol is the notable exception: it stabilizes microtubules *in vivo* and *in vitro* [e.g., Parness and Horwitz, 1981]. The vinca alkaloids, including vinblastine and vincristine, inhibit mitosis at substoichiometric concentrations but can also induce the formation of tubulin paracrystals *in vivo* [e.g., Starling and Burns, 1975] and spiral aggregates of protofilaments *in vitro* [e.g., Himes et al., 1976].

Despite their chemical heterogeneity, competition studies subdivide the ligands that bind to β-tubulin into two classes. The first includes colchicine, podophyllotoxin, and a variety of benzimidazoles. Colchicine binding to neuronal tubulin is competitively inhibited by podophyllotoxin [Cortese et al., 1977; Friedman et al., 1980], nocadazole [Hoebeke et al., 1976], and a variety of benzimidazole carbamates [Russell et al., 1992]. Similarly, high colchicine concentrations competitively inhibit the binding of [^{14}C]methyl(*N*-benzimidal-2-yl) carbamate (the active breakdown product of benomyl) to extracts from *Aspergillus nidulans* [Davidse and Flach, 1977], and mebendazole inhibits colchicine binding to partially purified *Ascaris* tubulin [Kohler and Bachman, 1981] and neuronal tubulin [Laclette et al., 1980]. Such competitive inhibition would be consistent with the ligands binding to topologically adjacent sites, although differences in the temperature dependencies of the binding of benzimidazole carbamates to neuronal and nematode tubulins indicate that the subtle sequence differences confer specific properties to the different tubulins [Russell et al., 1992]. Colchicine, podophyllotoxin, and a variety of competing analogues share a trimethoxyphenol ring; the benzimidazoles lack this moiety but have been proposed structurally to resemble parts of the colchicine A- and B-rings [Lin and Hamel, 1981].

The second class includes the vinca alkaloids, rhizoxin, dolastatin 10, maytansine, and phomopsin A and is characterized by the stabilization of the intrinsically labile colchicine-binding site [Wilson, 1970] and by the differential effects of the *N,N'*-ethylene*bis* (iodoacetamide) cross-linker. Colchicine, podophyllotoxin, and nocodazole each enhance the formation of the β-cys^{12} to β-cys^{201} or β-cys^{211} cross-link but inhibit the formation of the β-cys^{239} to β-cys^{354} cross-link [Roach and Luduena, 1984; Little and Luduena, 1985, 1987]. In contrast, vinblastine, matansine, and phomopsin A each has the opposite effects [Roach and Luduena, 1984; Luduena et al., 1990, 1992]; furthermore, each protects tubulin against the time-dependent sensitivity to alkylation by iodoacetamide in the presence of *bis* (1,8-anilinonaphthalene)-sulfonate and the loss of colchicine-binding activity. These ligands appear to bind at related sites, as dolastatin 10, phomopsin A, rhizoxin, and maytansine all inhibit the binding of [^{3}H]vinblastine and [^{3}H]vincristine [Bai et al., 1990a,b], while phomopsin A binding to two sites was inhibited by rhizoxin and ansamitocin P-3 (an analogue of maytansine) and, to a lesser extent, by vinblastine [Li et al., 1992a]. However, while maytansine, rhizoxin, and vinblastine competitively inhibits [^{3}H]vincristine binding, the inhibition by dolastatin 10 and phomopsin A is noncompetitive [Bai et al., 1990b]. Ligands of this second class therefore appear to bind to β-tubulin at proximally adjacent sites that are distinct from those that bind the colchicine–benzimidazole antimitotic ligands.

The stabilization of colchicine binding by ligands of this second class indicates that they probably optimize a specific conformational state. Colchicine binds slowly [Wilson, 1970; Garland, 1978; Lambeir and Engelborghs, 1981], indicating that it can only interact with a specific conformational state. Indeed, spectroscopic and fluorescence methods [Ray et al., 1981; Andreu and Timasheff, 1982a] indicate that colchicine binding requires β-tubulin to undergo an initial conformational change. By contrast, other ligands, including nocodazole [Hoebeke et al., 1976] and podophyllotoxin [Cortese et al., 1977], bind rapidly to neuronal tubulin, and mebendazole binds rapidly to *Haemonchus contortus* tubulin [Gill and Lacey, 1992]. These ligands therefore do not appear to require a prebinding conformational change. Consequently, conformational changes can occur both before and in response to the ligand binding. Indeed, the colchicine-induced change may be quite extensive as tubulin–colchicine complexes assemble into ribbons and sheets rather than *bona fide* microtubules [Andreu and Timasheff, 1982b]. Finally, specific β-tubulin isoforms, characterized by a limited number of amino acid changes (see details, this volume, and section X, below), bind colchicine at different rates [Banerjee and Luduena, 1991]. It is therefore probable that substitutions distant to the colchicine-binding site can markedly affect the kinetics of binding of this ligand and consequently that this conformational change is not restricted to a localized perturbation.

This conformational plasticity of tubulin, both before and after ligand binding, and that associated with the assembly-dependent GTP hydrolysis and phosphate release, seriously hamper the dissection of the ligand-binding sites. In particular, it is difficult when using either a mutational analysis or chemical modifications to distinguish between the direct effect of a specific residue on the ligand-binding site from indirect effects that result from perturbing the tertiary structure. For example, 2,4-dichlorobenzyl-thiocyanate, which reacts with β-cys^{239}, does not affect colchicine binding; colchicine does, however, reduce the alkylation by this reagent [Bai et al., 1989]. Finally, there are various drug-sensitive mutations that map to the α-subunit [Schatz et al., 1988], yet other evidence points to these drugs binding to the β-tubulin. It is therefore probable that the α-subunit mutations affect the ligand-binding site by perturbing the α/β-interface. Indeed, colchicine, but not other ligands that compete for this site, stabilize the α/β-dimer [Detrich et al., 1982], while colchicine, and a variety of tropolone analogues, potentiate the ultraviolet-induced dimerization and trimerization of tubulin [Wolff et al., 1992]. Such effects clearly demonstrates that the conformational changes associated with the binding of these ligands can affect the α/β-inter- and intradimer interfaces.

Mutational analysis, while having a very high spatial resolution, must therefore be treated with considerable caution. A further limitation is that appropriate genetic methods have, in general, only been developed for the fungi, yet most studies of microtubule assembly have used neuronal protein. It has, however, proved difficult to purify nonneuronal tubulin in sufficient quantities to study the *in vitro* assembly or other biochemical properties (but see Kilmartin [1981], Barnes et al., [1992] and Bellocq et al. [1992] for the limited success with fungal tubulins) such that there is a divergence between the availability of biochemical and genetic evidence. Indeed, as the fungal tubulins have markedly different drug sensitivities from those of the metazoa, the genetic analysis has concentrated on the effects of the sensitivity to various benzimidazoles while the biochemical studies have concentrated on those ligands that inhibit the *in vitro* assembly of neuronal tubulin.

Mutations can be subdivided into three classes: resistance, supersensitivity, and differing sensitivities to two related ligands. For example, colchicine resistance is conferred in *Chlamydomonas reinhardtii* by mutating β-lys^{350} to β-glu^{350} or β-met^{350} [Lee and Huang, 1990]. This might be due either to the altered amino acid directly affecting the colchicine-binding site or to a change elsewhere that perturbs the tertiary structure. The pleotropic effects of this mutation, including enhanced sensitivity to taxol and resistance to vinblastine and a variety of herbicides, has suggested that it affects the tertiary structure and, in particular, enhances microtubule stability [Schibler and Huang, 1991]. Conversely, mutations of *Aspergillus nidulans* (β-ala^{165} to β-val^{165}) confer enhanced sensitivity to benomyl, to its active breakdown product methyl (*N*-benzimidal-2-yl)carbamate, and to nocodazole but enhanced resistance to thiabendazole, indicating that β:165 may interact with the 4-thiazole side chain of thiabendazole [Jung and Oakley, 1990]. This is supported by the enhanced binding of [^{14}C]MBC to a strain containing the β-val^{165} change (K_a: 1.6 × 10^6 M) than to the wild type (0.45 × 10^6 M [Davidse and Flach, 1977; but see Jung and Oakley, 1990, for strain identification), by a point mutation (β-phe^{167} to β-tyr^{167}) conferring benomyl resistance in *Neurospora crassa* [Orbach et al., 1986], and by the implication of residues close to β:165 in conferring the benzimidazole sensitivity of *Plasmodium falciparum* [Delves et al., 1987] and *Pneumocystis carinii* [Edlind et al., 1992; Dyer et al., 1992].

A residue of particular interest is β:198, since a β-glu^{198} to β-gly^{198} confers sensitivity to diethofencarb in *Neurospora crassa* [Fujimura et al., 1992], varying benomyl sensitivities in *Venturia inaequalis* (β-lys^{198} > β-glu^{198} > β-gly^{198} > β-ala^{198}) and field strains of other fungi [Koenraadt et al., 1992], while in *A. nidulans* the changes β-glu^{198}, β-asp^{198}, β-gln^{198}, and β-lys^{198} have differential effects on the benomyl and thiabendazole sensitivities [Jung et al., 1992]. It is therefore likely that β:198 contributes directly to the benzimidazole-binding site, even though a β-glu^{198} mutation has opposing effects in *V. inaequalis* [Koenraadt et al., 1992] and *N. crassa* [Fujimura et al., 1992]. This involvement is reinforced by the fungal-specific difference from the metazoan benomyl-resistant tubulins at β:200 (Fig. 1B), by the different sensitivities to benomyl and thiabendazole of the β-phe^{200} to β-tyr^{200} mutation in *A. nidulans* [Jung et al., 1992], by a β-phe^{200} to β-tyr^{200} mutation in a benomyl-resistant strain of *V. inaequalis* and field strains of other fungi [Koenraadt et al., 1992], and by the β-met^{200} or β-ser^{200} substitution in the benomyl-insensitive plants (Fig. 1B).

Benomyl, nocodazole, and thiabendazole resistance is also conferred in *A. nidulans* by a β-his^6 to β-tyr^6 mutation [Jung et al., 1992], by a β-tyr^{50} to β-ser^{50} change, and by a β-gln^{134} to β-lys^{134} mutation, which also renders the microtubules hyperstable at high temperatures [Jung et al., 1987]. Benomyl resistance is conferred by a β-arg^{318} to β-trp^{318} mutation [Huffaker et al., 1988]. Finally, a β-arg^{241} to β-his^{241} mutant of *Sac. cerevisiae* requires benomyl for growth at low temperatures [Thomas et al., 1985], which is significant since ligands binding to the colchicine-site affect the alkylation of β-cys^{239} and its cross-linking to β-cys^{354} [Roach and Luduena, 1984; Roach et al., 1985; Bai et al., 1989].

The potential involvement of residues close to β-318 in binding the benzimidazole/colchicine/podophyllotoxin family of ligands is strongly supported by a comparison of the measured colchicine- and podophyllotoxin-binding constants with the known β-tubulin sequences [Burns, 1992]. This analysis highlighted the potential role of β:316 and proposed that this residue interacts with the trimethoxy-substituted (or A-) ring of colchicine and podophyllotoxin.

There has been relatively little attention paid to mapping mutations affecting the second class of antimitotic drugs. A β-asn^{100} to β-ile^{100} mutation in *A. nidulans* confers rhizoxin-resistance [Takahashi et al., 1990], consistent with the natural resistance of *Sac. cerevisiae* and *Sch. pombe*, which have β-val^{100} and β-ile^{100}, respectively. This has been directly confirmed by examining the binding of [^{14}C]rhizoxin, [^{14}C]phomopsin A, and [^{3}H]vinblastine to extracts derived from *Saccharomyces, Schizosaccharomyces,* and *Aspergillus* strains with differing residues at β:100. Rhizoxin binding was restricted to the β-asn^{100} strains and was inhibited by ansamitocin P-3. In contrast, there was no binding of either vinblastine or phomopsin A to any of the extracts, consistent with the competition studies using neuronal tubulin that these ligands bind to a family of spatially related sites. Finally, dolastatin 10 and phomopsin

A are both modified peptides, being modifications of tyr-val-val-pro-val-glu and val-val-ile-pro-phe, respectively. It is possible that the inhibition of microtubule assembly by these ligands results from them mimicking an interaction between two β-tubulin peptides. Most β-tubulins contain the peptide phe-ser-val-val-pro (β:167–171), with substitutions (usually a methionine) being restricted to β:170. This peptide is adjacent to a residue implicated in benomyl binding (β:165). It is also immediately C terminal to a peptide that can be photocross-linked to GTP [Hesse et al., 1987] (see section VI, above), and both of these modified peptide ligands inhibit the GTPase activity of tubulin dimers (see below).

The genetic evidence therefore implicates a variety of residues in the antimitotic ligand-binding sites. In common with the analysis of the GTP-binding site, these residues are distributed throughout much of the β-tubulin sequence, and in particular there is strong evidence that β:165, β:198–200, and β:316–318 contribute to the benzimidazole-binding site. Significantly, the residues at β:165, β:200, and β316 are characteristic of the different phylogenetic orders (Fig. 1B), consistent with the order-specific differences in the sensitivity to the benzimidazole/colchicine class of antimitotic ligands. The folding of the tertiary structure probably brings these residues into close proximity, but, unlike the coordinated amino acid substitutions (considered below), it is unlikely that they directly interact with each other. Finally, it is unlikely that benzimidazole sensitivity is solely defined by this cluster of residues. For example, the *C. elegans* ben-1, but not the tub-1 or mec-7, gene product is benomyl sensitive, yet none of the ben-1–specific residues (β:56, β:218, β:231, β:278, β:328, and β:331) [Driscoll et al., 1989] correspond to this cluster.

None of the residues implicated in drug binding map within the peptides implicated in GTP binding. There is, however, a structural relationship between both types of antimitotic ligand-binding sites and the GTP-binding site. In particular, members of the colchicine/benzimidazole family stimulate the GTPase activity of tubulin dimers [David-Pfeuty et al., 1979; Andreu and Timasheff, 1981; Lin and Hamel, 1981], although podophyllotoxin either fails to promote the uncoupled hydrolysis [Andreu and Timasheff, 1981] or only stimulates it at substoichiometric concentrations [Lin and Hamel, 1981]. Consequently, the conformational change(s) associated with the binding of the colchicine/benzimidazole class may mimic the changes induced during microtubule assembly which result in the assembly-dependent GTP hydrolysis. This is reinforced by the stimulation by this class of antimitotic drugs of the GTP-modulated cross-linking of β-cys^{12}–β-cys^{201} or -cys^{211} [Roach and Luduena, 1984; Little and Luduena, 1985, and 1987] and by the location of β-cys^{12} close to the guanosine-binding site [Shivanna et al., 1993].

The two spatially related interactions between β-tubulin and the maytansine/vinblastine class of antimitotic inhibitors are either probably also close to the GTP-binding site or can effect a conformational change that perturbs this site, since these ligands inhibit the formation of the β-cys^{12}–β-cys^{201} or -β:cys^{211} cross-link. Furthermore, vinblastine, maytansine, rhizoxin, dolastatin 10, and phomopsin A each inhibit the endogenous GTPase of tubulin dimers and the assembly-dependent GTP hydrolysis [David-Pfeuty et al., 1979; Lin and Hamel, 1981; Andreu and Timasheff, 1981; Bai et al., 1990a]. Maytansine, phomopsin A, and dolastatin 10 also markedly affect GTP binding at 0°C, while rhizoxin is less effective and prevents the release of pre-bound GDP [Huang et al., 1985; Bai et al., 1990b]. Significantly, in view of the evidence that the GTP-binding site has a "dual personality," higher concentrations, particularly of maytansine and rhizoxin, are required at 37°C than at 0°C [Huang et al., 1985; Bai et al., 1990b], indicating that the sensitivity to these ligands depends, in part, on the precise conformation of the GTP–β-tubulin interaction.

The binding sites for both classes of antimitotic drugs therefore relate to the GTP-binding site and/or mechanism of GTP hydrolysis, while the low solubilities of many of the antimitotic ligands strongly suggest that they bind within the presumed hydrophobic cleft. It is striking that most, if not all, of the antimitotic ligands bind to the β-subunit and that the evidence for conformational plasticity and for the "dual personality" of the GTP-binding site also specifically relates to β-tubulin. Such observations strongly suggest that the inhibition of microtubule assembly is due to these ligands perturbing the normal sequence of conformational changes associated with subunit addition and the consequential assembly-dependent GTP hydrolysis. Indeed, colchicine–tubulin complexes coassemble with carrier tubulin [Sternlicht and Ringel, 1979; Farrell and Wilson, 1980], while the kinetics of the substoichiometric, colchicine-induced inhibition of microtubule assembly indicate a sequence of transient states, with the inhibition being determined mainly by the formation of the final-state complex [Skoufias and Wilson, 1992]. Furthermore, various yeast β-tubulin mutants that are likely to have an altered tertiary structure, for example, from the truncation of the C terminus [Matsuzaki et al., 1988; Katz and Solomon, 1988], also have altered benomyl sensitivities. The susceptibility of this sequence of conformation changes of β-tubulin following subunit addition and GTP hydrolysis may underlie the independent evolution of such a diverse range of naturally occurring inhibitors, particularly since the high conservation of β-tubulin (Fig. 1B) indicates that "escape mutations" might have deleterious effects. For example, certain of the benomy supersensitive yeast strains have altered cold sensitivities [Thomas et al., 1985; Huffaker et al., 1988], while a colcemid-dependent mutant of a mammalan β-tubulin is, following transfection into tissue culture cells, both supersensitive to taxol and more rapidly degraded [Whitfield et al., 1986].

IX. SURFACE RESIDUES

The identification of surface residues provides a further constraint on an understanding of the tertiary structure. Such residues are likely to be mainly located on the external face of the assembled microtubule as both the intra- and interdimer interactions are probably highly conserved. The surface residues are identified by three analytical approaches: a comparison of the available α- and β-tubulin sequences, from the interaction with other proteins, and by genetic methods.

Despite the remarkable conservation of both the α- and β-tubulins, a small number of sequences contain significant insertions or deletions. In particular, the testis-specific *D. melanogaster* β_3 has an insertion of six residues at β:60–61, *D. melanogaster* α_4 has two peptides with markedly altered sequences and additional residues (α:33–38 and α:44–50), *Euglena gracilis* has a deletion/modified sequence (β:30–35), while *Stylonchia* α_1 has a deletion of residues α:275–280. Such differences are unlikely to be accommodated unless the peptides lie at or close to the subunit surface. The same applies to α:41, since up to 17 residues can be inserted at this position of the *Sac. cerevisiae* tub-3 α-tubulin gene without a significant effect *in vivo* [Schatz et al., 1987]. The presence of multiple amino acid substitutions, notably at α:35, α:43, α:117, α:309, β:30, β:35, β:43, and β:85 [Burns, 1992], indicates that these positions are not under significant selective pressure and probably lie superficially. Finally, a comparison of the primary sequences also shows that single insertions or deletions can be accommodated at a small number of residue positions (see Burns [1992] for details), several of which correspond to the hypervariable residue positions.

The second approach relates to the interaction with other proteins. For example, the proteolytic sensitivities of α-arg^{339} and β-tyr^{281} [Mandelkow et al., 1985] indicate that these residues must lie at the subunit surface, in agreement with the surface prediction and with the *Stylonchia* α:275–280 deletion and the alignment of the α- and β-sequences. Similarly, the various posttranslational modifications of α- and β-tubulin require the altered residues to be accessible to the modifying enzymes. For example, α-lys^{40} of assembled microtubules can be acetylated, and antibodies bind to assembled microtubules that specifically recognize the acetylated isoforms [LeDizet and Piperno, 1987]: indeed, a number of the insertions and deletions map close to this residue. Secondly, the C terminus of many, but not all, α-tubulin isoforms can be tyrosinylated [Idriss et al., 1991; Paturle-Lafanechere et al., 1991], and, while the unassembled dimer is the preferred substrate of tubulinyl tyrosine ligase [Raybin and Flavin, 1975], assembled microtubules are preferentially detyrosinylated by the tubulinyl tyrosine carboxypeptidase [Kumar and Flavin, 1981; Arce and Barra, 1985]. The C-terminal tyrosine must therefore lie at the exposed surface, and indeed mono- and polyclonal antibodies that recognize the tyrosinylated and detyrosinylated isoforms label assembled microtubules [Wehland et al., 1983; Gundersen et al., 1984; Gu et al., 1988]. Although the tyrosinylation status has either no or only a minimal effect on microtubule assembly [Kumar and Flavin, 1982; Idriss et al., 1991], phosphorylation of this residue inhibits microtubule assembly, indicating that it may lie close to an interdimer interface (see section IV, above). Similarly, the C-terminal peptides of both α- and β-tubulin can be phosphorylated by a calmodulin-dependent protein kinase, and the assembly of this neuronal tubulin is inhibited [Wandosell et al., 1986]. In contrast, phosphorylation by casein kinase II, at β-ser^{444} [Diaz-Nido et al., 1990], occurs on assembled microtubules and has no effect on microtubule assembly *in vitro* [Serrano et al., 1987]; this residue also appears to lie on the exposed surface and not close to a subunit–subunit interaction. Significantly, this residue (or β-ser^{446}) of the neuronal-specific β_2 is phosphorylated *in vivo* [Luduena et al., 1988]. Finally, both α- and β-tubulins can be ADP-ribosylated *in vitro* by turkey or chick erythrocyte mono-ADP-ribosyltransferase [Scaife et al., 1992; Raffaelli et al., 1992]. Comparison of the tubulin sequences with those of other ADP-ribosylated proteins has indicated that this modification may occur at α:318–321 and β:318–321 [Burns et al., 1993], peptides that may lie close to a subunit–subunit interface as ADP-ribosylation inhibits microtubule assembly [Scaife et al., 1992].

An alternative rationale also places the sites of the glutamylation of neuronal α-glu^{445} and β-glu^{435} [Edde et al., 1990, 1992; Rudiger et al., 1992] on the exposed surface. It is unlikely that the addition of up to five highly charged glutamate residues could be accommodated at any other position without affecting the tertiary structure, yet the neuronal protein commonly used to study microtubule assembly *in vitro* is polyglutamylated.

Most of these posttranslational modifications lie within the highly variable C-terminal peptides, and their relative exposure is further supported by immunological, proteolytic, genetic, and protein-binding studies. Antibodies raised against the variable C-terminal peptides [Lopata and Cleveland, 1987; Gu et al., 1988] show that the exposed peptide is not restricted to the C-terminal glutamate or tyrosine residue. Similarly, antibodies have been raised that recognize epitopes within α:415–443 and α:430–443 [de la Vina et al., 1988], α:414–422 and α:426–450 [Breitling and Little, 1986], β:412–431 [de la Vina et al., 1988], and β:416–430 [Breitling and Little, 1986]. While these antibodies stain assembled microtubules, it is only possible, because of the low resolution of this approach, to state that these peptides include an epitope that lies at or close to the subunit surface and not necessarily that the entire peptide is exposed. Analysis of the β:400–436 and β:400–445 peptides by one- and two-dimensional nuclear magnetic resonance and by circular dichroism does, however, indicate that these C-terminal peptides lack, at pH 7.0, any regular secondary structure [Reed et al., 1992]. It is therefore probable not only that they lie on the exposed surface but that any tertiary structure, particularly of the highly conserved β:400–430 peptide (Fig. 1B), is conferred by the interactions with the remainder of the tubulin subunit(s) and with the various associated proteins.

This location is supported by the sensitivity of the C-terminal α- and β-peptides to subtilisin digestion to yield assembly-competent products [e.g., Serrano et al., 1984; Sackett et al.,

1985]. Mass spectrometry shows that the cleavage sites are at α-asp^{438} and β-gln^{433} and, to a lesser extent, at α-his^{406} and at β-his^{396} [Redeker et al., 1992]. Finally, MAPs of the MAP2/tau/MAP4 class binds to this these C-terminal peptides: There is no binding to subtilisin-digested tubulin [Serrano et al., 1984], while synthetic peptides to the C-terminal peptides (α:430–441 and β:422–434) bind to MAP2 and tau [Littauer et al., 1986; Maccioni et al., 1988], and antiidiotypic antibodies against these peptides recognize MAP1, MAP2, and tau [Rivas et al., 1988]. Similarly, the β:400–436 and β:400–445 peptides inhibit the MAP-dependent microtubule assembly [Reed et al., 1992], while modification of the multiple carboxyl residues characteristic of the C-terminal peptide, as confirmed by subtilisin digestion, eliminated the MAP dependency of microtubule assembly [Mejillano and Himes, 1991].

The exposed location of the C-terminal peptide of β-tubulin, starting at β:430, is further indicated by its extreme sequence variability [see Burns and Surridge, 1991] and by the demonstration that truncating the *Sac. cerevisiae* β-tubulin by 12 residues has no apparent effect *in vivo* [Katz and Solomon, 1988; Gu and Cowan, 1989], although it did affect the benomyl sensitivity (see section VIII, above).

Various antibodies, which were considered above in the review of the subunit–subunit interactions, bind to assembled microtubules and therefore generally include accessible epitopes. In addition, a 37 residue synthetic peptide, corresponding to the highly conserved β:400–436, attenuates the inhibition of sperm motility induced by an antitubulin antibody, suggesting that this peptide includes a dynein-binding site [Goldsmith et al., 1991]. This contrasts with an earlier [Paschal et al., 1989], and disputed [Cleveland et al., 1990], proposal that dynein and MAP2 interact with β:437–440, although this work preceded the precise definition of the subtilisin cleavage site.

While the variable C-terminal peptide, and at least part of the flanking and highly conserved sequence, lies on the exposed surface of the microtubule, it is unclear how much of the N-terminal sequences of α- and β-tubulin are exposed. Iodinated tau, but not MAP2, binds to a site within α:1–75 [Littauer et al., 1986], while a monoclonal antibody raised against β:1–12, but including an epitope between β:7 and β:12, labels fixed microtubules *in vivo* [Theodorakis and Cleveland, 1992]. The N-terminal peptide (met-arg-glu-ile) of β-tubulin is highly conserved; the only exceptions, which may represent sequencing errors, are *Achyla klebsiana* and the *Glycine max* Sβ_2 genes. This sequence is essential for the mechanism governing the autoregulation of β-tubulin synthesis in tissue culture cells [Yen et al., 1988], while the synthesis of α-tubulin in *Tetrahymena* is apparently not autoregulated [Stargell et al., 1992]. The downregulation of β-tubulin synthesis may result from the binding of the nascent polypeptide to an unknown cellular factor that is not tubulin [Theodorakis and Cleveland, 1992], as it emerges from the ribosome [Yen et al., 1988]. Deletion of the codons for β-arg^2 and β-glu^3 from the mouse Mβ_1 gene, followed by transient transfection into tissue culture cells, only slightly altered the ability of the protein to assemble *in vivo,* and a similar result was observed following deletion of the totally conserved, apart from *Dictylostelium,* α-arg^2 residue [Gu and Cowan, 1989]. More extensive N-terminal deletions, however, markedly reduced the *in vivo* assembly competency, indicating that the extreme N terminus of both α- and β-tubulin may not contribute significantly to specifying either the tertiary structure or the intersubunit interactions and may therefore be exposed. Indeed, the *Dictylostelium* sequence contains a glycine insertion at α:2. Furthermore, the α-, β-, and γ-tubulins have differing N-terminal peptides: γ-Tubulin has an inserted proline at β:2/3, while α- differs from β- and γ-tubulins by having an insertion at β:3/4. Such insertions, together with the tau-binding and immunofluorescence studies, imply that parts of the N-terminal peptide lie on the exposed surface of the assembled microtubule.

Any prediction of the tertiary structure of tubulin can therefore be further constrained by identifying those residues located on the exposed surface of the assembled microtubule. Significantly, none of the identified residues or peptides correpond to those previously identified as contributing to the GTP or antimitotic drug-binding sites.

X. TERTIARY STRUCTURE OF THE TUBULIN SUBUNITS

The x-ray crystallographic determination of the tertiary structure of tubulin will require the growth of suitable crystals. This is limited by the isoform complexity, resulting from the expression in most systems of multiple genes and the various substoichiometric posttranslational modifications, from the intrinsic lability of most tubulins and by the evidence that tubulin can adopt a variety of conformational states. The determination of the crystallographic structure of individual monomers or the α/β-heterodimer will be a significant development. It is, however, unlikely to explain fully the mechanism of microtubule assembly, the mechanism of GTP hydrolysis, or the basis of the inhibition by antimitotic drugs, as specific conditions will have been used to grow the crystals and as a result the structure of a specific conformation will have been determined. It is increasingly evident that the conformational plasticity, particularly of β-tubulin, is an important characteristic of tubulin, as illustrated by the "dual personality" of the GTP-binding site and the conformational changes associated with both GTP hydrolysis and the binding of antimitotic drugs.

The synthesis of the biochemical, genetic, and sequence analysis approaches do, however, place some restrictions on the probable tertiary structure. This can be further constrained by analyzing the probable proximity of specific residues to each other, an approach that is aided by the remarkable conservations of the α- and β-tubulins. In particular, the C-terminal peptides of vertebrate β-tubulins are divergent between isotypes of the same organism, but are conserved between organisms, such that the isotypes can be subdivided into six classes, each with a characteristic pattern of cellular expression [Joshi et al., 1987; Lopata and Cleveland, 1987]. Inspection of the vertebrate β-tubulin sequences showed a remarkable correlation between the sequence of these isotype-specific C-terminal peptides (from β:431) and the residues at β:35–37, β55–57, and β:124. This strongly suggested that these disparate residues interact with each other in the tertiary structure [Burns and Surridge, 1990], and this is consistent with the evidence concerning both the GTP- and colchicine/benzimidazole-binding sites.

Extending the analysis to all β-tubulin sequences revealed an additional coordinated amino acid substitution: there was a remarkable correlation between whether this C-terminal peptide contained a tyrosine or a phenylalanine residue and the amino acids at β:217–218. This not only suggested that the C-terminal peptide makes an additional interaction, but also that it effectively spans two structural domains, one containing the β:35–37, β:55–57, and β:124 residues and the other the β:217–218 residues. This led to the proposal [Burns and Surridge, 1990] that the conformational change associated with GTP hydrolysis, and hence underlying dynamic instability, involved the relative displacement of these two domains and that the inhibition of dynamic instability by MAP2 and tau was due to their interaction with the C-terminal peptide. This model is strongly supported by the differing assembly kinetics of tubulin either enriched with the β_2 isotype [Banerjee et al., 1990] or containing specific β-tubulin isotypes [Banerjee et al., 1992]; by the subsequent demonstrations that chemical modification of the carboxyl groups, primarily within the C-terminal peptide, enhances microtubule assembly [Mejillano et al., 1991, 1992]; by the marked differences in the dynamic instability of neuronal and erythrocyte tubulins [Trinczek et al., 1991]; and by the anomalous behavior *in vivo* of constructs consisting of the N terminus of *Drosophila* β_2 and the C terminus of the testis-specific *Drosophila* β_3 (see Raff, this volume, for details).

It may be possible, as more α-, β-, and γ-tubulin sequences become available, particularly from phylogenetically related organisms, to identify additional coordinated amino acid substitutions. Indeed, many of the phylogenetic-specific differences (Fig. 1A,B) may represent such substitutions, reflecting the minor perturbations in the tertiary structure and resulting in the differing drug sensitivities. Further analysis of the tubulin sequences, coupled with parallel genetic and biochemical studies, will refine the current understanding of the structure of the tubulin subunit, but is unlikely to offer the resolution that will be available from the determination of its crystallographic structure.

XI. REFERENCES

Amos LA, Baker TS (1979): Three-dimensional image of tubulin in zinc-induced sheets, reconstructed from electron micrographs. Int J Biol Macromol 1:146–156.

Andreu JM (1982): Interaction of tubulin with non-denaturing amphiphiles. EMBO J 1:1105–1110.

Andreu JM, de la Torre J, Carrascosa JL (1986): Interaction of tubulin with octyl glucoside and deoxycholate. 2. Protein conformation, binding of colchicine ligands, and microtubule assembly. Biochemistry 25:5230–5239.

Andreu JM, Munoz JA (1986): Interaction of tubulin with octyl glucoside and deoxycholate. 1. Binding and hydrodynamic studies. Biochemistry 25:5220–5230.

Andreu JM, Timasheff SN (1981): The ligand- and microtubule assembly–induced GTPase activity of purified calf brain tubulin. Arch Biochem Biophys 211:151–157.

Andreu JM, Timasheff SN (1982a): Conformational states of tubulin liganded to colchicine, tropolone methyl ether, and podophyllotoxin. Biochemistry 21:6465–6476.

Andreu JM, Timasheff SN (1982b): Tubulin bound to colchicine forms polymers different from microtubules. Proc Natl Acad Sci USA 79:6753–6756.

Arce CA, Barra HS (1985): Release of C-terminal tyrosine from tubulin and microtubules at steady state. Biochem J 226:311–317.

Arevalo MA, Nieto JM, Andreu D, Andreu JM (1990): Tubulin assembly probed with antibodies to synthetic peptides. J Mol Biol 214:105–120.

Bai R-L, Lin CM, Nguyen NY, Liu T-Y, Hamel E (1989): Identification of the cysteine residue of β-tubulin alkylated by the antimitotic agent 2,4-dichlorobenzyl thiocyanate, facilitated by separation of the protein subunits of tubulin by hydrophobic column chromatography. Biochemistry 28:5606–5612.

Bai R, Pettit GR, Hamel E (1990a): Dolastatin 10, a powerful cytostatic peptide derived from a marine animal. Inhibition of tubulin polymerization mediated through the vinca alkaloid binding domain. Biochem Pharmacol 39:1941–1949.

Bai R, Pettit GR, Hamel E (1990b): Binding of dolastatin 10 to tubulin at a distinct site for peptide antimitotic agents near the exchangeable nucleotide and vinca alkaloid sites. J Biol Chem 265:17141–17149.

Banerjee A, Luduena RF (1991): Distinct colchicine binding kinetics of bovine brain tubulin lacking the type III isotype of β-tubulin. J Biol Chem 266:1689–1691.

Banerjee A, Roach MC, Trcka P, Himes RH (1990): Increased microtubule assembly in bovine brain tubulin lacking the type III isotype of β-tubulin. J Biol Chem 265:1794–1799.

Banerjee A, Roach MC, Trcka P, Luduena RF (1992): Preparation of a monoclonal antibody specific for the class IV isotype of β-tubulin. J Biol Chem 267:5625–5630.

Barnes G, Louie KA, Botstein D (1992): Yeast proteins associated with microtubules in vitro and in vivo. Mol Biol Cell 3:29–47.

Barton JS, Vandivort DL, Heacock DH, Coffman JA, Trygg KA (1987): Microtubule assembly kinetics. Changes with solution conditions. Biochem J 247: 505–511.

Beese L, Stubbs G, Cohen C (1987): Microtubule structure at 18Å resolution. J Mol Biol 194:257–264.

Bellocq C, Andrey-Tornare I, Paunier-Doret A-M, Maeder B, Paturle L, Job D, Haiech J, Edelstien SJ (1992): Purification of assembly-competent tubulin from *Saccharomyces cerevisiae*. Eur J Biochem 210:343–349.

Boggs BA, Minotti AM, Loeb LM, Cook R, Cabral F (1988): Mutations affecting assembly of β-tubulin localize to a region near the carboxyl terminus. J Biol Chem 263:14566–14573.

Bond JF, Fridovich-Keil JL, Pillus L, Mulligan RC, Solomon F (1986): A chicken-yeast chimeric β-tubulin protein is incorporated into mouse microtubules in vivo. Cell 44:461–468.

Breitling F, Little M (1986): Carboxy-terminal regions of the surface of tubulin and microtubules. Epitope locations of YOL1/34, DM1A, and DM1B. J Mol Biol 189:367–370.

Burns RG (1991): α-, β-, and γ-tubulins: Sequence comparisons and structural constraints. Cell Motil Cytoskel 20:181–189.

Burns RG (1992): Analysis of the colchicine-binding site of β-tubulin. FEBS Lett 297:205–208.

Burns RG, Farrell KW, Surridge CD (1993): Should the tubulins be members of the GTPase superfamily? CIBA Symp 176:248–267.

Burns RG, Surridge C (1990): Analysis of β-tubulin sequences reveals highly conserved, coordinated amino acid substitutions. Evidence that these "hot spots" are directly involved in the conformational change required for dynamic instability. FEBS Lett 271:1–8.

Burton PR, Himes RH (1978): Electron microscope studies of pH effects on assembly of tubulin free of associated proteins. J Cell Biol 77:120–133.

Burton PR, Hinkley RE, Pierson GB (1975): Tannic acid–stained microtubules with 12, 13 and 15 protofilaments. J Cell Biol 65:227–233.

Carlier M-F (1983): Kinetic evidence for a conformational change of tubulin preceding microtubule assembly. J Biol Chem 258:2415–2420.

Carlier M-F, Didry D, Melki R, Chabre M, Pantaloni D (1988): Stabilization of microtubules by inorganic phosphate and its structural analogues, the fluoride complexes of aluminum and beryllium. Biochemistry 27:3555–3559.

Carlier M-F, Didry D, Simon C, Pantaloni D (1989): Mechanism of GTP hydrolysis in tubulin polymerization: Characterization of the kinetic intermediate microtubule-GDP-P_i using phosphate analogues. Biochemistry 28:1783–1791.

Carlier M-F, Didry D, Valentin-Ranc C (1991): Interaction between chromium GTP and tubulin. Stereochemistry of GTP binding, GTP hydrolysis, and microtubule stabilization. J Biol Chem 266:12361–12368.

Chene P, Mazarguil H, Wright M (1992): Microtubule assembly protects the region 28–38 of the β-tubulin subunit. Cell Motil Cytoskel 22:25–37.

Cleveland DW, Joshi HC, Murphy DB (1990): Tubulin site interpretation. Nature 344:389.

Chretien D, Metoz F, Verde F, Karsenti E, Wade RH (1992): Lattice defects in microtubules: Protofilament numbers vary within individual microtubules. J Cell Biol 117:1031–1040.

Combeau C, Carlier M-F (1989): Characterization of the aluminum and beryllium fluoride species bound to F-actin and microtubules at the site of the γ-phosphate of the nucleotide. J Biol Chem 264:19017–19021.

Correia JJ, Baty LT, Williams RC (1987): Mg^{2+} dependence of guanine nucleotide binding to tubulin. J Biol Chem 262:17278–17284.

Cortese F, Bhattacharyya B, Wolff J (1977): Podophyllotoxin as a probe for the colchicine binding site of tubulin. J Biol Chem 252:1134–1140.

Crepeau RH, McEwen B, Edelstein SJ (1978): Differences in α- and β-polypeptide chains of tubulin resolved by electron microscopy with image reconstruction. Proc Natl Acad Sci USA 75:5006–5010.

Croom HB, Correia JJ, Williams RC (1986): The effects of elevated pH and high salt concentrations on tubulin. Arch Biochem Biophys 249:397–406.

David-Pfeuty T, Simon C, Pantaloni D (1979): Effect of anti-mitotic drugs on tubulin GTPase activity and self-assembly. J Biol Chem 254:11696–11702.

Davidse LC, Flach W (1977): Differential binding of methyl benzimidazol-2-yl carbamate to fungal tubulin as a mechanism of resistance to this antimitotic agent in mutant strains of *Aspergillus nidulans*. J Cell Biol 72:174–193.

de la Vina S, Andreu D, Medrano FJ, Nieto JM, Andreu JM (1988): Tubulin structure probed with antibodies to synthetic peptides. Mapping of three major types of limited proteolysis fragments. Biochemistry 27: 5352–5365.

Delves CJ, Ridley RG, Goman M, Holloway SP, Hyde JE, Scaife JG (1989): Cloning of a β-tubulin gene from *Plasmodium falciparum*. Mol Microbiol 3:1511–1519.

Detrich HW, Williams RC, Wilson L (1982): Effect of colchicine binding on the reversible dissociation of the tubulin dimer. Biochemistry 21:2392–2400.

Diaz-Nido J, Serrano L, Lopez-Otin C, Vandekerckhove J, Avila J (1990): Phosphorylation of a neuronal-specific β-tubulin isotype. J Biol Chem 265:13949–13954.

Doolittle RF (1992): Reconstructing history with amino acid sequences. Prot Sci 1:191–200.

Downing KH, Jontes J (1992): Projection map of tubulin in zinc-induced sheets at 4 Å resolution. J Struct Biol 109:152–159.

Driscoll M, Dean E, Reilly E, Bergholz E, Chalfie M (1989): Genetic and molecular analysis of a *Caenorhabditis elegans* β-tubulin that confers benzimidazole sensitivity. J Cell Biol 109:2993–3003.

Duanmu C, Lin CM, Hamel E (1986): Tubulin polymerization with ATP is mediated through the exchangeable GTP site. Biochim Biophys Acta 881:113–123.

Dyer M, Volpe F, Delves CJ, Somia N, Burns S, Scaife JG (1992): Cloning and sequence of a β-tubulin cDNA from *Pneumocystis carinii*: Possible implications for drug therapy. Mol Microbiol 6:991–1001.

Edde B, Rossier J, Le Caer J-P, Prome J-C, Desbruyeres E, Gros F, Denoulet P (1992): Polyglutamylated α-tubulin can enter the tyrosination/detyrosination cycle. Biochemistry 31:403–410.

Edde B, Rossier J, Le Caer J-P, Desbruyeres E, Gros F, Denoulet P (1990): Post-translational glutamylation of α-tubulin. Science 247:83–85.

Edlind TD, Bartlett MS, Weinberg GA, Prah GN, Smith JW (1992): The β-tubulin gene from rat and human isolates of *Pneumocystis carinni*. Mol Microbiol 6:3365–3373.

Eichenlaub-Ritter U, Tucker JB (1984): Microtubules with more than 13 protofilaments in the dividing nuclei of ciliates. Nature 307:60–62.

Farr GW, Sternlicht H (1992): Site-directed mutagenesis of the GTP-binding domain of β-tubulin. J Mol Biol 227:307–321.

Farr GW, Yaffe MB, Sternlicht H (1990): α-Tubulin influences nucleotide binding to β-tubulin: An assay using picomoles of unpurified protein. Proc Natl Acad Sci USA 87:5041–5045.

Farrell KW, Wilson L (1980): Proposed mechanism for colchicine poisoning of microtubules reassembled in vitro from *Strongylocentrotus purpuratus* sperm tail outer doublet tubulin. Biochemistry 19:3048–3054.

Friedman PA, Platzer EG, Carroll EJ (1980): Tubulin characterization during embryogenesis in *Ascaris suum*. Dev Biol 76:47–57.

Fridovich-Keil JL, Bond JF, Solomon F (1987): Domains of β-tubulin essential for conserved functions in vivo. Mol Cell Biol 7:3792–3798.

Frydman J, Nimmesgern E, Erdjument-Bromage H, Wall JS, Tempst P, Hartl F-U (1992): Function in protein folding of TRiC, a cytosolic ring complex containing TCP-1 and structurally related subunits. EMBO J 11:4767–4778.

Fujimura M, Oeda K, Inoue H, Kato T (1992): A single amino-acid substitution in the β-tubulin gene of *Neurospora* confers both carbendazim resistance and diethofencarb sensitivity. Curr Genet 21:399–404.

Garland DL (1978): Kinetics and mechanism of colchicine binding to tubulin: Evidence for ligand-induced conformational change. Biochemistry 17:4266–4272.

Geahlen RL, Haley BE (1977): Interactions of a photoaffinity analog of GTP with the proteins of microtubules. Proc Natl Acad Sci USA 74:4375–4377.

Geahlen RL, Haley BE (1979): Use of a GTP photoaffinity probe to resolve aspects of the mechanism of tubulin polymerization. J Biol Chem 254:11982–11987.

Geuens G, Hill AM, Levilliers N, Adouette A, de Brabander M (1989): Microtubule dynamics investigated by microinjection of *Paramecium* axonal tubulin: lack of nucleation but proximal assembly of microtubules at the kinetochore during prometaphase. J Cell Biol 108:939–953.

Gill JH, Lacey E (1992): The kinetics of mebendazole binding to *Haemonchus contortus* tubulin. Int J Parasitol 22:939–946.

Goldsmith M, Connolly JA, Kumar N, Wu J, Yarbrough LR, van der Kooy D (1991): Conserved β-tubulin binding domain for the microtubule-associated motors underlying sperm motility and fast axonal transport. Cell Motil Cytoskel 20:249–262.

Gu W, Cowan NJ (1989): Assembly properties of altered β-tubulin polypeptides containing disrupted autoregulatory domains. Mol Cell Biol 9:3418–3428.

Gu W, Lewis SA, Cowan NJ (1988): Generation of antisera that discriminate among mammalian α-tubulins: introduction of specialized isotypes into cultured cells results in their co-assembly without disruption of normal microtubule function. J Cell Biol 106:2011–2022.

Gundersen GG, Kalnoski MH, Bulinski JC (1984): Distinct populations of microtubules: Tyrosinated and nontyrosinated β-tubulin are distributed differently in vivo. Cell 38:779–789.

Hamaguchi Y, Toriyama M, Sakai H, Hiramoto Y (1985): Distribution of fluorescently labeled tubulin injected into sand dollar eggs from fertilization through cleavage. J Cell Biol 100:1262–1272.

Hamel E, Batra JK, Lin CM (1986): Direct incorporation of guanosine 5′-diphosphate into microtubules without guanosine 5′-triphosphate hydrolysis. Biochemistry 25:7054–7062.

Hamel E, Lin CM (1984): Inhibition of tubulin polymerization with ribose-modified analogs of GDP and GTP. Reduced inhibition with microtubule associated proteins and magnesium. Biochim Biophys Acta 797: 117–127.

Hamel E, Lin CM (1990): Reexamination of the role of nonhydrolyzable guanosine 5′-triphosphate analogues in tubulin polymerization: Reaction conditions are a critical factor for effective interactions at the exchangeable nucleotide site. Biochemistry 29:2720–2729.

Hamel E, Lin CM, Kenney S, Skehan P (1991): Highly variable effects of beryllium and beryllium fluoride on tubulin polymerization under different reaction conditions: Comparison of assembly reactions dependent on microtubule associated proteins, glycerol, dimethylsulphoxide, and glutamate. Arch Biochem Biophys 286:57–69.

Hamel E, Lustbader J, Lin CM (1984): Deoxyguanosine nucleotide analogues: Potent stimulators of microtubule nucleation with reduced affinity for the exchangeable nucleotide site of tubulin. Biochemistry 23:5314–5325.

Hamelin M, Scott IM, Way JC, Culotti JG (1992): The mec-7 β-tubulin gene of *Caenorhabditis elegans* is expressed primarily in the touch receptor neurons. EMBO J 11:2885–2893.

Hanssens I, Baert J, van Cauwelaert F (1990): Effect of guanine nucleotides on the hydrophobic interaction of tubulin. Biochemistry 29:5160–5165.

Hesse J, Thierauf M, Ponstingl H (1987): Tubulin sequence region β:155–174 is involved in binding exchangeable guanosine triphosphate. J Biol Chem 262:15472–15475.

Himes RH, Kersey RN, Heller-Bettinger I, Samson FE (1976): Action of vinca alkaloids vincristine, vinblastine, and desacetylvinblastine amide on microtubules in vitro. Cancer Res 36:3798–3802.

Hiraoka Y, Toda T, Yanagida M (1984): The NDA3 gene of fission yeast encodes β-tubulin: A cold-sensitive nda3 mutation reversibly blocks spindle formation and chromosome movement in mitosis. Cell 39:349–358.

Hoebeke J, van Nijen G, de Brabander M (1976): Interaction of oncodazole (R17934), a new anti-tumoral drug, with rat brain tubulin. Biochem Biophys Res Commun 69:319–324.

Huang AB, Lin CM, Hamel E (1985): Maytansine inhibits nucleotide binding at the exchangeable site of tubulin. Biochem Biophys Res Commun 128:1239–1246.

Huffaker TC, Thomas JH, Botstein D (1988): Diverse effects of β-tubulin mutations on microtubule formation and function. J Cell Biol 106:1997–2010.

Idriss H, Stammers DK, Ross CK, Burns RG (1991): The dynamic instability of microtubules is not modulated by α-tubulin tyrosinylation. Cell Motil Cytoskel 20:30–37.

Joshi HC, Yen TL, Cleveland DW (1987): In vivo coassembly of a divergent β-tubulin subunit (cβ6) into microtubules of different function. J Cell Biol 105: 2179–2190.

Jung MK, Dunne PW, Oakley BR (1987): Sequence alterations in β-tubulin mutations of *Aspergillus nidulans*. J Cell Biol 105:277a.

Jung MK, Oakley BR (1990): Identification of an amino acid substitution in the benA, β-tubulin gene of *Aspergillus nidulans* that confers thiabendazole resistance and benomyl supersensitivity. Cell Motil Cytoskel 17:87–94.

Jung MK, Wilder IB, Oakley BR (1992): Amino acid alterations in the benA (β-tubulin) gene of *Aspergillus nidulans* that confer benomyl resistance. Cell Motil Cytoskel 22:170–174.

Karecla P, Hirst E, Bayley P (1989): Polymorphism of tubulin assembly in vitro. J Cell Sci 94:479–488.

Katz WS, Solomon F (1988): Diversity among β-tubulins: A carboxy-terminal domain of yeast β-tubulin is not essential in vivo. Mol Cell Biol 8:2730–2736.

Kellogg DR, Mitchison TJ, Alberts BM (1988): Behaviour of microtubules and actin filaments in living *Drosophila* embryos. Development 103:675–686.

Kilmartin JV (1981): Purification of yeast tubulin by self-assembly in vitro. Biochemistry 20:3629–3633.

Kim H, Ponstingl H, Haley BE (1987): Identification of the guanosine interacting peptide of the GTP binding site of β-tubulin using $8N_3GTP$. Fed Proc 46:2229.

Kirchner K, Mandelkow E-M (1985): Tubulin domains responsible for assembly of dimers and protofilaments. EMBO J 4:2397–2402.

Koenraadt H, Somerville SC, Jones AL (1992): Characterization of mutations in the β-tubulin gene of benomyl-resistant field strains of *Venturia inaequalis* and other plant pathogenic fungi. Phytopathology 82:1348–1354.

Kohler P, Bachman R (1981): Intestinal tubulin as a possible target for chemotherapeutic action of mebendazole in parasitic nematodes. Mol Biochem Parasitol 4: 325–336.

Kumar N, Flavin M (1981): Preferential action of a brain detyrosinylating carboxypeptidase on polymerized tubulin. J Biol Chem 256:7678–7686.

Kumar N, Flavin M (1982): Modulation of some parameters of assembly of microtubules in vitro by tyrosinolation of tubulin. Eur J Biochem 128:215–222.

Laclette JP, Guerra G, Zetina C (1980): Inhibition of tubulin polymerization by mebendazole. Biochem Biophys Res Commun 92:417–423.

Lambeir A, Engelborghs Y (1981): A fluorescence stopped flow study of colchicine binding to tubulin. J Biol Chem 256:3279–3282.

LeDizet M, Piperno G (1987): Identification of an acetylation site of *Chlamydomonas* α-tubulin. Proc Natl Acad Sci USA 84:5720–5724.

Lee JC, Field DJ, Lee LLY (1980): Effects of nocodazole on structures of calf brain tubulin. Biochemistry 19:6209–6215.

Lee YC, Houston LL, Himes RH (1976): Inhibition of the self-assembly of tubulin by diethyl pyrocarbonate and photooxidation. Biochem Biophys Res Commun 70:50–57.

Lee VD, Huang B (1990): Missense mutations at lysine 350 in β2-tubulin confer altered sensitivity to microtubule inhibitors in *Chlamydomonas*. Plant Cell 2:1051–1057.

Li Y, Kobayashi H, Tokiwa Y, Hashimoto Y, Iwasaki S (1992a): Interaction of phomopsin A with porcine brain tubulin. Inhibition of tubulin polymerization and binding at a rhizoxin binding site. Biochem Pharmacol 43:219–224.

Li Y, Kobayashi H, Hashimoto Y, Iwasaki S (1992b): Binding selectivity of rhizoxin, phomopsin A, vinblastine, and ansamitocin P-3 to fungal tubulins: Differential interactions of these antimitotic agents with brain and fungal tubulins. Biochem Biophys Res Commun 187:722–729.

Lin CM, Hamel E (1981): Effects of inhibitors of tubulin polymerization on GTP hydrolysis. J Biol Chem 256:9242–9245.

Linse K, Mandelkow E-M (1988): The GTP-binding peptide of β-tubulin: Localization by direct photoaffinity labelling and comparison with nucleotide-binding proteins. J Biol Chem 263:15205–15210.

Littauer UZ, Giveon D, Thierauf M, Ginzburg I, Ponstingl H (1986): Common and distinct tubulin binding sites for microtubule associated proteins. Proc Natl Acad Sci USA 83:7162–7166.

Little M, Krauhs E, Ponstingl H (1981): Tubulin sequence conservation. BioSystems 14:239–246.

Little M, Luduena RF (1985): Structural differences between brain β1- and β2-tubulins: Implications for microtubule assembly and colchicine binding. EMBO J 4:51–56.

Little M, Luduena RF (1987): Location of two cysteines in brain β_1-tubulin that can be cross-linked after removal of exchangeable GTP. Biochim Biophys Acta 912:28–33.

Little M, Seehaus T (1988): Comparative analysis of tubulin sequences. Comp Biochem Physiol 90B:655–670.

Lopata MA, Cleveland DW (1987): In vivo microtubules are copolymers of available β-tubulin isotypes: Localization of each of six vertebrate β-tubulin isotypes using polyclonal antibodies elicited by synthetic peptide antigens. J Cell Biol 105:1707–1720.

Luduena RF, Roach MC, Horowitz P (1986): The effects of the anilonaphthalenesulphonates on the alkylation of tubulin: Correlation between the appearance of sulphydryl groups and apolar binding sites. Biochim Biophys Acta 873:143–146.

Luduena RF, Roach MC, Prasad V, Lacey E (1990): Effect of phomopsin A on the alkylation of tubulin. Biochem Pharmacol 39:1603–1608.

Luduena RF, Roach MC, Prasad V, Pettit GR (1992): Interaction of dolastatin 10 with bovine brain tubulin. Biochem Pharmacol 43:539–543.

Luduena RF, Zimmermann H-P, Little M (1988): Identification of the phosphorylated β-tubulin isotype in differentiated neuroblastoma cells. FEBS Lett 230: 142–146.

Maccioni RB, Rivas CI, Vera JC (1988): Differential interaction of synthetic peptides from the carboxyl-terminal regulatory domain of tubulin with microtubule associated proteins. EMBO J 7:1957–1963.

Maccioni RB, Seeds NW (1982): Involvement of tryptophane residues in colchicine binding and the assembly of tubulin. Biochem Biophys Res Commun 108:896–903.

Mandelkow E-M, Herrmann M, Ruhl U (1985): Tubulin domains probed by limited proteolysis and subunit-specific antibodies. J Mol Biol 185:311–327.

Mandelkow E, Mandelkow E-M (1983): Structure of tubulin rings studied by x-ray scattering using synchrotron radiation. J Mol Biol 167:179–196.

Mandelkow E-M, Mandelkow E (1985): Unstained microtubules studied by cytoelectron microscopy. Substructure, supertwist and disassembly. J Mol Biol 181:123–135.

Mandelkow E-M, Mandelkow E, Milligan RA (1991): Microtubule dynamics and microtubule caps: A time-resolved cryo-electron microscopy study. J Cell Biol 114:977–1003.

Mandelkow E, Schultheiss R, Mandelkow E-M (1984): Assembly and three-dimensional image reconstruction of tubulin hoops. J Mol Biol 177:507–529.

Mandelkow E, Thomas J, Cohen C (1977): Microtubule structure at low resolution by x-ray diffraction. Proc Natl Acad Sci USA 74:3370–3374.

Matsuzaki F, Matsumoto S, Yahara I (1988): Truncation of the carboxy-terminal domain of yeast β-tubulin causes temperature-sensitive growth and hypersensitivity to antimitotic drugs. J Cell Biol 107:1427–1435.

Mazumdar M, Parrack PK, Bhattacharyya B (1992): Interaction of prodan with tubulin. A fluorescence spectroscopic study. Eur J Biochem 204:127–132.

Mejillano MR, Barton JS, Nath JP, Himes RH (1990): GTP analogues interact with the tubulin exchangeable site during assembly and upon binding. Biochemistry 29:1208–1216.

Mejillano MR, Himes RH (1989): Tubulin dimer dissociation detected by fluorescence anisotropy. Biochemistry 28:6518–6524.

Mejillano MR, Himes RH (1991): Assembly properties of tubulin after carboxyl group modification. J Biol Chem 266:657–664.

Mejillano MR, Tolo ET, Williams RC, Himes RH (1992): The conversion of tubulin carboxyl groups to amides has a stabilizing effect on microtubules. Biochemistry 31:3478–3483.

Mogensen MM, Tucker JB, Stebbings H (1989): Microtubule polarities indicate that nucleation and capture of microtubules occurs at cell surfaces in *Drosophila*. J Cell Biol 108:1445–1452.

Nath JP, Himes RH (1986): Localization of the exchangeable nucleotide binding domain in β-tubulin. Biochem Biophys Res Commun 135:1135–1143.

Oakley CE, Oakley BR (1989): Identification of γ-tubulin: A new member of the tubulin superfamily encoded by *mipA* gene of *Aspergillus nidulans*. Nature 338: 662–664.

O'Brien ET, Erickson HP (1989): Assembly of pure tubulin in the absence of free GTP: Effect of magnesium, glycerol, ATP, and the nonhydrolyzable GTP analogues. Biochemistry 28:1413–1422.

Orbach MJ, Porro EB, Yanofsky C (1986): Cloning and characterization of the gene for β-tubulin from a benomyl-resistant mutant of *Neurospora crassa* and its use as a dominant selectable marker. Mol Cell Biol 6:2452–2461.

Panda D, Roy S, Bhattacharyya B (1992): Reversible dimer dissociation of tubulin S and tubulin detected by fluorescence anisotropy. Biochemistry 31:9709–9716.

Parness J, Horwitz SB (1981): Taxol binds to polymerized tubulin in vitro. J Cell Biol 91:479–487.

Paschal BM, Obar RA, Vallee RB (1989): Interaction of brain cytoplasmic dynein and MAP2 with a common sequence at the C-terminus of tubulin. Nature 342: 569–572.

Paturle-Lafanechere L, Edde B, Denoulet P, van Dorsselaer A, Mazarguil H, Le Caer JP, Wehland J, Job D (1991): Characterization of a major brain tubulin variant which cannot be tyrosinated. Biochemistry 30:10523–10528.

Prasad ARS, Luduena RF, Horowitz PM (1986): Detection of energy transfer between tryptophan residues in the tubulin molecule and bound bis(8-anilinonaphthalene-1-sulphonate), an inhibitor of microtubule assembly, that binds to a flexible region on tubulin. Biochemistry 25:3536–3540.

Prescott AR, Foster KE, Warn RM, Gull K (1989): Incorporation of tubulin from an evolutionarily diverse source, *Physarum polycephalum,* into the microtubules of a mammalian cell. J Cell Sci 92:595–605.

Raffaelli N, Scaife RM, Purich DL (1992): ADR Ribosylation of chicken red cell tubulin and inhibition of microtubule self-assembly in vitro by the NAD^+-dependent avian ADP ribosyl transferase. Biochem Biophys Res Commun 184:414–418.

Ray K, Bhattacharyya B, Biswas BB (1981): Role of B-ring of colchicine in its binding to tubulin. J Biol Chem 256:6241–6244.

Raybin D, Flavin M (1975): An enzyme tyrosylating α-tubulin and its role in microtubule assembly. Biochem Biophys Res Commun 65:1088–1095.

Redeker V, Melki R, Prome D, Le Caer J-P, Rossier J (1992): Structure of tubulin C-terminal domain obtained by subtilisin treatment. The major α- and β-tubulin isotypes from pig brain are glutamylated. FEBS Lett 313:185–192.

Reed J, Hull WE, Ponstingl H, Himes RH (1992): Conformational properties of the β(400–436) and β(400–445) C-terminal peptides of porcine brain tubulin. Biochemistry 31:11888–11895.

Rivas CI, Vera JC, Maccioni RB (1988): Anti-idiotypic antibodies that react with microtubule associated proteins are present in the sera of rabbits immunized with synthetic peptides from tubulin's regulatory domain. Proc Natl Acad Sci USA 85:6092–6096.

Roach MC, Bane S, Luduena RF (1985): The effects of colchicine analogues on the reaction of tubulin with iodo[^{14}C]acetamide and N, N′-ethylenebis(iodoacetamide). J Biol Chem 260:3015–3023.

Roach MC, Luduena RF (1984): Difference effects of tubulin ligands on the intrachain cross-linking of β_1-tubulin. J Biol Chem 259:12063–12071.

Rousset B, Wolff J (1980a): Effect of lactoperoxidase-induced α-β tubulin splitting on colchicine binding and polymerization of brain tubulin. FEBS Lett 115:235–238.

Rousset B, Wolff J (1980b): Lactoperoxidase-tubulin interactions. J Biol Chem 255:2514–2523.

Roychowdhury S, Gaskin F (1988): Stereoselectivity of the guanyl-exchangeable nucleotide-binding site of tubulin probed by guanosine 5′-*O*-(2-thiotriphosphate) diasteroisomers. Biochemistry 27:7799–7805.

Rudiger M, Plassman U, Kloppel K-D, Wehland J, Weber K (1992): Class II tubulin, the major brain β-tubulin isotype is polyglutamylated on glutamic acid residue 435. FEBS Lett 308:101–105.

Rudolf JE, Kimble M, Hoyle HD, Subler MA, Raff EC (1987): Three *Drosophila* β-tubulin sequences: A developmentally regulated isoform (β_3), the testis-specific isoform (β_2), and an assembly-defective mutation of the testis-specific isoform ($\beta 2t^8$) reveal both an ancient divergence in metazoan isotypes and structural constraints for β-tubulin function. Mol Cell Biol 7:2231–2242.

Russell GJ, Gill JH, Lacey E (1992): Binding of [^{3}H]benzimidazole carbamates to mammalian brain tubulin and the mechanism of selective toxicity of the benzimidazole anthelmintics. Biochem Pharmacol 43:1095–1100.

Sackett DL, Bhattacharyya B, Wolff J (1985) Tubulin subunit carboxyl termini determine polymerization efficiency. J Biol Chem 260:43–45.

Sackett DL, Knutson JR, Wolff J (1990): Hydrophobic surfaces of tubulin probed by time-resolved and steady-state fluorescence of Nile Red. J Biol Chem 265: 14899–14906.

Sackett DL, Lippoldt RE (1991): Thermodynamics of reversible monomer–dimer association of tubulin. Biochemistry 30:3511–3517.

Sage CR, Davis AS, Dougherty C, Sullivan K, Farrell KW (1991): Phenotypical and biochemical characterization of a mutant site-directed to a putative GTP binding/hydrolysis domain of β-tubulin in the budding yeast *S. cerevisiae*. J Cell Biol 115:36a.

Salmon ED, Leslie RJ, Saxton WM, Karow ML, McIntosh JR (1984): Spindle microtubule dynamics in sea urchin embryos: Analysis using fluorescein-labeled tubulin and measurements of fluorescence redistribution after laser photobleaching. J Cell Biol 99:2165–2174.

Savage C, Hamelin M, Culotti JG, Coulson A, Albertson DG, Chalfie M (1989): mec-7 is a β-tubulin gene required for the production of 15 protofilament microtubules in *Caenorhabditis elegans*. Genes Dev 3:870–881.

Scaife RM, Wilson L, Purich DL (1992): Microtubule protein ADP-ribosylation in vitro leads to assembly inhibition and rapid depolymerization. Biochemistry 31:310–316.

Schatz PJ, Georges GE, Solomon F, and Botstein D (1987): Insertions of up to 17 amino acids into a region of α-tubulin do not disrupt function in vivo. Mol Cell Biol 7:3799–3805.

Schatz PJ, Solomon F, Botstein D (1988): Isolation and characterization of conditional-lethal mutations in the TUB1 α-tubulin gene of the yeast *Saccharomyces cerevisiae*. Genetics 120:681–695.

Schibler MJ, Huang B (1991): The col^{R}4 and col^{R}15 β-tubulin mutations in *Chlamydomonas reinhardtii* confer altered sensitivities to microtubule inhibitors and herbicides by enhancing microtubule stability. J Cell Biol 113:605–614.

Serrano L, Avila J, Maccioni RB (1984): Controlled proteolysis of tubulin by subtilisin: Localization of the site for MAP2 interaction. Biochemistry 23:4675–4681.

Serrano L, Diaz-Nido J, Wandosell F, Avila J (1987): Tubulin phosphorylation by casein kinase II is similar to that found in vivo. J Cell Biol 105:1731–1739.

Sherman G, Rosenberry TL, Sternlicht H (1983): Identification of lysine residues essential for microtubule assembly. Demonstration of enhanced reactivity during reductive methylation. J Biol Chem 258:2148–2156.

Shivanna BD, Mejillano MR, Williams TD, Himes RH (1993): The GTP-binding site of β-tubulin: Identification of cysteine-12 as the major site of crosslinking by direct photoaffinity labeling. J Biol Chem 268: 127–132.

Simic-Krstic Y, Kelley M, Schneiker C, Krasovich M, McCuskey R, Koruga D, Hameroff S (1989): Direct observation of microtubules with the scanning tunneling microscope. FASEB J 3:2184–2188.

Skoufias DA, Wilson L (1992): Mechanism of inhibition of microtubule polymerization by colchicine: Inhibitory potencies of unliganded colchicine and tubulin-colchicine complexes. Biochemistry 31:738–746.

Stargell LA, Heruth DP, Gaertig J, Gorovsky MA (1992): Drugs affecting microtubule dynamics increase α-tubulin mRNA accumulation via transcription in *Tetrahymena thermophila*. Mol Cell Biol 12:1443–1450.

Starling D, Burns RG (1975): Ultrastructure of tubulin paracrystals from sea urchin eggs, with determination of spacings by electron and optical diffraction. J Ultrastruct Res 51:261–268.

Sternlicht H, Ringel I (1979): Colchicine inhibition of microtubule assembly via copolymer formation. J Biol Chem 254:10540–10550.

Sternlicht H, Yaffe MB, Farr GW (1987): A model of the nucleotide-binding site in tubulin. FEBS Lett 214: 226–235.

Szasz J, Yaffe MB, Elzinga M, Blank GS, Sternlicht H (1986): Microtubule assembly is dependent on a cluster of basic residues in α-tubulin. Biochemistry 25:4572–4582.

Takahashi M, Matsumoto S, Iwasaki S, Yahara I (1990): Molecular basis for determining the sensitivity of eukaryotes to the antimitotic drug rhizoxin. Mol Gen Genet 222:169–175.

Theodorakis NG, Cleveland DW (1992): Physical evidence for cotranslational regulation of β-tubulin mRNA degradation. Mol Cell Biol 12:791–799.

Thomas JH, Neff NF, Botstein D (1985): Isolation and characterization of mutations in the β-tubulin gene of *Saccharomyces cerevisiae*. Genetics 112:715–734.

Tilney LG, Bryan J, Bush DJ, Fujiwara K, Mooseker MS, Murphy DB, Snyder DH (1973): Microtubules: Evidence for 13 protofilaments. J Cell Biol 59:267–275.

Trinczek B, Obermann H, Mandelkow E-M, Marx A, Mandelkow E (1991): Dynamic instability of erythrocyte microtubules. J Cell Biol 115:175a.

Trivinos-Lagos L, Ohmachi T, Albrightson C, Burns RG, Ennis HL, Chisholm RL (1993): The highly divergent α- and β-tubulin genes from *Dictyostelium discoideum* are encoded by single genes. Submitted.

Umesono K, Toda T, Hayashi S, Yanagida M (1983): Two cell division cycle genes NDA2 and NDA3 of the fission yeast *Schizosaccharomyces pombe* control microtubular organization and sensitivity to anti-mitotic benzimidazole compounds. J Mol Biol 168:271–284.

Wandosell F, Serrano L, Avila J (1987): Phosphorylation of α-tubulin carboxyl-terminal tyrosine prevents its incorporation into microtubules. J Biol Chem 262: 8268–8273.

Wandosell F, Serrano L, Hernandez MA, Avila J (1986): Phosphorylation of tubulin by a calmodulin-dependent protein kinase. J Biol Chem 261:10332–10339.

Waxman PG, del Campo AA, Lowe MC, Hamel E (1981): Induction of polymerization of purified tubulin by sulphonate buffers. Marked differences between 4-morpholineethanesulphonate (Mes) and 1,4-piperazineethanesulphonate (Pipes). Eur J Biochem 120: 129–136.

Wehland J, Willingham MC, Sandoval IV (1983): A rat monoclonal antibody reacting specifically with the tyrosylated form of α-tubulin. I. Biochemical characterization, effects on microtubule polymerization in vitro and microtubule polymerization and organization in vivo. J Cell Biol 97:1467–1475.

Whitfield C, Abraham I, Ascherman D, and Gottesman MM (1986): Transfer and amplification of a mutant β-tubulin gene results in colcemid dependence: Use of the transformant to demonstrate regulation of β-tubulin subunit levels by protein degradation. Mol Cell Biol 6:1422–1429.

Wilson L (1970): Properties of colchicine binding protein from chick embryo brain. Interactions with vinca alkaloids and podophyllotoxin. Biochemistry 9:4999–5007.

Wolff J, Hwang J, Sackett DL, Knipling L (1992): Colchicine photosensitizes covalent tubulin dimerization. Biochemistry 31:3935–3940.

Yaffe MB, Farr GW, Miklos D, Horwich AL, Sternlicht ML, Sternlicht H (1992): TCP1 complex is a molecular chaperone in tubulin biogenesis. Nature 358: 245–248.

Yen TJ, Gay DA, Pachter JS, Cleveland DW (1988): Autoregulated changes in stability of polyribosome-bound β-tubulin mRNAs are specified by the first 13 translated nucleotides. Mol Cell Biol 8:1224–1235.

Xu S, Roychowdhury S, Gaskin F, Epstein DL (1992): Ethacrynic acid inhibition of microtubule assembly in vitro. Arch Biochem Biophys 296:462–467.

Zabala JC, Cowan NJ (1992): Tubulin dimer formation via the release of α- and β-tubulin monomers from multimolecular complexes. Cell Motil Cytoskel 23:222–230.

Zabrecky JR, Cole RD (1982): Binding of ATP to tubulin. Nature 296:775–776.

ABOUT THE AUTHORS

ROY G. BURNS is a lecturer in the Biophysics Section of the Blackett Laboratory, Imperial College of Science, Technology and Medicine in London, where he teaches molecular cell biology. After receiving a B.Sc. (hons) in biological sciences from the University of East Anglia, he studied for a Ph.D. in the Department of Botany at the University of Edinburgh, examining the nucleic acid changes that occur during the early development of fern gametophytes. This was followed by postdoctoral study in the Kewalo Marine Laboratory at the University of Hawaii and at the University of Edinburgh, and by Visiting Scientist positions at the Harvard Medical School and the Marine Biological Laboratory at Woods Hole, Massachusetts. He was appointed to his current position in 1977, where he has established a research group which studies the kinetics of microtubule assembly and the properties of tubulin and the microtubule-associated proteins.

CHRISTOPHER D. SURRIDGE is a research associate in biophysics at Imperial College of Science Technology and Medicine, London. After receiving a B.Sc. (hons.) in biophysics from Leeds University, England, he pursued doctoral research under Roy Burns at Imperial College, London, where he earned a Ph.D. in 1992 for work on endogenously occurring inhibitors of microtubule assembly. Since then he has been investigating the interaction between microtubule-associated proteins and a variety of phospholipids.

Microtubules: 33–45

γ-Tubulin

Berl R. Oakley

I. INTRODUCTION

It has been known for many years that the major structural components of microtubules are heterodimers composed of α- and β-tubulin. Sequence comparisons of tubulin genes from many organisms reveal that all α-tubulins share a great deal of amino acid (aa) identity and constitute a family of proteins. Likewise, all β-tubulins are closely related and constitute a second family of proteins. α- and β-Tubulins also share substantial sequence identity with each other and thus are members of a superfamily of proteins that share a common evolutionary origin [Krauhs et al., 1981]. Many years of biochemical and molecular genetic experiments gave no indication that the tubulin superfamily extended beyond α- and β-tubulins. Indeed, one of the few apparent certainties in microtubule research was that there were only two classes of tubulin. Recent experiments have revealed that even this fundamental dogma of microtubule research is incorrect. There is an additional family of tubulins, the γ-tubulins. Since γ-tubulin has been discovered only recently there is much less known about this family of proteins than is known about α- and β-tubulins. Nevertheless, the published data indicate that, like α- and β-tubulins, γ-tubulins are probably ubiquitous in eukaryotes and are essential for microtubule function.

It is the purpose of this chapter to summarize the data available on γ-tubulin and to discuss the possible functions of γ-tubulin. Since data on γ-tubulin are relatively sparse, I will indulge in some speculation, hoping to define fruitful directions for γ-tubulin research.

II. THE DISCOVERY OF γ-TUBULIN

γ-Tubulin was discovered in a genetic screen designed to identify proteins that interact with β-tubulin [Weil et al., 1986; Oakley and Oakley, 1989]. Oakley et al. [1985] isolated revertants of *ben*A33, a heat-sensitive β-tubulin mutation of the filamentous fungus *Aspergillus nidulans*. *ben*A33 has been characterized extensively. It is a glutamine to lysine substitution at aa 134 (M.K. Jung and B.R. Oakley, unpublished data). It causes a blockage of mitosis at high temperatures and an inhibition of nuclear migration, but it does not cause an inhibition of microtubule assembly [Oakley and Morris, 1981]. It also confers resistance to a variety of antimicrotubule agents. In addition, its heat-sensitive blockage of mitosis, hence growth, is partially suppressed by a variety of antimicrotubule agents and by an α-tubulin mutation that appears to reduce microtubule stability [Oakley and Morris, 1981; Gambino et al., 1984; Oakley et al., 1987]. These data have led to the conclusion that *ben*A33 probably inhibits microtubule function by hyperstabilizing microtubules [Oakley and Morris, 1981].

Among 2,604 revertants of *ben*A33, Oakley and colleagues [1985, 1987] found a number of α- and β-tubulin mutations and mutations in two previously unknown loci, *mip*A and *mip*B [Weil et al., 1986]. When the *mip*A gene was cloned and sequenced [Oakley and Oakley,

1989], it was found to encode a novel protein that shared a great deal of aa identity with α- and β-tubulin and was clearly a member of the tubulin superfamily of proteins. Oakley and Oakley [1989] designated this protein γ-*tubulin*.

III. γ-TUBULIN IS WIDESPREAD IN EUKARYOTES

The existence of a new member of the tubulin superfamily was unexpected and raised a number of questions. One of the most obvious was whether γ-tubulin was ubiquitous or merely an oddity, albeit an interesting one, found only in *Aspergillus* and related fungi.

Since α- and β-tubulin are highly conserved proteins, it is generally possible to find conditions under which tubulin genes of one species hybridize specifically to tubulin genes of another species, even if the species are phylogenetically distant [e.g., Neff et al., 1983]. Guessing that γ-tubulin might be highly conserved as well, Zheng et al. [1991b] were able to clone a γ-tubulin cDNA from *Drosophila melanogaster* by low stringency hybridization using an *A. nidulans* γ-tubulin cDNA as a probe. Using the *D. melanogaster* cDNA as a probe, they were then able to clone a human γ-tubulin cDNA. Horio et al. [1991] were also able to use low stringency hybridization to clone the γ-tubulin gene of the fission yeast *Schizosaccharomyces pombe,* and Joshi et al. [1992] were able to use this approach to clone a portion of a mouse γ-tubulin gene.

When sequences of the predicted products of the genes were compared (Fig. 1), it was possible to identify regions that are highly conserved among γ-tubulins. This made it possible to design degenerate primers to use polymerase chain reaction procedures to clone γ-tubulin cDNAs or genes. Stearns et al. [1991] used this approach to clone γ-tubulin genes from the clawed frog *Xenopus laevis* and from *Sch. pombe*. In addition, portions of γ-tubulin genes have been cloned from rat, maize, diatom, the budding yeast *Yarrowia lipolytica* (unpublished data from T. Stearns, M. Kirschner, and Z. Cande, cited in Stearns et al., 1991), the thermophilic red alga *Cyanidium caldarum* (R.G. Burns, Y. Akkari, C.E. Oakley, and B.R. Oakley, unpublished data), and the parasitic protozoan *Leishmania mexicana* (C.E. Oakley, B.R. Oakley, and D. Fong, unpublished data). Taken together, these data demonstrate that γ-tubulin is extremely widespread among eukaryotes and may be ubiquitous. Indeed, the only organism in which extensive effort has not led to identification of a γ-tubulin gene is the budding yeast *Saccharomyces cerevisiae* [Stearns et al., 1991].

IV. A COMPARISON OF γ-TUBULIN SEQUENCES

Although the number of γ-tubulin sequences that have been published and/or are present in databases is relatively small, a comparison of the available full-length sequences (Fig. 1) allows us to draw several conclusions. First, γ-tubulins are highly conserved. When sequences from extremely divergent organisms are compared, they share at least 66% aa identity. The γ-tubulins of humans and the clawed frog *X. laevis* share even greater identity, a remarkable 98%. Since *Homo sapiens* and *X. laevis* are quite phylogenetically distant vertebrates, this result indicates that γ-tubulins may be extraordinarily conserved in vertebrates.

Second, γ-tubulins are similar to α- and β-tubulins. Typically γ-tubulins share approximately 28%–35% aa identity with α- and β-tubulins, and when conserved substitutions are taken into consideration they share more than 60% similarity [Oakley and Oakley, 1989]. The subject of the similarities of α-, β-, and γ-tubulins and the functional significance of the conserved regions of the molecules has been examined recently [Burns, 1991], and it is beyond the scope of this chapter to reiterate the details of this analysis. Two major points are worth mentioning, however. First, α-, β-, and γ-tubulins are likely to have similar secondary and tertiary structures. Second, some of the regions that are highly conserved among α, β-, and γ-tubulins may be involved in the binding of GTP.

```
H.s. MPREIITLQLGQCGNQIGFEFWKQLCAEHGISPEAIVEEFATEGTDRKDVFFYQADDEHY  60
X.l. ----------------------------------G-------------------------  60
D.m1 --S--------------------R--L------SGVL-D--ND-L------------D--  60
D.m2 --S--------------------R--L------DGVL-D---D-Q------------N--  60
S.p. -G-------A--------SQ--Q---L----G-DGTL-S-----V---------S--TR-  60
A.n. -------I-A-----NV-SQ--Q---L-----QDGNL-------G---------S--TR-  60

H.s. IPRAVLLDLEPRVIHSILNSPYAKLYNPENIYLSEHGGGAGNNWASGFSQGEKIHEDIFD 120
X.l. -----------------------N------------------------------------ 120
D.m1 --------------NT-MG-V-S-------V---K------------Y-----LQ-EV-- 120
D.m2 ------I-------NN-MT---S----Q--VF--K------------------VQ-EV-- 120
S.p. ----I-I------VNN--SDT-GS-------LITKN---------N-Y-HA-R-F---M- 120
A.n. ----I--------LNG-QSG--KNI-----FFIGQQ-I------GA-YAA--VVQ-EV-- 120

H.s. IIDREADGSDSLEGFVLCHSIAGGTGSGLGSYLLERLNDRYPKKLVQTYSVFPNQDEMSD 180
X.l. -----------------------------------------------------------H 180
D.m1 ---------------I------------M--FIM---A-------I--F--------I-- 180
D.m2 -L--------------------------M---V----SE-F----I-----------I-- 180
S.p. M--------------S-L-------------F------------II--------SQSV-- 180
A.n. M--------------MFL-------------F----M---F----I-------DTQA.A- 179

H.s. VVVQPYNSLLTLKRLTQNADCVVVLDNTALNRIATDRLHIQNPSFSQINQLVSTIMSAST 240
X.l. ------------------------------------------------------------ 240
D.m1 --------M-------TA--S-------------C--------------N-------V-- 240
D.m2 --------I-------KC--S--------------E-----T-T-T---N-------L-- 240
S.p. ----------A-----L---S------A--AH--A----T---T-H-Q------V----- 240
A.n. ---N------AMR-------S------A--S--VA----V-E---Q-T-R----V----- 239

H.s. TTLRYPGYMNNDLIGLIASLIPTPRLHFLMTGYTPLTTDQSVAS.....VRKTTVLDVMR 295
X.l. --------------------------------------------.....----------- 295
D.m1 ------S----N----T-P-----Q------------S-SDIHTQQLVN----------- 300
D.m2 ------S----N----T-------Q-----------MS-C.E-.AKTS.----------- 296
S.p. -------------VSI------S--C---L-S---F-NQ-.-EEAKA..I---------- 297
A.n. ---------H---V-I---------S---L-S---F-G-N.IDQAKT..----------- 296

H.s. RLLQPKNVMVSTGRDRQTNHCYIAILNIIQGEVDPTQVHKSLQRIRERKLANFIPWGPAS 355
X.l. ------------------------------------------------------------ 355
D.m1 -------M-----P-KSNH----S----------------------D--M-Q------T- 360
D.m2 -------M---ALT-K-SRQ-FVS-----------S------------NV---------- 356
S.p. ---L---Q---VNPSKKS..-F-S--D-----A--AD-----L-----RY-S-------- 355
A.n. -------R---INPSKSS..---S--------A---D-----L-----R--S-------- 354

H.s. IQVALSRKSPYLPSAHRVSGLMMANHTSISSLFERTCRQYDKLRKREAFLEQFRKEDMFK 415
X.l. ---------------------------N-----------------------------I-- 415
D.m1 -------S---VQ-N-------L------C-----ALM---NV---G---D---R--I-- 420
D.m2 -----P-S---VQ---K----------G-----K-ALA--------N---DN--R-S--Q 416
S.p. ------K----IKTN-------L------A---K--LD---R----N-----YK--AI-E 415
A.n. -----TK----IQNT-------L-----VAT--K-IVQ---R----N-----YK--AP-Q 414

H.s. DNFDEMDTSREIVQQLIDEYHAATRPDYISWGTQEQ 451
X.l. -----L-N--------------------------DK 451
D.m1 -DLN-L-N---T-DC-VQ--E----E--MQFSVKRGNGPVDSKSEDSRSVTSAGS 478
D.m2 -DLT-L-IA-DT-DC-VQ--E---QI--PQ-SPAVEASKAG 457
S.p. -DLN-F-S--DV-AD--N--E-CED-N-L-L 446
A.n. -GL--F-EA-AV-MD-VG--E--E-EN-LDPDAGKDEVGV 454
```

Fig. 1. *A comparison of γ-tubulin amino acid sequences. The single amino acid codes are given for γ-tubulins of* Homo sapiens *(H.s.) [Zheng et al., 1991b],* Xenopus laevis *(X.l.) [Stearns et al., 1991],* Drosophila melanogaster *(D.m1 is the zygotically expressed isotype [Zheng et al., 1991b], and D.m2 is the maternally expressed isotype [Zheng et al., 1991a],* Schizosaccharomyces pombe *(S.p.) [Stearns et al., 1991; Horio et al., 1991], and* Aspergillus nidulans *(A.n.) [Oakley and Oakley, 1989]. Dashes indicate identity with the* H. sapiens *γ-tubulin, and periods indicate the absence of a corresponding amino acid.*

While α- and β-tubulins have long been known to bind GTP, the binding sites of GTP on α- and β-tubulins remain the subject of some controversy. The model of the GTP binding site proposed by Mandelkow et al. [1985] postulates that at least three regions of α- and β-tubulin molecules are involved in the interaction with GTP. These regions are highly conserved in γ-tubulins. The regions are (referring to the numbers of the human γ-tubulin aa sequence in Fig. 1) PRA-L-DLEP (aa 62–71), which is postulated to be involved in the binding of the guanine moiety of GTP; GGTGSG (aa 143–148), which is postulated to be involved in the binding of phosphoryl groups; and VV-PYN (aa 182–187), which is postulated to be involved in the binding of the ribose moiety. If the Mandlekow et al. model for GTP binding is substantially correct, it is likely that γ-tubulin is a GTP-binding protein.

In addition to these regions there are some that are conserved in all tubulins for which there is as yet no postulated function [see Fig. 1 in Burns, 1991]. Finally, there are some regions that are shared among α- and γ-tubulins or β- and γ-tubulins but not among all tubulins [see Fig. 1 in Burns, 1991]. One of several examples is aa 99–104 (GAGNNW), which is remarkably conserved in β-tubulins and γ-tubulins but is not conserved in α-tubulins. This region may be involved in the binding of the antimicroubule agent rhizoxin to β-tubulin [Takahashi et al., 1990].

Some regions are highly conserved in γ-tubulins but are not shared with α- and β-tubulins. Examples of six or more consecutive aa's completely conserved without substantial conservation in the equivalent region of α- or β-tubulin are aa 46–54 (DRKDVFFY), aa 123–135 (DREAD-GSDSLEGF), aa 202–207 (VVVLDN), aa 285–298 (RKTTVLDVMRRLL), aa 348–353 (FIPWGP), and aa 355–360 (SIQVAL). Since there is virtually no information as to the functions of different regions of the γ-tubulin molecule, any hypotheses as to the functions of these regions are premature. It seems likely, however, that these regions must be involved in γ-tubulin–specific functions since they are not conserved in α- or β-tubulins.

Finally, the lack of conservation of some regions is informative. The region around aa 280–285, for example, shows a great deal of sequence diversity, and there are inserts of various lengths in this region. Since this region is variable in length, it is unlikely that it is important to the structure of the molecule and may form a loop that extends from the surface of the molecule. Since a great deal of variation is tolerated in this region the sequence may be relatively unimportant.

Figure 1 also shows that γ-tubulins exhibit a great deal of diversity in the regions near their C termini. All of the γ-tubulin sequences are relatively similar through aa 443, but beyond this point only the two vertebrate γ-tubulins show identity. In addition, the C termini of γ-tubulins differ significantly in length, with the *Sch. pombe* γ-tubulin ending at aa 446 and *D. melanogaster* γ-tubulin extending to aa 478. Interestingly α- and β-tubulin also exhibit a great deal of variation at their C termini [Little and Seehaus, 1988]. The C termini of α- and β-tubulin, however, generally contain many acid residues, indicating that the charge of the region may be important even if the precise sequence is not. The C termini of γ-tubulins are not particularly acidic, and their variation in length suggests that the region may not be of particular importance. In *Sch. pombe* this region is clearly unimportant, because the γ-tubulin gene has been truncated such that only aa 1–437 are expressed and cells carrying only the truncated gene are viable and exhibit no obvious alterations of growth or any processes related to microtubule function (T. Horio and M. Yanagida, personal communication).

V. γ-TUBULIN FAMILIES

There are families of α- and β-tubulin genes in most organisms [Raff, 1984; Sullivan, 1988]. Different genes are expressed at different levels and in different tissues, although the functional significance of the differential expression is open to debate. It was thus logical to expect that there would be families of γ-tubulin genes. To date, however, the evidence for γ-tubulin families is sparse. The experiments that led to the cloning of γ-tubulin genes in *A. nidulans, Sch. pombe, X. laevis,* and *H. sapiens* (HeLa) [Oakley and Oakley, 1989; Stearns et al., 1991; Horio et al., 1991; Zheng et al., 1991b] gave no indication that there was more than one γ-tubulin gene in each of these organisms. Recently, however, Zheng et al. [1991a] have discovered a second γ-tubulin gene in *D. melanogaster.*

The aa sequence of the second *D. melanogaster* γ-tubulin is shown in Figure 1. The two *D. melanogaster* γ-tubulins are relatively divergent, sharing approximately 82% aa identity. This is considerably less identity than is shared by *X. laevis* and human γ-tubulins and suggests that the two *D. melanogaster* γ-tubulins might have different functions. As yet there is no evidence that they do in fact, have different functions, but it is clear that they have different patterns of expression. One of the genes, γ^{zy} (the first *D. melanogaster* γ-tubulin discovered [D.m.1. in Fig. 1]), is expressed zygotically in both males and females. The second γ-tubulin gene, y^{mat} (D.m.2. in Fig. 1), is expressed maternally during oogenesis and is contributed to eggs [Wilson et al., 1992].

VI. γ-TUBULIN IS A LOW ABUNDANCE PROTEIN LOCATED AT MICROTUBULE ORGANIZING CENTERS

In *A. nidulans,* quantitation of the γ-tubulin mRNA by Northern blotting revealed that it was present at only about 5% of the level of the major β-tubulin mRNA [Oakley and Oakley, 1989]. This suggested that γ-tubulin might be a relatively minor protein, and indeed Stearns et al. [1991] have demonstrated that this is the case. In an *X. laevis* (XTC) cell line, γ-tubulin is about 0.005% of total cellular protein, whereas α- and β-tubulin together comprise about 2.5% of total protein. In *A. nidulans,* γ-tubulin is about 0.011% of total protein (C.E. Oakley, T. Horio, Y. Yoon, and B.R. Oakley, unpublished data). While this is somewhat higher than the levels in XTC cells, it is still a relatively minor protein and is much less abundant than α- or β-tubulin.

The intracellular location of γ-tubulin was first determined in *A. nidulans* [Oakley et al., 1990]. Immunofluorescence microscopy with γ-tubulin–specific antibodies showed little or no staining of microtubules. Rather, γ-tubulin was located at the spindle pole bodies, the fungal mictotubule organizing centers (MTOCs). Subsequently γ-tubulin was shown to be located at the centrosomes (metazoan MTOCs) (Fig. 2) of human [Zheng et al., 1991b], fruit fly [Zheng et al., 1991b], mouse [Zheng et al., 1991b; Joshi et al., 1992], clawed frog [Stearns et al., 1991], rat neuron [Baas and Joshi, 1992], and various other cell lines [Stearns et al., 1991]. Thus γ-tubulin is located at MTOCs of phylogenetically divergent organisms and might be a ubiquitous component of MTOCs. Furthermore, Stearns et al. [1991] have shown by electron microscopic immunolocalization that γ-tubulin is a component of the pericentriolar material, the amorphous material that is essential for microtubule assembly from the centrosome.

In nocodazole-treated cells in which microtubules have completely disassembled, γ-tubulin remains located at the centrosome [Stearns et al., 1991]. γ-Tubulin thus appears to be an intrinsic component of centrosomes and is not merely located at the centrosome because of an association with microtubules. Interestingly, in these experiments, small bodies of γ-tubulin sometimes dissociated from centrosomes. When nocodazole was removed, microtubules assembled from these γ-tubulin–containing bodies [Stearns et al., 1991]. γ-Tubulin is thus part of, or closely associated with, microtubule nucleating material. Finally, staining of the centrosomes by anti- γ-tubulin antibodies shows a cell cycle–specific variation in intensity [Zheng et al., 1991b; Stearns et al., 1991]. Staining intensity is highest in mitosis (in particular in prophase and prometaphase) when the microtubule nucleating capacity of the centrosome is highest and lowest in interphase when the centrosome nucleates the assembly of relatively few microtubules.

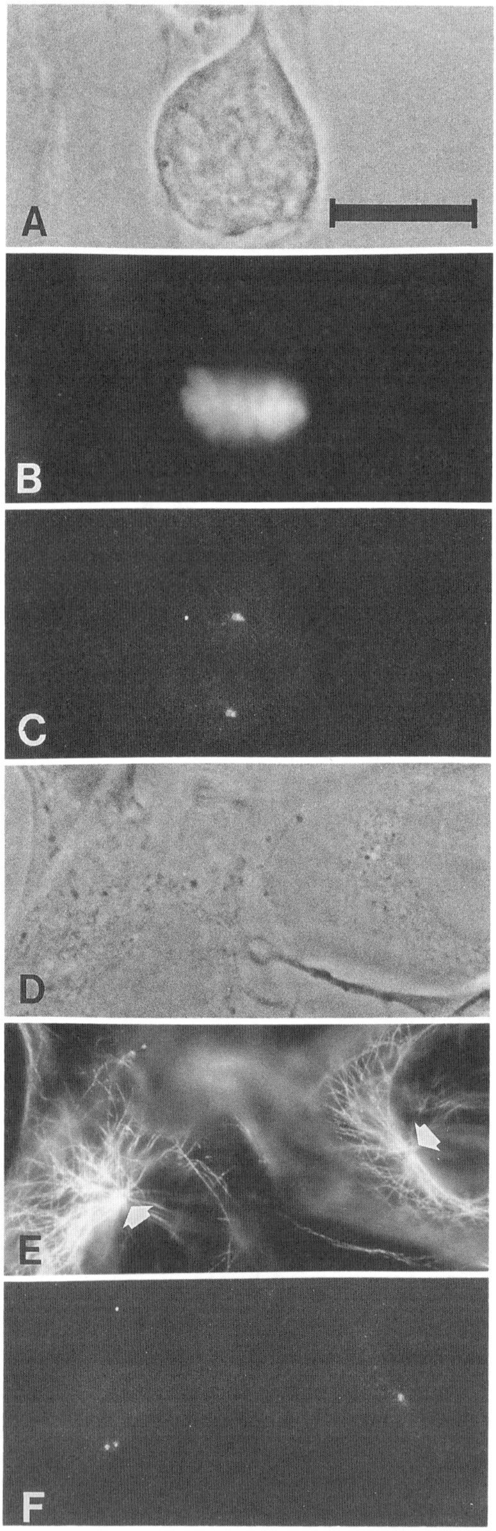

In the fission yeast *Sch. pombe,* γ-tubulin is found not only at the spindle pole body but also at a second type of MTOC that forms near the cytokinetic furrow after mitosis [Horio et al., 1991]. This result is important for two reasons. First it demonstrates that γ-tubulin is a component of MTOCs other than those that organize the mitotic spindle. This raises the possibility that γ-tubulin may be associated with many or all types of MTOCs. Second, since the MTOCs associated with the cytokinetic furrow in *Sch. pombe* arise *de novo* and there is no obvious γ-tubulin staining in the region before the MTOCs form, the γ-tubulin in these MTOCs must be transported to the MTOCs as they form. It may be transported from the spindle pole bodies or it may be dispersed in the cytoplasm before aggregating at the MTOCs as they form.

Though it is certain that much of the cell's γ-tubulin is located at MTOCs, the data available do not allow one to conclude that all of the γ-tubulin is located at MTOCs. In mammalian cells, Western blots indicate that the vast majority of the γ-tubulin is present in a cytoskeleton fraction that includes the centrosomes and many other structures [Joshi et al., 1992]. There appears to be little soluble γ-tubulin in these cells. The results from *Sch. pombe* [Horio et al., 1991], however, indicate that γ-tubulin can be transported through the cytoplasm, and in preliminary experiments in *A. nidulans* it appears that approximately 30% of the γ-tubulin in germlines is soluble and thus not MTOC

Fig. 2. *γ-Tubulin localization at the centrosomes of 3T6 (mouse fibroblast) cells.* **A–C:** *A metaphase cell: A, phase contrast micrograph; B, fluorescence micrograph showing chromosomes stained with the DNA-binding dye DAPI; and C, centrosomal staining with an anti-γ-tubulin antibody. The stained spot at the left of the upper centrosome is a precipitated complex due to preadsorption of the antibodies with bovine brain tubulin and does not reflect γ-tubulin location.* **D–F:** *Two interphase cells: D, phase contrast micrograph; E, microtubule stained with anti-β-tubulin antibody (arrows); and F, anti-γ-tubulin staining of centrosomes at the foci of the microtubule arrays. All magnifications are the same. Bar = 20 μm. [Reproduced from Zheng et al., 1991b, with permission of the publisher.]*

bound (C.E. Oakley, T. Horio, Y. Yoon, and B.R. Oakley, unpublished data).

If some γ-tubulin is not associated with MTOCs, it is reasonable to ask if some γ-tubulin might be a minor component of microtubules. While it is not possible to show that γ-tubulin is completely absent from microtubules, the few experiments that have been carried out suggest that microtubules contain very little, if any, γ-tubulin. Perhaps the most definitive experiments in this regard were carried out by Baas and Joshi [1992]. Using conditions that allowed the detection of very small amounts of γ-tubulin, Baas and Joshi found by immunoelectron microscopy and Western blotting that γ-tubulin was not detectable in axons and dendrites. Since axons and dendrites contain abundant microtubules, γ-tubulin must be a very minor component indeed.

VII. γ-TUBULIN IS ESSENTIAL FOR NUCLEAR DIVISION AND MICROTUBULE ASSEMBLY *IN VIVO*

Disruption of the γ-tubulin genes of *A. nidulans* and *Sch. pombe* have shown that γ-tubulin is essential for viability [Oakley et al., 1990; Stearns et al., 1991; Horio et al., 1991]. In *A. nidulans,* the gene disruption was carried out using the heterokaryon gene disruption procedure, developed by Osmani et al. [1988], that allows nuclei carrying recessive lethal gene disruptions to be maintained in heterokaryons. The phenotype of the disruption can then be determined using uninucleate spores (conidia) produced by the heterokaryon.

Disruption of the *A. nidulans* γ-tubulin gene results in a strong blockage of nuclear division and a weaker blockage of migration of nuclei through the germline [Oakley et al., 1990]. Examination of the microtubules in the disruptants by immunofluorescence microscopy revealed a virtually complete absence of mitotic spindles and a dramatic reduction in the lengths and numbers of cytoplasmic microtubules [Oakley et al., 1990]. Cytoplasmic microtubules were present in less than 5% of disruptant germlines, and when present they were often short and apparently disorganized. These results demonstrate that although γ-tubulin is apparently not a component of microtubules, it is essential for microtubule assembly *in vivo* in *A. nidulans.*

Interestingly, the disruption caused no significant increase in the mitotic index as judged by chromosomal condensation. This result is surprising in view of the fact that antimicrotubule agents and α- or β-tubulin mutations generally cause a substantial increase in mitotic index reflecting a blockage of mitosis [e.g., Oakley and Morris, 1981]. Apparently, although the γ-tubulin gene disruption strongly inhibits nuclear division, it does not cause a long blockage in the mitotic phase of the cell cycle. Rather, as disruptant nuclei enter mitosis, chromosomes condense but the spindle does not form and mitosis does not occur. Nuclei then enter G_1 after a normal mitotic period and proceed through the cell cycle. This interpretation would account for the fact that nuclei in the disruptants were much larger than normal and were obviously polyploid [Oakley et al., 1990]. While one must be careful not to overinterpret the data from a single experiment, one possibility is that γ-tubulin may be a part of the regulatory apparatus that controls the M to G_1 transition. If γ-tubulin were eliminated, as in the disruptant, the regulatory apparatus would no longer function and nuclei would transit readily from M to G_1 without having to complete mitosis successfully.

In *Sch. pombe,* the phenotype of a γ-tubulin disruption has been determined in plasmid loss experiments [Horio et al., 1991]. These experiments differ from the *A. nidulans* disruption experiments in an important way. In the *A. nidulans* procedure, there was apparently little carryover of γ-tubulin from the parental heterokaryon. γ-Tubulin was undetectable by immunofluorescence microscopy in 98% of germlings [Oakley et al., 1990]. The observed phenotype is thus the result of a complete or nearly complete absence of γ-tubulin. In the *Sch. pombe* plasmid loss experiments, there is a greater than normal amount of γ-tubulin at the beginning of the experiments because the functional γ-tubulin gene is on a multicopy plasmid vector. When selection pressure is removed the plasmid is lost, no more γ-tubulin is synthesized, and eventually γ-tubulin becomes growth limiting in the cell. The observed phenotype is caused by a deficit rather than an absence of γ-tubulin. As with *A. nidulans,* the plasmid loss experiments in *Sch. pombe* demonstrated that γ-tubulin is necessary for nuclear division. The phenotype of the arrest was somewhat different, however. In some arrested nuclei, normal mitotic spindles were absent but a tiny spot of microtubule staining was visible. In other nuclei, long mitotic spindles were present. In both cases chromosomes were condensed. A logical interpretation of these data was that in the cells lacking spindles γ-tubulin was essentially absent and in the cells with spindles γ-tubulin levels were subnormal such that mitotic spindles assembled but did not contain all of the microtubules necessary for the normal completion of mitosis [Horio et al., 1991]. Thus in both *A. nidulans* and *Sch. pombe* it appears that γ-tubulin is necessary for microtubule assembly.

There is also evidence that γ-tubulin is involved in microtubule assembly in mammalian cells. Joshi et al. [1992] carried out a number of experiments in which antibodies against γ-tubulin were injected into cultured mammalian cells. When the antibodies were injected into HeLa, mouse 3T3, and human pancreatic epithelial cells, they did not markedly affect the microtubule array. When the microtubules were caused to disassemble by cold or the antimicrotubule agent nocodazole, however, the antibodies prevented the reassembly of microtubules when cells were warmed or the nocodazole removed. When cells were injected with the antibodies in the G_2 phase of the cell cycle, cells entered mitosis normally but failed to form normal mitotic spindles. In many cases the spindles had reduced numbers of microtubules and in other cases there was a nearly complete absence of microtubules. When the antibodies were injected at metaphase, anaphase was blocked and mitotic microtubules diminished in number and length. Finally, the antibodies blocked the assembly of purified brain tubulin onto centrosomes *in vitro*.

One must be careful in interpreting such antibody injection experiments. It is clear that the antibodies inhibit microtubule nucleation, but they could do it in at least three ways. The simplest possibility is that γ-tubulin is essential for microtubule nucleation and the antibodies inhibit nucleation by binding to the nucleator. Alternatively, γ-tubulin could be near the actual nucleator so that antibodies binding to γ-tubulin could sterically inhibit the interaction of tubulin dimers with the real nucleator. Finally, the antibodies could bind to γ-tubulin and cause a disorganization or disturbance of the real nucleator such that nucleation is inhibited. While it is impossible to distinguish with certainty among these possibilities, the data from gene disruption experiments in *A. nidulans* and *Sch. pombe* strongly support the interpretation that γ-tubulin is directly involved in microtubule nucleation and the antibodies inhibit nucleation by binding to the nucleator.

Although γ-tubulin is important for microtubule assembly *in vivo,* it is apparently not required for microtubule assembly *in vitro*. Stearns et al. [1991] showed that when microtubules were assembled from frog egg extracts and purified by consecutive rounds of assembly and disassembly, γ-tubulin did not copurify with the microtubules and the purified microtubules lacked detectable γ-tubulin. γ-Tubulin is also undetectable in Western blots of microtubule protein purified from bovine brains by rounds of assembly and disassembly (Y. Yoon, T. Horio, and B.R. Oakley, unpublished).

VIII. A MODEL FOR γ-TUBULIN FUNCTION

The data I have summarized have suggested a working hypothesis for the functions of γ-tubulin *in vivo* (Fig. 3) [Oakley et al., 1991; Oakley, 1992]. γ-Tubulin is necessary for microtubule assembly *in vivo* [Oakley et al., 1991; Horio et al., 1991; Joshi et al., 1992] and probably for the assembly of microtubules from isolated centrosomes *in vitro* [Joshi et al., 1992], but it is not a requirement for the assembly of microtubule proteins into microtubules *in vitro*

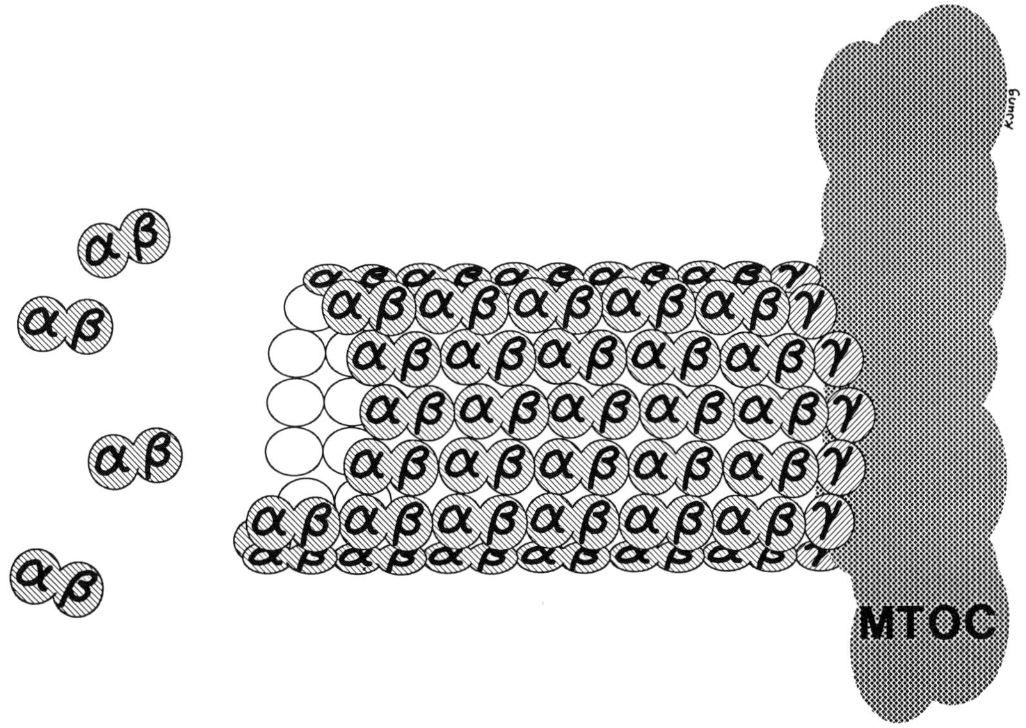

Fig. 3. *A model for γ-tubulin function. Circular aggregates of γ-tubulin at the microtubule-organizing center (MTOC) nucleate the assembly of microtubules. β-Tubulin binds to γ-tubulin. This establishes the orientation of tubulin dimers in protofilaments and thus defines microtubule polarity. [Courtesy of Dr. M. Katherine Jung. Reproduced with slight modifications from Oakley, 1992, with permission.]*

[Stearns et al., 1991]. It is located at MTOCs such as the centrosome, and, when γ-tubulin–containing material separates from the centrosomes after nocodazole treatment, this material has the ability to nucleate microtubule assembly [Stearns et al., 1991]. These data suggest strongly that γ-tubulin is important, probably essential, for the assembly of microtubules from MTOCs.

Although microtubule proteins have the ability to self-assemble into microtubules, assembly of tubulin dimers onto preexisting microtubules is strongly favored over formation of new microtubules [reviewed by Kirschner, 1978]. MTOCs have the ability to nucleate microtubule assembly *in vivo* and *in vitro* and in cells nearly all microtubules assemble onto MTOCs [reviewed by Brinkley, 1985]. Since γ-tubulin is quite similar to α- and β-tubulins, it is reasonable to hyopthesize that tubulin dimers bind to γ-tubulin. Aggregates of γ-tubulin at MTOCs, presumably rings or short cylinders, thus could serve as nucleation sites onto which tubulin dimers assemble. If γ-tubulin were eliminated by gene disruption [Oakley et al., 1990; Horio et al., 1991] or blocked with antibodies [Joshi et al., 1992] there would be fewer or no sites available for nucleation and microtubule assembly would be inhibited.

This hypothesis could also explain two other important functions of MTOCs. MTOCs apparently define the protofilament number of the microtubules that assemble from them. Brain microtubule proteins assemble into microtubules of various protofilament numbers in the absence of MTOCs but microtubules that assemble onto centrosomes *in vitro* have 13 protofilaments [Evans et al., 1985]. If γ-tubulin at the MTOC were in the form of 13mer rings, it could specify protofilament number.

MTOCs also define microtubule polarity. Microtubules are polar, and the ends of the microtubule have been designated plus and minus. In cells, the minus ends of microtubules are at MTOCs, or, more accurately, MTOCs nucleate microtubule assembly with the minus end proximal to and the plus end distal from the MTOC. If, as seems certain, γ-tubulin is firmly anchored to the MTOC and one subunit of the tubulin dimer bound specifically to γ-tubulin, the polarity of the microtubule would be established in a remarkably simple fashion. If, for example, the β-subunit of the tubulin dimer bound to γ-tubulin, the β-subunit of each dimer would be MTOC proximal and the α-subunit distal.

The genetic evidence available suggests that γ-tubulin may interact directly with β-tubulin. Weil et al. [1986] constructed a number of double mutant strains carrying mutant *mip*A (γ-tubulin) and heat-sensitive *ben*A (β-tubulin) alleles. Only three mutant *mip*A alleles exist, and each conferred the same phenotypes in the double mutants. The mutant *mip*A alleles were silent in a wild-type *ben*A background, but *mip*A/*ben*A double mutants had a variety of phenotypes each of which was different from the *ben*A mutation alone. Double mutants carrying mutant *mip*A alleles and *ben*A33 were much less heat sensitive than strains carrying *ben*A33 and the wild-type *mip*A allele, but they were partially cold sensitive. Double mutants carrying mutant *mip*A alleles and *ben*A11, *ben*A17, or *ben*A21 also showed greatly reduced heat sensitivity but were not cold sensitive. Double mutants carrying mutant *mip*A alleles and *ben*A30 or *ben*A31 were as heat sensitive as strains carrying *ben*A30 or *ben*A31 and the wild-type *mip*A allele but were partially cold sensitive. In addition, each of the *mip*A/*ben*A double mutants was slightly more sensitive to the antimicrotubule agent benomyl than strains carrying the mutant *ben*A allele in a wild-type *ben*A background. The simplest explanation for these allele-specific interactions between *mip*A and *ben*A mutations is that the γ-tubulin encoded by *mip*A and the β-tubulin encoded by *ben*A interact physically. Mutations in the *mip*A and *ben*A genes could thus alter the interactions and confer a variety of phenotypes. Subsequently we have constructed double mutant strains carrying mutant *mip*A alleles and all available mutant *tub*A (α-tubulin) alleles (M.A. Martin and B.R. Oakley, unpublished data). In no case have we found any evidence for interactions. The phenotype of the double mutants is indistinguishable from the phenotype conferred by the *tub*A allele in a wild-type *mip*A background. The simplest interpretation of these data is that γ-tubulin interacts physically with β-tubulin but not α-tubulin. This interpretation, in turn, suggests that the β-subunit of the tubulin dimer is MTOC proximal.

These genetic data raise two important questions. The first concerns the suppression of the heat sensitivity conferred by *ben*A33 by *tub*A and *mip*A mutations. Both *tub*A and *mip*A mutations suppress the heat sensitivity conferred by, and exhibit allele-specific interactions with, *ben*A33 [Oakley et al., 1987; Weil et al., 1986]. If, as postulated, allele specificity reflects physical interactions of the molecules encoded by these genes, these data suggest not only that both α- and γ-tubulins interact physically with β-tubulin but that they interact with the same region of the β-tubulin molecule. This might seem an unlikely possibility if α- and γ-tubulin are on opposite sides of β-tubulin, but, as Figure 3 shows, the working model we have postulated predicts that α- and γ-tubulin interact with the same region of β-tubulin.

The second question is more significant because it leads to the conclusion that γ-tubulin may influence the rate of tubulin dimer removal at MTOCs. Mutant *mip*A alleles suppress the heat sensitivity conferred by *ben*A11, *ben*A17, *ben*A21, and *ben*A33, and they reduce the benomyl resistance conferred by all heat-sensitive *ben*A alleles that have been tested. None of these *ben*A mutations prevents microtubule assembly [Oakley and Morris, 1981; M.K. Jung and B.R. Oakley, unpublished data], and *ben*A33 appears to increase microtubule stability [Oakley and Morris, 1981; Gambino et al., 1984]. The question that arises is that if γ-tubulin is simply a microtubule nucleator, how can it suppress the heat sensitivity conferred by these mutations and/or reduce their benomyl resistance? After all, γ-tubulin is apparently present only at the minus end of the microtubule but β-tubulin is present throughout the microtubule.

Two models suggests themselves. First, γ-tubulin mutations might alter the microtubule lattice such that the assembly and disassembly characteristics of the microtubule are altered. Second, γ-tubulin might be actively involved in the removal of tubulin dimers at the minus end of the microtubule. Formerly, the latter possibility would have seemed very unlikely, but Mitchison [1989] has demonstrated that tubulin subunits flux toward the pole in mammalian cells during metaphase and anaphase, and this can only occur if dimers are leaving the microtubule at the minus end.

If γ-tubulin is physically attached to β-tubulin at the MTOC, as we have hypothesized, how, then, can it be involved in, or even allow, subunit removal? A related question that may or may not concern γ-tubulin, but is important for our understanding of mitosis, is how can removal of subunits of at the minus ends of microtubules occur while the microtubules are anchored to MTOCs at or near their minus ends? The problem of the addition of subunits to or loss of subunits from attached ends of cytoskeletal elements is not unique to the minus ends of microtubules at centrosomes. Tubulin dimers are added to microtubules at kinetochores [Mitchison, 1989], and actin filaments that are linked to cell membranes incorporate actin monomers at their attached (barbed) ends [Tilney et al., 1981; Mooseker et al., 1982; Wang, 1985].

Ruhnau et al. [1989] and Gaertner and Wegner [1991] have proposed an intriguing model for the incorporation of actin monomers at attached ends that may be of value in explaining how tubulin dimers could be lost from the attached ends of microtubules at centrosomes. They have identified a protein, insertin [Ruhnau et al., 1989], that at concentrations substoichiometric to actin retards actin polymerization at the barbed end of microfilaments. Insertin has little effect on the critical concentration for assembly of microfilaments, however. An analysis of the kinetics of the reactions has led Wegner and colleagues to propose the "walking insertion" model for subunit addition. In this model insertin exists as a dimer at the barbed end of the microfilament. The two insertin molecules are staggered such that one (molecule A) is attached to one of the two strands of the double-helical microfilament (strand A) while the other (molecule B) is sufficiently removed from the end of the other strand (strand B) to allow the addition of an actin monomer to the strand. When an actin monomer is added, insertin molecule B binds to the end of strand B while insertin molecule A releases from strand A such that a new actin monomer can bind to the end of strand A. As this process is repeated cyclically, the microfilament lengthens while remaining attached to insertin molecules.

It is tempting to speculate that a similar process is involved in the loss of subunits at the minus ends of microtubules. Simple geometry dicates that subunits must be staggered at the ends of microtubules and thus that microtubule ends cannot be smooth. It is reasonable to hypothesize that a ring of γ-tubulin is attached to the minus end of a microtubule by one or a few protofilaments, allowing loss of subunits from other protofilaments. If attachment and release of γ-tubulin and protofilaments occurred in a cyclical fashion, subunits could be removed from all protofilaments of a microtubule while the microtubule remained attached to the ring of γ-tubulin. Thus the microtubule could disassemble while remaining firmly attached to the MTOC.

IX. DIRECTIONS FOR γ-TUBULIN RESEARCH

Since γ-tubulin has been discovered only recently, many fundamental questions about it remain unanswered. Since there are at least two expressed γ-tubulin genes in *D. melanogaster,* it is likely that there are two or more expressed γ-tubulin genes in higher plants and animals. One interesting focus of γ-tubulin research in the near future will be the determination of the numbers of expressed γ-tubulin genes in plants and animals, the determination of their patterns of expression, and the determination of whether sequence differences among γ-tubulins reflect the fact that they have different functions.

The hypothesis that γ-tubulin is a microtubule nucleator must be tested *in vitro,* and this means that functional γ-tubulin must be purified by some means. Purification of γ-tubulin would also allow one to determine some of the fundamental properties of the molecules such as whether it is a GTP-binding protein. It is likely that γ-tubulin purification will be a difficult task, since γ-tubulin is not an abundant protein and since it is likely to stick to MTOC proteins. If γ-tubulin really does influence the rate of tubulin dimer loss at the centrosome, its interaction with tubulin dimers must be more complex than simple nucleation of microtubule assembly and will require a great deal of study.

Since the ability of centrosomes to nucleate microtubule assembly changes in a cell cycle–specific fashion, if γ-tubulin is the centrosomal microtubule nucleator it must undergo changes during the cell cycle. The most obvious possibility is that it could be phosphorylated or otherwise posttranslationally modified, but it is possible that changes to the centrosome might occur such that accessibility of tubulin dimers to γ-tubulin might be altered. With regard to cell cycle regulation, the possible role of γ-tubulin in regulating the M to G_1 transition also needs to be investigated.

Finally, it is worth remembering that the tubulin dimer is the site of action for many agriculturally and medically important compounds, including cancer chemotherapy agents. Since γ-tubulin is an essential protein and is required for microtubule assembly *in vivo,* it is also a likely target for such compounds. Indeed, given the simlarities of γ-tubulin to α- and β-tubulin, it is possible that some of the antimicrotubule agents currently in use may bind to γ-tubulin as well as to α- or β-tubulin.

ACKNOWLEDGMENTS

I thank Liz Oakley for helpful criticisms of the manuscript and Tetsuya Horio, Kathy Jung, and Yisang Yoon for helpful discussions. Preparation of this manuscript was supported by grant GM31837 from the N.I.H.

X. REFERENCES

Baas PW, Joshi HC (1992): γ-Tubulin distribution in the neuron: Implications for the origins of neuritic microtubules. J Cell Biol 119:171–178.

Brinkley BR (1985): Microtubule organizing centers. Ann R Cell Biol 1:145–172.

Burns RG (1991): α-, β-, and γ-tubulins: Sequence comparisons and structural constraints. Cell Motil Cytoskel 20:181–189.

Evans L, Mitchison T, Kirschner M (1985): Influence of the centrosome on the structure of nucleated microtubules. J Cell Biol 100:1185–1191.

Gaertner A, Wegner A (1991): Mechanism of the insertion of actin monomers between the barbed ends of actin filaments and barbed end-bound insertin. J Muscle Res Cell Motil 12:27–36.

Gambino J, Bergen LG, Morris NR (1984): Effects of mitotic and tubulin mutations on microtubule architecture in actively growing protoplasts of *Aspergillus nidulans*. J Cell Biol 99:830–838.

Horio T, Uzawa S, Jung MK, Oakley BR, Tanaka K, Yanagida M (1991): The fission yeast γ-tubulin is essential for mitosis and is localized at two different microtubule organizing centers. J Cell Sci 99:693–700.

Joshi HC, Palacios MJ, McNamara L, Cleveland DW (1992): γ-Tubulin is a centrosomal protein required for cell cycle-dependent microtubule nucleation. Nature 356:80–83.

Kirschner MW (1978): Microtubule assembly and nucleation. Int Rev Cytol 54:1–71.

Krauhs E, Little M, Kempf T, Hofer-Warbinek R, Ade W, Ponstingl H (1981): Complete amino acid sequence of β-tubulin from porcine brain. Proc Natl Acad Sci USA 78:4156–4160.

Little M, Seehaus T (1988): Comparative analysis of tubulin sequences. Comp Biochem Physiol 90B: 655–670.

Mandelkow E-M, Kirschner K, Mandelkow E (1985): Domains, antigenicity, and substructure of tubulin. In De Brabander M, De Mey J (eds): Microtubules and Microtubule Inhibitors 1985. Amsterdam: Elsevier Science Publishers BV, pp 31–47.

Mitchison TJ (1989): Polewards microtubule flux in the mitotic spindle: Evidence from photoactivation of fluorescence. J Cell Biol 109:637–652.

Mooseker MS, Pollard TD, Wharton KA (1982): Nucleated polymerization of actin from the membrane-

associated ends of microvillar filaments in the intestinal brush border. J Cell Biol 95:223–233.
Neff NF, Thomas JH, Grisafi P, Botstein D (1983): Isolation of the β-tubulin gene from yeast and demonstration of its essential function in vivo. Cell 33:211–219.
Oakley (1992): γ-tubulin: The microtubule organizer? Trends Cell Biol 2:1–5.
Oakley BR, Oakley CE, Kniepkamp KS, Rinehart JE (1985): Isolation and characterization of cold-sensitive mutations at the *benA*, β-tubulin, locus of *Aspergillus nidulans*. Mol Gen Genet 201:56–64.
Oakley BR, Morris NR (1981): A β-tubulin mutation in *Aspergillus nidulans* that blocks microtubule function without blocking assembly. Cell 24:837–845.
Oakley BR, Oakley CE, Rinehart JE (1987): Conditionally lethal *tubA* α-tubulin mutations in *Aspergillus nidulans*. Mol Gen Genet 208:135–144.
Oakley BR, Oakley CE, Yoon Y, Jung MK (1990): γ-Tubulin is a component of the spindle-pole-body that is essential for microtubule function in *Aspergillus nidulans*. Cell 61:1289–1301.
Oakley CE, Oakley BR (1989): Identification of γ-tubulin, a new member of the tubulin superfamily encoded by *mip*A gene of *Aspergillus nidulans*. Nature 338: 662–664.
Osmani SA, Engle DB, Doonan JH, Morris NR (1988): Spindle formation and chromatin condensation in cells blocked at interphase by mutation of a negative cell cycle control gene. Cell 52:241–251.
Raff EC (1984): Genetics of microtubule systems. J Cell Biol 99:1–10.
Ruhnau K, Gaertner A, Wegner A (1989): Kinetic evidence for insertion of actin monomers between the barbed ends of actin filaments and barbed end-bound insertin, a protein purified from smooth muscle. J Mol Biol 210:141–148.
Stearns T, Evans L, Kirschner M (1991): γ-Tubulin is a highly conserved component of the centrosome. Cell 65:825–836.
Sullivan KF (1988): Structure and utilization of tubulin isotypes. Ann R Cell Biol 4:687–716.
Takahashi M, Matsumoto S, Iwasaki S, Yahara I (1990): Molecular basis for determining the sensitivity of eucaryotes to the antimitotic drug rhizoxin. Mol Gen Genet 222:169–175.
Tilney LG, Bonder EM, Derosier DJ (1981): Actin filaments elongate from their membrane-associated ends. J Cell Biol 90:485–494.
Wang Y-L (1985): Exchange of actin subunits at the leading edge of living fibroblasts: Possible role of treadmilling. J Cell Biol 101:597–603.
Weil CF, Oakley CE, Oakley BR (1986): Isolation of *mip* (microtubule-interacting protein) mutations of *Aspergillus nidulans*. Mol Cell Biol 6:2963–2968.
Wilson PG, Zheng Y, Oakley E, Oakley B, Fuller MT (1992): Differential expression and behavior of two γ-tubulin isoforms in *Drosophila*. Mol Biol Cell 3:346a.
Zheng Y, Oakley CE, Oakley BR (1991a): Identification of a second γ-tubulin gene in *Drosophila melanogaster*. J Cell Biol 115:382a.
Zheng Y, Jung MK, Oakley BR (1991b): γ-Tubulin is present in *Drosophila melanogaster* and *Homo sapiens* and is associated with the centrosome. Cell 65: 817–823.

ABOUT THE AUTHORS

BERL R. OAKLEY is Professor of Molecular Genetics at Ohio State University in Columbus, Ohio, where he teaches undergraduate and graduate courses in cell biology and molecular genetics. After receiving a B.S. magna cum laude from Duke University in North Carolina, he pursued his doctoral degree at Birkbeck College of the University of London, England, and earned a Ph.D. in 1974 for work on mitosis in microalgae. He carried out postdoctoral research at York University in Toronto, Canada, where he worked with I. Brent Heath on the structure of algal mitotic spindles, and at Rutgers Medical School (now the University of Medicine and Dentistry of New Jersey), where he worked with N. Ronald Morris on the genetics of mitosis in *Aspergillus*. Since 1982 his research has centered on the identification and characterization of proteins required for microtubule function. He is best known for the discovery of γ tubulin. His current research is centered on understanding the function of γ tubulin and other proteins involved in microtubule function. He is on the editorial board of *Cell Motility and the Cytoskeleton*.

Microtubules: 47–58

Regulation of Tubulin Synthesis

Don W. Cleveland and Nicholas G. Theodorakis

I. INTRODUCTION

Expression of α-and β-tubulin, the two major components of microtubules, is regulated in animal cells by events that operate on two levels. The first of these is the transcriptional activation of one or more members of the small multigene families (of up to six or seven functional genes in mammals) that encode either subunit. It is this first level that generates the diversity of tubulin gene products that, as reviewed by Raff (this volume) can in some cases confer special properties to the microtubules into which those isotypes assemble. Beyond transcriptional activation, a second level of gene control is one that sets the appropriate quantitative level of tubulin expression through what appears to be an autoregulatory pathway in which the stability of tubulin messenger RNA (mRNA) is governed by the concentration of its own translation product, the tubulin subunits themselves.

The initial observation, now nearly 14 years old, that suggested the existence of a feedback control system that established tubulin synthetic rates according to the concentration of unassembled tubulin subunits, emerged from analyzing tubulin synthetic rates in cells treated with drugs that interfere with normal microtubule assembly. Following colchicine-induced microtubule depolymerization (and a corresponding elevation in the unassembled subunit concentration), the only marked change in the pattern of newly synthesized proteins was a 5- to 10-fold lower level of synthesis of α- and β-tubulins [Ben Ze'ev et al., 1979]. A very different result emerged from treatment of cells with a second microtubule inhibitor, vinblastine, which causes precipitation of tubulin subunits into insoluble paracrystals: Tubulin synthesis was specifically elevated. This initial work was quickly confirmed using pulse labeling approaches and additional microtubule inhibitors to alter the level of unassembled tubulin subunits [Cleveland et al., 1981]. This effect is easily seen and an example of the specific alteration in tubulin synthetic rates is shown in Figure 1 following colchicine-induced elevation (Fig. 1B) or taxol-induced depletion (Fig. 1C) of the unassembled subunit concentration. In fact, all inhibitors that increase the concentration of unassembled subunits (e.g., colchicine and nocodazole) yield suppression of new synthesis whereas those that lower intercellular subunit levels (e.g., vinblastine and taxol) stimulate the level of new synthesis. This apparent autoregulation of synthesis has been documented in almost all animal cells [Cleveland et al., 1981], including those from species as divergent as mosquito and human. Furthermore, similar experiments in sea urchin embryos firmly support the view that autoregulation controls tubulin synthesis during early sea urchin development [Gong and Brandhorst, 1988].

Further evidence that tubulin synthesis is specified by the level of unassembled subunits came from a second, more direct approach that avoided the need for drugs (such as colchicine) whose absolute specificities cannot be known with certainty. By directly elevating tubulin content through microinjection of purified

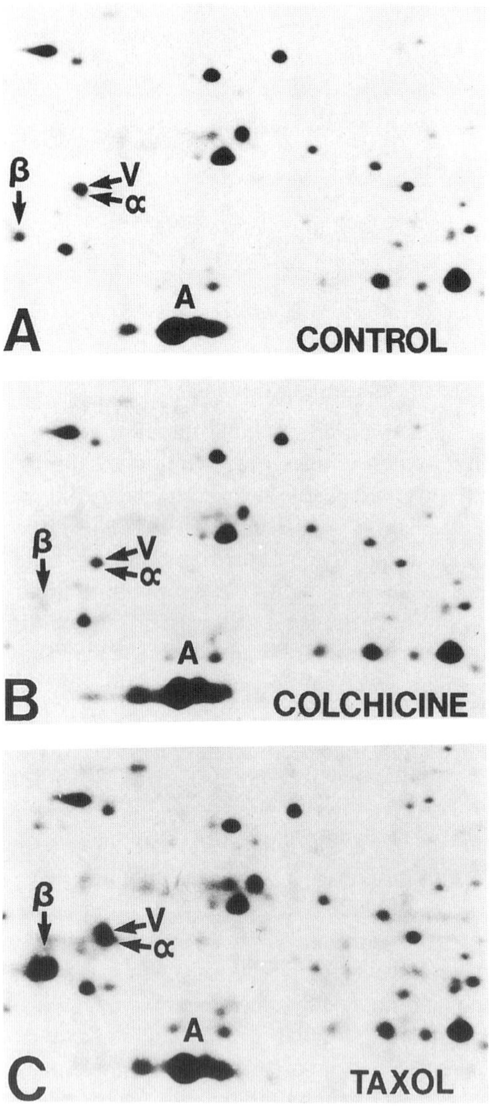

Fig. 1. *Specific alteration in synthesis of α- and β-tubulin polypeptides following alteration in the unpolymerized tubulin subunit concentration. Newly synthesized proteins in cultured hamster cells were pulsed labeled with [^{35}S]methionine and analyzed by two-dimensional gel electrophoresis.* **A:** *Protein synthesized in control cells with a normal level of unassembled tubulin subunits.* **B:** *Cells whose tubulin subunit concentration was elevated by colchicine-stimulated microtubule depolymerization.* **C:** *Cells with a taxol-induced depletion in unassembled tubulin concentration. The positions of α- and β-tubulin are indicated by the arrows; for reference, the positions of actin (A) and vimentin (V) are also indicated. (Data are unpublished observations from Pachter and Cleveland.)*

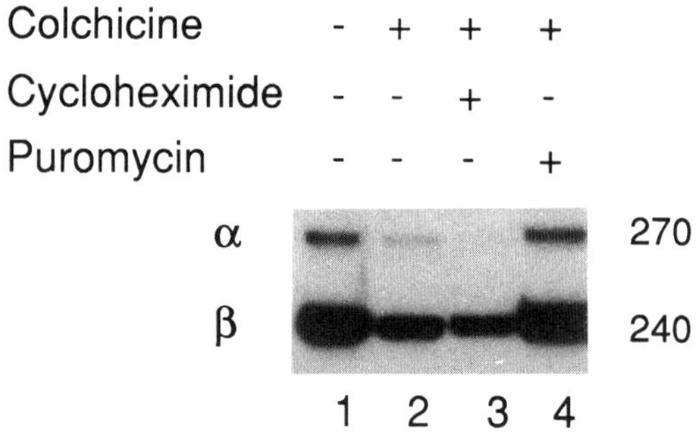

Fig. 2. *Increasing the unpolymerized tubulin subunit concentration mediates instability of tubulin mRNAs that are associated with ribosomes. Chinese hamster ovary (CHO) cells were treated for 4 hours with fresh medium (lane 1), medium with colchicine alone (lane 2), colchicine and low level cycloheximide (lane 3), or colchicine and puromycin (lane 4). α-Tubulin and β-tubulin mRNA were detected by S_1 nuclease protection of cytoplasmic RNA. (Data are unpublished observations from Bachurski, Theodorakis, Coulson, and Cleveland.)*

tubulin into living cells, it was found that a rapid and specific suppression of new tubulin synthesis was achieved after injecting enough tubulin to raise the original intracellular content by as little as 25%–50% [Cleveland et al., 1983].

II. TUBULIN SYNTHESIS IS REGULATED BY CHANGES IN THE STABILITY OF CYTOPLASMIC TUBULIN mRNAs

The changes in tubulin synthetic rates could result from changes in the efficiency of translation of tubulin mRNAs or from changes in the abundance of those mRNAs, thus leading to the measured changes in synthesis. Discrimination between those two possibilities emerged from measurement of tubulin mRNA levels [Cleveland et al., 1981]. This revealed that changes in the level of tubulin synthesis are not due to altered translation of tubulin mRNAs, but rather to specific changes in the abundance of tubulin mRNAs (see example in Fig. 2).

The autoregulated changes in tubulin mRNA levels seen after altering tubulin subunit levels could derive from three general pathways: transcriptional regulation, changes in RNA processing or transport, or regulation of cytoplasmic mRNA stability. Evidence that transcription does not play any role came from measuring the apparent rate of tubulin gene transcription using the nuclear "run-on" transcription assay. No changes in the transcription rates of α- and β-tubulin genes between nuclei isolated from colchicine-treated and untreated cells could be detected [Cleveland and Havercroft, 1983].

Even more convincing evidence for post-nuclear control of tubulin synthesis levels emerged from examination of selective changes in tubulin synthesis in enucleated cells [Caron et al., 1985; Pittenger and Cleveland, 1985]. Cytoplasts were prepared by fragmenting actin filaments with cytochalasin, followed by centrifugation to remove the nuclei under conditions in which most of the cytoplasm remained attached to the substratum. These studies demonstrated that the regulatory mechanism that controls tubulin mRNA levels must be cytoplasmic, since the cytoplasts were still able to respond to colchicine by decreasing the synthesis of β-tubulin. This finding not only ruled out a requirement for any nuclear event, but demonstrated that the decrease in β-tubulin mRNA levels in response to elevation of tubulin subunit levels must be due to altered stability of cytoplasmic tubulin mRNA.

A final piece of evidence that tubulin synthesis is controlled posttranscriptionally is that the abundance of RNAs encoded by hybrid gene constructs, which contain only a small portion of the β-tubulin coding region (the first exon) fused to the herpes simplex virus thymidine kinase gene, was still subject to changes in the level of tubulin subunits, even though these constructs contained no sequences from the tubulin promoter, introns, or downstream sequences [Gay et al., 1987].

III. THE MINIMAL REGULATORY SEQUENCE THAT CONFERS THE SELECTIVE INSTABILITY OF β-TUBULIN mRNA IS THE FIRST FOUR TRANSLATED CODONS

To address what sequences on a β-tubulin mRNA are responsible for its tubulin-dependent instability, hybrid genes bearing parts of a β-tubulin gene were transfected into cultured fibroblasts, and the levels of the resulting mRNAs were examined before and after altering the tubulin subunit levels with drugs that interfere with tubulin assembly. Sequences from β-tubulin mRNA that can confer onto a heterologous mRNA the ability to respond to changes in the level of tubulin subunits were initially localized to a small domain in the first exon that contained protein coding sequences; sequences in the 5′ untranslated region or other exons were not required [Gay et al., 1987; Yen et al., 1988a]. A series of constructs that fused progressively smaller regions of the β-tubulin coding sequence to a test gene (herpes simplex virus thymidine kinase) showed that the autoregulatory domain was contained within the first 13 translated nucleotides of β-tubulin mRNA, which encode the first four translated codons, met-arg-glu-ile (MREI) [Yen et al., 1988a]. That sequences within this 13 nucleotide domain were necessary for selective instability was further shown by deletion of either the second codon or the third and fourth codons of an otherwise authentic β-tubulin gene. This yielded an mRNA that was not subject to autoregulation. Combined with the earlier transfection results, these experiments demonstrated that this short, translated sequence was both necessary and sufficient for the autoregulated control of β-tubulin gene expression.

IV. DESTABILIZATION OF TUBULIN RNAs REQUIRES THEIR TRANSLATION

Concerning the issue of how cytoplasmic tubulin RNAs are degraded, several experiments have demonstrated that it is polyribosome bound RNAs that are selectively destabilized while RNAs not attached to ribosomes remain stable. The first of these experiments used drugs that block protein synthesis to show that inhibitors that remove RNAs from polysomes (e.g., pactamycin and puromycin) completely disrupt linkage of tubulin RNA content with the level of unassembled subunits [Pachter et al., 1987]. However, ongoing protein synthesis *per se* is not required since blockage of translation elongation with the inhibitor cycloheximide enhanced tubulin RNA instability in the presence of elevated subunit concentrations. An example of the dependency of RNA instability on translation is shown in Figure 2. In addition, by introducing premature translation termination codons into an authentic β-tubulin gene [Pachter et al., 1987] or into tubulin–thymidine kinase chimeras [Yen et al., 1988a], it was determined that RNAs transcribed from these genes were not autoregulated if translation terminated prior to codon 42. Thus, for an RNA to be a substrate for autoregulated instability, not only must it carry the 13 nucleotide coding sequence, but it must also be ribosome bound and its translation must proceed 3′ to codon 41.

V. DEGRADATION OF β-TUBULIN mRNA IS MEDIATED BY COTRANSLATIONAL BINDING OF A CELLULAR FACTOR TO THE β-TUBULIN NASCENT PEPTIDE

The preceding data were consistent with two possible models for the mechanism that establishes autoregulated instability of polyribosome-bound β-tubulin RNAs. In the first model, RNA degradation is facilitated by unpolymerized tubulin subunits binding (either directly or indirectly) to the 13 nucleotide recognition sequence present in β-tubulin mRNAs. That only ribosome-bound RNAs are substrates for degradation can be explained if translation is required to make the recognition domain accessible for binding. The second model arose from the realization that the sequence conferring regulating RNA instability actually encodes the first 4 N-terminal β-tubulin amino acids, suggesting that the true recognition event is a protein–protein interaction involving free subunits interacting (directly or indirectly) with the nascent tubulin polypeptide immediately after it emerges from the large ribosomal subunits.

Evidence supporting the polypeptide-recognition model first came from the observation that yeast, human, and chicken β-tubulin N-terminal sequences all specified autoregulation in mouse cells [Yen et al., 1988a]. Consistent with the polypeptide model, these genes encode the same N-terminal peptide, but the mRNA sequences have diverged. A more direct approach was to create a series of point mutations in the second and third codons of a human β-tubulin mRNA [Yen et al., 1988b]. Substitution of any of the other five arginine codons at the second codon resulted in mRNAs that were still substrates for autoregulated instability, whereas RNAs containing any of 11 mutated second codons that specified other amino acids were not. More recent work in which the arginine codon at the second position was altered to a codon for either of the two other basic amino acids lysine or histidine demonstrated that only arginine codons at the second position yielded mRNAs that were subject to autoregulated instability (C.J. Bachurski and D.W. Cleveland, unpublished results). A similar mutational analysis at the third codon (which encodes glutamate in all known β-tubulin mRNAs) revealed that only mutants that retain codons for glutamate (or the other acidic amino acid, aspartate) remain selectively regulated by tubulin subunit levels.

Other evidence that the regulatory sequence is the nascent peptide sequence rather than the mRNA sequence came from shifting the N-terminal peptide sequence out of its normal translated frame while leaving its position in the RNA nearly unchanged [Yen et al., 1988b]. A mutation that shifted the reading frame of the autoregulatory sequence after the first codon resulted in an mRNA that could not be regulated, although a similar mutation that shifted the reading frame after the fifth codon (thus preserving the autoregulatory sequence MREI) could still be autoregulated as long as the encoded peptide was greater than 40 amino acids long. These data, together with the observation that the mRNA must be translated for more than 40 codons (presumably to allow the nascent polypeptide to emerge from the ribosome), led to a model that an increase in the level of tubulin subunits allowed the cotranslational binding of a cellular factor, perhaps tubulin itself, to the nascent β-tubulin polypeptide. Presumably, this binding event alters the

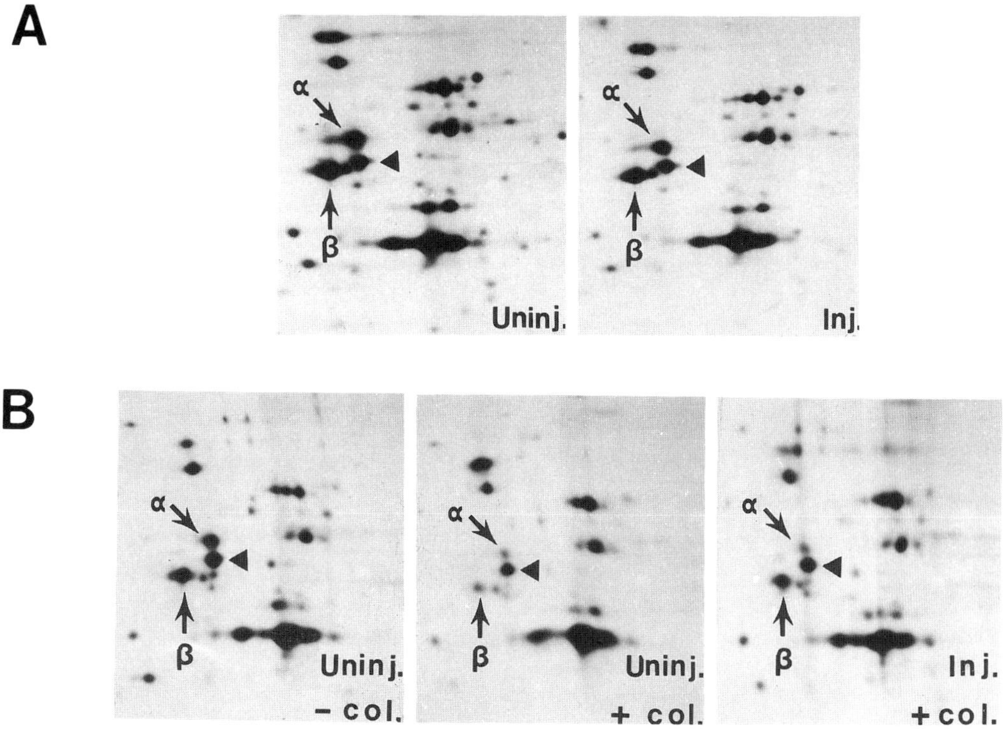

Fig. 3. *Microinjection of antibodies to the N terminus of β-tubulin disrupt the autoregulation of β-tubulin synthesis.* **A:** *CHO cells growing on glass chips were injected with a monoclonal antibody (18D6) that binds near the N terminus of β-tubulin* **(right)** *or left uninjected* **(left)** *and allowed to recover for 4 hours. The cells were then metabolically labeled with [^{35}S]-methionine, and labeled proteins were analyzed by two-dimensional gel electrophoresis and fluorography. The positions of α- and β-tubulin are indicated by the downward and upward-pointing arrows. Another spot that is not a tubulin but that migrates near the tubulins (probably vimentin) is indicated by the arrowhead.* **B:** *Cells that were untreated* **(left)**, *treated with colchicine for 4 hours* **(middle)**, *or microinjected with antibody and treated with colchicine for 4 hours* **(right)** *were analyzed as described above. [Reproduced from Theodorakis and Cleveland, 1992, with permission of the publisher.]*

polysomes to allow the binding or activation of a nuclease that can degrade tubulin mRNA.

Direct physical support for a required interaction of a cellular factor with the nascent β-tubulin polypeptide came from microinjecting into cultured cells monoclonal antibodies that bind to the N-terminal β-tubulin domain [Theodorakis and Cleveland, 1992]. Although injection of the antibody did not affect microtubule assembly, antibody-injected cells still suppressed the synthesis of α-tubulin in response to elevation of tubulin subunit levels after colchicine-induced microtubule disassembly. However, the presence of the antibody completely prevented measurable changes in β-tubulin synthesis (Fig. 3). Furthermore, since α-tubulin autoregulation was *not* affected, stabilization of β-tubulin mRNA could not be explained by an antibody-induced depletion of the pool of unassembled tubulin subunits. Rather, protection of β-tubulin mRNAs from regulated degradation must be due to a specific interaction of the antibody with the nascent β-tubulin polypeptide. Presumably, the binding of the antibody prevents the cotranslational binding of some cellular factor(s) to the nascent polypeptide, thereby disrupting the normal autoregulatory pathway. When combined with the

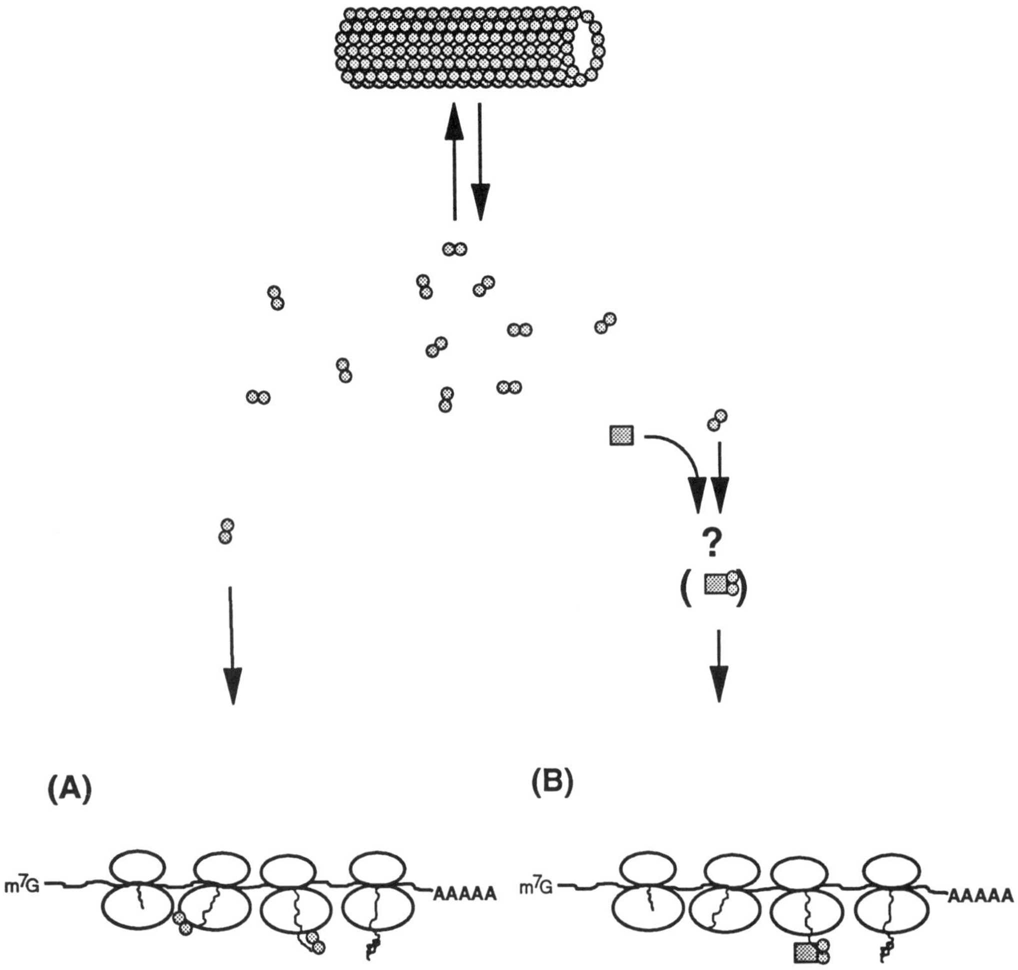

Fig. 4. *A model for autoregulated β-tubulin mRNA instability.* **A:** *Unassembled tubulin subunits (small circles) in the context of the ribosome (large ovals) bind to the nascent β-tubulin polypeptide as it emerges from the ribosome.* **B:** *Alternatively, another cellular factor (such as a molecular chaperone; small square) assist the binding of tubulin to the β-tubulin nascent peptide. This complex activates a nuclease that then degrades the polysome-bound mRNA.*

transfection experiments, the evidence is overwhelming that the regulatory target for β-tubulin synthesis is the recognition of the nascent polypeptide rather than of the mRNA itself (see Fig. 4).

Although the sequence that specifies autoregulation of tubulin mRNA stability in response to changes in the level of tubulin subunits is surprisingly short, two observations strengthen the claim that this domain can confer the required specificity. The first is that this regulatory sequence must lie at the N terminus. This was initially shown by translocation of a sequence encoding the 12 N-terminal amino acids of β-tubulin into the interior of a test gene; this resulted in an mRNA that was not responsive to changes in tubulin subunit levels [Yen et al., 1988b]. In fact, insertion of as few as two additional N-terminal codons (met-gly) abolishes the ability of an otherwise authentic mRNA to be autoregulated (Bachurski et al., unpublished data). Second, not only do all known β-tubulins except one begin with MREI, a search of current protein databases reveals that no other known eukaryotic protein has MREI at the N terminus.

VI. WHAT BINDS TO THE NASCENT β-TUBULIN PEPTIDE?

The autoregulatory model for controlling β-tubulin syntheis predicts that the cellular factor that interacts with the nascent β-tubulin polypeptide is the tubulin subunit itself. This idea is intellectually appealing for two reasons. First, this would represent a very simple way to couple the increase of tubulin subunits to the binding of the nascent peptide. Second, since protein–protein interactions between tubulin subunits are involved in tubulin heterodimer formation and microtubule assembly, it requires no stretch of imagination to suppose that similar interactions could also be used for cotranslational interactions between the tubulin subunit and the nascent peptide. Despite this, if there is a direct interaction between tubulin subunits and the nascent β-tubulin polypeptide, it must be of low affinity. Since the concentration of unassembled tubulin in cells with a normal complement of microtubules is about 10 μM [Hiller and Weber, 1978], if the affinity of tubulin for the nascent peptide were much greater than this (i.e., $K_d < 10$ μM), then all the nascent peptide would be saturated with tubulin subunits even before microtubules are depolymerized with colchicine.

Mindful of this,Theodorakis and Cleveland [1992] looked for such an interaction between tubulin and the N-terminal β-tubulin peptide by using two equilibrium methods. Using either gel filtration or equilibrium dialysis, no meaningful (i.e., physiologically relevant) binding could be detected for purified tubulin subunits and a synthetic peptide corresponding to the first 12 amino acids of β-tubulin. Therefore a simple, direct interaction between tubulin and the nascent β-tubulin polypeptide is probably not correct.

If not tubulin itself, what regulatory factor may bind to the nascent β-tubulin polypeptide? One possibility that comes to mind is that the tubulin–peptide interaction may require the presence of additional factors. Inasmuch as the degradation of β-tubulin mRNA occurs while the RNA is attached to ribosomes, one obvious possibility is that the productive complex is composed of tubulin, the nascent polypeptide, and the ribosome. (In this context, it is curious that ribosomes have been found in association with microtubules assembled from sea urchin oocytes [Suprenant et al., 1989].)

Alternatively, one might imagine the existence of another factor that monitors the level of free tubulin in the cell and that can bind to the nascent peptide, either alone or in a complex with tubulin. Although there is no direct experimental evidence that supports or refutes this notion, if such a factor exists it must be ubiquitous, since autoregulation of tubulin synthesis is found in all animals examined thus far [Cleveland et al., 1981; Lau et al., 1985]; furthermore, the factor must be abundant, since tubulin is an abundant translation product, and it must be able to bind to both tubulin subunits and to the nascent polypeptide. Possible candidates for suchs factors are molecular chaperones, proteins that can assist protein folding and/or assembly [Ellis, 1990] by binding to incompletely folded or assembled proteins. Two classes of ubiquitous and abundant chaperone proteins are the (hsp) heat shock protein 70 family of proteins and the chaperonin (hsp60, or groEL) class of proteins. The hsp70 family has recently been shown to be peptide binding [Flynn et al., 1989] that may interact with nascent polypeptides on polysomes [Beckmann et al., 1990]. Moreover, a cell line selected for resistance to microtubule inhibitors has alterations in at least one of these proteins, suggesting that it may interact with tubulin [Ahmad et al., 1990]. Another provocative observation is that this cell line also has a mutation in another protein, one of the family of hsp60 or groEL chaperonin class of molecular chaperones [Ahmad et al., 1990; Gupta, 1990]. Two observations demonstrate that chaperonins are likely to facilitate the folding of tubulin subunits. The first is that, following translation *in vitro,* newly made tubulin rapidly associates with a large complex that has many characteristics of chaperonins [Yaffe et al.,1992]. The second is that a purified chaperonin complex (along with two other, unidentified factors) can convert tubulin from an unfolded to a properly folded state [Gao et al., 1993]. Nevertheless, since chaperones of the hsp70 and/or hsp60/groEL class are believed to interact with a wide variety of substrates, it is difficult to imagine how they would specifically target tubulin mRNAs for degradation. Rather, we think it is more likely that their involvement in the more general phenomena of folding and assembly of newly synthesized proteins may be required for productive interaction of mature tubulin subunits with the nascent polypeptide.

VII. A MODEL FOR COTRANSLATIONAL TUBULIN RNA DEGRADATION

The sum of the present evidence suggests the model diagrammed in Figure 4 for the autoregulated degradation of β-tubulin mRNA mediated by the concentration of unassembled subunits. Free tubulin subunits (either alone or in association with another protein, such as a chaperonin) bind to the ribosome in such a way as to detect the presence of emerging polypeptides bearing the autoregulatory sequence. When this sequence is bound by the complex, a conformational change is triggered that alters the ribosomes translocation activity. This alteration either exposes or activates a latent nuclease that is present on the ribosome, allowing the degradation of the polysome-bound mRNA. Evidence that the ribosome is integrally involved in this β-tubulin mRNA degradation (apart from its obvious role of synthesizing the polypeptide recognition sequence) comes from studies using translation inhibitors [Pachter et al., 1987; Gay et al., 1989]. Low concentrations of cycloheximide or anisomycin slow ribosome translocation and accelerate tubule mRNA degradation. However, when either a single drug at a higher concentration or a combination of drugs added at low concentrations is used to abolish greater than 99% of protein synthesis and freeze ribosomes on the mRNA, autoregulated degradation of tubulin mRNA is prevented [Gay et al., 1989]. Emetine, another translation elongation inhibitor that freezes ribosomes on mRNA, also disrupts the autoregulated degradation of tubulin mRNA. The simplest interpretation of these data is that translational elongation is required for the degradation of tubulin mRNA.

Another unsolved problem is the identity of the nuclease that degrades tubulin mRNA. We think it is likely that this nuclease either is associated with the ribosome or is an integral part of the ribosome. Support for this view comes from the observation that nucleases can be identified and purified in polysome fractions from the cell [Ross and Kobs, 1986]. Association of the nuclease with polysomes also makes practical sense, since it would facilitate the localization of the nuclease to its substrate, ribosome-bound mRNA.

The demonstration that the binding of some cellular component to the nascent β-tubulin polypeptide must occur for the regulated degradation of β-tubulin mRNA raises the question of how this binding event is transduced into degradation of the mRNA. Obviously, the binding of the nascent polypeptide *per se* cannot be a sufficient stimulus for degradation of the polysome-bound mRNA because injection of an antibody that binds the nascent polypeptide does not cause degradation of the mRNA, but rather inhibits its decay (see Fig. 3).

VIII. REGULATION OF α-TUBULIN SYNTHESIS

α-Tubulin synthesis, like β-tubulin synthesis, is apparently controlled by selective changes in mRNA stability mediated by changes in the level of unassembled tubulin subunits (see Fig. 1) [Ben Ze'ev et al., 1979; Cleveland et al., 1981]. It is thus tempting to speculate that the mechanisms that regulate the two are similar. Indeed, like β-tubulin mRNA, only polysomal α-tubulin mRNAs are preferentially lost when tubulin subunit concentrations are elevated by treatment with colchicine (Fig. 2). Additionally, linkage of α-tubulin RNA stability with tubulin subunit levels is prevented when polysomes are disrupted by protein synthesis inhibitors [Pachter et al., 1987; Gong and Brandhorst, 1988] (see Fig. 2).

Inspection of the N-terminal sequences of α-tubulin polypeptides reveals that all vertebrate α-tubulins begin with tetrapeptide MREC, while α-tubulins from some simpler eukaryotes begin with MREV or MREI. This is reminiscent of the MREI autoregulatory domain identified for β-tubulin and suggests a common regulatory pathway for α- and β-tubulin. However, mutations to produce all possible amino acids at the fourth position of β-tubulin reveal that neither the α-tubulin N-terminal MREC nor the remaining 17 amino acids are efficient substrates for autoregulation (Bachurski et al., unpublished data). While the minimal β-tubulin sequence necessary for autoregulation is the tetrapeptide MREI, the regulatory domain controlling α-tubulin mRNA decay cannot be the analogous N-terminal domain.

Moreover, despite the similarities to β-tubulin, the domains on α-tubulin that confer autoregulated instability have proven to be difficult to identify. Many hybrid gene constructs that encode portions (or the entirety) of α-tubulin mRNA fused to heterologous test RNAs have been examined by transfection, but none was regulated by changes in the tubulin subunit concentration, even though the wild-type mRNAs in the host cells were autoregulated as expected. For reasons yet to be explained, almost all were poorly translated (Coulson, Sisodia, Yen, Theodorakis, and Cleveland, unpublished observations). Constructs that contained the full-length α-tubulin protein sequences with most of the α-tubulin 5′- and 3′-untranslated regions (but under the control of a heterologous promoter) were translated as efficiently as the endogenous α-tubulin mRNAs, but were not regulated by changes in the level of tubulin subunits (Bachurski et al., in preparation).

Finally, transfection of a complete wild-type rat α-tubulin gene into hamster cells yielded mRNAs that are identical to the authentic rat α-tubulin mRNA. Despite their efficient translation, even these wild-type mRNAs were not regulated by tubulin levels (Bachurski et al., in preparation). Why selective α-tubulin mRNA decay is not reproduced on RNAs encoded by transfected DNAs is perplexing; possibly there are species-specific, or isotype-specific regulatory domains in α-tubulin.

IX. WHAT *IN VIVO* ROLE DOES AUTOREGULATION SERVE?

An as yet unanswered question is the functional end that autoregulation serves. At first glance, autoregulation may appear to be largely redundant, since the polymerization reaction *in vitro* is self-buffering: Subunits continue to assemble until the level of free subunits is depleted to the critical concentration, the concentration at which subunits are at steady state with polymer [Oosawa and Asakura, 1975]. However, with the discovery of dynamic microtubules [Mitchison and Kirschner, 1984] that switch between growing and shrinking phases, Mitchison and Kirschner [1987] have demonstrated that the concept of a unique critical concentration for microtubule assembly in a closed system (i.e., in a cell with a fixed amount of tubulin and whose microtubules are nucleated from specific centrosomal sites) is not meaningful. Rather, the concentration of unassembled tubulin subunits at steady state is a variable determined by the number of microtubule ends and the total concentration of tubulin [Mitchison and Kirschner, 1987]. In this light, if a cellular mechanism existed to fix the tubulin subunit concentration (independent of assembly) and to regulate the number of microtubules (e.g., by controlling the nucleation capacity of centrosomes), then both the total tubulin concentration and the stability of individual microtubules would be uniquely determined. Thus autoregulation of tubulin synthesis may represent a portion of the cellular mechanism that ensures a nucleated array of dynamic microtubules.

X. OTHER MECHANISMS REGULATING TUBULIN EXPRESSION

Beyond transcriptional activation and RNA destabilization, protein stability represents a third level at which tubulin accumulation is controlled. A relevant example comes from studies using *Drosophila* tubulin mutants. In a *Drosophila* mutant that produces an unstable testis-specific β-tubulin subunit, the excess (wild-type) α-tubulin is degraded [Kemphues et al.,1982]. Similarly, in mammalian cells the expression of an unstable β-tubulin also leads to degradation of the excess α-tubulin [Boggs and Cabral, 1987]. Additionally, overexpression of normal β-tubulin leads to rapid degradation of the excess β-tubulin [Sisodia et al., 1990; Coulson and Cleveland, unpublished results]. Excess β-tubulin in mammalian cells is thus likely to be unstable [see also Sawada and Cabral, 1989]. However, despite many attempts, no cell lines have been isolated that constitutively overexpress α-tubulin to a great degree, perhaps indicating that α-tubulin overexpression is lethal to mammalian cells (Sisodia, Coulson, and Cleveland, unpublished results).

In budding yeast, however, overexpression of β-tubulin relative to α-tubulin is lethal, whereas overexpression of α-tubulin is not, instead resulting in a rapid loss of the excess accumulated α-tubulin protein [Burke et al., 1989; Katz et al., 1990; Weinstein and Solomon, 1990]. In yeast therefore it would seem that the cell need only control the level of β-tubulin, since excess α-tubulin does not accumulate. However, the relevance of these observations to animal cell tubulins is uncertain, given the much lower concentration of tubulin in budding yeast. Indeed, there is no evidence in budding yeast for the autoregulated mRNA destabilization mechanism described here. This is not surprising, since the properties of yeast tubulin are very different from animal tubulins; most of the tubulin in yeast does not appear to be cytoplasmic, but instead is used to form an intranuclear mitotic spindle. Therefore it is not surprising that budding yeast have not been observed to retain a *cytoplasmic* mechanism for regulating the level of tubulin. Together with the results obtained from *Drosophila*, these results indicate that, beyond the RNA degradation mechanism discussed here, there is also a posttranslational mechanism for balancing α- and β-tubulin levels.

XI. REFERENCES

Ahmad S, Ahuja R, Venner R, Gupta RS (1990): Identification of a protein altered in mutants resistant to microtubule inhibitors as a member of the major heat shock protein (hsp70) family. Mol Cell Biol 10: 5160–5165.

Beckmann RP, Mizzens LE, Welch WJ (1990): Interaction of hsp70 with newly synthesized proteins: Implications for protein folding and assembly. Science 246:850–854.

Ben Ze'ev A, Farmer SR, Penman S (1979): Mechanisms of regulating tubulin synthesis in cultured mammalian cells. Cell 17:319–325.

Boggs B, Cabral F (1987): Mutations affecting assembly and stability of tubulin: Evidence for a nonessential β-tubulin in CHO cells. Mol Cell Biol 7:2700–2702.

Burke D, Gasdaska P, Hartwell L (1989): Dominant effects of tubulin overexpression in *Saccharomyces cerevisiae*. Mol Cell Biol 9:1049–1059.

Caron JM, Jones AL, Ball LB, Kirschner MW (1985): Autoregulation of tubulin synthesis in enucleated cells. Nature 317:648–651.

Cleveland DW, Havercroft JC (1983): Is apparent autoregulatory control of tubulin synthesis nontranscriptionally controlled? J Cell Biol 97:919–924.

Cleveland DW, Lopata MA, Sherline P, Kirschner MW (1981): Unpolymerized tubulin modulates the level of tubulin mRNAs. Cell 25:537–546.

Cleveland DW, Pittenger MF, Feramisco JR (1983): Elevation of tubulin levels by microinjection suppresses new tubulin synthesis. Nature 305:738–740.

Ellis RJ (1990): Molecular chaperones: The plant connection. Science 250:954–959.

Flynn GC, Chappell TG, Rothman JE (1989): Peptide binding and release by proteins implicated as catalysts of protein assembly. Science 245:385–390.

Gao Y, Vainberg IE, Chow RL, Cowan NJ (1993): Two cofactors and cytoplasmic chaperonin are required for the folding of α and β tubulin. Mol Cell Biol 13:2478–2485.

Gay DA, Yen TJ, Lau JTY, Cleveland DW (1987): Sequences that confer β-tubulin autoregulation through modulated mRNA stability reside within exon 1 of a β-tubulin mRNA. Cell 50:671–679.

Gay DA, Sisodia SS, Cleveland DW (1989): Autoregulatory control of β-tubulin mRNA stability is linked to translational elongation. Proc Natl Acad Sci USA 86:5763–5767.

Gong Z, Brandhorst BP (1988): Stabilization of tubulin mRNA by inhibitors of protein synthesis in sea urchin embryos. Mol Cell Biol 8:3518–3525.

Gupta RS (1990): Mitochondria, molecular chaperone proteins, and the *in vivo* assembly of microtubules. Trends Biochem Sci 15:415–418.

Hiller G, Weber K (1978): Radioimmunoassay for tubulin: A quantitative comparison of the tubulin content of different established tissue culture cells and tissues. Cell 14:795–804.

Katz W, Weinstein B, Solomon F (1990): Regulation of tubulin levels and microtubule assembly in *Saccharomyces cerevisiae:* Consequences of altered tubulin gene copy number. Mol Cell Biol 10:5286–5294.

Kemphues KJ, Kaufman TC, Raff RA, Raff EC (1982): The testis-specific β-tubulin subunit of *Drosophila melanogaster* has multiple functions in spermatogenesis. Cell 31:655–670.

Lau Jt, Pittenger MF, Cleveland DW (1985): Reconstruction of appropriate tubulin and actin gene regulation after transient transfection of cloned β-tubulin and β-actin genes. Mol Cell Biol 5:1611–1620.

Mitchison T, Kirschner M (1984): Dynamic instability of microtubule growth. Nature 312:237–242.

Mitchison TJ, Kirschner MW (1987): Some thoughts on the partitioning of tubulin between monomer and polymer under conditions of dynamic instability. Cell Biophysics 11:35–55.

Oosawa R, Asakura S (1975): Thermodynamics of the Polymerization of Protein. New York: Academic Press.

Pachter JS, Yen TJ, Cleveland DW (1987): Autoregulation of tubulin expression is achieved through specific degradation of polysomal tubulin mRNAs. Cell 51: 283–292.

Pittenger MF, Cleveland DW (1985): Retention of autoregulatory control of tubulin synthesis in cytoplasts: Demonstration of a cytoplasmic mechanism that regulates the level of tubulin expression. J Cell Biol 101:1941–1952.

Ross J, Kobs G (1986): H4 histone mRNA decay in cell-free extracts initiates at or near the 3′ terminus and proceeds 3′ to 5′. J Mol Biol 188:579–593.

Sawada T, Cabral F (1989): Expression and function β-tubulin isotypes in Chinese hamster ovary cells. J Biol Chem 264:3103–3020.

Sisodia SS, Gay DA, Cleveland DW (1990): In vivo discrimination among β-tubulin isotypes: Selective degradation of a type IV β-tubulin isotype following overexpression in cultured animal cells. Nature New Biol 2:66–76.

Suprenant KA, Tempero LB, Hammer LE (1989): Association of ribosomes with *in vitro* assembled microtubules. Cell Motil Cytoskel 14:401–415.

Theodorakis NG, Cleveland DW (1992): Physical evidence for cotranslational regulation of β-tubulin mRNA degradation. Mol Cell Biol 12:791–799.

Weinstein B, Solomon F (1990): Phenotypic consequences of tubulin overproduction in *Saccharomyces cerevisiae:* Differences between alpha-tubulin and beta-tubulin. Mol Cell Biol 10:5295–5304.

Yaffe B, Farr GW, Miklos D, Horwich AL, Sternlicht ML, Sternlicht H (1992): TCP-1 complex is a molecular chaperone in tubulin biogenesis. Nature 358:245–248.

Yen TJ, Gay DA, Pachter JS, Cleveland DW (1988a): Autoregulated changes in stability of polyribosome-bound β-tubulin mRNAs are specified by the first thirteen translated nucleoides. Mol Cell Biol 8:1224–1235.

Yen TJ, Machlin PS, Cleveland DW (1988b): Autoregulated instability of β-tubulin mRNAs by recognition of the nascent amino terminus of β-tubulin. Nature 334: 580–585.

ABOUT THE AUTHORS

DON W. CLEVELAND is Professor of Biological Chemistry at the Johns Hopkins University School of Medicine in Baltimore, Maryland, where he oversees his research group focused on molecular biology and genetics of cytoskeletal proteins. After receiving a B.A. summa cum laude from New Mexico State University, he pursued doctoral research in the laboratory of Marc Kirschner at Princeton University, where he received a Ph.D. in 1977 for work on characterization of the microtubule associated protein *tau*. This was followed by a postdoctoral research period at the University of California at San Francisco, where he worked with Bill Rutter on defining the early molecular biological analysis of tubulin genes and their RNAs. Since 1981 Dr. Cleveland's research has been focused on defining the tubulin gene family, the polypeptides encoded by that family, and the underlying mechanisms controlling expression of individual members. This has led to the analysis of the post-transcriptional RNA control pathway described here. Other aspects of Dr. Cleveland's work involve defining the function of neurofilament proteins and identifying microtubule-associated components required for mitotic spindle assembly and chromosome transport. Dr. Cleveland has been an editor of *Molecular and Cellular Biology* and is now associate editor of the *Journal of Cell Biology.*

NICHOLAS G. THEODORAKIS is a postdoctoral fellow in Don Cleveland's group at the Johns Hopkins University School of Medicine. Dr. Theodorakis received his B.A. degree from Washington University in St. Louis, Missouri, and then pursued his doctoral training at Northwestern University in the laboratory of Rick Morimoto, where he earned a Ph.D. in 1989 for work on the heat shock response in vertebrate cells. In 1989 Dr. Theodorakis joined Dr. Cleveland's group, where he has been an active participant in deciphering the mechanism of autoregulated tubulin mRNA decay.

Microtubules: 59–83

Pharmacological Probes of Microtubule Function

Leslie Wilson and Mary Ann Jordan

I. INTRODUCTION

Studies on the actions of compounds that act on microtubules have played key roles in elucidating the functions and properties of microtubules in cells and *in vitro*. For example, during the 1970s and 1980s, colchicine, vinblastine, vincristine, nocodazole, and other such compounds were used extensively by investigators to determine whether specific cell processes involved the participation of labile (drug-sensitive) microtubules [reviewed by Soifer, 1975; Dustin, 1984]. Based on efforts by many investigators, a sophisticated understanding is now emerging of the mechanisms by which these compounds interact with tubulin and microtubules and perturb their functions. The early idea that these agents acted on microtubule-dependent cell functions only by destroying the microtubules or preventing their polymerization was simplistic. Microtubule functions in cells are highly complex. As described in Section 3 of this volume, microtubules not only serve as passive skeletal supports for the organization of the cytoplasm and for a variety of kinds of intracellular movement, but their remarkable polymerization dynamics are critical to many of their functions. For example, during mitosis, both dynamic spindle microtubules and molecular motor molecules work in concert to orchestrate a complex repertoire of chromosome movements as the spindle forms and functions. Drugs such as colchicine, the vinca alkaloids, and taxol interact with microtubule ends and surfaces and modulate the polymerization dynamics in ways that are far more interesting than previously imagined.

In the first two sections of this chapter, we describe the current state of knowledge regarding the binding of vinblastine, colchicine, and taxol to tubulin and microtubules and the effects of the drugs on the dynamics of tubulin addition and loss at microtubule ends. Then, with a focus on the action of these substances on mitosis, we describe how these agents are being used as tools for the elucidation of microtubule function in cells.

II. BINDING OF DRUGS TO TUBULIN AND MICROTUBULES: BINDING MECHANISMS

A. The Vinca Alkaloid–Binding Domain

1. Vinblastine binding to tubulin and to microtubules. Vinblastine (Fig. 1), a widely used antitumor drug obtained from the plant *Catharanthus roseus,* is the most extensively studied compound of a group of chemically related indole-dihydroindole compounds, including vincristine and vindesine, that inhibit mitosis by an action on spindle microtubules. The antimitotic activity of the vinca alkaloids was first described by Johnson et al. [1959] and Cutts et al. [1960] and has been described in detail recently by Jordan et al. [1991]. The antitumor activities of these agents have been described in recent reviews [Gerzon, 1980; Rowinsky and Donehower, 1991].

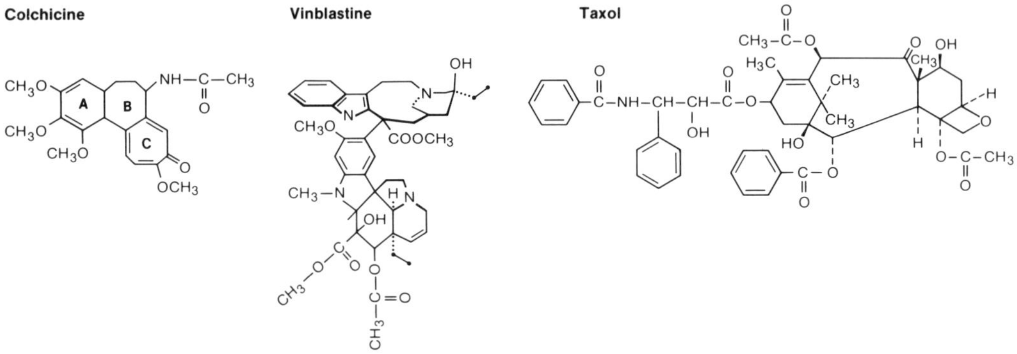

Fig. 1. *Structures of colchicine, vinblastine, and taxol.*

The binding of vinblastine to soluble and aggregated forms of tubulin including vinblastine–tubulin paracrystals has been studied extensively [reviewed by Wilson, 1975; Wilson et al., 1976; Luduena, 1979; Timasheff et al., 1991; Himes, 1991]. Vinblastine binding to tubulin occurs rapidly, and binding is rapidly reversible and relatively independent of temperature between 0° and 37°C. Confusion has existed regarding the number of vinblastine-binding sites per tubulin dimer and the affinity of the sites, because the binding affinity of vinblastine for one of the binding sites is strongly modulated by the state of tubulin self-association [Na and Timasheff, 1980a,b] and by solution conditions such as the Mg^{2+} concentration [Na and Timasheff, 1986]. There appear to be two vinblastine-binding sites per mole of tubulin dimer, and the intrinsic binding constants for the sites are weak, approximately $2 \times 10^4\ M^{-1}$ at 37°C. Vinca alkaloid binding to tubulin induces tubulin to self-associate into nonmicrotubule polymers and ordered aggregates through an isodesmic indefinite self-association pathway [Na and Timasheff 1980a, b]. Vinca alkaloid–induced self-association of tubulin in turn increases the affinity of one of the binding sites for the drug. For example, the apparent K_d for the binding of vinblastine to calf brain tubulin increased threefold as the tubulin concentration was raised from 16 to 78 μM [Na and Timasheff, 1986; reviewed by Timasheff et al., 1991; Himes, 1991].

The vinca alkaloid–induced self-association of tubulin can lead to formation of ordered hexagonal paracrystalline structures in vitro [Na and Timasheff, 1982], and the ability of vinca alkaloids to induce tubulin self-association most likely is responsible for the tubulin paracrystal formation in cells that occurs at high vinca alkaloid concentrations (e.g., > 10 μM vinblastine or vincristine) [see Fujiwara and Tilney, 1975]. Both the increased affinity of vinca alkaloids for self-associated tubulin and the increased binding affinity of tubulin for tubulin ligated to vinblastine probably play important roles in the perturbation of microtubule polymerization dynamics by the vinca alkaloids (discussed below).

Vinblastine binds to its tubulin-binding sites in intact microtubules with different affinities depending on whether the tubulin is located at the microtubule end (or ends) or is situated along the microtubule surface. Approximately 16–17 high-affinity vinblastine-binding sites per microtubule (K_d, 1–2 μM) are located at one or perhaps both ends of bovine brain microtubules [Wilson et al., 1982]. Between one and two low affinity vinblastine-binding sites per mole of tubulin dimer (K_d, 0.25–0.3 mM) are also located along the microtubule surface [Jordan et al., 1986; Singer et al., 1989]. The differences in affinities for vinblastine must be due either to conformational differences or to steric constraints imposed on the sites in tubulin due to the different locations of the tubulin dimers in the microtubule lattice. Beginning with concentrations higher than approximately 1–2 μM, vinblastine depolymerizes microtubules by causing splaying and peeling of protofilaments at both microtubule ends *in vitro* [Jordan et al., 1986; Himes, 1991]. The binding of the vinca alkaloids to the low affinity sites along the microtubule surface is probably responsible for the ability of relatively high vinca alkaloid concentrations to depolymerize microtubules *in vitro* and may play a role in tubulin paracrystal formation *in vivo*. The potent kinetic suppression of tubulin exchange that occurs at low vinblastine concentrations (<1 μM; to be described) is almost certainly due to vinblastine binding to the high affinity binding sites at the microtubule ends.

Unlike colchicine (described below), free vinblastine binds directly to microtubule ends without first forming a complex with soluble tubulin [Jordan et al., 1986]. Also, vinblastine does not become incorporated into the tubulin lattice of the microtubule, but incorporates strictly at the microtubule end or ends [Jordan and Wilson, 1990].

2. Interaction of other compounds with the vinblastine binding domain. Several natural products structurally unrelated to the vinca alkaloids have been identified that bind to tubulin at or close to the vinblastine site. The compounds maytansine and rhizoxin inhibit vinca alkaloid binding to tubulin competitively and are thought to bind to the vinblastine binding site [Bai et al., 1990b], whereas the peptide compounds dolastatin 10 and phomopsin A and a complex macrolide, halichondrin B, inhibit vinca alkaloid binding noncompetitively and are considered to bind at site(s) "near" the vinblastine-binding site [Bai et al., 1990b, 1991]. Interestingly, several of the drugs (maytansine, rhizoxin, dolastatin 10, and halichondrin B) inhibit cell proliferation with potencies from 10 to 100 times the potencies of vinblastine and vincristine [Bai et al., 1990a, 1991]. All of the compounds share the property with vinblastine of arresting cells in mitosis and inhibiting tubulin polymerization *in vitro* substoichiometrically (see below). However, the actions of these compounds on the polymerization dynamics of microtubules have not yet been described in detail. Binding of these structurally unrelated compounds to the vinca-binding domain of tubulin have similar effects on mitosis and microtubule polymerization, supporting the idea that the vinca-binding region may be a regulatory domain for controlling microtubule polymerization.

Some evidence has led to the suggestion that the vinca-binding domain may be adjacent to or overlap with the exchangeable GTP-binding site (E-site) in β-tubulin [Bai et al., 1990b]. Specifically, dolastatin 10, phomopsin A, maytansine, and halichondrin B inhibit E-site GTP exchange [Bai et al., 1990b, 1991; reviewed by Correia, 1991]. In addition, Luduena and co-workers [reviewed by Luduena and Roach, 1991] found that GTP shared the common ability with maytansine, rhizoxin, and vinblastine to induce formation of a specific intrachain sulfhydryl cross-link in β-tubulin with the bifunctional cross-linker *N,N′*-ethylene*bis*(iodoacetamide) (EBI), indicating that GTP and the drugs induce a similar conformational change in tubulin. However, inhibition of GTP exchange *per se* does not appear to be responsible for inhibition of microtubule polymerization because the relative inhibitory activities of the drugs on GTP exchange do not correlate with their relative abilities to inhibit microtubule polymerization [Bai et al., 1990a,b, 1991].

B. The Colchicine Binding Domain

1. Colchicine binding to tubulin and to microtubules. Colchicine (Fig. 1), isolated from the autumn crocus *Colchicum autumnale,* has played a fundamental role in elucidation of the properties and functions of tubulin and microtubules since it was first found to bind to "the subunits of microtubules" in cell extracts 25 years ago [Borisy and Taylor, 1967a,b; Wilson and Friedkin, 1967]. Indeed, tubulin was initially purified from bovine brain using bound [^{3}H]colchicine as a biochemical marker [Weisenberg et al., 1968].

The binding reaction between colchicine and tubulin is unusual. Beginning with the early work of Garland [1978] and Lambeir and Engelborghs [1981], studies on the kinetics of colchicine binding to tubulin have revealed that the binding reaction is a two-step process beginning with formation of a reversible, low-affinity pre-equilibrium complex. This step is followed by one or more slower steps in which conformational changes in tubulin lead to formation of a poorly reversible final-state tubulin–colchicine (TC) complex [reviewed by Hastie, 1991]. Conformational changes in tubulin upon binding to colchicine have been indicated by changes in the fluorescence of tubulin during TC complex formation [Garland, 1978; Lambeir and Engleborghs, 1981], by perturbation of the far ultraviolet-circular dichroism of tubulin [Andreu and Timasheff, 1982a], by an increase in the GTPase activity of tubulin on the binding of colchicine [David-Pfeuty et al., 1979; Andreu and Timasheff, 1981], by a decrease in the dissociation constant between α- and β-tubulin upon binding of colchicine [Detrich et al., 1982], and by the self-assembly of TC complexes into nonmicrotubule polymeric structures [Andreu et al., 1983].

Strong evidence exists in support of a two-subsite mechanism for the binding of colchicine to tubulin [Andreu and Timasheff, 1982a,b; Medrano et al., 1989; Andreu et al., 1991]. The data indicate that the trimethoxyphenol ring of colchicine (A-ring; Fig. 1) binds to one subsite and the methoxy tropolone moiety of colchicine (C-ring) binds to the second subsite. The B-ring does not appear to be involved in significant binding force between colchicine and tubulin. The initial binding of either the A- or C-ring to its subsite appears to induce a slow conformational change in the tubulin that brings the C- or A-ring binding subsite of tubulin into its proper position. While the nature of the slow conformational changes that occur during formation of the final state TC complex are unknown, it seems likely that the conformational changes are responsible for the powerful effects of TC complex on tubulin exchange at microtubule ends.

2. Interaction of other compounds with the colchicine-binding domain. Many other compounds have been isolated from natural sources or synthesized that bind to the colchicine site and inhibit microtubule polymerization substoichiometrically and inhibit mitosis. Among them, podophyllotoxin [Cornman and Cornman, 1951; Wilson, 1970; Wilson et al., 1976; Cortese et al., 1977], cornigerine [Hamel et al., 1988], combretastatins A-4 and A-2 [Lin et al., 1988, 1989], and MDL 27048 [Peyrot et al., 1989; Silence et al., 1992] all share structural *homology* with either the A- or C-rings of colchicine or both, whereas the benzimidazoles such as nocodazole [De Brabander et al., 1976; Hoebeke et al., 1976; Lee et al., 1980] and benomyl [Davidse and Flach, 1977; Friedman and Platzer, 1978; Ireland et al., 1979] have structural *similarity* to the A-ring of colchicine [Luduena and Roach, 1991].

Although there has been considerable information published on the binding of colchicine and related drugs to tubulin, the location of the colchicine-binding domain in tubulin remains unknown. Evidence has been obtained that the colchicine site resides on α-tubulin, on β-tubulin, or that both α-and β-tubulin are involved in the site [reviewed by Correia, 1991; Hastie, 1991]. Mutational analysis of the binding of benomyl to fungal tubulin clearly places important binding regions for the colchicine domain of fungal tubulin on the β-chain [Jung et al., 1992].

Several of the compounds that bind to the colchicine-binding domain (cornigerine, combretastatins A-2 and A-4, nocodazole, and a number of biphenyl analogues of colchicine, but not podophyllotoxin) share with colchicine the property of increasing the GTPase activity of tubulin in the absence of polymerization [Lin and Hamel, 1981; Hamel et al., 1988; Lin et al., 1989; Andreu et al., 1991]. This property is distinct from that of drugs that bind to the vinblastine domain, which in general appear to suppress GTPase activity in parallel with (probably causally related to) inhibition of microtubule polymerization [David-Pfeuty et al., 1979]. The relationship between the drug-induced conformational changes that increase tubulin GTPase activity and the drug-independent hydrolysis of GTP that accompanies tubulin polymerization is unknown.

There is still little information available on the effects of compounds that bind to the colchicine domain (other than colchicine; see below) on microtubule polymerization dynamics. Podophyllotoxin and nocodazole have been shown to inhibit microtubule treadmilling and/or dynamic instability [Caplow and Zeeberg, 1981; Jordan and Farrell, 1983; Schilstra et al., 1989; Toso RJ, Miller HP, Jordan MA, Wilson L, unpublished data]. In addition, some synthetic estrogens such as diethylstilbestrol that induce mitotic arrest, weakly inhibit microtubule assembly and appear to bind weakly in the colchicine domain [Hartley-Asp et al., 1985; Chaudoreille et al., 1991].

3. The binding of taxol to microtubules. Taxol (Fig. 1), a complex diterpene isolated from the bark of the western yew *Taxus brevifolia,* is a promising new antitumor drug [reviewed by Rowinsky et al., 1990; Rowinsky and Donehower, 1991]. It is a potent inhibitor of cell proliferation and appears to arrest cells in mitosis by a stabilizing action on microtubules [Fuchs and Johnson, 1978; Schiff and Horwitz, 1980]. In addition to stabilizing microtubules and greatly increasing the microtubule polymer mass in cells, taxol also induces microtubule "bundling" [Schiff and Horwitz, 1980; Turner and Margolis, 1984; Rowinsky et al., 1988; Roberts et al., 1989], and it has been suggested that the therapeutic and cytotoxic actions of taxol may be caused by the formation of persistent microtubule bundles during interphase rather than by mitotic arrest [Rowinsky et al., 1988].

Taxol binds specifically to microtubules in cells [Manfredi et al., 1982], and it binds reversibly to microtubules *in vitro* with an affinity of approximately 1 μM and a stoichiometry of 1 mol of taxol per 1 mol of tubulin in microtubules [Parness and Horwitz, 1981; Diaz and Andreu, 1993]. Either taxol does not bind to free tubulin subunits [Parness and Horwitz, 1981; Ringel and Horwitz, 1991a], or it binds to tubulin with much lower affinity than to microtubules [Carlier and Pantaloni, 1983]. Little is known about the location of the taxol-binding site in tubulin, although Rao et al. [1992] found that [^{3}H]taxol can be cross-linked to β-tubulin by irradiation with ultraviolet light. Taxol does not inhibit the binding to tubulin of colchicine, podophyllotoxin, or vinblastine [Kumar, 1981; Schiff and Horwitz, 1981]. Based on results of a study on the effects of taxol on polymerization of TC complex, Howard and Timasheff [1988] suggested that taxol may bridge laterally across adjacent tubulin subunits within a single microtubule at the solvent-exposed microtubule surface. Taxotere, a potentially important new semisynthetic and more water-soluble analog of taxol, has been introduced recently. Taxotere binds to the same site on microtubules as taxol and is slightly more potent than taxol in its effects on cells and on tubulin polymerization [Ringel and Horwitz, 1991b; Diaz and Andreu, 1993].

4. Other drug classes. Several compounds used as herbicides appear to be specific inhibitors of microtubule function in plant cells. Among these are the dinitroaniline compounds trifluralin and oryzalin and the phosphoric amide compounds amiprophos-methyl (APM) and butamiphos (Cremart) [reviewed by Morejohn and Fosket, 1991]. The compounds induce metaphase arrest, bind to plant and algal tubulin with relatively high affinity, inhibit taxol-induced microtubule polymerization of plant but not animal tubulins, and induce microtubule depolymerization in plant cells. It is not known whether these compounds bind to the colchicine domain, the vinblastine domain, or another location in plant tubulins. However, because the binding affinity of colchicine for plant tubulin is several orders of magnitude lower than its affinity for animal tubulins, plant tubulins do not appear to have a "colchicine-binding domain" *per se*. The actions of these herbicidal compounds on microtubule dynamics are not yet known. Griseofulvin, a compound used for the treatment of fungal infections in humans [reviewed by Hamel, 1990], blocks mammalian cells in mitosis at high concentrations [Weber et al., 1976]. Little is known about the interaction of griseofulvin with tubulin and microtubules. It appears to bind rapidly to tubulin at an unknown site, and it blocks the polymerization of mammalian brain microtubule protein at stoichiometric or superstoichiometric concentrations [Wehland et al., 1977; Keates, 1981; Sloboda et al., 1982].

III. MODULATION OF MICROTUBULE POLYMERIZATION DYNAMICS BY DRUGS *IN VITRO*

A. Overview of Microtubule Polymerization Dynamics

As described elsewhere in this volume and in several recent reviews, microtubules are not simple equilibrium polymers [e.g., Mitchison, 1988; Carlier, 1989; Gelfand and Bershadsky, 1991]. The energy provided by hydrolysis of tubulin-liganded GTP during addition of tubulin at the microtubule ends creates transitions at the ends between more stable and less stable states that give rise to two remarkable dynamic behaviors. One such behavior, called *dynamic instability* [Mitchison and Kirschner, 1984a,b], is a stochastic switching between shortening and growing phases at both ends of individual microtubules. The two ends of individual microtubules differ both structurally and kinetically, and the rates and durations of growing and shortening at one end, called the "plus" end, are much more extensive than the rates and durations of growing and shortening at the opposite or "minus" end [Horio and Hotani, 1986; Walker et al., 1988]. Growing and shortening behavior may be due to a stochastic gain and loss of a stabilizing "cap" at both microtubule ends, which is thought to consist of a short region of E-site GTP– or GDP–P_i-liganded tubulin at the ends of the microtubules [O'Brien et al., 1987; Carlier, 1989; Stewart et al., 1990; Melki et al., 1990]. However, the chemical nature of the stabilizing cap and the mechanisms governing the gain and loss of the cap are not yet known.

The dynamic instability behavior *in vitro* of microtubule-associated protein (MAP)–depleted microtubules assembled from vertebrate brain tubulin has been studied in several laboratories [e.g., Horio and Hotani, 1986; Walker et al., 1988; Gildersleeve et al., 1992; Toso et al., 1993]. Such microtubules are highly dynamic; their plus ends can grow and shorten for many micrometers at rapid rates. Dynamic instability is strongly suppressed by MAPs [Farrell et al., 1987; Hotani and Horio, 1988; Toso et al., 1993; Kowalski and Williams, 1993; Pryer et al., 1993]. However, even with microtubules containing a high concentration of MAPs such as bovine brain microtubules, consisting of 70% tubulin and 30% MAPs, dynamic instability occurs at both microtubule ends to a detectable extent [Toso et al., 1993].

A second dynamic behavior, called *treadmilling* or *flux* [Margolis and Wilson, 1978], which appears to be a consequence of dynamic instability at the microtubule ends [Farrell et al., 1987; Walker et al., 1988], occurs under conditions of polymer mass steady state. This is the net growing of microtubules at one end and net shortening at the opposite end [reviewed by Wilson and Margolis, 1981]. An imbalance in the extents of growing and shortening between the two ends must give rise to net growth at one end and net shortening at the opposite end. Tubulin exchange at microtubule ends due to treadmilling is readily distinguished from tubulin exchange due to dynamic instability in experiments using radiolabeled GTP to monitor tubulin exchange in MAP-rich microtubules at steady state [e.g., Wilson and Farrell, 1986; Jordan and Wilson, 1990]. The rate of flux, like the rates and extents of dynamic instability behavior at the microtubule ends, appears to be strongly suppressed by MAPs [Farrell et al., 1987].

Using radiolabeled GTP to monitor tubulin exchange at microtubule ends, the ends that grow at steady state operationally have been called the *net assembly* or *A-ends,* and the ends that shorten have been called the *net disassembly* or *D-ends*. The A ends appear to be the structurally defined plus ends of the microtubules. This assignment is based on analysis of the effects of low vinblastine concentrations on radiolabeled guanine nucleotide exchange with

MAP-rich bovine brain microtubules, which indicates that the drug acts predominantly or exclusively at microtubule A-ends [Jordan and Wilson, 1990], and analysis of the effects of low vinblastine concentrations on growing and shortening dynamics of individual microtubules by videomicroscopy, which indicates that vinblastine acts predominantly or exclusively at plus ends [Toso et al., 1993].

B. The Vinca Alkaloids

1. Substoichiometric inhibition of microtubule polymerization. Like colchicine (see below), vinblastine inhibits the polymerization of tubulin into microtubules *in vitro* at drug concentrations that are far below the concentration of tubulin in solution [Wilson et al., 1976; Owellen et al., 1976]. For example, 50% inhibition of microtubule polymerization in chick brain extracts occurred at 0.09 μM vinblastine while the free tubulin concentration present in the extract was 30 μM, a ratio of total tubulin to vinblastine of more than 300:1 [Wilson et al., 1976]. This and similar early work indicated that vinblastine does not inhibit microtubule polymerization by binding stoichiometrically to soluble tubulin, but rather, that it acts substoichiometrically by binding to the ends of the microtubules.

More recent studies on the binding of [^{3}H]vinblastine to steady-state MAP-rich bovine brain microtubules in relation to vinblastine-induced inhibition of tubulin exchange at microtubule ends have revealed that vinblastine inhibits tubulin exchange by 50% when an average of only 1.2 ± 0.3 molecules of vinblastine are bound per microtubule [Wilson et al., 1982]. In these experiments inhibition of net tubulin incorporation at A-ends was determined over a range of vinblastine concentrations, using [^{3}H]GTP to monitor tubulin addition; the vinblastine concentration that inhibited addition by 50% was 0.14 μM. In parallel fashion, [^{3}H]vinblastine binding to the microtubule ends was determined over a range of vinblastine concentrations, and the stoichiometry of vinblastine binding at 0.14 μM was determined. If all of the high affinity vinblastine-binding sites are located at the A-ends, the data would mean that binding of an average of one molecule of vinblastine per A-end reduces the rate of tubulin incorporation at the end by approximately 50%. If the 16–17 high affinity vinblastine-binding sites are evenly distributed at both microtubule ends, the results would mean that binding on average of one molecule of vinblastine per A-end is sufficient to reduce the rate of tubulin incorporation to near zero.

2. Kinetic stabilization of tubulin exchange at microtubule plus ends at low vinblastine concentration in the absence of significant microtubule depolymerization. Analysis of the effects of vinblastine at the ends of MAP-rich bovine brain microtubules at steady state using radiolabeled GTP to monitor tubulin incorporation and loss at both A- and D-ends has provided considerable information on the effects of the drugs on tubulin exchange. For example, the time course for net tubulin addition at the A-ends of the microtubules in the absence of vinblastine is shown in Figure 2. An

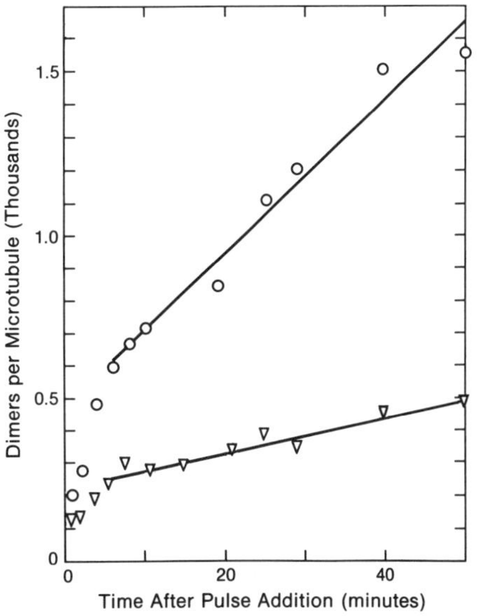

Fig. 2. *Time course for incorporation of tubulin dimers labeled with [^{3}H]GTP into MAP-rich bovine brain microtubules at steady state (circles, control microtubules; triangles, microtubules polymerized in the presence of 0.15 μM vinblastine). [Reproduced from Jordan and Wilson, 1990, with permission of the publisher.]*

initial burst of tubulin incorporation was followed by an approximately linear incorporation of tubulin that continued for the duration of the experiment (circles). The kinetically rapid burst of tubulin incorporation represented approximately 600 molecules of tubulin per microtubule and most likely was due to very short growing and shortening excursions at the microtubule ends extending into the polymer for a combined distance at both ends of approximately 0.36 μm [also see Toso et al., 1993]. The subsequent slower linear rate of incorporation reflects the treadmilling or flux rate (i.e., the net addition of tubulin at A-ends), which was approximately 0.4 tubulin dimers s^{-1}. Addition of 0.15 μM vinblastine (Fig. 2, triangles) reduced the magnitude of the initial burst of incorporation by 58% (to 0.15 μm of microtubule length) and reduced the flux rate by 78% (to 0.09 tubulin dimers s^{-1}). These and related data indicate that vinblastine suppresses both dynamic instability and treadmilling to approximately similar extents.

In the same study, 0.15 μM vinblastine was also found to inhibit the rate of tubulin dissociation at A-ends but not at D-ends. This was determined by analyzing simultaneously the effects of vinblastine on the rates of depolymerization at A- and D-ends after dilution of microtubules with ends differentially labeled with [^{3}H]GDP and [^{14}C]GDP [Jordan and Farrell, 1983]. The apparent dissociation rate constants for controls and microtubules incubated with 0.15 μM vinblastine are shown in Table I. The data indicate that vinblastine stabilizes (kinetically ''caps'') A-ends selectively. Significantly, inhibition of tubulin exchange by 0.15 μM vinblastine reduced the rate of tubulin exchange by 50% without inducing appreciable microtubule depolymerization (Table I). The critical tubulin subunit concentration was increased by only 20%, from 2.5 to 3.0 μM tubulin, and the polymer mass was reduced by less than 5% [Jordan and Wilson, 1990]. Thus low concentrations of vinblastine kinetically stabilize microtubule A- (or plus) ends without appreciably affecting the polymer mass.

3. Suppression of dynamic instability by low concentrations of vinblastine *in vitro*. Direct evidence that vinblastine suppresses the growing and shortening dynamics at plus ends of individual MAP-depleted and MAP-rich bovine brain microtubules has been obtained by videomicroscopy [Toso et al., 1993]. Suppression of dynamic instability of MAP-depleted microtubules by 0.5 μM vinblastine can be

TABLE I. Concentration of Antimitotic Drug That Inhibits the Treadmilling Exchange of Tubulin Into MAP-Rich Microtubules by 50% (K_i, Treadmilling), the Change in Mass of Microtubules Induced by Drug at the K_i, and the Inhibition of the Off-Rate Constants at the A- and D-Ends at This Drug Concentration

Drug	K_i, treadmilling (μM)	Change in mass of microtubules (%)	Dissociation rate constant (% inhibition)	
			A-ends	D-ends
Vinblastine	0.15	4	27	0
Colchicine	0.2	<5	78	5
Podophyllotoxin	0.5	5	—	—
Nocodazole	0.2	3	—	—
Taxol	6.5	5	76	79

Treadmilling was measured by the uptake of a pulse of [^{3}H]GTP into MAP-rich bovine brain microtubules at steady state. The off-rate constants were measured by diluting microtubules differentially radiolabeled at their A- and D-ends with [^{3}H]GTP and [^{14}C]GTP [Farrell and Jordan, 1982; Jordan and Farrell, 1983] and following the time course of loss of radiolabel from microtubules that were subsequently stabilized and collected on glass-fiber filters [Wilson et al., 1982]. The data are taken from Jordan and Wilson [1990], Wilson et al. [1985a,b], Jordan and Farrell [1983], and unpublished data of H. Miller and L. Wilson.

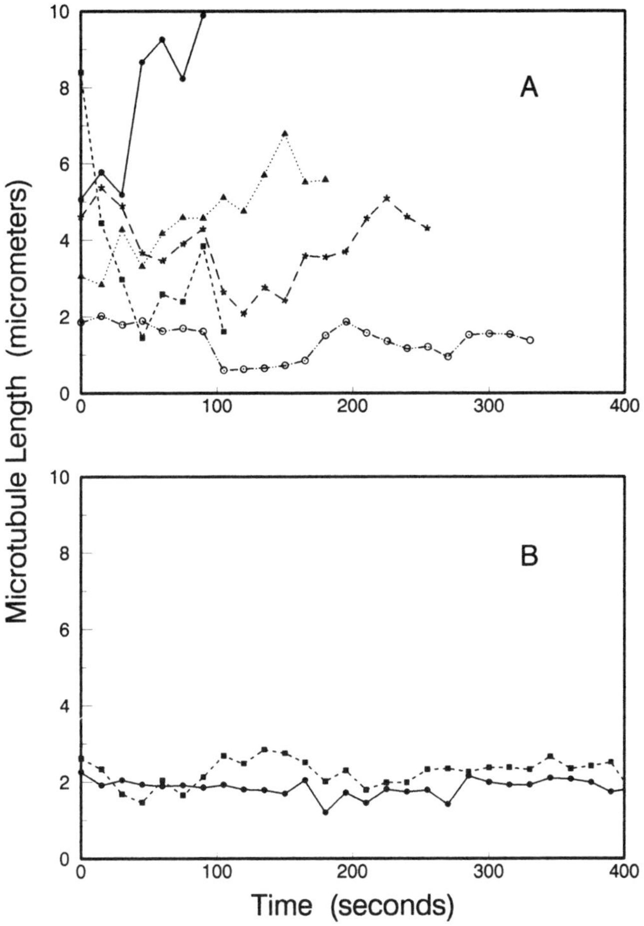

Fig. 3. *Changes in microtubule lengths in the absence* **(A)** *and presence* **(B)** *of 0.5 μM vinblastine. Each tracing shows the growing and shortening of an individual microtubule as studied with videomicroscopy. Microtubules were polymerized from MAP-depleted bovine brain tubulin (1.5 mg/ml, 30°C) at the plus ends of axonemal seeds in the absence or presence of the drug. [Reproduced from Toso et al., 1993, with permission of the publisher.]*

readily visualized by comparing the length changes of microtubules with time in the absence and presence of the drug (Fig. 3). Vinblastine strongly reduces both the growing and the shortening rates at plus ends in a concentration-dependent manner (Table II) [Toso et al., 1993]. In addition, both rates are suppressed to similar extents at any given vinblastine concentration. Such relatively similar degrees of inhibition of shortening and growing rates by vinblastine may in large part explain why at low concentrations the drug inhibits the rate of microtubule formation but does not induce appreciable microtubule depolymerization when it is added to preformed microtubules [Wilson et al., 1976; Jordan and Wilson, 1990].

In addition to suppressing growing and shortening rates, vinblastine strongly increases the percentage of time microtubules spend in a state of attenuated activity, neither growing nor shortening detectably. For example, the percent time spent in an attenuated state increased with increasing vinblastine concentration from 19%

TABLE II. Suppression of Dynamic Instability of Bovine Brain Microtubules *In Vitro* by Antimitotic Drugs*

Drug	Growing rate (dimers/s)	Shortening rate (dimers/s)	Percent time in attenuation**	Frequency of transition (events/s)	
				From shortening†	To shortening‡
Control	106 ± 13	104 ± 16	19 ± 3	0.050	0.026
Vinblastine (0.5 μM)	51 ± 4	45 ± 3	46 ± 7	0.053	0.020
Nocodazole (3 μM)	49 ± 5	41 ± 5	58 ± 2	0.059	0.014
Colchicine (0.2 μM TC)	35 ± 4	45 ± 4	69 ± 2	0.067	0.014
Taxol (0.5 μM)	48 ± 3	49 ± 5	53 ± 8	0.056	0.015

*Data are for plus ends of MAP-free bovine brain microtubules seeded with axonemal fragments, except for taxol, for which data are for both ends combined (plus and minus ends were indistinguishable). Data are from Toso et al. [1993], Jordan et al. [1993], and Toso RJ, Jordan MA, and Wilson L (unpublished).

**Attenuation is defined as a state in which there is no detectable length change in a 15 second interval.

†The frequency of transition from a shortening phase to either a growing phase or to a phase of attenuation; it equals the inverse of the mean phase duration of shortening.

‡The frequency of transition to shortening from a phase of either growing or attenuation equals the number of shortening events divided by the total time spent in the attenuation state or growing.

in control microtubules to 46% at 0.5 μM vinblastine (Table II). Vinblastine also increases the mean duration of the attenuation phase, reduces the mean duration of the growing phase, and reduces the mean duration of the shortening phase [Toso et al., 1993]. Thus the frequency of transition from shortening to growing or attenuation is increased by vinblastine (called *rescue* according to the terminology of Walker et al. [1988]) and the frequency of transition from growing or attenuation to shortening [*catastrophe,* Walker et al., 1988] is decreased by vinblastine (Table II).

4. How does vinblastine inhibit tubulin association and dissociation at plus ends? Vinblastine could inhibit tubulin addition at plus ends either by preventing new tubulin addition sterically when the drug is bound at the end in the vicinity of a tubulin association site or by inducing a conformational change in the tubulin that prevents new tubulin addition. When bound to tubulin at the plus end of a microtubule, vinblastine probably reduces the rate of tubulin dissociation by inducing a conformational change in the tubulin molecule to which it is bound, which increases the affinity of the tubulin for adjacent tubulin molecules in the microtubule wall. Such a mechanism would be consistent with the results of Na and Timasheff [1980a,b] indicating that vinblastine induces tubulin self-association. Because vinblastine bound to the microtubule end does not become buried in the microtubule by addition of new tubulin to the end [Jordan and Wilson, 1990], microtubule elongation must require dissociation of the vinblastine. Because 16–17 high affinity vinblastine-binding sites are at the microtubule ends [Wilson et al., 1982], it is possible that several vinblastine molecules may bind simultaneously to an individual microtubule end and bind to more than one tubulin addition site, but that binding of vinblastine to only one tubulin addition site is sufficient to block tubulin exchange. Such a mechanism is consistent with that described recently in a computer-simulation model of suppression of dynamic instability by podophyllotoxin and other antimitotic drugs [Bayley and Martin, 1991].

At microtubule plus ends, vinblastine decreases the transition frequency from the growing or attenuation phases into shortening, which may reflect increased retention of the stabilizing GTP or GDP–P_i cap, and it increases the transition frequency from shortening to growing or attenuation, which may reflect gain of the stabilizing cap [Toso et al., 1993]. Thus vinblastine may decrease the rate of cap loss by maintaining the tubulin-bound guanine nucleotide in the GTP or GDP–P_i state. Alternatively, the conformational changes that occur in tubulin when vinblastine binds to a microtubule end may mimic the conformation of tubulin–GTP or tubulin–GDP–P_i that may be responsible for microtubule stabilization.

C. Colchicine

1. Substoichiometric inhibition of microtubule polymerization by colchicine: Copolymer formation. Like vinblastine, colchicine inhibits microtubule polymerization *in vitro* at concentrations well below the concentration of tubulin free in solution [Olmsted and Borisy, 1973; Margolis and Wilson, 1977; Margolis et al., 1980], indicating that colchicine inhibits microtubule polymerization by interacting with the ends of the microtubules rather than with tubulin in soluble pools. However, in contrast with vinblastine, which binds directly to the ends of the microtubules, colchicine either cannot bind at all to microtubule ends or it does so very inefficiently. Instead, it first binds to soluble tubulin and forms a final-state TC complex, which then incorporates at the microtubule ends through a polymerization-dependent pathway [Margolis and Wilson, 1977; Skoufias and Wilson, 1992].

Early data indicated that, when tubulin is polymerized into microtubules *in vitro* in the presence of TC complex, the TC complex becomes incorporated into the microtubules along with unliganded tubulin resulting in formation of TC complex–tubulin copolymers [Sternlicht and Ringel, 1979; Sternlicht et al., 1983; Farrell and Wilson, 1984]. These data indicated that the binding of TC complex to microtubule ends does not completely prevent further tubulin addition to the ends, but rather, the ends remain competent to grow [see Sternlicht et al., 1983]. This has been shown directly in a recent study for microtubule A-ends. For example, in an experiment using 0.1 μM TC complex, the incorporation of only one to two molecules of colchicine per A-end inhibited the incorporation of tubulin by approximately 50% at the earliest measured time point. Then, small number of TC complexes continued to be incorporated, along with free tubulin, with very little change in the inhibited rate [Skoufias and Wilson, 1992].

Thus at low TC complex concentrations incorporation of TC complexes at the microtubule ends must disrupt the microtubule lattice at or near the end in a way that impairs the efficiency of new tubulin addition but does not completely destroy the ability of tubulin to be incorporated. When the ratio of TC complex to free tubulin is low relative to a possible critical transition ratio, the bound TC complex does not completely suppress tubulin addition, and new molecules of tubulin can add to the end and "bury" the initially exposed TC complex in the core of the microtubule. The cycle of TC complex addition and burial in the microtubule core probably repeats continuously under steady-state conditions. The mechanism appears to change at high TC complex concentrations, at which tubulin addition at the microtubule ends appears to be blocked completely [Farrell and Wilson, 1980].

Significantly, low concentrations of colchicine or TC complex, like low concentrations of vinblastine, inhibit tubulin exchange at microtubule ends in the absence of appreciable loss of polymer [Wilson and Farrell, 1986; Skoufias and Wilson, 1992](see Table I). For example, addition of 0.2 μM colchicine (Table I) or 0.1 μM TC complex to steady-state MAP-rich bovine brain microtubules inhibited tubulin exchange at the A-ends by 50% without exerting significant effects on the polymer mass [Skoufias and Wilson, 1992].

2. Kinetic stabilization of microtubule dynamics at both microtubule ends by TC complex. Similar to results with vinblastine [Jordan and Wilson, 1990], when low concentrations of colchicine are added to MAP-rich bovine brain microtubules at steady state, the drug inhibits both the rate of tubulin incorporation and tubulin loss at A-ends [*kinetic capping*, Margolis et al., 1980; Farrell and Wilson, 1984; Wilson and Farrell, 1986]. In contrast to the action of vinblastine, colchicine is able to cap both microtubule ends kinetically, but the A-ends are much more strongly affected than the D-ends at low drug concentrations [Farrell and Wilson, 1984]. For example, with 0.2 μM colchicine, the A-end dissociation rate constant was reduced by 78% compared with that for control microtubules, while that for the D-end was reduced by only 5% (Table I). Strong inhibition of the D-end dissociation rate constant occurs at high TC complex concentrations [Farrell and Wilson, 1984].

However, differences in the action of colchicine on the dynamics of tubulin addition and loss at microtubule ends have been obtained [Bergen and Borisy, 1983, 1986; Deery and Weisenberg, 1981] and remain to be explained. One possibility is that the varied responses of microtubules are due to the use of different microtubule proteins. In support of this idea, Banerjee and Luduena [1991] found that the kinetics of colchicine binding to tubulins composed of different β-tubulin isotypes differ significantly. In addition, the thermodynamic parameters governing the binding of colchicine to highly purified tubulins from unfertilized eggs and from flagella outer doublet microtubules of the sea urchin *Strongylocentrotus purpuratus* are substantially different [Wilson et al., 1984].

3. Suppression of dynamic instability *in vitro* by low concentrations of TC complex. Direct evidence that TC complex kinetically stabilizes plus ends of individual MAP-depleted bovine brain microtubules has been obtained by videomicroscopy (Toso RJ, Jordan MA, Farrell KW, Wilson L; unpublished data). For example, TC complex strongly reduces both the growing and the shortening rates (Table II). Like vinblastine, a specific concentration of TC complex inhibits the growing and shortening rates to a similar extent. In addition, TC complex strongly increases the percentage of time the microtubules remain in an attenuated state, neither growing nor shortening detectably (Table II). These data probably explain why addition of colchicine to microtubules already at steady state (especially those with stable minus ends) does not depolymerize the microtubules to a significant extent [see Skoufias and Wilson, 1992]. Like vinblastine, TC complex also increases the frequency of transition from shortening to growing or attenuation, and it decreases the frequency of transition from growing or attenuation to shortening (Table II). Thus the switching mechanism responsible for gain and loss of the stabilizing cap at plus ends appears to be affected by colchicine.

4. How does TC complex affect growing and shortening dynamics at microtubule ends? The mechanism by which incorporation of TC complex at microtubule ends inhibits tubulin addition and loss remains to be elucidated in detail. It is clear that the stabilizing effects of colchicine on the dynamics of tubulin addition and loss at microtubule ends are caused by the colchicine-liganded tubulin itself and appear to involve a conformational change in the tubulin induced by the binding of the drug. Inhibition of tubulin addition at the microtubule end could occur by steric hindrance due to the presence of the drug at an important interface between tubulin dimers in the microtubule lattice. However, such a simple steric blocking mechanism would not be expected to affect the rate of dissociation of tubulin at the microtubule ends. Therefore the colchicine-induced conformational change in tubulin must alter the tubulin so that once it becomes incorporated at a microtubule end it dissociates more slowly than unliganded tubulin.

An intriguing possibility is that the conformational changes induced in tubulin by the binding of colchicine are related in some way to the conformational changes that occur normally when unliganded tubulin adds to the end of a microtubule. However, the colchicine-induced conformational changes must differ from those that occur normally during tubulin addition. In this regard, it may be significant that the colchicine-induced conformational change in tubulin results in increased GTPase activity [David-Pfeuty et al., 1979; Andreu and Timasheff, 1981]. Thus the mechanism of GTP hydrolysis or the use of the energy released during hydrolysis might be altered when TC complex is incorporated at the ends. An interesting possibility is that the colchicine-induced conformational changes in tubulin may be mimicking the action of natural regulatory substances whose function is to modulate microtubule assembly dynamics by inducing "colchicine-like" conformational changes. Finally, it is curious that TC complex stabilizes microtubule plus ends in a manner that is remarkably similar to that of vinblastine, because the two drugs bind to different sites on tubulin, and the conformational changes induced in tubulin by the two drugs must be different based on the fact that colchicine stimulates the GTPase activity of tubulin whereas vinblastine does not or possibly inhibits it [see Lin and Hamel, 1981]. Therefore, vinblastine and colchicine appear to kinetically stabilize microtubule plus ends in a relatively similar way through different mechanisms.

D. Taxol

1. Stabilization of microtubules by taxol. Taxol profoundly affects the polymerization characteristics and stability of microtubules. It stimulates microtubule polymerization *in vitro*, promoting both the nucleation and elongation phases of the polymerization reaction, and it can reduce the critical tubulin subunit concentration (i.e., soluble tubulin concentration at steady state) to near zero [Schiff et al., 1979; Kumar, 1981; Howard and Timasheff, 1988]. Microtubules polymerized in the presence of taxol are extremely stable. They resist depolymerization by cold, calcium ions, dilution, and other antimitotic drugs [Schiff et al., 1979; Kumar, 1981; Howard and Timasheff, 1988; reviewed by Hamel, 1990]. Taxol induces the self-assembly of tubulin into microtubules at 0°C [Thompson et al., 1981], in the absence of exogenous GTP [Schiff and Horwitz, 1981; Hamel et al., 1981], without MAPs [Kumar, 1981], and at alkaline pH [Ringel and Horwitz, 1991a]. Howard and Timasheff [1988] have found that the linkage free energy provided by taxol binding to tubulin in the microtubule is −3 kcal/mol, which results in a stable polymer even at low temperature or in the absence of exogenous GTP. Taxol induces the formation of microtubules with predominantly 12 protofilaments rather than 13, both *in vitro* with pure tubulin [Andreu et al., 1992] and *in vivo* [Mogenson and Tucker, 1990]. Taxol also can induce the formation of hoops, ribbons, and other protofilamentous structures *in vitro* [Parness and Horwitz, 1981; Thompson et al., 1981, Bokros et al., 1993]. Because these aberrant assemblages of tubulin are ultimately replaced by microtubules, they appear to be less stable than microtubules [Thompson et al., 1981]; their formation can be minimized by conditions that reduce the rate of nucleation, such as gradually increasing the temperature in a stepwise manner [Bokros et al., 1993].

2. Stabilization of microtubule dynamics by taxol: A common mechanism shared with other drug classes. Although the detailed molecular mechanism is not yet elucidated, it is known that taxol inhibits the dynamic reactions at microtubule ends, suppressing both treadmilling and dynamic instability [Caplow and Zeeberg, 1982; Wilson et al., 1985a; Jordan et al., 1993]. A primary effect of the drug at concentrations that are stoichiometric or near-stoichiometric with respect to tubulin is to reduce the magnitudes of the dissociation rate constants equally at both microtubule ends, without exerting detectable effects on the apparent association rate constants (Table I) [Caplow and Zeeberg, 1982; Wilson et al., 1985a]. Analysis of the effects of substoichiometric taxol concentrations on dynamic instability (e.g., 1 molecule taxol to 40 molecules tubulin; Table II) indicates that the rate of shortening and the frequency of transition from growing or attenuation to shortening is decreased, and the percentage of time the microtubules spend in the attenuated state is greatly increased. Also, the rate of growing is decreased, possibly by reduction of the soluble tubulin concentration by the drug. Surprisingly, the frequency of transition from shortening to growing or attenuation is increased, indicating that, in ways that are not clear at the present time, taxol affects the switching between phases of growing and shortening similarly to vinblastine and colchicine.

Interestingly, the binding of taxol to microtubules at steady-state does not inhibit GTP hydrolysis as strongly as would be expected. For example, at taxol concentrations that are sufficient to bind stoichiometrically with tubulin in microtubules and that inhibit exchange of tubulin at the microtubule ends almost completely, the rate of GTP hydrolysis is reduced by less than 50% compared with microtubules incubated without taxol [Carlier and Pantaloni, 1983]. In addition, taxol does not appreciably affect the magnitude of the initial burst of tubulin exchange during a pulse of MAP-rich bovine brain microtubules at steady state with [^{3}H]GTP [Wilson et al., 1985a], which indicates that taxol may not inhibit dynamic instability at the extreme ends of the microtubules (<0.3–0.4 μm of length per microtubule), which is near the limit of detection by videomicroscopy [Walker et al., 1988; Toso et al., 1993].

IV. PHARMACOLOGICAL PROBES OF MICROTUBULE-BASED CELL FUNCTIONS

Colchicine, nocodazole, vinblastine, podophyllotoxin, and taxol have often been used at high concentrations in studies of microtubule behavior and function in cells. At such concentrations, the drugs either induce microtubule depolymerization or, as with taxol, induce microtubule polymerization and microtubule bundling. These studies are too numerous to review in this chapter; however, several will be mentioned briefly. For example, Cleveland et al. [1981] (see also Cleveland et al., this volume) used high concentrations of colchicine and nocodazole to reduce the microtubule polymer levels and taxol to increase the levels to demonstrate that the level of unpolymerized tubulin in cells modulates the synthesis of tubulin mRNA. In another recent study of microtubule-based processes in living cells, Lee et al. [1989] analyzed the reformation of the endoplasmic reticulum network in cells after inducing breakdown of the microtubules by nocodazole. By examining the reformation process after washing out nocodazole, and in the presence of taxol and the actin-specific drug cytochalasin B, they showed that formation of the endoplasmic reticulum network takes place through movement of the membranes along microtubules and not along actin filaments.

Finally, Cabral and Barlow [1991] and Minotti et al. [1991] used antimitotic drugs at high concentrations to isolate drug-resistant lines of Chinese hamster ovary cells that have altered α- or β-tubulins and altered microtubule function. Interestingly, the colcemid (a colchicine analogue)–resistant mutants have an increased fraction of polymerized tubulin, and Cabral and co-workers have suggested that the mutant cells have overcome their sensitivity to the microtubule destabilizing activity of colcemid by making their microtubules more stable. Similarly, taxol-resistant mutants have a decreased fraction of microtubule polymer, suggesting that these mutant cells responded to the microtubule stabilizing ability of taxol in a manner opposite that of colcemid by making their microtubules less stable than normal. To understand fully these results in terms of the drug-induced kinetic stabilization of microtubule dynamics that has been observed *in vitro* with all of these drugs may ultimately require characterizing the polymerization dynamics of microtubules in wild-type and mutant cells *in vivo*.

As described in the foregoing sections, very low concentrations of several antimitotic drugs can inhibit microtubule dynamics without appreciably altering the microtubule polymer mass. Thus it is possible to use these drugs at low concentrations to probe the functions of microtubule dynamics *in vivo*. Most of the work published to date using this methodology has focused on the role of dynamics in mitosis, which will be described in this section. However, vinblastine has recently been used in this manner to study the role of microtubule polymerization dynamics in the formation of new axons. Specifically, Baas and Ahmad [1993] found that, in the presence of vinblastine at concentrations that prevented microtubule polymerization, preassembled microtubules were transported into developing axons of cultured neuronal cells, with their plus ends leading, by a transport mechanism that did not require microtubule assembly.

A. The Role of Microtubule Dynamics in Mitosis: Evidence Obtained Using Pharmacological Probes

The most prominent perturbation of cell function induced by drugs that bind to tubulin and microtubules is to inhibit mitosis. The degree of mitotic block varies with the concentration of drug, with the length of incubation with drug, and with the type of cells studied. The exact molecular mechanism of mitotic inhibition has not been clear, although in general mitotic block by antimitotic drugs was thought to occur through an effect on microtubules. Historically, drug-induced mitotic block has been attributed to nearly total microtubule depolymerization with drugs such as colchicine or vinblastine and to enhanced microtubule polymerization with taxol.

However, recently Jordan et al. [1991, 1992, 1993] found that mitotic block in HeLa cells with many drugs is not associated with alteration in the levels of polymerized microtubules. In these experiments, the mass of polymerized microtubules was compared with the extent of mitotic block after incubation of the cells with antimitotic drugs for one cell cycle over a wide concentration range (Table III). Specifically, with low concentrations of vinblastine, vincristine, and nocodazole, mitotic block occurred in the presence of a normal complement of microtubules. For example, the relationship between accumulation of cells in mitosis and the levels of microtubule polymer in the cells after incubation with vinblastine is shown in Figure 4. It is clear that a 14-times higher concentration of vinblastine was required to depolymerize the microtubules than was required to block cells in mitosis (see Table III). Similarly, mitotic block occurs in the absence of any enhanced microtubule polymerization with taxol (Table III; Jordan et al., 1993).

TABLE III. The Concentration of Drug That Induces 50% Accumulation of HeLa Cells in Mitotic Metaphase (K_{met}) Compared With the Drug Concentration Required to Decrease (↓) or Increase (↑) the Microtubule Mass in Cells by 50% (K_{mass})

Drug	K_{met} (nM)	K_{mass} (nM)	K_{mass}/K_{met}
Vinblastine	0.80	11 ↓	14
Nocodazole	54	600 ↓	11
Podophyllotoxin	30	15 ↓	0.5
Colchicine	17	19 ↓	1.1
Taxol	5	13 ↑	2.6

HeLa cells were incubated with drug for 20 hours. Percent of cells in metaphase was determined by immunofluorescence microscopy, and microtubule polymer mass was determined by ELISA for tubulin in isolated cytoskeletons. Methods and results are from Jordan et al. [1991, 1992] and Thrower et al. [1993].

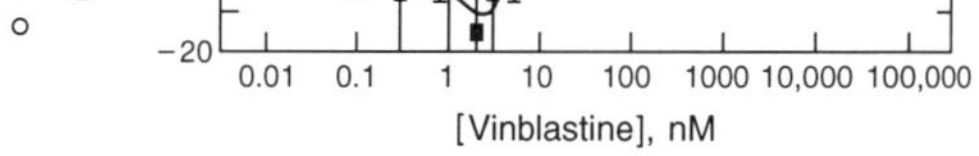

Fig. 4. *Relationship between metaphase arrest and microtubule depolymerization in HeLa cells incubated with vinblastine for one cell cycle (20 hours). Metaphase arrest was assayed by immunofluorescence microscopy of fixed cells and microtubule polymer mass was measured by ELISA for tubulin in stabilized cytoskeletons. At high concentrations of vinblastine (10–100 μM), the microtubule polymer mass increased as a result of vinblastine–tubulin paracrystal formation. [Reproduced from Jordan et al., 1992, with permission of the publisher.]*

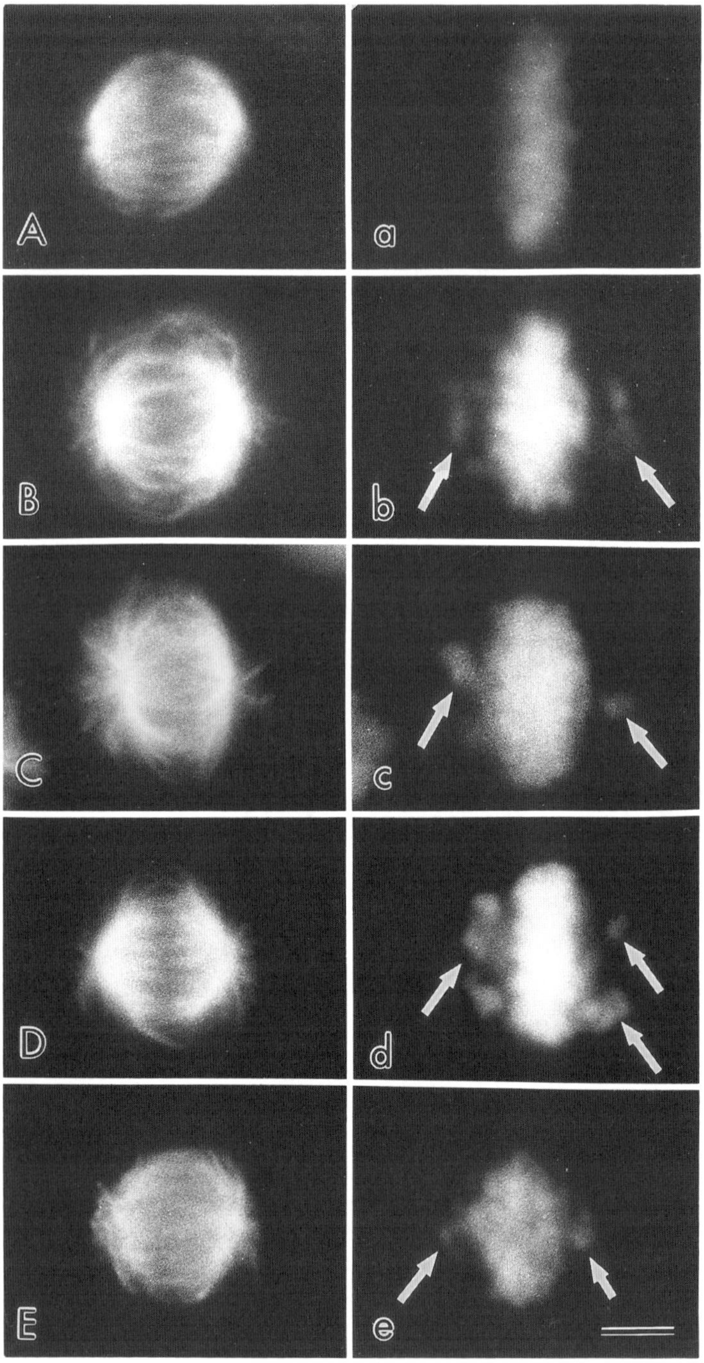

Fig. 5. *Microtubule and chromosome arrangements of HeLa cells blocked in metaphase with low concentrations of antimitotic drugs.* **A–E:** *Microtubules of the spindles.* **a–e:** *Chromosomes of the same spindles. A,a: No drug, normal spindle with a compact metaphase plate of chromosomes and very few astral microtubules. B,b: 33 nM nocodazole. C,c: 16 nM colchicine. D,d: 19 nM podophyllotoxin. E,e: 0.4 nM vinblastine. After incubation with the stated concentration of drug for 20 hours, bipolar spindles with chromosomes located near the spindle poles (arrows) and excluded from the otherwise compact metaphase plate were common. Methods are described by Jordan et al. [1992]. Bar = 5 μm.*

With colchicine and podophyllotoxin, however, mitotic block occurred at approximately the same drug concentrations that induced microtubule depolymerization (Table III).

Jordan et al. [1991, 1992, 1993] also found that all of the antimitotic drugs studied induce similar subtle rearrangements of spindle microtubule organization. For example, at the lowest effective drug concentrations, mitotically blocked spindles appear normal in most respects (Fig. 5). Blocked spindles are bipolar and have a compact equatorial plate containing most of the chromosomes, but the spindles are abnormal in that often from one to many chromosomes are situated near the poles or even outside the spindles at the ends of prominent astral microtubules rather than in the equatorial plate. In addition, the lengths of the central spindles are usually shortened as though the spindle poles are unable to separate to the full extent. It appears that there is a redistribution of microtubule polymer from the central spindles into the asters. Congression of chromosomes to the metaphase plate is inhibited. These spindles are called *type I* or *type II,* depending on whether they have only a few or many chromosomes near the poles (Figs. 5, 6). (At the lowest drug concentrations that induced mitotic block, a small percentage of blocked spindles appear to have completely normal chromosomal organization with no chromosomes located outside of the equatorial plate, although their central spindles are often slightly shorter than those of control spindles.) In addition to the presence of types I and II spindles, some spindles are monopolar and consist of a single star-shaped microtubule aster surrounded by a sphere of chromosomes enclosing one or two centrosomes. These spindles, called *type III* (Fig. 6), resemble the classic colchicine mitosis (C-mitosis) [see Dustin, 1984]. As the drug concentration is increased beyond the lowest effective concentration, the proportions of types I and II spindles decrease and type III increases (Fig. 6). With drugs such as vinblastine that induce microtubule depolymerization at relatively high concentrations, the size of the type III spindle diminishes as the drug concentration is increased further. At yet higher vinblastine concentrations, a type IV spindle forms, which contains no polymerized microtubules and consists only of a mass of condensed chromosomes surrounding the centrosomes.

Fig. 6. *Spindle organization induced in HeLa cells by increasing concentrations of antimitotic drugs. Normal bipolar or slightly abnormal bipolar spindles (types I and II) predominated at the lowest effective drug concentrations. The astral microtubules were longer and more numerous than in control cells, central spindles were shorter, centrioles pairs were disrupted, and a few chromosomes were found near the spindle poles rather than at the equatorial metaphase plate. Some spindles were extremely abnormal monopolar spindles (type III). The proportions of the more normal configurations decreased and the more abnormal configurations increased as the drug concentrations were increased. Drugs that induced this series of changes included vinblastine, vincristine, vindesine, vinepidine, vinrosidine, podophyllotoxin, nocodazole, colchicine, and taxol. The drawing summarizes results from the work of Jordan et al. [1991, 1992, 1993], and Wendell et al. [1993].*

Changes in spindle morphology that occurred with a low concentration of vinblastine (2 nM) were examined in detail to determine whether the bipolar spindles were truly blocked in mitosis. It was conceivable that cells incubated with low vinblastine concentrations simply required a long time to traverse prometaphase, to form normal spindles, and ultimately to enter anaphase. Another possibility was that the bipolar spindles simply may have been caught in the process of collapsing into classical C-mitotic or type III spindles. However, bipolar and monopolar spindles both formed within an hour of adding vinblastine to the cells, and the ratio of the two spindle forms remained approximately constant for the following 30 hours [Wendell et al., 1993]. Virtually no cells entered anaphase between 2 and 39 hours after addition of vinblastine (equivalent to approximately two cell cycles in untreated cells). Therefore abnormal bipolar spindles are truly blocked by vinblastine in a configuration that is nearly normal. They neither progress to anaphase nor collapse further to form monopolar spindles. Interestingly, bipolar spindles do lengthen during the course of drug incubation, and a very few normally organized spindles do appear at about 20 hours of vinblastine incubation, indicating that the spindles progress in the direction of normality in the presence of vinblastine (Jordan MA, Wendell K, and Wilson L, unpublished data). (After 28 hours in vinblastine, increasing numbers of cells revert to interphase without dividing.) The observations that 1) all of the antimitotic drugs thus far examined have very similar effects on mitotic spindle organization at the same concentrations at which they block mitosis (vinblastine, vincristine, vindesine, vinepidine, vinrosidine, nocodazole, podophyllotoxin, colchicine, and taxol) and 2) that all antimitotic drugs studied inhibit microtubule polymerization dynamics *in vitro* indicate strongly that the drugs all block mitosis by perturbing the polymerization dynamics of spindle microtubules.

B. Relationship Between the Drug Concentration Dependence for Inhibition of Microtubule Dynamics *In Vitro* and the Drug Concentration Dependence for Mitotic Block *In Vivo*

One puzzle that has existed in reconciling the mechanism of action of antimitotic drugs *in vivo* with their effects on microtubule dynamics *in vitro* concerns the disparity in the drug concentrations at which the effects occur. For example, 0.8 nM vinblastine induces 50% accumulation of HeLa cells in mitosis [Jordan et al., 1992], whereas vinblastine concentrations that are approximately 1–300-fold higher (0.1–0.3 μM), are required to inhibit treadmilling and dynamic instability *in vitro* [Jordan and Wilson, 1990; Toso et al., 1993]. However, vinblastine is concentrated in HeLa cells approximately 50–60-fold to between 0.1 and 0.4 μM when the drug concentration in the growth medium is between 2 and 10 nM [Jordan et al., 1991]. Thus the intracellular concentrations of vinblastine that are actually achieved are at the low end of the concentration range in which the drug kinetically stabilizes the ends of bovine brain microtubules *in vitro*. Interestingly, the intracellular tubulin concentration in HeLa cells is approximately 20 μM, indicating that in cells as well as *in vitro* vinblastine poisons microtubules substoichiometrically with ratios of approximately 1 vinblastine molecule to 200 tubulin molecules. Similar apparent disparities between the drug concentrations required to inhibit mitosis in HeLa cells and the concentrations required to inhibit tubulin exchange at microtubule ends *in vitro* have been observed for vincristine, colchicine, taxol, nocodazole, and podophyllotoxin. However, vincristine, colchicine, and taxol, like vinblastine, are extensively concentrated intracellularly to the same levels that stabilize growing and shortening dynamics of microtubules *in vitro* [Jordan et al., 1991, 1993; Thrower D, Wilson L, and Jordan MA, unpublished results].

C. Models for the Mechanism of Mitotic Block by Antimitotic Drugs

Mitotic spindle microtubules are extremely dynamic and exchange their tubulin with tubulin in the cytoplasmic pool with half-times of less than a minute [reviewed by McIntosh and Hering, 1991] (see also Wordemann and Mitchison, this volume; McIntosh, this volume). There are several stages during which suppression of spindle microtubule dynamics by antimitotic drugs could block mitotic progression. For example, during prometaphase, the plus ends of astral microtubules grow and shorten, effectively probing the cytoplasm until kinetochore attachment is attained [Rieder and Alexander, 1990; Hayden et al., 1990]. As discussed above (see Table II), vinblastine, colchicine, nocodazole, and taxol all suppress the growing and shortening rates at microtubule plus ends and increase the percentage of time that microtubules spend in a state of attenuated activity *in vitro*. The inhibition of chromosomal congression that occurs with low concentrations of antimitotic drugs may result from suppression of the dynamics of astral microtubules because only the microtubules emanating from one of the spindle poles may be sufficiently close to the chromosomes to achieve kinetochore attachment. In support of this idea, Wendell et al. [1993] found by serial-section electron microscopy that 27% fewer microtubules were attached to kinetochores of chromosomes in the metaphase plates of cells incubated with 2 nM vinblastine than were attached to chromosomes in the metaphase plates of control spindles.

The phenomenon of flux or treadmilling of mitotic spindle microtubules is another mitotic process that is vulnerable to poisoning by drugs. In the central spindle during metaphase and anaphase, tubulin is continually added at the plus ends of the microtubules and is lost at the minus (centrosomal) ends [Mitchison et al., 1986; Hamaguchi et al., 1987; Mitchison, 1989; Mitchison and Salmon, 1992]. All of the antimitotic drugs thus far examined (vinblastine, podophyllotoxin, nocodazole, colchicine, and taxol) inhibit the steady-state treadmilling of microtubules *in vitro* in the absence of significant change in the mass of polymerized microtubules (Table I). *In vivo* all of the drugs induce shortening of the interpolar distances of metaphase spindles at drug concentrations that have no effect on the mass of polymerized microtubules [Jordan et al., 1992, 1993]. This observation is perhaps most surprising with taxol, which one might initially expect to promote microtubule polymerization that might lead to a longer central spindle. One reasonable possibility that would account for the observations is that treadmilling of central spindle microtubules may be required for centrosome separation and that the reduced interpolar distances of spindles blocked in metaphase by low concentrations of antimitotic drugs might be causally related to inhibition of treadmilling.

Another possible mechanism by which inhibition of microtubule dynamics by drugs may block mitosis involves the signal for the transition from metaphase to anaphase. Microtubules appear to have some as yet unknown role in biochemical signaling of this transition. Several genetic and cell biological studies employing microtubule-disrupting drugs have implicated microtubules in the timing of the cell cycle, especially in the transition from metaphase to anaphase [Sluder, 1979; Dubel and Little, 1988; Alfa et al., 1990; Kung et al., 1990; Hoyt et al., 1991; Li and Murray, 1991]. While inhibition of the biochemical changes associated with this transition may result from failure to attain a structurally perfect spindle, the results of Jordan et al. [1991, 1992] suggest that dynamics of spindle microtubules may be in some way essential for the biochemical signal controlling the transition from metaphase to anaphase.

V. CONCLUSIONS

A detailed molecular understanding of the action of antimitotic drugs on microtubules *in vitro* is beginning to emerge. Interestingly, while the molecular interactions of vinblastine, colchicine, and taxol with tubulin and microtubules are all different, they suppress dynamic instability and treadmilling of microtubules in remarkably similar ways. In cells it is clear that antimitotic drugs not only affect the levels of microtubule polymer; at a more sensitive level, they also affect the dynamics of the microtubules. The precise characterization of the effects of such drugs on microtubule dynamics *in vitro* increasingly will allow antimitotic drugs to be used as fine probes of intracellular microtubule behavior and function.

ACKNOWLEDGMENTS

This work was supported by grants CH381D from the American Cancer Society (L.W.), NS13560 from the National Institute of Neurological Disorders and Stroke (L.W.), and CA57291 from the National Cancer Institute (M.A.J.). Its contents are solely the responsibility of the authors and do not necessarily represent the official views of the sponsoring organizations.

VI. REFERENCES

Alfa CE, Ducommun B, Beach D, Hyams JS (1990): Distinct nuclear and spindle pole body populations of cyclin-cdc2 in fission yeast. Nature 347:680–682.

Andreu JM, Bordas J, Diaz JF, Garcia de Ancos J, Gil R, Medrano FJ, Nogales E, Pantos E, Towns-Andrews E (1992): Low resolution structure of microtubules in solution. Synchrotron x-ray scattering and electron microscopy of taxol-induced microtubules assembled from purified tubulin in comparison with glycerol and MAP-induced microtubules. J Mol Biol 226:169–184.

Andreu JM, Gorbunoff MJ, Medrano FJ, Rossi M, Timasheff SN (1991): Mechanism of colchicine binding to tubulin. Tolerance of substituents in ring C′ of biphenyl analogs. Biochemistry 30:3777–3786.

Andreu JM, Timasheff SN (1981): The ligand and microtubule assembly induced GTPase activity of purified calf brain tubulin. Arch Biochem Biophys 211: 151–157.

Andreu JM, Timasheff SN (1982a): Conformational states of tubulin liganded to colchicine, tropolone methyl ether and podophyllotoxin. Biochemistry 21:6465–6476.

Andreu JM, Timasheff SN (1982b): Interaction of tubulin with single ring analogs of colchicine. Biochemistry 21:534–543.

Andreu JM, Wagenknecht T, Timasheff SN (1983): Polymerization of the tubulin-colchicine complex: Relation to microtubule assembly. Biochemistry 22: 1556–1566.

Baas PW, Ahmad FJ (1993): The transport properties of axonal microtubules establish their polarity orientation. J Cell Biol 120:1427–1437.

Bai RB, Pettit GR, Hamel E (1990a): Dolastatin 10, a powerful cytostatic peptide derived from a marine animal. Inhibition of tubulin polymerization mediated through the vinca alkaloid binding domain. Biochem Pharmacol 39:1941–1949.

Bai RB, Pettit GR, Hamel E (1990b): Binding of dolastatin 10 to tubulin at a distinct site for peptide antimitotic agents near the exchangeable nucleotide and vinca alkaloid sites. J Biol Chem 265:17141–17149.

Bai RB, Paull KD, Herald CL, Malspeis L, Pettit GR, Hamel E (1991): Halichondrin B and homohalichondrin B, marine natural products binding in the vinca domain of tubulin. Discovery of tubulin-based mechanism of action by analysis of differential cytotoxicity data. J Biol Chem 266:15882–15889.

Banerjee A, Luduena RF (1991): Distinct colchicine binding kinetics of bovine brain tubulin lacking the type III isotype of β-tubulin. J Biol Chem 266:1689–1691.

Bayley PM, Martin SR (1991): Microtubule dynamic instability: Some possible physical mechanisms and their implications. Biochemical Soc Trans 19:124–128.

Bergen LG, Borisy GG (1983): Tubulin–colchicine complex inhibits microtubule elongation at both plus and minus ends. J Biol Chem 258:4190–4194.

Bergen LG, Borisy GG (1986): Tubulin-colchicine complex (TC) inhibits microtubule depolymerization by a capping reaction exerted preferentially at the minus end. J Cell Biochem 30:11–18.

Bokros CL, Hugdahl JD, Hanesworth VR, Murthy JV, Morejohn LC (1993): Characterization of the reversible taxol-induced polymerization of plant tubulin in microtubules. Biochemistry 32:3437–3447.

Borisy GG, Taylor EW (1967a): The mechanism of action of colchicine. Binding of colchicine-^{3}H to cellular protein. J Cell Biol 34:525–533.

Borisy GG, Taylor EW (1967b): The mechanism of action of colchicine. Colchicine binding to sea urchin eggs and the mitotic apparatus. J Cell Biol 34:535–548.

Cabral F, Barlow SB (1991): Resistance to antimitotic agents as genetic probes of microtubule structure and function. Pharmacol Ther 52:159–171.

Caplow M, Zeeburg B (1981): Incorporation of radioactive tubulin into microtubules at steady state. Experimental and theoretical analysis of the effect of podophyllotoxin. J Biol Chem 256:5608–5611.

Caplow M, Zeeburg B (1982): Dynamic properties of microtubules at steady state in the presence of taxol. Eur J Biochem 127:319–324.

Carlier MF (1989): Role of nucleotide hydrolysis in the dynamics of actin filaments and microtubules. Int Rev Cytol 115:139–170.

Carlier MF, Pantaloni D (1983): Taxol effect on tubulin polymerization and associated guanosine 5′-triphosphate hydrolysis. Biochemistry 22:4814–4822.

Chaudoreille MM, Peyrot V, Braguer D, Codaccioni F, Crevat A (1991): Qualitative study of the interaction mechanism of estrogenic drugs with tubulin. Biochem Pharmacol 41:685–693.

Cleveland DW, Lopata MA, Sherline P, Kirschner MW (1981): Unpolymerized tubulin modulates the level of tubluin mRNAs. Cell 25:537–546.

Cornman I, Cornman ME (1951): The action of podophyllin and its fractions on marine eggs. Ann NY Acad Sci 51:1443–1488.

Correia JJ (1991): Effects of antimitotic agents on tubulin-nucleotide interactions. Pharmacol Ther 52:127–147.

Cortese F, Bhattacharyya B, Wolff J (1977): Podophyllotoxin as a probe for the colchicine binding site of tubulin. J Biol Chem 252:1134–1140.

Cutts JH, Beer CT, Noble RL (1960): Biological properties of vincaleukoblastine, an alkaloid present in *Vinca rosea* Linn, with reference to its anti-tumor action. Cancer Res 20:1023–1031.

David-Pfeuty T, Simon C, Pantaloni D (1979): Effect of antimitotic drugs on tubulin GTPase activity and self-assembly. J Biol Chem 254:11696–11702.

Davidse LC, Flach W (1977): Differential binding of methyl benzimidazol-2-yl carbamate to fungal tubulin as mechanism of resistance to this antimitotic agent in mutant strains of *Aspergillus nidulans*. J Cell Biol 72:174–193.

DeBrabander MJ, Van de Veire RML, Aerts FEM, Borgers M, Janssen PAJ (1976): The effects of methyl [5-(2-thienylcarbonyl)-1H-benzimidazol-2-yl]carbamate (R 17934; NSC 238159), a new synthetic antitumoral drug interfering with microtubules, on mammalian cells cultured *in vitro*. Cancer Res 36:905–916.

Deery WJ, Weisenberg RC (1981): Kinetic and steady-state analysis of microtubules in the presence of colchicine. Biochemistry 20:2316–2324.

Detrich HW III, Williams RC Jr, Wilson L (1982): Effect of colchicine binding on the reversible dissociation of the tubulin dimer. Biochemistry 21:2392–2400.

Diaz JF, Andreu JM (1993): Assembly of purified GDP-tubulin into microtubules induced by taxol and taxotere: Reversibility, ligand stoichiometry and competition. Biochemistry 32:2747–2755.

Dubel S, Little M (1988): Microtubule-dependent cell cycle regulation is implicated in the G_2 phase of *Hydra* cells. J Cell Sci 91:347–359.

Dustin P (1984): Microtubules, 2nd ed. Berlin: Springer-Verlag.

Farrell KW, Wilson L (1980): Proposed mechanism for colchicine poisoning of microtubules reassembled *in vitro* from *Strongylocentrotus purpuratus* sperm tail outer doublet tubulin. Biochemistry 19:3048–3054.

Farrell KW, Wilson L (1984): The differential kinetic stabilization of opposite microtubule ends by tubulin-colchicine complexes. Biochemistry 23:3741–3748.

Farrell KW, Jordan MA, Miller HP, Wilson L (1987): Phase dynamics at microtubule ends: The coexistence of microtubule length changes and treadmilling. J Cell Biol 184:1035–1046.

Fuchs DA, Johnson RK (1978): Cytologic evidence that taxol, an antineoplastic agent from *Taxus brevifolia*, acts as a mitotic spindle poison. Cancer Treat Rep 62:1219–1222.

Fujiwara K, Tilney LG (1975): Substructural analysis of the microtubule and its polymorphic forms. Ann NY Acad Sci 253:27–50.

Friedman PA, Platzer EG (1978): Interaction of anthelmintic benzimidazoles and benzimidazole derivatives with bovine brain tubulin. Biochim Biophys Acta 544:605–614.

Garland DL (1978): Kinetics and mechanism of colchicine binding to tubulin: Evidence for ligand induced conformational change. Biochemistry 17:4266–4272.

Gelfand V, Bershadsky AD (1991): Microtubule dynamics: Mechanism, regulation, and function. Annu Rev Cell Biol 7:93–116.

Gerzon K (1980): Dimeric catharanthus alkaloids. In Cassady JM, Douros JD (eds): Anticancer Agents Based On Natural Product Models. New York: Academic Press, pp 271–317.

Gildersleeve RF, Cross AR, Cullen KE, Fagan AP, Williams Jr RC (1992): Microtubules grow and shorten at intrinsically variable rates. J Biol Chem 267: 7995–8006.

Hamaguchi Y, Toriyama M, Sakai H, Hiramoto Y (1987): Redistribution of fluorescently labelled tubulin in the mitotic apparatus of sand dollar eggs and the effects of taxol. Cell Struct Funct 12:43–52.

Hamel E (1990): Interactions of tubulin with small ligands. In Avila J (ed): Microtubule Proteins. Boca Raton, FL: CRC Press, pp 89–191.

Hamel E, del Campo AA, Lowe MC, Lin CM (1981): Interactions of taxol, microtubule-associated proteins, and guanine nucleotides in tubulin polymerization. J Biol Chem 256:11887–11894.

Hamel E, Ho HH, Kang G-J, Lin CM (1988): Cornigerine, a potent antimitotic *Colchicum* alkaloid of unusual structure. Interactions with tubulin. Biochem Pharmacol 37:2445–2449.

Hartley-Asp B, Deinum J, Wallin M (1985): Diethylstilbestrol induces metaphase arrest and inhibits microtubule assembly. Mutat Res 143:231–235.

Hastie SB (1991): Interactions of colchicine with tubulin. Pharmacol Ther 512:377–401.

Hayden JJ, Bowser SS, Rieder C (1990): Kinetochores capture astral microtubules during chromosome attachment to the mitotic spindle: direct visualization in live newt lung cells. J Cell Biol 111:1039–1045.

Himes RH (1991): Interactions of the catharanthus (Vinca) alkaloids with tubulin and microtubules. Pharmacol Ther 51:257–267.

Hoebeke J, Van Nijen G, De Brabander M (1976): Interaction of oncodazole (R 17934), a new antitumoral drug, with rat brain tubulin. Biochem Biophys Res Commun 69:319–324.

Horio T, Hotani H (1986): Visualization of the dynamic instability of individual microtubules by dark-field microscopy. Nature 321:605–607.

Hotani H, Horio T (1988): Dynamics of microtubules visualized by darkfield microscopy: Treadmilling and dynamic instability. Cell Motil Cytoskel 10:229–236.

Howard WD, Timasheff SN (1988): Linkages between the effects of taxol, colchicine, and GTP on tubulin polymerization. J Biol Chem 263:1342–1346.

Hoyt MA, Totis L, Roberts BT (1991): *S. cerevisiae* genes required for cell cycle arrest in response to loss of microtubule function. Cell 66:507–517.

Ireland CM, Gull K, Gutteridge WE, Pogson CI (1979): The interaction of benzimidazole carbamates with mammalian microtubule protein. Biochem Pharmacol 28:2680–2682.

Johnson IS, Wright HF, Svoboda GH (1959): Experimental basis for clinical evaluation of anti-tumor principles derived from *Vinca rosea* Linn. J Lab Clin Med 54:830–837.

Jordan MA, Farrell KW (1983): Differential radiolabelling of opposite microtubule ends: Methodology, equilibrium exchange–flux analysis, and drug poisoning. Anal Biochem 130:41–53.

Jordan MA, Margolis RL, Himes RH, Wilson L (1986): Identification of a distinct class of vinblastine binding sites on microtubules. J Mol Biol 187:61–73.

Jordan MA, Wilson L (1990): Kinetic analysis of tubulin exchange at microtubule ends at low vinblastine concentrations. Biochemistry 29:2730–2739.

Jordan MA, Thrower D, Wilson L (1991): Mechanism of inhibition of cell proliferation by the vinca alkaloids. Cancer Res 51:2212–2222.

Jordan MA, Thrower D, Wilson L (1992): Effects of vinblastine, podophyllotoxin, and nocodazole on mitotic spindles: Implications for the role of microtubule dynamics in mitosis. J Cell Sci 102:401–416.

Jordan MA, Toso RJ, Thrower D, Wilson L (1993): Mechanism of mitotic block and inhibition of cell proliferation by taxol at low concentrations. Proc Natl Acad Sci USA (in press).

Jung MK, Wilder IB, Oakley BR (1992): Amino acid alterations in the benA (β-tubulin) gene of *Aspergillus nidulans* that confer benomyl resistance. Cell Motil Cytoskel 22:170–174.

Keates RAB (1981): Griseofulvin at low concentration inhibits the rate of microtubule polymerization *in vitro*. Biochem Biophys Res Commun 102:746–752.

Kowalski RJ, Williams RC Jr (1993): Microtubule-associated protein 2 alters the dynamic properties of microtubule assembly and disassembly. J Biol Chem 268:9847–9855.

Kumar N (1981): Taxol-induced polymerization of purified tubulin. J Biol Chem 256:10435–10441.

Kung AL, Sherwood SW, Schimke RL (1990): Cell line-specific differences in the control of cell cycle progression in the absence of mitosis. Proc Natl Acad Sci USA 87:9533–9557.

Lambeir A, Engelborghs Y (1981): A fluorescence stopped flow study of colchicine binding to tubulin. J Biol Chem 256:3279–3282.

Lee C, Ferguson M, Chen LB (1989): Construction of the endoplasmic reticulum. J Cell Biol 109:2045–2055.

Lee JC, Field DJ, Lee LLY (1980): Effects of nocadazole on structures of calf brain tubulin. Biochemistry 19:6209–6215.

Li R, Murray AW (1991): Feedback control of mitosis in budding yeast. Cell 66:519–531.

Lin CM, Hamel E (1981): Effects of inhibitors of tubulin polymerization on GTP hydrolysis. J Biol Chem 256:9242–9245.

Lin CM, Singh SB, Chu PS, Dempcy RO, Schmidt JM, Pettit GR, Hamel E (1988): Interactions of tubulin with potent natural and synthetic analogs of the antimitotic agent combrestatin: A structure–activity study. Mol Pharmacol 34:200–208.

Lin CM, Ho HH, Pettit GR, Hamel E (1989): Antimitotic natural products combrestatin A-4 and combrestatin A-2: Studies on the mechanism of their inhibition of the binding of colchicine to tubulin. Biochemistry 28:6984–6991.

Luduena RF (1979): Biochemistry of tubulin. In Roberts K, Hyams JS (eds): Microtubules. London: Academic Press, pp 65–116.

Luduena RF, Roach MC (1991): Tubulin sulfhydryl groups as probes and targets for antimitotic and antimicrotubule agents. Pharmacol Ther 49:133–152.

Manfredi JJ, Parness J, Horwitz SB (1982): Taxol binds to cellular microtubules. J Cell Biol 94:688–696.

Margolis RL, Wilson L (1977): Addition of colchicine–tubulin complex to microtubule ends: The mechanism of substoichiometric colchicine poisoning. Proc Natl Acad Sci USA 74:3466–3478.

Margolis RL, Wilson L (1978): Opposite end assembly and disassembly of microtubules at steady state *in vitro*. Cell 8:1–13.

Margolis RL, Rauch CT, Wilson L (1980): Mechanism of colchicine dimer addition to microtubule ends: Implications for the microtubule polymerization mechanism. Biochemistry 19:5550–5557.

McIntosh JR, Hering GE (1991): Spindle fiber action and chromosome movement. Annu Rev Cell Biol 7:403–426.

Melki R, Carlier M-F, Pantaloni D (1990): Direct evidence for GTP and GDP–P_i intermediates in microtubule assembly. Biochemistry 29:8921–8932.

Medrano FJ, Andreu JM, Gorbunoff MJ, Timasheff SN (1989): Roles of colchicine rings B and C in the binding process to tubulin. Biochemistry 28:5589–5599.

Minotti AM, Barlow SB, Cabral F (1991): Resistance to antimitotic drugs in Chinese hamster ovary cells correlated with changes in the level of polymerized tubulin. J Biol Chem 266:3987–3994.

Mitchison TJ (1988): Microtubule dynamics and kinetochore function in mitosis. Annu Rev Cell Biol 4:527–549.

Mitchison TJ (1989): Poleward microtubule flux in the mitotic spindle; evidence from photoactivation of fluorescence. J Cell Biol 109:637–652.

Mitchison TJ, Evans L, Schulze E, Kirschner M (1986): Sites of microtubule assembly and disassembly in the mitotic spindle. Cell 45:515–527.

Mitchison TJ, Kirschner M (1984a): Microtubule assembly nucleated by isolated centrosomes. Nature 312: 232–237.

Mitchison TJ, Kirschner M (1984b): Dynamic instability of microtubule growth. Nature 312:237–242.

Mitchison TJ, Salmon ED (1992): Poleward kinetochore fiber movement occurs during both metaphase and anaphase-A in newt lung cell mitosis. J Cell Biol 119:569–582.

Mogenson M, Tucker JB (1990): Taxol influences control of protofilament number at microtubule-nucleating sites in *Drosophila*. J Cell Sci 97:101–107.

Morejohn LC, Fosket DE (1991): The biochemistry of compounds with anti-microtubule activity in plant cells. Pharmacol Ther 51:217–230.

Na GC, Timasheff SN (1980a): Stoichiometry of the vinblastine-induced self-association of calf brain tubulin. Biochemistry 19:1347–1354.

Na GC, Timasheff SN (1980b): Thermodynamic linkage between tubulin self-association and the binding of vinblastine. Biochemistry 19:1355–1365.

Na GC, Timasheff SN (1982): *In vitro* vinblastine induced tubulin paracrystals. J Biol Chem 257:10387–10391.

Na GC, Timasheff SN (1986): Interaction of vinblastine with calf brain tubulin: Effect of magnesium ions. Biochemistry 25:6222–6228.

O'Brien ET, Voter WA, Erickson HP (1987): GTP hydrolysis during microtubule assembly. Biochemistry 26:4148–4156.

Olmsted JB, Borisy GG (1973): Characterization of microtubule assembly in porcine brain extracts by viscometry. Biochemistry 12:4282–4289.

Owellen RJ, Hartke CA, Dickerson RM, Hains FO (1976): Inhibition of tubulin–microtubule polyermization by drugs of the *Vinca* alkaloid class. Cancer Res 36: 1499–1502.

Parness J, Horwitz S (1981): Taxol binds to polymerized tubulin *in vitro*. J Cell Biol 91:479–487.

Peyrot V, Leynadier D, Sarrazin M, Briand C, Rodriguez A, Nieto JM, Andreu JM (1989): Interaction of tubulin and cellular microtubules with the new antitumor drug MDL 27048. J Biol Chem 264:21296–21301.

Pryer NK, Walker RA, Skeen VP, Bourns BD, Soboeiro MF, Salmon ED (1993): Brain microtubule-associated proteins modulate microtubule dynamic instability *in vitro*. J Cell Sci 103:965–976.

Rao S, Horwitz SB, Ringel I (1992): Direct photoaffinity labeling of tubulin with taxol. J Natl Cancer Inst 84:785–788.

Rieder CL, Alexander SP (1990): Formation of the astral mitotic spindle: Ultrastructural basis for the centrosome-kinetochore interaction. Electron Microsc Rev 3:269–300.

Ringel I, Horwitz SB (1991a): Effect of alkaline pH on taxol-microtubule interactions. J Pharmacol Exp Ther 259:855–860.

Ringel I, Horwitz SB (1991b): Studies with RP56976 (taxotere): A semisynthetic analogue of taxol. J Natl Cancer Inst 83:288–291.

Roberts JR, Rowinsky EK, Donehower RC, Robertson J, Allison DC (1989): Demonstration of the cell cycle positions of taxol-induced "asters" and "bundles" by sequential measurements of tubulin immunofluorescence, DNA content, and autoradiographic labeling of taxol-sensitive and resistant cells. J Histochem Cytochem 37:1659–1665.

Rowinsky EK, Donehower RC (1991): The clinical pharmacology and use of antimicrotubule agents in cancer chemotherapeutics. Pharmacol Ther 52:35–84.

Rowinsky EK, Cazenave LA, Donehower RC (1990): Taxol: A novel investigational antimicrotubule agent. J Natl Cancer Inst 82:1247–1259.

Rowinsky EK, Donehower RC, Jones RJ, Tucker RW (1988): Microtubule changes and cytotoxicity in leukemic cell lines treated with taxol. Cancer Res 48: 4093–4100.

Schiff PB, Fant J, Horwitz SB (1979): Promotion of microtubule assembly *in vitro* by taxol. Nature 277: 665–667.

Schiff PB, Horwitz SB (1980): Taxol stabilizes microtubules in mouse fibroblast cells. Proc Natl Acad Sci USA 77:1561–1565.

Schiff PB, Horwitz SB (1981): Taxol assembles tubulin in the absence of exogenous guanosine 5′-triphosphate or microtubule-associated proteins. Biochemistry 20:3247–3252.

Schilstra MJ, Martin SR, Bayley PM (1989): The effect of podophyllotoxin on microtubule dynamics. J Biol Chem 264:8827–8834.

Skoufias D, Wilson L (1992): Mechanism of inhibition of microtubule polymerization by colchicine: Inhibitory potencies of unliganded colchicine and tubulin-colchicine complexes. Biochemistry 31:738–746.

Silence K, D'Hoore A, Engelborghs Y, Peyrot V, Briand C (1992): Fluorescence stopped-flow study of the interaction of tubulin with the antimitotic drug MDL 27048. Biochemistry 31:11133–11137.

Singer WD, Jordan MA, Wilson L, Himes RH (1989): Binding of vinblastine to stabilized microtubules. Mol Pharmacol 36:366–370.

Sloboda RD, Van Blaricom G, Creasey WA, Rosenbaum JL, Malawista SE (1982): Griseofulvin: Association with tubulin and inhibition of *in vitro* microtubule assembly. Biochem Biophys Res Commun 105:882–888.

Sluder G (1979): Role of spindle microtubules in the control of cell cycle timing. J Cell Biol 80:674–691.

Soifer D (1975): The biology of cytoplasmic microtubules. Ann NY Acad Sci 253:1–848.

Sternlicht H, Ringle I (1979): Colchicine inhibition of microtubule assembly via copolymer formation. J Biol Chem 254:10540–10550.

Sternlicht H, Ringel I, Szasz J (1983): Theory for modelling the copolymerization of tubulin and tubulin–colchicine complex. Biophys J 42:255–267.

Stewart RJ, Farrell KW, Wilson L (1990): Role of GTP hydrolysis in microtubule polymerization: evidence for a coupled hydrolysis mechanism. Biochemistry 29: 6489–6498.

Timasheff SN, Andreu JM, Na GC (1991): Physical and spectroscopic methods for the evaluation of the interactions of antimitotic agents with tubulin. Pharmacol Ther 52:191–210.

Thompson WC, Wilson L, Purich DL (1981): Taxol induces microtubule assembly at low temperature. Cell Motil 1:445–454.

Thrower D, Jordan MA, Wilson L (1993): A quantitative solid-phase binding assay for tubulin. In Asai DJ (ed): Methods in Cell Biology, vol 37. San Diego: Academic Press (in press).

Toso RJ, Jordan MA, Farrell KW, Matsumoto B, Wilson L (1993): Kinetic stabilization of microtubule dynamic instability *in vitro* by vinblastine. Biochemistry 32: 1285–1293.

Turner PF, Margolis RL (1984): Taxol-induced bundling of brain-derived microtubules. J Cell Biol 99:940–946.

Walker RA, O'Brien ET, Pryer NK, Soboeiro MF, Voter WA, Erickson H, Salmon ED (1988): Dynamic instability of individual microtubules analyzed by video light microscopy: Rate constants and transition frequencies. J Cell Biol 107:1437–1448.

Weber K, Wehland J, Herzog W (1976): Griseofulvin interacts with microtubules both *in vivo* and *in vitro*. J Mol Biol 102:817–829.

Wehland J, Herzog W, Weber K (1977): Interaction of griseofulvin with microtubules, microtubule protein and tubulin. J Mol Biol 111:329–342.

Weisenberg RC, Borisy GG, Taylor WE (1968): The colchicine-binding protein of mammalian brain and its relation to microtubules. Biochemistry 7:4466–4478.

Wendell KL, Wilson L, Jordan MA (1993): Mitotic block in HeLa cells by vinblastine: ultrastructural changes in kinetochore-microtubule attachment and in centrosomes. J Cell Sci 104:261–274.

Wilson L (1970): Properties of colchicine binding proteins from chick embryo brain. Interactions with vinca alkaloids and podophyllotoxin. Biochemistry 9:4999–5007.

Wilson L (1975): Microtubules as drug receptors: pharmacological properties of microtubule protein. Ann NY Acad Sci 253:213–231.

Wilson L, Anderson K, Chin D (1976): Nonstoichiometric poisoning of microtubule polymerization: A model for the mechanism of action of the vinca alkaloids, podophyllotoxin, and colchicine. In Goldman R, Pollard T, Rosenbaum JL (eds): Cell Motility. Cold Spring Harbor, New York; Cold Spring Harbor Laboratory, pp 1051–1064.

Wilson L, Farrell KW (1986): Kinetics and steady-state dynamics of tubulin addition and loss at opposite microtubule ends: The mechanism of action of colchicine. Ann NY Acad Sci 466:690–708.

Wilson L, Friedkin M (1967): The biochemical events of mitosis II. The *in vivo* and *in vitro* binding of colchicine in grasshopper embryos and its possible relation to inhibition of mitosis. Biochemistry 6: 3126–3135.

Wilson L, Margolis RL (1981): Microtubule treadmills—possible molecular machinery. Nature 293: 705–711.

Wilson L, Jordan MA, Morse A, Margolis RL (1982): Interaction of vinblastine with steady-state microtubules *in vitro*. J Mol Biol 159:129–149.

Wilson L, Miller HP, Pfeffer TA, Sullivan KF, Detrich HW, III (1984): Colchicine-binding activity distinguishes sea urchin egg and outer doublet tubulins. J Cell Biol 99:37–41.

Wilson L, Miller HP, Farrell KW, Snyder KB, Thompson WC, Purich DL (1985a): Taxol stabilization of microtubules *in vitro:* Dynamics of tubulin addition and loss at opposite microtubule ends. Biochemistry 24:5254.

Wilson L, Miller HP, Jordan MA, Farrell KW (1985b) Dynamics of tubulin addition and loss at opposite ends of bovine brain microtubules *in vitro:* Analysis using double-labelled microtubules. In DeBrabander M, DeMey J (eds): Microtubules and Microtubule Inhibitors. Amsterdam: Elsevier Science Publishers BV, pp 71–78.

ABOUT THE AUTHORS

LESLIE WILSON is Professor of Biochemistry and Pharmacology in the Department of Biological Sciences at the University of California, Santa Barbara, where he teaches undergraduate and graduate courses in molecular pharmacology and cell biology. After receiving a B.S. in pharmacy from the Massachusetts College of Pharmacy and Allied Health Sciences in Boston, he pursued doctoral research in the laboratory of Morris Friedkin in the Department of Pharmacology at Tufts University School of Medicine, where he earned a Ph.D. in 1967 for work on the identification and characterization of tubulin as the receptor for colchicine. This was followed by postdoctoral research at the University of California, Berkeley, where he worked in Daniel Mazia's laboratory on the heterodimeric subunit structure of tubulin and on the mechanisms of action of vinblastine and colchicine. He was Assistant Professor of Pharmacology at Stanford University School of Medicine before joining the faculty at the University of California, Santa Barbara, in 1975. Dr. Wilson's research has involved elucidation of the polymerization dynamics of microtubules and determination of the mechanisms by which drugs and microtubule-associated proteins modulate microtubule dynamics. He is co-editor of *Methods in Cell Biology* and an associate editor of *Biochemistry.*

MARY ANN JORDAN is Associate Research Biologist at the University of California, Santa Barbara. After studying mathematics at Carleton College, receiving a B.A. magna cum laude in mathematics from the University of Minnesota, and participating in an NSF-sponsored marine biological expedition to the South Pacific from Stanford University, she pursued doctoral research at the University of Rochester in the laboratory of Dr. Jerome Kaye and Dr. Rachel McMaster-Kaye, where she earned a Ph.D. in 1969 for ultrastructural studies of spermiogenesis. This was followed by postdoctoral research in the ultrastructure of spermatogenesis in the laboratory of Dr. David Phillips at Washington University in St. Louis, and research and teaching positions in structural and marine biology at the University of Wyoming, the University of Michigan, and Utah State University. In 1978 she joined the laboratory of Dr. Leslie Wilson at the University of California, Santa Barbara, to pursue postdoctoral research in microtubule biochemistry, ultrastructure, and the effects of vinblastine on tubulin exchange at microtubule ends. She continues there as a research biologist, pursuing and directing research in microtubule dynamics and function in mitosis and in the mechanism of action of antimitotic drugs, particularly vinblastine and taxol, in blocking mitosis and inducing cell death.

Microtubules: 85–109

The Role of Multiple Tubulin Isoforms in Cellular Microtubule Function

Elizabeth C. Raff

J. B. S. Haldane is reputed to have once remarked that the heavenly father must have an inordinate fondness for beetles, since he created so many of them. Perhaps the same is true of tubulin genes?

I. INTRODUCTION

Microtubules are evolutionarily ancient organelles used in cells of all eukaryotic organisms to perform diverse functions involved in cell division, cell motility, and the generation and maintenance of cell shape. In different cells, or at different times or places in the same cell, many different kinds of microtubule-based structures are assembled, each unique in architecture and mode of function. Early in the history of tubulin biochemistry, the assembly into microtubules *in vitro* of purified tubulin in the absence of any other proteins definitively established that formation of a tubular polymer is an intrinsic property of the α/β-tubulin heterodimer [Lee and Timasheff, 1975; Erickson and Voter, 1976; Herzog and Weber, 1977]. However, it has also long been clear that microtubule assemblages *in vivo* are comprised not only of the α- and β-tubulin subunits themselves, but also of a complex and diverse set of permanently or transiently associated additional proteins.

The morphology and functional properties in the cell of any given microtubule structure will therefore be determined in a complex way by a multiplicity of factors, including the total spectrum of constitutent proteins and their interaction with cellular regulatory signals. Just one of the factors that must be considered as we try to work out how the specificity of microtubule assembly and function is controlled *in vivo* is the fact that different microtubule arrays are constructed from different α- and β-tubulins. This is the specific part of the question that I will consider here: What are the roles played in cellular microtubule function by the structural diversities of α- and β-tubulins?

II. TUBULIN GENE FAMILIES AND PATTERNS OF ISOFORM EXPRESSION AND USAGE

Most species express multiple isoforms of both α- and β-tubulin that are distinct but related proteins encoded in multigene families [Raff, 1984; Sullivan and Cleveland, 1986; Raff et al., 1987; Sullivan, 1988; Joshi and Cleveland, 1990; Murphy, 1991; Ludueña et al., 1992; Chasan, 1992]. The size of the gene families ranges from the small families in fungi consisting of one or two genes each for α- and β-tubulin, intermediate-sized gene families such as the *Drosophila melanogaster* family of four members each, to larger gene families typified by the vertebrate families of five to seven genes each for α- and β-tubulin. The record for functional tubulin genetic complexity has recently been taken by *Arabidopsis thaliana*, with a total of 16 expressed tubulin genes—seven α and nine β [Marks et al., 1987; Oppenheimer et al., 1988; Kopczak et al., 1992; Snustad et al., 1992]; thus, as the authors point out, this otherwise rather modest green plant devotes about 0.05% of its genome to tubulin genes.

The expression and usage of tubulin isoforms varies widely across phylogeny; examples of the patterns of tubulin gene expression in several species are given in the following discussion. However, in spite of the complexity of detail, one can make some generalizations about the mode of utilization of different members of a given tubulin gene family. The typical pattern of expression is that one or more genes encode "major" isoforms that are expressed at high levels at many developmental times and/or in many tissues or cell types, while the remaining genes encode "minor" isoforms expressed in lower levels and/or fewer cell types. Frequently more than one gene from each family may be expressed in a single cell type. In some cases one or more major isoforms may be expressed nearly constitutively, but the temporal and spatial regulation of expression is often complex for both major and minor isoforms. Note that the designation of an isoform as *minor* may in some cases be misleading, since an isoform that contributes only a minor part of the overall tubulin repertoire of an organism may be the predominant form in a specific cell type, as, for example is the case for the divergent vertebrate hematopoietic β-tubulin [Murphy et al., 1987; Monteiro and Cleveland, 1987]. Another general feature of tubulin gene families in higher eukaryotes is that the minor isoforms represent the more diverged protein sequences, while the major isoforms are more highly conserved in sequence both within a given lineage and when compared to major isoforms in other species, as for example, in the vertebrate [Lewis et al., 1985a,b; Monteiro and Cleveland, 1987] and *Drosophila* [Biajolin et al., 1984; Theurkauf et al., 1986; Rudolph et al., 1987] families. All species in which the tubulin gene families have been studied have similar numbers of α- and β-tubulin genes, but there is no evidence in higher eukaryotes for coordinate expression of obligate α/β-gene pairs. Rather, expression of the α- and β-tubulin gene families appears to be independently regulated. Thus, given that multiple genes for both subunits may be expressed in a single cell, the potential for genetic diversity at the heterodimer level may considerably exceed the number of individual tubulin genes.

Finally, in addition to genetic complexity, additional diversity within the cell may be generated at the protein level. Extensive and complex posttranslational modifications of α-tubulins exist, including detyrosination/tyrosination, acetylation, and glutamylation [L'Hernault and Rosenbaum, 1985; Piperno and Fuller, 1985; Bulinski et al., 1988; Baas and Black, 1990; Eddé et al., 1990, 1991]. Protein modifications are much less extensive for β-tubulins, but modification of specific minor β-tubulin isoforms occurs, including a phosphorylated *Drosophila* isoform [Diaz et al., 1990, 1991] and both phosphorylation and glutamylation of certain vertebrate isoforms [Gard and Kirchner, 1985; Diaz-Nido et al., 1990; Lee et al., 1990; Alexander et al., 1991; Rudiger et al., 1992]. Posttranslational modification of α-tubulin may be the primary or only source of tubulin heterogeneity in species that lack genetic tubulin diversity, for example, in ciliated and flagellated protists [Youngblom et al., 1984; Silflow et al., 1985; reviewed by Silflow, 1991], but also in at least one metazoan, the brine shrimp *Artemia salina* [Langdon et al., 1991a,b]. The precise functional roles played by tubulin protein modifications have remained tantalizingly elusive, although detyrosination and perhaps also acetylation of α-tubulin may have a role, albeit indirect, in microtubule stability [Bulinski et al., 1988; Khawaja et al., 1988; Webster et al., 1990].

III. DO STRUCTURALLY DIFFERENT TUBULINS HAVE INTRINSICALLY DIFFERENT BIOCHEMICAL PROPERTIES THAT DETERMINE MICROTUBULE FUNCTION *IN VIVO*?

A priori, one can posit two nonmutually exclusive "reasons" why it would be advantageous to an organism to have multiple tubulin genes. One possibility is that the existence of multiple tubulin genes might simply provide an organism a means by which differential temporal and spatial demands for tubulin can be precisely modulated. In this case differences in sequences of different isoforms might simply reflect the time of separation of the sequences and would thus represent accumulation of allowable changes in protein-coding sequence compatible with the basic requirements for microtubule assembly, including dimerization of the α- and β-subunits and end-to-end and side-to-side associations of dimers within the polymer (see Burns, this volume).

However, given what we know about structure–function relationships in proteins, the existence of diverged tubulin structural genes suggests the possibility that there has been evolutionary selection for proteins with different functional features. This indeed is the classic general picture of one way in which protein function can be modulated during evolution: Duplication of a gene ensures that the organism can both retain the original function of the gene and at the same time allows the possibility of divergence of the two copies of the gene, with subsequent selection for different functional capacities of the two gene products. The proposition that structurally different tubulin subunits each possess capabilities for assembly of different microtubule arrays was the crux of the multitubulin hypothesis stated by Fulton and Simpson [1976] at the time when microtubule-philes first began to realize that it was not just "tubulin" but rather "the tubulins" that we had to think about.

Two kinds of observations support the view that structurally distinct tubulin isoforms may also be functionally distinct. One, the spatially and temporally restricted patterns of expression of different genes that encode structurally distinct isoforms, together with the regulated patterns of covalent modifications at the protein level, are in themselves suggestive of functional specialization. Second, the evolutionary conservation of both the organization and function of the vertebrate tubulin gene families supports this hypothesis. The β-tubulin families in different vertebrate species, from man to chicken (i.e., representing about three hundred million years since a last common ancestor), are very similar to each other, exhibiting a high degree of conservation both of specific features of protein sequence and of patterns of expression and utilization for both the major and minor isoforms [Sullivan and Cleveland, 1986; Sullivan, 1988]. Thus the similarity between corresponding β-tubulin isoforms in two different vertebrate species is greater than the similarity among different isoforms within a species. A particular feature that is conserved in the corresponding isoforms in different vertebrate species is the sequence of the 15 amino acids at the C terminus, acidic in character but otherwise highly variable in sequence between different isoforms in the same species. The C-terminal sequence therefore has been used to designate sequence classes, or isotypes, within the vertebrate β-tubulins; an internal variable region near position 56 in the N-terminal region of the molecule also exhibits isotype-defining features [Sullivan and Cleveland, 1986]. Burns and Surridge [1990] have shown that the isotype-specific C terminus and the internal variable region exhibit evolutionary covariation. In sum, the concomitant conservation of both sequence and pattern of expression of the vertebrate β-tubulins strongly suggests that selection has operated to maintain isotype-specific functional features.

Accumulated data on microtubule function in many different organisms has in fact revealed that both kinds of mechanism operate in biological systems—Both regulatory and functional pressures have driven selection for tubulin heterogeneity. It is clearly not always the case that a tubulin is tubulin is a tubulin; on the other hand, there is by no means a tubulin for every tubule. It is precisely because there is not a single answer that the question of the functional role of structural differences in isoforms has been so surprisingly difficult to settle.

There have thus far been four unambiguous experimental comparisons of the functional capacities of two isoforms by molecular and genetic manipulations that allow one tubulin isoform to be replaced with another. These experiments allow direct assessment of whether two structurally distinct isoforms also possess distinct functional properties that are significant in the context of microtubule assembly in a cell. The score thus far is even: In two cases, for fungal tubulins, the answer is no; in two cases, for *Drosophila* tubulins, the answer is yes.

The genome of the budding yeast *Saccharomyces cerevisiae* contains only one β-tubulin gene, *TUB2*, but has two α-tubulin genes, *TUB1* and *TUB3*, that encode proteins that differ by 10% in amino acid sequence [Schatz et al., 1986a]. The *TUB1* protein is essential for normal growth; null mutations in the *TUB1* gene are lethal. The *TUB3* gene, on the other hand, is not essential for viability, but null mutations in the gene result in poor viability of spores, showing that the *TUB3* protein is normally utilized in this stage of the life cycle (and that strains without *TUB3* would almost certainly be at an enormous selective disadvantage in the wild). By genetic and molecular manipulation of gene copy number, Schatz and his colleagues [1986b] compared the relative ability of the two α-tubulin isoforms to function at different stages in the life cycle. They found that, if present at sufficient level, either protein alone could support all of the microtubule functions required for all aspects of normal growth, including assembly of meiotic and mitotic spindles and nuclear movement during cell division and mating. They therefore concluded that the genetic difference between the two genes (i.e., one is essential for viability and one is not) is simply a reflection of gene expression levels and that there is no significant qualitative functional difference between the two α-tubulin isoforms, in spite of their difference in amino acid sequence.

The filamentous mold *Aspergillus nidulans* has two genes each for α- and β-tubulin, whose expressions and functions have been studied by Morris and May and their colleagues. The situation for the *Aspergillus* β-tubulins is quite similar to that of the yeast α-tubulins. The genes *benA* and *tubC* encode β-tubulins that differ in sequence by 17% [May et al., 1987]. Similarly to the yeast α-tubulin pair, the *Aspergillus benA* gene is essential for viability, while *tubC* normally functions in asexual sporulation but is not essential for this process [Oakley and Morris, 1980; May et al., 1985; Weatherbee et al., 1985; May and Morris, 1988]. May [1989] generated strains that express the *tubC* protein as their sole β-tubulin by replacing the *benA* gene with a chimeric gene construct in which the *tubC* coding sequences are driven by the *benA* promoter. These strains were fully viable and did not exhibit any abnormalities in microtubule function, demonstrating that the two β-tubulins of *Aspergillus* are functionally interchangeable in spite of what for tubulins is a really considerable divergence in amino acid sequence. I should note that the overall *Aspergillus* tubulin situation is more complicated, since there is additional genetic heterogeneity to account for in the two α-tubulin isoforms, the relative functional capabilities of which have not been established. The α-tubulins are even more divergent than the two β-tubulins, differing by 28% in sequence, and both genes are essential, *tubA* for vegetative growth and *tubB* for sexual development [Kirk and Morris, 1991; Doshi et al., 1991].

At any rate, for both the yeast α-tubulins and the *Aspergillus* β-tubulins, it is reasonable to conclude that the minor isoform of each gene pair serves to ensure a sufficient tubulin pool during a specific stage of the life cycle but does not provide any crucial specialized functional properties. But perhaps there are still lingering doubts: There is indeed an unanswerable caveat to the interpretation of these experiments. The findings of Schatz et al. [1986b] and May [1989] are elegant demonstrations that the isoform pairs in question are functionally interchangeable for growth of the organism under normal laboratory growth conditions. We can never be sure that the much longer and more subtle experiment performed by evolution has not selected for functional differences significant over the long haul but too slight to detect in the scale of the lab experiment. In fact, I suspect that this is almost certainly so. Second, the laboratory experiments cannot measure all the contingencies in the selective history of the organism represented in the present tubulin gene sequences. For example, *S. cerevisiae* can also grow in a filamentous mode [Gimeno et al., 1992], for which tubulin requirements have not been established. Finally, while the fungal isoform-swap experiments show that each isoform in the pairs compared is equally capable of supplying required cellular function for growth in the laboratory, these data do not mean that the isoform pairs are identical proteins biochemically. The *Aspergillus* β-tubulins are in fact clearly not identical, since the two isoforms exhibit different sensitivity to the antimicrotubule drug benomyl [May, 1989]. Drug binding affinity provides a useful probe for aspects of tubulin molecular structure (see Burns, this volume), but it is not clear how—or whether—this property reflects features relevant to cellular function. (Note that differences in drug sensitivity of growth were also observed in the yeast experiments but could be attributed to α-tubulin pool levels rather than to isoform composition [Schatz et al., 1986b].)

In sum, considering all of the possible ramifications of the fungal isoform-swap data, I would taken the position that the simplest interpretation of these gene-swap experiments really does give us the crux of the answer: For these two fungal species, the differences in the coding sequences in the main reflect time of divergence of the genes. In these two cases, differential expression of the tubulin gene pairs in question thus primarily reflects regulatory requirements for tubulin usage during specific parts of the life cycle.

A comparison of two β-tubulin isoforms in *Drosophila* gave us quite a different picture, and finally provided a definitive example of intrinsically different functional properties *in vivo* exhibited by structurally different tubulins. The *Drosophila* genome contains four β-tubulin genes. Two of the genes, β1 and β2, encode highly conserved proteins that are about 95% similar in sequence not only to each other but also to other major metazoan β-tubulins [Rudolph et al., 1987; Michiels et al., 1987]. The other two *Drosophila* isoforms, β3 and β4, on the other hand, are considerably diverged not only from β1 and β2 (to which they are only 85%–88% similar in sequence), but from β-tubulins in other species as well [Rudolph et al., 1987; Raff et al., 1987; Diaz et al., 1990, 1991]. We compared the function of one of the major isoforms, β2, with that of the developmentally regulated minor isoform, β3 [Hoyle and Raff, 1990].

The normal contexts of expression of *Drosophila* β2- and β-tubulins are quite different. The β2-gene is expressed solely in the postmitotic cells of the male germline and is the only β-tubulin expressed in these cells [Kemphues et al., 1979, 1982; Michiels et al., 1989]. However, although it is restricted in site of expression to a single tissue, β2-tubulin has multiple microtubule functions in spermatogenesis and is required for assembly of all microtubules in the postmitotic germ cells, including the meiotic spindles, two different classes of cytoskeletal tubules, and the motile sperm tail axoneme [Kemphues et al., 1979, 1980, 1982, 1983; Raff and Fuller, 1984; Fuller et al., 1987, 1988]. The developmental pattern for β3-tubulin utilization presents an inverse situation. The β3-gene is expressed at several times in development and in many cell types [Raff et al., 1982; Michiels et al., 1987; Gasch et al., 1988; Kimble et al., 1989] and is essential for viability of the fly [Kimble et al., 1990]. However, our analysis of the cellular sites of β3 function leads us to conclude that this isoform is utilized primarily in one general category of microtubules—cytoskeletal microtubule arrays, often transient, that are involved in mediating cell shape changes or tissue arrangements [Kimble et al., 1989, 1990].

Using *Drosophila* stocks transformed with a hybrid gene construct in which β2 regulatory sequences drive the β3-coding sequence, we examined the ability of β3 to support the multiple functions normally provided in the male germ cells by β2 [Hoyle and Raff, 1990]. β3 alone could not replace β2, even when present

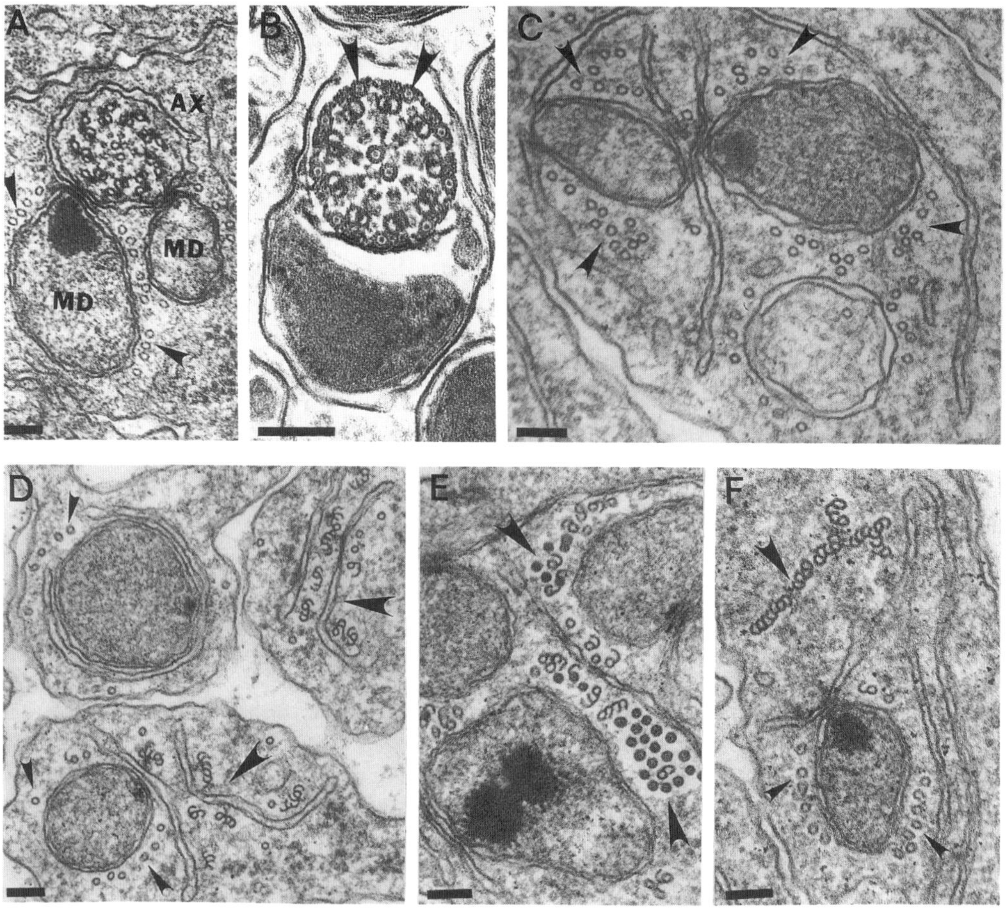

Fig. 1. β-*Tubulin usage in* Drosophila *spermatogenesis. Cross sections through developing spermatids.* **A,B:** *Spermatids in wild-type fertile males illustrating some of the microtubule classes supported by the testis-specific isoform β2-tubulin. A, an intermediate stage spermatid showing the developing axoneme (AX) with the accompanying mitochondrial derivatives (MD) and the associated cytoplasmic tubules (arrowheads) required for elongation of the mitochondrial derivatives. At this stage in axoneme morphogenesis, the typical "9 + 2" structure of nine doublet tubules surrounding a central pair of microtubules is present; the nine outer accessory tubules associated with the 'B' tubule of each doublet are forming. At this stage, the central pair and accessory tubules do not yet contain the luminal filament present in the singlet axoneme tubules of mature* Drosophila *sperm. Note that the mitochondrial associated tubules are disassembled subsequent to mitochondrial elongation. B, a mature spermatid in which axoneme morphogenesis and mitochondrial derivative development are complete. The mature axoneme consists of the "9 + 2" structure of nine doublet tubules surrounding a central pair of microtubules, plus the nine outer accesory tubules associated with the 'B' tubule of each doublet. Each of the central pair and accessory tubules contains a luminal filament (arrowheads). The major mitochondrial derivative is filled with the paracrystalline material (just beginning to accumulate in the stage shown in A); the minor mitochondrial derivative is now much smaller than the major derivative (compare with A). The mitochondrial derivative-associated tubules are no longer present.* **C:** *An intermediate stage spermatid in a transgenic male in which β3-tubulin is the sole β-tubulin available in the postmitotic male germ cells [see discussion of the transgenic constructs by Hoyle and Raff, 1990]. In such males the mitochondrial derivative-associated tubules (arrowheads) are the only functional microtubules assembled. Mitochondrial elongation, mediated by this associated cytoplasmic microtubule array, occurs almost normally. However, no axonemes are assembled.* **D–F:** *Spermatids in transgenic males in which carboxy-truncated β2-tubulin, designated β2ΔC, is the sole β-tubulin available in the*

in the testes in levels equivalent to β2 levels in fertile males. When expressed in the absence of the endogenous isoform (i.e., in testes of males carrying a null mutation in the β2 gene), β3-tubulin supported assembly and wild-type function of only one class of cytoskeletal microtubules, those that mediate elongation of the mitochondrial derivative that in the *Drosophila* sperm extends aong the axoneme (see Fig. 1A–C). However, β3-tubulin could not support axoneme assembly or function of spindles or the cytoskeletal tubules required for shaping of the spermatid nucleus. In testes of males in which both β3 and the endogenous testis-specific isoform β2 were coexpressed, the meiotic spindles and all classes of cytoskeletal microtubules functioned normally, but, if β3 exceeded 20% of the total β-tubulin pool, it disrupted normal axoneme assembly, resulting in dominant male sterility. We can draw two important conclusions from these findings. First, the two *Drosophila* isoforms β2 and β3 are not functionally equivalent. Second, within a single cell, assembly of different microtubule arrays exerts different requirements with regard to the structure of the component β-tubulins. This is true within different categories of single-walled microtubules (e.g., spindles and different kinds of cytoskeletal arrays), but our data also suggest that assembly of the doublet tubules of the axoneme poses the most severe structural constraints on microtubule assembly.

The differences in the relative functional capacities of β2 and β3 raise yet another question. That is, relative to the conserved β2 protein, is the diverged β3 protein specialized or merely limited? Selection must operate on the β2 sequence to maintain the capacity of the β2 protein for assembly of multiple morphological classes of microtubules; in particular, in addition to assembly of several different functional classes of singlet tubules, β2 is utilized to assemble the doublet tubules of the motile axoneme. β3, however, appears to be utilized only for singlet tubules, primarily transient cytoskeletal arrays. The diverged β3 sequence may thus simply represent drift away from structural elements required for other classes of microtubules. Alternatively, the β3 sequence may represent "fine tuning" for a particular aspect of structure required for the microtubules in which β3 normally functions. Although at present unanswered, this question could be experimentally addressed by the reciprocal experiment to that described above: Can β2 replace β3?

The situation in the postmitotic cells in the *Drosophila* male germline provides the simplest kind of context for experimental study of tubulin function, since in these cells several different kinds of microtubules are assembled, but all utilize a single species of heterodimer composed of the testis-specific β2-tubulin isoform in combination with the predominant *Drosophila* 84B α-tubulin isoform (discussed

---

postmitotic male germ cells [see discussion of the transgenic contruct by Fackenthal et al., 1993a]. D, Parts of two early stage spermatids containing functional mitochondrial derivative-associated tubules (small arrowheads) and abortive axoneme attempts (large arrowheads) consisting of disorganized arrays of doublet tubules, many of which are morphologically aberrant. E, A later stage spermatid; the abortive axoneme contains both aberant doublet tubules and singlet tubules that can be identified as having axonemal identity because they contain the luminal filament normally typical of the central pair and accessory tubules. F, An intermediate stage spermatid illustrating the failure of β2ΔC to support correct organization of the supramolecular architecture of the axoneme; the normal mitochondrial derivative-associated tubules (small arrowheads) as well as an abortive axoneme attempt (large arrowhead) are both present, the typical β2ΔC phenotype.

Within the limits we have measured, when β3 or β2ΔC are the sole β-tubulins in the postmitotic male germ cells, the qualitative phenotypes of microtubule assembly in spermatogenesis are not sensitive to the testis tubulin pool level. In the examples shown here, the β3-tubulin level in the male in C is equivalent to approximately 60% the β2-tubulin level in a normal wild-type male; the levels of β2ΔC in the males in D and E are equivalent to approximately 25% the β2-tubulin level in a normal wild-type male, and the β2ΔC level in the male shown in F is approximately equal to the β2-tubulin level in a normal wild-type male. Note that in males with the normal testis-specific isoform, β2-tubulin, fertility is supported by β2-tubulin levels from 50% to four times the normal β2-tubulin level in a normal wild-type male (i.e., diploid for the β2 gene). Bars = 100 nm [Micrographs were taken by F.R. Turner; F is from Fackenthal et al., 1993, with permission of the publisher.]

in the following section). The more general situation is that multiple α- and/or β-tubulins are present in the total cellular tubulin pool, as indeed is the situation in both yeast and *Aspergillus,* in which the major and minor α- and β-tubulin isoforms, respectively, are both present in cells at a specific stage in the life cycle.

The remaining case in which two structurally distinct tubulin isoforms have been shown also to be functionally distinct is provided by the study of Matthews and her colleagues [1993] on the role of the two α-tubulin isoforms present in the early *Drosohila* embryo. These data also provide the best example to date of a situation in which joint participation of more than one isoform of a given tubulin family is required for wild-type microtubule function. Before looking at this work in detail, therefore, I want briefly to address the general questions that have to be considered in thinking about the functional role played by the presence in the cell of more than one class of α- and/or β-tubulin.

IV. THE ROLE OF MULTIPLE TUBULIN ISOFORMS IN A SINGLE CELL

Given a multicomponent tubulin pool, two kinds of utilization are possible. Assembly of different microtubules may depend on selective utilization of different subsets of isoforms; alternatively, all isoforms available may be uniformly incorporated into all microtubules. Subcellular sorting of tubulin isoforms has been observed in a relatively small number of instances, primarily by immunocytochemical localization techniques. Isoform sorting has been observed in several cultured neuronal cell types [Burgoyne et al., 1988; and Remolona, 1989; Denoulet et al., 1989; Joshi and Cleveland, 1989]. Falconer et al. [1992] reported tubulin isoform sorting during differentiation of both neuronal and muscle cells in culture. We observed differential localization of the predominant *Drosophila* isoform β1 and the divergent isoform β3 in the developing wing blade cells [Kimble et al., 1989]. Of particular note is the observation by Oka et al. [1990] of differential localization of tubulins even within a single structure, the mitotic spindle of the sea urchin embryo. An experimental situation that suggests the existence of cellular mechanisms to distinguish between isoform classes are the results of Sisodia et al. [1990]. These authors obtained high transcriptional and translational levels of a class IV β-tubulin supported by a chicken gene overexpressed in mammalian cells in culture, but found that the heterologous protein failed to accumulate to significant levels, suggesting selective degradation of the chicken β-tubulin and hence selective omission from incorporation into microtubules relative to the endogenous isoforms. These results also point out that different cell types may have very different mechanisms for controlling both the size and structural composition of their tubulin pools: For example, in the *Drosophila* oocyte [Matthews et al., 1993] and testis [Fackenthal et al., 1993b] there do not seem to be mechanisms similar to the autologous control of tubulin pools at the level of message stability observed in cultured mammalian cells by Cleveland and his co-workers (see Cleveland [1988] and this volume).

The mechanisms by which subcellular sorting may be achieved are not known, but, taken together, observations such as described above provide suggestive evidence for functional sorting of tubulins during assembly of morphologically distinct microtubule arrays. The relative paucity of observations of subcellular sorting may reflect the actual case, i.e., that isoform sorting may occur in only a few cell types, particularly in neurons. Alternatively, it may be that subcellular sorting is difficult to document. Certainly, however, the bulk of evidence available at present suggests that the more general case is that all the tubulin subunits present in a cell are uniformly incorporated into all the mictotubule arrays assembled in the cell, as is suggested by the numerous instances of incorporation of heterologous tubulins into functional microtubules *in vivo* (see the discussion below) and as has also been directly demonstrated for the wild-type isoform mixes in several vertebrate cell types in culture [Lopata and Cleveland, 1987; Lewis et al, 1987; Sawada and Cabral, 1989] and in the *Drosophila* embryo [Theurkauf, 1992; Matthews et al., 1993].

Given uniform incorporation into microtubules of all members of the cellular soluble tubulin pool, there are two possible explanations for the selective advantage for expression of multiple isoforms in one cell— exactly analogous to the possible roles for multiple tubulin genes themselves. That is, expression of multiple tubulins in a single cell might simply serve to ensure a sufficient pool of tubulin available to meet the requirements for microtubules, and it would not really matter which of the isoforms were present, as long as a tubulin pool of sufficient size were maintained. This is the finding in the fungal isoform-swap experiments: In each case the role of the minor isoform appears to be to ensure a sufficient tubulin pool for a particular developmental process. Alternatively, the specific mix of isoforms might be required to generate the properties of the resultant polymer. In this case each isoform would contribute unique biochemical properties necessary in a specific combination to generate a hybrid polymer with the required properties. The α-tubulins in the *Drosophila* embryo provide an example of just such a situation, that is, in which microtubule function *in vivo* not only requires more than one isoform, but is also dependent on the relative ratio of different isoforms in the total tubulin pool.

D. melanogaster has four α-tubulin genes, designated by their chromosomal location, which encode three structurally distinct α-tubulin isoforms [Theurkauf et al., 1986]. The predominant *Drosophila* 84B α-tubulin isoform is highly similar in sequence to the major α-tubulins in many other species. The 84B α-tubulin gene is expressed constitutively, and the 84B protein constitutes the majority of α-tubulin present in cells of all tissues at all stages of development [Raff, 1984; Matthews and Kaufman, 1987; Matthews et al., 1989]. A second gene, 84D, probably represents a relatively recent duplication of the 84B gene and encodes a protein that differs in sequence from 84B by only two conservative amino acid substitutions [Theurkauf et al., 1986]. 84D is expressed in most but not all of the tissues in which 84B is expressed, but at very much lower levels, the major exception being that 84D is not expressed at all in the adult male [Matthews et al., 1989]. Expression of the 84D gene thus enhances the pool of the major isoform in most cells in females, but its absence in the male shows that it is dispensible for viability. The remaining two *Drosophila* α-tubulin genes are co-expressed with the 84B gene in developmentally restricted patterns. 85E α-tubulin, 95% similar in sequence to the 84B and 84D proteins, is accumulated only in a small subset of tissues, in cells that are lineally unrelated but that all appear to have a morphological requirement for extensive cell shape changes during development [Matthews et al., 1990]. Expression of the fourth gene, 67C, occurs only during oogenesis [Theurkauf et al., 1986; Matthews et al., 1989]. Differing in amino acid sequence from the 84B and 84D isoforms by 33%, the *Drosophila* 67C isoform is the most divergent of all known α-tubulins.

Studies by Matthews and colleagues show that normal spindle function in early *Drosophila* development not only requires both the predominant 84B/D isoforms and the extraordinarily divergent maternal 67C isoform, but is also sensitive to the relative proportions of the two proteins in the tubulin pool. Matthews et al. [1989, 1993] found that the tubulin pool of mature *Drosophila* oocytes is made up of a mix of these two α-tubulin isoforms, approximately 80% the predominant 84B/D isoforms and 20% the divergent maternal-specific 67C isoform (in combination with a single β-tubulin, the predominant β1-isoform). Both the 84B and 67C α-tubulin genes are essential for early development. 84B is required at all stages in the *Drosophila* life cycle; mutations in the 84B gene result in a variety of phenotypes, including defects in oogenesis and embryogenesis [Matthews and Kaufman, 1987]. Loss-of-function mutations in the maternally expressed 67C gene result in defects both in meiosis and in subsequent cleavage stage mitoses, resulting in failure of embryos produced by homozygous mutant mothers to develop [Rees, 1990; Matthews et al., 1993].

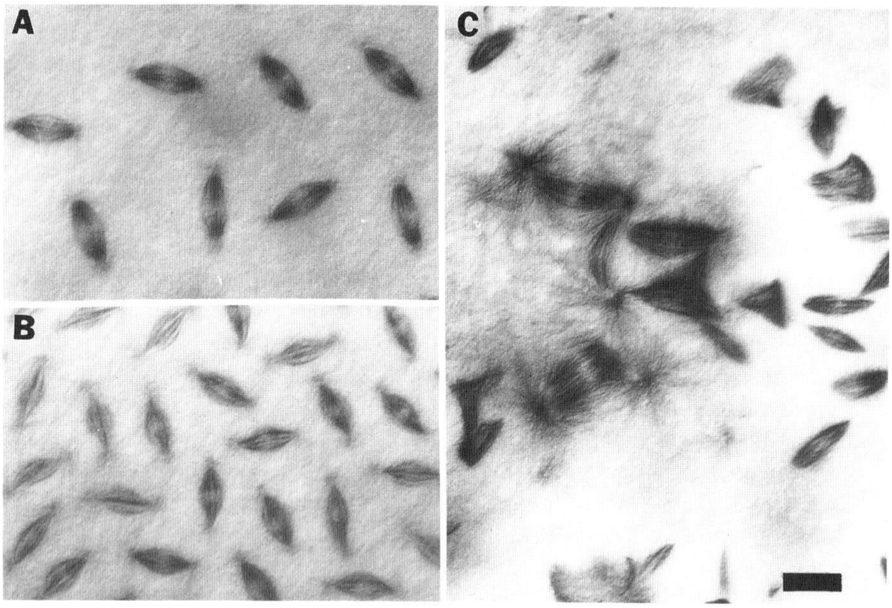

Fig. 2. α-*Tubulin usage in* Drosophila *development. Immunostained spindles in early* Drosophila *embryos.* **A,B:** *Spindles in wild-type embryos stained with an antiserum specific for the divergent maternal 67C α-tubulin isoform [see Rees, 1990; Matthews et al., 1993].* **C:** *The aberrant mitotic spindle morphology that results when the divergent maternal 67C α-tubulin isoform is overexpressed in an otherwise wild-type animal. The embryo shown is from a transgenic female carrying four copies of the 67C α-tubulin gene (the two endogenous copies plus two transgenic copies; see Matthews et al. [1993] for description of the 67C transgene constructs); spindles are stained with an antiserum that recognizes β-tubulin. Note that while some spindles are normal in morphology, many spindles are markedly abnormal. As illustrated here, the most striking feature of mitotic spindles assembled in embryos in which the ratio of the 67C to 84B/D α-tubulin isoforms is too high is the enormous size of some of the spindles. These mitotic defects result in lethality of most embryos with excess 67C α-tubulin. Note that an excess of 67C α-tubulin relative to the 84B/D isoforms affects only mitosis; meiosis occurs normally. Magnification is the same for all panels. Bar = 10 μm. [The micrographs were provided by K.A. Matthews and are from Matthews et al., 1993, with permission of the publisher.]*

The maternal effect embryonic lethality resulting from mutations in the 67C gene is due to a requirement for the 67C protein and not simply to a decrease in α-tubulin pool size, since the lethality could be rescued by introduction of a wild-type copy of the 67C gene in transgenic flies, but could not be rescued by increasing the gene dosage of 84B. Quite unexpectedly, it was found that overexpression of 67C in otherwise wild-type flies caused severe defects in mitosis in the early embryo, resulting in lethality (see Fig. 2). However, the mitotic defects resulting from the presence of excess 67C α-tubulin could be compensated for by a concomitant increase in 84B levels; thus the effect of excess 67C is due to the altered ratio of the 67C and 84B/D isoforms, not to an increase in α-tubulin level *per se*. Excess levels of 67C did not affect meiosis, however. Furthermore, increased 84B in otherwise wild-type flies had no developmental consequences.

Taken together, these results reveal complex α-tubulin requirements for microtubule function in early *Drosophila* development. The 67C and the 84B α-tubulin isoforms are not functionally equivalent, and both isoforms must be present in the tubulin pool of both the oocyte and the early embryo in order for normal spindle function. There is both a lower limit and an upper limit on the ratio of the 67C and 84B/D α-tubulin isoforms that will support normal development, but function of the meiotic and mitotic spindles are differentially sensitive to perturbations in the normal wild-type ratio: Meiosis fails in the absence of functional 67C α-tubulin, but can tolerate an increase in either 67C or 84B relative to wild type. Mitotic spindle function requires 67C but cannot tolerate an increase in the ratio of 67C to 84B/D above some critical threshold.

The simplest interpretation of the observed phenotypes is that copolymerization of heterodimers containing the 67C and 84B/D α-tubulin isoforms generates a mixed polymer with functional properties different from a uniform polymer containing either isoform alone. Presumably the differences in the relative α-tubulin structural requirements of the meiotic and mitotic spindles reflect in some way the differences in their morphology and function. In addition to the basic difference in the nature of the chromosome separations in the two kinds of division, another distinction between the meiotic and mitotic spindles is the contrast in their required stability in the cell. The mature oocyte remains arrested in metaphase of meiosis I until fertilization. On the other hand, probably the most noteworthy characteristic of early *Drosophila* development is the rapidity of the syncytial mitotic cleavages in the early embryo, which, with karyokinesis occurring synchronously approximately every 10 minutes, is one of the fastest eukaryotic cell cycle times known.

Given the remarkably large sequence divergence between the 67C and 84B/D isoforms, it is not possible with certainty to discern which aspects of structure are most relevant to the differences in functionality of the two isoforms. However, I am tempted to crawl out onto the limb of speculation and note two structural features that in particular demarcate the 67C protein from virtually all other α-tubulins. First, the 84B sequence contains the C-terminal tyrosine required for the enzymatic detryrosination/tyrosination cycle [Bulinski et al., 1987] and the lysine at position 40 demonstrated to be the site of acetylation in *Chlamydomonas* and mouse [L'Hernault and Rosenbaum, 1985; Eddé et al., 1991]. The α-tubulin detyrosination cycle has not been studied in *Drosophila,* but acetylation of the 84B/D isoforms has been demonstrated using an antiserum specific for acetylated α-tubulin [Piperno and Fuller, 1985], and the 84B isoform undergoes extensive and complex developmentally regulated modifications [Matthews et al., 1989]. However, the divergent 67C sequence lacks both of the relevant residues, having an acidic residue, glutamatic acid, at position 40 and phenyalanine at the C terminus [Theurkauf et al., 1986]. Thus 67C α-tubulin cannot undergo either of the virtually universal posttranslational modifications observed for most other α-tubulins. It should be noted that nondetyrosinable α-tubulin is not unique to *Drosophila;* vertebrates also have at least one such isoform [Paturle-Lafanechere et al., 1991]. It is possible that α-tubulins that cannot undergo specific posttranslational modifications may represent a general selective mechanism for generating functionally distinct isoforms. At present this remains highly speculative, a tantalizing but unplaceable piece in the puzzle of the relationship between tubulin isotype usage and specification of the functional properties of microtubules.

Second, the 67C protein is structurally unique from other α-tubulins in that this protein has an insertion of 11 additional amino acids following residue 50 compared with the other two *Drosophila* isoforms or seven additional amino acids at this position compared with all other α-tubulins [Theurkauf et al., 1986]. Now at the end of my speculative limb (perhaps already ominously bending), I will note that the divergent *Drosophila* β3-tubulin has a similar unique feature, an insertion of six amino acids at position 56, in the corresponding internal variable region of the β-tubulin molecule [Rudolph et al., 1987]. As discussed further below, in the β-tubulin molecule this region has been identified as an isotype-defining feature and also may physically interact with the isotype-specific C terminus [Cleveland and Sullivan, 1986; Burns and Surridge, 1991] (see also Burns, this volume); our molecular and genetic analyses of variant alleles of *Drosophila* β2-tubulin are consistent with this model [Fackenthal et al., 1993b]. Perhaps the corresponding region in α-tubulin also plays a role in determining specific functional properties of microtubules.

The situation in the early *Drosophila* embryo, in which two α-tubulin isoforms are present together with a single β-tubulin, is relatively amenable to experimental manipulation, particularly because study of microtubule function in *Drosophila* can be approached genetically. It may be a much more daunting proposition to disentangle the relationship of the structures of the individual members of more complex isoform mixes to the biochemical and physiological properties of the microtubules into which they are incorporated, given the possible combinatorial generation of multiple heterodimers containing different α- and β-tubulin partners. For example, a single vertebrate cell may express as many as five different isoforms from each tubulin family [Sullivan and Cleveland, 1986; Lewis et al., 1985b, 1987; Lewis and Cowan, 1988; Lopata and Cleveland, 1987; Joshi and Cleveland, 1989].

A suggestive piece of evidence for the role of the structural diversity of proteins in complex isoform mixes in determining microtubule assembly properties is provided by the work of Banerjee and colleagues [1990], who demonstrated the influence of a single isoform on the assembly properties of a complex isoform mix. They showed that the assembly properties of the native mix of tubulins in bovine brain homogenates differed markedly from those of preparations that had been depleted, by antibody-affinity chromatography, of a class III neuronal-specific β-tubulin isoform that normally constitutes about 25% of the total β-tubulin in native preparations. In subsequent work they showed that each β-tubulin isoform purified and studied individually has unique biochemical parameters for assembly *in vitro,* different from the properties of the mix as a whole [Banerjee et al., 1992]. In these experiments, β-tubulins were subtracted from the total mix using isotype-specific antibodies and then the properties of assembly *in vitro* examined for preparations containing a single form of β-tubulin in combination with the total native set of α-tubulins present in the homogenates. Since the isoform mix represented in a total brain homogenate may not necessarily reflect the specific mix present in any one given cell type, these experiments address the general question of how the assembly properties of an isoform mix may differ from the assembly properties intrinsic to an individual isoform but do not necessarily reflect the exact isoform mix for any specific cell type. However, these results provide strong support for the hypothesis that assembly properties of an isoform mix, and perhaps the properties of the resultant polymer, may be controlled *in vivo* by differential combinatorial expression of different α- and β-tubulins.

There are still many unanswered questions. It is not known, for example, whether there is ever preferential association of specific α- and β-tubulin pairs or whether, as has been the implicit assumption, the heterodimers present in the soluble pool reflect all the possible combinations resulting from random dimerization of all of the α- and β-subunits in the pool. Likewise, another implicit assumption is that when multiple heterodimers are coassembled, they are randomly incorporated into the growing microtubule, resulting in a uniform polymer reflecting all of the components of the soluble tubulin pool. It is not clear whether these are good general assumptions or not. Certainly the observations of isoform sorting described above show that there are mechanisms that at least in some cell types allow isoform classes to be distinguished and handled differently in the cell. Baker et al. [1990] examined at the ultrastructural level the location along individual microtubules of different vertebrate β-tubulin isoforms coassembled *in vitro,* with specific reference to the behavior of the extremely divergent hematopoietic isoform. They observed nonuniform incorporation of different isoforms along the length of individual microtubules. However, they concluded that while this phenomenon clearly reflected the different assembly properties of the isoforms examined, it might not be directly relevant to microtubule assembly *in vivo*.

V. ISOFORM MIXES, THRESHOLD EFFECTS, AND THE PERSISTENCE OF TUBULIN PROTEINS *IN VIVO*

Microtubule assembly displays extraordinary variety both *in vitro* and *in vivo*. Certainly the plasticity—or one could even say promiscuity—of tubulin assembly *in vitro* does not necessarily reflect physiological conditions in a cell, but it is nonetheless absolutely clear that cells are extremely tolerant of variations in the components of the tubulin pool. Incoporation of heterologous tubulin into functional mictotubules arrays *in vivo* is in fact the very feature of microtubule assembly that has allowed experimenters a means of visualizing and thus analyzing the mechanism of function of microtubule-based structures such as the spindle. Thus, to cite only a few of many possible examples, purified bovine brain tubulin introduced by microinjection into cultured mammalian cells [Keith et al., 1981; Saxton et al., 1984], sea urchin embryos [Salmon et al., 1984], or *Drosophila* embryos [Kellog et al., 1988] is coassembled with endogenous tubulin into functional spindles and cytoplasmic microtubules. A reciprocal experiment by Prescott et al. [1989] showed that tubulin from an evolutionarily distant organism, *Physarum polycephalum,* was coincorporated with the endogenous tubulin into functional spindles and cytoplasmic microtubules when injected into cultured mammalian cells. Furthermore, these authors demonstrated that the heterologous tubulin conferred resistance to colchicine-induced microtubule disassembly *in vivo,* normally a property of microtubules in *Physarum* but not in mammalian cells, providing further evidence for modulation of the properties of the polymer by specific members of an isoform mix. Similarly, heterologous tubulins expressed from genes introduced in transient transfection experiments or by stable transformation into the genome are coincorporated into microtubules with the endogenous tubulins [see, for example, Bond et al., 1986; Fridovich-Keil et al., 1987; Gasch et al., 1988].

Typically, however, the heterologous tubulins constitute a relatively small part of the total tubulin pool. This brings up the question of threshold effects. The importance to function of relative amounts of different isoforms in the tubulin pool *in vivo* is demonstrated by the role of α-tubulins in *Drosophila* development discussed above. A related question is that of the potential importance of the quantity of tubulin *per se*. A cell type in which tubulin level is important is illustrated by experiments in *S. cerevisiae* to alter expression level of the endogenous isoforms; yeast cells can function on 50% lower than normal tubulin levels, but a significant excess of tubulin is lethal [Burke et al., 1989; Katz et al., 1990; Weinstein and Solomon, 1990]. In *Aspergillus,* the quantity of tubulin may also be important; in this organism level of expression appears to be regulated at gene transcription [May et al., 1990]. Tubulin level *per se,* however, may not be not crucial in all cell types; *Drosophila* embryos and male germ cells tolerate tubulin levels from 50% to fourfold over normal levels [Matthews et al., 1993; Hoyle and Raff, 1990; Fackenthal et al., 1993a].

Because of the possibility of effects due to tubulin levels and to threshold effects, experiments in which relative function of different isoforms is tested by coexpression must be interpreted with caution. Thus coassembly of a given heterologous tubulin subunit into functional microtubules may not reveal crucial biological differences in the functional properties of the test tubulin from those of the endogenous isoform unless a situation is achieved in which the heterologous subunit constitutes a significant proportion of the total pool, or, better, completely replaces the endogenous isoform—since it may be hard to know what level constitutes "significant." For example, the phenotypes resulting from mutant forms of β2-tubulin in *Drosophila* spermatogenesis show that even reaching a level of 50% of the pool may not be enough to allow conclusions about the intrinsic functional capacity of the test subunit. Some recessive mutations in the β2-tubulin gene encode β2 variants that are fully stable and in fertile heterozygous males are coincorporated in equal amounts with wild-type β2-tubulin into functional mature motile sperm, even though in homozygous males the mutant protein alone cannot support normal microtubule function, resulting in male sterility due to failure of some or all of the microtubule-dependent events required to complete spermatogenesis [Raff and Fuller, 1984; Fuller et al., 1987, 1988]. For these mutant β2 proteins, the threshold at which defective function can be discerned therefore lies somewhere between 50% and 100% of the pool.

Second, in an experimental situation, both the threshold for effects on microtubule function, as well as the apparent intrinsic assembly capacity of the test tubulin alone, will depend on the cellular context in which it is asked to function. Thus, although the divergent *Drosophila* β3-tubulin isoform disrupted axoneme assembly when it comprised 20% of the total β-tubulin pool, we could not readily define a threshold for functional disruption of other classes of microtubules; for example, we did not observe disruption of spindles at any level of β3 relative to the normal β2 isoform that we were able to achieve, even though β3 alone could not support meiosis [Hoyle and Raff, 1990].

Finally, another important consideration in understanding the biological role of structurally diverse tubulins, related to the question of the threshold effects, is the persistence of tubulin proteins *in vivo*. Patterns of gene expression in many instances are very sharply restricted, but the tubulins synthesized during one developmental time may persist well after the end of gene expression and thus constitute part of the tubulin pool for a much longer time than the period of gene action or may persist in the later daughter cell progeny of the cells in which the tubulin was originally synthesized. This fact has perhaps not been well enough appreciated in consideration of patterns of isoform usage.

Persistence of the protein products of tubulin genes whose expressions are delimited in time or space was demonstrated by Gull and Dove and their co-workers for tubulin isoform usage in *P. polycephalum*. This microbial eukaryote has a life style at least as complex as that of any yuppy, having alternate plasmodial, amoeboid, and flagellate forms. *Physarum* has a moderately complex tubulin gene family of at least five α-tubulin and three β-tubulin genes [Schedl et al., 1984]. The transition between different forms is accompanied by multiple changes in gene expression, including tubulin genes [Burland et al., 1983]. However, residual tubulins are built into the architecturally distinct microtubule arrays typical of the new cell type [Paul et al., 1987; Diggins-Gilicinski et al, 1989]; moreover, developmental changes in microtubule morphology are also not abrupt, implying flexibility at the level of microtubule suprastructure as well as at the level of isoform usage [Solnica-Krezel et al., 1990, 1991].

In addition to providing an example of the persistence of tubulins in the cytoplasm, the *Physarum* tubulin gene switches at life cycle changes appear to represent a case in which regulatory pressures have provided the primary selective force for multiple genes rather than expression of functionally diverged proteins. The best argument for this is the tubulin gene switch during the amoeba to flagellate transition, which represents a change in gene expression but not a change in protein. Paul et al. [1992] found that, in flagellates, expression of the *betA* β-tubulin gene increases 100-fold over that of the *betB* β-tubulin gene in amoebae. However, the proteins encoded by the *betA* and *betB* genes are at least 99% similar. The quite plausible conclusion drawn by the authors is that expression of the *betA* gene provides a means to meet the tubulin levels required for assembly of the flagellar axoneme, but that the *betA* and *betB* gene proucts are probably equivalent proteins. It is also a reasonable conclusion that in this case selection has directly acted to maintain similar protein sequence, given the 15% divergence of the two genes in nucleotide sequence [Paul et al., 1992].

In *Drosophila* development, described above, persistence of the functionally unique divergent

67C α-tubulin isoform is essential for later microtubule function. 67C α-tubulin is loaded into the egg during oogenesis, where it is required for function; subsequently the gene is neither transcribed nor is "left-over" maternal message translated in the embryo, but the protein synthesized earlier is required for function of the embryonic cleavage stage spindles [Matthews et al., 1987, 1993]. The mutant phenotypes do not reflect a requirement for 67C α-tubulin in later developmental stages. However, the 67C isoform does not disappear rapidly from the embryonic tubulin pool, but rather persists throughout the remainder of embryogenesis and is incorporated into all microtubule arrays, although the immunolocalization studies indicate that loss of the 67C isoform from the tubulin pool may not be uniform in all cell types [Theurkauf, 1992].

Similarly, our observations suggest persistence of β-tubulins in *Drosophila* development. The predominant *Drosophila* isoform, β1-tubulin, is synthesized during oogenesis and throughout development and is the only β-tubulin present in the early embryo [Raff et al., 1982; Raff, 1984], but the β1-gene is not zygotically expressed in the mesoderm cells in which expression of the β3-gene occurs [Gasch et al., 1988; Buttgereit et al., 1991]. However, we found that even in mutant embryos homozygous for a deletion that removes the β3-gene, some microtubules appear in the developing muscle cells, suggesting that sufficient levels of β1-tubulin persist in the cytoplasm to support microtubule assembly at the correct time preceding sarcomere differentiation [Dettman and Raff, unpublished observations; Dettman, 1993]. Nevertheless, retention of tubulins made at one time in development by later stages is not necessarily the general case. For example, during spermatogenesis in *Drosophila,* β1-tubulin is the only β-tubulin present in early-stage cells after formation of the primary spermatocytes, synthesis of the testis-specific β2-tubulin takes over and β2 supports all subsequent microtubule-dependent functions. In the absence of β2-tubulin, in mutant animals, no postmitotic microtubule-dependent functions take place, nor can any microtubules be detected at the ultrastructural level [Kemphues et al., 1982]. Thus if any β1-tubulin persists from earlier stages in the tubulin pool of mature spermatocytes or spermatids, it is not sufficient to support any microtubule assembly.

VI. DO EXPERIMENTS TO TEST TUBULIN FUNCTIONAL EQUIVALENCE IN FUNGI AND IN THE FRUIT FLY REALLY ADDRESS EXACTLY THE SAME QUESTION?

The experiments discussed above to compare the relative functional capacities of different tubulin isoforms provide examples of isoforms that are experimentally interchangeable and isoforms that are not. Thus far, however, the interchangeable isoforms are found in fungi and the functionally distinct isoforms in *Drosophila*. This raises some hard questions. Do the two kinds of results tell us primarily about different kinds of selective pressures operating in the evolution of tubulin genes? Or do these different results also suggest that there are fundamental differences in the cellular requirements for microtubule assembly in "simple" single-celled eukaryotes versus in a more complex multicellular eukaryote? In addition to the isoform-swap data, there are other indications that the tubulin structural requirements for microtubule assembly in yeast may be qualitatively different from those in a complex metazoan (again, as represented by the situation in the *Drosophila* male germline). As described below, two different structural alterations of the β-tubulin molecule that cause profound disruption of the function of the testis-specific *Drosophila* β2 isoform cause only subtle effects on microtubule function when introduced into the single β-tubulin of *S. cerevisiae*. Clearly in yeast as well as in the fly, the wild-type β-tubulin would have selective advantages in evolution over either of the two variant forms. The problem is how to interpret the differences in both the nature and degree of functional disruption in the two organisms in the laboratory situation.

The first difference observed in structure–function relationships between yeast and *Drosophila* β-tubulins indeed suggests the somewhat surprising possibility that there may be considerable difference in the stringency of structural requirements for fundamental aspects of microtubule assembly in fungi and metazoans. The amino acid at position 288 is acidic in all known β-tubulins, from the human isoforms to the β-tubulin in *Giardia lamblia,* the most primitive known eukaryote (in 52 of the 54 presently known β-tubulin sequences this residue is glutamic acid; it is aspartic acid in the remaining two isoforms). For comparisons of β-tubulin primary structures see Burns and Surridge (this volume). Conservation of an acidic residue at this position across the entirety of eukaryotic phylogeny is powerful evidence of very strong selective constraint. The most obvious conclusion is that this acidic residue plays a required functional role common to all β-tubulins. However, experimental tests of the effect of alterations at this conserved position reveal that the situation is more complex.

Substitution of lysine for glutamic acidic at residue 288 in the *Drosophila* β2 isoform indeed disrupts assembly and function of all β2-tubulin–based microtubules in *Drosophila* spermatogenesis [Rudolph et al., 1987; Fuller et al., 1987, 1988]. The mutant β2-tubulin with lysine at position 288 is defective in assembly of all classes of spermatogenic microtubules, forming primarily microtubules of S-shaped cross section rather than normal closed microtubules; we interpret this aberrant microtubule morphology to reflect inability of dimers to form correct side-to-side associations in the polymer [Fuller et al., 1988]. Surprisingly, however, the same substitution introduced into the single β-tubulin of *S. cerevisiae* results in little effect on microtubule function, the sole phenotype being marginally increased drug sensitivity of growth [Praitis et al., 1991]. The logical inference from this result is that the glutamic acid to lysine change at residue 288 in the yeast β-tubulin did not cause a fundamental alteration of assembly properties. However, microtubules supported by the yeast variant were not examined at the ultrastructural level; thus one possibility is that the microtubules assembled from the yeast lys^{288} variant are morphologically altered as in *Drosophila* but that spindle and cytoskeletal microtubule function can still occur. If this is so, then the different phenotypes may reflect differences between yeast and *Drosophila* cells at the level of interactions of tubulin-based suprastructure with other cellular components rather than basic folding properties of the β-tubulin molecule. At present, the interpretation of the strict evolutionary conservation of an acidic amino acid at position 288 contrasted with the disparate effects of changes in this residue on β-tubulin function in yeast and fly must remain speculative.

A second way in which β-tubulin functions in yeast and flies appear to differ is in the functional role of the C-terminal domain, known to be a surface feature of microtubules and implicated in binding of certain microtubule-associated proteins [Serrano et al., 1984; Breitling and Little, 1986; Littauer et al., 1986; Dráber et al., 1989]. For the C-terminus, the *S. cerevisiae* and *Drosophila* examples may serve as an indication of the evolutionary route for selection for tissue-specific functions. Matsuzaki et al. [1988] found that removal of the region of the C terminus of the single β-tubulin of *S. cerevisiae* corresponding to the variable C-terminal domain of β-tubulin isoforms in higher eukaryotes resulted in relatively little functional disruption. Relative to wild-type cells, yeast cells with the C-terminal truncated β-tubulin as their only source of β-tubulin showed increased sensitivity to antimicrotubule drugs and to elevated growth temperatures, but were fully viable and grew and divided normally under standard laboratory culture conditions.

We have constructed the corresponding truncation of *Drosophila* testis-specific β2-tubulin isoform, to produce a β2 variant missing only the final variable 15 amino acid acidic C terminus [Fackenthal et al., 1993a]. In contrast to the situation in yeast, the truncated *Drosophila* β2 is markedly different in properties from the wild-type β2 and cannot support wild-type microtubule function in spermatogenesis. Males that have the C-terminal truncated β2-tubulin as their only source of β-tubulin in the postmitotic germline cannot produce sperm and thus are sterile.

Taken together, these experiments suggest that the β-tubulin C terminus may have one function common to all organisms and another class of functions that may be unique to metazoans. First, it is clear that the β-tubulin C terminus is not required for microtubule assembly *per se,* as was shown by previous studies of microtubule assembly *in vitro* [Serrano et al., 1984; Maccioni et al., 1985; White et al., 1987] and confirmed by the studies of the truncated β-tubulins in yeast and fly *in vivo*. However, we found that the truncated *Drosophila* β2 isoform exhibits decreased stability relative to the wild-type protein such that the level of protein accumulated in the tubulin pool per gene copy is only about half that of wild-type β2[Fackenthal et al., 1993a]. Thus the presence of the acidic C terminus is required either for the intrinsic stability of β-tubulin itself or of the heterodimer into which it is incorporated. Note that the *Drosophila* β3 isoform ectopically expressed in the testis is as stable as the endogenous β2 isoform [Hoyle and Raff, 1990]; thus the stability function in the testis does not require the specific sequence of the β2 C terminus. Perhaps this general function is the only role of the β-tubulin C terminus in fungi. Stability of the yeast truncated β-tubulin was not examined directly, but slightly decreased cellular tubulin pools might plausibly explain the observed growth phenotype.

It is important to note that the failure in microtubule function of the truncated *Drosophila* β2-tubulin was not caused by the decreased stability *per se*. It is in fact the microtubule assembly phenotype of the truncated β2 that suggests a second role of the C terminus that may be specific to metazoans. The sequence of the variable C terminus of the β-tubulin molecule has been shown to be the primary isotype-defining feature among vertebrate β-tubulin families [Sullivan and Cleveland, 1986; Sullivan, 1988]. This observation suggests that the unique sequence of the variable C terminus of each isoform in a metaozoan β-tubulin family may play a role in determining the functional specificity of the isoform. The phenotype for microtubule function that we observed for the *Drosophila* truncated β2-tubulin provides strong support for this hypothesis [Fackenthal et al., 1993a]. Truncated β2 does not exhibit any intrinsic defects in microtubule assembly properties and can be assembled into both cytoplasmic microtubules and into presumptive axonemal doublet tubules of normal morphology. However, in males that express only the truncated β2, the organization of microtubules into several of the testis-specific microtubule-based suprastructures is defective. The most remarkable defect of the truncated β2 is the failure to form organized axonemes, even though both doublet and singlet microtubules of presumptive axonemal identity are assembled (see Fig. 1D–F). The failure in axoneme morphogenesis thus lies not in failure of microtubule assembly but in failure to achieve the interactions of microtubules with other components, such as the spokes and linkers, required for the "9 + 2" axonemal architecture. These results show that the variable C-terminal domain is important for determining isoform-specific function; however, it is also important to note that the properties of a chimeric β3 protein, in which the normal β3 carboxy region has been replaced by the corresponding sequences from the β2 molecule, show that the C-terminal domain does not alone determine the specific functional features of β2-tubulin [Hoyle et al., 1991].

A reasonable hypothesis is that the β-tubulin acidic C terminus has a role in mediation of protein stability common to all organisms but also plays a second role in metazoans. The general stability function of the acidic C-terminal domain may involve intrinsic folding properties of the β-tubulin molecule, interactions with α-tubulin, or interactions with general microtubule-associated proteins that mediate stability in the cell. In metazoans, it appears that in addition there have evolved sequence-specific interactions of the C terminus specific of a given isoform with other cellular components required for formation of microtubule-based suprastructures unique to a given tissue or cell type. These specific protein–protein interactions governed by the unique sequence of the C terminus of each isoform thus act to mediate some aspects of the isoform-specific functions of different members of the β-tubulin family.

A final point that is relevant to the consideration of differences in the structural requirements for microtubule function between cells of species from widely separated phyla is that even within a single cell type in *Drosophila* different sets of microtubules display disparate β-tubulin structural requirements. For example, in developing *Drosophila* spermatids, the assembly and function of the cytoskeletal microtubules required for elongation of the mitochondrial derivatives can be fully supplied both by the ectopically expressed wild-type β3 isoform and the carboxy truncated β2 variant, even though axoneme morphogenesis in the same cell cannot be supported by either of these β-tubulins [Hoyle and Raff, 1990; Fackenthal et al., 1993a]. The mitochondrial-associated tubules are the most robust to all the perturbations in tubulin structure examined here and thus for the spermatid cells appear to represent the "ground state" or most minimal requirements for microtubule assembly. If β3 and β2 were compared under conditions in which only assembly of microtubules equivalent to the mitochondrial elongation tubules were required, it would be concluded that the two isoforms were interchangeable. The substantial differences between the intrinsic functional capacities of β3 and β2 could be defined precisely because β3 function could be tested in several different categories of microtubule assembly.

Given the disparate structural requirements of different sets of microtubules in a single metazoan cell type, it is perhaps not too surprising after all that the stringency of structural requirements may differ between simpler and more complex eukaryotes. As stated earlier above, the relatively minor differences in growth phenotype obtained in the tubulin substitution experiments in fungi probably really do indicate that in *S. cerevisiae* and *Aspergillus nidulans* the primary role for multiple tubulins is in cellular control over timing and amount of tubulin synthesis. However, functional differences betwen isoforms that are subtle in the laboratory setting may nonetheless reflect meaningful selective advantage in evolution. Furthermore, I think that, taken together, the different results obtained with fungi and flies can indeed be taken to indicate that there are real differences in the nature of even very fundamental cellular requirements for microtubule function in different kinds of organisms.

VII. IS SELECTION FOR FUNCTIONALLY SPECIALIZED TUBULINS THE GENERAL RULE FOR THE ROLE OF MULTIPLE TUBULINS IN COMPLEX EUKARYOTES?

I think the answer is yes, but . . . We have argued previously that the primary selective force driving the formation and maintenance of multiple tubulin genes may be the need for regulation of timing and amount of tubulin but that subsequent selection will drive toward functional specialization among isoforms [Raff et al., 1987]. Although this is likely to be true for all organisms, it may be that the complexity of demands for differentiated cell function in multicellular eukaryotes is such that the functional differences between isoforms are greater, and therefore more readily experimentally demonstrable, among different members of the larger tubulin gene families of higher eukaryotes. Another feature of tubulin gene families in multicellular species is that the most prevalently utilized isoforms represent the most highly conserved sequences. Comparisons of isoform utilization patterns suggest that it is a general rule that the more conserved sequence isoforms are utilized in situations in which multiple morphological classes of microtubules must be assembled. I would argue that the structural requirements necessary for assembly of the multiple-walled tubules of the centriole, basal body, and axoneme may constitute the greatest selective constraint on tubulin sequence. Thus the predominant *Drosophila* isoform, β1, which is utilized to assemble the centriole and basal body and the testis-specific isoform, β2, utilized to assemble the motile sperm tail axoneme, are about 95% similar both to each other and to the major vertebrate isoforms. Considering the lack of tubulin genetic diversity in flagellated and ciliated protists, Silflow [1991] has likewise argued that the requirement for axoneme serves as a major constraint on tubulin gene evolution. Consistent with this hypothesis, the single β-tubulin of *Chlamydomonas,* utilized for assembly of all morphological classes of microtubules both singlet and multiple walled, is recognizably more like the major conserved isoforms of vertebrates and *Drosophila* than any divergent minor isoform in these families [see discussion by Raff et al., 1987]. The sequence conservation of the flagellate-specific *Physarum* β-tubulin [Paul et al., 1992] is also consistent with this idea. The more divergent isoforms in the gene families of higher eukaryotes probably reflect in a complicated way the sum of allowed divergence concomitant with the absence of this particular aspect of selective constraint, together with selection for specialized functional features.

In many ways we are left with the problem we started out with still intact. That is, it is very hard to unravel which features of the expression of any given isoform reflect which aspect of its evolutionary history. Evidence for functional differences between different tubulin isoforms is the demonstration that structurally different proteins exhibit different biochemical or assembly properties *in vitro* [Rothwell et al., 1986; Baker et al., 1990; Banerjee et al., 1990, 1992]. These data are comforting to a protein chemist and indeed constitute compelling evidence for intrinsic biochemical differences among the proteins in question. However, the fungal isoform swap experiments described above are a potent reminder that neither restricted patterns of expression nor primary sequence divergence *per se* are necessarily sufficient to allow inference of differences in intrinsic functional capacity that are significant within the cell, at least under experimental conditions.

Nonetheless, expression of a structurally distinct isoform correlated with unique microtubule function is quite strongly suggestive of functional selection. A particularly good example is the *Caenorhabditis elegans* β-tubulin isoform encoded by the *mec7* gene. The *mec7* β-tubulin is expressed in only 6 of the 959 somatic cells in the worm— the six neurons that mediate touch sensitivity [Savage et al., 1989; Hammelin et al., 1992]. The physiological response of touch sensitivity depends on a unique microtubule-based cytoskeletal array of specialized 15-protofilament microtubules (all other microtubules in the worm have 11 protofilaments), the assembly of which depends on the expression of functional *mec7* β-tubulin; in the absence of *mec7* function, a smaller number of 11 protofilament microtubules are assembled in the touch neurons, which are sufficient to mediate neuron extension but that do not mediate touch sensitivity [Chalfie and Sulston, 1981; Chalfie and Thomson, 1982]. The simplest hypothesis is that the specialized microtubule architecture required for touch sensitivity is directed by the *mec7* isoform. However, while the correlation of the unique *mec7* β-tubulin sequence with a unique microtubule architecture is compelling, alternative possibilities have not been entirely ruled out. The most crucial possibility that needs to be eliminated is the possibility that loss of *mec7* simply results in loss of a sufficient tubulin pool; it has not been directly experimentally tested whether another β-tubulin could supply the function required for touch sensitivity if expressed at a sufficient level. The *mec7* sequence has several amino acid substitutions not present in other known β-tubulins [Savage et al., 1989]. However, these sequence features may not be not an absolute prerequisite for 15 protofilament architecture. In *Drosophila* most microtubules have the more usual 13 protofilament substructure, but the elegant studies of Tucker and colleagues have shown that the elongating cells of the developing wing blade have a complex microtubule-based cytoskeleton utilizing 15 protofilament microtubules [Tucker et al., 1986; Mogensen and Tucker, 1987; Mogensen et al., 1989]. Assembly of the 15 protofilament tubules in *Drosphila* does not appear to be directed by the β-tubulin components, since the β-tubulins expressed in the wing cells (β1 and β3) [Kimble et al., 1989] are utilized in other cell types for assembly of microtubules of standard architecture. Moreover, Mogensen and Tucker [1990] found that nucleation of 15 protofilament tubules was extremely labile; if cultured wing discs were treated with taxol before microtubule assembly, the subsequent wing cell tubules were composed of 12 protofilaments (not previously seen in *Drosophila)* or 13 protofilaments (the normal for other *Drosophila* tubules).

One final note: An alternative possibility for the worm touch cell microtubules is that it is the rest of the microtubules in the worm that require specialized tubulin functional features—It could be that assembly of the unusual (but standard for the worm) 11 protofilament microtubules requires structural features not necessary for the 15 protofilament tubules. In this case, we have gone through the looking glass, and the *mec7* sequence might represent allowed divergence given release from the constraints imposed by the requirement for the assembly of 11 protofilament microtubules.

VIII. CONCLUSIONS: ARE THERE ANY?

Generality of principle but complexity of detail is a frustrating but real feature of all biological systems, by no means a problem only for understanding microtubule assembly. Cellular processes that seem momentarily clear after a superficial level of study almost always get more puzzling when just that one more experiment is done. And this keeps happening at each subsequent level of explication—the puzzles just move down to a deeper level of detail. The accompanying sense of frustration felt by the experimenter is exemplified in Twitty's description [1966] of the embryologist Holtfreter in an earlier part of the century, as he attempted to interpret his studies of the neural organizer. After the first exciting demonstration of neural tube induction, "for the next several years similar results followed thick and fast—and confusingly. In fact, many of them were so puzzling and so paradoxical that Holtfreter was to remark despairingly that the analysis of organizer action was rapidly 'bringing chaos out of order.' "

The cell is a machine that was built while it was running, and the evolutionary history of each adjustment is built into the present mode of function. For the question of the role of multiple tubulin isoforms, the evidence at hand clearly demonstrates functional specialization of different tubulin isoforms, but also shows that functional specialization is only part of the answer and that each part of the answer brings up more questions. Note that precisely the same questions—and with the same set of possible multiple answers—can be asked with regard to the diversity that exists in virtually all of the cytoskeletal proteins. For example, compare the question of the role of actin isoform usage with that posed here for tubulin isoforms: In a discussion of the question of to what degree actin isoforms are functionally distinct, Rubenstein [1990] concluded that it is hard to decide. And not just for each cytoskeletal protein, but also for each associated set of interacting proteins there is a complex story emerging showing that specificity of function *in vivo* results from combinations of generic, specialized, overlapping, and redundant functions.

Taken together, the evidence available in the literature suggests that tubulin heterogeneity provides both regulatory and functional requirements but that the two kinds of requirements may be complexly interrelated. The precise and exquisite specificity of microtubule function includes differential tubulin structural requirements for different microtubule arrays, but is also dependent on the plasticity and adaptibility of the system, as indicated by the wide tolerance of microtubule assembly *in vivo* for variations in subunit composition and the fact that intrinsically defective tubulins may be coincorporated into functional polymer; by the existence of threshold effects that may differ for different microtubule-based structures; and by the potential for multiple cellular controls over tubulin levels and isoform usage. All of these features reveal the underlying biological complexity of the mechanisms to control microtubule-based cellular processes. Perhaps the chaos *is* the order.

ACKNOWLEDGMENTS

I thank K.A. Matthews for sharing data on α-tubulin usage in *Drosophila* development prior to publication and R.A. Raff for critical reading of the manuscript. The work from my laboratory described in this chapter was supported by USPHS grant HD 16739.

IX. REFERENCES

Alexander JE, Hunt DF, Lee MK, Shabanowitz J, Michel H, Berlin SC, MacDonald TL, Sundberg RJ, Rebhun LI, Frankfurter A (1991): Characterization of posttranslational modifications in neuron-specific class III β-tubulin by mass spectroscopy. Proc Natl Acad Sci USA 88:4685–4689.

Asai DJ, Remolona NM (1989): Tubulin isotype usage *in vivo:* A unique spatial distributin of the minor neuronal-specific β-tubulin isotype in *Pheochromocytoma* cells. Dev Biol 132:398–409.

Baas PW, Black MM (1990): Individual microtubules in the axon consist of domains that differ in both composition and stability. J Cell Biol 111:495–509.

Baker HN, Rothwell SW, Grasser WA, Wallis KT, Murphy DB (1990): Copolymerization of two distinct tubulin isotypes during microtubule assembly *in vitro*. J Cell Biol 110:97–104.

Banerjee A, Roach MC, Trcka P, Ludueña RF (1990): Increased microtubule assembly in bovine brain tubulin lacking the type III isotype of β-tubulin. J Biol Chem 265:1794–1799.

Banerjee A, Roach MC, Trcka P, Ludueña RF (1992): Preparation of a monoclonal antibody specific for the class IV isotype of β-tubulin. J Biol Chem 267: 5625–5630.

Biajolan S, Falkenberg D, Renkawitz-Pohl R (1984): Characterization and developmental expression of β tubulin genes in *Drosophila*. EMBO J 3:2543–2548.

Bond J, Fridovich-Keil JL, Pillus L, Mulligan RC, Solomon F (1986): A chicken-yeast chimeric β-tubulin protein is incorporated into mouse microtubules *in vivo*. Cell 44:461–468.

Breitling F, Little M (1986) Carboxy-terminal regions on the surface of tubulin and microtubules. J Mol Biol 189:367–370.

Bulinski JC, Richards JE, Piperno G (1988): Posttranslational modifications of α-tubulin: Detyrosination and acetylation differentiate populations of interphase microtubules in cultured cells. J Cell Biol 106:1213–1220.

Burgoyne RD, Cambray-Deakin MA, Lewis SA, Sarkar S, Cowan NJ (1988): Differential distribution of β-tubulin isotypes in cerebellum. EMBO J 7:2311–2319.

Burke D, Gasdaska P, Hartwell L (1989): Dominant effects of tubulin overexpression in *Saccharomyces cerevisiae*. Mol Cell Biol 9:1049–1059.

Burland TG, Gull K, Schedl T, Boston RS, Dove WF (1983): Cell-type dependent expression of tubulins in *Physarum*. J Cell Biol 97:123–141.

Burns RG, Surridge C (1990): Analysis of β-tubulin sequences reveals highly conserved, coordinated amino acid substitutions. FEBS Lett 27:1–8.

Buttgereit D, Leiss D, Michiels F, Renkawitz-Pohl R (1991): During *Drosophila* embryogenesis the β1 tubulin gene is specifically expressed in the nervous system and the apodemes. Mech Dev 33:107–118.

Chalfie M, Sulston J (1981): Developmental genetics of the mechanosensory neurons of *Caenorhabditis elegans*. Dev Biol 82:358–370.

Chalfie M, Thomson JN (1982): Structural and functional diversity in the neuronal microtubules of *Caenorhabditis elegans*. J Cell Biol 93:15–23.

Chasan R (1992): Multitudinous microtubules. Plant Cell 4:505–506.

Cleveland DW (1988): Autoregulated instability of tubulin mRNAs: A novel eukaryotic regulatory mechanism. TIBS 13:339–343.

Denoulet P, Filliatreau G, De Nechaud B, Gros F, Di Giamferardino L (1989): Differential axonal transport of isotubulins in the motor axons of the rats sciatic nerve. J Cell Biol 108:965–971.

Dettman RW (1993): A genetic analysis of a developmentally regulated *beta-tubulin* in *Drosophila melanogaster.* PhD thesis, Indiana University, Bloomington, Indiana.

Diaz HB, Conine MA, Raff EC (1990): Identification of the *Drosophila melanogaster* β4-tubulin isoform and its pattern of expression. J Cell Biol 111:412a.

Diaz HB, Hutchens JA, Conine MA, Raff EC (1991): Evolutionary relationships among metazoan β-tubulin genes. J Cell Biol 115:150a.

Diaz-Nido J, Hernandez MA, Avila J (1990): Microtubule proteins in neuronal cells. In Avila J (ed): Microtubules. Boca Raton, FL: CRC Press, pp 193–257.

Diggins-Gilicinski M, Solnica-Krezel L, Burland TG, Paul ECA, Dove WF (1989): The localization of the divergent β2-tubulin isotype in the microtubule arrays of *Physarum polycephalum*. J Cell Sci 94:217–226.

Doshi P, Bossie CA, Doonan JH, May GS, Morris NR (1991): Two α-tubulins genes of *Aspergillus nidulans* encode divergent proteins. Mol Gen Genet 225:129–141.

Dráber P, Dráberová E, Linhartová I, Viklick'y (1989): Differences in the exposure of C- and N-terminal tubulin domains in cytoplasmic microtubules detected with domain-specific monoclonal antibodies. J Cell Sci 92:519–528.

Eddé B, Rossier J, Le Caer J-P, Desbruyeres E, Gros F, Denoulet P (1990): Posttranslational glutamylation of α-tubulin. Science 247:83–85.

Eddé B, Rossier J, Le Caer JP, Berwald-Netter Y, Koulakoff A, Gros F, Denoulet P (1991): A combination of posttranslational modifications is responsible for the production of neuronal alpha-tubulin heterogeneity. J Cell Biochem 46:134–142.

Erickson HP, Voter WA (1976): Polycation-induced assembly of purified tubulin. Proc Natl Acad Sci USA 78:2813–2817.

Fackenthal JD, Turner FR, Raff EC (1993a): Tissue-specific microtubule functions in *Drosophila* spermatogenesis require the β2-tubulin isotype-specific carboxy terminus. Dev Biol 58:213–227.

Fackenthal JD, Hutchens JA, Turner FR, Raff EC (1993b): Structural analysis of mutations in the Drosophila β2-tubulin isoform reveal regions in the β-tubulin molecule required for general and for tissue-specific functions. Submitted for publication.

Falconer MM, Echeverri CJ, Brown DL (1992): Differential sorting of beta tubulin isotypes into colchicine-stable microtubules during neuronal and muscle differentiation of embryonal carcinoma cells. Cell Motil Cytoskel 21:313–325.

Fridovich-Keil J, Bond JF, Solomon F (1987): Domains of β-tubulin essential for conserved functions *in vivo*. Mol Cell Biol 7:3792–3798.

Fuller MT, Caulton JH, Hutchens JA, Kaufman TC, Raff EC (1987): Genetic analysis of microtubule structure: A β-tubulin mutation causes the formation of aberrant microtubules *in vivo* and *in vitro*. J Cell Biol 104: 385–394.

Fuller MT, Caulton JH, Hutchens JA, Kaufman TC, Raff EC (1988): Mutations that encode partially functional β2-tubulin subunits have different effects on structurally different microtubule arrays. J Cell Biol 107: 141–152.

Fulton C, Simpson PA (1976): Selective synthesis and utilization of flagellar tubulin. The multi-tubulin hypothesis. In Goldman R, Pollard T, Rosenbaum J (eds): Cell Motility. Cold Spring Harbor, NY: Cold Spring Harbor Press, pp 987–1005.

Gard D, Kirschner M (1985): A polymer-dependent increase in phosphorylation of β-tubulin accompanies differentiation of a mouse neuroblastoma cell line. J Cell Biol 100:764–774.

Gasch A, Hinz U, Leiss D, Renkawitz-Pohl R (1988): The expression of β1 and β3 tubulin genes of *Drosophila melanogaster* is spatially regulated during embryogenesis. Mol Gen Genet 211:8–16.

Gimeno CJ, Ljungdahl PO, Styles CA, Fink GR (1992): Unipolar cell divisions in the yeast *S. cerevisiae* lead to filamentous growth: Regulation by starvation and RAS. Cell 68:1061–1076.

Hamelin M, Scott IA, Way JC, Culotti JG (1992): The *mec-7* β-tubulin gene of *Caenorhabditis elegans* is expressed primarily in the touch receptor neurons. EMBO J 11:2885–2893.

Herzog W, Weber K (1977): *In vitro* assembly of purified tubulin into microtubules in the absence of microtubule-associated proteins and glycerol. Proc Natl Acad Sci USA 74:1860–1864.

Hoyle HD, Raff EC (1990): Two *Drosophila* beta tubulin isoforms are not functionally equivalent. J Cell Biol 111:1009–1026.

Hoyle HD, Turner FR, Raff EC (1991): Function of the carboxy terminus in a *Drosophila* beta tubulin chimera. J Cell Biol 115:382a.

Joshi H, Cleveland DW (1990): Diversity among tubulin subunits: Toward what functional end? Cell Motil Cytoskel 16:159–163.

Joshi HC, Cleveland DW (1989): Differential utilization of β-tubulin isotypes in differentiating neurites. J Cell Biol 109:663–673.

Katz W, Weinstein B, Solomon F (1990): Regulation of tubulin levels and microtubule assembly in *Saccharomyces cerevisiae:* Consequences of altered tubulin gene copy number. Mol Cell Biol 10:5286–5304.

Keith CH, Feramisco JR, Shelanski M (1981): Direct visualization of fluorescein-labelled microtubules in vitro and in microinjected fibroblasts. J Cell Biol 88:234–240.

Kellog DR, Mitchison TJ, Alberts BM (1988): Behavior of microtubules and actin filaments in living *Drosophila* embryos. Develoment 103:675–686.

Kemphues KJ, Kaufman TC, Raff RA, Raff EC (1982): The testis-specific β-tubulin subunit in *Drosophila melanogaster* has multiple functions in spermatogenesis. Cell 31:655–670.

Kemphues KJ, Raff EC, Kaufman TC (1980): Mutation in a testis-specific β-tubulin in *Drosophila:* Analysis of its effects on meiosis and map location of the gene. Cell 21:445–451.

Kemphues KJ, Raff EC, Kaufman TC (1983): Genetic analysis of *B2t*, the structural gene for a testis-specific β-tubulin subunit in *Drosophila melanogaster*. Genetics 105:345–356.

Kemphues KJ, Raff RA, Kaufman TC, Raff EC (1979): Mutation in a structural gene for a β-tubulin specific to testis in *Drosophila melanogaster*. Proc Natl Acad Sci USA 76:3991–3995.

Khawaja S, Gunderson GG, Bulinski JC (1988): Enhanced stability of microtubules enriched in detyrosinated tubulin is not a direct function of detyrosination level. J Cell Biol 106:141–149.

Kimble M, Dettman RW, Raff EC (1990): The β3-tubulin gene of *Drosophila melanogaster* is essential for viability and fertility. Genetics 126:991–1005.

Kimble M, Incardona JP, Raff EC (1989): A variant β-tubulin isoform of *Drosophila melanogaster* (β3) is expressed primarily in tissues of mesodermal origin in embryos and pupae, and is utilized in populations of transient microtubules. Dev Biol 131:415–429.

Kirk KE, Morris NR (1991): The *tubB* α-tubulin gene is essential for sexual development in *Aspergillus nidulans*. Genes Dev 5:2014–2023.

Kopczak SD, Haas NA, Hussey PJ, Silflow CD, Snustad DP (1992): The small genome of *Arabodopsisc* contains at least six expressed α-tubulin genes. Plant Cell 4:539–547.

Langdon CM, Rafiee P, MacRae TH (1991a): Synthesis of tubulin during early postgastrula development of *Artemia:* Isotubulin generation and translational regulation. Dev Biol 148:138–146.

Langdon CM, Freeman JA, MacRae TH (1991b): Posttranslationally modified tubulins in *Artemia:* Prelarval development in the absence of detyrosinated tubulin. Dev Biol 148:147–155.

Lee JC, Timasheff SN (1975): The reconstitution of microtubules from purified calf brain tubulin. Biochemistry 14:5183–5187; erratum, (1976):15:934.

Lee MK, Tuttle JB, Rebhun LI, Cleveland DW, Frankfurter A (1990): The expression and posttranslational modification of a neuron-specific β-tubulin isotype during chick embryogenesis. Cell Motil Cytoskel 17:118–132.

Lewis SA, Cowan NJ (1988): Complex regulation and functional versatility of mammalian α- and β-tubulin isotypes during the differentiation of testis and muscle cells. J Cell Biol 106:2023–2033.

Lewis SA, Gilmartin ME, Hall JL, Cowan NJ (1985a): Three expressed sequences within the human β-tubulin multigene family each define a distinct isotype. J Mol Biol 182:11–20.

Lewis SA, Lee MG-S, Cowan NJ (1985b): Five mouse tubulin isotypes and their regulated expression during development. J Cell Biol 101:852–861.

Lewis SA, Gu W, Cowan NJ (1987): Free intermingling of mammalian β tubulin isotypes among functionally distinct microtubules. Cell 49:37–46.

L'Hernault SW, Rosenbaum JL (1985): *Chlamydomonas* alpha-tubulin is posttranslationally modified by acetylation on the epsilon-amino group of lysine. Biochemistry 24:473–478.

Littauer UZ, Giveon D, Thierauf M, Ginzburg I, Ponstingl H (1986): Common and distinct tubulin binding sites for microtubule-associated proteins. Proc Natl Acad Sci USA 83:7162–7166.

Lopata MA, Cleveland DW (1987): *In vivo* microtubules are copolymers of available β-tubulin isotypes: Localization of each of six vertebrate β-tubulin isotypes using polyclonal antibodies elicited by synthetic peptide antigens. J Cell Biol 105:1707–1720.

Ludueña RF, Banerjee A, Khan IA (1992): Tubulin structure and biochemistry. Curr Opin Cell Biol 4:53–57.

Maccioni RB, Serrano L, Avila J (1985): Structural and functional domains of tubulin. BioEssays 2:165–169.

Marks DM, West J, Weeks D (1987): The relatively large beta-tubulin gene family of *Arabidopsis* contains a member with an unusual transcribed 5′ noncoding sequence. Plant Mol Biol 10:91–104.

Matsuzaki F, Matsumoto S, Yahara I (1988): Truncation of the carboxy-terminal domain of yeast β-tubulin causes temperature-sensitive growth and hypersensitivity to antimitotic drugs. J Cell Biol 107:1427–1435.

Matthews KA, Kaufman TC (1987): Developmental consequences of mutations in the 84B α-tubulin gene of *Drosophila melanogaster.* Dev Biol 119:100–114.

Matthews KA, Miller DFB, Kaufman TC (1989): Developmental distribution of RNA and protein products of the *Drosophila* α-tubulin gene family. Dev Biol 132:45–61.

Matthews KA, Miller DFB, Kaufman TC (1990): Functional implications of the unusual spatial distribution of a minor α-tubulin isotype in *Drosophila:* A common thread among chortotonal ligaments, developing muscle, and testis cyst cells. Dev Biol 137:171–183.

Matthews KA, Rees D, Kaufman TC (1993): A functionally specialized α-tubulin is required for oocyte meiosis and cleavage mitosis in *Drosophila*. Development 117:977–991.

May GS (1989): The highly divergent β-tubulins of *Aspergillus nidulans* are functionally interchangeable. J Cell Biol 109:2267–2274.

May GS, Gambino J, Weatherbee JA, Morris NR (1985): Identification and functional analysis of β-tubulin genes by site-specific integrative transformation in *Aspergillus nidulans*. J Cell Biol 101:712–719.

May GS, Morris NR (1988): Developmental regulation of a conidiation specific β-tubulin in *Aspergillus nidulans*. Dev Biol 128:406–414.

May GS, Tsang ML-S, Smith H, Fidel S, Morris NR (1987): *Aspergillus nidulans* β-tubulin genes are unusually divergent. Gene 55:231–243.

May GS, Waring RB, Morris NR (1990): Increasing *tub* C β-tubulin synthesis by placing it under the control of a *ben* A β-tubulin upstream sequence causes a reduction in *ben* A β-tubulin level but has no effect on microtubule function. Cell Motil Cytoskel 16:214–220.

Michiels F, Falkenburg D, Müller AM, Hinz U, Otto U, Bellmann R, Glätzer KH, Brand R, Bialojan S, Renkawitz-Pohl R (1987): Testis-specific β2 tubulins are identical in *Drosophila melanogaster* and *D. hydei* but differ from the ubiquitous β1 tubulin. Chromosoma 95:387–395.

Michiels F, Gasch A, Kaltschmidt B, Renkawitz-Pohl R (1989): A 14 bp promoter element directs the testis specificity of the *Drosophila* β2 tubulin gene. EMBO J 8:1559–1565.

Mogensen MM, Tucker JB (1987): Evidence for microtubule nucleation at plasma membrane-associated sites in *Drosophila*. J Cell Sci 88:95–107.

Mogensen MM, Tucker JB (1990): Taxol influences control of protofilament number at microtubule-nucleating sites in *Drosophila*. J Cell Sci 97:101–107.

Mogensen MM, Tucker JB, Stebbings H (1989): Microtubule polarities indicate that nucleation and capture of microtubules occurs at cell surfaces in *Drosophila*. J Cell Biol 108:1445–1452.

Monteiro MJ, Cleveland DW (1987): Sequence of chicken cβ7 tubulin: Analysis of a complete set of vertebrate β-tubulin isotypes. J Mol Biol 199:439–446.

Murphy DB (1991): Functions of tubulin isoforms. Curr Opin Cell Biol 3:43–51.

Murphy DB, Wallis KT, Machlin PS, Ratrie H III, Cleveland DW (1987): The sequence and expression of the divergent β-tubulin in chicken erythrocytes. J Biol Chem 262:14305–14312.

Oakley BR, Morris NR (1980): Nuclear movement is β-tubulin dependent in *Aspergillus nidulans*. Cell 19:255–262.

Oka MT, Arai T, Hamaguchi Y (1990): Heterogeneity of microtubules in dividing sea urchin eggs revealed by immunofluorescence microscopy: Spindle microtubules are composed of tubulin isotypes different from those of astral microtubules. Cell Motil Cytoskel 16:239–250.

Oppenheimer DG, Haas N, Silflow C, Snustad DP (1988): The β-tubulin gene family of *Arabidopsis thaliana:* Preferential accumulation of the β1 transcript in roots. Gene 63:87–102.

Paturle-Lafanechere L, Eddé B, Denoulet P, Van Dorsselaer A, Mazarquil H, Le Caer JP, Wehland J, Job D (1991): Characterization of a major brain tubulin variant which cannot be tyrosinated. Biochemistry 30:10;523–528.

Paul ECA, Buchschacher GL Jr, Cunningham DB, Dove WF, Burland TG (1992): Preferential expression of one β-tubulin gene during flagellate development in *Physarum*. J Gen Microbiol 138:229–238.

Paul ECA, Roobol A, Foster KE, Gull K (1987): Patterns of tubulin isotype synthesis and usage during mitotic spindle morphogenesis in *Physarum*. Cell Motil Cytoskel 7:272–281.

Piperno G, Fuller MT (1985): Monoclonal antibodies specific for an acetylated form of α-tubulin recognize the antigen in cilia and flagella from a variety of organisms. J Cell Biol 101:2085–2094.

Praitis V, Katz WS, Solomon F (1991): A codon change in β-tubulin which drastically affects microtubules structure in *Drosophila melanogaster* fails to produce a significant phenotype in *Saccharomyces cerevisiae*. Mol Cell Biol 11:4726–4731.

Prescott AR, Foster KE, Warn RM, Gull K (1989): Incorporation of tubulin from an evolutionarily diverse source, *Physarum polycephalum,* into the microtubules of a mammalian cell. J Cell Sci 92:595–605.

Raff EC (1984): Genetics of microtubule systems. J Cell Biol 99:1–10.

Raff EC, Diaz HB, Hoyle HD, Hutchens JA, Rudolph JE, Subler MA (1987): Origin of multiple gene families: Are there both functional and regulatory constraints? In Raff RA, Raff EC (eds): Development as an Evolutionary Process. New York: Alan R. Liss, Inc., pp 203–238.

Raff EC, Fuller MT (1984): Genetic analysis of microtubule function in *Drosophila*. In Borisy GG, Cleveland DW, Murphy DB (eds): Molecular Biology of the Cytoskeleton. Cold Spring Harbor, NY: Cold Spring Harbor Laboratory, pp 293–304.

Raff EC, Fuller MT, Kaufman TC, Kemphues K, Rudolph JE, Raff RA (1982): Regulation of tubulin gene expression during embryogenesis in *Drosophila melanogaster.* Cell 28:33–40.

Rees D (1990): A Genetic and Developmental Analysis of Mutations in *αTUB67C:* A Highly Divergent, Maternally Expressed α-Tubulin in *Drosophila melanogaster.* Ph.D. thesis, Indiana University, Bloomington, Indiana.

Rothwell SW, Grasser WA, Murphy DB (1986): Tubulin variants exhibit different assembly properties. Ann NY Acad Sci 466:103–110.

Rubenstein PA (1990): The functional importance of multiple actin isoforms. BioEssays 12:309–315.

Rudiger M, Plessman U, Kloppel KD, Wehland J, Weber K (1992): Class II tubulin, the major brain beta tubulin isotypes, is polyglutamylated on glutamic acid residue 435. FEBS Letts 308:101–105.

Rudolph JE, Kimble M, Hoyle HD, Subler MA, Raff EC (1987): Three *Drosophila* beta-tubulin sequences: A developmentally regulated isoform (β3), the testis-specific isoform (β2), and an assembly-defective mutation of the testis-specific isoform ($\beta 2t^8$) reveal both an ancient divergence in metazoan isotypes and structural constraints for beta-tubulin function. Mol Cell Biol 7:2231–2242.

Salmon ED, Leslie RJ, Saxton WM, Karow ML, McIntosh JR (1984): Spindle microtubule dynamics in sea urchin embryos: Analysis using a fluorescein-labeled tubulin and measurements of fluorescence redistribution after laser photobleaching. J Cell Biol 99:2165–2174.

Savage C, Hamelin M, Culotti JG, Coulson A, Albertson DG, Chalfie M (1989): *mec-7* is a β-tubulin gene required for the production of 15-protofilament microtubules in *Caenorhabditis elegans*. Genes Dev 3:870–881.

Sawada T, Cabral F (1989): Expression and function of β-tubulin isotypes in Chinese hamster ovary cells. J Biol Chem 264:3013–3020.

Saxton WM, Stemple DW, Leslie ED, Salmon ED, Zavortink M, McIntosh JR (1984): Tubulin dynamics in cultured mammalian cells. J Cell Biol 99:2175–2186.

Schatz PJ, Solomon F, Botstein (1986a): Genetically essential and nonessential α-tubulin genes specify functionally interchangeable proteins. Mol Cell Biol 6:3722–3733.

Schatz PJ, Pillus L, Grisafi P, Solomon F, Botstein D (1986b) Two functional α-tubulin genes of the yeast *Saccharomyces cerevisiae* encode divergent proteins. Mol Cell Biol 6:3711–3721.

Schedl T, Owens J, Dove WF, Burland TG (1984): Genetics of the tubulin gene families of *Physarum*. Genetics 108:143–164.

Serrano L, de la Torre J, Maccioni RB, Avila J (1984): Involvement of the carboxyl-terminal domain of tubulin in the regulation of its assembly. Proc Natl Acad Sci USA 81:5989–5993.

Silflow CD (1991): Why do tubulin gene families lack diversity in flagellate/ciliate protists? Protoplasma 164:9–11.

Silflow CD, Chisholm RL, Conner TW, Ranum LP (1985): The two alpha-tubulin genes of *Chlamydomonas reinhardi* code for slightly different proteins. Mol Cell Biol 5:2389–2398.

Sisodia SS, Gay DA, Cleveland DW (1990): *In vivo* discrimination among β-tubulin isotypes: Selective degradation of a type IV β-tubulin isotype following overexpression in cultured animal cells. New Biologist 2:66–76.

Snustad DP, Haas NA, Kopczak SD, Silflow CD (1992): The small genome of *Arabodipsis* contains at least nine expressed β-tubulin genes. Plant Cell 4:549–556.

Solnica-Krezel L, Burland TG, Dove WF (1991): Variable pathways for developmental changes of mitosis and cytokinesis in *Physarum polycephalum*. J Cell Biol 113:591–604.

Solnica-Krezel L, Diggins-Gilicinski M, Burland TG, Dove WF (1990): Variable pathways for developmental changes in composition and organization of microtubules in *Physarum polycephalum*. J Cell Sci 96: 383–393.

Sullivan KF (1988): Structure and utilization of tubulin isotypes. In Palade GE, Alberts BM, Spudich JA (eds): Annual Review of Cell Biology. Palo Alto, CA: Annual Reviews, Inc., pp 687–716.

Sullivan KF, Cleveland DW (1986): Identification of conserved isotype-defining variable region sequences for four vertebrate β-tubulin polypeptide classes. Proc Natl Acad Sci USA 83:4327–4331.

Theurkauf WE (1992): Behavior of structurally divergent α-tubulin isotypes during *Drosophila* embryogenesis: Evidence for post-translational regulation of isotype abundance. Dev Biol 154:205–217.

Theurkauf WE, Baum H, Bo J, Wensink PC (1986): Tissue-specific and constitutive α-tubulin genes of *Drosophila melanogaster* code for structurally distinct proteins. Proc Natl Acad Sci USA 83:8477–8481.

Tucker JB, Milner MJ, Currie DA, Muir JW, Forrest DA, Spencer MJ (1986): Centrosomal microtubule-organizing centres and a switch in the control of protofilament number for cell surface-associated microtubules during *Drosophila* wing morphogenesis. Eur J Cell Biol 41:279–289.

Twitty VC (1966): Of Scientists and Salamanders. San Francisco: W.H. Freeman and Co. pp 42–43.

Weatherbee JA, May GS, Gambino J, Morris NR (1985): Involvement of a particular species of β-tubulin (β3) in conidial development in *Aspergillus nidulans*. J Cell Biol 101:706–711.

Webster DR, Wehland J, Weber K, Borisy GG (1990): Detyrosination of alpha tubulin does not stabilize microtubules *in vivo*. J Cell Biol 111:113–122.

Weinstein B, Solomon F (1990): Phenotypic consequences of tubulin overproduction in *Saccharomyces cerevisiae*. Mol Cell Biol 10:5296–5304.

White EA, Burton PR, Himes RH (1987): Polymorphic assembly of subtilisin-cleaved tubulin. Cell Motil Cytoskel 7:31–38.

Youngblom J, Schloss JA, Silflow CD (1984): The two β-tubulin genes of *Chlamydomonas reinhardtii* code for identical proteins. Mol Cell Biol 4:2686–2696.

ABOUT THE AUTHOR

ELIZABETH C. RAFF is Professor of Biology at Indiana University in Bloomington, Indiana, where she teaches graduate and undergraduate courses in cell biology, development, and biochemistry. After receiving a B.S. in biochemistry with highest distinction from the Pennsylvania State University, she pursued doctoral research with J.J. Blum at Duke University, earning a Ph.D. degree in biochemistry in 1968 for work on mechanisms of ciliary motility. Her postdoctoral work with F.W. Putnam at Indiana University focused on questions concerning protein structure. In 1974 Dr. Raff returned to the study of microtubules, and in 1979 established genetic analysis in *Drosophila* as an experimental system for study of tubulin function *in vivo*. Genetic analysis has proven to be a powerful tool for approaching problems in cell biology. Dr. Raff's laboratory has demonstrated functional specialization of tubulin isoforms and, moreover, has established that there are hierarchical tubulin structural requirements for different microtubule assemblages, even within the same cell. Dr. Raff was a member of the Cellular and Molecular Basis of Disease Review Committee, N.I.G.M.S., 1984–1987, and an elected member of the Publications Committee of the American Society of Cell Biology, 1987–1989. She is currently a member of the Molecular Cytology Study Section, N.I.H.

Microtubules: 111–137

Microtubule Dynamics *In Vitro*

Peter M. Bayley, Kam K. Sharma, and Stephen R. Martin

I. INTRODUCTION

The availability of a simple experimental system for the assembly of microtubules from purified tubulin dimer *in vitro* has stimulated extensive studies on the fundamental kinetic and dynamic properties of microtubules under well-defined solution conditions. An individual microtubule undergoes transitions between periods of growth and shortening, so-called microtubule dynamic instability. This apparently unique property is related to the utilization of GTP in microtubule assembly and is directly observable both *in vivo* and *in vitro*. In this chapter we review the current understanding of the mechanisms of this phenomenon in terms of experimental observations of microtubule dynamics and biochemical and kinetic characterization of the underlying processes. A numerical model is discussed, the lateral cap model, that allows computer simulation of microtubule dynamic properties for different microtubule lattices under a wide range of conditions. Finally, the proposed molecular mechanism for dynamic instability is extended to examine how small regulator molecules, such as nucleotides, drugs, and accessory proteins, may contribute, in conjunction with other factors, to control mechanisms governing the extent and rates of changes in the distribution of dynamic microtubules in a cellular environment.

II. MICROTUBULE ASSEMBLY

Much of the fundamental work in the purification and characterization of tubulin as the principal component of the microtubule system was described in various chapters in the earlier edition of this volume [Amos, 1979; Jacobs, 1979; Luduena, 1979; Scheele and Borisy, 1979]. Since then a monograph [Dustin, 1984] and a number of reviews have described various physical and biochemical aspects of tubulin and microtubule assembly, including thermodynamics [Timasheff and Grisham, 1980; Correia and Williams, 1983], kinetics [Hill and Kirschner, 1982; Engelborghs, 1990], and nucleotide requirements [Carlier, 1989]. Most recently, reviews of different aspects of the dynamic properties have appeared [Bayley, 1990; Gelfand and Bershadsky, 1991; Erickson and O'Brien, 1992; Bayley and Martin, 1992b; Mandelkow and Mandelkow, 1992]. From these and other works can be derived the following general principles governing microtubule assembly.

A. Guanine Nucleotides

Tubulin αβ-heterodimer (M_r = 100,000) binds two molecules of guanine nucleotide: The α-subunit binds one molecule of GTP irreversibly at the nonexchangeable site (N-site), and this is presumed to have a structural role, whereas the β-subunit binds one molecule of nucleotide reversibly at the exchangeable site (E-site). At this E-site, GTP binds with a Mg^{2+}-dependent affinity of $5 \times 10^7\ M^{-1}$; alternatively, GDP binds in a Mg^{2+}-independent fashion with threefold lower affinity [Zeeberg and Caplow, 1979; Fishback and Yarbrough, 1984; Mejillano and Himes, 1991]. Many GTP analogues, including nonhydrolyzable analogues, can also bind at the E-site.

B. Self-Assembly and Seeded Assembly

At elevated temperatures (e.g., 37°C) and at typical concentrations $[C_{tot}] > 10\ \mu M$, tubulin–GTP will spontaneously assemble into microtubules, and the E-site GTP is hydrolyzed stoichiometrically with the extent of microtubule assembly. Self-assembly exhibits a concentration-dependent lag-phase during which microtubule nucleation occurs. Alternatively, the assembly proceeds more efficiently if it is "seeded" by the addition of preformed nuclei in the form of short microtubules or subcellular components such as axonemal fragments or centrosomes.

C. The Steady State and Critical Concentration

Assembly continues to a steady state, at which the microtubule mass remains constant, provided a source of GTP is present; if the GTP becomes exhausted, the microtubules eventually disassemble. At the steady state, there remains a concentration of tubulin–GTP in pseudoequilibrium with the assembled microtubules. By analogy with equilibrium polymerization [Oosawa and Asakura, 1975], this is termed the *steady state* or *critical concentration,* C_c. The condition for tubulin assembly, under a given set of solution and environmental conditions, is that the total concentration of tubulin–GTP, $[C_{tot}]$, is greater than C_c.

The assembly process is generally exponential over most of its amplitude, and the observed rate constant, k_{obs}, is related to the microtubule number concentration (i.e., the concentration of nuclei present, $[C_N]$) and the bimolecular rate constant for addition to the microtubule end, k_+, such that $k_{obs} = k_+ \cdot [C_N]$. For $k_+ = 10^7\ M^{-1} \cdot s^{-1}$ and $[C_N] = 1$ nM, $k_{obs} = 10^{-2}\ s^{-1}$, and the characteristic time for assembly, $1/k_{obs}$, is 100 seconds. The amount of polymer formed is $[C_{pol}] = [C_{tot}] - C_c$. For $[C_{tot}] = 20\ \mu M$, and $C_c = 10\ \mu M$, $[C_{pol}] = 10\ \mu M$; and for $[C_N] = 1$ nM, the average number of tubulin molecules per microtubule is 10^4, and the mean length $\approx 6\ \mu m$ (a 13 protofilament microtubule has about 1,625 tubulins/μm).

D. Assembly Promoters

C_c and k_+ are dependent on buffer conditions, including cations such as Mg^{2+}, and on the presence of assembly promoters, such as glycerol, sucrose, dimethylsulfoxide (DMSO) or polyethylene glycol. In general, such promoters also increase the diversity of assembled structures [Carlier and Pantaloni, 1981; Karecla et al., 1989], characteristic of the tendency of tubulin to form polymorphic structures. Microtubule-associated proteins (MAPs) also promote assembly, lowering the C_c.

E. Polarity and Protofilament Number

In aqueous buffer systems, the microtubules assembled from tubulin–GTP are markedly similar at the ultrastructural level to those observed *in vivo*. They have a cylindrical structure of effectively parallel protofilaments in which the α/β-heterodimer is arranged in a head to tail orientation, producing a polar microtubule with one end "α-out" and the other β-out." However, the exact specification of the geometrical lattice of tubulin molecules within the microtubule is not known conclusively. Microtubules assembled from tubulin (by self-assembly) frequently contain protofilament number n in the range 12 to 16, though $n = 13$ and $n = 14$ are more common [Wade et al., 1990]. Microtubules grown onto seeding structures, such as axonemal fragments, and centrosomes, appear to retain the protofilament number dictated by the nucleating structure [Allen and Borisy, 1974; Evans et al., 1985; Scheele et al., 1982]. This situation is further complicated by the recent observation [Chrétien et al., 1992] that changes in protofilament number within a single microtubule are observed (albeit rather infrequently) for microtubules self-assembled *in vitro*.

F. Microtubule Lattices

A further degree of uncertainty exists in terms of the symmetry properties of the lattice. For $n = 13$, there exists only one perfectly sym-

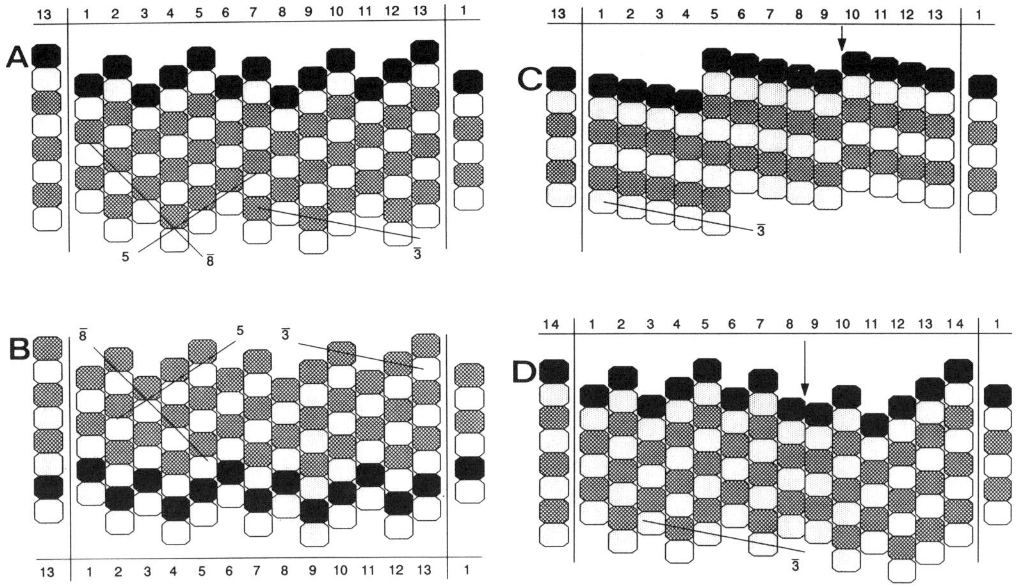

Fig. 1. *Microtubule lattices, shown in two-dimensional representation [Amos, 1979]; three-dimensional representations are given in McEwen and Edelstein [1977]; see also Fig. 6). Closed symbols denote β-subunit-containing GTP; cross-hatched symbols, β-subunit-containing GDP; and stippled symbols, α-subunit of tubulin. The microtubule end shows the single layer of T–GTP, according to the lateral cap model (see section IV).* **A,B:** *The two ends of the 13 protofilament A-type lattice, showing the $\bar{3}$-start alternating α–β–α monomer helix and the 5- and $\bar{8}$-start dimer helices (the bar denotes a left-handed helix). A is the β-out and B is the α-out end.* **C:** *The 13 protofilament B-lattice: the $\bar{3}$-start monomer helix is continuously α-subunit or continuously β-subunit, apart from the α–β "seam" (arrow).* **D:** *The 14 protofilament A-lattice: The $\bar{3}$-start monomer helix is alternating α–β–α, apart from the α–α/β–β "seam" (arrow). The 5-and $\bar{8}$-start helices are present but not continuous. (The 14 protofilament B-lattice resembles C but with one more protofilament.)*

metrical helical lattice of the tubulin α/β-heterodimer, namely, the A-type lattice, in which the α-subunit and the β-subunit alternate along the $\bar{3}$-start monomer helix (see Fig. 1a) [Amos, 1979; Mandelkow et al., 1986]. An alternative arrangement, the B-type lattice, has homologous contacts (α–α or β–β) along the $\bar{3}$-start helix, except for one position, a "seam," at which a single heterologous, α–β contact occurs (see Fig. 1b). For n = 14, the A-type lattice contains a single homologous α–α, β–β contact (see Fig. 1c). Structural studies have not yet reached the resolution necessary to distinguish between the α-subunit and the β-subunit; hence the question of monomer arrangement is currently open. Tubulin assembles into a variety of structures other than microtubules, including sheets, in some of which the protofilaments are arranged in antiparallel orientation. Hence polymorphism is a characteristic of tubulin assembly, and caution must be exercised in designating a single exact lattice for the microtubule.

G. Microtubule Disassembly

Observations of the kinetics of microtubule assembly/disassembly showed that, under conditions in which microtubules are unstable (e.g., low tubulin concentration, removal of GTP, or low temperature), the disassembly is surprisingly fast [Karr et al., 1980; Farrell et al., 1983; Carlier et al., 1984]. Seen retrospectively, the interpretation of some of the earlier work is complicated by the use of microtubule protein. The presence of MAPs has a pronounced effect on C_c and assembly kinetics [Scheele and

Borisy, 1979; Bayley et al., 1985]. Taken together, these experiments introduced the concept that disassembly principally involves the rapid loss of tubulin–GDP from the end of an assembled microtubule. The rates, in the range 100 to 400 s^{-1} per microtubule end, can be increased more than 10-fold by the presence of [Mg^{2+}] >5 mM and [Ca^{2+}] >1 mM [Gal et al., 1988]. The suggestion was made that residual tubulin–GTP could comprise a terminal structure [Karr et al., 1979; Carlier and Pantaloni, 1981] and that this "cap" could stabilize the microtubule against fast disassembly of tubulin–GDP [Carlier et al., 1984; Mitchison and Kirschner, 1984b]. Although still controversial, this mechanism has become a central tenet of subsequent interpretations.

H. The Coupling of GTP Hydrolysis and Tubulin–GTP Addition

A serious limitation to detailed understanding of how such a mechanism might work lies in the considerable uncertainty about the exact relationship in time between the addition of a molecule of tubulin–GTP to the end of a growing microtubule and the hydrolysis of one molecule of GTP to GDP. If hydrolysis is a slow process relative to the addition, the assembled microtubule should contain substantial amounts of tubulin–GTP. Original observations [Carlier and Pantaloni, 1981] supported this idea and led to an extensive theoretical treatment of the ability of this tubulin–GTP "cap" to stabilize the microtubule [Chen and Hill, 1985; reviewed by Carlier, 1989]. Subsequent analytical measurements of the guanine nucleotide compositions of microtubules attempting to quantify the extent of any GTP-containing structure in steady-state microtubules have generally concluded that it is below the limits of detection [O'Brien et al., 1987; Schilstra et al., 1987; Stewart et al., 1990]. One recent estimate of ^{32}P present on steady-state microtubules assembled in γ-^{32}P-GTP suggests that the average number could be on the order of 12 $\pm$ 3 molecules per microtubule end [Wilson et al., 1990]. These results indicate that, at steady state, where dynamic instability is most readily observable, GTP hydrolysis is relatively fast compared with the (bimolecular) addition reaction of tubulin– GTP, and any stabilizing region is extremely small.

I. Kinetic Intermediates in Assembly

Extensive work has suggested a possible role for the complex of tubulin–GDP–P_i as a stabilizer of microtubules, in analogy with the kinetic role of the myosin–ADP–P_i complex as a kinetically important intermediate species in ATP hydrolysis by actomyosin in skeletal muscle. Suggestive evidence lies in the ability of phosphate analogues to stabilize microtubules [Carlier et al., 1989], but this behavior equally resembles the low critical concentration observed with certain nonhydrolyzable GTP analogues such as GMP–PNP [Carlier et al., 1984], and the relationship to microtubule dynamics is only indirect. Tubulin–GDP–P_i must, of course, occur at least as a transient intermediate in the conversion of tubulin–GTP to tubulin–GDP in the microtubule lattice. This species has in fact been detected in a single turnover experiment using taxol-stabilized microtubules in which microtubule dynamics and disassembly are excluded [Melki et al., 1990]. However, during microtubule assembly, GTP hydrolysis appears to be irreversible [Osei et al., 1988], i.e., effectively excluding tubulin–GDP–P_i as an intermediate from which GTP might be formed. Furthermore, phosphate itself was not able to stabilize individual microtubules against disassembly by loss of tubulin–GDP [Caplow et al., 1989], and phosphate does not inhibit steady-state tubulin exchange into microtubules [Schilstra et al., 1991]. Thus the stabilizing species does not appear at this stage to be kinetically or analytically distinguishable from the tubulin–GTP molecule that has added to a terminal position of the microtubule lattice.

III. EXPERIMENTAL OBSERVATIONS OF MICROTUBULE DYNAMICS *IN VITRO*

The concept of microtubule dynamic instability derives from experimental observations by Mitchison and Kirschner [1984a,b] that showed that microtubules have unusual dynamic properties. The length distribution of a microtubule population at steady state can change relatively quickly, with the average length increasing as a function of time. This work was followed by the direct observation by Horio and Hotani [1986] of substantial changes in length of individual microtubules at, or close to, steady state, using dark-field microscopy. Microtubules show both growing and shortening phases that can interconvert. Several reviews discuss the important biological implications of these observations [Kirschner, 1988; Mitchison and Kirschner, 1988], and the topic of microtubule dynamics *in vivo* is further detailed by Wordeman and Mitchison in this volume.

Experimental observations are broadly divided into two groups: In the first the "bulk" properties of a microtubule population are monitored. More recently, microscopic methods have been developed to observe the properties of individual microtubules in real time. In earlier work [Margolis and Wilson, 1978; Wilson et al., 1985] pulse-chase techniques were used to quantitate the dynamic property of microtubule "treadmilling." This phenomenon was inferred [Wegner, 1976] from the observation of the existence of different critical concentrations for actin assembly at the two ends of an F-actin polymer. Recent measurements [Farrell et al., 1987] suggest that the end-dependent incorporation of tubulin into microtubules due to treadmilling phenomena is generally considerably less than that due to the substantial length excursions of dynamic instability. A radioactive labeling technique has also been used [Schilstra et al., 1988] to estimate the incorporation of tubulin into microtubules at steady state, due to dynamic instability. It was shown [Schilstra et al., 1991] that dynamics can be significantly influenced by solution conditions and can be suppressed by substoichiometric concentrations of antimitotic drugs [Schilstra et al., 1989] acting in an end-dependent fashion [Margolis and Wilson, 1977] (see also section IV.C.5., below). Other work examining GTP utilization during microtubule assembly is referred to in section II.H., above. The phenomenon of microtubule oscillations has recently been reviewed [Mandelkow and Mandelkow, 1992].

A. Measurement of Microtubule Dynamic Instability *In Vitro*

Regarding population versus direct methods, most quantitative measurements on microtubule populations depend on the evaluation of mean microtubule length or microtubule number concentration. Such measurements present the problems of chemical fixation, the identification of short microtubules, and the preservation against breakage of long microtubules. Electron microscopy, optical microscopy, and immunofluorescence have all been applied successfully.

The alternative approach of direct observation of individual microtubules has become possible through the use of video enhancement techniques and image processing. In addition to dark-field techniques, differential interference contrast (DIC) microscopy has potentially the highest depth resolution (being close to optical resolution, ≈ 0.2 μm) and has proved extremely successful both *in vitro* and *in vivo* [see Salmon et al., 1989, for a review of the technique].

The experimental basis for most modeling is now *microscopic* rather than *macroscopic*. However, there is a serious problem of scale in relating the two regimes. The reader is invited to consider how many observations would be necessary to determine accurately the rate constants for the spontaneous interconversion of A↔B simply by measuring the lifetimes of individually observed molecules of A and B. Simulation suggests, in the analogous case of microtubule transitions, that extensive observations would be required to define the lifetime adequately for each set of conditions [Bayley et al., 1991]. This is difficult to achieve experimentally owing to the length of observation required for individual events at $C > C_c$ and $C < C_c$ and to the limits of reproducibility and stability of the protein system under observation conditions. Thus the quantitation of

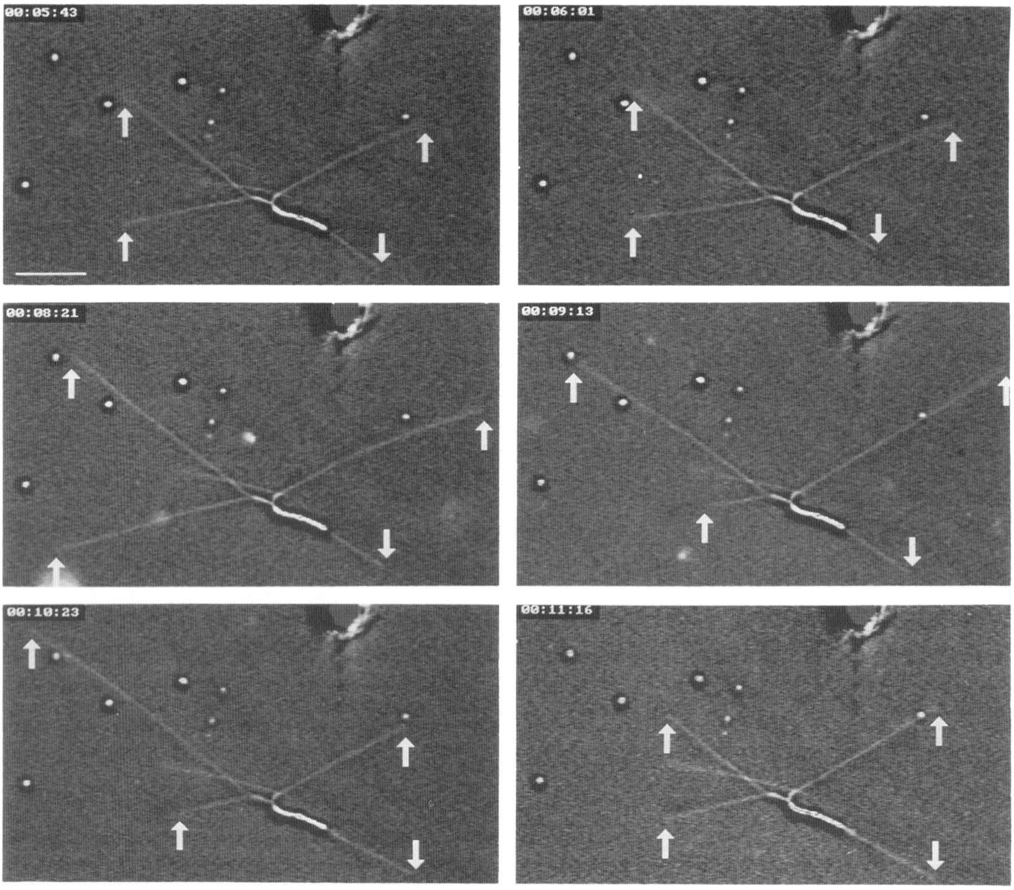

Fig. 2. *Bipolar dynamic microtubule growth* in vitro *seeded from a* Chlamydomonas *axoneme. T–GTP (15 μM) incubated with a seed at 37–40°C, observed by video-enhanced dark-field microscopy at time = 5 minutes 43 seconds to 11 minutes 16 seconds. The two ends of the axoneme are distinguished by their morphology and the microtubule distribution. Both ends show dynamic transitions within this short time period. Bar = 10 μm.*

microtubule dynamic instability necessarily involves some intrinsic limitations.

B. Results

1. Microtubule length versus time. Using DIC observation, Walker et al. [1988] provided the first quantitation of microtubule transitions as a function of tubulin–GTP (T–GTP) concentration. They demonstrated the typical sawtooth behavior of length versus time [cf. Horio and Hotani, 1986], the dependence of growth rate on [T-GTP], and the limited concentration range about C_c in which transition behavior was observable. The end dependence of the kinetics was derived from the visual identification of the ends of axonemes, from which microtubule growth was initiated.

Some of these properties are illustrated from our own dark-field observations. Figure 2 shows the behavior of several microtubules attached to both ends of an axoneme from a *Chlamydomonas* flagellum. The behavior of these microtubules is independent and uncorrelated. The polarity of the microtubules is identified by the characteristic fraying of the distal end of the isolated axoneme into numerous thinner fibers

[Allen and Borisy, 1974]. This identification is confirmed by measurements that show that microtubules grow from the frayed end at a rate distinctly faster than from the opposite end.

Figure 3 shows simultaneous multiple microtubule growth from one end of an axonemal fragment from *Tetrahymena* cilia. The microtubules exhibit independent and markedly different behavior, although two ends may be closer in space than 1 μm. The plot of length versus time for one of these microtubules (Fig. 4) shows frequent transitions between periods of steady growth and rapid shortening. Points are digitized from the video record at approximately 3 second intervals. The rates of individual periods of growth appear approximately linear ($\leqslant$ S.D. 10%) over time intervals of 170 to 1,000 seconds. Periods 3 and 4, possibly separated by a brief pause, differ in rate by a factor of 1.22. Periods of rapid shortening appear intrinsically less linear (e.g., periods 2 and 9), but the quantitation is more difficult. The constancy of growth rates is discussed in the next section.

To investigate the end-dependence of these properties, observations have been made of the growth properties of microtubules initiated from the two ends of a seed derived from chemically cross-linked tubulin microtubules [Koshland et al., 1988] absorbed onto a coverslip (see Fig. 5). Increasing concentrations of [T–GTP] are introduced to the same seeds, and the growth behavior can be compared unambiguously for the opposite ends of a single microtubule. In dark-field optics, the new growth appears brighter; the extremities of the microtubule are unattached and show limited radial movement relative to the seed. At high concentration of T–GTP, microtubules grow at both ends, whereas at lower concentrations, growth appears to be limited to the faster growing end. Measurements of growth rates (at both ends) suggest that this is an intrinsic property of the ethylene glycol bis-succinimidylsuccinate (EGS) seeds.

2. Transition behavior. The work of Walker et al. [1988] suggested that both transition frequencies depended linearly on [T–GTP]. More recent observations on the growth to shortening transitions of microtubules subjected to rapid dilution indicate that the frequency of this transition is, in fact, significantly nonlinearly dependent on [T–GTP] (see Fig. 6 in Walker et al. [1991]; note logarithmic y-axis). This is an important point in subsequent modeling, since it shows that transitions between states are highly cooperative. (This property is also essential for the mechanism of microtubule oscillations [Chen and Hill, 1987; Bayley et al., 1989; Mandelkow and Mandelkow, 1992].

This work showed that the lifetime of the growing state was not related to the speed of growth before dilution to low [T–GTP]. This result appears inconsistent with any mechanism other than the tight-coupling of GTP hydrolysis with T–GTP addition (see section II.H., above). This is further evidence in support of the smallness of the region conferring stability on the growing state, even at concentrations significantly above the C_c. Likewise, stop-flow experiments on (bulk) microtubule disassembly also failed to detect a significant lag in microtubule disassembly on dilution, consistent with a short capping region [Voter et al., 1991].

3. The constancy of microtubule growth. O'Brien et al. [1990] observed that shortening rates of individual microtubules are markedly enhanced by conditions such as elevated [Mg^{2+}] and that disassembly can be discontinuous. Extensive observations showing that individual microtubules may grow and shorten at variable rates were recently reported by Gildersleeve et al. [1992], using DIC optics at high resolution. The basic pattern of growth-shortening transitions was observed (cf. Fig. 4). However, careful statistical analysis showed that, while both growth and shortening occurred in apparently linear phases, significant variations in rate occurred over short and intermediate time scales ($\approx$10–100 seconds). These discontinuities appear not to be related to specific structural locations, but may be due to local, transient structural variations. These authors note a number of reports of observations *in vivo* as well as *in vitro,* showing variations in rates, as well as temporary "pauses" in length changes. Further interesting information has come from observations [Dye et al., 1992] of

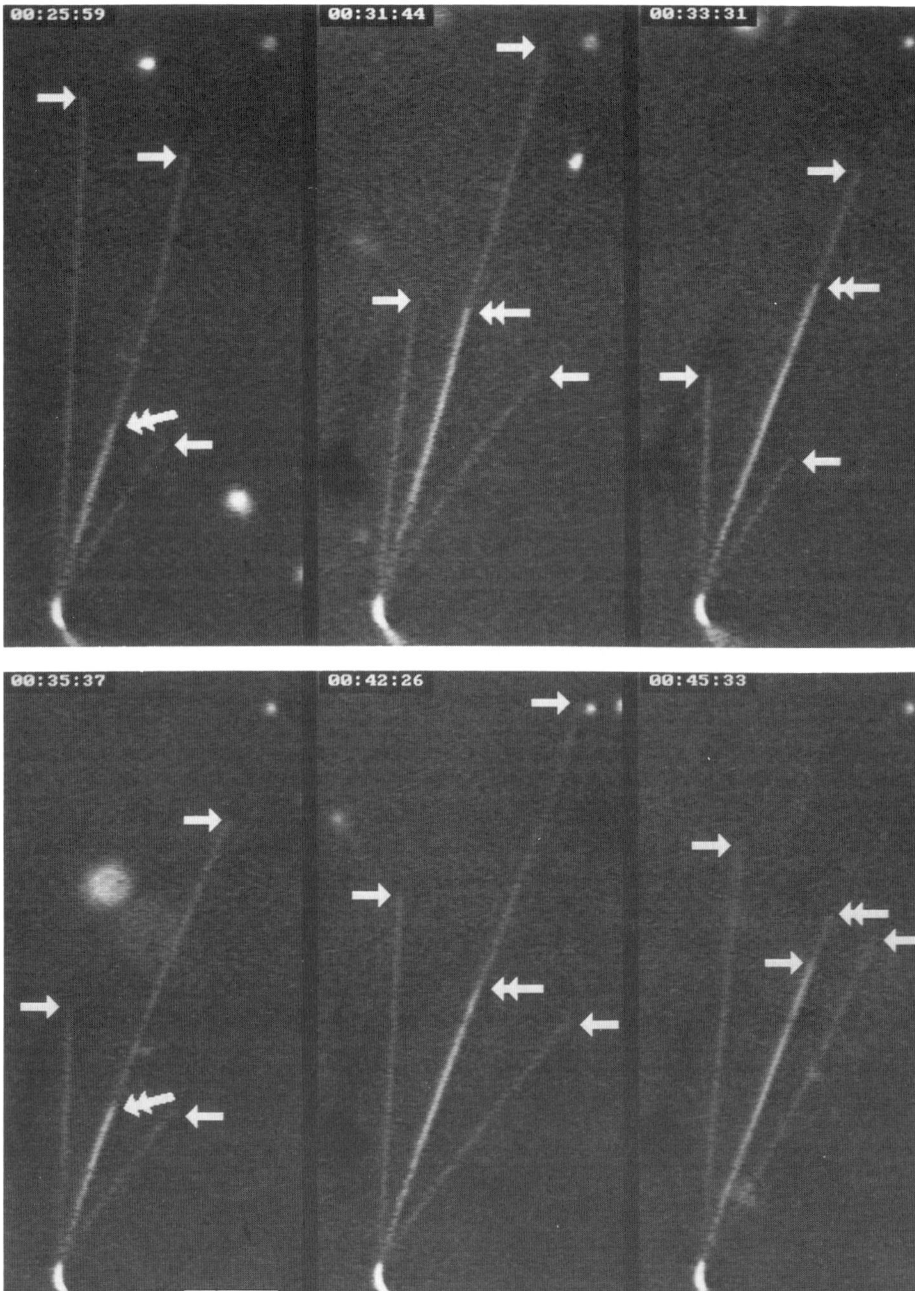

Fig. 3. *Dynamic microtubule growth* in vitro *seeded from an axoneme of* Tetrahymena *cilia. T–GTP (≈25 μM) incubated with a seed at 30°C, observed by video-enhanced dark-field microscopy at time = 25 minutes 59 seconds to 45 minutes 33 seconds. The four individual microtubules behave independently with unrelated periods of growth and shortening (see also Fig. 4). The more intense microtubule (end marked with a double-headed arrow) was resolvable earlier in the sequence, but having once aligned with its longer neighbor it remained so. At t = 45 minutes 33 seconds, this neighboring microtubule shortens past its growing end.*

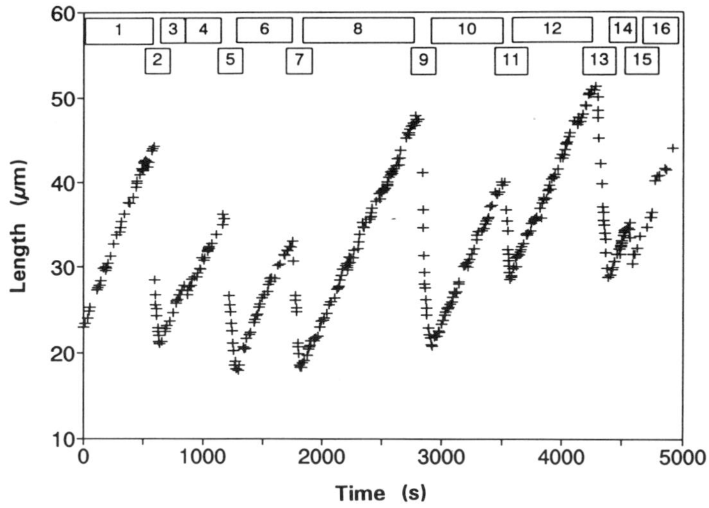

Fig. 4. *Observed microtubule dynamic instability. Analysis of length versus time for one microtubule from Figure 3 (the rightmost, single microtubule). Rates of growth and shortening were obtained by linear regression for 16 periods. The mean rate was calculated for both phases from* $Rate_{GorS} = \Sigma(r_i \cdot t_i)/\Sigma(t_i)$ *for* $i = G$ *or* S.

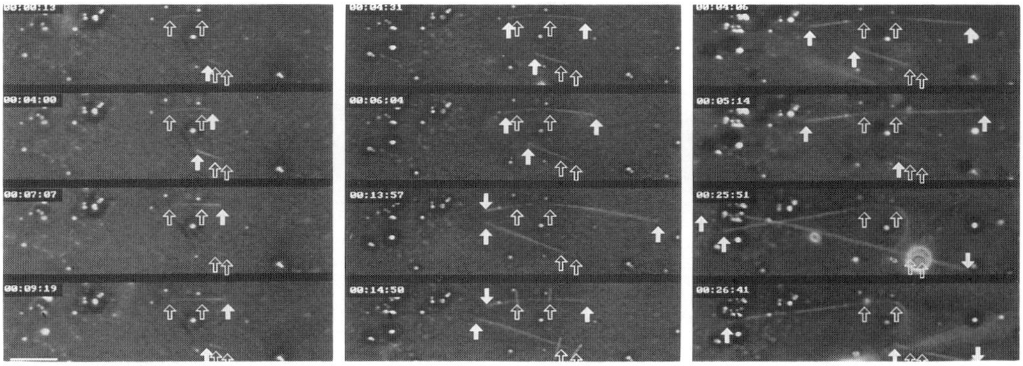

Fig. 5. *Bipolar dynamic microtubule growth* in vitro *seeded from fragments of single EGS cross-linked microtubules. The ends of the seeds are marked throughout with open arrows. Three experiments are shown using the same two seeds in a flow-cell into which tubulin at three different concentrations is introduced. Temperature is 37°–40°C.* **Left** *(0 minutes 13 seconds to 9 minutes 19 seconds): at 6 μM T–GTP, growth (closed arrows) is found on one end of each seed (top seed, right-hand end; bottom seed, left-hand end) both presumed the plus end, with the minus end initiating growth less effectively. The bottom seed (plus end) shows a G→S and an S→G transition.* **Middle** *(4 minutes 31 seconds to 14 minutes 50 seconds): At 10 μM T–GTP, growth occurs at both ends of the top seed, but only at the plus end of the bottom seed. The top seed (plus end) shows a G→S transition.* **Right** *(4 minutes 6 seconds to 26 minutes 41 seconds): At 15 μM T–GTP, growth occurs at both ends of both seeds. The bottom seed shows both G→S and S→G transitions (left hand, plus end). Total experiment duration was 3 hours. Bar = 12 μm. EGS, ethylene glycol* bis-*succinimidylsuccinate (see Koshland et al., 1988).*

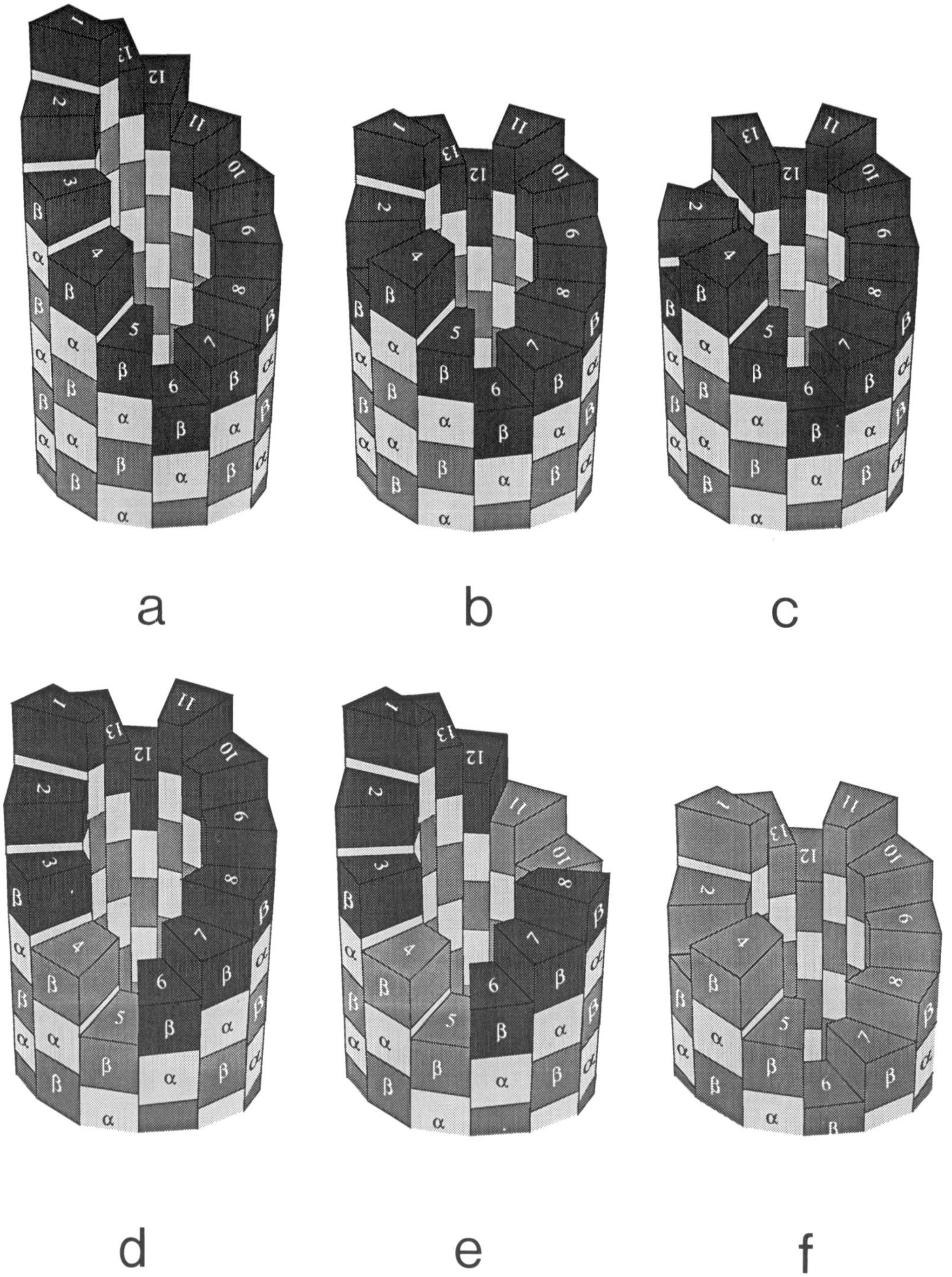

Fig. 6. *The lateral cap model for microtubule dynamics, showing different end configurations for a 13 protofilament A-type lattice (β-out end). See text for application of rate constants for the association and dissociation processes involved in the numerical simulation of dynamic instability. Black = β-tubulin-GTP; medium grey = β-tubulin–GDP; light grey = α-tubulin.* **a–c:** *The T–GTP cap is shown as a terminal layer, seen in different configurations of the lattice end. For example, in a there is one binding site, protofilament 6,*

microtubules, stabilized against disassembly by being "tethered" to stable seeds at both ends. The microtubule wall was seen to break down, apparently by loss of tubulin molecules. This is apparently a very rare event (for a 20 μm microtubule, once in 40 minutes on average) and the loss of T–GDP from an intact wall occurs with rate $\approx 10^{-8} s^{-1}$. While this process appears unlikely to contribute significantly to microtubule turnover in realistic conditions, it provides a further parameter characterizing the stability of the microtubule lattice.

It is clear therefore that the microtubule, composed only of tubulin plus its associated guanine nucleotide, has remarkable dynamic properties. The next section presents a survey of current work addressing the problem of rationalizing this unusual physical behavior, based on relatively simple structural and kinetic concepts.

IV. MODELS OF MICROTUBULE DYNAMICS *IN VITRO*

There are many different ideas about what constitutes a satisfactory model for a biological or other scientific process. Most accessible are structural models, and several such models for microtubules, have been alluded to already (see Fig. 1). In the absence of a three-dimensional structure for tubulin, and the potential diversity of microtubule lattice structures, only a limited schematic representation can be given, and the reader is cautioned against implications of diagrammatic models that may be more apparent than real.

By contrast, numerical models seek to provide something closer to the dynamic working system, to account for changes in experimental observables as a function of a controlling variable, such as time, concentration, temperature, and so forth. Such models clearly require a structural basis (e.g., ideal structures, as above) and also the introduction of physical parameters (e.g., rate constants) that may not be obtainable by direct observation. The converse is that an observable parameter (e.g., the lifetime of a state of a system) is often a complicated function of the intrinsic properties of that system.

Two different numerical approaches may be distinguished—both relevant to the treatment of microtubule properties. In one, an analytical solution is sought, that is, the behavior of the physical property is (eventually) expressed as an equation involving combinations of the intrinsic parameters. Thus the time dependence of an intermediate in a complex kinetic process can be expressed in terms of the intrinsic rate constants of the system. Analytical models may not incorporate much structural information and often have the disadvantage of mathematical complexity and unfamiliarity to the nonspecialist.

In the other approach, the behavior of the physical property is *simulated* in terms of an idealized structure, with kinetic events usually represented by a series of individual steps, repeated successively (by computer), the results of one step being the starting point for the next. Monte Carlo simulation offers a means of assessing the result of a large number of competing processes whose relative rates and hence probabilities can be stipulated and whose contribution is treated on a stochastic basis. Effective programs can often be implemented with modest computing power.

---

and one dissociation site, protofilament 1. In b, binding sites are on protofilaments 3, 6, and 12; dissociation sites are on protofilaments 1, 4, and 11. In c, binding sites are on protofilaments 1, 3, 6, and 12; dissociation sites are on protofilaments 2, 4, 11, and 13. **d–f:** *Progressive loss of tubulin–GTP (αβ-dimer) exposes T–GDP: a possible sequence starting from b is that d is generated by spontaneous loss of T–GTP from protofilament 4 in b, facilitating subsequent loss from 5; e is generated from d by loss of T–GTP from protofilament 11, facilitating subsequent loss from protofilaments 10 and 9; f is generated from e by loss of remaining T–GTP, leaving a microtubule with an all T–GDP end, i.e., the typical shortening state.*

Both types of numerical modeling were introduced to the field of microtubules by Hill and coworkers, [Hill and Kirschner, 1982; Hill and Carlier, 1983; Hill and Chen, 1984; Chen and Hill, 1985]. However, the initial basis of this work derived from experimental findings of a substantial fraction of tubulin-GTP within the assembled microtubule, i.e., somewhat different from those now accepted as typical (see section II.H.). The interested reader is referred to this literature for further details, [Hill, 1987; Carlier, 1989]. We follow the methodology of Chen and Hill [1985] in applying a Monte Carlo approach to the stochastic processes of tubulin addition and dissociation, but with a significantly different formulation of the properties and behavior of the ends of the microtubule.

A. Nomenclature

The nomenclature that has grown up around dynamic instability is somewhat confusing. We take the term *dynamics* to refer to the time-dependent changes in microtubule length, from which all other parameters derive. It should be noted that this dynamic behavior is most clearly observed close to the C_c. At [T–GTP] $>>$ C_c, microtubules grow continuously; at [T–GTP] $<<$ C_c, they shorten continuously. Visually, the dynamics are associated with the ability to see both phases of behavior, i.e., growing (G) and shortening (S). The phenomenon of "switching" from the state of growing to shortening was termed *catastrophe*. We prefer to term this process the *G→S transition*. Similarly for a microtubule initially in shortening phase, the "switching" from shortening to growth was termed *rescue;* we refer to this as the *S→G transition*. In fact the term *switch* itself has a mechanistic implication that may not be justifiable, since it suggests an "all-or-nothing" transition, and the reality is likely to be more complicated.

We use subscripts *G* and *S* to designate the kinetic parameters associated with either phase. Thus the system can be described phenomenologically in terms of four observable parameters, rates of growth and shortening, R_G, R_S, and the lifetimes of either state, T_G, T_S; the "transition frequencies" are then $k_{GS} = 1/T_G$ and $k_{SG} = 1/T_S$.

We note also that there is an ambiguity in the term *dynamic:* If the microtubule transitions become kinetically more "dynamic," i.e., increased rates of transitions, the length excursions necessarily are reduced. This state may well appear as a suppression of length changes.

B. Objectives of the Modeling

The basic question requiring a mechanistic answer is how can a microtubule undergo transitions, apparently at random, between a slowly growing state and a relatively rapidly shortening state? This behavior appears to be unique among all protein assembly systems. As discussed above, the *in vitro* experimental system contains only tubulin plus GTP under suitable experimental conditions. Thus it is well defined biochemically.

Also as discussed above (see section II.G.), the prevalent concept is that the microtubule is stabilized in growth by a "cap" located at the microtubule end. By definition, this region is required to differ in some way from the tubulin molecules in the main body of the microtubule, but its biochemical nature is not established unequivocally. Since T–GTP is the species adding to the microtubule end, it is logical to assign this as the initial terminal species. We have argued (in section II.I) that there is insufficient evidence, at this time, for a species intermediate between T–GTP and T–GDP acting as a *kinetically significant* species. We shall return to this point later. For the moment, the stabilizing effect of terminal T–GTP is taken as a postulate, to be rejected if it proves inadequate. Since any terminal structure appears to be so small as to be at the limit of analytical detection (see section II.H), the working hypothesis can be stated: *The growing microtubule is stabilised in growth by the presence of a cap composed of a terminal layer of molecules of T–GTP, whose loss causes the microtubule to shorten rapidly and whose reestablishment causes the microtubule to return to its state of growth.*

The objectives of the modeling are to test this hypothesis in terms of a specific question: Can a model for microtubule dynamic instability with a cap composed of a single terminal layer of T–GTP quantitatively reproduce the characteristic growing–shortening transition behavior? The different steps in seeking to answer this question by modeling microtubule dynamic instability are thus:

1. To define a structural model representing a suitable idealized lattice geometry, (cf. Fig. 1).
2. To derive kinetic rate constants for the association and dissociation processes, based, where possible, on experimental values.
3. With the minimum elaboration, to devise a mechanism whereby tubulin addition can preserve the basic condition that the cap is limited in size, but that addition and loss can occur as competing processes on a stochastic or random basis.
4. Hence to simulate the dynamic behaviour in terms of the characteristic property, microtubule length vs. time, (the L(t) plot) for an individual microtubule, and the growth rate vs [tubulin-GTP] (the J(C) plot) for a bulk microtubule solution.
5. To characterise the transition behaviour, using extended plots of L(t) at different [tubulin-GTP] to assess the predicted dependence of state lifetimes on [tubulin-GTP].
6. To examine the implications and predictions of the model, in terms of the effects on the dynamic properties of additional factors such as GDP, drugs, and other small molecules.

C. The Lateral Cap Model

The approach initially considers the 13 protofilament A-lattice, since the simulation is simplified on account of the lattice symmetry. Recently the treatment has been extended to the other microtubule lattice types discussed in section II.F. above. Dynamic instability is shown to be a potential property of all of them [Bayley and Martin, 1992a; Martin et al., 1993].

Irrespective of lattice type, the model is characterized by a microtubule end with a *terminal layer of T–GTP*. To distinguish this from the arrangement [Chen and Hill, 1985] involving an extensive longitudinal array of T–GTP, we call this the *lateral cap model*. This is illustrated in Figure 6.

The basic treatment has been described elsewhere [Bayley et al., 1989, 1990; Bayley and Martin, 1992b]. Briefly, it comprises the principle that the addition of T–GTP to a terminal site in the lattice is directly coupled to the hydrolysis of a previously terminal T–GTP molecule. This satisfies the 1:1 stoichiometry of addition and hydrolysis and also maintains the cap structure as being no more than one molecule of T–GTP deep, consistent with available experimental evidence (see sections II.H. and III.B.2.). This tight coupling corresponds to the *forced* hydrolysis described by O'Brien et al. [1987]; we avoid this term since the hydrolysis process is evidently spontaneous. Tight coupling is to be distinguished from the uncoupled addition reaction as previously formulated by Carlier and Pantaloni [1981] and Chen and Hill [1985].

Coupled hydrolysis also suggests an interpretation of the need for GTP hydrolysis during assembly of tubulin-GTP. With 1:1 stoichiometry, one can write the microtubule elongation step formally as

$$\mathrm{MT(T{-}GDP_x : T{-}GTP_y) + T{-}GTP} \\ \rightarrow \mathrm{MT(T{-}GDP_{x+1} : T{-}GTP_y)} + \Delta G_h,$$

where x is the number of T–GDP molecules in the lattice and y is the number of T–GTP molecules in the cap. The net effect is to increase the microtubule lattice by one molecule of T–GDP, a process that would be energetically unfavorable but that becomes possible through the availability of ΔG_h, the free energy of hydrolysis of GTP. If the cap is lost, the intrinsically unstable T–GDP lattice spontaneously disassembles. In this view the capped microtubule is *metastable* and is maintained only by the availability of GTP. Loss of the cap, or exhaustion of the GTP, allows the system to return to its lowest free energy, i.e., the state of dissociated T–GDP.

The stipulation of tight coupling is in fact a statement of the kinetic condition that hydrolysis proceeds faster than the addition of further T–GTP. It has the practical advantage of reducing the number of computational steps, as well as being consistent with the experimental evidence for a small cap under steady-state conditions, where dynamic instability is most pronounced.

Figure 6a–c illustrates the lateral cap of T–GTP as the terminal layer on a 13 protofilament A-type microtubule lattice. It may be noted that different end configurations are possible, with different numbers of equivalent sites (range 1–5) determined by the geometrical properties of the lattice. Figure 6 illustrates the "β-out" end of the microtubule; treatment of the opposite ("α-out") is given by Martin et al. [1991].

The observations of Walker et al. [1988] showed that the growth rate of either end of individual microtubules was apparently a linear function of [T–GTP], but with a negative intercept at [T–GTP] = 0, implying that T–GTP also dissociates from the growing end. With the lateral cap, this exposes a T–GDP. Hence subsequent steps may involve a microtubule end with sites containing a variable number of T–GDP molecules (Fig. 6d–f). In the optimal shortening state, the end is composed entirely of T–GDP, whereas in the optimal growing state it is composed entirely of T–GTP.

1. Numerical treatment. Rate constants are derived assuming a single value for the site association rate constant for T–GTP, k(+T) (such that the bimolecular rate constant for growth corresponds to the experimental value of $k_+ \approx 10^7\ M^{-1} \cdot s^{-1}$ per end). Values for the dissociation rate constants reflect the calculated affinities of T–GTP at different sites according to the following reasoning. The affinity for T–GTP at a growing end involves a site with two neighboring T–GTP molecules and is designated $K_{T/TT}$. As the proportion of T–GDP in the site increases, the affinity of the site for T–GTP is postulated to decrease so that the affinity for T–GTP at a shortening end (all T–GDP) is considerably lower than for a growing end (all T–GTP). (Without this property, the shortening phase would barely exist.) The lower limit is the affinity of T–GDP at a shortening end, i.e., given by the critical concentration for assembly of T–GDP, generally inaccessible experimentally, but taken to be several orders of magnitude greater than that for T–GTP. Thus the affinities are ordered as

$$K_{T/TT} > K_{T/TD} \approx K_{T/DT} > K_{T/DD} > K_{D/DD}.$$

The corresponding dissociation rate constants are derived from these via $k(-T/TT) = k(+T)/K_{T/TT}$, and so forth. Thus a set of six constants, comprising the association rate constant k(+T), and maximally five dissociation rate constants, designated k(−T/xy) (where xy = TT, TD, DT, and DD) and k(−D/DD) are sufficient to define the kinetic model. (This simplified description is used here for illustrative purposes: A more general treatment has been developed that allows inclusion of all possible binding sites [Martin et al., 1993].

The Monte Carlo method (see section IV, above) allows a mechanism containing a number of competing pathways to be simulated on a probabilistic basis; its application to this formulation is described in detail elsewhere [Bayley et al., 1990]. For a given value of [C] = [T–GTP], the association rate (k[+T].[C]) and the five dissociation rates are defined. At any step in the simulation, and for the microtubule end configuration at that point, the relative probability is assessed of n possible addition reactions and n possible dissociation reactions, where n is the current number of possible sites (maximum n = 5 for the 13 protofilament A-lattice; see Fig. 6). One of these reactions is selected on the basis of a random choice. The lattice configuration is then adjusted by addition or subtraction as appropriate, generating a new end configuration and set of sites, and the process is repeated from a new time origin.

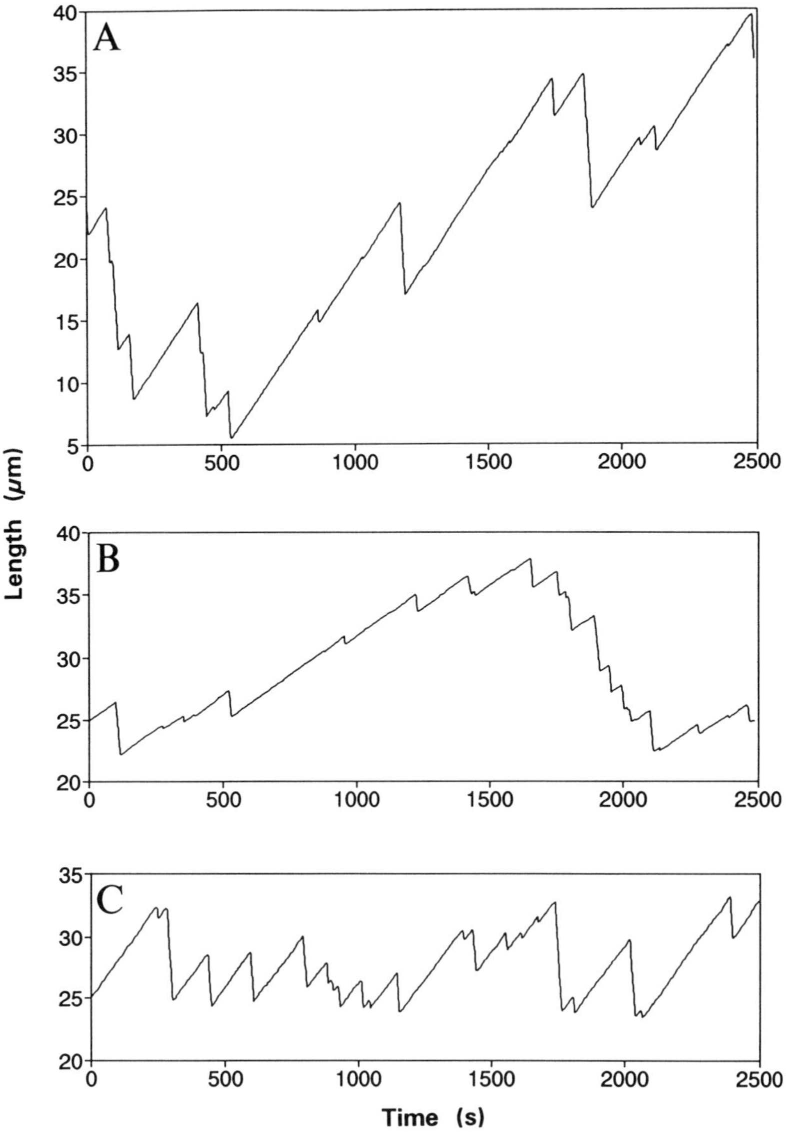

Fig. 7. *Microtubule length as a function of time (L(t) plot) computed for the β-out ends of three microtubule lattices: 13 protofilament A-type* (**A**), *13 protofilament B-type* (**B**), *and 14 protofilament A-type* (**C**).

2. Results: Microtubule length versus time. The change of length with time, L(t), is computed (see Fig. 7). Over long periods of time (Δt) the net change in length (ΔL) gives the total microtubule population mean growth rate (ΔL/Δt) at this value of [C]. Repeating this for different [C] values allows the J(C) plot to be obtained and the steady-state condition to be determined ($\Delta L = 0$ at $[C] = C_c$, the "critical" concentration). Typical plots are shown in Figure 8. The C_c is affected by the values of several rate constants, of which k(−T/TT) is most important. Note also in Figures 7 and 8 that the characteristic properties of the transitions

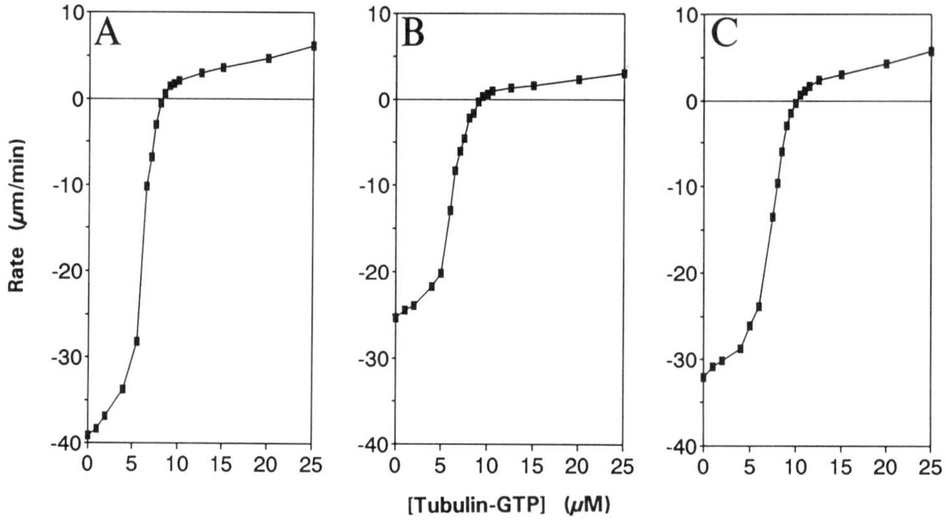

Fig. 8. *Net growth rate as a function of [T–GTP] (J(c) plot) computed for the β-out ends of three microtubule lattices: 13 protofilament A-type* (**A**), *13 protofilament B-type* (**B**), *and 14 protofilament A-type* (**C**).

between growth and shortening in L(t), and the form of the J(c) plot, are not confined to the 13 protofilament A-lattice. Both A- and B-type microtubule lattices, irrespective of protofilament number, can have the same behavior but with significantly different growth and shortening rates and state lifetimes.

Examination of the L(t) plots generated for different values of [T–GTP] shows that the dynamic transitions are found to occur in a range of concentrations spanning the C_c. Assigning a minimum excursion distance of 0.2 μm (corresponding to some 300 tubulin molecules), the L(t) plot is analyzed to obtain the average lifetime of the periods of either growth or shortening. It is found that the lifetime of the growing phase, T_G, increases nonlinearly above C_c as a high power of the T–GTP concentration; likewise, the lifetime of the shortening phase, T_S, increases nonlinearly below the C_c [Bayley et al., 1990].

The results thus show that a simple stochastic model does indeed produce G→S and S→G transitions and predicts a nonlinear dependence on [T–GTP] for the transition lifetimes. Examination in detail of the progress of the simulation (e.g., for the 13 protofilament (A-lattice) shows, as expected, that each kinetic step is independent, i.e., stochastic, but that a number of patterns can be discerned. The growing state, G, is typically associated with an end of 13 T–GTP molecules, (T = 13, D = O). However, a mixed end (T = 12, D = 1) will usually revert to (T = 13, D = O). Likewise, the shortening state S has an end that is typically (T = 0, D = 13); the configuration (T = 1, D = 12) usually reverts to (T = O, D = 13). This means that, at steady state, the numerous intermediate states are present only to a small extent (see Fig. 9). They behave as true intermediates in being liable to pass in either direction at a given time. Thus while the system behaves predominantly in a two-state, all or nothing fashion on the macroscopic scale, the possibility exists for periods of small excursions (i.e., a few additions and subtractions) below the observable limit. Such behavior can, in principle, give the effect of an apparent short "pause" in the L(t) function. More important, growth and shortening rates in the region of the C_c are less than their optimal values owing to the presence of microtubule end configurations other than exclusively T–GTP or T–GDP [Bayley et al., 1990].

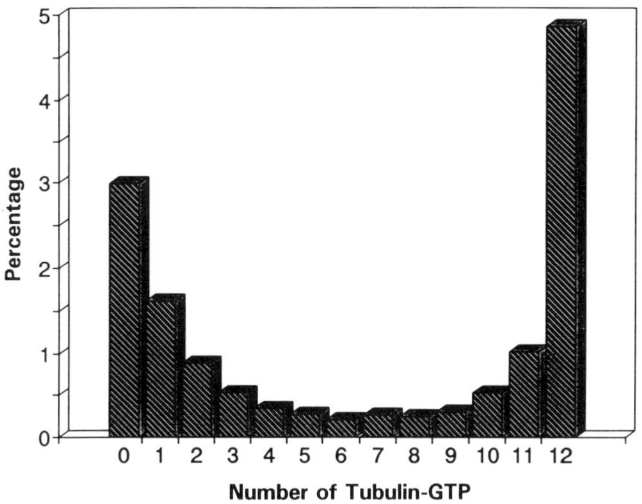

Fig. 9. *Computed microtubule end configurations during dynamic instability. The percentage of time a microtubule spends with different numbers of T-GTP in the terminal layer compared with 86% for the 13 T-GTP state (not plotted). Simulated for 13 protofilament A-type lattice, β-out end, at the calculated C_c (8.25 μM).*

3. Results: Transition behavior. The essentially two-state behavior also implies that the transitions between states have a strongly nonlinear, or cooperative, quality. This results from the feature, on this model, that the affinity for adding T–GTP is greater for sites with a greater proportion of T–GTP. This helps to promote cooperative addition reactions of T–GTP to the S state to favor the S→G transition. Likewise, progressive exposure of T–GDP in terminal sites favors the cooperative loss of T–GTP from the lattice to promote the G→S transition. It is predicted in the lateral cap model that transition behavior must be cooperative, and the dependence of the affinity of different sites for T–GTP on the nucleotide composition of the binding site provides a possible structural basis of this prediction. Nonlinearity is essential for the mechanism of microtubule oscillations (see Section III.B.2.).

Figure 10 shows the available experimental data and the calculations of the model for the dependence of transition frequencies on [T–GTP]. Note the logarithmic scale for the y-axis. Both experimental data and the calculated behavior show a strong dependence of the transition rates on T–GTP concentration. Although there are some absolute differences, it is clear that the lateral cap model is indeed capable of reproducing the general form of most of the experimental data.

In addition to the relative affinities mentioned above, there is one further important factor that appears to be essential to the process of microtubule dynamic instability. This is the structural or geometric property of the microtubule lattice. If the tubulin molecule, like tobacco mosaic virus protein, assembled in a one-start helix, then for a lattice comprising n_p protofilaments there would be one unique "lateral" binding site and $n_p - 1$ identical "longitudinal or axial" binding sites. Loss of the n_p terminal molecules (in a G→S transition) or regain of n_p terminal molecules (in an S→G transition) would be extremely cooperative, and the range for significant dynamic activity (spontaneous G→S and S→G transitions) would be extremely small. With the idealized microtubule lattices (or with mixed lattices), the cooperative unit is clearly less than n_p because the tubulin molecule symmetry, and hence the lattice symmetry, dictates (for the 13 protofilament

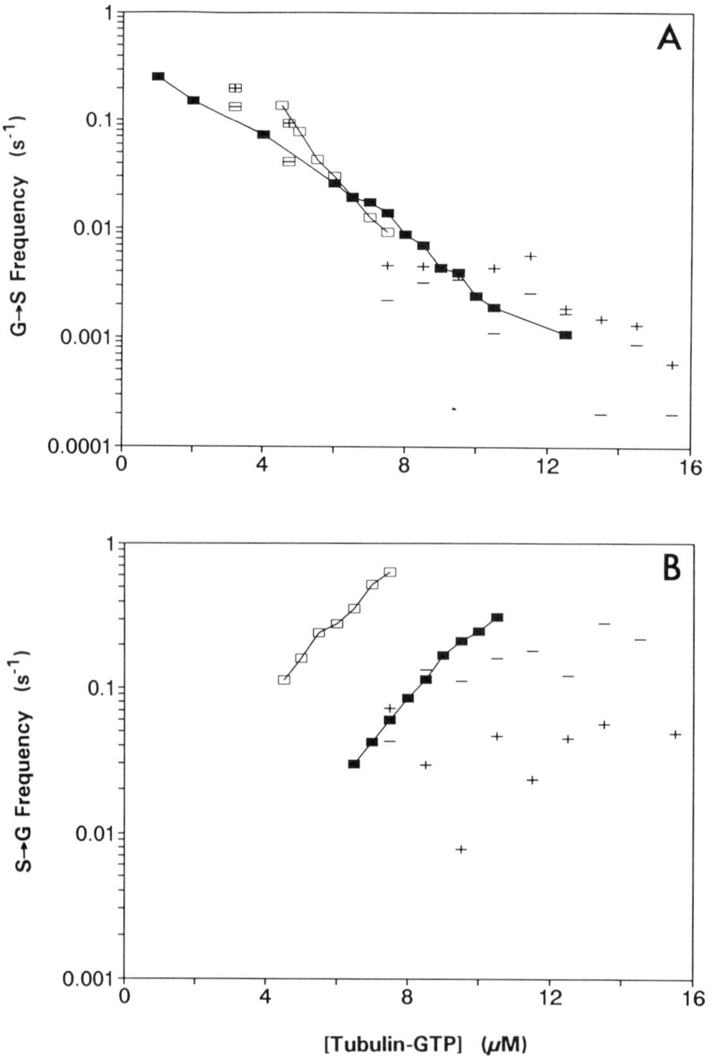

Fig. 10. *Frequencies of G→S* (**A**) *and S→G* (**B**) *transitions as a function of [T–GTP]. Comparison of computed values for the β-out (■) and α-out (□) ends of the 13 protofilament A-lattice with experimental values taken from Walker et al. [1988] (+, plus end; −, minus end) and from Walker et al. [1991] (⊞, plus end; ⊟, minus end).*

A-lattice) a five-start (dimer) lattice. Inspection shows that the end of such a microtubule is composed potentially of a series of five-start "runs" of terminal tubulin molecules of average length approximately four. Hence the model predicts an intermediate degree of cooperativity that allows dynamic instability to occur in a significant concentration range around the C_c.

The results presented in Figures 7–10 show calculations for the β-out end. The treatment of the α-out end has been described for the 13 protofilament A-lattice [Martin et al., 1991]. Dynamic instability, with characteristic differences, is predicted to occur at the opposite ends of the microtubule.

4. Other properties: The effect of T–GDP on dynamics. In general, a departure from the basic helically symmetrical lattice of the 13 protofilament A-type structure requires a modification in the formulation to incorporate the new types of subunit–subunit contacts. This has been achieved, in a general treatment, by considering affinities as deriving from the sums of explicitly defined subunit interaction energies [Bayley and Martin, 1992a]. A similar approach was introduced for the actin assembly system by Erickson [1989]. The assignment of free energies is based on the ranges of affinities deduced for the treatment as described above (section IV.C.1.). This enables "seams" of α–α plus β–β interactions to be treated as well as the full B-type lattice, in which these interactions predominate. The results, as illustrated in Figures 7 and 8, show that the characteristic properties of microtubule dynamic instability are not confined to a single lattice type, such as the helically symmetrical 13 protofilament A-type lattice. The striking general finding is that relatively small intersubunit effects can have surprisingly large cumulative effects on lattice properties.

These energy values also allow treatment of intrinsic lattice stability. For example, using the terms derived for the description of microtubule dynamic instability, the rate constant for dissociation of T–GDP from a buried position in the microtubule wall is calculated to be on the order of $1.5 \times 10^{-8} s^{-1}$, in good agreement with a similar order of magnitude value deduced by Dye et al. [1992] from observations of tethered microtubules. By contrast, the propagation of a "defect" appears to be orders of magnitude more favorable.

Other developments include the definition of self-consistent parameters for the binding of T–GDP to typical sites at the microtubule end. This allows treatment of the complicated effects of introducing into the system different relative amounts of T–GTP and T–GDP. The general result is that the treatment predicts regions of behavior where T–GDP could suppress microtubule dynamics, but without necessarily causing extensive microtubule disassembly (see next section). It thus appears possible that microtubule dynamic behavior could be controlled, in part, by the metabolic state of a cell, which would be sensitive to this balance of GTP–GDP.

5. Substoichiometric effects of antimitotic drugs. Margolis and Wilson [1977] described the "poisoning" of microtubule growth by concentrations of the antimitotic drug colchicine, substoichiometric to that of tubulin. They deduced that the effect operated through the microtubule ends. (The action of antimitotic drugs on the tubulin system is reviewed by Correia [1991].)

Because of the formulation of the lateral cap model, it is relatively simple to adapt it to accommodate the inclusion of an additional species of tubulin–nucleotide–drug complex (or tubulin-X). We have examined the consequences for the calculated dynamic properties of microtubules of including such a species. The minimum assumption is that it should bind to the lattice with (site-dependent) affinities identical to those of T–GTP but that the inhibitory effect is exerted by reducing the affinity for addition of further T–GTP molecules specifically in the adjacent sites.

A pictorial representation of this mechanism is shown in Figure 11. The presence of tubulin-X inhibits addition in either 1) the 5-start direction, 2) the $\bar{8}$-start direction, or c) in both 5-start and $\bar{8}$-start directions together. The structural effect is similar in all three cases: The microtubule adopts an end configuration of one dissociation site (such as protofilament 1, 13, or 13, respectively, in Figure 11a–c) and one or more inhibited addition sites (protofilaments 6, 5, or 4+6, respectively). The microtubule remains effectively inactive while the binding is inhibited, with activity restricted to limited shortening and recovery occurring in the region of the dissociation site. This is a natural consequence of the fact that the microtubule end is a closed system whose extension (microtubule growth) can be arrested by the presence of as little as one inhibitory molecule

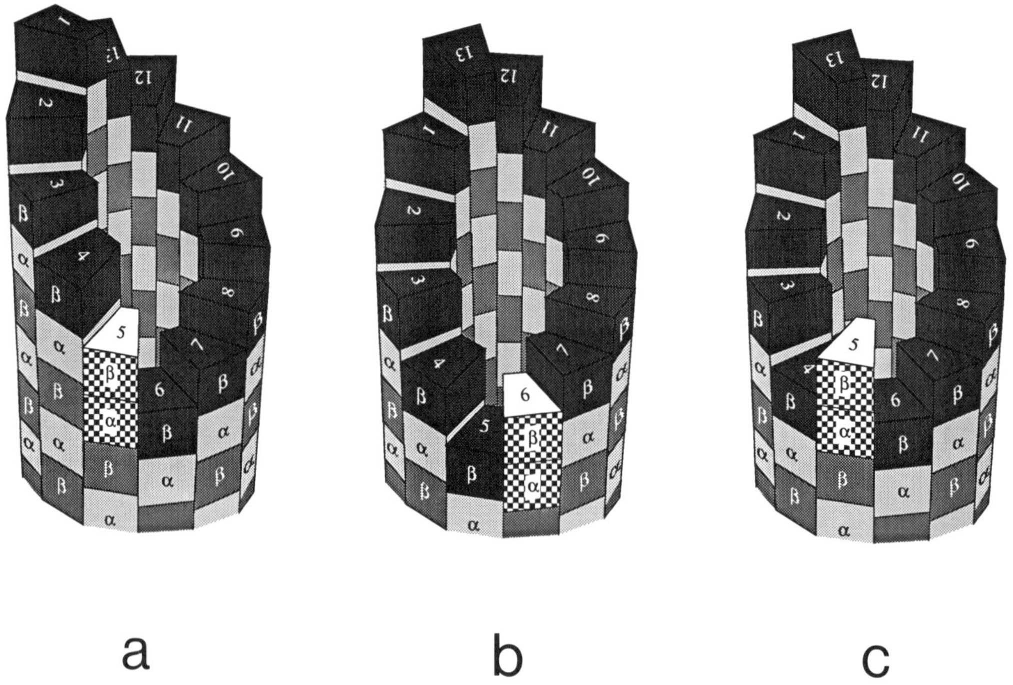

Fig. 11. *Possible interactions of substoichiometric concentrations of a drug molecule with dynamic microtubules. Black = β-tubulin–GTP; medium grey = β-tubulin–GDP; light grey = α-tubulin; checkered = tubulin–drug complex. The tubulin–drug complex is visualized as binding to the lattice, but inhibiting additional growth in* **a:** *the right-handed 5-start direction;* **b:** *the left-handed 8-start direction; or* **c:** *both directions. The effect is suppression of dynamic length excursions; the microtubule adopts the configuration with a single dissociation site (protofilament 13) and a single inhibited binding site (protofilaments 6 or 5 for a or b, respectively). In c, protofilaments 13 and 5 are dissociation sites, and protofilaments 4 and 6 are inhibited binding sites.*

per microtubule end. Further net growth is suspended until *either* an addition reaction occurs, with low probability, at an inhibited site (creating a new and normal addition site and including the species tubulin-X in the lattice), *or* until sufficient dissociation occurs to allow release of tubulin-X and restore the possibilities of normal dynamic growth.

The consequences for dynamics depend on the magnitude of the inhibitory effect, as well as the ratio of [tubulin-X] to [T–GTP]. Figure 12 shows a simulated L(t) plot for a dynamic microtubule exposed to different amounts of the species tubulin-X, assuming only a 10-fold decrease in affinity in the adjacent sites in the 5- and $\bar{8}$-start directions. Even with a low ratio of tubulin-X to T–GTP, the treatment predicts that there will be a strong suppression of the amplitude of dynamic length excursions. Under these conditions, there is little effect on the steady-state concentration. This result is therefore interesting both from the point of view of the substoichiometric action of antimitotic drugs and for the more general mechanism that is implied for the control of microtubule dynamic instability.

For the drugs, a mechanism such as this would suggest that the primary action at extremely low concentrations is to control microtubule dynamics. Experimentally, podophyllotoxin was shown preferentially to reduce the dynamic incorporation of tubulin into microtubules [Schilstra et al., 1989]. Jordan and Wilson [1990] showed the effects of low vinblastine

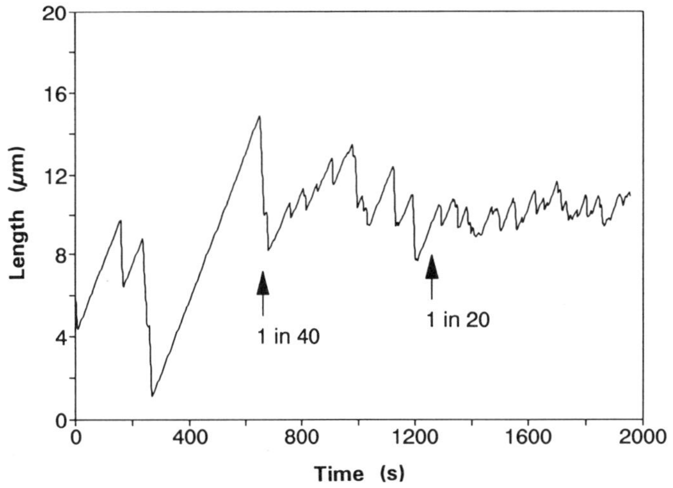

Fig. 12. *Microtubule length as a function of time. The computed effect of added tubulin–GTP–drug. The simulation was performed for the β-out end of the 13 protofilament A-lattice, and tubulin–GTP–drug was added at the level indicated and at times marked by the arrows. The* total *tubulin concentration used in the simulation was adjusted to the new C_c after each addition of tubulin–GTP–drug. The C_c values are 8.25, 9.2, and 10.1 μM for zero, 1 in 40, and 1 in 20 tubulin–GTP–drug, respectively. Addition of tubulin–GTP–drug to the microtubule end inhibits addition of further tubulin molecules in* both *the 5- and $\tilde{8}$-start directions (see Fig. 11).*

concentration on tubulin exchange into microtubules, and the direct inhibition of dynamic length excursions has been reported for nocodazole [Jordan et al., 1991] and colchicine or vinblastine [Toso et al., 1991]. Colchicine (as the tubulin–colchicine complex) was found by Hays and Salmon [1986] to stabilize isolated mitotic spindles, again consistent with suppression of the end-dependent dynamic processes of microtubules. Also, Skoufias and Wilson [1992] recently estimated the average level of incorporation of colchicine into microtubules under conditions of substoichiometric poisoning and found it to be on the order of 13 to 26 molecules per microtubule. This would be consistent with a mechanism favoring dissociation of the tubulin-X complex to enable further growth to occur.

It may be mentioned here that the formulation for the effects of T–GDP follows rather closely that used for the tubulin–drug complexes. If T–GDP can add directly to the lattice, its effect can be treated in terms of the parameters already set out in the simple formulation of the problem, again as a reduction in affinity in the adjacent site. In this case, the cap will have received an additional destabilizing effect by the presence of the T–GDP, which will tend to favor the G→S transition. Thus T–GDP will favor the shortening state and increase the C_c. This effect is capable of changing dynamic properties quite markedly, particularly if it is assumed that the affinity for tubulin–GDP at a predominantly T–GTP site is similar to that for T–GTP at the same site. T–GDP thus becomes a good competitor with T–GTP, with the added consequence that, once added at the end, it favors the G→S transition. The fact that it is often difficult to exclude the presence or accumulation of GDP during experiments of tubulin assembly means that GDP could be an unwelcome source of experimental variability in dynamics experiments. But, more importantly, GDP does appear capable of acting as a regulator of dynamic activity.

One consequence of the simple formulation of Figure 11 is that the process that causes reduction of T–GTP affinity at one end of the microtubule (e.g., β-out as illustrated) would imply reduced affinity of the tubulin-X species at the opposite end, i.e., the action of the tubulin–drug complex would be end dependent. On the whole, past experimental evidence does not show strong end dependence for the substoichiometric action of drugs like colchicine on pure tubulin microtubules, though tubulin–colchicine (at higher concentrations) was shown to cap minus end microtubules preferentially in the presence of MAPs [Bergen and Borisy, 1986]. The direct observational methods now available should permit a closer examination of the polarity of action of tubulin–drug complexes under controlled conditions.

In terms of more general mechanisms of the control of microtubule length changes, these considerations suggest that a specific protein molecule, with high affinity for the microtubule end, could act similarly. Such a species might well act, unidirectionally, as a nucleator, as does gelsolin in the case of F-actin polymerization. Most significantly, it could readily exert control of microtubule growth in an end-dependent manner by interfering with the subsequent addition of T–GTP molecules, even though it might only be present at a concentration substoichiometric relative to that of tubulin.

Two recent reports on the dynamic properties of microtubules grown from centrosomes in the presence of cytoplasmic extracts identify an increased G→S transition frequency as characteristic of the effect of mitosis-specific phosphorylation of proteins interacting with microtubules [Belmont et al., 1990], and a cyclin B-dependent kinase has been implicated [Verde et al., 1992]. It is interesting that a similar effect could be produced *in vitro* by an increase in the effective C_c, for example, by GDP. *In vivo,* the level of GDP might itself be influenced indirectly by the activity of a protein kinase or, more directly, by the effect of the nucleoside diphosphate kinase that would normally be required in its conversion back to GTP. This mechanism, operating on the effective C_c, appears potentially more sensitive than one that uses the concentration of free tubulin dimer as a controlling factor.

V. CRITIQUE OF THE *STATUS QUO*

The phenomenon of microtubule dynamic instability is now well established and marks one of the highlights of microtubule activity of the past decade. It appears to be a property unique to microtubules. This uniqueness apparently derives not only from the spontaneous GTP hydrolysis associated with the assembly process but also from the geometrical or spatial properties of the microtubule lattice(s). These in turn reflect symmetry properties of the tubulin αβ-heterodimer. The current uncertainty about structure will, it is hoped, be resolved in due course. Whether this resolution will favor a single type of microtubule structure or, more intriguingly the general occurrence of mixed A-type and B-type lattices [McEwen and Edelstein, 1977] remains to be seen. The simulations reported here suggest that the dynamic properties do not depend on a single specific lattice. Once the possibility of multiple symmetry breaking "seams" is established, a large and diverse family of dynamic lattice structures could be possible.

The experimental observation of microtubule dynamics is now relatively easy, but the quantitation of the parameters that define the behavior is intrinsically difficult. This is due to the limited range of T–GTP concentration over which the dynamic transitions are observable. A small change in conditions can cause a shift in C_c that alters dynamics considerably. Above the C_c, G→S transitions are less frequent, and their stochastic nature requires observations over periods of hours to obtain lifetime data suitable for averaging. Similarly below the C_c, even long microtubules (e.g., 100 μm), shortening typically at 20 μm per minute, have relatively low probability of experiencing a transition into the growing state. Even the observation of steadily growing or shortening microtubules is not without problems: The constraints of optical resolution still impose a lower limit on the magnitude of a discontinuity that can be effectively registered. It is possible that many minor transitions go undetected.

In spite of these limitations, sufficient is known to justify attempts to seek a quantitative representation of the phenomenon of microtubule dynamic instability, using modeling based on idealized structures, with a limited set of assumptions about mechanisms and the assignment of reasonable kinetic parameters. The results given here do indicate that a model based on a single terminal layer of T–GTP as the stabilizing factor between microtubule assembly and disassembly is able to reproduce a substantial amount of the currently available experimental evidence.

Recent reviews have also addressed the ability of the concept of a single terminal layer of T–GTP to explain kinetic and biochemical evidence [Erickson and O'Brien, 1992; Caplow, 1992]. There appears to be a general consensus that any species located at the microtubule ends to which the stabilizing effect of a cap may be attributed is either at, or less than, the limits set by present analytical and kinetic methods. These limits appear to be approximately in the same range as the protofilament number, i.e., 12 to 17, implying at most a single layer region or structure. The lateral cap model represents the simplest view, namely, that, in the absence of convincing evidence for a specific intermediate species, the logical proposition is to see whether T–GTP (which is the initial adduct) can indeed provide the explanation. With the principle of tightly coupled hydrolysis (which is less an assumption than a defined kinetic limit), the mechanism of dynamic transitions follows directly from the initial postulates regarding relative affinities. In terms of quantitation, any idealized model is likely to diverge somewhat from experimental results. However, these divergences appear to be due not only to the necessary simplifications of the process of modeling, but may also owe something to the experimental difficulties alluded to above.

The review by Erickson and O'Brien [1992] expresses concern that the concentration dependence of transition frequencies predicted by the lateral cap model is significantly greater than was shown in the apparently linear dependence of such frequencies on T–GTP concentration in the original work of Walker et al. [1988]. A similar point has recently been made by Caplow [1992]. Both comments appear to ignore the requirement for a nonlinear mechanism in microtubule oscillation. Laying aside the difficulties of obtaining sufficient observations over a range of tubulin concentrations, and specific questions of data analysis, as discussed above, the more recent work of Walker et al. [1991] has extended the concentration range of the data and demonstrated the nonlinear dependence on [T–GTP] of the mean rate of the transition from G to S state. The comparison between experiment and calculation is given in Figure 10. Within the limitations of experiment, analysis, and simulation, the agreement is surprisingly good, probably better than could reasonably be expected.

As regards the biochemical implications of the lateral cap model, Caplow [1992] makes a further point. This concerns the utlization of GTP by a seeded microtubule assembly system and is based on a novel interpretation of the terminal layer concept. He suggests that GTP hydrolysis should increase as the square of [T–GTP], since it is necessary to lay down one layer of T–GTP before addition to the second layer causes GTP hydrolysis to occur. This view appears hard to justify: It is dependent on the properties attributed to the terminal layer of the seeds and would evidently cease to operate once further T–GTP molecules were added beyond the depth of two. This question appears capable of experimental resolution but requires measurements of GTP usage by dynamic microtubules at very low tubulin concentrations. Such systems are, in principle, available [Caplow and Shanks, 1990; Symmons et al., 1991], but decisive data appears still to be lacking at this stage.

The recent observations of the variable nature of microtubule growth and shortening reported by Gildersleeve et al. [1992] has opened up a new area of enquiry. These results appear experimentally well founded and pose a new level of problem for the quantitative description of microtubule properties. The modeling described here necessarily will produce a steady growth rate (for a given lattice formulation), within the limitation previously described [Bayley et al., 1990] that microscopic variations in rate will be expected, particularly close to the C_c, due to the occurrence of "imperfect" or "mixed" microtubule ends containing both T–GTP and T–GDP. This effect does not appear sufficient to explain the different, approximately linear portions of growth (or shortening) seen in the results of Gildersleeve et al. [1992]. A more likely explanation could be that, when observed at sufficiently high spatial and temporal resolution, microtubule growth is indeed less regular than conceived in idealized models. The possibility must exist for defects in lattice structure, which could involve changes in A/B-lattice type, with inclusion of more AB or BB "seams," as well as changes in protofilament number, as reported by Chrétien et al., [1992]. The simulations suggest that all of these factors could introduce changes in growth and shortening rates. These might persist for considerable periods, i.e., as long as the new configuration survived. They would not need to be permanently sited dislocations in view of the possibility that defects could migrate axially, acting themselves as extra addition or dissociating sites. One further idea that deserves consideration as a source of variation in observable properties is that microtubule elongation may involve a (possibly small) number of "abortive" additions, for example, tubulin molecules added wrong-way round (i.e., antiparallel to the normal axial direction). Formation of tubulin sheets appears to involve antiparallel protofilaments. Multiple structures, hooks, junctions, and so forth, are well known under assembly-forcing conditions. Only a small number of such reactions might suffice to affect dynamic observations quite dramatically. Their frequency might depend on specific solution conditions, as well as on small chemical or genetic differences in tubulin preparations.

These appear to be matters requiring predominantly experimental clarification. The theoretical model exists to be improved as data become available. At the present stage, it appears reasonable to conclude that the working hypothesis elaborated above in section IV.B. remains valid. The lateral cap model deriving from it provides a useful, if simplified, means of interpreting the phenomenon of microtubule dynamic instability as an intrinsic property of microtubules, irrespective of exact lattice type and operating at both ends of the polar structure.

Undoubtedly, in the integrated biological environment, as experienced by microtubules in the cytoplasm, there will be additional factors potentially influencing their behavior. Nonetheless, it appears that the remarkable phenomenon of microtubule dynamic instability as demonstrated *in vitro* derives essentially from physical properties attributable ultimately to the lattice interactions of the tubulin dimer and the assembly-dependent hydrolysis of GTP. The intrinsic metastability of the growing microtubule lattice, resulting from these properties, appears well-suited to the biological requirement to provide a cytoskeletal structure sufficiently stable to contribute to supporting and directing cellular processes but also able to respond rapidly with changes in its spatial organization in the course of different types of cellular function.

ACKNOWLEDGMENTS

The authors thank Martyn Symmons for advice on the use of EGS seeds; Maria Schilstra for her contributions to the work described from this laboratory; several colleagues for useful comments on the manuscript; and Joe Brock for Figures 6 and 11. The work was supported in part by the E.C. Science Programme (twinning grants 852/00255 and 902/00203).

VI. REFERENCES

Allen C, Borisy GG (1974): Structural polarity and directional growth of microtubules of *Chlamydomonas* flagella. J Mol Biol 90:381–402.

Amos LA (1979): Structure of microtubules. In Roberts K, Hyams JS (eds): Microtubules. London: Academic Press, pp 1–64.

Bayley PM (1990): What makes microtubules dynamic? J Cell Sci 95:329–334.

Bayley PM, Martin SR (1992a): Modelling microtubule dynamics for different lattice types. Biophys J 61:a27.

Bayley PM, Martin SR (1992b): Microtubule dynamic instability: Basic mechanisms and numerical modelling by computer simulation. Comm Theor Biol 2: 1–24.

Bayley PM, Butler FMM, Clark DC, Manser EJ, Martin SR (1985): The assembly of microtubule protein *in vitro:* The kinetic role in microtubule elongation of oligomeric fragments containing microtubule-associated proteins. Biochem J 227:439–455.

Bayley PM, Schilstra MJ, Martin SR (1989): A lateral cap model of microtubule dynamic instability. FEBS Lett 259:181–184.

Bayley PM, Schilstra MJ, Martin SR (1990): Microtubule dynamic instability: Numerical simulation of microtubule transition properties using a lateral cap model. J Cell Sci 95:33–48.

Bayley PM, Martin SR, Sharma KK (1991): Dynamic microtubules: Experimental observation and computer simulation of polar microtubule behaviour with lateral cap model mechanisms. Am Inst Physics Conf Proc 226:187–199.

Belmont LD, Hyman AA, Sawin KE, Mitchison TJ (1990): Real-time visualization of cell cycle dependent changes in microtubule dynamics in cytoplasmic extracts. Cell 62:579–589.

Bergen LG, Borisy GG (1986): Tubulin-colchicine complex (TC) inhibits microtubule depolymerisation by a capping reaction exerted preferentially at the minus end. J Cell Biochem 30:11–18.

Caplow M (1992): Microtubule dynamics. Curr Opin Cell Biol 4:58–65.

Caplow M, Shanks J (1990): Mechanisms of the microtubule GTPase reaction. J Biol Chem 265:8935–8941.

Caplow M, Ruhlen R, Shanks J, Walker RA, Salmon ED (1989): Stabilisation of microtubules by tubulin–GDP–Pi subunits. Biochemistry 28:8136–8141.

Carlier M-F (1989): Role of nucleotide hydrolysis in the dynamics of actin filaments and microtubules. Int Rev Cytol 115:139–170.

Carlier M-F, Pantaloni D (1981): Kinetic analysis of guanosine-5′-triphosphate hydrolysis associated with tubulin polymerisation. Biochemistry 20:1918–1924.

Carlier M-F, Hill TL, Chen Y-D (1984): Interference of GTP hydrolysis in the mechanism of microtubule assembly: An experimental study. Proc Natl Acad Sci USA 81:771–775.

Carlier M-F, Didry D, Simon C, Pantaloni D (1989): Mechanism of GTP hydrolysis in tubulin polymerisation: Characterisation of the kinetic intermediate microtubule-GDP-Pi using phosphate analogues. Biochemistry 28:1783–1791.

Chen Y-D, Hill TL (1985): Monte Carlo study of the GTP cap in a five-start helix model of a microtubule. Proc Natl Acad Sci USA 82:1131–1135.

Chen Y-D, Hill TL (1987) Theoretical studies on oscillations in microtubule polymerisation. Proc Natl Acad Sci USA 84:8419–8423.

Chrétien D, Metoz F, Verde F, Karsenti E, Wade RH (1992): Lattice defects in microtubules: Protofilament numbers vary within individual microtubules. J Cell Biol 117:1031–1040.

Correia JJ (1991): Effects of anti-mitotic agents on tubulin-nucleotide interactions. Pharmacol Ther 52:127–141.

Correia JJ, Williams RC (1983): Mechanisms of assembly and disassembly of microtubules. Annu Rev Biophys Bioeng 12:211–235.

Dustin P (1984): Microtubules, 2nd ed. Berlin: Springer-Verlag.

Dye RB, Flicker PF, Lien DY, Williams RC (1992): End-stabilised microtubules observed in vitro: Stability, subunit interchange, and breakage. Cell Motil Cytoskel 21:171–186.

Engelborghs Y (1990): Dynamic aspects of microtubule assembly. In Avila J (ed): Microtubule proteins. Boca Raton, FL: CRC Press, pp 1–35.

Erickson HP (1989): Co-operativity in protein-protein association. The structure and stability of the actin filament. J Mol Biol 206:465–474.

Erickson HP, O'Brien ET (1992): Microtubule dynamic instability and GTP hydrolysis. Annu Rev Biophys Biomol Struct 21:145–166.

Evans T, Mitchison TM, Kirschner MW (1985): Influence of the centrosome on the structure of nucleated microtubules. J Cell Biol 100:1185–1191.

Farrell KW, Himes RH, Jordan MA, Wilson L (1983): On the non-linear relationship between the initial rates of dilution-induced microtubule disassembly and the initial free subunit concentration. J Biol Chem 258: 14148–14156.

Farrell KW, Jordan MA, Miller HP, Wilson L (1987): Phase dynamics at microtubule ends: The coexistence of microtubule length changes and treadmilling. J Cell Biol 104:1035–1046.

Fishback JL, Yarbrough LR (1984): Interaction of 6-mercapto-GTP with bovine brain tubulin. Equilibrium aspects. J Biol Chem 259:1968–1973.

Gal V, Martin SR, Bayley PM (1988): Fast disassembly of microtubules induced by Mg^{2+} or Ca^{2+}. Biochem Biophys Res Commun 155:1464–1470.

Gelfand V, Bershadsky AD (1991): Microtubule dynamics: Mechanism, regulation and function. Annu Rev Cell Biol 7:93–116.

Gildersleeve RF, Cross AR, Cullen RE, Fagen AP, Williams RC (1992): Microtubules grow and shorten at intrinsically variable rates. J Biol Chem 267:7995–8006.

Hays TS, Salmon ED (1986): The stabilization of microtubules in isolated spindles by tubulin-colchicine complex. Cell Motil Cytoskel 6:282–290.

Hill TL (1987): Linear Aggregation Theory in Biology. New York: Springer-Verlag.

Hill TL, Carlier M-F (1983): Steady-state theory of the interference of GTP hydrolysis in the mechanism of microtubule assembly. Proc Natl Acad Sci USA 80:7234–7238.

Hill TL, Chen YD (1984) Phase changes at the end of a microtubule with a GTP cap. Proc Natl Acad Sci USA 81:5772–5776.

Hill TL, Kirschner MW (1982): Bioenergetics and kinetics of microtubule and actin filament assembly-disassembly. Int Rev Cytol 78:1–125.

Horio T, Hotani H (1986): Visualisation of the dynamic instability of individual microtubules by dark field microscopy. Nature 321:605–607.

Jacobs M (1979): Tubulin and nucleotides. In Roberts K, Hyams JS (eds): Microtubules. London: Academic Press, pp 255–277.

Jordan MA, Wilson L (1990): Kinetic analysis of tubulin exchange at microtubule ends at low vinblastine concentrations. Biochemistry 29:2730–2739.

Jordan MA, Toso RJ, Thrower D, Miller HP, Wilson L (1991): Inhibition of microtubule dynamics *in vitro* and *in vivo* by nocodazole. J Cell Biol 115:36a.

Karecla P, Hirst E, Bayley P (1989): Polymorphism of tubulin assembly *in vitro*. J Cell Sci:94:479–488.

Karr TL, Podrasky AE, Purich DL (1979) Participation of guanine nucleotides in nucleation and elongation steps of microtubule assembly. Proc Natl Acad Sci USA 76:5475–5479.

Karr TL, Kristofferson D, Purich DL (1980): Mechanism of microtubule depolymerization. Correlation of rapid induced disassembly experiments with a kinetic model for endwise depolymerisation. J Biol Chem 255: 8560–8566.

Kirschner MW (1988): Biological implications of microtubule dynamics. Harvey Lect 83:1–20.

Koshland DE, Mitchison TJ, Kirschner MW (1988): Polewards chromosome movement driven by microtubule depolymerization *in vitro*. Nature 331:499–504.

Luduena RF (1979): Biochemistry of tubulin. In Roberts K, Hyams JS (eds): Microtubules. London: Academic Press, pp 65–116.

Mandelkow E-M, Mandelkow E (1992): Microtubule oscillations. Cell Motil Cytoskel 22:235–244.

Mandelkow E-M, Schultheiss R, Rapp R, Muller M, Mandelkow E (1986): On the surface lattice of microtubules: Helix starts, protofilament number, seam, and handedness. J Cell Biol 102:1067–1073.

Margolis RL, Wilson L (1977): Addition of colchicine-tubulin complex to microtubule ends: The mechanisms of substoichiometric colchicine poisoning. Proc Natl Acad Sci USA 74:3466–3470.

Margolis RL, Wilson L (1978): Opposite end assembly and disassembly of microtubules at steady state *in vitro*. Cell 13:1–8.

Martin SR, Schilstra MJ, Bayley PM (1991): Opposite-end behaviour of dynamic microtubules. Biochim Biophys Acta 1073:555–561.

Martin SR, Schilstra MJ, Bayley PM (1993): Dynamic instability of microtubules: Monte Carlo simulation and application to different types of microtubule lattice. Biophys J (in press).

McEwen B, Edelstein SJ (1977): Evidence for a mixed lattice in microtubules reassembled in vitro. J Mol Biol 139:123–145.

Mejillano MR, Himes RH (1991): Binding of guanine nucleotides and Mg^{2+} to tubulin with a nucleotide-depleted exchangeable site. Arch Biochem Biophys 291:356–362.

Melki R, Carlier M-F, Pantaloni D (1990): Direct evidence for GTP and GDP-Pi intermediates in microtubule assembly. Biochemistry 29:8921–8932.

Mitchison T, Kirschner MW (1984a): Microtubule assembly nucleated by isolated centrosomes. Nature 312: 232–237.

Mitchison T, Kirschner MW (1984b): Dynamic instability of microtubule growth. Nature 312:237–242.

Mitchison T, Kirschner MW (1988): Cytoskeletal dynamics and nerve growth. Neuron 1:761–772.

O'Brien ET, Voter WA, Erickson HP (1987): GTP hydrolysis during microtubule assembly. Biochemistry 26:4148–4156.

O'Brien ET, Salmon ED, Walker RA, Erickson HP (1990) Effects of magnesium on the dynamic instability of individual microtubules. Biochemistry 29:6648–6656.

Oosawa F, Asakura S (1975): Thermodynamics of the Polymerisation of Protein. London: Academic Press.

Osei AA, Everett GW, Himes RH (1988): GTP hydrolysis by tubulin occurs with water oxygen incorporation into inorganic phosphate. Biochim Biophys Acta 955:269–271.

Salmon ED, Walker RA, Pryer NK (1989): Video-enhanced differential interference contrast light microscopy. Biotechniques 7:624–633.

Scheele RB, Borisy GG (1979): Structure of microtubules. In Roberts K, Hyams JS (eds): Microtubules. London: Academic Press, pp 175–254.

Scheele RB, Bergen LG, Borisy GG (1982): Control of the structural fidelity of microtubules by initiation sites. J Mol Biol 154:485–500.

Schilstra MJ, Martin SR, Bayley PM (1987): On the relationship between nucleotide hydrolysis and microtubule assembly: Studies with a GTP regenerating system. Biochem Biophys Res Commun 147:588–595.

Schilstra MJ, Martin SR, Bayley PM (1988) Dynamic processes in steady state microtubules studied by tubulin exchange kinetics. In Rousset BAF (ed): Structure and Functions of the Cytoskeleton. Colloque Inserm 171. London, Paris: John Libbey, pp 43–49.

Schilstra MJ, Martin SR, Bayley PM (1989): The effect of podophyllotoxin on microtubule dynamics. J Biol Chem 264:8827–8834.

Schilstra MJ, Bayley PM, Martin SR (1991): The effect of solution composition on microtubule dynamic instability. Biochem J 277:839–847.

Skoufias DA, Wilson L (1992): Mechanism of inhibition of microtubule polymerisation by colchicine: Inhibitory potencies of unliganded colchicine and tubulin-colchicine complexes. Biochemistry 31:738–746.

Stewart RJ, Farrell KW, Wilson L (1990): Role of GTP hydrolysis in microtubule polymerization: Evidence for a coupled hydrolysis mechanism. Biochemistry 29:6489–6498.

Symmons MF, Martin SR, Bayley PM (1991): Biochemical studies of microtubule dynamic instability. J Cell Biol 115:37a.

Timasheff SN, Grisham LM (1980): *In vitro* assembly of cytoplasmic microtubules. Annu Rev Biochem 49: 565–591.

Toso RJ, Jordan MA, Wilson L (1991): Colchicine and vinblastine suppress dynamic instability of individual microtubules. J Cell Biol 115:36a.

Verde F, Dogterom M, Stelzer E, Karsenti E, Leibler S (1992): Control of microtubule dynamics and length by cyclin A- and cyclin B-dependent kinases in *Xenopus* egg extracts. J Cell Biol 118:1097–1108.

Voter WA, O'Brien ET, Erickson HP (1991): Dilution-induced disassembly of microtubules: Relation to dynamic instability and the GTP cap. Cell Motil Cytoskel 18:55–62.

Wade RH, Chrétien D, Job D (1990): Characterisation of microtubule protofilament numbers. How does the surface lattice accommodate? J Mol Biol 212:775–786.

Walker RA, O'Brien ET, Pryer NK, Soboeiro MF, Voter WA, Erickson HP, Salmon ED (1988): Dynamic instability of individual MAP-free microtubules analysed by video light microscopy: Rate constants and transition frequencies. J Cell Biol 107:1437–1448.

Walker RA, Pryer NK, Salmon ED (1991): Dilution of individual microtubules observed in real time *in vitro:* Evidence that the cap size is small and independent of elongation rate. J Cell Biol 114:73–81.

Wegner A (1976): Head-to-tail polymerization of actin. J Mol Biol 108:139–150.

Wilson L, Miller HP, Jordan MA, Farrell KW (1985): Dynamics of tubulin addition and loss at opposite ends of bovine brain microtubules in vitro: Analysis using double-labelled microtubules. In De Brabander M, De Mey J (eds): Microtubules and Microtubule Inhibitors. Amsterdam: Elsevier, pp 61–69.

Wilson L, Miller HP, Farrell KW (1990): Measurements of ^{32}P at microtubule ends: Evidence for an extremely small GTP or P_i cap. J Cell Biol 111:28a.

Zeeberg M, Caplow M (1979): Determination of free and bound microtubular protein and guanine nucleotide under equilibrium conditions. Biochemistry 18: 3880–3886.

ABOUT THE AUTHORS

PETER M. BAYLEY is a member of the MRC scientific staff at the National Institute for Medical Research, London. He received his doctorate in protein chemistry from Oxford University in 1965, and did postdoctoral work with John Schellman in Eugene, Oregon, and with Lubert Stryer at Stanford University. His present work focuses on optical spectroscopy of protein conformation, calcium–calmodulin interactions, the kinetics of conformational change, and cytoskeletal self-assembly systems. He has been closely involved with the British Biophysical Society, and is an officer of the European Biophysical Societies' Association. Since 1983 he has been the managing editor of the *European Biophysics Journal*.

KAM K. SHARMA is a member of the staff at the National Institute for Medical Research. He received his degree from the University of Liverpool in 1979. His present work focuses on dark-field video microscopy applied to studies of microtubule dynamics and the quantitation of dynamic properties *in vitro*.

STEPHEN R. MARTIN is a member of the staff at the National Institute for Medical Research. He received his doctorate in kinetics from the University of Kent at Canterbury in 1973, and did postdoctoral work with Hartmut Diebler at the Max Planck Institute for Biophysical Chemistry, Göttingen, Germany. His present work focuses on optical spectroscopy of protein and nucleic acid conformation, calcium–calmodulin interactions, and computer simulation of biochemical processes.

PROTEINS ASSOCIATED WITH MICROTUBULES

Microtubules: 141–154

The MAP1 Family

Reinhold Müller, Stefan Kindler, and Craig C. Garner

I. INTRODUCTION

Recent progress in understanding the functional relationship between different microtubule-associated proteins (MAPs) in both dividing cells and differentiating neurons has been made through the molecular characterization of these proteins. We focus on two members of the MAP1 family, MAP1A and MAP1B, whose structural and functional relationships have long been debated. The characterization of both proteins at the molecular level has revealed that they are structurally related MAPs performing complementary functions during neuronal maturation. We describe biochemical, ultrastructural, and functional studies and discuss the primary structure of these two proteins.

Historically, MAP1 has been defined as a high molecular weight heat-labile brain protein of about 350 kD that remains bound to microtubules through multiple purification cycles of polymerization and depolymerization [Sloboda et al., 1975]. MAP1 is distinct from a second high molecular weight MAP, called MAP2, that is heat stable [Herzog and Weber, 1978] (see Oakley, section II, this volume). MAP1 has been found in several laboratories to be resolvable into two to four bands, depending on the source of material used for preparation and on electrophoretic conditions [Wiche et al., 1991]. The apparent variation in resolution resulted in uncertainties concerning the relationship between various MAP1 species as is evident from the diversity of names used: MAP1A, -1B, and -1C [Bloom et al., 1984a; Wiche et al., 1986], MAP1.1 and -1.2 [Greene et al., 1983], MAP1 and MAP5 [Riederer et al., 1986], MAP1x [Calvert and Anderton, 1985], and MAP1 and MAPX [Binder et al., 1984]. Monoclonal antibodies against individual MAP1 subcomponents and cDNA analysis have helped to resolve the relationship between some of these proteins. It would appear that MAP1A, MAP1.1, and MAP1 [Bloom et al., 1984a; Greene et al., 1983; Riederer et al., 1986; Binder et al., 1984] refer to one protein, while MAP1B, MAP1.2, MAP5, MAP1x and MAPX [Bloom et al., 1985; Greene et al., 1983; Riederer et al., 1986; Calvert and Anderton, 1985; Binder et al., 1984] refer to a second. A third MAP1 protein with distinct biochemical properties, called MAP1C, is a cytoplasmic dynein [Paschal and Vallee, 1987] and will not be discussed further (see Holtzbauer Cral and Collins, this volume). In this discussion we will adopt the nomenclature that named the MAP1 subcomponents with the letters A, B, and C.

II. SPATIAL AND TEMPORAL PATTERNS OF EXPRESSION

MAP1A, the largest MAP1 species, has a molecular mass of 350 kD as measured by SDS-PAGE [Riederer and Matus, 1985], while MAP1B is slightly smaller, about 320 kD [Riederer et al., 1986]. These two proteins exhibit a complementary pattern of expression during rat brain development (Fig. 1) [Matus, 1988]. The level of MAP1B is highest during the first few postnatal days of rat brain devel-

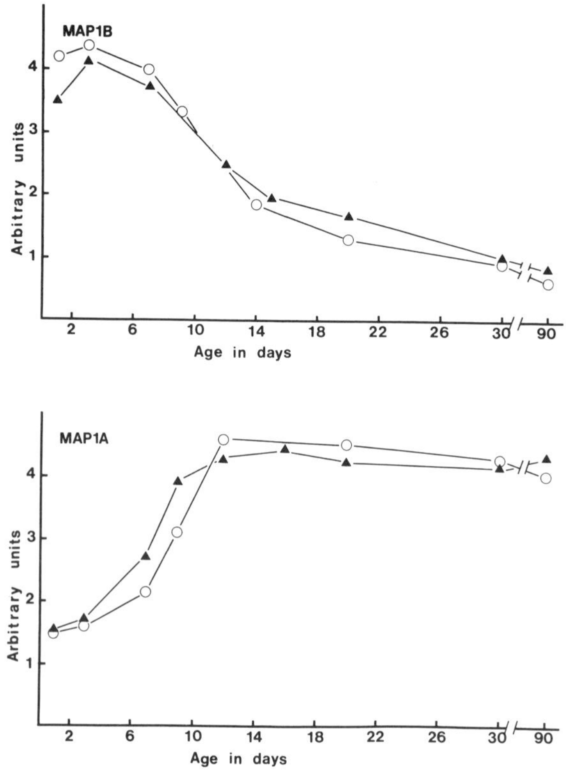

Fig. 1. *Expression of MAP1A* **(bottom)** *and MAP1B* **(top)** *during rat cerebral cortical development. The relative amounts of mRNA and protein encoding MAP1A and MAP1B expressed during postnatal days 1, 3, 7, 9, 12, 15, 20, 30, and 90 were determined by RNA slot blot analysis and by an ELISA. (○), protein levels; (▲), mRNA levels. [Reproduced from Garner et al., 1990, with permission from the publisher.]*

opment when most neurons are differentiating and sending out processes. At around postnatal day 10 (P10), the amount of MAP1B begins to decline, reaching the low basal level seen in the mature brain by P30 (Fig. 1) [Riederer et al., 1986; Garner et al., 1990]. MAP1A, in contrast, is virtually absent during early postnatal development; then, just after P10, as neurons begin to obtain their final form, the MAP1A level increases, reaching a maximum in the mature brain (Fig. 1) [Riederer and Matus 1985; Garner et al., 1990].

MAP1B is predominantly axonal, though it is also found in dendrites in both juvenile and adult neurons [Bloom et al., 1985; Calvert and Anderton, 1985; Riederer et al., 1986; Schoenfeld et al., 1989]. While isoforms of MAP2, tau, and MAP3 are also expressed during early brain development [Matus, 1988], only MAP1B is consistently found in extending neurites [Tucker et al., 1988]. MAP1B is the first of the MAPs to appear in cell bodies and neurites of avian motor neurons [Tucker et al., 1988], the retinal ganglion cells [Tucker and Matus, 1987], as well as rat cerebellar parallel fibers [Calvert et al., 1987], suggesting a role in neurite outgrowth [Tucker, 1990]. This assumption is supported by the persistent expression of MAP1B in several adult brain areas in which neurons continue to proliferate and to extend neurites. Examples include the photosensitive cells of the mammalian retina [Tucker and Matus, 1988], the olfactory epithelium and the nerve layer in the olfactory bulb

[Viereck et al., 1989], and the mossy fibers in the hippocampus [Schoenfeld et al., 1989]. In cultured cells that have been induced to extend neurites by nerve growth factor (NGF), for example, PC12 cells, MAP1B is the most prominent MAP [Greene et al., 1983; Brugg and Matus, 1988]. MAP1B's dramatic temporal changes [Drubin et al., 1985; Aletta et al., 1988; Brugga nd Matus, 1988] and its association with microtubules in neurites and growth cones [Brugg and Matus, 1988] has led to the proposal that MAP1B plays a vital role in regulating the dynamic assembly of microtubules during neurite outgrowth. This central role of MAP1B as a neurite-promoting factor is supported by molecular studies using antisense oligonucleotides that block not only the MAP1B expression but also neurite outgrowth in cultured PC12 cells [Matus, personal communication).

MAP1A, in contrast, is only weakly expressed in developing axonal fibers, moderately in mature axons, but intensely in developing and mature dendrites [Bloom et al., 1984a; Huber and Matus, 1984; Riederer and Matus, 1985; Schoenfeld et al., 1989]. MAP1A's appearance seems always correlated with the stabilization of the mature neuronal form. For example, it is strongly and specifically expressed in mature retinal ganglion cells and their axons connected to the optic tectum [McKerracher et al., 1989; Okabe et al., 1989]. In the cerebellum, MAP1A is initially found in the cell soma and proximal segments of extending Purkinje cell dendrites. As the cerebellum matures and Purkinje cells attain their final form, MAP1A is found throughout these dendrites [Schoenfeld et al., 1989]. Interestingly, MAP1A is conspicuously absent from the olfactory nerve layer throughout development [Schoenfeld et al., 1989], where MAP1B continues to be present even in the adult [Schoenfeld et al., 1989; Viereck et al., 1989]. Thus the late expression of MAP1A, its presence in mature axonal and dendritic processes, and its absence from immature axons suggest that this protein plays a morphological role in the stabilization of mature axons and dendrites.

III. BIOCHEMICAL PROPERTIES OF MAP1A

Adult rat MAP1A can be separated into two proteins by SDS-PAGE with 5 M urea [Wiche et al., 1991]. In early postnatal development the larger polypeptide is primarily present (Müller and Garner, unpublished observations). Although MAP1A can be phosphorylated by endogenous protein kinases [Herrmann et al., 1985], in particular by a casein kinase II–like activity [Diaz-Nido et al., 1990], there is no indication at present that one MAP1A species results from the phosphorylation of the other as has been shown for MAP1B (see below).

Early studies on purified MAP1 from bovine brain showed that it could stimulate the assembly of tubulin *in vitro* [Kuznetsov et al., 1981]. Immunopurified MAP1A from bovine and rat brain are stoichiometric complexes of a heavy chain (>300 kD) and three light chains: LC1 (34 kD), LC2 (30 kD), and LC3 (18 kD) [Vallee and Davis, 1983; Kuznetsov et al., 1986; Kuznetsov and Gelfand, 1987; Schoenfeld et al., 1989]. The molar ratios of the light chains to the heavy chain are about 1.4 for LC1, 1.9 for LC2, and 2.0 for LC3 [Kuznetsov and Gelfand, 1987; Schoenfeld et al., 1989]. The light chains were found to bind a 60 kD chymotryptic fragment that remains associated with the microtubule surface [Schoenfeld et al., 1989]. What role do the light chains play during the binding of MAP1A to microtubules? One set of experiments has addressed this question but did not distinguish between MAP1A and MAP1B. Vallee and Davis [1983] found that MAP1 complexes containing the light chains not only bind microtubules *in vitro* but also stimulate the assembly of microtubules. MAP1 stripped of light chains also binds taxol-stabilized microtubules but no longer stimulates the assembly of microtubules [Vallee and Davis, 1983]. It is, however, unclear whether the light chains are essential for stimulating microtubule assembly or if this activity of MAP1 was destroyed during the preparation of the heavy chain [Vallee and Davis, 1983].

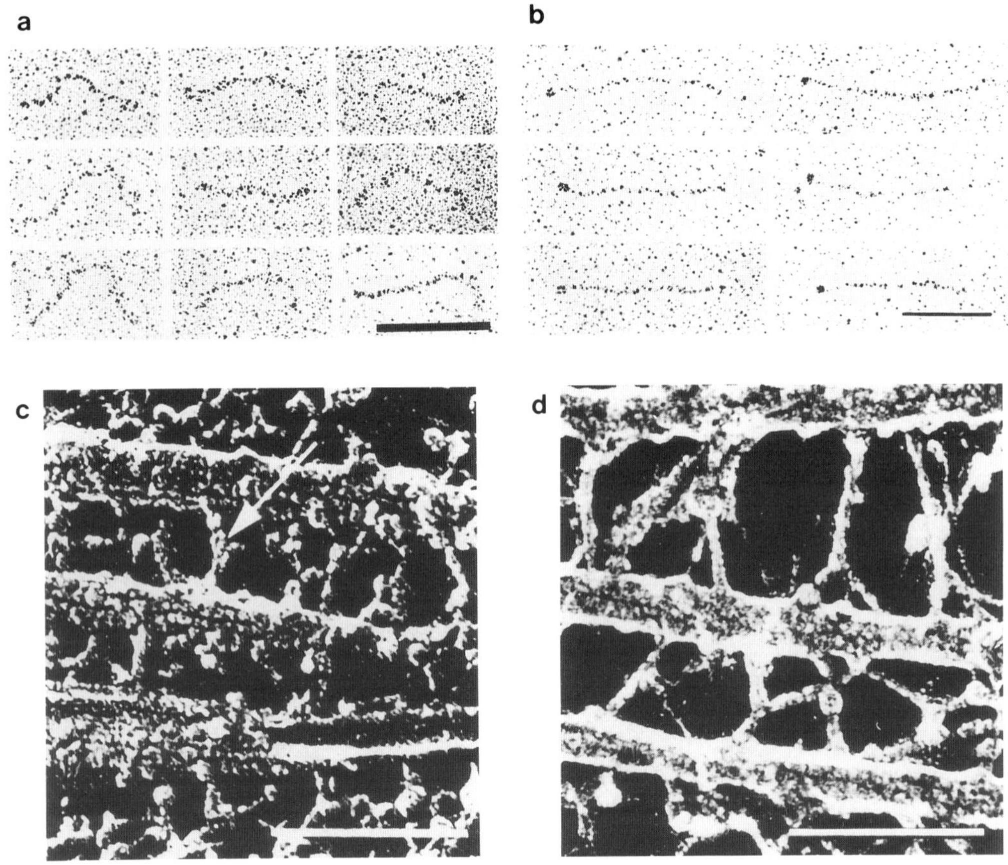

Fig. 2. *Affinity-purified MAP1A* (**a**) *and MAP1B* (**b**) *were absorbed onto mica and examined by the low-angle rotary shadowing technique. Molecules appear as extended, long, thin, filamentous structures. MAP1B also has a small spherical portion at one end. When pellets of microtubules polymerized with MAP1A* (**c**) *or MAP1B* (**d**) *are examined by the quick-freeze, deep-etch technique, MAP1A and MAP1B can be seen to form thin long filamentous projections that appear to cross-bridge adjacent microtubules. Bar = 100 nm. (Courtesy of N. Hirokawa.)*

IV. ULTRASTRUCTURE OF MAP1A

Rotary shadowing of affinity-purified MAP1A indicates that it is a highly elongated, flexible molecule (100–150 nm in length) [Shiomura and Hirokawa, 1987a] (Fig. 2a). When purified MAP1A bound to *in vitro* polymerized microtubules was examined at the ultrastructural level, long filamentous material was seen cross-linking microtubules (Fig. 2c) [Shiomura and Hirokawa, 1987a]. Such cross-links suggested that MAP1A attaches to the microtubule surface via at least one domain and that the remaining part of the molecule is involved in forming bridges between microtubules. Hirokawa and his colleagues have used gold-labeled monoclonal antibodies against MAP1A to visualize *in situ* filamentous projections on the surface of microtubules in Purkinje cell dendrites using immunoelectron microscopy [Hirokawa et al., 1985; Shiomura and Hirokawa, 1987a,b]. These labeled filamentous projections were primarily observed between neighboring microtubules, occasionally between microtubules and membrane organelles, and seldom between microtubules and neurofilaments [Hirokawa et al., 1985; Shiomura and Hirokawa,

1987a]. Could these filamentous projections perform a spacing function to regulate the distance between microtubules in mature axons and dendrites? Or is there a cross-linking function via specific binding partners to the surrounding cytoskeleton? To date, no specific partners have been identified, though some studies suggest that MAP1A might bind MAP2, neurofilaments, and perhaps plectin (Furtner and Wiche, 1987; Herrmann and Wiche, 1987]. Clearly, these are some of the more important questions to be addressed in future experiments.

V. BIOCHEMICAL PROPERTIES OF MAP1B

MAP1B can also be separated into two proteins by SDS-PAGE. The higher molecular mass species of 320 kD is a highly phosphorylated form of the 300 kD protein [Viereck et al., 1989; Fischer and Romano-Clarke, 1990; Riederer et al., 1990]. MAP1B, like MAP1A, is phosphorylated by endogenous protein kinases [Herrmann et al., 1985; Diaz-Nido et al., 1990]. One potential enzyme is a casein kinase II–like protein [Diaz-Nido et al., 1990], and a second appears to be an NGF-activated protein kinase [Tsao et al., 1990]. During brain development the amount of phosphorylated MAP1B (MAP1B-P) decreases much faster than the "unphosphorylated" form (MAP1B-U) [Riederer et al., 1990; Viereck et al., 1989]. Also MAP1B-P is apparently the major form in axons [Riederer et al., 1990; Sato-Yoshitake et al., 1989], while MAP1B-U is concentrated in dendrites [Riederer et al., 1990; Schoenfeld et al., 1989; Sato-Yoshitake et al., 1989]. The phosphorylated form appears to have biochemical properties different from the "unphosphorylated" form, since cat brain MAP1B-P partitioned to the pellet during high speed centrifugation while MAP1B-U remained soluble. This result implies that MAP1B-P is associated with cold-stable microtubules or with the membrane cytoskeleton, while MAP1B-U is not [Riederer et al., 1990].

Studies on cultured PC12 cells are also providing clues to the role of phosphorylation in the function of MAP1B. Uninduced PC12 cells contain basal levels of MAP1B-U that are not associated with cytoplasmic microtubules [Brugg and Matus, 1988]. When PC12 cells are stimulated by NGF to differentiate and to extend neurites, not only do the levels of MAP1B increase [Greene et al., 1983; Drubin et al., 1985, 1988; Aletta et al., 1988; Brugg and Matus, 1988], but there is also a concomitant increase in MAP1B-P [Greene et al., 1983; Aletta et al., 1988]. Interestingly, MAP1B in induced PC12 cells becomes increasingly associated with the microtubules in the outgrowing neurite and in the growth cones [Brugg and Matus, 1988]. These data suggest that the phosphorylated form of MAP1B is the species that not only binds microtubules but also perhaps stimulates their assembly or stabilizes assembled filaments. These suggestions are supported by *in vitro* studies in which casein kinase II–phosphorylated MAP1B binds microtubules more efficiently than the unphosphorylated form [Diaz-Nido et al., 1988]. Thirty phosphorylation sites are predicted for casein kinase II, based on the primary sequence of MAP1B, including potential sites situated around the microtubule binding and the projection domains [Noble et al., 1989]. Since MAP1B has been shown to be phosphorylated at multiple sites [Herrmann et al., 1985; Diaz-Nido et al., 1990], the role of phosphorylation in the function of the protein has become a challenging problem that may be easier to address now that cDNA clones for MAP1B are available.

Like MAP1A, MAP1B, in its immunopurified form, is also an oligomeric protein composed of a heavy chain (300 kD) and two light chains, LC1 and LC3 [Vallee and Davis, 1983; Kuznetsov et al., 1986; Kuznetsov and Gelfand, 1987]. The stoichiometry of heavy to light chain is 2.6 for LC1 and 1.1 for LC3 [Schoenfeld et al., 1989]. LC1 and LC3 were found to bind a 120 kD chymotryptic fragment of MAP1B containing the microtubule-binding domain [Kuznetsov et al., 1986; Hammarback et al., 1991]. Even before the molecular structure of MAP1A and MAP1B were known (see below), it was this common feature of associated light chains that suggested to Vallee and his collaborators that MAP1A and MAP1B must be structurally similar, in spite of the fact that other biochemical and molecular data (for example, affinity for microtubules, spatial distribution, and the lack of any cross-reacting monoclonal antibodies) indicated that the two proteins were unrelated.

VI. ULTRASTRUCTURE OF MAP1B

Affinity-purified MAP1B appears in rotary shadowing experiments as a thin and elongated molecule (186 ± 38 nm long) that in contrast to MAP1A possesses a spherical portion at one end (Fig. 2b) [Sato-Yoshitake et al., 1989]. As seen for MAP1A, MAP1B that is bound to *in vitro* taxol-assembled microtubules exhibits filamentous projections decorating the microtubule surface (Fig. 2d) [Sato-Yoshitake et al., 1989]. The spherical portion of MAP1B appears to bind to the microtubule surface, suggesting that it is, or lies very close to, the microtubule-binding domain of MAP1B. It is not clear if this spherical structure is an intrinsic feature of the heavy chain or the result of light chain–heavy chain interactions. *In situ,* MAP1B, like MAP1A, was found by immunoelectron microscopy to form long filamentous cross-bridges primarily between microtubules [Hirokawa et al., 1985; Sato-Yoshitake et al., 1989]. These data imply that the primary function of MAP1A and MAP1B in developing and mature axons and dendrites is to interact along and between microtubules and not with neurofilaments or the membrane cytoskeleton.

VII. MOLECULAR CHARACTERIZATION OF MAP1A and MAP1B

The first cDNA clones for MAP1 were isolated from mouse brain mRNA by Lewis et al. [1986] using expression cloning. The identity of the cloned MAP1 species was determined by generating a polyclonal antiserum against a fusion/protein encoded by one of the cDNA clones. This antiserum was found to cross-react with a protein of the same mobility as MAP1B from rat brain [Noble et al., 1989] and PC12 cells [Aletta et al., 1988]. Together with immunocytochemical studies demonstrating that this antiserum stained structures in the developing and mature cerebellum identical to MAP1B (Matus and Cowan unpublished data), these data permitted Cowan and his colleagues to conclude that the MAP1 cDNA cloned was for MAP1B [Noble et al., 1989]. Partial cDNA clones for rat MAP1B have also been isolated in several laboratories [Garner et al., 1988; Rienitz et al., 1989; Safaei and Fischer, 1989; Hammarback et al., 1991; Zauner et al., 1992]. Both rat and mouse MAP1B are encoded by an 11 kb mRNA [Lewis et al., 1986; Garner et al., 1988; Safaei and Fischer, 1989] and are transcribed from a single gene [Lewis et al., 1986; Garner et al., 1990]. The MAP1B gene is located on mouse chromosome 13 [Garner et al., 1990]. The 11 kb mRNA was found to be more prevalent in developing than in adult brain (Fig. 1) [Lewis et al., 1986; Safaei and Fischer, 1989; Garner et al., 1990] in agreement with the pattern seen for the protein [Riederer et al., 1986].

The isolation of cDNA clones encoding rat brain MAP1A was first reported by Matus and colleagues in 1988. The mRNA encoding MAP1A is slightly smaller than that for MAP1B (10 kb) [Garner et al., 1988] and is transcribed from a single gene located on mouse chromosome 2 [Garner et al., 1990]. This mRNA, in contrast to that for MAP1B, is more abundant in adult than in developing brain (Fig. 1). The near overlap in the relative levels of MAP1A and MAP1B with their mRNAs in developing rat cerebral cortex (Fig. 1) suggests that one mechanism controlling the expression of MAP1A and MAP1B during development may be at the transcriptional level [Garner et al., 1990]. Other mechanisms including protein/mRNA turnover rate or translational control have yet to be reported.

The mRNAs encoding MAP1A and MAP1B have both been detected in tissues other than brain, albeit mostly at low levels [Lewis et al., 1986; Langkopf et al., 1992], except for heart and muscle, where MAP1A mRNA levels are higher than in other nonneuronal tissues [Langkopf et al., 1992]. Whether these elevations are caused by contaminating peripheral neurons was not examined. The presence of MAP1A and MAP1B mRNAs in nonneuronal cells is consistent with earlier studies describing a broader distribution of these proteins outside the brain [Bloom et al., 1984b; Wiche et al., 1984]. In cultured cells, MAP1A-like proteins have been identified associated with spindle fibers, microtubule-organizing centers, stress fibers, and nuclear flecks [Bloom et al., 1984b; Wiche et al., 1984; De Mey et al., 1987; Asai et al., 1985; Sato et al., 1985]. Though mRNAs of the same size as brain MAP1A mRNA have been found in these tissues and nonneuronal cell lines [Lewis et al., 1986; Langkopf et al., 1992], little is known about the structure or function of these nonneuronal MAP1 species.

The full length nucleotide sequences for MAP1B and MAP1A cDNAs were first reported by Noble et al. [1989] and Langkopf et al. [1992], respectively. The MAP1B cDNA contains an open reading frame encoding 2,464 amino acid residues, while the MAP1A cDNA contains 2,774. The calculated molecular weights for MAP1B (M_r 255,534) and MAP1A (M_r 299,221) are clearly smaller than those observed by SDS-PAGE (320 and 350 kD, respectively). However, the smaller values are consistent with the anomalous mobilities in SDS-PAGE of filamentous proteins [Goedert et al., 1991]. A comparison of the deduced amino acid sequences of MAP1A and MAP1B via dot matrix analysis revealed that these two proteins contain distinct regions of similarity flanked by apparently unrelated regions (Fig. 3A) [Langkopf et al., 1992]. Eight segments of similarity could be identified, six in the amino-terminal portion and two at the C terminus. In addition to sequence similarities, the two proteins also exhibit a similar charge distribution (Fig. 3B). As observed for other MAPs [Goedert et al., 1991], MAP1A and MAP1B have positively charged domains flanked by negatively charged regions. The positively charged domains in MAP2, MAP4, and tau have been shown to delineate their microtubule-binding domains, while the negatively charged regions span their projection domains [Goedert et al., 1991]. In MAP1A and MAP1B, one positively charged domain overlaps the first three segments of similarity, and another is situated in the C terminus, overlapping similarity segments 7 and 8 (Fig. 3B) [Langkopf et al., 1992]. One microtubule-binding domain in MAP1B has been localized to a small region in the N-terminal half of the molecule [Noble et al., 1989]. This region overlaps the N-terminal positively charged domain and contains the short motif (KKEX), where X is either an acidic or a hydrophobic residue, repeated 21 times. This region was shown to be essential for MAP1B binding to microtubules in transfected cells and has the ability to bind *in vitro* polymerized microtubules [Noble et al., 1989]. A similar region containing 11 repeats of this motif is present in MAP1A and is situated in the second segment of similarity between MAP1A and MAP1B (Fig. 3). By analogy to MAP1B, this region is likely to be at least part of the MAP1A microtubule-binding domain. Interestingly, protein sequences flanking the segment containing the KKEX repeats exhibit far more sequence similarity between MAP1A and MAP1B than the KKEX repeat containing segments themselves [Langkopf et al., 1992]. Whether these conserved segments are also involved in microtubule binding remains to be determined.

VIII. LIGHT CHAINS: MAP1A AND MAP1B ARE POLYPROTEINS

Vallee and his colleagues utilized a polyclonal antiserum generated against a mixture of bovine light chains to isolate a cDNA clone encoding LC1 from a rat brain cDNA expression library. Confirmation of the nucleotide sequence was established by sequencing several internal peptides of bovine LC1 [Hammarback et al., 1991]. Surprisingly, a comparison of the nucleotide sequence of the rat LC1 cDNA with that of mouse MAP1B cDNA revealed a 94.4% identity with the C-terminal end of the MAP1B cDNA nucleotide sequence [Hammarback et al., 1991]. The amino acid sequence of bovine LC1 is colinear with the single open read-

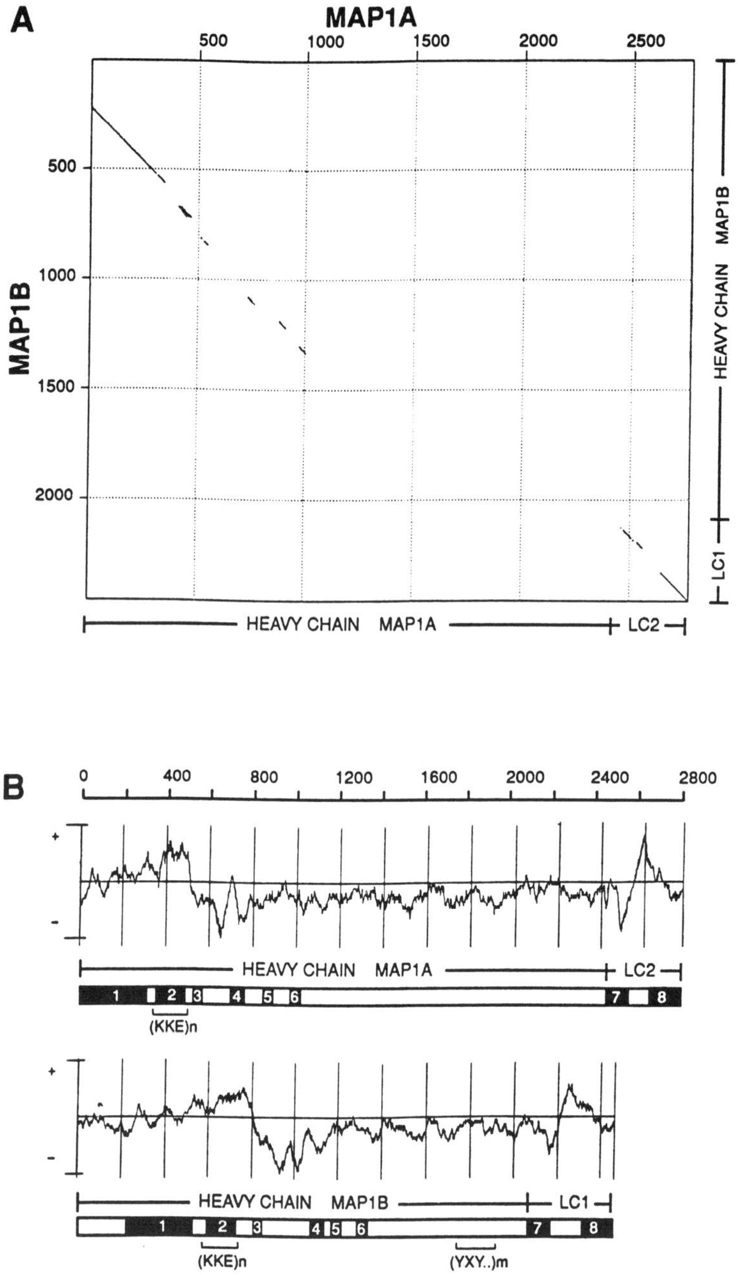

Fig. 3. A: *Diagonal dot matrix comparison between the deduced MAP1A/LC2 and MAP1B/LC1 polyprotein sequences. The comparison was performed using a DOTPLOT-DNASTAR (Madison, WI) program: Parameter limits were set at 50% homology with a window of 30 amino acids. Approximate borders between heavy and light chains are delineated to the left and across the bottom of the plot.* **B:** *Charge distribution and schematic comparison of primary structure similarities of MAP1A/LC2 and MAP1B/LC1. The charge distribution from the deduced amino acid sequences from MAP1A/LC2 and MAP1B/LC1 were determined with the PROTEIN-DNASTAR program, window 50. Shaded and numbered segments in the linear diagrams indicate the relative locations of similar sequences. The (K/R)(K/R)(D/E)(X = acidic or hydrophobic residue) domains (seg. 2) are 26% identical while other segments vary between 50% and 88% identity. KKEX represents this repeated motif, found in a portion of MAP1B to bind microtubules YXY, the 12 imperfect 15 amino acid repeats, and LC1 and LC2 are the sequences for the corresponding two light chain polypeptides. [Reproduced from Langkopf et al., 1992, with permission of the publisher.]*

ing frame present in the MAP1B and LC1 cDNA sequences, suggesting that LC1 is encoded in the last ~300 codons of the MAP1B cDNA. Further proof that these sequences encode LC1 was obtained by demonstrating that light chain antibodies reacted with a fusion protein spanning the tail region of MAP1B. LC1 antibodies were specific for bovine LC1 and did not cross-react with MAP1B heavy chain [Hammarback et al., 1991]. The N terminus of LC1 was predicted to lie between amino acid 2,201 and 2,216 of the MAP1B cDNA gene product [Hammarback et al., 1991].

Analogous experiments on LC2 demonstrated that this protein is situated at the C-terminal end of the MAP1A cDNA product (Fig. 3) [Langkopf et al., 1992]. A comparison of the sequences predicted to span LC1 and LC2 revealed that they are 55% identical, account for the seventh and eighth segment of similarity between the deduced MAP1A and MAP1B sequences, and possess the second positively charged domain (Fig. 3). In an *in vitro* binding assay tubulin was shown to bind directly protein sequences spanning LC1 [Rienitz et al., 1989; Zauner et al., 1992], perhaps because of LC1's positively charged region. Therefore, light chains have two presumptive binding sites, one for the heavy chain and one for tubulin. It is not yet known whether the light chains constitute part of the heavy chain microtubule binding domain and/or if they have a modulatory role regulating the binding of MAP1A and MAP1B to microtubules.

In rat and calf brain homogenates as well as in several cell lines, the light chains are found as separate proteins from the heavy chains [Vallee and Davis, 1983; Kuznetsov et al., 1986; Schoenfeld et al., 1989]. How are heavy and light chains generated? It would appear that they are not created by alternative splicing of a primary transcript, since both the LC1 and LC2 coding sequences are only found on mRNAs of the same size as the mRNAs for the corresponding heavy chains [Hammarback et al., 1991; Langkopf et al., 1992]. Reinitiation at a methionine near the predicted start of LC1 also seems unlikely [Hammarback et al., 1991]. The favored mechanism involves a specific proteolytic processing of a precursor protein. While a precursor for MAP1B/LC1 has not been identified [Hammarback et al., 1991], an affinity-purified antibody against LC2 did react with a protein of slower mobility on SDS-PAGE than either of the previously identified MAP1A species [Langkopf et al., 1992]. This finding would support a model (Fig. 4) in which both LC1 and LC2 are processed from their respective precursors by a specific not yet identified protease. Such a model would also allow the released light chains to associate freely with either heavy chain as exemplified by the association of both LC1 and LC2 with immunoprecipitated MAP1A [Schoenfeld et al., 1989].

The model does not account for the stoichiometry measured between the light chains and heavy chains [Schoenfeld et al., 1989]. For example, the translation of the MAP1B/LC1 mRNA followed by processing should lead to a 1:1 stoichiometry for LC1 to MAP1B. How-

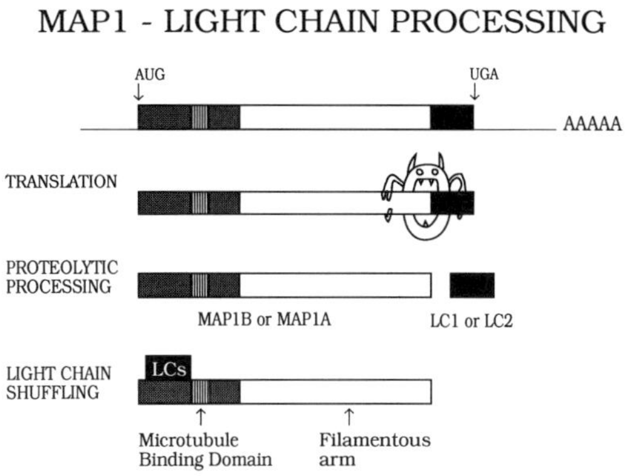

Fig. 4. *Proposed processing model of light chains from heavy chains. Spliced and processed transcripts encoding either MAP1A/LC2 or MAP1B/LC1 are translated to produce a precursor that is subsequently cleaved specifically by a putative protease. The released light chains (black boxes) are then free to bind to the N-terminal portion of the heavy chains (grey stippled boxes) near the microtubule-binding domain (KKEX repeats; striped box). The light chains could either partake in or regulate the binding of the heavy chains to the microtubules. The filamentous arms (white box) extend from the microtubule surface to interact with the surrounding milieu.*

ever, biochemical studies predict a ~3:1 ratio. In addition, it is not clear by what mechanism LC1 levels remain elevated in mature brain to become associated with MAP1A when the transcript levels for the MAP1B/LC1 encoding mRNA have already declined (Fig. 1). Perhaps MAP1B has a higher turnover rate than LC1.

With respect to the origin of LC3, it would appear not to be coencoded with a third MAP1-like molecule (J. Hammarback, personal communication).

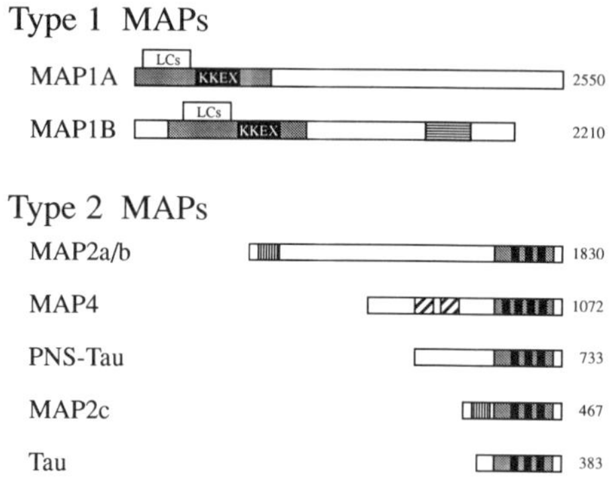

Fig. 5. *Schematic representation of the MAP1A and MAP1B family of microtubule-associated proteins (type 1 MAPs) to the MAP2, MAP4, and tau family (type 2 MAPs). Type 1 MAPs carry their microtubule-binding domain (KKEX) as well as light chain binding sites in the N-terminal region of the molecule (grey stippled box), whereas type 2 MAPs bind microtubules via their C-terminal portion (gray stippled box) that contain three or four imperfect 18 amino acid repeats (black boxes). The rest of the protein serves as a projection domain (white box) variable in length that extends from the microtubule surface and possesses different characteristics. Horizontal, diagonal, and vertical striped boxes indicate the localization of 12 imperfect 15 amino acid repeats in MAP1B, acidic domains in MAP4, and a binding region for the regulatory subunit of the cAMP-dependent protein kinase on MAP2b and -2c, respectively. The numbers refer to the amino acid residues in each protein. Values for MAP1A and MAP1B are approximate, because the exact location of the proteolytic polyprotein cleavage sites are unknown.*

IX. DOMAIN STRUCTURE OF MAP1A/LC2 AND MAP1B/LC1

Functional studies on recombinant MAP1A and MAP1B cDNA products have demonstrated that these two proteins have three principle domains. The C terminus of each precursor is a light chain that is rapidly cleaved by a presumptive protease. Once the light chain is removed, the heavy chain thus generated has two domains also found in other brain MAPs. At one end (the N terminus of MAP1A and MAP1B and the C terminus of MAP2, MAP4, and tau) (Fig. 5) is a positively charged microtubule-binding domain. The remaining portion of each molecule exhibits an overall negative charge (Fig. 3B) [Goedert et al., 1991; Langkopf et al., 1992] and is predicted to be the projection domain that extends from the microtubule surface and forms filamentous bridges with neighboring microtubules [Hirokawa et al., 1985; Shiomura and Hirokawa, 1987b; Goedert et al., 1991]. In the case of MAP1A and MAP1B, the light chains may associate with the heavy chain near the microtubule-binding domain [Kuznetsov et al., 1986; Schoenfeld et al., 1989].

Although all brain MAPs thus far described possess a positively charged domain that binds the negatively charged microtubule surface, MAPs can be divided into two families based on the structure of their microtubule-binding domains. MAP1A and MAP1B form one family, each one containing a multiply repeated KKEX motif. MAP2, MAP4, and tau form a second family in which each member possesses three to four 18-amino-acid-residue repeats in the microtubule-binding domain [Goedert et al., 1991]. Based on these characteristics, we propose that MAP1A and MAP1B be referred to as *type I MAPs* and the MAP2, MAP4, and tau family as *type 2 MAPs*. Based on *in vivo* and *in vitro* studies, one function for MAPs is to promote and stabilize the assembly of microtubules [Goedert et al., 1991]. Variations in the structure of the microtubule-binding domains could permit MAPs to bind either at separate sites along the microtubules and/or with different affinities. *In vitro* studies with MAP1 and MAP2 showed that both bind a common sequence within the variable C-terminal region of different β-tubulin isoforms [Cross et al., 1991], suggesting that members of both families bind the same region of the microtubule surface. Functional differences between the type I and type 2 MAPs do exist; for example, MAP2 and tau are better promoters of microtubule assembly *in vitro* than MAP1A or MAP1B [Olmsted, 1986]. In addition, cells transfected with MAP2 and tau tend to reorganize microtubules [Lewis et al., 1988; Kanai et al., 1989], whereas MAP1B only binds microtubules [Noble et al., 1989].

The second major functional difference between type I and type 2 MAPs is thought to lie in their structurally different projection domains. Based on the molecular weight of each protein and their length measured by rotary shadowing experiments, each should project arms of different lengths from the microtubule surface. These arms could act to cross-link microtubules to the surrounding cytoskeleton or provide spacing. Clear differences observed between the spacing of microtubules in axons versus dendrites are thought to be due to the MAP that is present [Matus, 1990]. Mature axons, for example, have closely spaced microtubules and possess predominantly the short-armed tau, while mature dendrites have more widely spaced microtubules and contain high levels of the long-armed MAP2a and -2b and MAP1A [Matus, 1990].

The primary structure of the proposed projection domains of MAP1A and MAP1B are unrelated. Therefore, one would expect to find here structural domains responsible for the functional differences between MAP1A and MAP1B. Specifically, there should be a domain in MAP1B involved in the plastic events associated with neurite outgrowth and in MAP1A a domain involved in the stabilization of neuronal form. In the projection domain of MAP1B is a region that contains 15 imperfect 12 amino acid repeats [Noble et al., 1989; Rienitz et al., 1989]. This domain also contains 9 of the 30 potential casein kinase II phosphorylation sites [Noble et al., 1989]. Although no repeats are found in MAP1A, the MAP1A projection domain does contain 20 of 29 putative casein kinase II phosphorylation sites, suggesting that the interaction of the MAP1A projection domain with its surrounding may be regulated via phosphorylation. Future experiments will clearly need to focus on the binding partners.

X. CONCLUSIONS

The structural, biochemical, and functional properties of two MAPs, MAP1A and MAP1B, have been described. These two proteins are translated from mRNAs of different sizes, expressed at different times during brain development, and transcribed from genes located on separate chromosomes. Though dissimilar along most of their length, MAP1A and MAP1B are structurally related proteins exhibiting several discrete segments of high similarity. They are similarly organized with potential microtubule-binding domains at their N termini and light chains encoded in the last 300 aa residues of each protein. The mRNAs may be translated into precursors that are apparently proteolytically cleaved into heavy and light chains. Heavy and light chains reassociate to form a multisubunit microtubule- binding domain. These common features suggest that MAP1A and MAP1B are derived from an ancestral gene that diverged and acquired specialized functions to provide the nervous system with MAPs that are either required for the dynamic outgrowth of neurites or for the stabilization of neuronal form. Future research should focus on defining the structural and functional domains in these proteins and on deciphering the role of the various phosphorylation sites.

XI. REFERENCES

Aletta JM, Lewis SA, Cowan NJ, Greene LA (1988): Nerve growth factor regulates both the phosphorylation and steady-state levels of microtubule-associated protein 1.2 (MAP1.2). J Cell Biol 106:1573–1581.

Asai DJ, Thompson WC, Wilson L, Dresden CF, Schulman H, Purich DL (1985): Microtubule-associated proteins (MAPs): A monoclonal antibody to MAP1 decorates microtubules in vitro but stains stress fibers and not microtubules in vivo. Proc Natl Acad Sci USA 82:1434–1438.

Binder LI, Frankfurter A, Kim H, Caceres A, Payne MR, Rebhun LI (1984): Heterogeneity of microtubule-associated protein 2 during rat brain development. Proc Natl Acad Sci USA 81:5613–5617.

Bloom GS, Schoenfeld TA, Vallee RB (1984a): Widespread distribution of the major polypeptide component of MAP1 (microtubule-associated protein 1) in the nervous system. J Cell Biol 98:320–330.

Bloom GS, Luca FC, Vallee RB (1984b): Widespread cellular distribution of MAP1A in the mitotic spindle and in interphase microtubules. J Cell Biol 98:331–340.

Bloom GS, Luca FC, Vallee RB (1985): Microtubule-associated protein 1B: Identification of a major component of the neuronal cytoskeleton. Proc Natl Acad Sci USA 82:5404–5408.

Brugg B, Matus A (1988): PC12 cells express juvenile microtubule-associated proteins during nerve growth factor-induced neurite outgrowth. J Cell Biol 107: 643–650.

Calvert R, Anderton BH (1985): Microtubule-associated protein MAP1 which is expressed at elevated levels during development of the rat cerebellum. EMBO J 4:1171–1176.

Calvert RA, Woodhams PL, Anderton BH (1987): Localization of an epitope of a microtubule-associated protein 1 x in outgrowing axons of the developing rat central nervous system. Neurosci 23:131–141.

Cross D, Dominnguez J, Maccioni RB, Avila J (1991): MAP1 and MAP2 binding sites at the C-terminus of β-tubulin. Studies with synthetic tubulin peptides. Biochemistry 30:4362–4366.

De Mey J, Aerts F, De Raeymaeker M, Daneels G, Moeremans M, De Wever B, Vandre DD, Vallee RB, Borisy RB, De Brabander M (1987): Microtubule-associated protein (MAP) 1A: A phosphoprotein associated with spindle fibers and microtubule organizing centers. Prog Zool 34:187–205.

Diaz-Nido J, Serrano L, Mendez E, Avila J (1988): A casein kinase II–related activity is involved in phosphorylation of microtubule-associated protein MAP-1B during neuroblastoma cell differentiation. J Cell Biol 106:2057–2065.

Diaz-Nido J, Serrano L, Hernandez MA, Avila J (1990): Phosphorylation of microtubule proteins in rat brain at different developmental stages: Comparison with that found in neuronal cultures. J Neurochem 54:211–222.

Drubin DG, Feinstein SC, Shooter EM, Kirschner MW (1985): Nerve growth factor-induced neurite outgrowth in PC12 cells involves the coordinate induction of microtubule assembly and assembly-promoting factors. J Cell Biol 101:1799–1807.

Drubin D, Kobayashi S, Kellogg D, Kirschner M (1988): Regulation of microtubule protein levels during cellular morphogenesis in nerve growth factor-induced PC12 cells. J Cell Biol 106:1583–1591.

Fischer I, Romano-Clarke G (1990): Changes in microtubule-associated protein MAP1B phosphorylation during rat brain development. J Neurochem 55:328–333.

Furtner R, Wiche G (1987): Binding specificities of purified porcine brain α- and β-tubulin subunits and of microtubule-associated proteins 1 and 2 examined by electron microscopy and solid-phase binding assays. Eur J Cell Biol 45:1–8.

Garner CC, Farmer L, Matus A (1988): Single-shot cloning of multiple cDNAs coding for a set of related microtubule-associated proteins. Gene 71:483–490.

Garner CC, Garner A, Huber G, Kozak C, Matus A (1990): Molecular cloning of microtubule-associated protein 1 (MAP1A) and microtubule-associated protein 5 (MAP1B): Identification of distinct genes and their differential expression in developing brain. J Neurochem 55:146–154.

Goedert M, Crowther RA, Garner CC (1991): Molecular characterization of microtubule-associated proteins tau and MAP2. TINS 14:193–199.

Greene LA, Liem RKH, Shelanski ML (1983): Regulation of a high molecular weight microtubule-associated protein in PC12 cells by nerve growth factor. J Cell Biol 96:76–83.

Hammarback JA, Obar RA, Hughes SM, Vallee RB (1991): MAP1B is encoded as a polyprotein that is processed to form a complex N-terminal microtubule-binding domain. Neuron 7:129–139.

Herrmann H, Dalton JM, Wiche G (1985): Microheterogeneity of microtubule-associated proteins. MAP-1 and MAP-2, and differential phosphorylation of individual subcomponents. J Biol Chem 260:5797–5803.

Herrmann H, Wiche G (1987): Plectin and IFAP-300 kDa are homologous proteins binding to microtubule-associated proteins 1 and 2 and to the 240 kDa subunit of spectrin. J Biol Chem 262:1320–1325.

Herzog W, Weber K (1978): Fractionation of brain microtubule-associated proteins. Isolation of two different proteins which stimulate tubulin polymerization in vitro. Eur J Biochem 92:1–8.

Hirokawa N, Bloom G, Vallee RB (1985): Cytoskeletal architecture and immunocytochemical localization of microtubule-associated proteins in regions of axons associated with rapid axonal transport. The β,β′-iminodipropionitril-intoxicated axon as a model. J Cell Biol 101:227–239.

Huber G, Matus A (1984): Immunocytochemical localization of microtubule-associated protein 1 in rat cerebellum using monoclonal antibodies. J Cell Biol 98:777–781.

Kanai Y, Takenura R, Oshima T, Mori H, Ihara Y, Yanagisawa M, Masaki T, Hirokawa N (1989): Expression of multiple tau isoforms and microtubule bundle formation in fibroblasts transfected with a single tau cDNA. J Cell Biol 109:1173–1184.

Kuznetsov SA, Rodionov VI, Gelfand VI, Rosenblat VA (1981): Microtubule-associated protein MAP-1 promotes microtubule assembly in vitro. FEBS Lett 135:241–244.

Kuznetsov SA, Rodionov VI, Nadezhdina ES, Murphy DB, Gelfand VI (1986): Identification of a 34-kD polypeptide as a light chain of microtubule-associated protein-1 (MAP-1) and its association with a MAP-1 peptide that binds to microtubules. J Cell Biol 102: 1060–1066.

Kuznetsov SA, Gelfand VI (1987): 18 kDa microtubule-associated protein: Identification as a new light chain (LC3) of microtubule-associated protein 1 (MAP-1). FEBS Lett 212:145–148.

Langkopf A, Hammarback JA, Müller R, Vallee RB, Garner CC (1992): Microtubule-associated proteins 1A and LC2: Two protein encoded in one messenger RNA. J Biol Chem 267:16561–16566.

Lewis SA, Sherline P, Cowan NJ (1986): A cloned cDNA encoding MAP1 detects a single copy gene in mouse and a brain-abundant mRNA whose level decreases during development. J Cell Biol 102:2106–2114.

Matus, A (1988): Microtubule-associated proteins: Their potential role in determining neuronal morphology. Annu Rev Neurosci 11:29–44.

Matus A (1990): Microtubule-associated proteins and the determination of neuronal form. J Physiol P 84:134–137.

McKerracher L, Vallee RB, Aguayo AJ (1989): Microtubule-associated protein 1A (MAP1A) is a ganglion cell marker in adult rat retina. Vis Neurosci 2:349–356.

Noble M, Lewis SA, Cowan NJ (1989): The microtubule binding domain of microtubule-associated protein MAP1B contains a repeated sequence motif unrelated to that of MAP2 and tau. J Cell Biol 109:3367–3376.

Okabe S, Shiomura Y, Hirokawa N (1989): Immunocytochemical localization of microtubule-associated proteins 1A and 2 in rat retina. Brain Res 483:335–346.

Olmsted JB (1986): Microtubule-associated proteins. Annu Rev Cell Biol 2:421–457.

Paschal BM, Vallee RB (1987): Retrograde transport by the microtubule associated protein MAP1C. Nature 330:181–183.

Riederer B, Matus A (1985): Differential expression of distinct microtubule-associated proteins during brain development. Proc Natl Acad Sci USA 82:6006–6009.

Riederer B, Cohen R, Matus A (1986): MAP5: A novel microtubule-associated protein under strong developmental regulation. J Neurocytol 15:763–775.

Riederer BM, Guadano-Ferraz A, Innocenti GM (1990): Difference in distribution of microtubule-associated proteins 5a and 5b during the development of cerebral cortex and corpus callosum in cats: dependence on phosphorylation. Dev Brain Res 56:235–243.

Rienitz A, Grenningloh G, Hermans-Borgmeyer I, Kirsch J, Littauer UZ, Prior P, Gundelfinger EG, Schmitt B, Betz H (1989): Neuraxin, a novel putative structural protein of the rat central nervous system that is immunologically related to microtubule-associated protein 5. EMBO J 8:2879–2888.

Safaei R, Fischer I (1989): Cloning of a cDNA encoding MAP1B in rat brain: Regulation of mRNA levels during development. J Neurochem 52:1871–1879.

Sato C, Tanabe K, Nishizawa K, Nakayama T, Kobayashi T, Nakamura H (1985): Localization of 350K molecular weight and related proteins in both the cytoskeleton and nuclear flecks that increase during G1 phase. Exp Cell Res 160:206–220.

Sato-Yoshitake, Shiomura, Miyasaka, Hirokawa (1989): Microtubule-associated protein 1B: Molecular structure, localization, and phosphorylation-dependent expression in developing neurons. Neuron 3:229–238.

Schoenfeld TA, McKerracher L, Obar R, Vallee RB (1989): MAP 1A and MAP1B are structurally related microtubule associated proteins with distinct developmental patterns in the CNS. J Neurosci 9:1712–1730.

Shiomura Y, Hirokawa N (1987a): The molecular structure of microtubule-associated protein 1A (MAP1A) in vivo and in vitro. An immunoelectron microscopy and quick-freeze, deep-etch study. J Neurosci 7: 1461–1469.

Shiomura Y, Hirokawa N (1987b): Colocalization of microtubule-associated protein 1A and microtubule-associated protein 2 on neuronal microtubules in situ revealed with double-label immunoelectron microscopy. J Cell Biol 104:1575–1578.

Sloboda RD, Rudolph SA, Rosenbaum JL, Greengard P (1975): Cyclic AMP-dependent endogenous phosphorylation of a microtubule-associated protein. Proc Natl Acad Sci USA 72:177–181.

Tsao H, Aletta JM, Greene LA (1990): Nerve growth factor and fibroblast growth factor selectively activate a protein kinase that phosphorylates high molecular weight microtubule-associated proteins. J Biol Chem 265:15471–15480.

Tucker RP, Matus A (1987): Developmental regulation of two microtubule-associated proteins (MAP2 and MAP5) in the embryonic avian retina. Development 101:535–546.

Tucker RP, Matus A (1988): Microtubule-associated proteins characteristic of embryonic brain are found in the adult mammalian retina. Dev Biol 130:423–434.

Tucker RP, Binder LI, Matus A (1988): Neuronal microtubule-associated proteins in the embryonic avian spinal cord. J Comp Neurol 271:44–55.

Tucker RP (1990): The roles of microtubule-associated proteins in rain morphogenesis: A review. Brain Res Rev 15:101–120.

Vallee RB, Davis SE (1983): Low molecular weight microtubule-associated proteins are light chains of microtubule-associated protein 1 (MAP 1). Proc Natl Acad Sci USA 80:1342–1346.

Viereck C, Matus A (1990): The expression of phosphorylated and non-phosphorylated forms of MAP5 in the amphibian CNS. Brain Res 508:257–264.

Viereck C, Tucker RP, Matus A (1989): The adult rat olfactory system expresses microtubule-associated proteins found in the developing brain. J Neurosci 9:3547–3557.

Wiche G, Briones E, Koszka U, Artlieb U, Koepler R (1984): Widespread occurrence of polypeptides related to microtubule-associated proteins (MAP1 and MAP2) in non-neuronal cells and tissues. EMBO J 3:991–998.

Wiche G, Herrmann H, Dalton JM, Foisner T, Leichtfried FE, Lassmann H, Koszka C, Briones E (1986): Molecular aspects of MAP1 and MAP2: microheterogeneity, in vitro localization and distribution in neuronal and nonneuronal cells. Ann NY Acad Sci 466:180–198.

Wiche G, Oberkanins C, Himmler A (1991): Molecular structure and function of microtubule-associated proteins. Int Rev Cytol 124:217–273.

Zauner W, Kratz J, Stanton J, Feick P, Wiche G (1992): Identification of two distinct microtubule binding domains on recombinant rat MAP1B. Eur J Cell Biol 57:66–74.

ABOUT THE AUTHORS

REINHOLD MÜLLER is scientific assistant of the Faculty of Biology/Chemistry of the University of Bremen, Germany, in the group of Dietmar Blohm. He teaches undergraduate and graduate courses dealing with topics in neurobiology and molecular biology. After receiving a diploma in chemistry from the Ruhr-University in Bochum, Germany, he pursued his doctoral research at the same university in the laboratory of Olaf Pongs, where he earned a Ph.D. in 1990 for work on the molecular biology of a family of mRNA species encoding voltage-dependent potassium channels in the *Shaker* locus of *Drosophila melanogaster.* This was followed by postdoctoral research in the Center for Molecular Neurobiology in Hamburg, in the laboratory of Craig C. Garner, on mRNA-transport and the structure and function of microtubule-associated proteins in the rat brain. Since the spring of 1993 he has worked on the role of these and other proteins during neuronal differentiation in different cell culture systems.

STEFAN KINDLER is currently a research fellow at the California Institute of Technology in Pasadena, where he is studying postsynaptic density proteins in rat brain in the laboratory of Mary B. Kennedy. After receiving his diploma in biology from the University of Hamburg, Germany, he carried out his doctoral research at the Center for Molecular Neurobiology in Hamburg in the laboratory of Craig C. Garner, where he earned a Ph.D. in 1991 for cloning and characterizing cDNAs encoding MAP2b and 2c from rat brain. During the following year, as a postdoctoral fellow in the same research group, he studied the transport of mRNAs and proteins into the dendrites of neuronal cells.

CRAIG C. GARNER has recently joined the faculty of the University of Alabama in Birmingham, Alabama, as Associate Professor of Cell Biology in the Neurobiology Research Center, where he will teach undergraduate and graduate courses dealing with topics in developmental neurobiology and cell biology. After receiving a B.A. in biochemistry from Rutgers University in New Jersey, he pursued his doctoral research at Purdue University in the laboratory of Klaus Herrmann, where he earned a Ph.D. in 1984 for work on the molecular biology of 3-deoxy-D-*arabino*-heptulosonate 7 phosphate synthase, the first enzyme in the shikimate pathway in *Escherichia coli*. This was followed by postdoctoral research with Andrew Matus at the Friedrich Miescher Institute in Basel, Switzerland, on the molecular characterization of several microtubule-associated proteins from rat brain. In 1988 he accepted a 5-year position as group leader at the Center for Molecular Neurobiology in Hamburg, Germany, where his research has and is now focused on the identification and characterization of structural proteins of synapses in the developing rat brain. In 1989 he was co-recipient of the Gian Töndury Prize for distinguished research.

Microtubules: 155–166

MAP2

Andrew Matus

I. INTRODUCTION

MAP2 is a member of the family of microtubule-associated proteins that includes tau (see Goedert, this volume) and MAP4 (see Bulinski, this volume), which bind to microtubules through similar multimeric binding domains. It was first discovered as a high molecular weight (M_r ~280,000) component of brain microtubules that lowers the critical concentration of tubulin needed for polymerization, and this aroused interest in it as a potential regulator of microtubule assembly in cells [Murphy and Borisy, 1975; Sloboda et al., 1975]. MAP2 is unusually heat stable for so large a protein, remaining soluble when microtubule preparations are boiled [Fellous et al., 1977]. This combination of properties made it easy to purify, and its major biochemical properties were rather quickly determined. It proved to be a long, fibrous molecule made up of a relatively short microtubule-binding domain and a long "tail," or projection domain, that extends freely from the microtubule surface [Murphy and Borisy, 1975; Amos, 1977; Kim et al., 1979; Vallee, 1980]. It was also found to be multiply phosphorylated [Murphy and Borisy, 1975; Vallee, 1980] and, remarkably, could be purified with a cyclic AMP-dependent protein kinase bound to its projection domain [Vallee et al., 1981].

The use of brain as a starting material for MAP2 isolation is a convenience based on its very high microtubule content. Because of this it was not initially appreciated that MAP2 is unique to neural tissue. Once specific antibodies were obtained it became clear that throughout the brain MAP2 is almost exclusively expressed in neurons [Matus et al., 1981], the few known exceptions involving MAP2 expression in glial cells [Papasozomenos and Binder, 1986; Tucker et al., 1988]. In addition it was found that the high molecular weight form (then the only form known) is selectively associated with dendrites [Bernhardt and Matus, 1984; De Camilli et al., 1984]. Its presence there in very high quantities suggested that MAP2 is involved in determining the characteristic arrangement of dendritic microtubules as longitudinal bundles [Matus, 1988]. The filamentous projection domain of MAP2 bore an equally suggestive resemblance to the fine filaments that can be seen connecting microtubules in electron micrographs of dendrites [Hirokawa et al., 1989]. Together, these observations have led to the current view of MAP2 as a stabilizer and cross-linker of neuronal microtubules.

II. MAP2 AND NEURONAL MORPHOGENESIS

Several features of MAP2 suggest that it is involved in the molecular mechanism of neuronal morphogenesis, of which the most conspicuous are its pattern of expression during development and its polarized distribution within the neuronal cytoplasm. MAP2 exists in two high molecular weight forms, MAP2a and MAP2b, and a low molecular weight form, MAP2c [Burgoyne and Cumming, 1984; Binder et al., 1984; Garner et al., 1988a]. Of these,

MAP2b is expressed continuously throughout life, MAP2c is expressed at high levels only during development, and MAP2a is found only in mature neurons of the adult brain [Binder et al., 1984; Burgoyne and Cumming, 1984; Riederer and Matus, 1985]. In the rat the appearance of MAP2a and the disappearance of MAP2c occur at the same time, between postnatal days 10 and 20, which corresponds to the period of synaptogenesis when neuronal circuitry becomes consolidated. The same changeover takes place in the amphibian *Xenopus*, where it occurs at metamorphosis [Viereck et al., 1988]. Thus a developmentally regulated exchange in the expression of these two forms seems to be a common feature of the circuit stabilization in many vertebrates. The association of MAP2c with the growth of neuronal processes is also suggested by its continued expression in the adult mammalian olfactory system [Viereck, et al., 1989], where new circuitry continues to be formed during adult life.

III. PRIMARY STRUCTURE OF MAP2

Complete amino acid sequences have been published for MAP2b from mouse [Lewis et al., 1988] and rat [Kindler et al., 1991] and for MAP2c from rat [Doll et al., 1990]. No independent sequence has yet been reported for MAP2a, and it remains possible that the difference between these two forms represents a posttranslational modification. Rat MAP2b consists of 1,830 amino acids, giving a calculated molecular mass of 200,000 Da, while MAP2c has 470 amino acids, yielding a molecular mass of 42,300 Da. The most striking feature of the MAP2 sequence is a region near the C-terminus that contains three similar repeats of 18 amino acids joined by less conserved "spacers" of 13 amino acids (Fig. 1). The sequence of this region, which bears overall homology to the C-terminal regions of tau [Lewis et al., 1988] and MAP4 [Aizawa et al., 1991; Chapin and Bulinski, 1991; West et al., 1991], is essential for MAP2 binding to tubulin [Lewis et al., 1988].

The primary sequence of MAP2c consists of two continuous stretches of amino acids from the extreme ends of MAP2b joined together with 1,372 amino acids from the center of MAP2b, corresponding to most of the projection domain missing (Fig. 1) [see Papandrikopoulou et al., 1989]. Corresponding to these

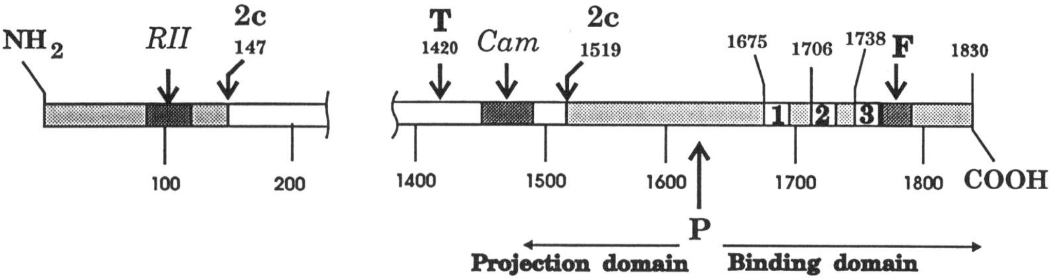

Fig. 1. *A diagrammatic representation of the major primary sequence features of MAP2. Note that a gap of approximately 1,100 amino acids has been introduced in the projection domain of MAP2b where there are currently no known functional elements. The N-and C-terminal regions that are common to MAP2b and MAP2c are shaded pale grey, and the boundaries of the 2c sequence, at residues 147 and 1,519, are marked above. The interface between the microtubule-binding domain and the projection domain are indicated from a unique thrombin cleavage site [Joly et al., 1989] at "P" below. The site of transition from net negative to net positive charge, around residue 1,420, is marked by "T" above. The site at which cyclic AMP–dependent protein kinase binds through its regulatory subunit is marked by* "RII" *and a putiative calmodulin-binding motif by* "Cam." *The three similar 18 amino acid sequence motifs that make up the core of the tubulin-binding domain are represented by the numbered white bars near the C-terminus, and the short sequence required for microtubule bundling is marked by "F" above.*

forms there are two developmentally regulated mRNAs, of 6 and 9 kb [Garner and Matus, 1988]. Since there is only one MAP2 gene the change in their expression appears to involve alternative splicing.

Molecular cloning experiments have also led to the identification of the region of MAP2 responsible for its bound kinase activity. This is located near the N terminus of the molecule, between residues 83 and 113 (Fig. 1), and is mediated via the regulatory subunit of the type II cyclic AMP dependent kinase [Rubino et al., 1989; Obar et al., 1989]. Isolated microtubule preparations also contain type II calcium–calmodulin-activated protein kinase [Vallano et al., 1985; Schulman et al., 1985], and a putative calmodulin-binding motif, based on homology with calmodulin-binding domains in other proteins, has been identified in the projection domain of MAP2b [Kindler et al., 1991] (see Fig. 1). However, there may be additional sites because calmodulin appears to bind to MAP2c as well as MAP2b (F. Leitch and A. Matus, unpublished observations).

Beyond these rather meager details little is known about the structure of MAP2. On the basis of electron microscopy [Voter and Erickson, 1982] and physical chemical measurements [Hernandez et al., 1986] it appears that MAP2 exists in solution as an extended filament, with little secondary structure, and this is in agreement with predictions based on its largely hydrophilic amino acid sequence [Lewis et al., 1988]. It is noteworthy that whereas the projection domain of the molecule bears a net negative charge, the C-terminus is positively charged. There is a distinct transition between these two charge zones at around residue 1,420 (site "T" in Fig. 1). The latter is appropriate to the attachment of MAP2 to negatively charged sequences in tubulin, since its binding to microtubules is ionic in nature.

IV. THE MECHANISM OF MAP2 BINDING TO MICROTUBULES

In one approach to defining the tubulin-binding domain in more detail, the three 18 amino acid repeats were synthesized separately as oligopeptides and their capacity for binding to tubulin and promoting its polymerization measured [Joly et al., 1989]. The second repeat, termed *m2*, was found to be the most potent, but its activity was more than 1,000-fold less than that of the intact MAP2 molecule. This is consistent with the results of experiments with genetically engineered deletion constructs of MAP2, which show that sequences flanking the three-repeat domain make a significant contribution to tubulin binding [Lewis et al., 1989; Lewis and Cowan, 1990]. Further insight into the combined function of the three domains may be gleaned from studies of the similarly structured tubulin-binding region of tau (see Goedert, this volume). One of these studies indicates that despite their close linkage within the primary sequence, each 18 amino acid repeat binds to microtubules with a high degree of autonomy, and it was proposed that the relatively low affinity of each individual repeat is a device for allowing rearrangement and even translational movement of the molecule on the microtubule surface to occur after binding [Butner and Kirschner, 1991]. The second study gives further evidence for the necessity of sequences neighboring the repeat region in determining tau binding and also makes the important point that MAP sequences adequate for binding to isolated microtubules *in vitro* may be inadequate for mediating binding to cellular microtubules in transfected cells [Lee and Rook, 1992].

Recent experiments suggest that the binding of successive MAP2 molecules to the microtubule surface may be cooperative (D. Murphy, personal communication). This was most noticeable in electron micrographs showing clustering of MAP2 molecules on the microtubule surface when added to preformed tubulin polymers at low molar ratios. In addition, quantitative binding data for MAP2 binding show characteristics of cooperativity.

V. MAP2 AND MICROTUBULE CROSS-LINKING

The high concentrations of MAP2 in neuronal processes, together with its filamentous structure, suggest an obvious relationship to the filaments that connect microtubules in the neuronal cytoplasm [Heuser and Kirschner, 1980; Hirokawa et al., 1988]. Two views have been expressed concerning the nature of this linkage. One of these proposes that the MAPs actively cross-link microtubules, while the other favors the idea that these proteins act as struts that determine a minimum spacing between neighboring microtubules. Currently, the weight of evidence seems to favour the latter interpretation. Although MAP2 forms fine filaments between microtubules *in vitro* when they are pelleted, MAP2-containing microtubules do not form bundles spontaneously, even in concentrated suspensions. This is not because the MAP2 of isolated microtubules is "unavailable" to form cross-links or because of any physical hindrance to bundling, because when bivalent antibodies to MAP2 are added to microtubule suspensions, bundles rapidly form [Matus et al., 1987]. Physical measurements comparing pure tubulin polymers with tubulin plus MAP2 have failed to show any of the differences in properties that would be expected if there were interactions between the MAP2 projection domains that cover the microtubule surface [Sato et al., 1988; Friden et al., 1988]. It thus seems that any binding of microtubule-attached MAP2 molecules to one another must be of extremely low affinity. It is by no means implausible that this is the case under physiological conditions. Within the confines of a neuronal process the density of microtubules is very high [Peters et al., 1976], and electron microscopy indicates that their surfaces are densely covered with MAP2 molecules [Hirokawa et al., 1988]. Under these conditions even extremely weak binding between pairs of MAP2 molecules could be sufficient to bundle the densely packed microtubules. This scheme has the added attraction that a multiplicity of weak cross-links could accommodate local parting of adjacent microtubules to allow for the passage of transported organelles and for the parting of microtubules at branch points in processes.

If MAP2 does forge cross-links between microtubules it might be expected that particular sites in the molecule mediate the formation of dimers. Recently, electron microscopic evidence for the formation of antiparallel dimers has been presented [Wille et al., 1992], although in earlier ultrastructural studies purified MAP2 molecules were found to be mainly monomeric [Voter and Erickson, 1982; Gottlieb and Murphy, 1985]. The biochemical properties of MAP2 in solution also favor the monomeric state [Hernandez et al., 1986].

A. Bundling of Microtubules in MAP2-Transfected Cells

The availability of cDNA clones for MAP2 means that its influence on cellular microtubules can now be evaluated by transfecting it into nonneuronal cells where it is not normally expressed. The first study of this kind, with MAP2b, showed that it not only binds to nonneuronal microtubules but also induces their bundling [Lewis et al., 1989]. Further experiments using deletion mutants produced the surprising result that microtubules were still bundled after most of the N-terminal projection domain had been removed. Instead a short sequence near the C-terminus, between residues 1,758 and 1,780 in MAP2b, was shown to be essential [Lewis and Cowan, 1990]. This seemed to suggest that this short C-terminal sequence, which is juxtaposed to the tubulin-binding domain, makes the filamentous cross-links between microtubules in neuronal processes. However, this idea is contradicted by the fact that it is the long N-terminal projection that forms the filamentous side-arms on microtubules *in vitro* [Amos, 1977; Kim et al., 1979; Vallee, 1980]. Amother problem is that the C-terminal extension is too short to account for the intermicrotubule spacing found in neuronal processes, leading to the suggestion that perhaps not all the MAP2 repeats bind to the microtubule surface or that a third, unknown molecule bridges the gap [Lewis and Cowan, 1990].

All of these ideas are based on MAP2 being an active cross-linker of microtubules. If instead it plays the role of passive spacer then alternative explanations for the necessity of the C-terminal sequence become possible. For example, it might be required to potentiate the binding of MAP2 to microtubules or to organize the presentation of the molecule in a manner necessary for bundling to occur by other means. Another argument against MAP2 being actively involved in cross-linking is that microtubule bundling can be induced in cells that do not contain neuronal MAPs by treating them with a variety of chemical agents that stabilize microtubules. On this basis it has been suggested that bundling depends on as yet unidentified ubiquitous cross-linking molecules that can only operate when microtubules have been stabilized by either chemicals or molecules like MAP2 [Chapin et al., 1991]. However, it is worth considering whether the long rods that stabilized microtubules form inside neuronal processes and MAP-transfected cells require specific cross-linking to become bundled. As suggested above, very weak interactions, originating from bound MAPs or even from the stabilized tubulin polymers themselves, might be sufficient to produce adherence between microtubules if frequently repeated along their length. Favoring this interpretation is the observation that microtubules in nonneuronal cells are even more tightly bundled when stabilized by taxol than they are by the expression of MAP2 [Weisshaar et al., 1992].

B. Is the Projection Domain of MAP2 a Flexible Spacer Element?

Some very recent evidence strongly implicates the projection arm of MAP2 as a passive spacer element between microtubules. In these experiments the baculovirus transfection system was used to express very high levels of either MAP2b or MAP2c in the insect recipient cell line SF9. This induced these cells to grow long processes, and examination of these in the electron microscope showed that the microtubules they contain were more than 50 nm apart in cells expressing MAP2b but less than 20 nm apart in those expressing MAP2c [Chen et al., 1992]. Currently available evidence thus suggests that the length of the MAP2 projection domain determines microtubule spacing. This in turn suggests that the concentration of the long projection arm forms of MAP2 (MAP2a and MAP2b) dendrites is associated with wider spacing between microtubules than can be achieved by the short-arm form, MAP2c, which occurs in axons. A further possibility is that this wider microtubule spacing might contribute to the larger cross-sectional area of dendrites compared with axons (Fig. 2A). Similarly, the switch in expression from short-arm MAP2c to the additional long-arm form MAP2a that occurs during the maturation of neuronal processes might be related to the increase in thickness of dendrites that has been observed to occur during their growth [Morest, 1969].

If the projection arm determines microtubule spacing then any change in its state of extension could alter the spacing between dendritic microtubules. It has been suggested that such changes could be mediated by phosphorylation [Friedrich and Aszodi, 1991], and indeed the projection arm of MAP2 is phosphorylated by a variety of protein kinases [Theurkauf and Vallee, 1983; Hernandez et al., 1987; Tsuyama et al., 1986; Hoshi et al., 1992]. MAP2 in brain is very highly phosphorylated [Tsuyama et al., 1987], raising the possibility that electrostatic repulsion between closely spaced charged phosphate moieties in the projection arm region of the molecule would favor the extended form. Such a phenomenon has been demonstrated for tau, which becomes long and stiff when phosphorylated [Hagestedt et al., 1989].

Friedrich and Aszodi [1991] also suggested the possibility that changes in the length of the projection arm may mediate dendritic branching by pushing apart microtubule bundles at branch points. This is an intriguing idea, and there is evidence that in some circumstances MAP2 may indeed accumulate at branch points in dendrites [Matus et al., 1990]. However, such a scheme demands that MAP2 could mediate different spacing between individual microtubules within the cross-sectional area of a

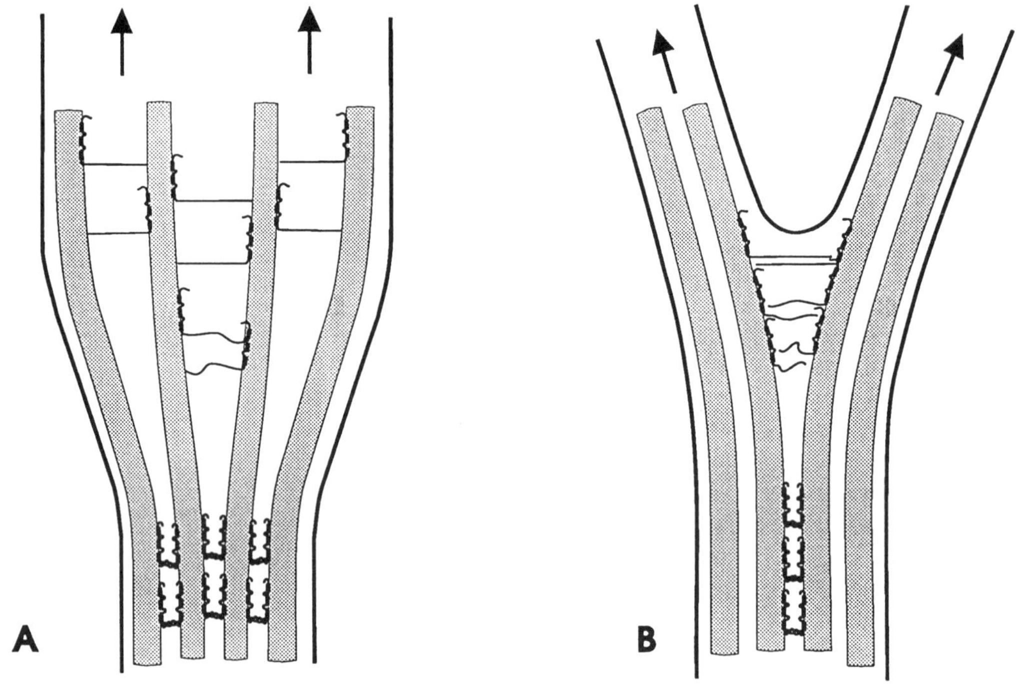

Fig. 2. *Two views of microtubule spacing by the projection arm of MAP2.* **A:** *The longer sidearm of MAP2b (above) leads to more widely spaced microtubules than does the shorter sidearms of MAP2c (below). This could play a part in determining the diameter of the process as indicated by the drawn outline. Another possibility is that changes in the state of extension of the long MAP2b sidearm, possibly mediated by phosphorylation [Friedrich and Aszodi, 1991], could regulate the spacing between microtubules. This is illustrated by the extension of the sidearms of the MAP2 molecules between the two central microtubules, which are drawn as going from randomly arranged (below) through partially extended (middle) to fully extended (top).* **B:** *Another suggested scheme in which extension of the MAP2b sidearms, again mediated by phosphorylation, leads to divergence of the microtubules at branch points in dendrites. As the diagram makes clear, this would require very precise regulation of MAP2 phosphorylation within local areas of cytoplasm.*

single process (Fig. 2B), and at present there is no known mechanism that shows the promise of conferring such extreme precision.

VI. TARGETING OF HIGH MOLECULAR WEIGHT MAP2 TO DENDRITES

A variety of mechanisms can be envisaged to account for the selective association of high molecular weight MAP2 with dendrites. These fall into two groups, those involving the specific targeting of the high molecular weight forms to dendrites and those involving the exclusion of the high molecular weight forms from axons. The latter scheme suggests the existence of a filter mechanism located at the entry to the axon. It has the added attraction that it could also apply to other components of the neuronal cytoplasm that are excluded from axons, such as ribosomes and Golgi membranes. However, the presence of high molecular weight MAP2 in the proximal axons of cerebellar basket cells [Bernhardt and Matus, 1984] and motor neurons [Papasozomenos et al., 1985] argues against such a scheme. Another difficulty for an axonal exclusion mechanism is that in cultured hippocampal cells high molecular weight MAP2 is initially present in the axon from which it subsequently disappears [Caceres et al., 1984].

A. Is Axonal MAP2 Targeted for Proteolysis?

An alternative proposal is that the absence of high molecular weight MAP2 from axons is the result of its destruction by proteases located there. Evidence in favor of this idea comes from experiments in which biotin-labeled high molecular weight MAPs was injected into cultured motor neurons and found first to spread into all processes but subsequently disappear from axons while persisting in dendrites [Okabe and Hirokawa, 1989]. Unfortunately these experiments cannot be considered definitive because of the possibility that pressure injection of high molecular weight MAP2 forces it into the axonal compartment, overriding the normal sorting mechanisms. However, these results confirm that proteolysis of high molecular weight MAP2 in axons does occur, at least as a fail-safe mechanism. It has been pointed out that the primary amino acid sequence of MAP2 contains multiple repeats of the motif proline–glutamic acid–serine–threonine (PEST [see Friedrich and Aszodi, 1991], which is a feature of many proteins that are programmed for proteolytic destruction. Certainly, high molecular weight MAP2 is notoriously sensitive to brain proteases. However, the low molecular weight form, MAP2c, which is present in axons also contains PEST sequences [Friedrich and Aszodi, 1991]. Thus, while these proteolytic mechanisms are undoubtedly important in regulating MAP2 function and perhaps its cytoplasmic distribution, they are as yet poorly understood.

B. Is There a Dendritic Targeting Signal for High Molecular Weight MAP2?

The alternative to the exclusion hypothesis is that some mechanism exists for the specific accumulation of high molecular weight MAP2 in dendrites. This most obvious way in which this might occur is through the specific export of protein from the cell body into dendrites. However, the 9 kb mRNA that encodes MAP2b has been shown to be present in dendrites [Garner et al., 1988b], and the transport of newly synthesized RNA into the dendrites of cultured neurons has been demonstrated [Davis et al., 1990],so the concentration of high molecular weight MAP2 in dendrites could also be the results of its local synthesis there. Either scheme would depend on a signal which links protein or mRNA to a dendrite-oriented transport mechanism. While there is currently no information regarding the nature of such a mechanism, a potential clue exists in the orientation of microtubules in dendrites, where approximately half of them have their plus ends toward the cell body [Burton, 1988; Baas et al., 1988]. This mean that in half the population of dendritic microtubules a retrograde-directed motor system, such as that mediated by dynein, could operate away from the cell body, whereas in axons, in which all the microtubules have their plus ends distally, dynein transport would be uniformly toward the cell body [Black and Baas, 1989]. Thus if high molecular weight MAP2 protein or mRNA were transported by a ''retrograde'' motor system it would be constantly moved toward the cell body in axons but would be transported distally by the reverse-oriented microtubules in dendrites.

VII. CELLULAR FUNCTIONS OF MAP2

To date there is relatively little direct evidence regarding the function of MAP2 in living neurons. This is beginning to change, particularly with the availability of DNA sequence and the consequent possibility of utilizing manipulations involving molecular genetics.

One effective means of judging the contribution of a protein to cellular function is to attenuate its expression by introducing into the cell nucleic acid sequences that are complimentary (antisense) to part of the mRNA sequence that encodes the target protein. This may be achieved either by introducing a plasmid from which antisense RNA is transcribed by cellular enzymes or by adding to the medium short, synthetic antisense oligonucleotides that can enter the cells. In many cases this has been shown specifically to block synthesis of a target protein, either by interfering with translation or by forming a heteroduplex with the target mRNA that leads to destruction by scavenger ribonucleases. Whatever the exact mechanism may be, two studies have been published showing the feasibility of using this approach for MAP2. In the first, the plasmid technique was used in conjunction with a teratocarcinoma cell line (P19) that, when treated with retinoic acid, differentiates into a neuroblastoma-like cell that expresses MAP2 [Dinsmore and Solomon, 1991]. The antisense MAP2 RNA produced in these cells effectively lowered MAP2 levels and dramatically reduced their ability to grow processes. In addition, even though these cells expressed other neuronal differentiation markers they did not cease dividing, as they normally do when differentiating into the neuronal pathway. This led to the suggestion that MAP2 itself may be involved in the control of cell division in neurons [Dinsmore and Solomon, 1991].

In another recent study antisense MAP2 oligonucleotides were shown to inhibit neurite outgrowth from cultured cerebellar neurons [Caceres et al., 1992]. The cells appeared to be blocked at the stage where they had elaborated lamellipodial veils but had not yet grown the minor processes that are characteristic of the next phase of differentiation [stage II of Dotti et al., 1988]. Previously, antisense oligonucleotides had been used to reduce tau expression in the same cerebellar cells [Caceres et al., 1991]. In that case minor processes were produced but they did not progress to the next stage, where a single axon grows rapidly. It thus seems that suppression of these two related microtubule-associated proteins has distinct effects on neuronal differentiation. It is also interesting that even though the anti-MAP2–treated cells still synthesized tau, they did not elaborate processes, suggesting that some aspects of MAP2 function may be absolutely required for process initiation.

VIII. RIGIDIFICATION OF MICROTUBULES VIA THE MULTIMERIC BINDING DOMAIN OF MAP2

One of the most striking aspects of MAP2 structure is the presence of three repeats of the same motif in its tubulin-binding domain. Recent experiments suggest that the purpose of this repetition is to impart stiffness to the microtubules. Nonneuronal cells do not normally express MAP2. Under these circumstances the microtubules emanate as single filaments from a centrosomal microtubule-organizing center (MTOC). When stabilized by transfection with MAPs the microtubules become independent of the MTOC and form bundles [Lewis et al., 1989; Weisshaar et al., 1992]. Both of these phenomena occur in the absence of MAP2 when cells are treated with taxol [Schiff and Horwitz, 1980; De Brabander et al., 1981] so neither appears to depend on any feature of MAP2 beyond its ability to stabilize tubulin in the polymerized state. However, MAP2-stabilized microtubules have two properties that are not produced by taxol. First, the MAP2 microtubules are much longer than the taxol microtubules. This is probably because although MAP2 is an effective stabilizer of microtubules it is a relatively weak initiator of microtubule formation compared with taxol [De Brabander et al., 1981]. Second, MAP2-stabilized microtubules are stiff. This shows itself in the conformation that the MAP2-microtubules adopt when they are longer than the cell diameter. Where they encounter the cell surface they bend, typically forming rings around the periphery of the cell. At the same time the cell surface is pushed out into a rounded shape, suggestive of the microtubules pushing outward [Weisshaar et al., 1992].

Further evidence in favor of MAP2 imparting stiffness to microtubules comes from experiments in which the cortical actin network of MAP2-transfected cells was depolymerized with the drug cytochalasin B. When this is done the cells grow stiff processes containing MAP2-decorated microtubules [Edson et al., 1993]. The origin of this ability to make microtubules rigid is suggested by the structure of MAP2

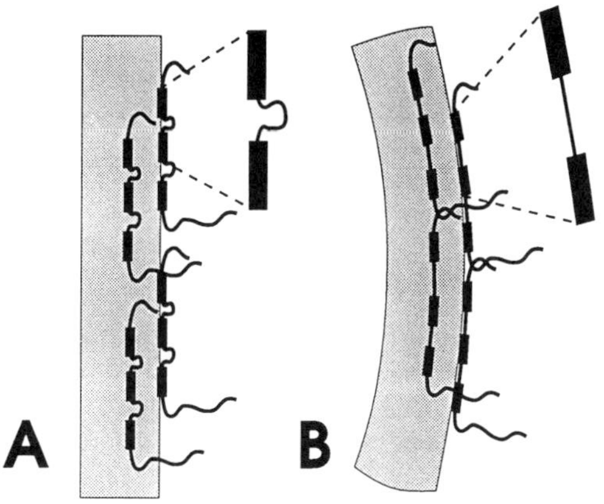

Fig. 3. *Binding of MAP2 to microtubules limits their flexibility. The consequences of attachment to microtubules of the multimeric tubulin-binding domain of MAP2 is illustrated for several molecules of MAP2c on the surface of a short segment of a microtubule. Each MAP2 molecule contains three tubulin-binding motifs joined by short stretches of 13 amino acids. Each motif can bind to a separate tubulin subunit in the wall of the microtubule, and this, repeated all over the microtubule surface, limits the flexibility of the microtubule because the subunits are closely linked together [Edson et al., 1993].*

itself. It binds to microtubules through a domain that contains three repeats of similar 18 amino acid motifs that are linked together by short, flexible stretches of 13 amino acids. The three 18 amino acid motifs are so positioned that they can bind to neighboring tubulin subunits in the wall of the microtubule (Fig. 3A). As a result the movement of the subunits relative to one another is limited by the relatively short 13 amino acid linker sequence (Fig. 3B), and this limits the flexibility of the microtubule [Edson et al., 1993]. These results indicate that by producing long, rigid microtubules MAP2 can mediate the formation of axial cytoskeletal elements inside cells that are capable of supporting processes. Such a combination of properties could be of great importance in producing neuronal processes that are capable of supporting the load imposed on them by the surrounding mass of neural tissue in the central nervous system and in resisting physical disturbances induced by movement.

IX. POSSIBLE CONTRIBUTIONS OF MAP2 TO NEURONAL PLASTICITY

The combination of properties of MAP2 suggest that it may contribute in several ways to the plasticity of neuronal structure, particularly in the developing nervous system. One possibility that has already been mentioned arises from the extensive phosphorylation of MAP2, which may contribute to the state of extension and rigidity of the projection side arm. Phosphorylation also influences the efficacy of MAP2 in promoting tubulin polymerization *in vitro* [Jameson and Caplow, 1981; Burns et al., 1984], and microinjection experiments have shown that depending on its state of phosphorylation MAP2 in living cells can either be entirely bound to microtubules or soluble in the cytoplasm [Brugg and Matus, 1991]. Several other aspects of MAP2 function suggest that phosphorylation effects of this kind may be important in determining neuronal plasticity, especially during development. It has been shown that the state of MAP2 phosphorylation changes during the critical period of visual system plasticity in cats and that manipulations of visual system maturation (e.g., dark-rearing) that extend the critical period, result in postponement of the change in phosphorylation [Aoki and Siekevitz, 1985]. MAP2 is a substrate for a wide variety of protein kinases [Hernandez et al., 1987]. These include mitogen-activated kinases [Silliman and Sturgill, 1989; Hoshi et al., 1992] and those that are activated by cyclic AMP [Sloboda et al., 1975; Theurkauf and Vallee, 1983], and calcium–calmodulin [Schulman, 1984; Tsuyama et al., 1986]. Levels of both the latter second-messenger systems in neurons are influenced by activation of neurotransmitter receptors, suggesting possible pathways for the coupling of synaptic transmitter and MAP2 phosphorylation [see Matus, 1991, for further discussion]. Glutamate receptor activation has indeed been shown to have a pronounced effect on MAP2 phosphorylation [Halpain and Greengard, 1990].

Another aspect of MAP2 regulation, relating to the high molecular weight dendritic forms, is the presence of the 9 kb MAP2 mRNA in dendrites. This suggests the possibility that during development local concentrations of MAP2 in the dendritic cytoplasm are regulated by the level of protein synthesis [Matus, 1991]. This in turn could be regulated by changes in the levels of second messengers, such as calcium and cyclic AMP, that are known to influence protein synthesis. An initial study has indeed suggested that stimulation of glutamate receptors leads to the elevation of dendritic MAP2 in cultured neuron concentrations as measured by immunofluorescence staining intensity [Bigot et al., 1991]. Thus glutamate receptor activation may influence MAP2 function via both phosphorylation and net protein synthesis.

X. REFERENCES

Aizawa H, Emori Y, Mori A, Murofushi H, Sakai H, Suzuki K (1991): Functional analyses of the domain structure of microtubule-associated protein-4 (MAP-U). J Biol Chem 266:9841–9846.

Amos L (1977): The rearrangement of high molecular weight microtubule-associated proteins. J Cell Biol 72:642–654.

Aoki C, Siekevitz P (1985): Ontogenetic changes in the cyclic adenosine 3′,5′-monophosphate-stimulable phosphorylation of cat visual cortex proteins, particularly and microtubule-associated protein 2 (MAP2): Effects of normal and dark rearing and the exposure to light. J Neurochem 5:2465–2483.

Baas PW, Deitch JS, Black MM, Banker GA (1988): Polarity orientation of microtubules in hippocampal neurons: Uniformity in the axon and nonuniformity in the dendrite. Proc Natl Acad Sci USA 85:8335–8339.

Bernhardt R, Matus A (1984): Light and electron microscopic studies of the distribution of microtubule-associated protein 2 n rat brain: A difference between dendritic and axonal cytoskeletons. J Comp Neurol 226:203–221.

Bigot D, Hunt S, Matus A (1991): Reorganization of the cytoskeleton in rat neurons following stimulation with excitatory amino acids in vitro. Eur J Neurosci 3:551–558.

Binder LI, Frankfurter A, Kim H, Caceres A, Payne MR, Rebhun LI (1984): Heterogeneity of microtubule-associated protein 2 during rat brain development. Proc Natl Acad Sci USA 81:5613–5617.

Black MM, Baas PW (1989): The basis of polarity in neurons. Trends Neurosci 12:211–214.

Brugg B, Matus A (1991): Phosphorylation determines the binding of microtubule-associated protein 2 (MAP2) to microtubules in living cells. J Cell Biol 114:735–743.

Burgoyne RD, Cumming R (1984): Ontogeny of microtubule-associated protein 2 in rat cerebellum: differential expression of the doublet polypeptides. Neurosci 11:156–167.

Burns RG, Islam K, Chapman R (1984): The multiple phosphorylation of the microtule-associated protein MAP2 controls the MAP2:tubulin interaction. Eur J Biochem 141:609–615.

Burton PR (1988): Dendrites of mitral cell neurons contain microtubules of opposite polarity. Brain Res 473:107–115.

Butner KA, Kirschner MW (1991): Tau protein binds to microtubules through a flexible array of distributed weak sites. J Cell Biol 115:717–730.

Caceres A, Banker G, Steward O, Binder L, Payne M (1984): MAP2 is localized to the dendrites of hippocampal neurons which develop in culture. Dev Brain Res 13:314–318.

Caceres A, Potrebic S, Kosik KS (1991): The effect of tau antisense oligonucleotides on neurite formation of cultured cerebellar macroneurons. J Neurosci 11: 1515–1523.

Caceres A, Mautino J, Kosik KS (1992): Suppression of MAP2 in cultured cerebellar macroneurons inhibits minor neurite formation. Neuron 9:607–618.

Chapin SJ, Bulinski JC, Gundersen GG (1991): Microtubule-bundling in cells. Nature 349:24.

Chapin SJ, Bulinski JC (1991): Non-neuronal 210 $\times$ 10^3 M_r microtubule-associated protein (MAP4) contains a domain homologous to the microtubule-binding domain of neuronal MAP2 and tau. J Cell Sci 98:27–36.

Chen J, Kanai Y, Cowan N, Hirokawa N (1992): Differential functions of MAP2, MAP2c and tau on the organisation of microtubules in the cells. Mol Biol Cell 3:259a.

Davis L, Burger B, Banker GA, Steward O (1990): Dendritic transport: Quantitative analysis of the time course of somatodendritic transport of recently synthesized rna. J Neurosci 10:3056–3068.

De Brabander M, Geuens G, Nuydens R, Willebrords R, De Mey J (1981): Taxol induces the assembly of free microtubules in living cells and blocks the organizing capacity of the centrosomes and kinetochores. Proc Natl Acad Sci USA 78:5608–5612.

De Camilli P, Miller PE, Navone F, Theurkauf WE, Vallee RB (1984): Distribution of microtubule-associated protein 2 in the nervous system of the rat studied by immunofluorescence. Neuroscience 11:817–846.

Dinsmore JH, Solomon F (1991): Inhibition of MAP2 expression affects both morphological and cell division phenotypes of neuronal differentiation. Cell 64:817–826.

Doll T, Papandrikopoulou A, Matus A (1990): Nucleotide and amino acid sequences of embryonic rat MAP2c. Nucleic Acid Res 18:361.

Dotti CG, Sullivan CA, Banker GA (1988): The establishment of polarity by hippocampal neurons in culture. J Neurosci 8:1454–1468.

Edson K, Weisshaar B, Matus A (1993): Actin depolymersation induces process formation in MAP2-

transfected non-neuronal cells. Development 117: 689–700.

Fellous A, Francon J, Lennon AM, Nunez J (1977): Microtubule assembly in vitro. Eur J Biochem 78:167–174.

Friden B, Nordh J, Wallin M, Deinum J, Norden B (1988): Effects of proteolysis of the extending parts of high-molecular-weight microtubule-associated proteins on interactions between microtubules. Biochim Biophys Acta 955:135–142.

Friedrich P, Aszodi A (1991): MAP2: A sensitive cross-linker and adjustable spacer in dendritic architecture. FEBS Lett 295:5–9.

Garner CC, Brugg B, Matus A (1988a): A 70-kilodalton microtubule-associated protein (MAP2c), related to MAP2. J Neurochem 50:609–615.

Garner CC, Tucker RP, Matus A (1988b): Selective localization of messenger RNA for cytoskeletal protein MAP2 in dendrites. Nature 336:674–677.

Garner CC, Matus A (1988): Different forms of microtubule-associated protein 2 are encoded by separate mRNA transcripts. J Cell Biol 106:779–783.

Gottlieb RA, Murphy DB (1985): Analysis of the microtubule-binding domain of MAP-2. J Cell Biol 101: 1782–1789.

Hagestedt T, Lichtenberg B, Wille H, Mandelkow EM, Mandelkow E (1989): Tau protein becomes long and stiff upon phosphorylation: Correlation between paracrystalline structure and degree of phosphorylation. J Cell Biol 109:1643–1651.

Halpain S, Greengard P (1990): Activation of NMDA receptors induces rapid dephosphorylation of the cytoskeletal protein MAP2. Neuron 5:237–246.

Hernandez MA, Avila J, Andreu JM (1986): Physicochemical characterization of the heat-stable microtubule-associated protein MAP2. Eur J Biochem 154:41–48.

Hernandez MA, Wandosell F, Avila J (1987): Localization of the phosphorylation sites for different kinases in the microtubule-associated protein MAP2. J Neurochem 48:84–93.

Heuser JE, Kirschner M (1980): Filament organization revealed in platinum replicas of freeze-dried cytoskeletons. J Cell Biol 86:212–234.

Hirokawa N, Hisanaga SI, Shiomura Y (1988): MAP2 is a component of crossbridges between microtubules and neurofilaments in the neuronal cytoskeleton: Quick-freeze, deep-etch immunoelectron microscopy and reconstitution studies. J Neurosci 8:2769–2779.

Hoshi M, Ohta K, Gotoh Y, Mori A, Murofushi H, Sakai H, Nishida E (1992): Mitogen-activated-protein-kinase-catalyzed phosphorylation of microtubule-associated proteins, microtubule-associated protein 2 and microtubule-associated protein 4, induces an alteration in their function. Eur J Biochem 203:43–52.

Jameson L, Caplow M (1981): Modification of microtuble steady-state dynamics by phosphorylation of the microtuble-associated proteins. Proc Natl Acad Sci USA 78:3413–3417.

Johnson GVW, Litersky JM, Jope RS (1991): Degradation of microtubule-associated protein 2 and brain spectrin by calpain: A comparative study. J Neurochem 56:1630–1638.

Joly JC, Flynn G, Purich DL (1989): The microtubule-binding fragment of microtubule-associated protein-2: Location of the protease-accessible site and identification of an assembly-promoting peptide. J Cell Biol 109:2289–2294.

Kim H, Binder LI, Rosenbaum JL (1979): The periodic association of map2 with brain microtubules in vitro. J Cell Biol 80:266–276.

Kindler S, Schulz B, Goedert M, Garner CC (1991): Molecular structure of microtubule-associated protein 2b and 2c from rat brain. J Biol Chem 265:19679–19684.

Lee G, Rook SL (1992): Expression of tau protein in non-neural cells: Microtubule binding and stabilization. J Cell Sci 102:227–237.

Lewis SA, Wang D, Cowan NJ (1988): Microtubule-associated protein MAP-2 shares a microtubule binding motif with tau protein. Science 242:936–939.

Lewis SA, Ivanov IE, Lee GH, Cowan NJ (1989): Organization of microtubules in dendrites and axons is determined by a short hydrophobic zipper in microtubule-associated and tau. Nature 342:498–505.

Lewis SA, Cowan N (1990): Microtubule bundling. Nature 345:674.

Matus A, Bernhardt R, Hugh Jones T (1981): High molecular weight microtubule-associated proteins are preferentially associated with dendritic microtubules in brain. Proc Natl Acad Sci USA 78:3010–3014.

Matus A, Riederer B, Huber G (1987): Influence of monoclonal antibodies on microtubule assembly. J Neurochem 49:714–720.

Matus A (1988): Microtubule-associated proteins: Their potential role in determining neuronal morphology. Annu Rev Neurosci 11:29–44.

Matus A, Delhaye-Bouchaud N, Mariani J (1990): Microtubule-associated protein 2 (MAP2) in Purkinje cell dendrites: Evidence that factors other than binding to microtubules are involved in determining its cytoplasmic distribution. J Comp Neurol 297:435–440.

Matus A (1991): Microtubule-associated proteins and neuronal morphogenesis. J Cell Sci 15:(suppl)61–67.

Morest DK (1969): The growth of dendrites in the mammalian brain. Z Anat Entwickl-Gesch 128:290–317.

Murphy DB, Borisy GG (1975): Association of high-molecular-weight proteins with microtubules and their role in microtubule assembly in vitro. Proc Natl Acad Sci USA 72:2696–2700.

Obar RA, Dingus J, Bayley H, Vallee RB (1989): The RII subunit of cAMP-dependent protein kinase binds to a common amino-terminal domain on microtubule-associated proteins 2A, 2B, and 2C. Neuron 3:639–645.

Okabe S, Hirokawa N (1989): Rapid turnover of microtubule-associated protein MAP2 in the axon revealed by microinjection of biotinylated MAP2 into cultured neurons. Proc Natl Acad Sci USA 86:4127–4131.

Papandrikopoulou A, Doll T, Tucker RP, Garner CC, Matus A (1989): Embryonic MAP2 lacks the cross-linking sidearm sequences and dendritic targeting signal of adult MAP2. Nature 340:650–652.

Papasozomenos SC, Binder LI, Bender PK, Payne MR (1985): Microtubule-associated protein 2 without axons of spinal motor neurons: associations with microtubules and neurofilaments in normal and beta, beta '-iminodipropionitrile-treated axons. J Cell Biol 100:74–85.

Papasozomenos SC, Binder LI (1986): Microtubule-associated protein 2 (MAP2) is present in astrocytes of the optic nerve but absent from astrocytes of the optic tract. J Neurosci 6:1748–1756.

Peters A, Palay SL, Webster HDF (1976): The Fine Structure of the Nervous System. Philadelphia: W.B. Saunders.

Riederer B, Matus A (1985): Differential expression of distinct microtubule-associated proteins during brain development. Proc Natl Acad Sci USA 82:6006–6009.

Rubino HM, Dammerman M, Shafit-Zagardo B, Erlichman J (1989): Localization and characterization of the binding site for the regulatory subunit of type II cAMP-dependent protein kinase of MAP2. Neuron 3:631–638.

Sato M, Schwartz WH, Selden SC, Pollard TD (1988): Mechnical properties of brain tubulin and microtubules. J Cell Biol 106:1205–1211.

Schiff PB, Horwitz SB (1980): Taxol stabilizes microtubules in mouse fibroblast cells. Proc Natl Acad Sci USA 77:1561–1565.

Schulman H (1984): Differential phosphorylation of MAP-2 stimulated by calcium-calmodulin and cyclic AMP. Mol Cell Biol 4:1175–1178.

Schulman H, Kuret J, Jefferson AB, Nose PS, Spitzer KH (1985): Ca^{2+}/calmodulin-dependent microtubule-associated protein 2 kinase: Broad substrate specificity and multifunctional potential in diverse tissues. Biochemistry 24:5320–5327.

Silliman CC, Sturgill TW (1989): Phosphorylation of microtubule-associated protein 2 by MAP kinase primarily involves the projection domain. Biochem Biophys Res Commun 160:993–998.

Sloboda RD, Rudolph SA, Rosenbaum JL, Greengard P (1975): Cyclic AMP-dependent endogenous phosphorylation of a microtubule-associated protein. Proc Natl Acad Sci USA 72:177–181.

Theurkauf WE, Vallee RB (1983): Extensive cAMP-dependent and cAMP-independent phosphorylation of microtubule-associated protein 2. J Biol Chem 258:7883–7886.

Tsuyama S, Bramblett GT, Huang K-P, Flavin M (1986): Calcium/phospholipid-dependent kinase recognizes sites in microtubule-associated protein 2 which are phosphorylated in living brain and are not accessible to other kinases. J Biol Chem 261:4110–4116.

Tsuyama S, Terayama Y, Matsuyama S (1987): Numerous phosphates of microtubule-associated protein 2 in living rat brain. J Biol Chem 262:10886–10892.

Tucker RP, Binder LI, Viereck C, Hemmings BA, Matus A (1988): The sequential appearance of low- and high-molecular weight forms of MAP2 in the developing cerebellum. J Neurosci 12:4503–4512.

Vallano ML, Goldenring JR, Buckholz TM, Larson RE, DeLorenzo RJ (1985): Separation of endogenous calmodulin- and cAMP-dependent kinases from microtubule preparations. Proc Natl Acad Sci USA 82:3202–3206.

Vallee RB (1980): Structure and phosphorylation of microtubule associated protein 2 (MAP2). Proc Natl Acad Sci USA 77:3206–3210.

Vallee RB, DiBartolomeis MJ, Theurkauf WE (1981): A protein kinase bound to the projection portion of MAP2 (microtubule-associated protein 2). J Cell Biol 90:568–576.

Viereck C, Tucker RP, Binder LI, Matus A (1988): Phylogenetic conservation of brain microtubule-associated proteins MAP2 and tau. Neurosci 26:893–904.

Viereck C, Tucker RP, Matus A (1989): The adult rat olfactory system expresses microtubule-associated proteins found in the developing brain. J Neurosci 9:3547–3557.

Voter WA, Erickson HP (1982): Electron microscopy of MAP 2 (microtubule-associated protein 2). J Ultrastruct Res 80:374–382.

Weisshaar B, Doll T, Matus A (1992): Reorganisation of the microtubular cytoskeleton by embryonic microtubule-associated protein 2 (MAP2c). Development 116:1151–1161.

West RR, Tenbarge KM, Olmsted JB (1991): A model for microtubule-associated protein 4 structure. Domains defined by comparisons of human, mouse, and bovine sequences. J Biol Chem 266:21886–21896.

Wille H, Mandelkow EM, Dingus J, Vallee RB, Binder LI, Mandelkow E (1992): Domain structure and antiparallel dimers of microtubule-associated protein 2 (MAP2). J Struct Biol 108:49–61.

ABOUT THE AUTHOR

ANDREW MATUS is Neurobiology Group Leader at the Friedrich Miescher Institute in Basel, Switzerland, and teaches cellular neurobiology at the Zoological Institute of the University of Basel. He completed a B.Sc. in chemistry and an M.Sc. in cell biology at the University of Auckland, New Zealand, and transferred to University College London, England, to study anatomy and embryology for the Ph.D. degree. He was subsequently a research fellow and then Lecturer in Anatomy at University College before moving to his present position in 1978. His research interests lie in exploring the way that structural molecules contribute to neuronal development and plasticity. For some years his group has concentrated on role of microtubules in neuronal morphogenesis. Dr. Matus and his associates are currently using gene cloning techniques to investigate the function of regulatory microtubule-associated proteins.

Microtubules: 167–182

MAP4

Jeannette Chloë Bulinski

I. INTRODUCTION

Microtubules (MTs) play key roles in a variety of cellular functions, including chromosome separation at mitosis, trafficking of vesicles, and establishment and maintenance of cell shape and organization of the Golgi, endoplasmic reticulum, and other organelles. Studies using pharmacological inhibitors of MTs have demonstrated the MT dependence of these processes; however, they do not explain the mechanisms by which MTs contribute to disparate cellular functions. To understand the *in vivo* workings of an MT, we must understand the function and interactions of each of its molecular components. Although MTs are composed chiefly of the α- and β-polypeptides of tubulin, a variety of microtubule-associated proteins (MAPs) are imputed to diversify the function of MTs. For example, MT-dependent ATPases are responsible for bidirectional movement of MTs, chromosomes, and vesicles along MTs (see Scholey and Vale; Collins, Earnshaw, and McIntosh, this volume). Enzymes that posttranslationally modify MTs can also be considered to be MAPs. Other MAPs promote assembly of MTs and are bound along the MT wall. These assembly-promoting MAPs (AP-MAPs) are thought to influence MT dynamics and MT interactions with organelles or other cytoplasmic components *in vivo*.

The focus of this discussion is on MAP4, which is the predominant nontubulin component of MTs in a wide variety of vertebrate tissues, cell types, and MT organelles. As described below, MAP4 has been isolated from a number of different mammals thus far, in which its *in vivo* distribution and developmental expression have been examined. MAP4 has also been purified and characterized biochemically. More recently, several MAP4 species have been cloned and subjected to extensive molecular analysis, and their sequences, predicted molecular structure, and relationship to other AP-MAPs have been ascertained. The ubiquitous expression and developmental regulation of MAP4 suggest that it is functionally important in MT-dependent events. Now that molecular tools have been brought to bear on this problem, a determination of the function(s) of MAP4 *in vivo* should be within our purview.

II. IDENTIFICATION AND PURIFICATION OF MAP4

The original self-assembly schemes devised to purify MTs from vertebrate brain tissue [e.g., Shelanski et al., 1973; Borisy et al., 1975] resulted in the copurification of several species, called MAP1, MAP2, and tau protein, that were believed to be functionally important components of MTs [Murphy and Borisy, 1975; Dentler et al., 1975; Weingarten et al., 1975]. Because MTs are abundant and important in neuronal cell function and in brain development, further study of these MAPs promises to aid in our understanding of MT function within the nervous system. In fact, more recent studies demonstrated that the brain MAPs are either absent or are present in trace amounts outside of the nervous system [Weatherbee et al., 1982; Bloom et al., 1984]. This restricted distribution suggested that the neuronal MAPs perform specialized functions in neuronal cells. Indeed, the unique morphology of neurons suggested the need for a unique MT cytoskeleton. It seemed unlikely that extrapolation of the properties and functions of neuronal MAPs to the MT-dependent events of proliferating and differentiating cells outside the nervous system would be fruitful. Instead, an isolation of MTs, along with any MAPs, from a nonneuronal source loomed as a necessary step in the investigation of MT function in nonneuronal cells.

The original investigations of nonneuronal MTs exploited cells of the rapidly dividing cell line HeLa, because it is a pure cell type and is amenable to large-scale culture [Bulinski and Borisy, 1979; Bulinski, 1986]. Self-assembly protocols much like those utilized in the study of brain MTs were used for the selective purification of assembly-competent tubulin and proteins whose association with MTs might be expected to be physiologically relevant. The choice of HeLa cells for MT purification experiments was important in order to have a tubulin concentration sufficient to allow self-assembly in the extract at the start of the procedure and to have sufficient material to carry out the multiple cycles of assembly–disassembly that are required to identify MAPs and to distinguish them from trace quantities of impurities that may linger through the early steps of the purification.

Accordingly, the first purifications of nonneuronal MTs, including MAP4 (then called 210 kD MAP because of its electrophoretic mobility) were accomplished by Bulinski and Borisy [1979] and by Weatherbee et al. [1980]. Actually, these preparations served to identify two species of MAPs, a triplet of ~210 kD (MAP4) and a doublet of 125 kD, as quantitatively copurifying MAPs. All other extract proteins were either present in trace amounts in purified material or did not appear to copurify with MTs. These early studies also demonstrated that chimeric MTs could be reconstituted in which both the 210 kD MAP (MAP4) and 125 kD MAP associated with and sedimented with brain as well as HeLa tubulin, suggesting that both MAPs were capable of a specific interaction with MTs.

Many polycationic substances are capable of binding nonspecifically to MTs [Murphy et al., 1977]; therefore, the isolation of any putative MAPs, such as those isolated from HeLa cells, immediately raises questions about the specificity of the protein's interaction with MTs. This question can be best addressed by demonstrating that the putative MAP is localized on MTs *in vivo*. To test the significance of the MT-binding capacity of the 210 kD MAP (MAP4), specific antibodies were raised and used in immunofluorescence of multiple types of cultured cells [Bulinski and Borisy, 1980a,b]. Figure 1 shows a portion of these results, in which a fibrous pattern of anti-210 kD MAP (MAP4) immunofluorescence can be seen. Double staining with antibodies against tubulin, as well as dissolution of the MTs with pharmacological agents or cold treatment, demonstrated that the MAP4 fluorescence was associated with the MT cytoskeleton. Figure 1 shows a variety of primate cultured cells, including cells from a number of bodily tissues and from all stages of the cell cycle (see Fig. 1). In fact,

---

Fig. 1. *Indirect immunofluorescence staining of primate cells with 210 kD MAP (MAP4) antibody.* **a:** *Interphase human foreskin fibroblast.* ×*430.* **b:** *Interphase human foreskin fibroblast.* ×*370.* **c:** *Mitotic human fetal kidney cells, metaphase* **(left)** *and anaphase* **(right)**. ×*675.* **d:** *Colony of interphase human bladder primary cells.* ×*260.* **e:** *Interphase tetraploid bladder cells.* ×*460.* **f:** *Glial cell, human fetal brain explant culture.* ×*770.* **g:** *Metaphase human liver cell (Chang cell line).* ×*670.* **h:** *Human liver (Chang line) undergoing cytokinesis; midbody staining is shown.* ×*670.* **i:** *Colony of interphase African Green Monkey kidney cells (CV-1 line).* ×*460.* **j:** *Interphase African Green monkey kidney cell (BSC-1 line).* ×*530.* **k:** *Telophase African green monkey kidney cell (BSC-1 line).* ×*700.* **l:** *Interphase Abyssinian colobus brain (glial) cell.* ×*580.* **m:** *Mitotic stumptail macaque spleen cells; metaphase,* **top**, *prometaphase,* **bottom.** ×*350.* **n:** *Prometaphase owl monkey kidney cell.* ×*920.* **o:** *Interphase cotton-topped Marmoset leukocyte.* ×*1,160. For procedures used and origin of cell lines and cultures, see Bulinski and Borisy [1980b]. [Reproduced from Bulinski and Borisy, 1980b, with permission of the publisher.]*

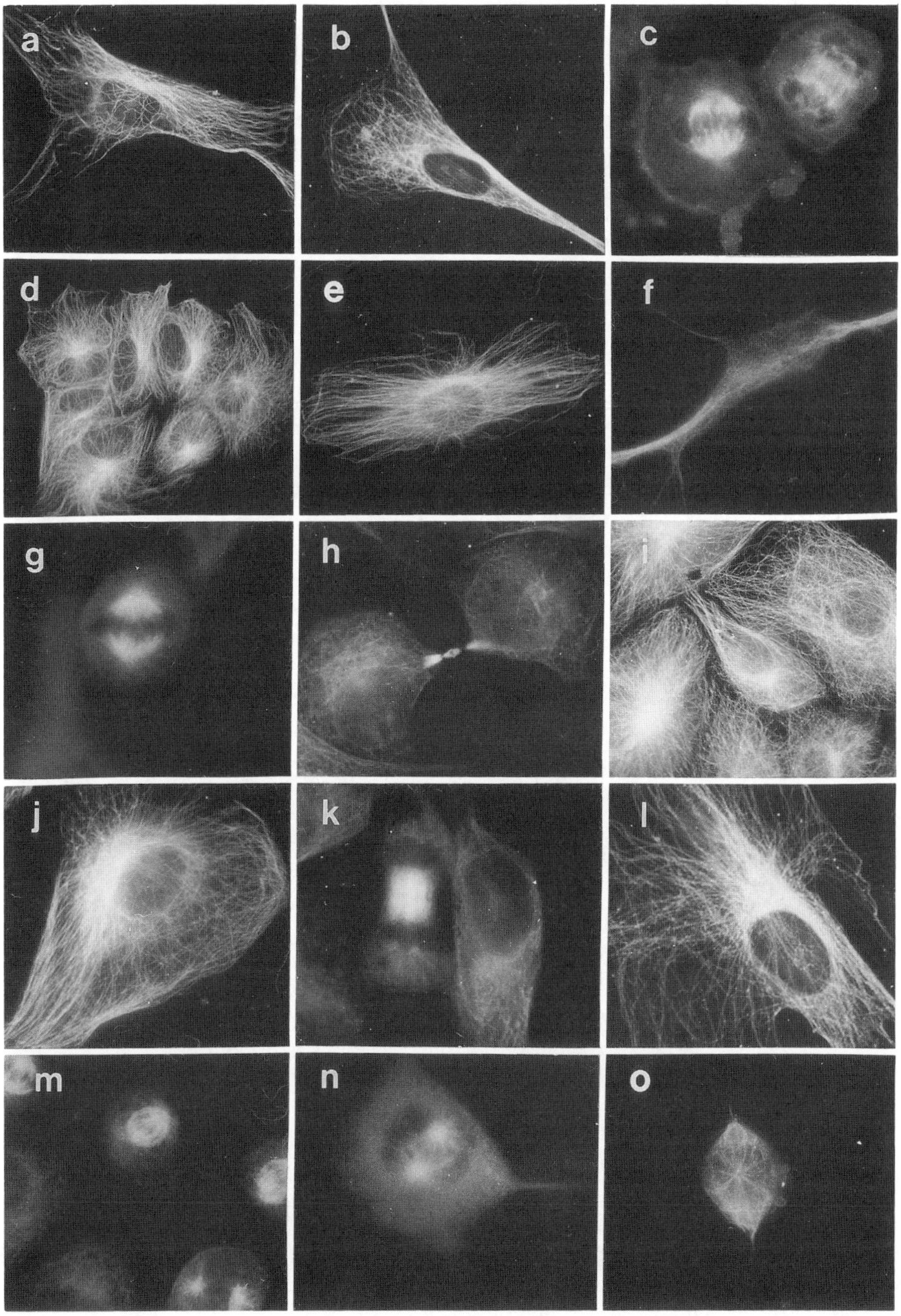
a
b
c
d
e
f
g
h
i
j
k
l
m
n
o

the 210 kD MAP (MAP4) was detected in all types of tissue (epithelial, connective, muscle, and nervous tissues); MAP4 immunofluorescence patterns appeared to be identical to those of tubulin in all cell types except mature sperm, in which no MAP4 was detected. Further insights into the subcellular distribution of MAP4 were gained from immunoelectron microscopy with the anti-210 kD MAP (MAP4) antisera [DeBrabander et al., 1981]. At this level of resolution MAP4 was detected along the length of apparently all cellular MTs, on all classes of MTs within the mitotic spindle, on MTs formed after mitosis or drug-induced polymerization, and on aberrant MT structures such as vinblastine-induced paracrystals and taxol-induced MT bundles.

Anti-HeLa 210 kD MAP (MAP4) antibodies were also used to immunostain gels; MAP4 was detected in every instance, except that in adult human brain the expression appeared to be diminished. Curiously, the antibodies prepared against the human MAP isolated from HeLa cells showed a marked species specificity; cross-reactivity with a species of ~210 kD was observed in material from any type of primate (human, New World and Old World monkeys, and prosimians) but showed no cross-reactivity with proteins from lower animals. Although one would expect that homologues of the MAP would exist in association with MTs in many organisms, the species specificity of antibodies that were produced significantly obstructed efforts to identify homologous species in other animals [Bulinski and Borisy, 1980b; Parysek et al., 1984].

Putative homologues of the 210 kD MAP (MAP4) were rapidly identified, though they were not necessarily immediately recognized as such. For example, Olmsted and Lyon [1981] identified a 215 kD protein as one of several species apparent on two-dimensional gels of material that copurified with MTs from mouse neuroblastoma cells. The relationship of this protein to the protein isolated from HeLa cells could not readily be ascertained; not only was antibody cross-reaction precluded by species differences, but the protein was reported to be differentiation specific in these neuronal cells, a feature that seemed to be inconsistent with the cell and tissue distribution of the human protein. Actually, the neuroblastoma MAP seems to be expressed in constant stoichiometry to assembled tubulin, which was shown to increase during the differentiation of mouse neuroblastoma [Olmsted, 1981]. Further studies on the mouse neuroblastoma MAP revealed that it is widely distributed among tissues and developmental stages [Parysek et al., 1984; Olmsted et al., 1986], and it shares certain biochemical properties with the HeLa protein. Because of evidence suggesting that the neuroblastoma MAP is the mouse homologue of the HeLa 210 kD MAP, Parysek et al. [1984] proposed the name *MAP4* for the neuroblastoma MAP and the entire family of similar MAPs.

The development of an MT purification procedure in which MT assembly is assisted by the MT-stabilizing drug taxol [Vallee, 1982], as well as the observation that the AP-MAPs exhibit thermostability [Vallee, 1985], allowed more efficient strategies of purifying MAPs from sources that were not especially enriched in tubulin. Members of the MAP4 family identified in this way include a doublet of 200–220 kD species from rat pituitary tissue and cultured cells [Bloom et al., 1985], a 190 kD MAP from bovine adrenal medulla (also called MAP-U) [Murofushi et al., 1986] and a 200 kD MAP from rat liver [Kotani et al., 1988]. The latter two species were shown to be related to HeLa MAP4 on the grounds of immunological cross-reactivity and the pattern of proteolytic fragments generated from each [Kotani et al., 1988].

MAPs of mass similar to MAP4 were also identified in *Drosophila* [Goldstein et al., 1986], in *Xenopus* [Gard and Kirschner, 1987], and in diatoms [Wordeman and Cande, 1987]. Of these, only the *Drosophila* 205 kD MAP has been studied in detail [Irminger-Finger et al., 1990]. Although the *Drosophila* 205 kD MAP shares some properties with AP-MAPs, protein sequence information demonstrates that it is unrelated to MAP4. As yet, the existence of MAP4 homologues in nonmammalian species has not been demonstrated.

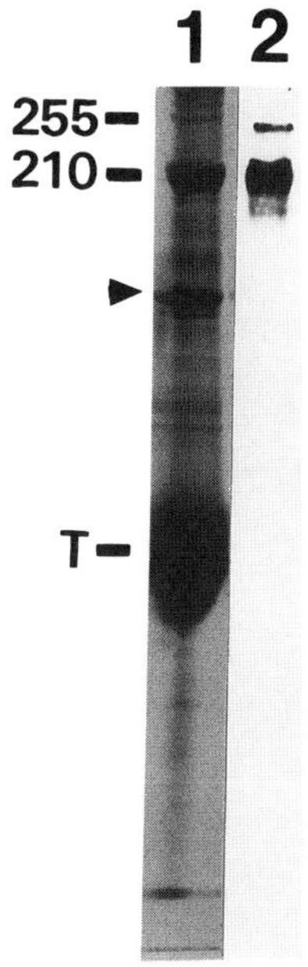

Fig. 2. *Western blot of anti-210 kD MAP antibody.* **Lane 1**, *Coomassie blue staining;* **lane 2,** *anti-210 kD MAP4 antibody staining of twice-cycled MT protein from HeLa cells purified, electrophoresed, and immunostained according to Chapin and Bulinski [1991]. 255, 210, and T mark the positions of the 255 kD species, the 210 kD MAP4 species, and tubulin, respectively, while the arrowhead marks the position of the 125 kD MAP. Note that antibodies prepared against the 210 kD MAP4 species are cross-reactive with the 255 kD species, as well as the 210 kD triplet of polypeptides to which antisera were prepared.*

Antibodies prepared against the HeLa 210 kD MAP exhibited an unexpected and informative cross-reactivity with another species, of approximately 255 kD, which is at least 10- to 50-fold less abundant than the 210 kD MAP4 species (Fig. 2). In addition to the antibody cross reaction, this protein seems to be a member of the MAP4 family because it shows biochemical behavior similar to that of the 210 kD MAP4. Also, its one-dimensional peptide maps are almost identical to the peptide maps of the 210 kD MAP4 (Fig. 3). To date, little is known about the origin of this MAP, which we call the *255 kD MAP4;* it is most likely translated from alternatively spliced mRNA transcribed from the MAP4 gene. To date, this MAP4 species has only been detected in primate material; presumably its low abundance and the species specificity of available antibodies have hampered detection of this minor MAP4 species in other organisms.

Reactivity of a monoclonal antibody prepared against rat brain MT preparations was used to identify a MAP whose reported electrophoretic behavior (M_r ~180 kD) was similar to that of MAP4 [Huber et al., 1985]. This MAP was named *MAP3;* the anti-MAP3 antibody was used to demonstrate the presence of MAP3 in glia at all stages of development, in axons of immature neuronal cells [Bernhardt et al., 1985], and in differentiating PC-12 cells [Brugg and Matus, 1988]. Later examination of nonneuronal tissues uncovered a much more widespread distribution of MAP3 throughout the rat [Huber and Matus, 1990] and suggested that MAP3 might actually be a MAP4 family member. In support of this hypothesis, the monoclonal antibody that defines MAP3 was shown to cross-react with all MAP4 family members identified in human (including 255 kD MAP), rat, and to a lesser extent mouse [Chapin et al., 1991]. Recombinant MAP4 fusion proteins demonstrated that anti-MAP3 was immunoreactive with a region near the N terminus of all MAP4 family members identified to date. The most logical interpretation of these findings, as well as the similar tissue distribution and biochemical properties of MAPs 3 and 4, is that MAP3 is synonymous with MAP4.

III. BIOCHEMICAL PROPERTIES OF MAP4

A. MAP4 Is an Assembly-Promoting MAP

Originally biochemical studies of MAP4 were performed with a HeLa MAP fraction that contained MAP4 as well as a 125 kD MAP [Bulinski and Borisy, 1980c]. In these studies, MAP4 was found to stimulate both the rate and extent of tubulin assembly, actions similar to those of the neuronal AP-MAPs, MAP2, and tau. These results suggested that MAP4 acts by a mechanism equivalent to that of other AP-MAPs, which promote MT nucleation and elongation by binding specifically to MT poly-

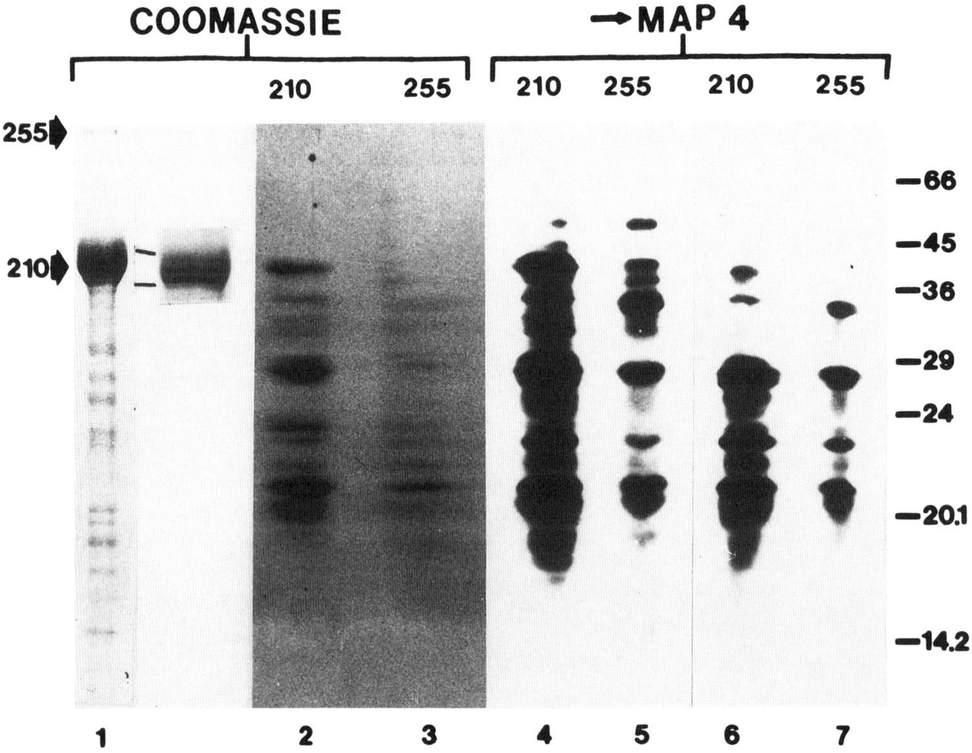

Fig. 3. *One-dimensional peptide maps of human 210 kD MAP (MAP4) and 255 kD MAP.* **Lane 1**, *heat-stable fraction of HeLa MAPs, electrophoresed on a 5% polyacrylamide gel.* **Inset** *at right shows a higher resolution electropherogram demonstrating heterogeneity of the MAP4 triplet. MAP4* **(lanes 2, 4, 6)** *and 255 kD MAP* **(lanes 3, 5, 7)** *bands were excised from a preparative gel and subjected to V8 proteolysis during electrophoresis on a 15% polyacrylamide gel, as described by Cleveland et al. [1977]. Two different amounts of V8 protease (0.3 μg, for lanes 2–5; 1.2 μg for lanes 6 and 7) were used in order to yield different efficiencies of cleavage. Electropherograms were stained with Coomassie brilliant blue (COOMASSIE) or electrophoretically transferred to nitrocellulose and immunostained with anti-210 kD MAP4 antibody (→MAP4), as described by Chapin and Bulinski [1991]. Electrophoretic positions of intact 255 kD MAP (255) and 210 kD MAP4 (210) are indicated at left.*

mer [Murphy et al., 1977]. In additional experiments, the ability of MAP4 to stimulate MT assembly was deduced from the decrement in the capacity of MAP fractions to augment MT assembly after MAP4 had been immunodepleted from the fractions [Huber et al., 1986]. Finally, the ability of MAP4 to stimulate MT assembly was tested in the most definitive manner, using both purified MAP4 [Murofushi et al., 1986; Aizawa et al., 1987] and recombinant MAP4 [Aizawa et al., 1991]. In some of these experiments, MAP4 was heat treated during its purification; therefore, in case heat treatment alters the properties of MAP4, the type of preparation used should be kept in mind when comparing the results of these biochemical experiments.

In every kind of preparation that has been examined, the binding of MAP4 to MTs is nucleotide insensitive, and MAP4 exhibits saturable binding. These properties are similar to those observed for other AP-MAPs; in fact, tau protein has been shown to compete with MAP4 for binding to MTs, suggesting that MAP4 and this neuronal MAP occupy similar or identical sites along the MT lattice [Aizawa et al.,

1987]. Although accurate calculations are complicated by uncertainties in the native molecular mass of MAP4 (see below) and by the inevitable proteolysis of the MAP4 used in MT-binding studies, the saturation level of MAP4 that can bind to MTs has been found to range between 1:2.3 and 1:20 (mol MAP4:mol tubulin) [Murofushi et al., 1986; Chapin and Bulinski, 1993]. These values are similar to the level (1:4.3) at which the neuronal MAP, MAP2, was reported to saturate its MT-binding sites [Kim et al., 1986].

B. MAP4 is a Fibrous Molecule Possessing at Least Two Functional Domains

To gain a better understanding of the function and nature of the interaction between MAP4 and MTs, the size and shape of the native MAP4 molecule were determined by classic biophysical measurements of sedimentation and diffusion coefficient [Bulinski and Borisy, 1980c]. These results demonstrated that MAP4 is a highly elongated molecule, and its native molecular weight, calculated from the sedimentation and diffusion data, suggested that the molecular weight of the native MAP4 molecule in solution is ~440 kD. Thus MAP4 was proposed to be a dimeric species, comprised of two ~220 kD species. Naturally, if MAP4 is a dimer, it might contain more than one MT-binding site per molecule. However, two concerns about the original study suggest that further studies are needed to determine the true molecular size of MAP4. First, the original measurements were made with a MAP mixture; although it appeared that the MAP4 present did not interact with other components in the mixture, this is still a formal possibility. Second, the molecular mass calculated from MAP4's deduced amino acid sequence is approximately 125 kD [e.g., West et al., 1991]. Therefore the native molecular weight attributed to MAP4 in the original report would require a trimeric complex of MAP4 polypeptides; this seems unlikely and would be unprecedented among cytoskeletal proteins. Probably the molecular characterization of MAP4 should be reinvestigated with pure MAP4 (of course, purified without heat treatment), a task that should be quite easily accomplished now that large amounts of recombinant MAP4 have been purified [Aizawa et al., 1991].

The fibrous structure of MAP4 has been corroborated by some of its other properties. First, its extreme sensitivity to proteases suggests an extended conformation. Second, MAP4 is thermostable [Murofushi et al., 1986]; the capability to renature after heat denaturation is the hallmark of fibrous proteins such as the actin-binding protein tropomyosin and several of the other AP-MAPs [Vallee, 1985]. In addition, direct visualization of MAP4-containing MTs in electron microscopy has revealed an elongated extension from the MT surface [Murofushi et al., 1986].

The structural data that suggested that MAP4 is a highly elongated protein akin to other AP-MAPs also suggested that it might consist of two distinct domains: one domain that binds to the MT wall and another that projects from the MT surface. This kind of structure was originally demonstrated for the high molecular mass AP-MAP, MAP2 [Vallee and Borisy, 1977]. For MAP4, the two domains could be proteolytically separated; a 27 kD MT-binding domain could be cleaved from the remainder of the molecule (the so-called projection domain) by mild proteolysis [Aizawa et al., 1987]. The former domain is capable of binding to MTs and stimulating their assembly, independent of the remainder of the molecule. The projection domain, on the other hand, is the extended portion of the molecule that has been visualized projecting from the MT wall [Murofushi et al., 1986; Aizawa et al., 1991]. Obviously, there are distinct similarities between MAP4 and the neuronal AP-MAPs with regard to domain structure, heat stability, and mode of MT binding.

C. MAP4 Is a Heterogeneous Protein

As discussed above, all MAP4 preparations studied to date consist of multiple species of M_r ~200,000. In studies of mouse MAP4, the three polypeptides MAP4A, -4B, and -4C were shown to be similar by one-dimensional proteolytic mapping [Olmsted et al., 1986]. In addition to the three closely spaced bands, the electrophoretic pattern of primate MAP4 also exhibits a species of M_r ~255,000 that is immunologically and proteolytically related to the 210 kD MAP4 forms (see Figs. 2 and 3). It is not clear how the heterogeneous species of MAP4 are derived. All of the neuronal AP-MAPs identified thus far are phosphoproteins; *in vitro,* the level of phosphorylation of MAPs has been shown to change their structure [Hagestedt et al., 1989] and their MT-stabilizing activity [Jameson and Caplow, 1981; Lindwall and Cole, 1984; Brugg and Matus, 1988]. At least some of the electrophoretic heterogeneity exhibited by the neuronal AP-MAP tau stems from disparate levels of phosphorylation *in vivo* [Kanai et al., 1989].

No direct evidence has yet shown that MAP4 exists as a phosphoprotein in interphase cells. However, MAP4 becomes phosphorylated at the onset of mitosis, such that MAP4 derived from mitotic cells is reactive with the mitosis-specific antiphosphoprotein antibody MPM-2 [Vandre et al., 1991]. During mitosis, MAP4 is a preferred substrate of a spindle-associated kinase [Tombes et al., 1991]. Since MAP4 is the predominant MAP in many types of dividing cells, its phosphorylation could conceivably contribute to the dramatic increase in dynamic behavior and change in organization of MTs that has been observed during mitosis.

D. Molecular Studies of MAP4

Recently, three different laboratories have begun to apply the tools of molecular biology to begin to decipher the function of nonneuronal MAP4. Aizawa et al. [1990, 1991], West et al. [1991], as well as studies in our laboratory [Chapin and Bulinski, 1991, 1992; Chapin et al., 1993] have identified the full-length coding sequences of MAP4 derived from human, cow, and mouse, respectively, and have begun to provide structural and functional information concerning this family of molecules.

Figure 4 shows a simple model for the structure of MAP4 that is suggested from its primary sequence. Although simplistic, this model shows several notable features of the MAP. First, the projection domain of MAP4 is located in the N-terminal two-thirds of the molecule. This region possesses a series of 13 amino acid repeats of unknown functional significance, including, in human MAP4, a tandem repeat of 26 amino acids that may represent a DNA sequence duplication [Chapin and Bulinski, 1991; West et al., 1991]. The projection domain has been identified through *in vitro* studies [Aizawa et al., 1991; Chapin and Bulinski, 1991]; not surprisingly, it has a net acidic charge. Among the three species examined thus far, the projection domain is not highly conserved, except at the extreme N terminus [West et al., 1991].

In contrast, the MT-binding domain of MAP4 has a net basic charge and is more highly conserved among species, especially within its repeat-bearing region, and at the extreme C terminus of the molecule. This C-terminal one-third of the MAP4 molecule is known from studies of recombinant MAP4 fragments to constitute its single MT-binding domain [Chapin

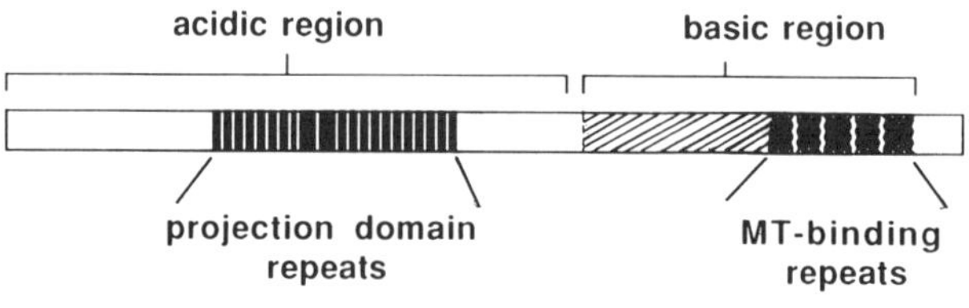

Fig. 4. *Primary structure of MAP4. MAP4 is depicted as a linear sequence, with an acidic region and a basic region outlined in brackets. Imperfect repeated segments are shown within the N-terminal projection domain and the C-terminal MT-binding domain; the cross-hatched area (▨) indicates the proline-rich region that is located within the MT-binding, basic region of the MAP's structure.*

and Bulinski, 1991; Aizawa et al., 1991]. It is striking that, in comparison to the neuronal MAPs MAP2 and tau, the MT-binding domain of MAP4 is situated at the same approximate place in the molecule, it contains repeat sequences almost identical in sequence (see below), and it carries the same charge. These similarities argue strongly that the mode of MT binding is conserved among these MAPs. In any case, there are some differences between MAP4 and the neuronal MAPs within the MT-binding domain. For example, the neuronal MAPs exhibit sequence homology in the regions flanking the MT-binding repeats, which MAP4 does not share. Furthermore, although all three MAPs possess a proline-rich basic region just N-terminal to the repeats, the proline-rich region of MAP4 (cross-hatched region in Fig. 4) is more extensive than that of the other two MAPs. Also, this proline-rich region in MAP4 is capable of binding to MTs in the absence of the repeat region, a capacity that the proline-rich regions of each of the neuronal MAPs lack. Finally, in the C-terminal one-third of the proline-rich region of MAP4, there are consensus sites for phosphorylation by several kinases, including protein kinases A and C [West et al., 1991]; a decreased positive charge in this region, brought about by phosphorylation, could possibly regulate the binding of MAP4 to MTs *in vivo*.

Isolation of MAP4-encoding cDNA clones resulted in the unexpected identification of four structurally distinct MAP4 species; although each clone encodes a MAP4 protein virtually identical in all regions outside of the MT-binding repeats (excepting species differences and minor discrepancies), the MT-binding repeats of each of the four differ. Figure 5 details the sequence of the repeat-bearing region of the most widely expressed MAP4 species, isoform IV. As shown, four of the five imperfect repeats and the spacer sequences intervening show sequence similarities when compared with the analogous repeat sequences of tau and/or MAP2. The second repeat, however, is markedly different. With its curiously long spacer sequence and its divergence from the consensus repeat sequence, one could easily discount the homology it does have with other imperfect repeats [West et al., 1991; Chapin and Bulinski, 1992]. However, the net positive charge of repeat two, its position between MT-binding repeats one and three, and its 10 homologous amino acid residues all indicate that repeat two is a functional MT-binding segment. The effect of this repeat on the efficacy of the MT-binding domain is not yet known.

Figure 5 documents the MT-binding domain structure of the four imperfect repeats and spacer sequences found within MAP4, and Figure 6 depicts the arrangement of each of the repeats and spacers into the four MAP4 isoforms that have been identified. Isoforms I and III were identified by the isolation of multiple cDNA clones from a human fetal brain library [Chapin and Bulinski, 1991], form II was cloned from a bovine kidney (MDBK) cell line [Aizawa et al., 1991], and form IV was cloned from both human and mouse cell lines and tissues [West et al., 1991].

Thus far, definitive evidence has been gathered documenting the *in vivo* expression of only two MAP4 forms, isoforms III and IV (see below). It is significant that isoform IV, whose MT-binding domain is quite unlike the MT-binding domains of the neuronal MAPs, is widely expressed in most proliferating cells and differentiated tissues [Chapin et al., 1993]. Isoform III, whose structure within the repeat region resembles that of a differentiation-specific isoform of tau, shows a more limited expression pattern in certain differentiated tissues (e.g., see Fig. 7).

The expression of MAP4 species with structurally distinct MT-binding domains may contribute to functional differences in binding of MAP4 to MTs. Given differences in MT-binding properties, one could hypothesize that isoforms of MAP4 differentially affect MT dynamics. It is possible that forms containing more MT-binding repeat sequences might bring about more potent MT-binding and stabilization; however, this tidy model might be collapsed once the MT-binding contribution of the second (pseudo) repeat is known. Because of

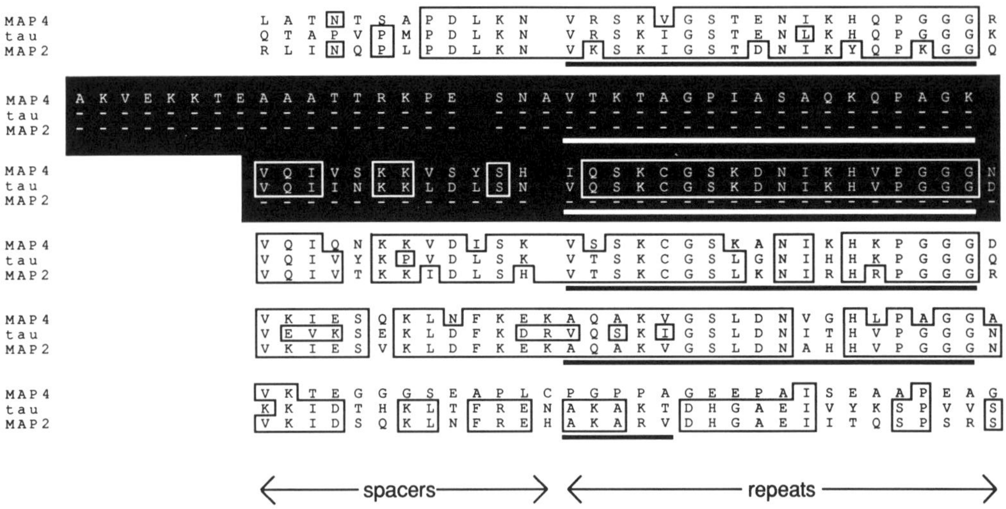

Fig. 5. *Amino acid sequence of the five imperfect repeats comprising part of the MAP4 MT-binding domain. Where applicable, each is aligned with repeat segments found within the AP-MAPs tau and MAP2; the repeat region is situated at the same approximate position of MAP4 and tau, as well, about 100 residues from their respective C termini. As illustrated, the MT-binding region is quite well conserved among these three proteins; this conserved region is characterized by a repeated 18 amino acid motif (bold underline), separated by less highly conserved spacer sequences of 13–14 amino acids.*

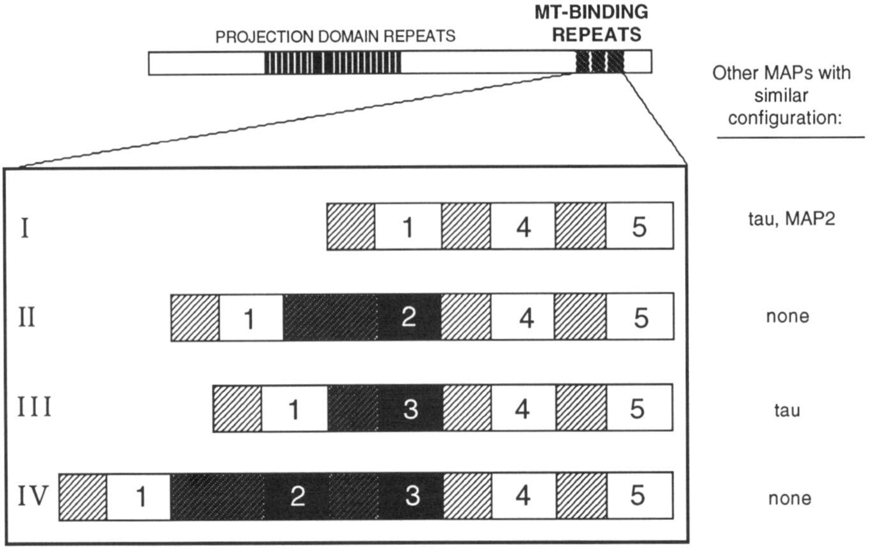

Fig. 6. *Schematic diagram of the repeat-bearing portion of MAP's four isoforms. Each of the four alternatively spliced isoforms of MAP4 that has been identified thus far contains three, four, or five imperfect repeats and spacers. Arrangement of the repeat-bearing motifs in each isoform is shown. Sequences of each of the five repeats is shown in Figure 5.*

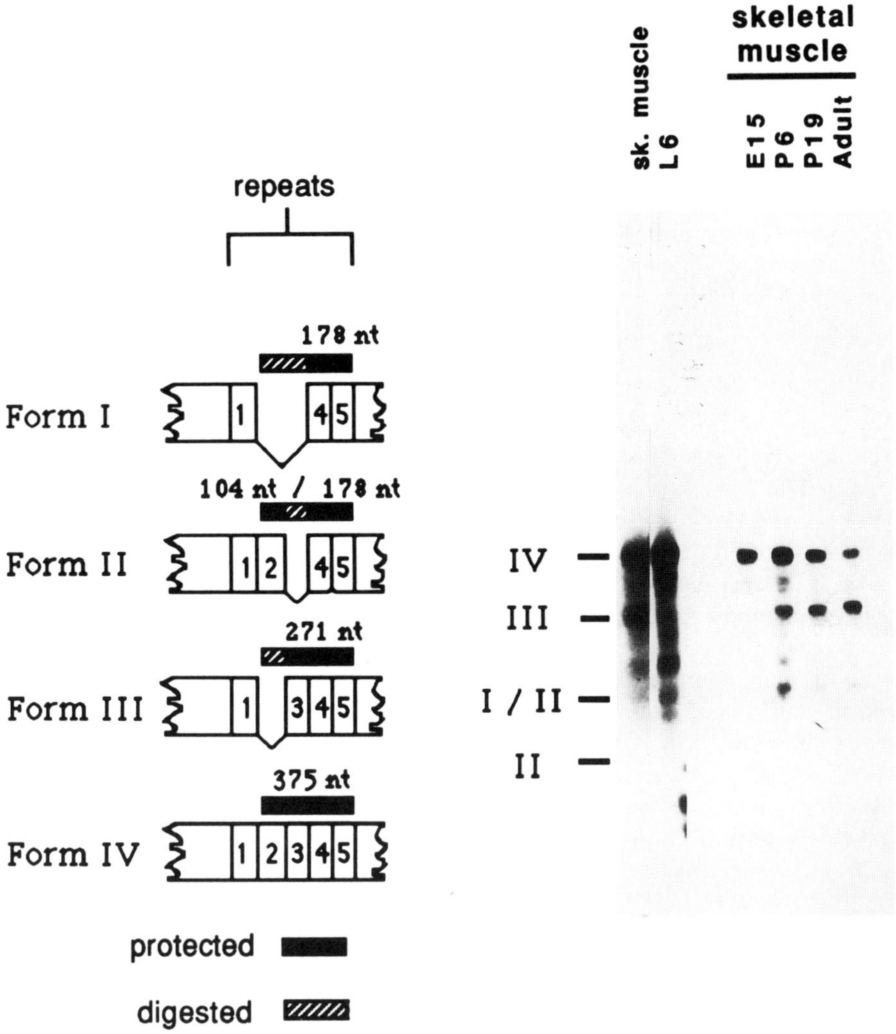

Fig. 7. *RNAse protection analysis of alternatively spliced MAP4 forms. A cloned PCR product from rat L6 cells, encoding the five-repeat form (isoform IV) of the MT-binding domain of MAP4, was used to make a radiolabeled RNA probe. This probe was used in RNAse protection assays with total RNA from rat cell lines and tissues, and the protected fragments were analyzed on nondenaturing gels. The schematic diagram at left indicates the protected fragments expected for each isoform, and the expected position of each fragment is shown on the electropherogram. The corresponding human antisense RNA probe used in assays with sense RNA transcribed from human cDNA clones encoding isoforms I, III, and IV yielded the expected fragments (data not shown). The right-hand portion shows electrophoretic bands obtained from RNA from undifferentiated L6 myoblasts (5 μg) and skeletal muscle (30 μg each) from adult rats and from various ages as indicated. Isoform IV predominates in myoblasts and in muscle at early stages of development, while isoform III appears and increases in abundance during later development.*

the relatively poor conservation of the second repeat sequence itself, as well as the unusual length of the spacer sequence with repeat two, effects of this repeat on overall MT-binding activity of MAP4 cannot be predicted. Butner and Kirschner [1991] tested the MT-binding capacity of *in vitro* translated peptides, each of which corresponded to the tau MT-binding repeats, except with spacer regions of varying length. Their results demonstrate that for the well-conserved tau repeat region the spacer length is not as important as diffuse charge interaction. However, the effect of MAP4's "pseudo" repeat two and its overlong spacer segment on overall MT-binding ability has yet to be tested adequately. In any case, the existence of this peculiar repeat within the ubiquitously expressed isoform of MAP4 is another of its unique features, which may functionally distinguish MAP4 from its neuron-specific counterparts.

E. Developmental Regulation of MAP4 Species

The observation that MAP4 is developmentally regulated further supports the notion that members of this MAP family are of functional importance in the morphogenetic events of development. The most obvious regulation observed is the decrease in the expression of MAP4 protein that accompanies neural development in rat brain [Bernhardt et al., 1985], in mouse brain [Olmsted et al., 1986], and in human neuroblastoma cells [Chapin and Bulinski, 1993]. In fact, in immunostained human neuroblastoma cells, individual MTs and MT-filled processes depleted in MAP4 arise during neurite outgrowth; this correlates temporally with the overall decrease in MAP4 expression. On the other hand, *increased* expression of MAP4 has been documented during mouse neuroblastoma [Olmsted and Lyon, 1981] and rat PC-12 cell [Brugg and Matus, 1988] differentiation. This apparent up- and downregulation is not really contradictory; Matus [1991] has classified MAP4 as a "juvenile" MAP form that is likely to function during more plastic stages of neuronal development, with decreased expression in more mature neurons. This assessment is consistent with the finding that, in the mature nervous system, MAP4 expression is pronounced only in nonneuronal cells [Parysek et al., 1985].

Expression of the heterogeneous species that make up the MAP4 triplet observable by polyacrylamide gel electrophoresis is also temporally regulated during brain development in rat [Riederer and Matus, 1985] and in mouse [Olmsted et al., 1986]. Expression of these MAP4 species is also regulated in different tissues of an adult organism, as demonstrated in rat tissues by Huber and Matus [1990]. The nature of this regulation, or even whether it is transcriptional, translational, or posttranslational, is not known, though.

MAP4 species with different MT-binding domains have been observed to undergo developmentally regulated expression. For example, most tissues and all types of proliferating cells examined thus far express MAP4 isoform IV. In muscle tissue of developing animals (Fig. 7) and also in heart muscle and brain [Chapin et al., 1993], MAP4 transcripts encoding isoform III are also expressed. It is likely that the other two isoforms of MAP4, whose expression has not yet been detected, are expressed in tissues and/or developmental stages not yet carefully scrutinized in these sensitive nuclease protection experiments.

Southern blot experiments [West et al., 1991] as well as *in situ* hybridization studies [Chapin et al., 1993] have demonstrated that MAP4 is transcribed from a single gene, which is localized, in humans, at the 3p21 chromosomal locus. Therefore, alternative splicing of mRNA from this gene must give rise to the heterogeneous forms of MAP4 that possess different MT binding domains. The gene encoding tau protein is also subject to alternative splicing that results in the expression of a tau form that contains four repeats [Himmer, 1989]. The tau four-repeat form, whose MT-binding domain virtually mimics that of isoform III of MAP4, has been shown, through *in vitro* assays, to be a more potent stabilizer of MTs than its three-repeat counterpart [Goedert and Jakes, 1990; Scott et al., 1991; Butner and Kirschner, 1991]. Expression of the four-repeat tau isoform in transfection studies [Lee and Rook, 1992] and in differentiating cells [Hanemaaijer and Ginzburg, 1991] have both provided *in vivo* evidence of its capacity to enhance MT stabilization greatly. Expression of different MAP4 forms may contribute analogously to regulation of MT dynamics in other cell types during development. For example, if MAP4 isoform III binds unusually tenaciously to MTs, it may contribute to the high degree of stability that is characteristic of MTs in differentiated brain and muscle tissue. It will be important in future studies to perform comparisons of the MT stabilization properties of MT-binding domains containing each of the structures that expressed forms of MAP4 are known to possess.

MAP4 regulation during development may involve another level of complexity that has not yet been investigated. Northern blot analysis of proliferating and differentiated cells and tissues showed a doublet of transcripts of ~5.5 and 6.5 kb, as well as one or two very large species (8–10 kb) [West et al., 1991, Chapin et al., 1993]. However, during muscle development, the large transcripts have been observed to increase in relative abundance, becoming as abundant as the similar transcripts in heart, and coming to predominante in skeletal muscle. In testis, West et al. [1991] have reported an even larger messenger RNA transcript (~11 kb), in addition to the 5.5–6.5 kb pair. The significance of these large mRNAs is not known. They may encode high molecular mass forms of MAP4, such as the 255 kD MAP4 species. A similarly large tau transcript encodes a large tau protein that is expressed in the nervous system [Georgieff et al., 1993]. However, the size of the transcript cannot necessarily be assumed to correlate with the size of the protein. Further studies of alternative splicing, alternative translational start sites, and mRNA processing may reveal other members of the MAP4 family whose expression is developmentally regulated.

F. *In Vivo* Function of MAP4

Because MAP4 is a nearly ubiquitous component of the MTs of proliferating and differentiating nonneuronal cells and is subject to developmental regulation, it is likely to be important in a number of MT functions. MAP4 could possibly be involved in mitosis, cell motility, organelle distribution or movements, or cellular morphogenetic events. Thus far, several clues to MAP4 functions exist. Since MAP4 is an AP-MAP capable of stabilizing MTs *in vitro* [Bulinski and Borisy, 1980c; Murofushi et al., 1986], its immunolocalization along MTs *in vivo* [Bulinski and Borisy, 1980a,b; Parysek et al., 1984; Kotani et al., 1987] implies that its normal function may be to stabilize all or a subset of the MTs in a particular cell. Indeed, since dynamic instability of MTs *in vitro* is abrogated by MAPs [Wilson et al., 1985], the dynamics of MTs *in vivo* would be expected to be dampened by the presence of MAP4 along the MTs. Tau protein, another AP-MAP, has been shown to stabilize MTs *in vivo* when microinjected [Drubin and Kirschner, 1986] or transfected [Kanai et al., 1989] into fibroblasts, and transfection of the MT-binding domain of tau is sufficient to effect the MT stabilization observed [Lee and Rook, 1992]. Studies in which the cellular amount of MAP4 is manipulated, using approaches similar to those used for tau, will doubtless establish whether MAP4 plays an important role in controlling microtubule dynamics *in vivo* and whether MAP4's MT-binding domain carries out this function independent of its projection domain.

In contrast to the MT-binding domain, there are currently no compelling hypotheses explaining the function of MAP4's projection domain. Two possibilities are that the projection domain exerts a regulatory effect on the MT-binding domain, or it participates in binding molecules or organelles, thus tethering them to MTs. MAP2's projection domain has been shown to contain a binding site for the regulatory subunit of type II cAMP-dependent protein kinase [Theurkauf and Vallee, 1982]; however, MAP4 contains no similar site, and it does not exhibit this binding interaction [Obar et al., 1989]. MAP4's sequence contains consensus sequences for phosphorylation by several protein kinases, most notably cdc2, protein kinase A, and protein kinase C [Chapin and Bulinski, 1991; West et al., 1991; Aizawa et al., 1991]. The projection domain of MAP4 could provide binding sites for these kinases or their regulatory subunits within the MT cytoskeleton. However, even if binding interactions of this type were measurable *in vitro*, one would need to establish their *in vivo* significance.

MAP4's projection domain could possess binding sites for elements of other filament systems or for organelles. In this scenario, MAP4's extended conformation might contribute to higher order interaction between MTs and other structures in the cytoplasm. Now that transfections and anti-sense knockouts can be performed in tissue culture cells, monitoring the behavior of cells overexpressing full-length MAP4 or its domains, or cells lacking MAP4, may provide hints of functions involving either MAP4's projection domain or the whole molecule.

Only when experimental evidence outlining the basic function(s) of MAP4 is in hand can we hope to understand the differential functions of phosphorylated MAP4 species or of the disparate members of the MAP4 family. Many important questions remain to be investigated and answered before the functional role of MAP4 and MTs can be understood in proliferating cells and during cell differentiation and transformation.

NOTE ADDED IN PROOF

Results obtained while this volume was in preparation have provided dramatic insights into MAP4 function, and have implicated MAP4 as a regulator of MT dynamics at the interphase-to-mitosis transition. In a manuscript entitled, "Cyclin B interaction with microtubule-associated protein 4 (MAP4) targets p34^{cdc2} kinase to microtubules and is a potential regulator of M-phase microtubule dynamics," which was submitted in mid-1993, Ookata, Hisanaga, Bulinski, Murofushi, Hotani, Okumura, Tachibana, and Kishimoto have demonstrated that MAP4 interacts specifically with the mitotic cyclin, cyclin B, both *in vitro* and *in vivo*. Since a complex of cyclin B and p34^{cdc2} kinase constitutes M-phase promoting factor, or MPF, and since MAP4 is capable of interacting not only with free cyclin B, but also with the MPF complex, the interaction of MAP4 with cyclin B serves to target MPF to interphase and mitotic MTs. Furthermore, MAP4 readily undergoes phosphorylation by p34^{cdc2} kinase with the MAP4/MPF complex, and this phosphorylation significantly alters MAP4's functional properties. Ookata et al. have shown that, although phosphorylation of MAP4 by mitotically active cyclin B/p34^{cdc2} complex does not abrogate MAP4's binding to MTs, the phosphorylated MAP4 is no longer capable of performing the MT-stabilizing function characteristic of unphosphorylated MAP4. Thus, recent work provides an optimistic closing note to this discussion of MAP4; one can now predict that regulatory functions of MAP4 during the interphase-to-mitosis transition and throughout mitosis will be elucidated in the very near future.

ACKNOWLEDGMENTS

This chapter was written during the tenure of grant BE-8 from the American Cancer Society. The author thanks Steve Chapin, Ron Liem, and Chia-Man Lue for helpful discussions.

IV. REFERENCES

Aizawa H, Emori Y, Mori A, Mirofushi H, Sakai H, Suzuki K (1991): Functional analysis of the domain structure of microtubule-associated protein 4 (MAPU). J Biol Chem 266:9841–9846.

Aizawa H, Emori Y, Mirofushi H, Kawasaki H, Sakai H, Suzuki K (1990): Molecular cloning of a ubiquitously distributed microtubule-associated protein with M_r 190,000. J Biol Chem 265:13849–13855.

Aizawa H, Mirofushi H, Kotani S, Hisanaga S, Hirokawa N, Sakai H (1987): Comparison of a major heat-stable microtubule-associated protein in HeLa cells and 190-kDa microtubule-associated protein in bovine adrenal cortex. J Biol Chem 262:3782–3787.

Bernhardt R, Huber G, Matus A (1985): Differences in the developmental patterns of three microtubule-associated proteins in the rat cerebellum. J Neurosci 5:977–991.

Bloom GS, Luca FC, Vallee RB (1985): Identification of high molecular weight microtubule-associated proteins in anterior pituitary tissue and cells using taxol-dependent purification combined with microtubule-associated protein specific antibodies. Biochemistry 24:4185–4191.

Bloom GS, Luca FC, Vallee RB (1984): Widespread cellular distribution of MAP1A (microtubule-associated protein 1A) in the mitotic spindle and on interphase microtubules. J Cell Biol 98:331–340.

Borisy GG, Marcum JM, Olmsted JB, Murohy DB, Johnson KA (1975) Purification of tubulin and associated high molecular weight proteins from porcine brain and characterization of microtubule assembly *in vitro*. Ann NY Acad Sci 253:107–132.

Brugg B, Matus A (1988): PC12 cells express juvenile microtubule-associated proteins during nerve growth factor-induced neurite outgrowth. J Cell Biol 107: 643–650.

Bulinski JC (1986): Purification and assay of primate 210k microtubule-associated protein. Methods Enzymol 134:147–156.

Bulinski JC, Borisy GG (1979): Self-assembly of HeLa tubulin and the identification of HeLa microtubule-associated proteins. Proc Natl Acad Sci USA 76:293–297.

Bulinski JC, Borisy GG (1980a): Immunofluorescence localization of HeLa cell MAPs on microtubules *in vitro* and *in vivo*. J Cell Biol 87:792–801.

Bulinski JC and Borisy GG (1980b): Widespread distribution of a 210,000 mol wt microtubule-associated protein in cells and tissue of primates. J Cell Biol 87:802–808.

Bulinski JC, Borisy GG (1980c): Microtubule-associated proteins from cultured HeLa cells: Analysis of molecular properties and effects on microtubule polymerization. J Biol Chem 255:11570–11576.

Brugg B, Matus A (1991): Phosphorylation determines the binding of microtubule-associated protein 2 (MAP2) to microtubules in living cells. 114:735–743.

Butner KA, Kirschner MW (1991): Tau protein binds to microtubules through a flexible array of distributed weak sites. J Cell Biol 115:717–730.

Chapin SJ, Bulinski JC (1991): Non-neuronal 210kD microtubule-associated protein (MAP4) contains a domain homologous to the microtubule-binding domains of neuronal MAP2 and tau. J Cell Sci 98:27–36.

Chapin SJ, Bulinski JC (1992): Microtubule stabilization by assembly promoting microtubule-associated proteins: A repeat performance. Cell Motil Cytoskel 4:27–36.

Chapin SJ, Bulinski JC (1993): Cellular microtubules heterogeneous in their content of MAP4 (210 kDMAP). Cell Motil Cytoskel (in press).

Chapin SJ, Bulinski JC, Matus A (1991): MAP3 and MAP4 are closely related members of a microtubule-associated protein family. J Cell Biol. 115:339a.

Chapin S, Lue C-M, Yu M-T, Bulinski JC (1993): Tissue-specific expression and developmental regulation of distinct forms of MAP4 generated by alternative RNA splicing. Manuscript submitted.

DeBrabander M, Bulinski JC, Geuens G, DeMey J, Borisy GG (1981): Ultrastructural immunocytochemical localization of the 210K microtubule-associated protein in cultured cells of primates. J Cell Biol 91:438–445.

Dentler WL, Granett S, Rosenbaum JL (1975): Ultrastructural localization of the high molecular weight proteins with *in vitro* assembled brain microtubules. J Cell Biol 65:237–241.

Drubin DG, Kirschner MW (1986): Tau protein function in living cells. J Cell Biol 103:2739–2746.

Gard DL, Kirschner MW (1987): A microtubule-associated protein from *Xenopus* eggs that specifically promotes assembly at the plus end. J Cell Biol 105:2203–2216.

Georgieff IS, Liem RKH, Mellado W, Nuñez J, Shelanski ML (1991): High molecular weight tau: Preferential localization in the peripheral nervous system. J Cell Sci 100:55–60.

Goedert M, Jakes R (1990): Expression of separate isoforms of tau protein: Correlation with the tau pattern in brain and effects on tubulin polymerization. EMBO J 9:4225–4230.

Goldstein LSB, Laymon RA, McIntosh JR (1986): A microtubule-associated protein in *Drosophila melanogaster:* Identification, characterization and isolation of coding sequences. J Cell Biol 102:2076–2087.

Hagestedt T, Lichtenberg B, Willie H, Mandelkow E-M, Mandelkow E (1989): Tau protein becomes long and stiff upon phosphorylation: Correlation between paracyrstalline structure and degree of phosphorylation. J Cell Biol 109:1643–1651.

Hanemaaijer R, Ginzburg I (1991): Involvement of mature tau isoforms in the stabilization of neurites in PC12 cells. J Neurosci Res 30:163–171.

Himmler A (1989): Structure of the bovine tau gene: Alternatively spliced transcripts generate a protein family. Mol Cell Biol 9:1389–1396.

Huber G, Matus A (1990): Microtubule-associated protein 3 (MAP3) expression in non-neuronal tissues. J Cell Sci 95:237–246.

Huber G, Alaimo-Beuret D, Matus A (1985): MAP3: Characterization of a novel microtubule-associated protein. J Cell Biol 100:496–507.

Huber G, Pehling G, Matus A (1986): The novel microtubule-associated protein MAP3 contributes to the *in vitro* assembly of brain microtubules. J Biol Chem 261:2270–2273.

Irminger-Finger I, Laymon RA, Goldstein LSB (1990): Analysis of the primary sequence and microtubule-binding region of the *Drosophila* 205K MAP. J Cell Biol 111:2653–2572.

Jameson L, Caplow M (1981): Modification of microtubule steady-state dynamics by phosphorylation of the microtubule associated proteins. Proc Natl Acad Sci USA 78:3413–3417.

Kanai Y, Takemura R, Oshima T, Mori H, Ihara Y, Yanangisawa M, Masaki T, Hirokawa N (1989): Expression of multiple isoforms and microtubule bundle formation in fibroblasts transfected with a single tau cDNA. J Cell Biol 109:1173–1184.

Kim H, Jensen CG, and Rehbun LI (1986): The binding of MAP-2 and tau on brain microtubules in vitro: Implications for microtubule structure. Ann NY Acad Sci 466:218–239.

Kotani S, Murofushi H, Aizawa H, Kaji K, Sakai H (1987): Identification of the 190 kD microtubule-associated proteins in cultured fibroblasts and its association with interphase and mitotic microtubules. Cell Struct Func 12:1–9.

Kotani S, Murofushi H, Maekawa S, Aizawa H, Sakai H (1988): Isolation of rat liver microtubule-associated proteins: Evidence for a family of microtubule-associated proteins with molecular mass of around 200,000 which distribute widely among mammalian cells. J Biol Chem 263:5385–5389.

Lee G, Rook SL (1992): Expression of tau protein in non-neuronal cells: Microtubule binding and stabilization. J Cell Sci 102:227–237.

Lindwall G, Cole D (1984): Phosphorylation affects the ability of tau protein to promote microtubule assembly. J Biol Chem 259:5301–5305.

Matus A (1991): Microtubule-associated proteins and neuronal morphogenesis. J Cell Sci Suppl 15:61–67.
Murphy DB, Borisy GG (1975): Association of high molecular weight proteins with microtubules and their role in microtubule assembly *in vitro*. Proc Natl Acad Sci (USA) 72:2696–2700.
Murphy DB, Vallee RB, Borisy GG (1977): Identity and polymerization-stimulatory activity of the nontubulin proteins associated with microtubules. Biochemistry 16:2598–2605.
Murofushi H, Kotani S, Aizawa H, Hisanaga S, Hirohawa N, Sakai H (1986): Purification and characterization of a 190-kD microtubule-associated protein from bovine adrenal cortex. J Cell Biol 103:1911–1919.
Obar RA, Dingus J, Bayley H, Vallee RB (1989): The RII subunit of cAMP-dependent protein kinase binds to a common amino-terminal domain in microtubule-associated proteins 2A, 2B, and 2C. Neuron 3:639–645.
Olmsted JB (1981): Tubulin pools in differentiating neuroblastoma cells. J Cell Biol 89:418–423.
Olmsted JB, Lyon HD (1981): A microtubule-associated protein specific to differentiated neuroblastoma cells. J Biol Chem 256:3507–3511.
Olmsted JB, Asnes CF, Parysek LM, Lyon HD, Kidder GM (1986): Distribution of MAP4 in cells and in adult and developing mouse tissues. Ann NY Acad Sci 466:292–305.
Parysek LM, Asnes CF, Olmsted JB (1984): MAP4: occurrence in mouse tissues. J Cell Biol 99:1309–1315.
Parysek LM, DelCerro M, Olmsted JB (1985): Microtubule-associated protein 4 antibody: A new marker for astroglia and oligodendroglia. Neuroscience 15:869–876.
Riederer B, Matus A (1985): Differential expression of distinct microtubule-associated proteins during brain development. Proc Natl Acad Sci USA 82:6006–6009.
Scott CW, Blowers DP, Barth PT, Lo MMS, Salama AI, Caputo CB (1991): Differences in the abilities of human tau isoforms to promote microtubule assembly. J Neurosci Res 30:154–162.
Shelanski ML, Gaskin F, Cantor CR (1973): Assembly of microtubules in the absence of added nucleotide. Proc Natl Acad Sci USA 70:765–768.
Theurkauf WE, Vallee RB (1982): Molecular characterization of the cAMP-dependent protein kinase bound to microtubule-associated protein 2. J Biol Chem 257:3284–3290.
Tombes RM, Peloquin JG, Borisy GG (1991): Specific association of an M-phase kinase with isolated mitotic spindles and identification of two of its substrates as MAP4 and MAP1B. Cell Regulation 2:861–874.
Vallee RB (1982): A taxol-dependent procedure for the isolation of microtubules and microtubule-associated proteins. J Cell Biol 92:435–442.
Vallee RB (1985): On the use of heat-stability as a criterion for the identification of microtubule-associated proteins. Biochem Biophys Res Commun 133:128–133.
Vallee RB, GG Borisy (1977): Removal of the projections from cytoplasmic microtubules *in vitro* by digestion with trypsin. J Biol Chem 252:377–382.
Vandre DD, Centonze VE, Peloquin J, Tombes RM, Borisy GG (1991): Proteins of the mammalian mitotic spindle: Phosphorylation/dephosphorylation of MAP4 during mitosis. J Cell Sci 98:577–588.
Weatherbee JA, RB Luftig, Weihing RR (1980) Purification and reconstitution of HeLa microtubules. Biochemistry 19:4116–4123.
Weatherbee JA, P Sherline, RN Mascardo, JG Izant, RB Luftig, RR Weihing (1982): Microtubule-associated proteins of HeLa cells: Heat stability of the 200,000 mol wt HeLa MAPs and detection of the presence of MAP2 in HeLa cell extracts and cycled microtubules. J Cell Biol 92:155–163.
Weingarten M, Lockwood A, Hwo S, Kirschner M (1975): A protein factor essential for microtubule assembly. Proc Natl Acad Sci USA 72:1858–1862.
West RR, Tenbarge KM, Olmsted JB (1991): A model for microtubule-associated protein 4 structure. J Biol Chem 266:21886–21896.
Wilson L, Miller HP, Farrell KW, Snyder KB, Thompson WC, Purich DL (1985): Taxol stabilization *in vitro:* Dynamics of tubulin addition and loss at opposite microtubule ends. Biochemistry 24:5254–5262.
Wordeman L, and Cande ZW (1987): Reactivation of spindle elongation *in vitro* is correlated with the phosphorylation of a 205 kD spindle associated protein. Cell 50:535–543.

ABOUT THE AUTHOR

JEANNETTE CHLOË BULINSKI is currently Associate Professor of Cell Biology, Anatomy, and Pathology at Columbia University, College of Physicians and Surgeons, in New York, where she teaches cell biology courses to both graduate students and medical students. After receiving a B.A. in chemistry from the University of Colorado, she did predoctoral work in the laboratory of Gary Borisy at the University of Wisconsin, studying microtubules in non-neuronal cells (in which she discovered MAP4, the topic of her chapter). After receiving her Ph.D. in molecular biology in 1980, Dr. Bulinski studied myogenic differentiation, in postdoctoral research with Steve Hauschka at the University of Washington. In 1982 she assumed a faculty position at the University of California, where she examined the role of actin isoforms and posttranslational modifications of tubulin during myogenesis and other differentiative processes. Currently Dr. Bulinski is continuing to investigate the function of the cytoskeleton during cell differentiation and the cell cycle. In 1985 she was named a NSF Presidential Young Investigator, and in 1987 she received a Career Development Award from UCLA. Presently Dr. Bulinski serves as Co-Director of the multi-disciplinary Graduate Program in Cell, Molecular, and Biophysical Studies at Columbia University.

Microtubules: 183–200

Tau Protein and Alzheimer's Disease

M. Goedert, R. Jakes, M.G. Spillantini, and R.A. Crowther

I. INTRODUCTION

Tau was one of the first microtubule-associated proteins to be identified. It was purified from pig brain by Weingarten et al. [1975], who also demonstrated its association with microtubules and showed that it promotes microtubule assembly *in vitro*. It was subsequently found that tau protein extracted from adult pig brain is heterogenous in that it runs as several bands of 55–65 kD apparent molecular weight on denaturing gels [Cleveland et al., 1977a,b]. By contrast, when extracted from immature rodent brain, it was found to run as one broad band of 55 kD [Mareck et al., 1980; Couchie and Nunez, 1985], indicating a developmental switch in its expression. Part of this heterogeneity was shown to result from phosphorylation, as alkaline phosphatase treatment simplifies the pattern of tau bands [Lindwall and Cole, 1984]. A further level of complexity was introduced with the finding that a protein of 110 kD apparent molecular weight from peripheral ganglia reacts with anti-tau antibodies [Drubin et al., 1985; Peng et al., 1986].

Within the past few years the molecular diversity of neuronal tau proteins has been explained, and several novel functional aspects have emerged. This has mainly resulted from the isolation of cDNA clones encoding various tau isoforms [Lee et al., 1988; Goedert et al., 1988, 1989a,b; Kosik et al., 1989; Himmler et al., 1989; Mori et al., 1989; Kanai et al., 1989; Goedert et al., 1992a; Couchie et al., 1992] and the use of the proteins expressed from these clones in functional experiments [Kanai et al., 1989; Goedert and Jakes, 1990; Knops et al., 1991; Lee and Rook, 1992; Gallo et al., 1992; Bramblett et al., 1993; Lo et al., 1993]. An additional impetus came from the direct demonstration that abnormally phosphorylated tau is a major component of the paired helical filament of Alzheimer's disease [Goedert et al., 1988; Wischik et al., 1988a,b; Kondo et al., 1988; Lee et al., 1991], one of the two defining neuropathological abnormalities of that disease. The demonstration that filaments like those found in Alzheimer's disease can be formed *in vitro* from recombinant tau protein fragments [Wille et al., 1992; Crowther et al., 1992] further underlines the central role that tau plays in the neurofibrillary pathology of Alzheimer's disease.

II. TAU ISOFORMS AND THEIR EXPRESSION

All known tau isoforms are produced from a single gene by alternative mRNA splicing. The human tau gene is located on the long arm of chromosome 17 at band position 17q21 [Neve et al., 1986]. The isolation of cDNA and genomic clones has led to the determination of tau sequences from mouse [Lee et al., 1988; Couchie et al., 1992], man [Goedert et al., 1988, 1989a,b; Andreadis et al., 1992], rat [Kosik et al., 1989; Mori et al., 1989; Kanai et al., 1989; Goedert et al., 1992a], and cow [Himmler et al., 1989]. The corresponding cDNA libraries were made from the nervous system, reflecting the predominant tissue source

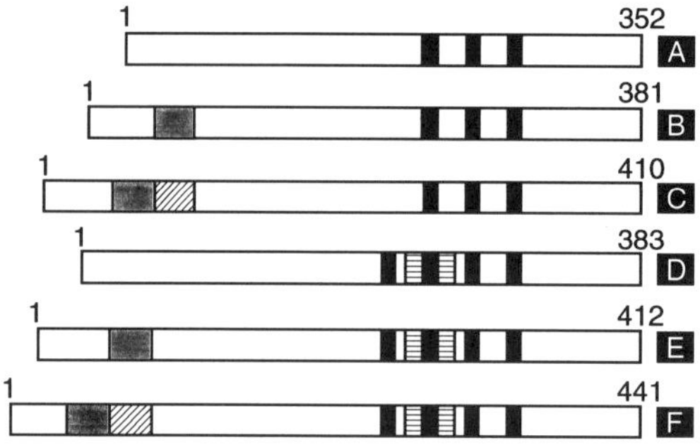

Fig. 1. *Schematic representation of six human brain tau isoforms. The three or four tandem repeats are marked by black bars, and the three inserts that distinguish the isoforms are shaded. Isoform A is expressed in fetal human brain, whereas all six isoforms are expressed in adult human brain.*

of tau. Tau protein has also been identified in chicken erythrocytes [Murphy et al., 1985], human monocytes [Kim et al., 1991], and mammalian testis [Ashman et al., 1992]. However, no sequences from these nonneuronal tissues have been reported.

A. The Central Nervous System

The six tau isoforms from human brain that have been isolated as full-length cDNA clones have sequences that range from 352 to 441 amino acids in length and differ from each other by the presence or absence of three inserts [Goedert et al., 1989b] (Fig. 1). The most striking feature of the tau sequences is the presence of three or four tandem repeats of 31 or 32 amino acids located in the C-terminal half, each containing a characteristic pro–gly–gly–gly motif. The extra repeat in the isoforms with four repeats is inserted within the first repeat of the isoforms with three repeats in a way that preserves the periodic pattern. Twelve residues are identical in the four repeats, and a further six residues are identical in three of the four repeats (Fig. 2). Experiments with synthetic peptides and with tau proteins produced by expression in *Escherichia coli* show that the repeats constitute microtubule-binding domains [Ennulat et al., 1989; Aizawa et al., 1989; Lee et al., 1989; Maccioni et al., 1989; Butner and Kirschner, 1991]. Microtubules assembled in the presence of tau show arms projecting from the surface [Hirokawa et al., 1988].

Tau thus consists of a C-terminal microtubule-binding domain and an N-terminal projection domain, with the latter determining the spacing between adjacent microtubules [Chen et al., 1992]. Microtubule polymerization is promoted by synthetic 18-mer peptides from the repeat region, although the molar concentration of peptide required is two orders of magnitude higher than that of full-length tau protein [Ennulat et al., 1989; Aizawa et al., 1989; Maccioni et al., 1989]. Recombinant tau isoforms with four repeats induce faster microtubule assembly than those with three repeats, as might be expected if one tau molecule links a number of tubulin subunits [Goedert and Jakes, 1990]. Similar microtubule-binding domains are found in the otherwise structurally unrelated high molecular weight microtubule-associated proteins MAP2 and MAP4 [Lewis et al., 1988; Aizawa et al., 1989]. Besides being distinguished by the presence of three or four tandem repeats, some tau isoforms contain 29 or 58 amino acid inserts located near

D	L	K	N	V	K	S	K	I	G	S	T	E	N	L	K	H	Q	P	G	G	G	K	V	Q	I	I	N	K	K	L
D	L	S	N	V	Q	S	K	C	G	S	K	D	N	I	K	H	V	P	G	G	G	S	V	Q	I	V	Y	K	P	V
D	L	S	K	V	T	S	K	C	G	S	L	G	N	I	H	H	K	P	G	G	G	Q	V	E	V	K	S	E	K	L
D	F	KDR		V	Q	S	K	I	G	S	L	D	N	I	T	H	V	P	G	G	G	N	K	K	I	E	T	H	K	L

Fig. 2. *Four tandem repeats in the human tau sequence. Twelve residues (boxed) are identical in the four repeats, while six residues (boxed) are identical in three out of the four repeats. These tandem repeats are located near the C-terminus and constitute the microtubule-binding domains of tau. Tau can exist with three or four tandem repeats.*

the N-terminus [Goedert et al., 1989b; Himmler et al., 1989; Kosik et al., 1989] (Fig. 1). These sequences are rich in proline and charged amino acid residues and share some similarity with the C-terminus of the middle neurofilament subunit. Their presence might well modulate the unknown function of the projection domain, the sequence of which shows a striking charge separation, with acidic residues grouped toward its N-terminus (residues 1–125) and basic residues toward its C-terminus (residues 126–250) (see Fig. 10B, below).

When individual human tau isoforms that have been expressed in *E. coli* are mixed together, they give a characteristic set of six bands, ranging from 48 to 67 kD in apparent molecular weight [Goedert and Jakes, 1990]. True molecular weights of the different isoforms range from 37 to 46 kD; therefore tau is anomalously retarded on denaturing gels, as has been clear since the sequencing of the first tau isoforms [Lee et al., 1988; Goedert et al., 1988]. Comparison of the pattern of recombinant tau isoforms with that of native tau shows that the four major tau isoforms in adult human brain correspond to isoforms with three and four repeats without N-terminal inserts and to isoforms with three and four repeats containing the first N-terminal insert of 29 amino acids (isoforms A, B, D, and E in Fig. 1). Isoforms with three and four repeats with the 58 amino acid N-terminal insert are also found (isoforms C and F in Fig. 1), albeit at lower levels [Goedert and Jakes, 1990]. These findings reflect the levels of mRNA transcripts encoding the various tau inserts [Goedert et al., 1989a,b]. Transcripts encoding isoforms with three and four repeats are found throughout the adult human brain, and their levels do not vary much between different regions; the same is true of transcripts encoding isoforms with the first N-terminal insert. Transcripts encoding isoforms with the 58 amino acid insert are found at only low levels. By contrast, in fetal brain, although transcripts encoding isoforms with three repeats are abundant, no transcripts encoding isoforms with four repeats, nor the N-terminal inserts, can be detected. The same is true of tau protein extracted from immature brain. The developmental shift of tau bands from a simple fetal pattern to a more complex adult pattern thus involves the transition from the expression of the isoform with three repeats and no inserts to the expression of all six isoforms. As with all microtubule-associated proteins, the fetal expression pattern of tau persists in tissues in which neuronal growth continues in the adult, such as retinal photoreceptors [Tucker and Matus, 1988] and the olfactory system [Viereck et al., 1989].

Transcripts encoding the various tau isoforms are expressed not only in a manner that is stage specific but also in a manner that is cell type specific. *In situ* hybridization has shown that tau protein mRNAs occur in nerve cells [Goedert et al., 1989a; Couchie et al., 1990; Takemura et al., 1991] (Fig. 3) and that different neuronal cell types express different isoforms [Goedert et al., 1989a]. Transcripts encoding isoforms with three or four repeats are found in pyramidal cell bodies throughout all layers of the adult human cerebral cortex. In the hippocampal formation transcripts encoding isoforms with three repeats are present in granule cells throughout the cornu ammonis (CA) layers and pyramidal cells in the subiculum. Transcripts encoding isoforms with four repeats are also found in pyramidal cells in the subiculum and hippocampus, but not in granule cells of the dentate gyrus.

No information is at present available about the cellular distribution of the various tau protein isoforms. Antibodies that recognize all tau isoforms label axons and only a small number of neuronal cell bodies and dendrites in adult brain from a variety of species [Binder et al., 1985; Brion et al., 1988; Migheli et al., 1988; Trojanowski et al., 1989; Shin et al., 1991; Riederer and Innocenti, 1991]. By contrast, in juvenile cat brain, a larger number of cell bodies and dendrites is labeled [Riederer and Innocenti, 1991]. Immunohistochemistry using anti-tau antibodies has also shown staining of a small number of glial cells, suggesting that tau expression may not be entirely limited to nerve cells [Migheli et al., 1988; Shin et al., 1991].

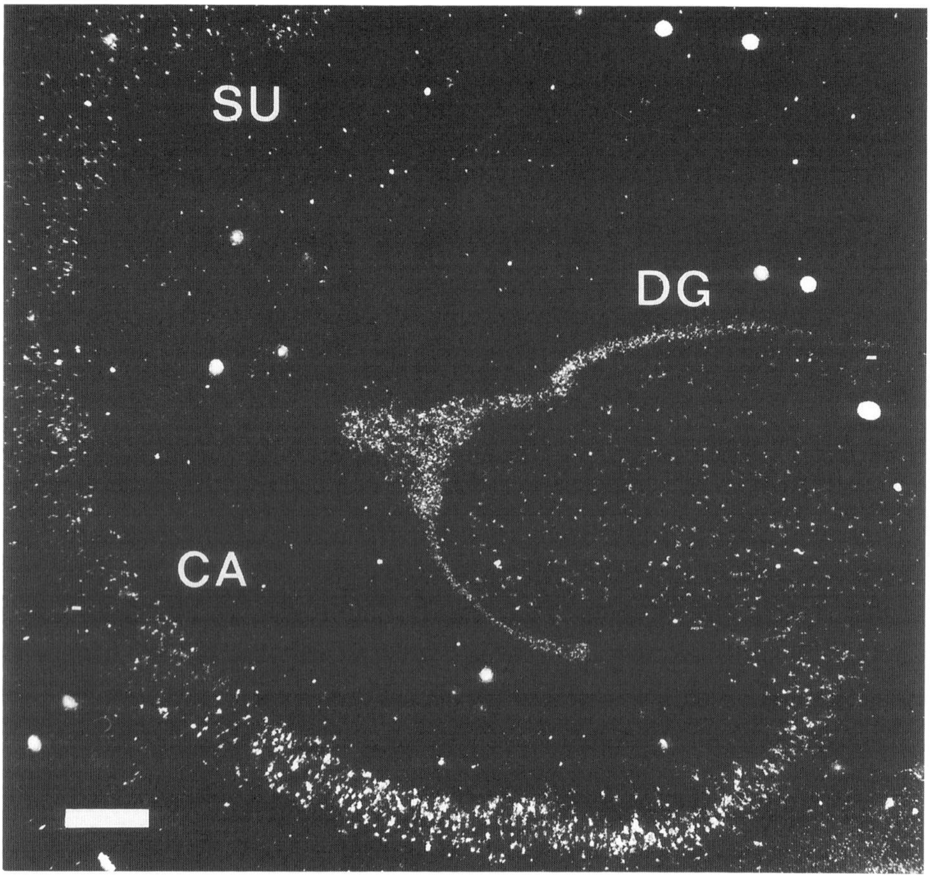

Fig. 3. *Cellular localization of tau protein mRNAs in adult human hippocampal formation. A probe specific for transcripts encoding isoforms with three repeats labels numerous nerve cell bodies. CA, cornu ammonis; DG, dentate gyrus; SU, subiculum. Bar = 350 μm. [Reproduced from Goedert et al., 1989a, with permission of the publisher.]*

The human cDNA clones isolated to date account for most, if not all, tau isoforms present in adult human brain. The screening of mouse and cow cDNA and genomic libraries has suggested the existence of other forms of tau [Lee et al., 1988; Himmler, 1989; Himmler et al., 1989]. These include bovine isoforms with an alternative N-terminus or an alternative extended C-terminus and a mouse isoform with an extension of 25 amino acids at the C-terminus. The status of these reported clones is unclear, since the existence of the corresponding transcripts or proteins has not been documented and since the complexity of tau isolated from bovine or mouse brain is no greater than that of tau from human brain.

B. The Peripheral Nervous System

As mentioned above, the peripheral nervous system and some cultured cell lines appear to express a larger form of tau [Drubin et al., 1985; Peng et al., 1986]. One such "big tau" isoform, which was isolated as a full-length cDNA clone from a rat PC12 pheochromocytoma cDNA library, is 686 amino acids in length [Goedert et al., 1992a]. It contains a sequence identical to the longest brain tau isoform but with an additional 254 amino acid

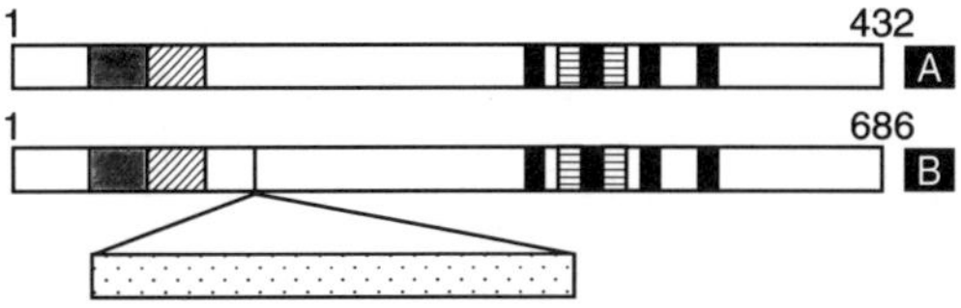

Fig. 4. *Schematic representation of the largest rat brain tau isoform* **(upper)** *and the peripheral nervous system-specific rat big tau isoform* **(lower)**. *The brain isoform contains four microtubule-binding repeats and the 58 amino acid N-terminal insert. Big tau contains an additional 254 amino acid sequence (*⬚*), which is inserted as shown.*

insert in the N-terminal half (Fig. 4). This insert, which is encoded by a single exon, shows no significant homology with known proteins. Big tau identified by this clone is probably identical with the 110 kD tau-like protein previously reported in peripheral ganglia and in cell lines with peripheral neuron-like characteristics [Drubin et al., 1985, 1986, 1988; Peng et al., 1986; Oblinger et al., 1991; Georgieff et al., 1991]. It should bind to microtubules with the same affinity as other isoforms with four repeats. However, it provides a projection domain double the size of that found in brain tau. A clone for big tau has also been isolated from a mouse neuroblastoma cDNA library [Couchie et al., 1992]. It encodes a 733 amino acid protein that contains a 237 amino acid insert that is highly homologous to the rat insert. In addition, it contains a 66 amino acid sequence homologous to a sequence previously identified in some bovine brain tau clones [Himmler et al., 1989]. At present, the significance of the latter sequence is unclear, as corresponding transcripts have not been identified. The sequence of a human big tau insert of 253 amino acids has been derived from the sequencing of genomic clones [Andreadis et al., 1992]. Unlike the remainder of the tau-coding region, the big tau insert is only 50% homologous between man, mouse, and rat [Goedert et al., 1992; Couchie et al., 1992; Andreadis et al., 1992], suggesting that it may be required to add length to the tau sequence. Further studies are needed to provide a detailed description of the tau isoform composition in sensory and sympathetic ganglia.

Big tau is encoded by an 8 kb mRNA that is derived from the same gene as the set of 6 kb mRNAs that encodes the smaller tau isoforms [Goedert et al., 1992a; Couchie et al., 1992]. The 8 kb big tau mRNA is found in sensory and sympathetic ganglia in newborn and adult rats, but not in brain or spinal cord of either newborn or adult rats. By contrast, the 6 kb tau mRNA is found in brain and spinal cord in newborn and adults rats. However, the 6 kb mRNA is also present in sensory and sympathetic ganglia in newborn but not adult rats. It follows that in developing sensory and sympathetic ganglia the smaller tau isoforms are expressed in conjunction with big tau isoform(s), whereas in adult ganglia only big tau isoform(s) are present. This adds a further level of complexity to the developmental expression of tau isoforms in the nervous system. In sensory and sympathetic ganglia, big tau mRNA is localized in nerve cell bodies (Fig. 5), as is tau mRNA in the central nervous system.

Big tau is thus an isoform that is specifically expressed in the peripheral nervous system, implying the existence of alternative splicing between peripheral and central nervous systems. The topographical division into peripheral and central nervous systems is thus paralleled by molecular differences that may result in a distinct organization of the cytoskeleton between the two major divisions of the vertebrate nervous system.

III. TAU AND ITS FUNCTIONS

Ever since its discovery, tau has been known to be a potent promoter of tubulin polymerization *in vitro* [Weingarten et al., 1975; Cleveland et al., 1977a,b]. Consequently, tau proteins are frequently ascribed a role in tubulin polymerization during neurite formation. In PC12 cells the levels of tau mRNA and protein increase during neurite outgrowth, and this increase correlates with an increase in the levels of polymerized tubulin [Drubin et al., 1985, 1988]. Moreover, in experiments in which tau

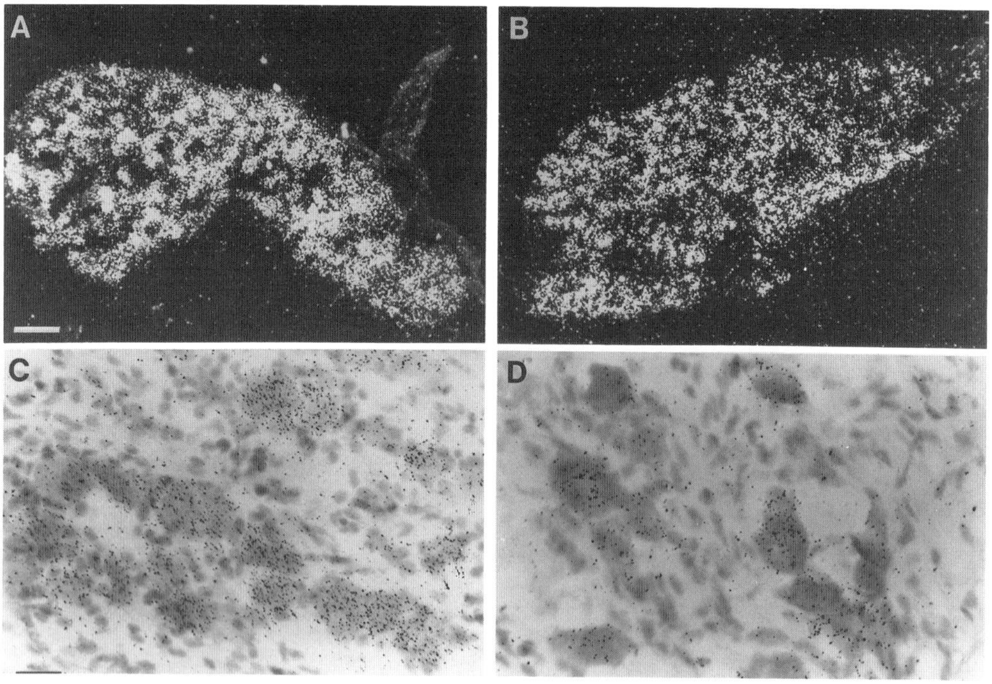

Fig. 5. *Cellular localization of big tau mRNA in adult rat sensory and sympathetic ganglia. A probe specific for the sequence encoding the big tau insert labels numerous nerve cell bodies.* **A,B:** *Dark-field photomicrographs of dorsal root ganglion (A) and superior cervical ganglion (B) sections.* **C,D:** *Bright-field photomicrographs of dorsal root ganglion (C) and trigeminal ganglion (D) sections. Bars = 105 μm for A and B, 25 μm for C and D. [Reproduced from Goedert et al., 1992a, with permission of the publisher.]*

was injected into a fibroblast cell line, an increase in microtubule mass was observed [Drubin et al., 1986].

Microinjection of tau into fibroblasts also leads to an increased resistance of microtubules to depolymerizing agents, implying a role for tau in the stabilization of microtubules once they have formed [Drubin et al., 1986]. Similar conclusions were reached in studies on developing cerebellar nerve cells and adult olfactory cells, in which the appearance of large amounts of tau protein was found to correlate with the appearance of a stable cytoskeleton [Ferreira et al., 1989; Viereck et al., 1989]. The binding of tau to microtubules reduces their dynamic instability. Tau has been shown to reduce the frequency of transitions between growing and shrinking phases of microtubules nucleated by centrosomes *in vitro* [Bré and Karsenti, 1990]. By direct observation of single microtubules tau was shown to increase the rate of microtubule polymerization, to inhibit the rate of depolymerization, and to inhibit the rate of transition into the catastrophic shrinking phase [Drechsel et al., 1992]. Adult tau isoforms with four repeats are better at promoting microtubule assembly and at binding to microtubules than fetal isoforms with three repeats [Goedert and Jakes, 1990; Butner and Kirschner, 1991]; this is consistent with the need for a more dynamic cytoskeleton during development.

Several laboratories have transfected various tau constructs into mammalian cells that do not normally express tau [Kanai et al., 1989; Knops et al., 1991; Lee and Rook, 1992; Gallo et al., 1992; Bramblett et al., 1993; Lo et al., 1993]. In transfected cells tau bound to microtubules, enhanced microtubule stability and, in some cases, induced microtubule bundling. The thickness of the bundles was 10–100 times that of normal microtubules, with an average distance of about 20 nm between adjacent microtubules [Kanai et al., 1989]. Bundling is probably a consequence of the microtubule stabilization effected by tau [Lee and Rook, 1992]. No bundles are observed in the absence of the microtubule-binding repeats. Deletion of the proline-rich region upstream of the repeats also prevents bundle formation, suggesting an important role for this region in microtubule binding and bundling *in vivo*. In addition, the length of the N-terminal region and the presence of the first half of the C-terminal tail contribute to bundling efficiency [Kanai et al., 1992]. It is at present unclear whether microtubule bundling has a physiological correlate or whether it results from the nonphysiological overexpression of tau in some cell types. In Sf9 insect cells, the overexpression of tau resulted not only in microtubule bundles but also in the extension of long and thin neurite-like processes [Knops et al., 1991; Baas et al., 1991; Chen et al., 1992]. When stably transfected with cDNAs encoding each of the six human brain tau isoforms 3T3 cells developed distinct morphologies in the absence of microtubule bundles [Lo et al., 1993]. Thus, cell lines expressing three-repeat tau isoforms developed large flat cell bodies, while cells expressing four-repeat isoforms had small and round cell bodies. All lines, except those expressing the shortest tau isoform, displayed very long, thin, neurite-like processes. Conversely, the introduction of antisense oligonucleotides into developing cerebellar nerve cells to block tau expression prevented the differentiation of the short neurites, one of which would otherwise have developed into an axon [Caceres and Kosik, 1990; Caceres et al., 1991]. Similarly, tau antisense oligonucleotides produced a retraction of neurites in PC12 cells treated with nerve growth factor [Hanemaaijer and Ginzburg, 1991]. These results suggest that axonal morphology may be dependent on tau expression, although it is unclear whether tau plays a crucial role in the establishment of neuronal polarity.

In vitro reconstruction studies using tau and tubulin have indicated that tau molecules can form cross-bridges of about 18 nm length between adjacent microtubules [Hirokawa et al., 1988]. Similar cross-bridges are observed *in vivo* in Sf9 insect cells overexpressing tau, as well as in axons from cerebellar Purkinje cells [Chen et al., 1992]. When viewed by electron microscopy individual tau molecules are rod-like with a mean length of 35 nm and a width of 3–5 nm [Wille et al., 1992]. Surprisingly, the microtubule-binding domain alone measures 20–25 nm; thus the N- and C-termini extend the total length by only 10–15 nm. Available structural data suggest that tau is an extended molecule without much secondary structure. Binding studies indicate that the relatively weak affinity of the repeat region of tau for microtubules is concentrated in three or four 18 amino acid binding elements, which are separated by flexible linker sequences [Butner and Kirschner, 1991]. The binding energy appears to derive from a series of weak interactions contributed by small groups of amino acids, suggesting that tau might adopt multiple conformations and possibly migrate on the surface of the microtubule with important consequences for its function.

IV. TAU PHOSPHORYLATION

Phosphorylation is the only known posttranslational modification of tau [Cleveland et al., 1977a,b; Butler and Shelanski, 1986]. It is of functional importance, since phosphorylated tau is less effective than nonphosphorylated tau at promoting microtubule assembly [Yamamoto et al., 1983; Lindwall and Cole, 1984]. Besides the alternative mRNA splicing that results in the expression of different tau isoforms in immature and adult nerve cells, phosphorylation thus represents a second mechanism that modulates the affinity of tau for microtubules. Only little is known about the residues in tau that are phosphorylated *in vivo* or about the relevant protein kinases and phosphatases.

Tau protein can be phosphorylated *in vitro* by cAMP-dependent protein kinase [Pierre and

Nuñez, 1983], multifunctional Ca^{2+}–calmodulin-dependent protein kinase II (CaM kinase) [Yamamoto et al., 1983], protein kinase C [Hoshi et al., 1987; Baudier et al., 1987], casein kinase I [Pierre and Nuñez, 1983], casein kinase II [Steiner et al., 1990], v-fms kinase [Kim et al., 1991], proline-directed serine/threonine protein kinase [Vulliet et al., 1992], cdc2 kinase [Drewes et al., 1992; Ledesma et al., 1992; Mawal-Dewan et al., 1992], glycogen synthase kinase-3 [Yang et al., 1991; Hanger et al., 1992; Mandelkow et al., 1992], and mitogen-activated protein kinase [Drewes et al., 1992; Ledesma et al., 1992; Goedert et al., 1992c]. CaM kinase phosphorylates a single serine residue that is located 26 amino acids from the C-terminus (serine 416, using the nomenclature of the longest human brain tau isoform; see Fig. 10, below) [Steiner et al., 1990]. Protein kinase C phosphorylates serine 324 and probably serines 262, 293, and 356 [Correas et al., 1992]; these serines are located in the microtubule-binding domain, accounting for the reduced ability of tau to bind to microtubules when phosphorylated by protein kinase C [Hoshi et al., 1987; Correas et al., 1992]. Mitogen-activated protein kinase phosphorylates most of the 17 Ser–Pro and Thr–Pro sites (see Fig. 10A, below) [Drewes et al., 1992; Gustke et al., 1992; Goedert et al., 1992c], and glycogen synthase kinase-3 phosphorylates some Ser–Pro sites [Hanger et al., 1992; Mandelkow et al., 1992]. Two protein kinases (called tau protein kinase 1 and tau protein kinase 2) were purified from brain by virtue of their ability to phosphorylate tau proteins [Ishiguro et al., 1991, 1992]; like mitogen-activated protein kinase, they phosphorylate some Ser–Pro and Thr–Pro sites. Proline-directed protein kinase phosphorylates serines 202, 205 and 235 and threonine 231; all these residues are followed by a proline [Vulliet et al., 1992]. Two protein kinases (PK36 and PK40) were isolated from bovine brain on the basis of their ability to phosphorylate intermediate and heavy neurofilament subunits [Roder and Ingram, 1991]. These kinases also phosphorylate tau at Ser–Pro and Thr–Pro sites. Protein phosphatases 2A, 2B, and 2C have been shown to dephosphorylate tau phosphorylated by cAMP-dependent protein kinase and CaM kinase [Yamamoto et al., 1985, 1990], while tau protein phosphorylated by the 42 kD isoform of mitogen-activated protein kinase (also known as *ERK2)* is selectively dephosphorylated by protein phosphatase 2A [Goedert et al., 1992c]. At present, the significance of these findings *in vitro* for the phosphorylation and dephosphorylation *in vivo* of tau is unclear.

Recently, antibodies were produced that recognize only phosphorylated tau [Lee et al., 1991; Greenberg et al., 1992; Biernat et al., 1992; Mercken et al., 1992]; they have been used to uncover the developmental regulation of tau phosphorylation in brain [Kanemaru et al., 1992; Bramblett et al., 1993; Goedert et al., 1993]. Antiserum T3P recognizes Ser 396 (see Fig. 10A, below) in a phosphorylated state, and antibody AT8 recognizes Ser 202 (see Fig. 10A, below) in a phosphorylated state. Neither antibody recognizes adult tau, but they both recognize fetal human brain tau (Fig. 6). Tau is partially phosphorylated at these sites until birth, at which time it is abruptly dephosphorylated, indicating a tight developmental regulation of tau phosphorylation. Dephosphorylation occurs before the isoform switch, suggesting a dual regulation of the affinity of tau for microtubules during development. Serines 202 and 396 are never again phosphorylated in normal human brain; however, their phosphorylation is one of the characteristics of the tau that is present in the paired helical filaments of Alzheimer's disease. Other antibodies recognize only dephosphorylated tau. Thus antibodies tau 1 and E-2 see adult tau but fail to recognize tau from paired helical filaments, indicating phosphorylation at or near their epitopes in Alzheimer's disease [Grundke-Iqbal et al., 1986; Wood et al., 1986; Dickson et al., 1992].

V. TAU IN ALZHEIMER'S DISEASE

Alzheimer's disease is characterized neuropathologically by the presence of abundant amyloid plaques and neurofibrillary changes in

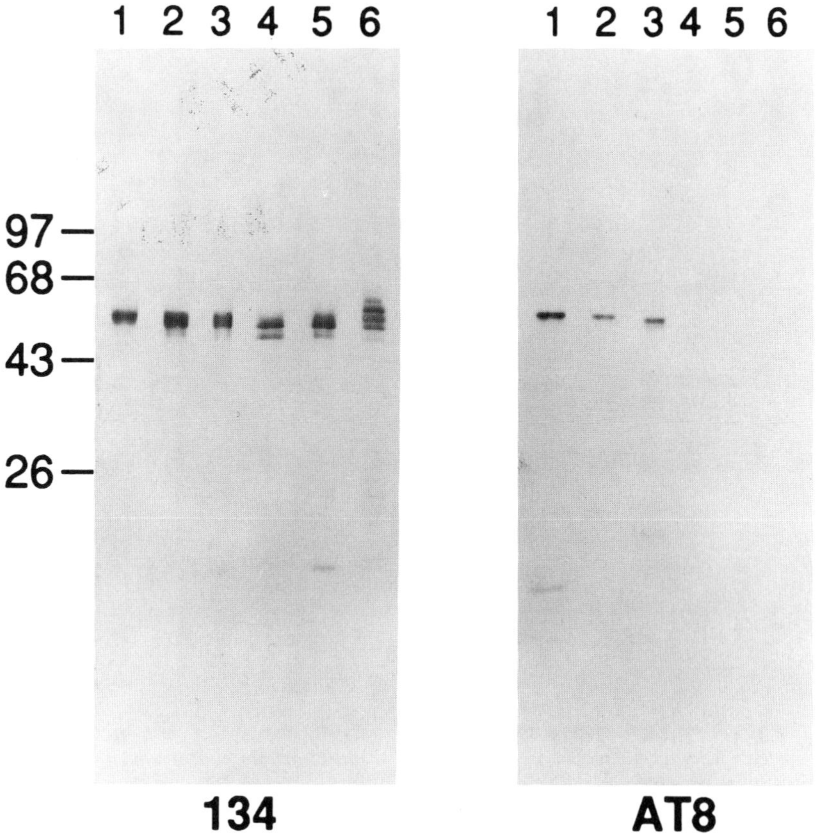

Fig. 6. *Developmental regulation of tau protein phosphorylation in human brain. Immunoblots of tau extracted from human brain at various stages using antiserum 134 and the phosphorylation-dependent antibody AT8; 134 recognizes all tau isoforms irrespective of phosphorylation, whereas AT8 requires serine 202 to be phosphorylated.* **Lanes 1,** *20 weeks gestation;* **2,** *23 weeks gestation;* **3,** *39 weeks gestation;* **4,** *postnatal day 2;* **5,** *postnatal day 5;* **6,** *adult. [Reproduced from Goedert et al., 1993, with permission of the publisher.]*

cerebral cortex and hippocampus, as well as in some subcortical nuclei. Neurofibrillary changes are found in nerve cell bodies as neurofibrillary tangles, in dendrites as neuropil threads, and in the abnormal neurites associated with senile plaques. The presence of neurofibrillary changes correlates with the presence of dementia [McKee et al., 1991; Arriagada et al., 1992] and their topographical distribution can be used to stage the progression of the disease [Braak and Braak, 1991].

The principal fibrous component of all three neurofibrillary lesions is a structure not normally seen in the cytoskeleton, known as the paired helical filament (PHF) [Kidd, 1963] (Fig. 7). The PHF consists of two strands of subunits that twist around one another in a helical fashion. When viewed in the electron microscope the helical twist and relative disposition of the two strands give rise to images in which the width alternates between 8 and 20 nm, with an apparent period of about 80 nm [Crowther and Wischik, 1985]. Pronase treatment of tangle preparations removes a fuzzy coat and leaves behind a pronase-resistant core [Wischik et al., 1988b]. From the electron micrographs it is possible to compute a map of the cross-sectional density in the core. Such maps show two

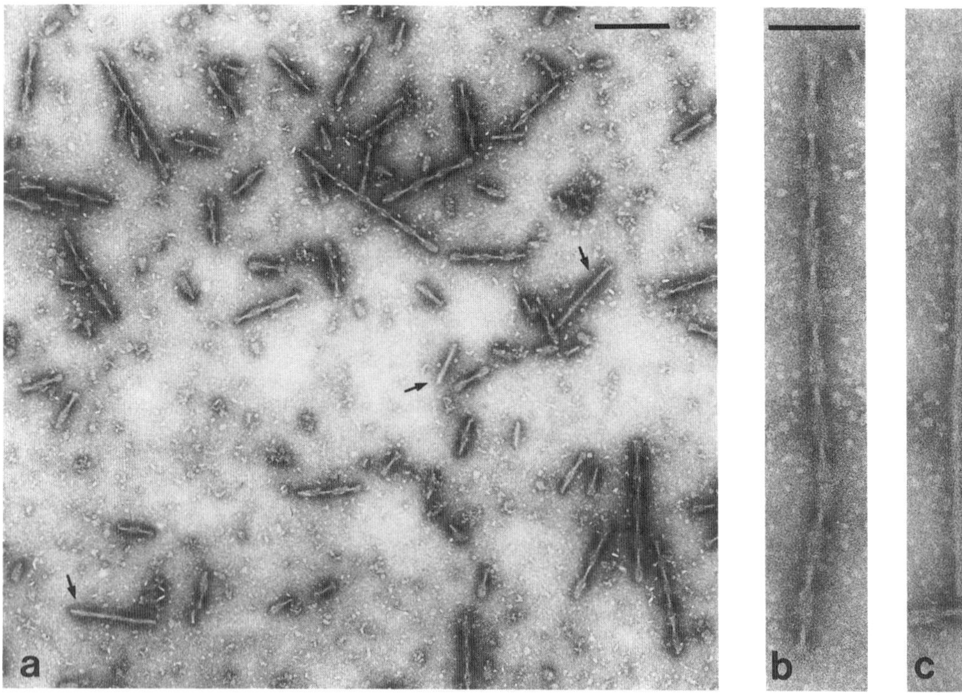

Fig. 7. *Electron micrographs of negatively stained abnormal filaments from an Alzheimer's disease brain.* **a:** *Low power view showing predominantly paired helical filaments, but with a few straight filaments (arrows). Bar = 200 nm.* **b,c:** *High power view of a paired helical filament and a straight filament, respectively. Bar = 100 nm. [Reproduced from Goedert et al., 1992b, with permission of the publisher.]*

C-shaped morphological units, which correspond to the two strands of the PHF, each with three domains [Crowther and Wischik, 1985] (Fig. 8a). The straight filament (SF), which constitutes a minor species, contains identical or closely related subunits that are differently arranged [Crowther, 1991]. Thus the computed cross section of the SF shows two C-shaped subunits very similar to those seen in the PHF but arranged back to back rather than base to base (Fig. 8b).

Biochemical studies have shown that tau forms a major component of the PHF [Goedert et al., 1988; Wischik et al., 1988a; Kondo et al., 1988; Lee et al., 1991]. Proteolytic digestion of PHFs monitored by labeling with antibodies specific for various regions of the protein indicates that the N-terminal half and the C-terminal region form the fuzzy coat around the filament, while the repeat region forms the core of the structure [Ksiezak-Reding and Yen, 1991; Goedert et al., 1992b]. Sequence analysis of tau fragments recovered from pronase-treated tangle PHFs has shown that both three- and four-repeat isoforms contribute to the core of the PHF but that in both cases a length of protein containing only three repeats is protected [Jakes et al., 1991]. These results suggest that the self-association of tau through the microtubule-binding region may lead to PHF formation. This has been confirmed with the demonstration that recombinant tau fragments comprising three microtubule-binding repeats can form filaments like those found in Alzheimer's disease [Wille et al., 1992; Crowther et al., 1992]. These findings show that tau is the only component necessary to form PHFs and end a long-standing controversy regarding the molecular nature of the PHF. At present, it is not known in detail how the packing of the three microtubule-

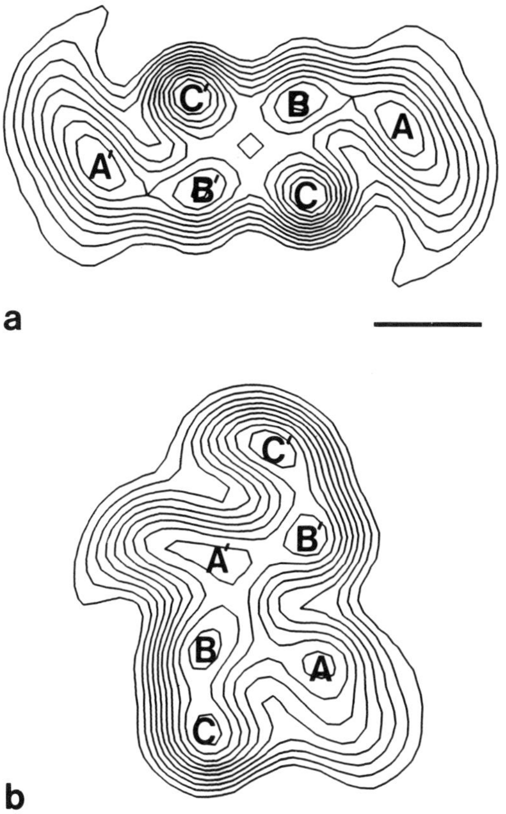

Fig. 8. *Computed cross sections through a paired helical filament* **(a)** *and a straight filament* **(b)**. *The contour maps denote stain exclusion. Both cross sections show two C-shaped morphological units, each displaying three peaks of density, equivalently designed A, B, and C and A′, B′ and C′. Each C-shaped unit corresponds to the density within one strand of the two strands that constitute each filament type. Bar = 4 nm. [Reproduced from Crowther, 1991, with permission of the publisher.]*

binding repeats of tau produces the C-shaped subunits of the PHF. However, the fold of the repeated sequences is likely to produce equivalent surfaces on the subunit, leading to the alternative packings seen in PHFs and SFs.

Tau protein extracted from Alzheimer PHFs runs as three bands of 60, 64 and 68 kD apparent molecular weight [Flament et al., 1989; Greenberg and Davies, 1990; Ksiezak-Reding and Yen, 1991; Lee et al., 1991; Brion et al., 1991] (Fig. 9). These bands, which contain the whole of tau, run more slowly on gels than normal or recombinant tau. However, after extensive dephosphorylation, they align with the six recombinant brain isoforms, indicating that PHF tau consists of all six isoforms in an abnormally phosphorylated state [Goedert et al., 1992b] (Fig. 9). The relative amounts of the different isoforms recovered from PHFs are similar to those observed in normal human brain. PHF tau has a greatly reduced ability to bind to microtubules; this functional impairment results entirely from abnormal phosphorylation, since dephosphorylated PHF tau binds as well as does normal tau to microtubules [Bramblett et al., 1993]. The generation of PHF tau leads to reduced levels of microtubule-binding competent tau in those regions of Alzheimer's disease brain that contain PHF tau and abundant neurofibrillary lesions [Bramblett et al., 1992]. The reduced binding of PHF tau to microtubules, coupled with reduced levels of normal tau, probably destabilizes microtubules in Alzheimer's disease, resulting in the impairment of vital cellular processes, such as rapid axonal transport, and leading to the degeneration of the affected nerve cells.

Several sites that are abnormally phosphorylated in PHF tau have been identified through mass spectrometry and the use of phosphorylation-dependent antibodies; they include, in the numbering of the longest human brain tau isoform, threonine 231 and serines 202, 235, 262, and 396 [Lee et al., 1991; Hasegawa et al., 1992; Goedert et al., 1993]. The phosphorylation-dependent antibodies do not recognize normal adult brain tau, but recognize fetal tau, the three PHF tau bands, and isolated PHFs, indicating that in Alzheimer's disease all six adult brain tau isoforms are likely to be abnormally phosphorylated in a way similar to the normal phosphorylation of the single fetal tau isoform [Kanemaru et al., 1992; Bramblett et al., 1993; Goedert et al., 1993]. It will be interesting to see whether this holds for all the sites that are abnormally phosphorylated in PHF tau. With the exception of serine 262, which is only phosphorylated in a fraction of PHF tau [Hasegawa et al., 1992], the known sites

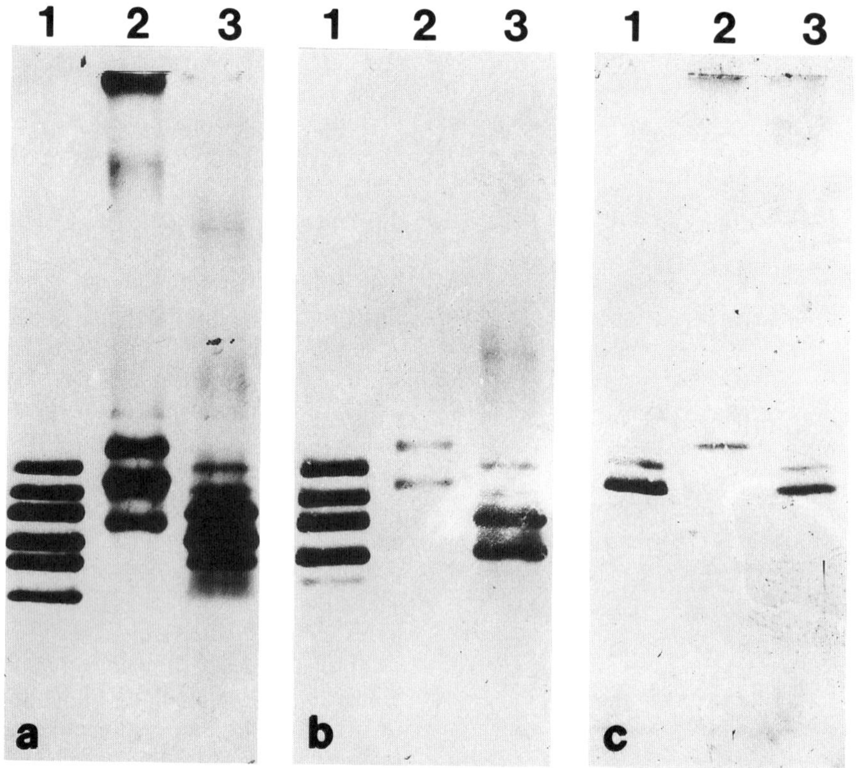

Fig. 9. *Immunoblots before and after dephosphorylation of paired helical filament tau compared with recombinant tau isoforms. In each panel,* **lane 1** *represents the six recombinant human tau isoforms,* **lane 2** *represents paired helical filament tau before dephosphorylation, and* **lane 3** *represents paired helical filament tau after dephosphorylation.* **a:** *Blotted with an antiserum raised against a peptide corresponding to the N-terminus of tau, a region common to all isoforms.* **b:** *Blotted with an antiserum raised against a peptide corresponding to the first N- terminal insert, a region found in four out of the six isoforms.* **c:** *Blotted with an antiserum raised against a peptide corresponding to the second N-terminal insert, a region found in two out of the six isoforms. Note the alignment of the paired helical filament tau bands after dephosphorylation (lane 3 in each panel) with the recombinant tau isoforms. [Reproduced from Goedert et al., 1992b, with permission of the publisher.]*

are serine or threonine residues followed by a proline [Lee et al., 1991; Biernat et al., 1992; Hasegawa et al., 1992; Goedert et al., 1993]. The longest human brain tau isoform contains 17 Ser–Pro and Thr–Pro sites that are distributed through the N-terminal half and the C-terminal region but that are absent from the microtubule-binding region (Fig. 10A).

Exactly how phosphorylation inhibits binding of tau to microtubules and subsequently promotes its assembly into PHFs is unclear. It is likely that phosphorylation of a few specific sites impairs binding to tubulin. It is less likely that phosphorylation of the same specific sites is necessary to induce the conformation of tau favoring PHF assembly, since the repeat region without any flanking sequences and their phosphorylation sites can assemble into filaments [Wille et al., 1992; Crowther et al., 1992]. Perhaps more likely, based on the striking overlap of potential phosphorylation sites (Fig. 10A) and the cluster of basic residues just upstream of the repeat region (Fig. 10B), is that general phosphorylation of these sites lead to charge neutralization. Self-aggregation is then no longer inhibited, and the phosphorylated tau assembles into PHFs.

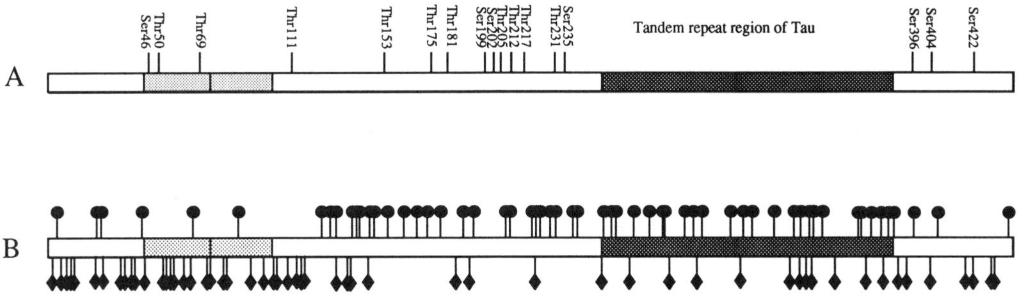

Fig. 10. **A:** *Distribution of 17 Ser–Pro and Thr–Pro sites in the longest human brain tau isoform. The tandem repeat region and the N-terminal inserts are shaded. Note the absence of Ser–Pro and Thr–Pro sites in the repeat region and a cluster of these sites just N-terminal of the repeats.* **B:** *Distribution of basic (●) and acidic (◆) residues in the longest human brain tau isoform. The N-terminal region up to the repeats shows a striking charge distribution, with the first 125 amino acids having a strong net negative charge and the second 125 amino acids having a strong net positive charge. Note the overlap between the Ser–Pro and Thr–Pro cluster upstream of the repeats and the positively charged region.*

The abnormal phosphorylation of tau that occurs in Alzheimer's disease could result from increased activity of a tau kinase or from decreased activity of a tau phosphatase. The phosphorylation of Ser–Pro and Thre–Pro sites in PHF tau suggests that a protein kinase with that specificity may be involved. Known candidates are PK36 and PK40 [Roder and Ingram, 1991], tau protein kinase 1 [Ishiguro et al., 1992], glycogen synthase kinase-3 [Yang et al., 1991; Hanger et al., 1992; Mandelkow et al., 1992], proline-directed protein kinase [Vulliet et al., 1992] and mitogen-activated protein kinase [Drewes et al., 1992; Ledesma et al., 1992; Goedert et al., 1992c]. The latter, which consists of several isoforms that are found in adult brain [Boulton et al., 1991], has been shown to phosphorylate recombinant tau at most Ser–Pro sites and Thr–Pro sites, including serines 202 and 396 [Drewes et al., 1992; Goedert et al., 1992c]. Moreover, recombinant tau phosphorylated by mitogen-activated protein kinase has a 10-fold reduced ability to bind to microtubules when compared with nonphosphorylated tau [Drechsel et al., 1992]. Mitogen-activated protein kinase is activated by a variety of growth factors and neurotransmitters through several intermediate steps consisting of a protein kinase cascade [Pelech and Sanghera, 1992]. In Alzheimer's disease any one of these steps could be affected.

Recombinant tau protein that has been phosphorylated at Ser–Pro and Thr–Pro sites by the 42 kD isoform of mitogen-activated protein kinase is selectively dephosphorylated by protein phosphatase 2A [Goedert et al., 1992c]. This enzyme exists *in vivo* as a 36 kD catalytic subunit that is complexed to a 60 kD A-subunit, which in turn is complexed to one of several B-subunits [Cohen, 1989]. The latter exist as several isoforms produced from distinct genes, at least one of which is nervous system specific [Mayer et al., 1991]. Interestingly, the catalytic subunit of protein phosphatase 2A dephosphorylates tau very poorly, whereas the trimeric form of the enzyme is much more effective than the dimeric form consisting of the A-subunit and the catalytic subunit. This raises the possibility that a defect in a B-subunit might underlie the abnormal phosphorylation of tau in Alzheimer's disease.

The challenge ahead lies in identifying the protein kinases and/or protein phosphatases that lead to the abnormal phosphorylation of adult tau. This may in turn permit the development of effective strategies aimed at preventing or delaying the appearance of the neurofibrillary pathology of Alzheimer's disease.

ACKNOWLEDGMENTS

We thank Mr. D.J. Crowther for drawing Figures 2 and 10 and Dr. L.A. Amos for helpful comments on the manuscript.

VI. REFERENCES

Aizawa H, Emori Y, Murofushi H, Kawasaki H, Sakai H, Suzuki K (1990): Molecular cloning of a ubiquitously distributed microtubule-associated protein with M_r 190,000. J Biol Chem 265:13849–13855.

Andreadis A, Brown MW, Kosik KS (1992): Structure and novel exons of the human tau gene. Biochemistry 31:10626–10633.

Arriagada PV, Growdon JH, Hedley-White ET, Hyman BT (1992): Neurofibrillary tangles but not senile plaques parallel duration and severity of Alzheimer's disease. Neurology 42:631–639.

Ashman JB, Hall ES, Eveleth J, Boekelheide K (1992): Tau, the neuronal heat-stable microtubule-associated protein, is also present in the cross-linked microtubule network of the testicular spermatid manchette. Biol Reprod 46:120–129.

Baas PW, Pienkowski TP, Kosik KS (1991): Processes induced by tau expression in Sf9 cells have an axon-like microtubule organization. J Cell Biol 115: 1333–1344.

Baudier J, Lee S-H, Cole RD (1987): Separation of the different microtubule-associated tau protein species from bovine brain and their mode II phosphorylation by Ca^{2+}/phospholipid-dependent protein kinase C. J Biol Chem 262:17584–17590.

Biernat J, Mandelkow EM, Schröter C, Lichtenberg-Kraag B, Steiner B, Berling B, Meyer H, Mercken M, Vandermeeren A, Goedert M, Mandelkow E (1992): The switch of tau protein to an Alzheimer-like state includes the phosphorylation of two serine-proline motifs upstream of the microtubule-binding region. EMBO J 11:1593–1597.

Binder LI, Frankfurter A, Rebhun LI (1985): The distribution of tau in the mammalian central nervous system. J Cell Biol 101:1371–1378.

Boulton TG, Nye S, Robbins DJ, Ip NY, Radziejewska E, Morgenbesser SD, DePinho RA, Panayotatos N, Cobb MH, Yancopoulos GD (1991): ERKs: A family of protein serine/threonine kinases that are activated and phosphorylated in response to insulin and NGF. Cell 65:663–675.

Braak H, Braak E (1991): Neuropathological stageing of Alzheimer-related changes. Acta Neuropathol 82: 239–259.

Bramblett GT, Trojanowski JQ, Lee VM-Y (1992): Regions with abundant neurofibrillary pathology in human brain exhibit a selective reduction in levels of binding-competent tau and accumulation of abnormal tau isoforms (A68 proteins). Lab Invest 66:212–222.

Bramblett GT, Goedert M, Jakes R, Merrick SE, Trojanowski JQ, Lee VM-Y (1993): Abnormal tau phosphorylation at Ser^{396} in Alzheimer's disease recapitulates development and contributes to reduced microtubule binding. Neuron 10:1089–1099.

Bré MH, Karsenti E (1990): Effects of brain microtubule-associated proteins on microtubule dynamics and the nucleating activity of centrosomes. Cell Motil Cytoskel 15:88–98.

Brion JP, Guilleminot J, Couchie D, Flament-Durand J, Nuñez J (1988): Both adult and juvenile tau microtubule-associated proteins are axon specific in the developing and adult rat cerebellum. Neuroscience 25:139–146.

Brion JP, Hanger DP, Couck AM, Anderton BH (1991): A68 proteins in Alzheimer's disease are composed of several tau isoforms in a phosphorylated state which affects their electrophoretic mobilities. Biochem J 279:831–836.

Butler M, Shelanski ML (1986): Microheterogeneity of microtubule-associated tau proteins is due to differences in phosphorylation. J Neurochem 47: 1517–1522.

Butner KA, Kirschner MW (1991): Tau protein binds to microtubules through a flexible array of distributed weak sites. J Cell Biol 115:717–730.

Caceres A, Kosik KS (1990): Inhibition of neurite polarity by tau antisense oligonucleotides in primary cerebellar neurons. Nature 343:461–463.

Caceres A, Potrebic S, Kosik KS (1991): The effect of tau antisense oligonucleotides on neurite formation of cultured cerebellar macroneurons. J Neurosci 11: 1515–1523.

Chen J, Kanai Y, Cowan NJ, Hirokawa N (1992): Projection domains of MAP2 and tau determine spacings between microtubules in dendrites and axons. Nature 360:674–677.

Cleveland DW, Hwo S-Y, Kirschner MW (1977a): Purification of tau, a microtubule-associated protein that induces assembly of microtubules from purified tubulin. J Mol Biol 116:207–225.

Cleveland DW, Hwo S-Y, Kirschner MW (1977b): Physical and chemical properties of purified tau factor and the role of tau in microtubule assembly. J Mol Biol 116:227–247.

Cohen P (1989): The structure and regulation of protein phosphatases. Annu Rev Biochem 58:453–508.

Correas I, Diaz-Nido J, Avila J (1992): Microtubule-associated protein tau is phosphorylated by protein kinase C on its tubulin binding domain. J Biol Chem 267:15721–15728.

Couchie D, Nuñez J (1985): Immunological characterization of microtubule-associated proteins specific for the immature brain. FEBS Lett 188:331–335.

Couchie D, Legay F, Guilleminot J, Lebargy F, Brion JP, Nuñez J (1990): Expression of tau protein and tau

mRNA in the cerebellum during axonal outgrowth. Exp Brain Res 82:589–596.
Couchie D, Mavilia C, Georgieff IS, Liem RKH, Shelanski ML, Nuñez J (1992): Primary structure of high molecular weight tau present in the peripheral nervous system. Proc Natl Acad Sci USA 89:4378–4381.
Crowther RA, Wischik CM (1985): Image reconstruction of the Alzheimer paired helical filament. EMBO J 4:3661–3665.
Crowther RA (1991): Straight and paired helical filaments in Alzheimer disease have a common structural unit. Proc Natl Acad Sci USA 88:2288–2292.
Crowther RA, Olesen OF, Jakes R, Goedert M (1992): The microtubule binding repeats of tau protein assemble into filaments like those found in Alzheimer's disease. FEBS Lett 309:199–202.
Dickson DW, Ksiezak-Reding H, Liu WK, Davies P, Crowe A, Yen S-HC (1992): Immunocytochemistry of neurofibrillary tangles with antibodies to subregions of tau protein: Identification of hidden and cleaved tau epitopes and a new phosphorylation site. Acta Neuropathol 84:596–605.
Drechsel DN, Hyman AA, Cobb MH, Kirschner MW (1992): Modulation of the dynamic instability of tubulin assembly by the microtubule-associated protein tau. Mol Biol Cell 3:1141–1154.
Drewes G, Lichtenberg-Kraag B, Döring F, Mandelkow EM, Biernat J, Dorée M, Mandelkow E (1992): Mitogen activated protein (MAP) kinase transforms tau protein into an Alzheimer-like state. EMBO J 11: 2131–2138.
Drubin DG, Feinstein SC, Shooter EM, Kirschner MW (1985): Nerve growth factor-induced neurite outgrowth in PC12 cells involves the coordinate induction of microtubule assembly and assembly-promoting factors. J Cell Biol 101:1799–1807.
Drubin DG, Kirschner MW (1986): Tau protein function in living cells. J Cell Biol 103:2739–2746.
Drubin DG, Kobayashi S, Kellogg D, Kirschner M (1988): Regulation of microtubule protein levels during cellular morphogenesis in nerve growth factor-treated PC12 cells. J Cell Biol 106:1583–1591.
Ennulat DJ, Liem RKH, Hashim GA, Shelanski ML (1989): Two separate 18-amino acid domains of tau promote the polymerization of tubulin. J Biol Chem 264:5327–5330.
Ferreira A, Busiglio J, Caceres A (1989): Microtubule formation and neurite growth in cerebellar macroneurons which develop *in vitro:* Evidence for the involvement of the microtubule-associated proteins MAP-1a, HMW-MAP2 and tau. Dev Brain Res 49:213–228.
Flament S, Delacourte A, Hémon B, Défossez A (1989): Characterization of two pathological tau protein variants in Alzheimer's disease. J Neurol Sci 92: 133–141.
Gallo JM, Hanger DP, Twist EC, Kosik KS, Anderton BH (1992): Expression and phosphorylation of a three-repeat isoform of tau in transfected non-neuronal cells. Biochem J 286:399–404.
Georgieff IS, Liem RKH, Mellado W, Nuñez J, Shelanski ML (1991): High molecular weight tau: Preferential localization in the peripheral nervous system. J Cell Sci 100:55–60.
Goedert M, Wischik CM, Crowther RA, Walker JE, Klug A (1988): Cloning and sequencing of the cDNA encoding a core protein of the paired helical filament of Alzheimer disease. Proc Natl Acad Sci USA 85: 4051–4055.
Goedert M, Spillantini MG, Potier MC, Ulrich J, Crowther RA (1989a): Cloning and sequencing of the cDNA encoding an isoform of microtubule-associated protein tau containing four tandem repeats: Differential expression of tau protein mRNAs in human brain. EMBO J 8:393–399.
Goedert M, Spillantini MG, Jakes R, Rutherford D, Crowther RA (1989b): Multiple isoforms of human microtubule-associated protein tau: Sequences and localization in neurofibrillary tangles of Alzheimer's disease. Neuron 3:519–526.
Goedert M, Jakes R (1990): Expression of separate isoforms of human tau protein: Correlation with the tau pattern in brain and effects on tubulin polymerization. EMBO J 9:4225–4230.
Goedert M, Spillantini MG, Crowther RA (1992a): Cloning of a big tau microtubule-associated protein characteristic of the peripheral nervous system. Proc Natl Acad Sci USA 89:1983–1987.
Goedert M, Spillantini MG, Cairns NJ, Crowther RA (1992b): Tau proteins of Alzheimer paired helical filaments: Abnormal phosphorylation of all six brain isoforms. Neuron 8:159–168.
Goedert M, Cohen ES, Jakes R, Cohen P (1992c): p42 MAP kinase phosphorylation sites in microtubule-associated protein tau are dephosphorylated by protein phosphatase $2A_1$. Implications for Alzheimer's disease. FEBS Lett 312:95–99.
Goedert M, Jakes R, Crowther RA, Six J, Lübke U, Vandermeeren M, Cras P, Trojanowski JQ, Lee VM-Y (1993): The abnormal phosphorylation of tau protein at serine 202 in Alzheimer's disease recapitulates phosphorylation during development. Proc Natl Acad Sci USA 90:5066–5070.
Goto S, Yamamoto H, Fukunaga K, Iwasa T, Matsukado Y, Miyamoto E (1985): Dephosphorylation of microtubule-associated protein 2, tau factor, and tubulin by calcineurin. J Neurochem 45:276–283.
Greenberg SG, Davies P (1990): A preparation of Alzheimer paired helical filaments that displays distinct tau proteins by polyacrylamide gel electrophoresis. Proc Natl Acad Sci USA 87:5827–5831.
Greenberg SG, Davies P, Schein JD, Binder LI (1992): Hydrofluoric acid-treated tau PHF proteins display the

same biochemical properties as normal tau. J Biol Chem 267:564–569.

Grundke-Iqbal I, Iqbal K, Tung Y-C, Zaidi MS, Wisniewski HM, Binder LI (1986): Abnormal phosphorylation of the microtubule-associated protein tau in Alzheimer cytoskeletal pathology. Proc Natl Acad Sci USA 83:4913–4917.

Gustke N, Steiner B, Mandelkow EM, Biernat J, Meyer HE, Goedert M, Mandelkow E (1992): The Alzheimer-like phosphorylation of tau protein reduces microtubule binding and involves Ser–Pro and Thr–Pro motifs. FEBS Lett 307:199–205.

Hanemaaijer R, Ginzburg I (1991): Involvement of mature tau isoforms in the stabilization of neurites in PC12 cells. J Neurosci Res 30:163–171.

Hanger DP, Hughes K, Woodgett JR, Brion JP, Anderton BH (1992): Glycogen synthase kinase-3 induces Alzheimer's disease-like phosphorylation of tau: Generation of paired helical filament epitopes and neuronal localization of the kinase. Neurosci Lett 147:58–62.

Hasegawa M, Morishima-Kawashima M, Takio K, Suzuki M, Titani K, Ihara Y (1992): Protein sequence and mass spectrometric analyses of tau in the Alzheimer's disease brain. J Biol Chem 267:17047–17054.

Himmler A (1989): Structure of the bovine tau gene: Alternatively spliced transcripts generate a protein family. Mol Cell Biol 9:1389–1396.

Himmler A, Drechsel D, Kirschner MW, Martin DW Jr (1989): Tau consists of a set of proteins with repeated C-terminal microtubule-binding domains and variable N-terminal domains. Mol Cell Biol 9:1381–1388.

Hirokawa N, Shiomura Y, Ogabe S (1988): Tau proteins: The molecular structure and mode of binding on microtubules. J Cell Biol 107:1449–1459.

Hoshi M, Nishida E, Miyata Y, Sakai H, Miyoshi T, Ogawara H, Akiyama T (1987): Protein kianse C phosphorylates tau and induces its functional alterations. FEBS Lett 217:237–241.

Ishiguro K, Omori A, Kato K, Tomizawa K, Imahori K, Uchida T (1991): A serine/threonine proline kinase activity is included in the tau protein kinase fraction forming a paired helical filament epitope. Neurosci Lett 128:195–198.

Ishiguro K, Takamatsu M, Tomizawa K, Omori A, Takahashi M, Arioka M, Uchida T, Imahori K (1992): Tau protein kinase I converts normal tau protein into A68-like component of paired helical filaments. J Biol Chem 267:10897–10901.

Jakes R, Novak M, Davison M, Wischik CM (1991): Identification of 3- and 4-repeat tau isoforms within the PHF in Alzheimer's disease. EMBO J 10:2725–2729.

Kanai Y, Takemura R, Oshima T, Mori H, Ihara Y, Yanagisawa M, Masaki T, Hirokawa N (1989): Expression of multiple tau isoforms and microtubule bundle formation in fibroblasts transfected with a single tau cDNA. J Cell Biol 109:1173–1184.

Kanai Y, Chen J, Hirokawa N (1992): Microtubule bundling by tau proteins *in vivo:* Analysis of functional domains. EMBO J 11:3953–3961.

Kanemura K, Takio K, Miura R, Titani K, Ihara Y (1992): Fetal-type phosphorylation of the tau in paired helical filaments. J Neurochem 58:1667–1675.

Kidd M (1963): Paired helical filaments in electron microscopy of Alzheimer's disease. Nature 197:192–193.

Kim H, Strong TV, Anderson SJ (1991): Evidence for tau expression in cells of monocyte lineage and its *in vitro* phosphorylation by v-fms kinase. Oncogene 6:1085–1087.

Knops J, Kosik KS, Lee G, Pardee JD, Cohen-Gould L, McColongue L (1991): Overexpression of tau in a nonneuronal cell induces long cellular processes. J Cell Biol 114:725–733.

Kondo J, Honda T, Mori H, Hamada Y, Miura R, Ogawara H, Ihara Y (1988): The carboxyl third of tau is tightly bound to paired helical filaments. Neuron 1:827–834.

Kosik KS, Orecchio LD, Bakalis S, Neve RL (1989): Developmentally regulated expression of specific tau sequences. Neuron 2:1389–1397.

Ksiezak-Reding H, Yen S-H (1991): Structural stability of paired helical filaments requires microtubule-binding domains of tau: A model for self-association. Neuron 6:717–728.

Ledesma MD, Correas I, Avila J, Diaz-Nido J (1992): Implication of brain cdc2 and MAP2 kinases in the phosphorylation of tau protein in Alzheimer's disease. FEBS Lett 308:218–224.

Lee G, Cowan N, Kirschner M (1988): The primary structure and heterogeneity of tau protein from mouse brain. Science 239:285–288.

Lee G, Neve RL, Kosik KS (1989): The microtubule binding domain of tau protein. Neuron 2:1615–1624.

Lee G, Rook SL (1992): Expression of tau protein in non-neuronal cells: Microtubule binding and stabilization. J Cell Sci 102:227–237.

Lee VM-Y, Balin BJ, Otvos L, Trojanowski JQ (1991): A68—a major subunit of paired helical filaments and derivatized forms of normal tau. Science 251: 675–678.

Lewis SA, Wang D, Cowan NJ (1988): Microtubule-associated protein MAP2 shares a microtubule binding motif with tau protein. Science 242:936–939.

Lindwall G, Cole RD (1984): Phosphorylation affects the ability of tau protein to promote microtubule assembly. J Biol Chem 259:5301–5305.

Lo MMS, Fieles AW, Norris TE, Dargis DG, Caputo CB, Scott CW, Lee VM-Y, Goedert M (1993): Human tau isoforms confer distinct morphological and functional properties to stably transfected fibroblasts. Mol Brain Res (in press).

Maccioni RB, Vera JC, Dominguez J, Avila J (1989): A discrete repeated sequence defines a tubulin binding domain on microtubule-associated protein tau. Arch Biochem Biophys 275:568–579.

Mandelkow EM, Drewes G, Biernat J, Gustke N, Van Lint J, Vandenheede JR, Mandelkow E (1992): Glycogen synthase kinase-3 and the Alzheimer-like state of microtubule-associated protein tau. FEBS Lett 314: 315–321.

McKee AC, Kosik KS, Kowall NW (1991): Neuritic pathology and dementia in Alzheimer's disease. Ann Neurol 30:156–165.

Mareck A, Fellous A, Francon J, Nunez J (1980): Changes in composition and activity of microtubule-associated proteins during development. Nature 284:353–355.

Mawal-Dewan M, Sen PC, Abdel-Ghany M, Shalloway D, Racker E (1992): Phosphorylation of tau protein by purified $p34^{cdc28}$ and a related protein kinase from neurofilaments. J Biol Chem 267:19705–19709.

Mayer RE, Hendrix P, Cron P, Matthies R, Stone SR, Goris J, Merlevede W, Hofsteenge J, Hemmings BA (1991): Structure of the 55-kDa regulatory subunit of protein phosphatase 2A: Evidence for a neuronal-specific isoform. Biochemistry 30:3589–3597.

Mercken M, Vandermeeren M, Lübke U, Six J, Boons J, Van de Voorde A, Martin JJ, Gheuens J (1992): Monoclonal antibodies with selective specificity for Alzheimer tau are directed against phosphatase-sensitive epitopes. Acta Neuropathol 84:265–272.

Migheli A, Butler M, Brown K, Shelanski ML (1988): Light and electron microscope localization of the microtubule-associated tau protein in rat. J Neurosci 8:1846–1851.

Mori H, Hamada Y, Kawaguchi M, Hona T, Kondo J, Ihara Y (1989): A distinct form of tau is selectively incorporated into Alzheimer's paired helical filaments. Biochem Biophys Res Commun 159:1221–1226.

Murphy DB, Wallis KT (1985): Erythrocyte microtubule assembly *in vitro*. J Biol Chem 260:12293–12301.

Neve RL, Harris P, Kosik KS, Kurnit DM, Donlon TA (1986): Identification of cDNA clones for the human microtubule-associated protein tau and chromosomal localization of the genes for tau and microtubule-associated protein 2. Mol Brain Res 1:271–280.

Oblinger MM, Argasinski A, Wong J, Kosik KS (1991): Tau gene expression in rat sensory neurons during development and regeneration. J Neurosci 11:2453–2459.

Pelech SL, Sanghera JS (1992): Mitogen-activated protein kinases: Versatile transducers for cell signaling. Trends Biochem Sci 17:233–238.

Peng I, Binder LI, Black MM (1986): Biochemical and immunological analyses of cytoskeletal domains of neurons. J Cell Biol 102:252–262.

Pierre M, Nunez J (1983): Multisite phosphorylation of tau proteins from rat brain. Biochem Biophys Res Commun 115:212–219.

Riederer BM, Innocenti GM (1991): Differential distribution of tau proteins in developing cat cerebral cortex and corpus callosum. Eur J Neurosci 3:1134–1145.

Roder HM, Ingram VM (1991): Two novel kinases phosphorylate tau and the KSP site of heavy neurofilament subunits in high stoichiometric ratios. J Neurosci 11:3325–3343.

Shin R-W, Iwaki T, Kitamoto T, Tateishi J (1991): Hydrated autoclave pretreatment enhances tau immunoreactivity in formalin-fixed normal and Alzheimer's disease brain tissues. Lab Invest 64:693–702.

Steiner B, Mandelkow EM, Biernat J, Gustke N, Meyer HE, Schmidt B, Mieskes G, Söling HD, Drechsel D, Kirschner MW, Goedert M, Mandelkow E (1990): Phosphorylation of microtubule-associated protein tau: Identification of the site for Ca^{2+}-calmodulin dependent kinase and the relationship with tau phosphorylation in Alzheimer tangles. EMBO J 9:3539–3544.

Takemura R, Kanai Y, Hirokawa N (1991): *In situ* hybridization of tau mRNA in developing rat brain. Neuroscience 44:393–407.

Trojanowski JQ, Schuck T, Schmidt ML, Lee VM-Y (1989): Distribution of tau proteins in the normal human central and peripheral nervous system. J Histochem Cytochem 37:209–215.

Tucker RP, Matus AI (1988): Microtubule-associated proteins characteristic of embryonic brain are found in the adult mammalian retina. Dev Biol 130:423–434.

Viereck C, Tucker RP, Matus A (1989): The adult rat olfactory system expresses microtubule-associated proteins found in the developing brain. J Neurosci 9:3547–3557.

Vulliet R, Halbran SM, Braun RK, Smith AJ, Lee G (1992): Proline-directed phosphorylation of human tau protein. J Biol Chem 267:22570–22574.

Weingarten MD, Lockwood AH, Hwo S-H, Kirschner MW (1975): A protein factor essential for microtubule assembly. Proc Natl Acad Sci USA 72:1858–1862.

Wille H, Drewes G, Biernat J, Mandelkow EM, Mandelkow E (1992): Alzheimer- like paired helical filaments and antiparallel dimers formed from microtubule-associated protein tau *in vitro*. J Cell Biol 118:573–584.

Wischik CM, Novak M, Thogersen HC, Edwards PC, Runswick MJ, Jakes R, Walker JE, Milstein C, Roth M, Klug A (1988a): Isolation of a fragment of tau derived from the core of the paired helical filament of Alzheimer disease. Proc Natl Acad Sci USA 85:4506–4510.

Wischik CM, Novak M, Edwards PC, Klug A, Tichelaar W, Crowther RA (1988b): Structural characterization of the core of the paired helical filament of Alzheimer disease. Proc Natl Acad Sci USA 85:4884–4888.

Wood JG, Mirra SS, Pollock NJ, Binder LI (1986): Neurofibrillary tangles of Alzheimer disease share antigenic determinants with the axonal microtubule-associated protein tau. Proc Natl Acad Sci USA 83:4040–4043.

Yamamoto H, Fukunaga K, Tanaka E, Miyamoto E (1983): Ca^{2+}- and calmodulin- dependent phosphorylation of microtubule-associated protein 2 and tau factor, and inhibition of microtubule assembly. J Neurochem 41:1119–1126.

Yamamoto H, Fukunaga K, Goto S, Tanaka E, Miyamoto E (1985): Ca^{2+}- and calmodulin-dependent regulation of microtubule formation via phosphorylation of microtubulin-associated protein 2, tau factor, and tubulin, and comparison with the cyclic AMP-dependent phosphorylation. J Neurochem 44:759–768.

Yamamoto H, Saitoh Y, Yasugawa S, Miyamoto E (1990): Dephosphorylation of tau factor by protein phosphatase 2A in synaptosomal fractions, and inhibition by aluminum. J Neurochem 55:683–690.

Yang SD, Yu JS, Lai YG (1991): Identification and characterization of the ATP Mg-dependent phosphatase activator (F_A) as a microtubule protein kinase in the brain. J Prot Chem 10:171–181.

ABOUT THE AUTHORS

M. GOEDERT is a member of the scientific staff of the Medical Research Council Laboratory of Molecular Biology, Cambridge, England. A native of Luxembourg, he obtained an M.D. degree at the University of Basel (Switzerland) in 1980 and a Ph.D. at the University of Cambridge in 1984, before joining the MRC Laboratory of Molecular Biology. His work is in the field of molecular neurobiology.

R. JAKES is a Senior Scientific Officer in the Structural Studies Division of the MRC Laboratory of Molecular Biology, Cambridge, U.K. In 1970, while working with Professor Brian Hartley, he was first to demonstrate the crystalizability of tryptophanyl t-RNA synthetase. He later worked on proteins involved in cell motility and cell structure, purifying, characterizing and sequencing striated and smooth muscle myosin light chains, profilin from profilactin, villin, and cingulin. More recently he joined work to elucidate the composition and structure of paired helical filaments derived from neurofibrillary tangles (Alzheimer's disease). Consequently, he is interested in tubulin-associated and other brain proteins. He has held visiting scientist positions at the Department of Biochemistry, University of São Paulo, Brazil, and the Department of Zoological Cell Biology, University of Stockholm, Sweden.

M.G. SPILLANTINI holds an MRC postdoctoral fellowship at the Laboratory of Molecular Biology of the Medical Research Council in Cambridge, England, where she works on the molecular neuropathology of Alzheimer's disease. She is on leave from the Institute of Cell Biology of the Italian National Research Council, Rome, where she is a member of the scientific staff. After receiving a degree in biology summa cum laude in 1981 from the University of Florence, Italy, she pursued research in the Department of Clinical Pharmacology of the University of Florence and then at the Unité de Neurobiologie of the INSERM in Paris. She continued research at the Medical Research Council in Cambridge, first at the Molecular Neurobiology Unit, and then, beginning in 1988, at the Laboratory of Molecular Biology, where, working in Dr. M. Goedert's group, she obtained a Ph.D. in molecular biology from Cambridge University. She is a member of Peterhouse, Cambridge.

R.A. CROWTHER is a member of the scientific staff of the Medical Research Council Laboratory of Molecular Biology, Cambridge, England, and fellow of Peterhouse, Cambridge, where he teaches mathematics to science undergraduates. After taking the mathematical tripos and the postgraduate diploma in computing at Cambridge University, he pursued doctoral research at the Laboratory of Molecular Biology, where he was concerned with new methods in protein crystallography. Following postdoctoral work in Edinburgh, he returned to the Laboratory of Molecular Biology to develop methods for computer image processing of electron micrographs. Since then, he has applied these techniques to studying the structures of various macromolecular assemblies, including viruses, muscle, coated vesicles, and, most recently, the paired helical filaments of Alzheimer's disease. In 1984 he taught a course on electron microscopy and image processing in the Graduate Biophysics Program at Harvard. He was elected to membership of the European Molecular Biology Organization in 1985 and to a fellowship of the Royal Society in 1993.

Microtubules: 201–220

Tektins

Jan M. Norrander and Richard W. Linck

I. INTRODUCTION

Tektins are a set of proteins that form extended filaments in the walls of ciliary and flagellar microtubules. Three tektins have been isolated and studied: tektin A (nominally 53–57 kD), tektin B (51–52 kD), and tektin C (46–47 kD). The three tektins are present in the basal bodies and all nine doublet microtubules of sperm flagella; at least one tektin appears to be present in the central singlet microtubules. Though originally isolated from invertebrate flagellar microtubules, there is immunological evidence indicating a more universal presence of tektins or tektin-like proteins in other types of microtubules, specifically the centrioles, centrosomes, and spindles of mammalian cells. Biochemical and immunological studies of the tektin proteins reveal some similarities between tektins and intermediate filament (IF) proteins. The recently obtained cDNA sequence of tektin A has allowed for a more detailed analysis of the similarities, differences, and possible evolutionary relationship between tektins and IF proteins. In this discussion we present biochemical, immunological, and molecular biological characterizations of flagellar tektins, detail immunological evidence for the presence of tektins in other microtubule systems, compare and contrast tektins and IF proteins, and review possible biological functions of tektins.

II. FRACTIONATION OF FLAGELLAR MICROTUBULES

The structure of microtubules can be studied by the fractionation of microtubules into smaller but stable components. By comparing the structures of these components with electron microscopy (EM) and by determining their polypeptide composition by SDS-PAGE, the possible role of various polypeptides within microtubules can be partially elucidated.

Ciliary and flagellar axonemes can be fractionated into stable components by a variety of methods. For example, doublet microtubules can be purified that, depending on the species, contain one or only remnants of the central pair microtubules [Gibbons, 1965; Linck, 1973; Linck and Langevin, 1981; Witman et al., 1972a], and incubation of flagellar doublet microtubules at 40°C causes the solubilization of the B-subfiber, resulting in the production of pure A-tubules composed of 13 protofilaments and an adlumenal component [Stephens, 1970; Witman et al., 1972a; Linck and Langevin, 1981; Tilney et al., 1973]. The discovery of tektin proteins resulted from the extraction of axonemal microtubules with 0.5% Sarkosyl (sodium dodecyl sarcosinate), which solubilizes most of the doublet microtubule wall but leaves behind relatively stable ribbons consisting of approximately three protofilaments (pfs) [Meza et al., 1972; Witman, 1970; Witman et al., 1972a,b]. Electron microscopy studies indicate that this ribbon component originates not from the "partition," as proposed by earlier workers, but from the region where the inner B-subfiber wall and the radial spokes make contact with the A-tubule; however, the precise origin of these ribbons has not yet been determined [Linck, 1976]. More recently, a similar fractionation has been achieved with ciliary axonemes of marine invertebrates, using a thermal treatment [Stephens et al, 1989]; in this case nine stable pf ribbons are obtained, which are still linked in their ninefold axonemal arrangement, presumably by the continued presence of nexin linkages. This observation is consistent with a model wherein the stable pf ribbon occupies a location in the A-tubule wall near the attachment of the radial spokes and the inner B-tubule wall. Analysis of *Strongylocentrotus purpuratus,* Sarkosyl-insoluble pf ribbons by SDS-PAGE showed them to consist primarily of tubulin, three proteins subsequently identified as tektins A, B and C, and a 77 and 83 kD pair of polypeptides (Fig. 1). There are a number of other components present in lesser amounts but these have not been studied.

Because these Sarkosyl-insoluble pf ribbons represent such a stable component of doublet microtubules, a further study of their polypeptide components was warranted. These ribbons could be further extracted by the use of chaotropic agents such as KI, NaSCN, or urea, which left behind relatively stable filaments [Linck et al., 1982; Linck and Langevin, 1982]. Extraction of axonemal microtubules with the combination of 0.5% Sarkosyl and 2M urea yields stable 2 nm filaments that are primarily composed of equimolar amounts of the three tektin proteins found in Sarkosyl-insoluble pf ribbons [Linck et al., 1982; Linck and Langevin, 1982]. These tektin filaments often aggregate into sheets and bundles of six or more filaments. Extraction with 0.5% Sarkosyl/3M urea selectively removes tektin C and results in filaments composed of only tektins A and B. The name *tektin* is derived from the Greek *tektonos,* meaning architecture or builder and was an early term used by Mazia [1968] for cytoskeletal and microtubule proteins.

III. BIOCHEMISTRY OF TEKTIN PROTEINS

One of the most abundant sources of tektin proteins is the sperm flagellar axonemes of sea urchins. The sea urchins *S. purpuratus, Strongylocentrotus droebachiensis,* and *Lytechinus pictus* have all been used as tektin sources; however, the majority of experiments has been performed with the tektins of *S. purpuratus,* and as a result these are the most thoroughly characterized. *L. pictus* tektins have also been used with the only notable difference being a slight shift in molecular weights on SDS-PAGE [Steffen and Linck, 1988].

Tektins from *S. purpuratus* were compared against each other and to tubulin by high-resolution, two-dimensional tryptic peptide mapping and by amino acid analysis [Linck and Stephens, 1987]. There is a 63%–67% coincidence in the number and position of tryptic peptides from tektin B as compared with tektins A and C and >70% coincidence between A and C. None of the tektins has tryptic peptides with any significant degree of similarity to those of α- or β-tubulin. The amino acid compositions of tektins A, B, and C are all similar but distinctly different from those of α- and β-tubulin; however, the amino acid compositions of the tektins are very similar to those of the IF proteins: a family of highly α-helical, coiled-coil proteins that form filamentous networks in most mammalian cells. Tektins are also similar to IF proteins in terms of secondary structure. Circular dichroism studies of filaments composed of tektins A and B show them to be approximately 70% α-helical [Linck and Langevin,

Fig. 1. *Polypeptide composition of* S. purpuratus, *Sarkosyl-insoluble pf ribbons. Doublet microtubules were extracted with 0.5% Sarkosyl and the resulting protofilament ribbons resolved by two-dimensional IEF/SDS-PAGE [Linck and Langevin, 1982]. The gel shown at the top was stained with Serva Blue to reveal the major polypeptide components of the pf ribbons; molecular weights (* $\times 10^{-3}$*) are indicated along the one-dimensional SDS-PAGE profile at the left, and the pI range is shown at the top. The three immunoblots below were stained with affinity-purified anti-*S. purpuratus *tektins. Anti-55 (antibody against tektin A) stains three 55 kD proteins (a, b, and c). Anti-51 (antibody against tektin B) stains two 51 kD proteins (a' and b'). Anti-47 (antibody against tektin C) stains four 47 kD proteins (a'', b'', c'', and d'') and crossreacts with the 55 kD protein a. None of the tektin antibodies crossreact with the tubulins or the ribbon components of 77 and 83 kD (indicated by arrows).*

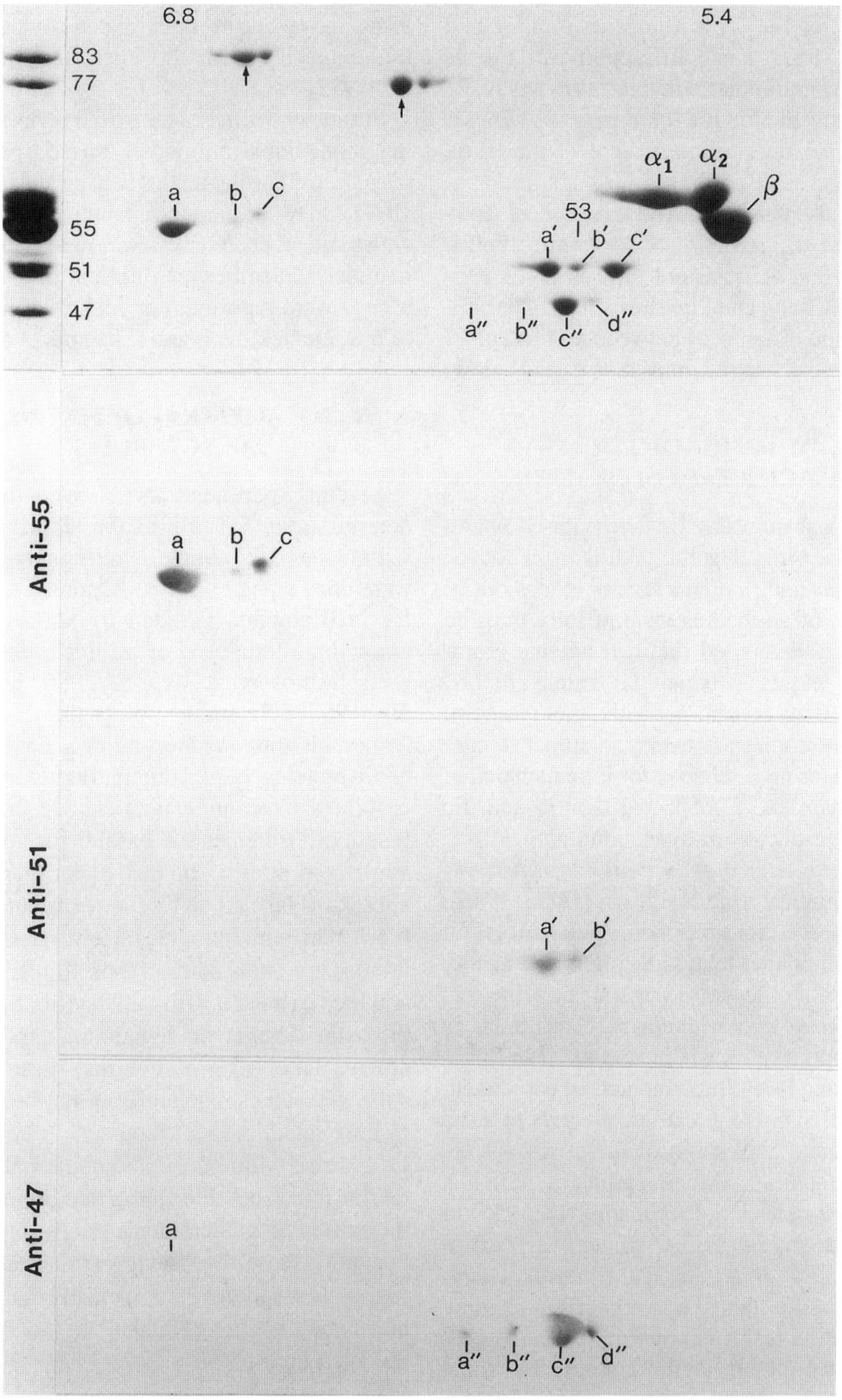
6.8
5.4
83
77
55
51
47
a
b
c
α_1
α_2
β
53
a′
b′
c′
a″
b″
c″
d″
Anti-55
Anti-51
Anti-47

1982]; other estimates of 55%–60% have been obtained for the same filaments [Chang and Piperno, 1987]. X-ray diffraction of filaments containing equimolar amounts of tektins A, B, and C showed strong α-type patterns [Beese, 1984]. The low proline content of the tektin proteins revealed in their amino acid compositions is also consistent with an α-helical structure, as is the presence of large polypeptide domains that are resistant to proteolytic degradation [Chang and Piperno, 1987]. The similarities and differences between tektins and IF proteins are discussed more thoroughly below.

IV. IMMUNOLOGICAL CHARACTERIZATION OF TEKTINS

Polyclonal antibodies have been raised against each of the three flagellar tektins from both *S. purpuratus* and *L. pictus* [Linck et al., 1987]. Both sets of antibodies were affinity purified using SDS-denatured tektin filaments (composed of tektins A, B, and C) from *L. pictus*. The specificity and cross-reactivity of these two sets of antibodies are nearly identical. Under SDS-denaturing conditions for immunoblotting, antitektins A and B are primarily monospecific for their respective antigens, and antitektin C cross-reacts only slightly with tektin A; however, under renaturing conditions [Dunn, 1986], there is some cross-reaction of one antitektin for the other two tektins [Steffen and Linck, 1992]. Antibodies raised against flagellar tektins also cross-react with ciliary tektins from sea urchins and molluscs [Linck et al., 1987; Stephens et al., 1989]. Immunoblots of two-dimensional IEF/SDS-PAGE carried out on both tektin filaments and Sarkosyl-insoluble pf ribbons are identical; antitektin A stains primarily a 55 kD/pI 6.9 spot, antitektin B primarily two 51 kD/pI 6.15 spots, and antitektin C primarily a 47 kD/pI 6.15 spot (Fig. 1). Antitektin C also weakly stains the major spot stained by antitektin A (55 kD/pI 6.9); this result is consistent with the results of one-dimensional immunoblots, where antitektin C weakly stains the 55 kD tektin protein. Antitektins A and C also stain several minor spots shifted by 0.1 pI units. These may represent isoelectric variants, phosphorylated forms, novel tektins or unrelated contaminants that copurify with the antigens [Linck et al., 1987].

Similar experiments have been performed using monoclonal antibodies raised against *S. purpuratus* flagellar tektins [Chang and Piperno, 1987]. On two-dimensional immunoblots stained with these monoclonals, several additional polypeptides and somewhat different pI values for tektins were reported; e.g., tektin A had a pI of 6.8, and tektins B and C had pIs of 5.8.

V. LOCALIZATION OF TEKTINS IN MICROTUBULES

Several experiments have been conducted to determine the location of the tektin proteins within the microtubule. Preliminary studies were done using an antibody prepared against tektin filaments, isolated by Sarkosyl-urea extraction, composed of all three tektin proteins [Amos et al., 1985, 1986; Linck et al., 1985]. On immunoblots of whole axonemes this antibody preparation stains all three tektin proteins. Using immunofluorescence microscopy, these antitektins stained the entire length of axonemes and basal bodies of methanol-fixed sperm. In EM experiments, the antitektins labeled tektin filaments but not unfixed, native microtubules. However, if microtubules are attached to carbon film EM grids and then extracted with Sarkosyl-urea to solubilize the tubulin, the remaining, exposed filaments label with antitektins; these results suggest that the tektin antigens may be masked by microtubule-associated proteins or by tubulin, i.e., buried within the microtubule wall. Antitektins also label thin fibrils that project from the ends of both microtubules and stable pf ribbons. All these results suggest that the tektin filaments are integral components of microtubules and that they extend along the length of the microtubule wall.

Other experiments have been carried out with affinity-purified, polyclonal antibodies specific for each tektin [Linck et al., 1987]; the characterization of these antibodies is described above. *L. pictus* sperm axonemes can be splayed into nine doublet microtubules that are held together at one end by the basal body; in this species the central pair singlet microtubules are lost in the preparation [Steffen and Linck, 1988]. Each tektin antibody stains the entire length of all nine doublets and their A-tubule extensions, suggesting that tektins A, B, and C are all present in each doublet of sperm axonemes. Using another species, the bat star *Patiria miniata,* it was possible to retain a part of the central pair; a faint staining of these microtubules is seen with antitektin C.

The partitioning of tektins between the soluble and insoluble fractions of extracted axonemes was examined. In one study, monoclonal antibodies were used to show that all three *S. purpuratus* flagellar tektins are exclusively associated with the Sarkosyl-insoluble pf ribbons; none are present in the Sarkosyl-soluble fraction [Chang and Piperno, 1987]. In another study, ciliary axonemes of both molluscan gill tissue and sea urchin embryos were fractionated by a 40°C heat treatment or by Sarkosyl-urea into a "skeleton" of the former axoneme, consisting of nine outer filaments/pf ribbons surrounding one or two central filaments/pf ribbons [Stephens et al., 1989]; tektins A and B were found to be associated with the "skeleton," and only tektin C was partially solubilized.

Finally, tektins are also present in basal bodies. By immunofluorescence, all three polyclonal antitektins intensely stain the basal bodies of sea urchin sperm [Steffen and Linck, 1988]. On immunoblots of isolated basal bodies from *Chlamydomonas*, monoclonal antibody 3-7-1 to tektin C cross-reacts with ~46–50 kD polypeptides [Chang and Piperno, 1987]. The possible presence of tektin proteins in basal bodies is reasonable, given that tektins are present in the A-tubules of axonemes and that the A-tubules and basal bodies are continuous structures.

VI. EVIDENCE FOR TEKTINS OR TEKTIN-LIKE PROTEINS IN NONAXONEMAL CYTOSKELETAL STRUCTURES

There are numerous immunological findings that provide evidence for the presence of tektin or tektin-like proteins in nonaxonemal cytoskeletal structures. An abbreviated nomenclature has been developed for the polyclonal antibodies; for example, anti-(L.p.)-tektin A and anti-(S.p.)-tektin A refer to the antibodies raised against tektin A from *L. pictus* and *S. purpuratus,* respectively.

A. Centrioles and Centrosomes

In interphase cells of CHO (Chinese hamster ovary), HeLa (human cervical epithelioid carcinoma), and LLC-PK_1 (pig kidney epithelia) at least one antitektin stains paired dots; the number of dots is cell cycle dependent, with one pair typically seen in interphase cells and two pairs in mitotic cells [Steffen and Linck, 1988, 1989b]. The number, size, and localization of these structures indicate that they correspond to centrioles. Centrioles were stained by all antitektins in LLC-PK_1 cells, by anti-(L.p.)-tektin C in CHO and HeLa cells, and by anti-(L.p. and S.p.)-tektin A in CHO cells. Centriole staining was not observed when antitektins were preabsorbed with either whole tektin filaments or the respective purified tektin. Because Fc-specific interaction with the intermediate filament protein vimentin had been described [Hansson et al., 1984], Fab fragments for each of the antitektins were used for staining and gave the same results as uncleaved antitektins. In 3T3 (mouse embryonic fibroblast), CHO, GF (gerbil fibroma), and IM (Indian mundjacshin) cells, antitektins stain larger juxtanuclear structures about 1 μm in diameter or polar structures in mitotic spindles; the size of these structures clearly differs from that of centrioles, but their size, location, and cell-cycle dependency indicate they are centrosomes.

B. Midbody

By immunofluorescence, staining of the midbody region is seen in 3T3 and CHO cells with antitektin C and in LLC-PK_1 cells with all three antitektins [Steffen and Linck, 1989b]. As the cells proceed through mitosis, changes in the tektin staining pattern are seen that correspond to the development and disassembly of the midbody. In late anaphase, when the chromosomes have almost reached the spindle poles, a diffuse staining of filamentous material is seen in the equator region that is aligned parallel to the cell axis. This staining pattern becomes more compact in early telophase. As telophase continues, the filaments appear to shorten and move closer together until they form an intensely stained midbody. As the daughter cells begin to separate, the staining gradually decreases.

C. Mitotic and Meiotic Spindles

Isolated mitotic spindles from sea urchin eggs, when treated with 1 mM free calcium, lose almost all their tubulin but still retain a spindle shape [Hays and Salmon, 1983; Leslie et al., 1987; Rebhun and Palazzo, 1988]. The remnant left after microtubule depolymerization is referred to as the *spindle matrix*. In oocytes from the clam *Spisula solidissima,* using a calcium/cold treatment (0.5 mM free Ca^{2+} at 4°C), 95% of the tubulin can be removed from meiotic spindles; under these conditions the spindle shape is retained [Steffen and Linck, 1992]. On SDS-PAGE immunoblots, the following cross-reactions were observed between flagellar antitektins and *S. solidissima* spindle matrix components: anti-(L.p.)-tektin A stains a 48 kD protein; anti-(L.p.)-tektin C stains 64, 100, and >200 kD proteins; and anti-(S.p.)-tektin C stains a 52 kD and faintly a 97 kD protein. However, by immunofluorescence microscopy, only anti-(L.p.)-tektin C revealed a distinct labeling of isolated spindles and artificially activated *Spisula* eggs.

Immunofluorescence staining of the spindle area is also observed in mammalian cells; e.g., with anti-tektin B in CHO cells, with anti-tektin C in HeLa cells, and with anti-tektins A and C in LLC-PK_1 cells. The staining appears diffuse and fibrous with slightly higher intensities at the spindle poles, the latter corresponding to centrioles and centrosomes. Staining is observed from prometaphase until early anaphase [Steffen and Linck, 1989b]. When isolated spindles of LLC-PK_1 cells were treated with Ca^{2+} to destabilize tubulin, anti-(S.p.)-tektin C still labeled the Ca^{2+} resistant spindle remnant, and a strong immunofluorescence signal was observed at the spindle poles. On immunoblots, anti-(S.p.)-tektin C recognized a >100 kD protein. By immunoelectron microscopy of LLC-PK_1 cells, anti-(S.p.)-tektin C appears to stain linear tracts within the spindles; however, the resolution of the technique is not high enough to determine whether the staining is of microtubules or of some other independent fibrous structures [Steffen and Linck, 1992].

In all of these cases of invertebrate and mammalian spindle matrix staining by immunofluorescence and by immuno-EM, the resolution of the technique is not high enough to determine whether the antitektin label is associated with spindle microtubules or with filamentous, nonmicrotubule entities.

VII. SIMILARITIES BETWEEN TEKTINS AND IF PROTEINS

Before describing the similarities between tektins and IF proteins, a brief summary of IF proteins and their structure is indicated [Goldman et al., 1986; Steinert and Roop, 1988; Wang et al., 1985; Parry and Steinert, 1992]. Intermediate filaments are cytoskeletal elements in most mammalian cells and are composed of a group of related proteins; IFs have also been found in some invertebrates [Weber et al., 1988, 1989; Dodemont et al., 1990]. More than 40 IF proteins encoded by a multigene family have been identified. IF and IF-like proteins (i.e., lamins) have been grouped into six or more classes, based on sequence data:

1. I. Acidic keratins
2. II. Neutral–basic keratins
3. III. Vimentin, desmin, glial fibrillary acid protein, peripherin
4. IV. Neurofilaments, α-internexin
5. V. Lamins
6. VI. Nestin

The basic structural subunit of an IF is a dimer. Most IF subunits are homopolymers but keratins are obligate heteropolymers. IF proteins differ in molecular weight and immunological properties but are structurally related. All IF proteins have a highly α-helical central rod domain of ~37 kD that forms the backbone of intermediate filaments. This rod domain is composed of α-helical regions separated by nonhelical linker regions. The rod domain can be divided into two halves that are structurally similar, segment 1 (α-helical regions 1A and 1B separated by nonhelical linker L1) and segment 2 (α-helical regions 2A and 2B separated by nonhelical linker L2); segments 1 and 2 are separated by the nonhelical linker L12. The placement and size of the various α-helical and nonhelical linker regions is highly conserved among all IF proteins; an important exception is found in lamins and invertebrate IF proteins, which have an extra 42 amino acids in region 1B. The α-helical domains are composed of nonidentical heptad repeats in which usually more than 75% of the first and fourth residues are nonpolar; these hydrophobic residues line up to form a helical stripe along the outside of the α-helix, and they interact with a corresponding hydrophobic stripe in a second IF protein to form and stabilize a coiled-coil dimer. In the central rod domain, IF proteins within a class share strong primary sequence homology (70%–95%); however, between classes the homology is only 30% or less [Steinert and Roop, 1988]. The only consistent regions of sequence homology found in all IF proteins are two regions of consensus sequence: nine residues [LNDRL(or F)AXYI] at the start of domain 1A, and 18 residues [LD(or E)XEIA-XYRKLLEGEEXR(or K)] at the end of domain 2B. In this latter region, the sequence YRKLLEGEE represents the immunological epitope for the IFA monoclonal antibody that recognizes most forms of IF proteins [Pruss et al., 1981; Magin et al., 1987]. The N- and C-terminal domains of IF proteins are random coil structures that are highly variable in size and sequence; these end domains account for most of the molecular weight and immunological differences between IF proteins and presumably provide functional properties to the IFs.

Initial studies of the tektin proteins revealed an intriguing similarity to IF proteins. A monoclonal antibody against tektin filaments was used to detect a ~48 nm periodicity along the axis of tektin filaments [Amos et al., 1986]; similar structural repeats have been detected in IFs [Ip et al., 1985; Steven et al., 1985], and theoretically correspond to the ~37 kD, highly α-helical central rod domain. Both tektin and intermediate filaments are relatively insoluble and are composed of smaller, 2 nm fibrils; these fibrils presumably represent coiled coils [Amos et al., 1986; Linck and Langevin, 1982; Linck et al., 1985; Steinert et al., 1985]. Biochemical studies revealed that tektins and IF proteins are similar in terms of their pI range (5.8–7.0), molecular weights (46,000–68,000), and amino acid composition [Linck and Stephens, 1987]. Finally, there are several studies that reveal an immuno cross-reactivity between tektin antibodies and IF proteins. Various antitektin antibodies stain IF patterns in mammalian cells, and there is some evidence that antitektins cross-react with lamins in the nuclear envelopes of mammalian cells and sperm heads, as reviewed below.

A. HeLa Cells

Anti-(L.p.)-tektin B shows an immunofluorescence staining pattern virtually identical to anti-keratin in HeLa cells [Steffen and Linck, 1989a]. Stained filaments are distributed homogeneously throughout the interphase cytoplasm with no major focal point; during cell division they form a cage-like structure around the mitotic apparatus. Treatment of HeLA cells with cytochalasin B will alter the antikeratin staining pattern, and anti-(L.p.)-tektin B staining is altered in the same manner. On SDS-PAGE immunoblots of whole cells, 0.5% Triton X-100 extracted cells, and 0.5% Sarkosyl/2M urea extracted cells, anti-(L.p.)-tektin B stains an ~54 kD polypeptide band and faintly cross-reacts with an ~52 kD band. On two-dimensional SDS-PAGE immunoblots of whole cells, anti-

(L.p.)-tektin B stains isoforms or different proteins of the 54 kD band with pIs of 6.1 and 6.2; these peptides are also recognized by a commercially available monoclonal anti-keratin. The specificity of this cross-reaction was verified by affinity purification of the antitektin antibodies with HeLa keratins bound to nitrocellulose [Olmsted, 1986] and then demonstrating that these keratin-purified antitektin antibodies still cross-reacted with sea urchin tektins [Steffen and Linck, 1989a].

Initially, it was assumed that this cross-reaction of antitektins with IF proteins occurred due to antibodies against tektin α-helical structure staining the α-helical central rod domain of IF proteins. To test this hypothesis, IFs isolated from HeLa cells were digested with α-chymotrypsin which removes the end domains, leaving an ~39 kD fragment that accounts for most of the central rod; subsequently the rod domain is cleaved into smaller fragments of ~18 and ~21 kD [Geisler et al., 1982; Steinert and Roop, 1988]. All these polypeptides were stained by polyclonal antiserum raised against HeLa IFs. However, staining by anti-(L.p.)-tektin B was reduced by more than 98% within 2 minutes. These results could suggest that tektins have epitopes in common with the end domains of certain keratins [Steffen and Linck, 1989a].

B. LLC-PK1 Cells

Immunofluorescence staining of LLC-PK1 cells with anti-(L.p.)-tektin B also reveals a filamentous network; however, this differs from that seen with HeLa cells in terms of size and distribution of the filamentous material [Steffen and Linck, 1989a]. EM immunogold labeling with anti-(L.p.)-tektin B reveals a filamentous pattern that colocalizes with IFs. On SDS-PAGE immunoblots of whole cells, 0.5% Triton X-100 extracted cells, 0.5% Sarkosyl/2M urea extracted cells, and isolated IFs, anti-(L.p.)-tektin B stains an ~55 kD polypeptide and faintly cross-reacts with an ~52 kD band. On two-dimensional IEF/SDS-PAGE immunoblots of whole cells, anti-(L.p.)-tektin B stains two isoforms or different proteins of 55 kD with pIs of 6.1 and 6.3. Anti-(S.p.)-tektin A and anti-(S.p.)-tektin C also show immunofluorescence staining of filament patterns. Anti-(S.p.)-tektin A labels 8–10 nm filaments by EM immunogold labeling.

C. BHK-21 Cells

A monoclonal antibody against flagellar tektin C from *S. purpuratus* (3-7-1) stains filamentous structures in BHK-21 cells; similar structures were seen when BHK-21 cells were stained with antivimentin antibodies [Chang and Piperno, 1987]. Filaments were also observed when 3T3, HeLa, and PtK_2 cells were stained with this antibody, and staining of myotubes from chick embryos revealed desmin-like striated patterns. On immunoblots of cytoskeletal proteins isolated from BHK-21 cells, the antibody cross-reacts with a 55 kD band that is also stained by antivimentin and antidesmin antibodies.

D. Nuclear Envelopes and Sperm Heads

By immunofluorescence microscopy, a second monoclonal antibody against flagellar tektin C from *S. purpuratus* (3-10-1) stains nuclear envelopes of BHK-21 cells and chick embryo monocytes. On immunoblots, this antibody cross-reacts with nuclear lamins A and C from rat hepatocytes [Chang and Piperno, 1987]. Anti-(L.p.)-tektin B stains the nuclear envelope in LLC-PK_1 cells [Steffen and Linck, 1989b], and anti-(L.p.)-tektin C appears to stain the nuclear envelope of GF cells [Steffen and Linck, 1989b] and chick embryo neurons [Edson et al., 1991]. Possible nuclear envelope staining is also seen with sea urchin sperm [Steffen and Linck, 1989b]. Anti-(L.p.)-tektin B affinity purified with *L. pictus* tektins and non-affinity-purified anti-(S.p.)-tektin B intensely stain the outline of the sperm head. All anti-tektins also stain the acrosome regions, and anti-tektin C stains areas coincident with the mitochondria.

E. Cross-Reaction of Anti-IF Protein Antibodies With Tektin Proteins

A monoclonal antibody, IFA, which has been shown to stain most IF proteins [Pruss et al.,

1981], cross-reacts weakly with tektins A and B and less so with tektin C on immunoblots of purified tektins [Amos et al., 1986]. Commercially available, nonaffinity-purified polyclonal antidesmin and antivimentin also cross-react with tektins A and C [Chang and Piperno, 1987]. However, the possibility that this reaction is due to preimmune activity cannot be eliminated.

VIII. PRIMARY STRUCTURE OF TEKTIN A1

A. Isolation and Identification of Tektin cDNA Clones

While the above results would appear to indicate that tektin proteins are related to IF proteins, it was necessary to obtain their amino acid sequences to determine definitively whether tektins are another class of IF proteins. The first step in obtaining primary sequence data was to construct an expression library [Norrander et al., 1992]. Since it had been shown that tektins of the same size and immunospecificity were present in the cilia of sea urchin embryos [Linck et al., 1987], an expression library was constructed from *S. purpuratus* blastula mRNA. Immunoscreening of this library has provided cDNA clones for all three tektin proteins. Two clones for tektin A have been completely sequenced; the cDNA clone tekA10-2 contains approximately 70% of tektin A's coding sequence while tekA5-8 contains the entire coding sequence (Fig. 2). The sequences of the two clones have >97% nucleotide identity and >99% amino acid identity within the protein coding region but have considerable differences in their 5′- and 3′-untranslated regions, indicating the possibility that tektins are encoded by a multigene family. Allowing for this, the tektin A amino acid sequence obtained from clone tekA5-8 has been named *tektin A1*. TekA5-8 was identified as a tektin A clone by several criteria. First, the molecular weight, acidic pI, amino acid composition, and predicted α-helical content of the protein coded for by tekA5-8 all compare favorably with previous experimental measurements for tektin A protein from sea urchin sperm. Second, the pattern of expression of tektin A1 in sea urchin embryo development is consistent with it being a ciliary/flagellar protein (see below). Finally, definitive evidence of tekA5-8's identity was provided by partial amino acid sequencing of *S. purpuratus* flagellar tektin A (Table I). Tektin A1 was cleaved with N-chlorosuccinimide (NCS), the cleavage products separated by SDS-PAGE and transferred to Immobilon [Norrander et al., 1992]. Two of the resulting peptide bands (26 and 12.5 kD) were excised and partially sequenced. The sequences of these two bands is identical, probably because the 12.5 kD band arose from a partial cleavage of the 26 kD band at His-374 or -377. A secondary sequence was also obtained from the 12.5 kD band that matches a predicted, comigrating fragment generated by NCS cleavage at Trp-128 and Trp-230.

B. Expression of Tektin A1 mRNA

TekA5-8 hybridizes to a 2,750 base embryonic mRNA. The developmental expression of tektin A1 was studied by fertilizing and growing *S. purpuratus* embryos *in vitro,* removing aliquots at various points of development, isolating mRNA, and blotting and hybridizing with radiolabeled tekA5-8 [Norrander et al., 1992]. Tektin A1 mRNA reaches peak levels of expression at the blastula stage when ciliogenesis occurs. These observations support the conclusion that tekA5-8 codes for tektin A rather than a cytoplasmic IF protein or nuclear lamin. Several minor bands are visible at earlier points after fertilization; these may represent mRNA for tektin A components of centrioles, centrosomes, or mitotic spindles.

C. Predicted Secondary Structure of Tektin A1

The predicted structure of tektin A1 is a series of α-helical domains separated by nonhelical linkers. Tektin A1 has a central rod segment (residues 87–432) that is predicted to be highly α-helical (70% by the criteria of Robson and Suzuki [1976]). These results compare favorably with the 70% α-helix predicted by cir-

```
                                                                                   TGAAGCGTTCAAGTGACCCGCGGACC  26

     CACCATTTACTTCGTACCGCAGCAACTGTTTAGATTTGAGATTTTATTACCGACCAAAACTACTCAAGTCTATATTTTTAATACAATCACTGGCAAACAATTCCATC  133

     ATG GAT GCT GGT GCA ACT CTC CTT TCC AGG AGC TAT GCT CCA ACG ATC CCT GTC TAT CCA ACA CAG ACC ACA GTA GGA ACC  214
1    met asp ala gly ala thr leu leu ser arg ser tyr ala pro thr ile pro val tyr pro thr gln thr thr val gly thr

     AAG ACA GAC CAG GCC CTG TCT CAA GAT CTT GCC AAG ATG AGC GGC CTA GGT GAA ACC GGA GTC TAC GGT GTT CCC ACT GGG  295
28   lys thr asp gln ala leu ser gln asp leu ala lys met ser gly leu gly glu thr gly val tyr gly val pro thr gly

     GCA CCG GCC GCA CAG GGT TTC CGC TCC GGC AAG CAC ACC ACT CAG GAA TGG CAC GAA TCC AAC TAC AAC AAA TAT TTC CAG  376
55   ala pro ala ala gln gly phe arg ser gly lys his thr thr gln glu trp his glu ser asn tyr asn lys tyr phe gln

     AGT TTC ACA GAC CGC GAT AAC GCT GAG AGG CTA TGC CAC GAG TCC AAG CAG CTC TCA AAC GAG ACC CAC GCT CTG ACC ATG  457
82   ser phe thr asp arg asp asn ala glu arg leu cys his glu ser lys gln leu ser asn glu thr his ala leu thr met

     CGC ACA CAG GCC GAC GTC ACG AAG AAG CTC GGG GAC AGG ATG AAT GAC ATC AAC TTC TGG AAG TTT GAG CTC AAC CGT GAG  538
109  arg thr gln ala asp val thr lys lys leu gly asp arg met asn asp ile asn phe trp lys phe glu leu asn arg glu

     ATT GAG GAG ATG ATC GAG GAG ACA GAC CTG CTA TGT GCA CAG AAG AAG CGG TTA GAG AAT GCG CTG GAT GCA ACA GAA GTG  619
136  ile glu glu met ile glu glu thr asp leu leu cys ala gln lys lys arg leu glu asn ala leu asp ala thr glu val

     CCA CTG AAG ATC GCT AGG GAT AAT CTG ACC TGC AGG TCA AGG AGG CAA GAT ATT GAT CTT GTT GGA GAC AGG GTA GAG ATG  700
163  pro leu lys ile ala arg asp asn leu thr cys arg ser arg arg gln asp ile asp leu val gly asp arg val glu met

                                                                                 asp leu val gly asp arg val glu met
                                                                                 GAT CTT GTT GGA GAC AGG GTA GAG ATG

     GCT CTA AAT AAG GAG GTA GAT ATC ATC ACA AAA GTT CAA GAT CTA CTC AAG AGA ACA TTA GAA CAA TCT GAT AGA CAA ATC  781
190  ala leu asn lys glu val asp ile ile thr lys val gln asp leu leu lys arg thr leu glu gln ser asp arg gln ile

     ala leu asn lys glu val asp ile ile thr lys val gln asp leu leu lys arg thr leu glu gln ser asp arg gln ile
     GCT CTA AAT AAG GAG GTA GAT ATC ATC ACA AAA GTT CAA GAT CTA CTC AAG AGA ACA TTA GAA CAA TCT GAT AGA CAA ATC

     AAG CTC AAC CGT GGT TCA AAG CAC AAA TTG ACC ATG GAC TGG TCG GAT AAA CTC TCC GCC TTC AAG ATT GAT GAG AAA TGT  862
217  lys leu asn arg gly ser lys his lys leu thr met asp trp ser asp lys leu ser ala phe lys ile asp glu lys cys
                      +                                                       +
     lys leu asn arg gly ser lys his lys leu thr met asp trp ser asp lys leu ser ala phe lys ile asp glu lys cys
     AAG CTC AAC CGT GGG TCA AAG CAC AAA TTG ACC ATG GAC TGG TCG GAT AAA CTC TCT GCC TTC AAG ATT GAT GAG AAA TGT

     ACT GGA CTT AAC AAC AAT TCA ACT GAG ATC CAG TAT AAG GAA GGA TCG GCC AAA TTT GAG GCA GTT CAA ACT AAT CCA CAA  943
244  thr gly leu asn asn asn ser thr glu ile gln tyr lys glu gly ser ala lys phe glu ala val gln thr asn pro gln
                                                  +               +                                       +
     thr gly leu asn asn asn ser thr glu ile gln tyr lys glu gly ser ala lys phe glu ala val gln thr asn pro gln
     ACT GGA CTT AAC AAC AAT TCA ACT GAG ATC CAG TAC AAG GAA GGA TCA GCC AAA TTT GAG GCA GTT CAA ACT AAC CCA CAA

     TCC TGG GCA GAG TTT TCG CAT GAC AAT GTC GTC CGT GCT GAG CAT GAG CGT CTT GCC TCC CAG CAG CTC CGC AAC CTG ATA  1024
271  ser trp ala glu phe ser his asp asn val val arg ala glu his glu arg leu ala ser gln gln leu arg asn leu ile
                                                      +       +       +                                       +
     ser trp ala glu phe ser his asp asn val val arg ala glu his glu arg leu ala ser gln gln leu arg asn leu ile
     TCC TGG GCA GAG TTT TCG CAT GAC AAT GTC GTC CGT GCC GAG CAC GAG CGC CTT GCC TCC CAG CAG CTC CGC AAC CTG ATC

     GAC CAG ATC CTC ACA GAC ACC TCC AAC GAC ATG AGG GAA CAG TGC AAC ACC GTC AAC ACA GAG TTT GCT CGC CGC ATC GAA  1105
298  asp gln ile leu thr asp thr ser asn asp met arg glu gln cys asn thr val asn thr glu phe ala arg arg ile glu
                          +
     asp gln ile leu thr asp thr ser asn asp met arg glu gln cys asn thr val asn thr glu phe ala arg arg ile glu
     GAC CAG ATC CTC ACA GAT ACC TCC AAC GAC ATG AGG GAA CAG TGC AAC ACC GTC AAC ACA GAG TTT GCT CGC CGC ATC GAA

     GAG ATG AAT GAC GCC AAG ACG AAG ATG GAA AAC CAC TTG CTG AAG ACC GTT GAA GAT ATC GCT GGC ATG GAG AAG AAC ATC  1186
325  glu met asn asp ala lys thr lys met glu asn his leu leu lys thr val glu asp ile ala gly met glu lys asn ile
                  +           +   +                       +       +
     glu met asn asp ala lys thr lys met glu asn his leu leu lys thr val glu asp ile ala gly met glu lys asn ile
     GAG ATG AAT GAT GCC AAG ACA AAA ATG GAA AAC CAC TTG CTC AAG ACT GTT GAA GAT ATC GCT GGC ATG GAG AAG AAC ATC
```

Figure 2.

```
    AAG GAT CTG ACC CAG GCC GTC AAG GAC AAA GAG GCC CCA ATG AAG GTG GCC CAG ACC AGG TTG GAC CAC AGA ACA CAC AGG 1267
352 lys asp leu thr gln ala val lys asp lys glu ala pro met lys val ala gln thr arg leu asp his arg thr his arg

    lys asp leu thr gln ala val lys asp lys glu ala pro met lys val ala gln thr arg leu asp his arg thr his arg
    AAG GAT CTG ACC CAG GCC GTC AAG GAC AAA GAG GCC CCA ATG AAG GTG GCC CAG ACC AGG TTG GAC CAC AGA ACA CAC AGG

    CCC AAT GTG GAA CTC TGC AGG GAC CCA GCT CAG TAC AGG ATG GTC CAA GAG GTC GGT GAG ATC CAG GAC TCC ATC GAC AAG 1348
379 pro asn val glu leu cys arg asp pro ala gln tyr arg met val gln glu val gly glu ile gln asp ser ile asp lys
                                         +                                           +
    pro asn val glu leu cys arg asp pro ala gln tyr arg met val gln glu val gly glu ile gln asp ser ile asp lys
    CCC AAT GTG GAA CTC TGC AGG GAC CCA GCC CAG TAC AGG ATG GTC CAA GAG GTC GGT GAG ATT CAG GAC TCC ATC GAC AAG

    CTC CAG CAG AAG CTG GCC GAG TCC AAG GCC TCG CTC AAG GAC CTG ATG GAC ACC CGC ATG GCT CTG GAG AAG GAG ATT GCC 1429
406 leu gln gln lys leu ala glu ser lys ala ser leu lys asp leu met asp thr arg met ala leu glu lys glu ile ala

    leu gln gln lys leu ala glu ser lys ala ser leu lys asp leu met asp thr arg met ala leu glu lys glu ile ala
    CTC CAG CAG AAG CTG GCC GAG TCC AAG GCC TCG CTC AAG GAC CTG ATG GAC ACC CGC ATG GCT CTG GAG AAG GAG ATT GCC

    CTT AAG AAG AAC AGC ATC TTC GTT GAC CGC GAC AAG TGC CTC AAG TTC CGC ACC CGT TAC CCA TCC ACC TCC AAG CTG GTC 1510
433 leu lys lys asn ser ile phe val asp arg asp lys cys leu lys phe arg thr arg tyr pro ser thr ser lys leu val
     *                                                                                                  +   +
    pro lys lys asn ser ile phe val asp arg asp lys cys leu lys phe arg thr arg tyr pro ser thr ser lys leu val
    CCT AAG AAG AAC AGC ATC TTC GTT GAC CGC GAC AAG TGC CTC AAG TTC CGC ACC CGT TAC CCA TCC ACC TCC AAG CTT GTT

    GGC TAT CAG TAA ACTGTTCTGATGTGCTCGGTTTCTCAAAATTTGTAATACCACGACTACAAAAGTGAGCATGTTCAGCATAGCTTCATCAGTCCATGTAGGA 1613
460 gly tyr gln

    gly tyr gln
    GGC TAT CAG TAA ACTGTTCTCAGTTGTACTCAGTTTCTCACAAGTTCTAAAATAATGACTACAGAAGTGAGCATGTTCAGTTCAGTATAGTTTCATCAGTCCA

    AACATGTCACAAATTGTAATACCATGGCTACAAAAGTGAGCATGTTCAGTATATCTTCATCAGTCCATGTAGGCAACAAAGATTTGTTTTCTAAACTTATGGCAACG 1720
    TGTAGGCAACACGAACGTTAACTGTATTGAAAACAAAATACAGAGACCTCCAAGCTTGTTCGGTATCAGTAAACTGTTCTCAATTGTGCTCAGTTTGTCACAAATTG

    AGAATCGATATCATTTGACTATCTGATATTATATGTTTTATGGTAATTAGTTCAAAGGCAAATAGTGAAATATGAACTTGTATTGAAAACAAAATACAGCATTTTTG 1827
    TAAAACCATGACTACAAAAGTGAGCATGTTCATTATAGTTTCATCATTCCATGTAGGCAACGAAGATGTGTTCTCTAAACTTATCTAAATTGATATCATTTGACTAT

    AAGGAAAAAAAAATACAAAACCTTTTGCTCACTTTTGTAGTTGTTGAAAAACAGCAAGCCCTGAGTTTTATTTATTTATCTAATATCTGAAACAGTATTGGCCATAC 1934
    CTGATATTTTATGTTTTAGGGTAATTAGTTCAAAGGCAAATAGTGAAATATTAACCTGTATTGAAAACAAAATACAGCATTTTTGAAAAGGAGATACCAAACCTTTT

    AAGCATGAGCTGGATTGAAGGAGTGATAAACAAATCTACATGAAAGTACCATGTCAGTTGACCCCTTGAGAAAATGAGGACTGCTAGATATTTGGACTCGGTTGGAA 2041
    GCTCACTTTTGTAGTTGTTAAAAAACAGCAAGCCCTTAGTTTTATTTATTTATCTAATATCTGAAACAGTATTGGCCATTCAAGCATGAGCTGGAACGAAGGATCGA

    TGTAGTAATATGCATGAAATGGATATGCTGAAATGGGCGCTAACATGCGAATCCACACAGCGGCATATTGCATGTGTTAAACTTTGCATCCAAGAATGCAGGACTAT 2148
    TAAACAAATCTACATGAAAGTACCATGTCAGTTGACCCCTTGAGAAAATGAGGACTGCTAGATATTTGGACTTGGTTAGAATGTAGTAATATGCGTAAAATGGATAT

    AGATCGTTTAGATTTATCTCAGACATACATGTAAGAGAAATTTGTTGTAATAGGATAGATATGTTTGTATATCATGTCAAAACAGCAATTCAAAGATGTGCATGTTT 2255
    GCTGGAATGTGCGCTAACTCGCATATCCATGTGTCCAGGAATGCAGGACTATCGATTGTTTAGATTTATCTTAGACATGAGAGAAATTTGTATTACTAGGATGGATA

    TTGGTGTTTTTTTAGTATTAAATTGCTTGAGTACTCAATACCGATCGGTATCTTTGTATCAATAAAATATACTTTTTGTCATTCCACACTTTCATTAGGAAAAAAAT 2362
    TGTTTGTATATCATGTCAACAGCAAATCAGAGATGTGCATGTTTTTTTTAGTATTGACTTGCTTGAGTACTTAATACTGATAGTTTTATTTGTATCAATGAAATATA

    ACATCCGTCCTTTGATGCACATAATCTATCTCGTCGATTCATGAACAATATATATTTTCTAGTTTCATGGTATAGCACATTCTTGTGAGTTTAGCTGCAATTTCTTA 2469
    GTTTTTGTCATTCCAAAGTTTCATTAGGAAATGAATACATCCGTCCTTTGATGCACATAATCTATCTCGTCGTCGATTCATGAACAATATATATTTTCTAGTTTCAT

    CTTGCAATAACATCTTCTTTAGGTATATATTCATTAGGTTAAATGTTCCCCTATCTATTCTATTATCAAGATTTTTTCTATATATTTGTGCCATGTATTATAGTGTG 2576
    GGTATAGCACATTCTTGTGAATTTAGCTGCAATTTCTTTCTTGCAATAACATCTTTTTTAGGTATATATTCATTAGGTTAAATGTTCCCCTATCTATTCTATTTCCA

    TATATATTTTGAGAAGCCGATTAATTTGTACCTTCGTGAAATTGTGAATCATGCCTTGACTTGAGTGATCATAATAAAAA                           2656
    AGATGTTTTCTATATATTTGTGCCATGTATTATAGTGTGTATATATTTTGAGAAGAGTCATTCAATTATGCCGATTAATTTGTACCTTCGTGAAATTGTGAATCATG

    CCTTGACTTGAGTGATCATAATAAAATTATATAAAAAAAAAAA
```

Fig. 2. *Nucleotide and deduced amino acid sequences of tektin A cDNA clones tekA5-8 and tekA10-2 (shaded sequence). Amino acid residue numbers are listed on the left, and nucleotide residue numbers for tekA5-8 are on the right.* +, *Differences in nucleotide sequences between tekA5-8 and tekA10-2 within the coding region;* *, *differences in amino acid sequence.*

TABLE I. Amino Acid Sequences of Flagellar Tektin A Peptide Fragments*

26 kD fragment main sequence	12.5 kD Fragment Main sequence	cDNA sequence	Secondary sequence	cDNA sequence
A,E,D	A	A273	K	K129
E	E	E	F	F
F	F	F	—	E
SV	SV	S	L	L
—	N(H)	H	N(H)	N
D	D	D	R	R
N	N	N	E	E
V	V	V	I	I
V	V	V	E	E
R	R	R	(D)	E
A	A	A	M	M
E	E	E	I(D)	I
—	—	H		
—	E	E		
—	—	R		
L	DL	L		
A	A	A		
—	D	S290		
QV	Q	Q		
QD	Q	Q		
	DL	L		
	—	R		
	—	N		
	L	L		
	I	I		
	D	D		
	—	Q		
	IA	I		
	L	L		
	T	T		
	D	D		

*Amino acid sequences of the N-terminal ends of fragments from *S. purpuratus* flagellar tektin A, compared with equivalent sequences deduced from cDNA tekA5-8 coding for *S. purpuratus* embryonic ciliary tektin A1. The only discrepancy (aspartic vs. serine) at position 290 could represent a difference between ciliary versus flagellar tektin A1, or it could represent sequence differences between members of the tektin A family. In some positions alternative residues are given; gaps indicate undetermined residues.

cular dichroism of tektin filaments [Linck and Langevin, 1982] and their strong α-type X-ray patterns [Beese, 1984]; see Figure 4 and discussion below.

IX. RELATIONSHIP BETWEEN TEKTIN A1 AND IF PROTEINS

A search of sequence data banks revealed that tektin A1 does not show a high degree of amino acid homology with any other protein [Norrander et al., 1992]. At the most, tektin A1 shares a 13%–15% sequence identity with α-helical, coiled-coil proteins such as keratins, vimentins, nuclear lamins A and C, myosins, and tropomyosins. The various similarities previously noted between tektins and IF proteins (α-helical structure, amino acid composition, pI, MW, filament structure, solubility properties and immunocross-reactivity) suggested that a more detailed comparison between these two protein classes was warranted. The only consistent homology revealed between tektin A1 rod sequence versus various IF protein rods occurs near their C termini, where one of the two IF protein consensus sequences occurs. However, comparison of sequence data reveals that tektin A1 contains only part of this consensus sequence and is lacking the important last nine residues (YRKLLEGEE). Tektin A1 lacks the entire other IF protein consensus sequence located at the beginning of domain 1A.

Despite the low amino acid homology between tektin A1 and IF proteins, tektin A1 does show a long-range structural similarity to IF proteins (Fig. 3) [Norrander et al., 1992]. To facilitate the comparison between tektin A1 and IF proteins, conventional IF protein terminology will be used to refer to the various regions of tektin A1 secondary structure. Like IF proteins, tektin A1 is predicted to contain a central rod domain that is largely α-helical; major portions of the α-helical domains contain the heptapeptide repeats necessary to form coiled coils. The homologous region near the C terminus of tektin A1 was used as a guide to align tektin A1 sequence with IF protein sequences so as to maximize amino acid identities and heptad repeats. Tektin A1's predicted α-helical and nonhelical linker domains are similar in size and location to those of IF proteins, but there are some important differences. Tektin A1 contains two additional linkers whose locations correspond to sites of IF protein structural

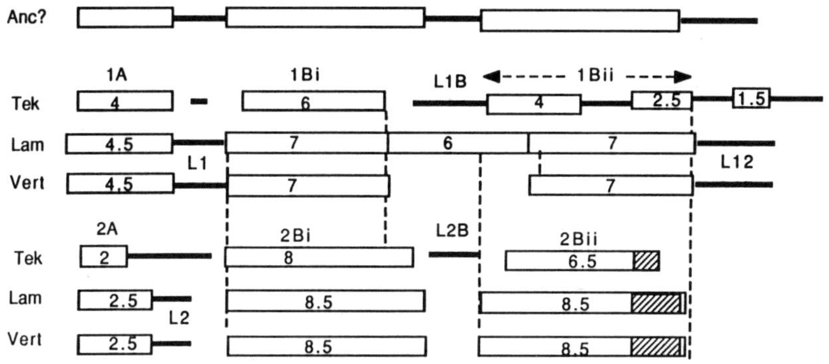

Fig. 3. *Schematic illustration of the structural relationship between the rod domains of tektin A1 (tek), lamins and invertebrate IF proteins (Lam), and vertebrate IF proteins (Vert). Segments 1 and 2 of the rod domains have been aligned vertically to highlight the structural similarity between these two regions. Open boxes represent sequence strongly predicted to be α-helical; numbers within boxes indicate the number of heptad repeats. Solid lines represent nonhelical linkers. Hatched regions indicate the location of the IF protein concensus sequence in region 2B; tektin A1 contains approximately half of this sequence. The length and position of α-helical and nonhelical linker regions is similar among all three protein types. Two additional nonhelical linker regions found in tektin A1 occur at sites of IF protein structural anomalies; linker L1B at the site of a 42 amino acid addition in region 1B of lamins and invertebrate IF proteins and linker L2B at the site of a stutter in heptad phasing. (Anc?) represents the structure of a possible common ancestral gene that may have duplicated to produce tektins and IF proteins.*

anomalies. One of these linkers appears in region 1B at the site where 42 additional amino acids are located in nuclear lamins and invertebrate IF proteins [McKeon et al., 1986; Fisher et al., 1986; Wolin et al., 1987; Weber et al., 1988, 1989; Dodemont et al., 1990]. Adapting standard IF protein terminology, this tektin A1 linker has been named *L1B*, and the resulting halves of rod 1B are referred to as *rods 1Bi* and *1Bii*. The second linker, L2B, is located at the site where a stutter in heptad phasing occurs in IF proteins [Steinert and Roop, 1988; Conway and Parry, 1988]; this linker splits rod 2B into rods 2Bi and 2Bii.

Although tektin A1 and IF proteins are not sufficiently homologous for tektins to be included as a new member of the IF protein family, there are aspects of the tektin sequence that raise the possibility that tektins and IF proteins might have arisen from a common ancestral gene (Fig. 3). For IF proteins, the structural similarity between segments 1 and 2 led to speculation that these arose via a gene duplication event. The low primary sequence homology between segments 1 and 2 did not support this hypothesis [Quax-Jeuken et al., 1983]. In tektin A1, there is considerably more evidence to support a gene duplication hypothesis. Not only do segments 1 and 2 of tektin A1 have a very similar secondary structure, but when tektin A1 was searched for internal repeats, a significant primary sequence homology was found between the two halves (residues 60–250 and 260–450) of the central rod domain (35% homologous with 22% identity and 13% conservative substitutions). This homology is much greater than that of any IF protein. The probability of this degree of internal homology occurring by chance is 10^{-5}. Aligning tektin A1 and IF protein sequences as described above, then using tektin A1 sequence as a guide, segments 1 and 2 of IF proteins can also be aligned to produce the best match, if breaks are inserted at the point of the stutter in IF protein rod domain 2B and in the middle of rod 1B where the extra 42 amino acids of lamins and invertebrate IF proteins are inserted.

The overall structural similarity between tektin and IF proteins could have occurred by convergent evolution; however the possibility cannot yet be excluded that tektins and IF proteins arose instead from a common ancestral gene, structurally similar to one-half of the central rod domains of tektins and IF proteins. In such a case, this gene may have undergone a gene duplication event to produce a protein precursor to tektins and IF proteins [Norrander et al., 1992]. There are three additional pieces of data that are consistent with this evolutionary scheme. First, there are consistent differences between the aligned regions of segments 1 and 2 that are found in both tektin A1 and IF proteins, such as the size difference between rod 1A and 2A. This suggests that the possible gene duplication event occurred before tektins and IF proteins diverged. Second, there is a strong homology between lamins and invertebrate IF proteins that tends to indicate that these protein classes are closer to the putative ancestral protein than vertebrate IF proteins [Weber et al., 1989]. Third, segments 1 and 2 of tektin A1 show greater sequence homology than those of IF proteins, suggesting that tektins have diverged even less than IF proteins from the putative ancestral protein [Chen et al., 1993; Norrander et al., 1992].

IF proteins are related to each other not only by their protein structure but also by their gene structure. The placement of introns is essentially identical for all type III IF proteins; there is some variation in the intron positions of type I and II keratins, but they are clearly related to type III genes [for review, see Steinert and Roop, 1988], while neurofilament genes have a totally different gene structure [Lewis and Cowan, 1986]. Lamin and invertebrate IF protein gene structure is nearly identical and closely related to type III IF proteins [Weber et al., 1988, 1991; Döring and Stick, 1990; Dodemont et al., 1990]. Intron placement differences between invertebrate IF proteins and lamins arise at the sites of two signal sequences present in lamins that are essential for the compartmentalization of lamin proteins in the nuclear envelope; these sequences are absent in both vertebrate and invertebrate IF proteins. These facts suggest that vertebrate IF proteins evolved from a lamin-like ancestor [Weber et al., 1991]. Obviously, more sequence data of both IF proteins and tektins, including tektin genomic sequences, are required before we understand whether tektins and IF proteins are evolutionary related or whether their protein structural similarities have occurred by chance.

X. MODEL FOR TEKTIN ASSOCIATION WITH TUBULIN

The similarities between tektins and IF proteins aside, the clearest definition of the tektins is as microtubule-associated proteins. The biochemical, immunological, and structural evidence reviewed here suggest that tektin filaments exist as extended polymers within the walls of ciliary and flagellar microtubules. This evidence includes the following: 1) antitektins stain filaments that project axially from the ends of microtubules [Linck et al., 1985; Steffen and Linck, 1988]; 2) all three antitektins stain the entire length of all nine doublet microtubules [Steffen and Linck, 1988]; and 3) X-ray diffraction data indicate that α-helices are oriented axially in isolated tektin filaments [Beese, 1984]. Based on these results, we expect tektin molecules to associate with several tubulin dimers within the microtubule wall. A search for possible tubulin-binding sites was conducted by comparing the sequence of tektin A1 to those of microtubule-associated proteins (MAPs). Only weak similarities were found [Norrander et al., 1992]; however, tektin A1 does contain sites that are structurally similar to putative tubulin-binding sites found in MAP2, tau, and a trypanosome MAP [Schneider et al., 1988; Lewis et al., 1989; Goedert et al., 1991]. The tubulin-binding sites in these MAPs consist of a cluster of positively charged residues apparently located on a loop between two helical segments; in tektin A1 the two additional linker regions L1B and L2B both have this feature and may represent novel tubulin-binding sites. Other possible tubulin sites are suggested by a pattern of two to three consecutive positively charged residues whose locations roughly correspond to the spacing of tubulin monomers and dimers when tektin A1 is aligned with tubulin as shown in Figure 4 [Chen et al., 1993; Norrander et al., 1992].

XI. CONCLUSIONS

The structural organization and function of tektins in ciliary and flagellar microtubules have not been resolved. The various data presented above suggest two possibilities for the structure of tektin filaments within microtubules. Tektin filaments may exist as 2–3 nm fibrils located between groves of tubulin protofilaments. Alternatively, they may themselves comprise one or more of the 13 protofilaments of the A-microtubule wall (Fig. 5). The composition of Sarkosyl-insoluble pf ribbons is consistent with both proposals. Within pf ribbons, the molar ratio of tubulin (α and β) to tektins (A, B, and C) to the 77/83 kD pair of polypeptides is consistent with models in which two of the three protofilaments in the pf ribbon are composed of tubulin and the remaining protofilament is composed of either tubulin or tektins, with the 77/83 kD proteins being associated periodically along the ribbon [Linck, 1990]. In the proposed B lattice structure of microtubules [Linck and Langevin, 1981; Linck et al., 1981; Mandelkow et al., 1977, 1986; Mandelkow and Mandelkow, 1989; McEwen and Edelstein, 1980; Woodrum and Linck, 1980], there is a "seam" or helical discontinuity, where two protofilaments are joined differently. It has been proposed that this discontinuity in the microtubule wall and the Sarkosyl-insoluble pf ribbon structure are the same, with the tektin protofilament comprising the "seam" [Linck, 1990].

Possible functional roles of tektins include microtubule stability, microtubule assembly, ciliary length determination, and as a scaffold for the arrangement of axonemal components such as dynein arms, radial spokes, and nexin filaments. The fact that tektin proteins are associated with pf ribbons, a relatively stable component of flagellar doublet microtubules, suggests that tektins might be a factor in microtubule stability. Tektins also appear to be present in relatively stable structures such as basal bodies and centrioles [Steffen and Linck, 1988, 1989b]. In addition, certain classes of cytoplasmic microtubules also have a greater stability to drugs, cold, and Ca^{2+} treatment; included in this group are certain spindle microtubules and midbody microtubules [Brinkley and Cartwright, 1975]. Tektin polyclonal antibodies cross-react with both spindles and midbodies, suggesting the presence of tektins or tektin-like proteins in these structures [Steffen and Linck, 1989b, 1992].

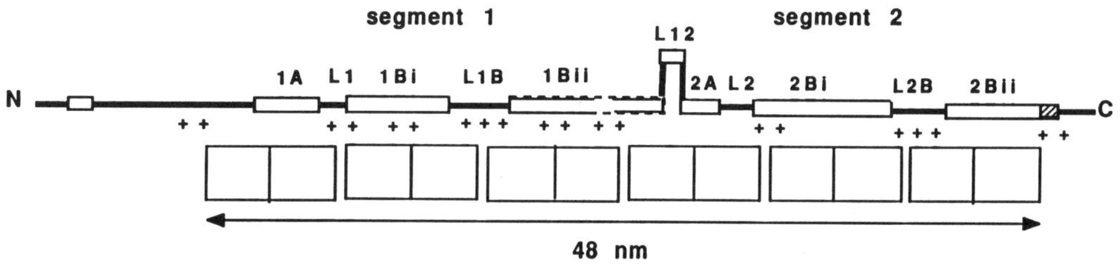

Fig. 4. *A model of the proposed structure of tektin A1, showing how it may associate with tubulin along a microtubule protofilament. Open boxes represent regions of tektin A1 predicted to be α-helical and solid lines represent nonhelical linker regions. The hatched region indicates the tektin A1 sequence that matches approximately half of the IF protein consensus sequence found in region 2B. The double boxes represent 8 nm long tubulin dimers. The legnths of individual tektin molecules have not been directly measured; however, by analogy with lamins, each tektin molecule is probably long enough to line up beside a row of 12 tubulin monomers in the microtubule lattice. Pluses indicate two or three consecutive positively charged residues in the tektin sequence that may represent tubulin-binding sites. The location of these possible binding sites roughly aligns with the spacing of tubulin dimers and monomers.*

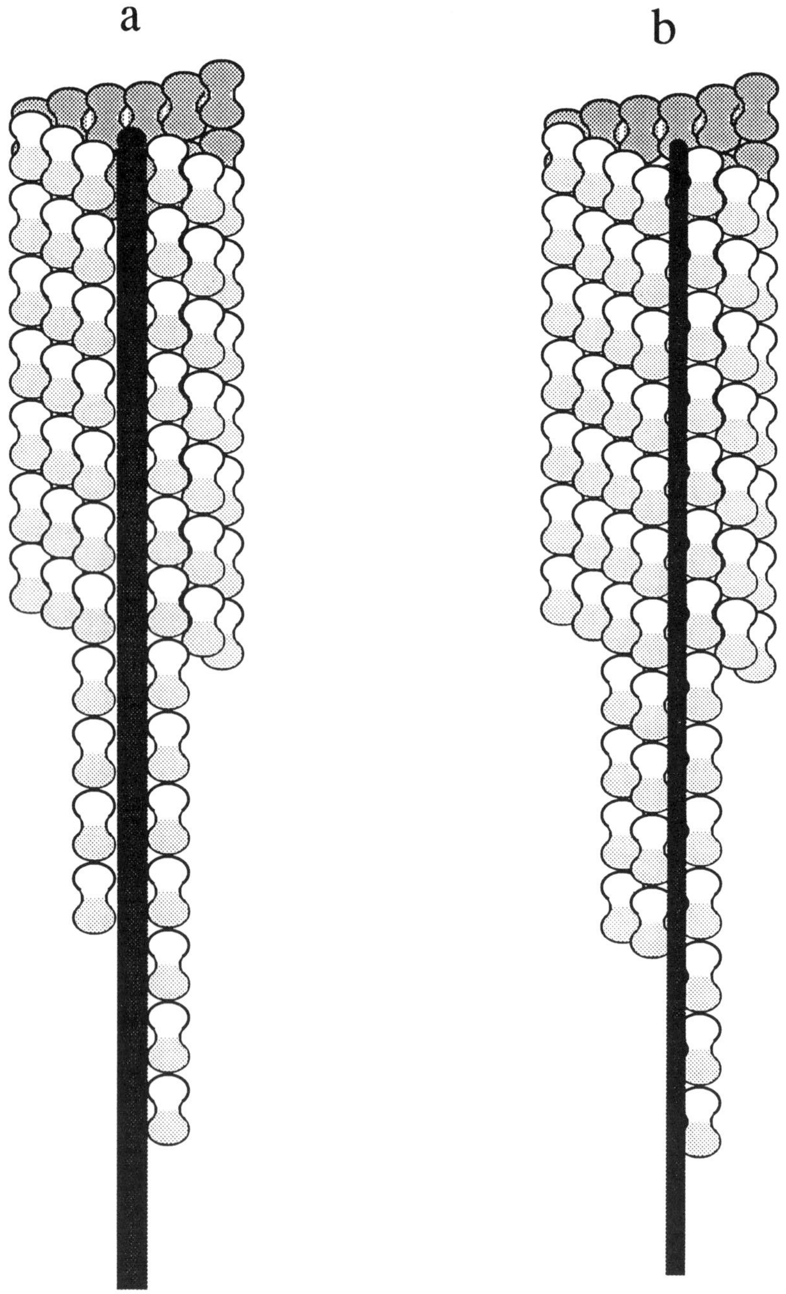

Fig. 5. *Alternative models for the arrangement of tektins in microtubules. The dumbbell shapes represent αβ-tubulin dimers aligned in a B-type lattice to form the protofilaments of the microtubule wall [Harrison et al., 1993; Linck and Langevin, 1981; Mandelkow et al., 1986; Song et al., 1993]. Tektin polymers (in black) are shown in a, as one of the 13 protofilaments of the microtubule wall, and in b, as a 2-3 nm fibril located in the groove between two tubulin protofilaments. In either case, the tektins could be involved in forming a "seam", i.e., the stagger in the match-up of tubulin dimers that occurs in a microtubule with the B-type lattice. Adapted from the authors' work.*

Although the coassembly of tubulin and tektin has not been demonstrated *in vitro,* it could be important in the regulation of microtubule assembly [Stephens, 1977, 1989]. The formation of a tektin tubulin complex (i.e., the pf ribbon or "seam") may trigger microtubule assembly [Linck, 1990] or act to form microtubules with the correct structure.

Another possible function of tektins is indicated by experiments that suggest that ciliary microtubule length in sea urchin embryos is limited by the amount of tektin A synthesized [Stephens, 1989]. After deciliation of *Strongylocentrotus droebachiensis* sea urchin embryos, tektin A is synthesized *de novo* and in limited amounts; additional synthesis of tektin A is required for a second round of ciliary regeneration. Nearly all of the tektin A synthesized in response to deciliation is incorporated into the ciliary axoneme [Stephens, 1977, 1989]. There is some evidence that suggests that the control of tektin A synthesis during ciliary regeneration may occur at the transcriptional level [Norrander et al., 1988]. Based on these observations, it has been proposed that tektin A is a major factor in the elongation and/or length determination of ciliary microtubules [Stephens, 1989].

Finally, tektins may be important factors in determining the arrangement of axonemal com ponents [Amos and Klug, 1974; Amos et al., 1976]. Preliminary evidence indicates tektin filaments may associate with dynein-like polypeptides [Linck, 1990] and nexin filaments [Stephens et al., 1989] and possibly radial spoke components. Other lines of investigation suggest that outer dynein arms determine their own 24 nm axial periodicity as observed in reassociation experiments between *Chlamydomonas* dynein and mammalian brain microtubules [Haimo et al., 1979]; however, the more complex spacings of inner dynein arms may be specified instead by positional information on the doublet microtubule [Smith and Sale, 1992]. The axial arrangement and complex charge distribution of tektin A1 could provide such positional information [Chen et al., 1993; Norrander et al., 1992].

All of the suggested roles for tektin proteins presented above represent exciting possibilities. However, it is clear that much more experimental work is needed to elucidate the part tektin proteins play in microtubule structure and function.

ACKNOWLEDGMENTS

This work was supported by Public Health Services grant GM35648 and Minnesota Medical Foundation grant SMF-527-86 to R.W.L. and by National Research Service Award 1F32-GM11669 to J.M.N.

XII. REFERENCES

Amos WB, Amos LA, Linck RW (1985): Proteins closely similar to flagellar tektins are detected in cilia but not in cytoplasmic microtubules. Cell Motil 5:239–249.

Amos WB, Amos LA, Linck RW (1986): Studies of tektin filaments from flagellar microtubules by immunoelectron microscopy. J Cell Sci Suppl 5:55–68.

Amos LA, Klug A (1974): Arrangement of subunits in flagellar microtubules. J Cell Sci 14:523–549.

Amos LA, Linck RW, Klug A (1976): Molecular structure of flagellar microtubules. In Goldman R, Pollard T, Rosenbaum J (eds): "Cell Motility." Cold Spring Harbor, NY: Cold Spring Harbor Laboratory, pp 847–867.

Beese L (1984): Microtubule Structure: An Analysis by X-Ray Diffraction. Ph.D. thesis. Waltham, MA: Brandeis University.

Brinkley BR, Cartwright J (1975): Cold-labile and cold-stable microtubules in the mitotic spindle of mammalian cells. Ann NY Acad Sci 253:428–439.

Chang XJ, Piperno G (1987): Cross-reactivity of antibodies specific for flagellar tektin and intermediate filament subunits. J Cell Biol 104:1563–1568.

Chen R, Perrone CA, Amos LA, Linck RW (1993): Tektin B1 from ciliary microtubules: Primary structure as deduced from the cDNA sequence and comparison with tektin A1. J Cell Sci 106 (in press).

Conway JF, Parry DAD (1988): Intermediate filament structure: 3. Analysis of sequence homologies. Int J Biol Macromol 10:79–98.

Dodemont H, Riemer D, Weber K (1990): Structure of an invertebrate gene encoding cytoplasmic intermediate filament (IF) proteins: Implications for the origin and diversification of IF proteins. EMBO J 9:4083–4094.

Döring V, Stick R (1990): Gene structure of nuclear lamin LIII of *Xenopus laevis:* A model for the evolution of IF proteins from a lamin-like ancestor. EMBO J 9:4073–4081.

Dunn SD (1986): Effect of the modification of transfer buffer composition and the renaturation of proteins in gels on the recognition of proteins on Western blots by monoclonal antibodies. Anal Biochem 157: 144–153.

Edson KJ, Linck RW, Letourneau PC (1991): Identification of a tektin-like protein associated with neurofilaments in the developing chick nervous system. J Neurosci Res 30:105–115.

Fisher DZ, Chaudhary N, Blobel G (1986): cDNA sequencing of nuclear lamins A and C reveals primary and secondary structural homology to intermediate filament proteins. Proc Natl Acad Sci USA 83:6450–6454.

Geisler N, Kaufmann E, Weber K (1982): Protein chemical characterization of three structurally distinct domains along the protofilament unit of desmin 10 nm filaments. Cell 30:277–286.

Gibbons IR (1965): Chemical dissection of cilia. Arch Biol 76:317–352.

Goedert M, Crowther RA, Garner CC (1991): Molecular characterization of microtubule-associated proteins tau and MAP2. Trends Neurosci 14:193–199.

Goldman RD, Goldman A, Green KJ, Jones JCR, Jones SM, Yang HY (1986): Intermediate filament networks: Organization and possible functions of a diverse group of cytoskeletal elements. J Cell Sci Suppl 5:69–97.

Haimo LT, Telzer BR, Rosenbaum JL (1979): Dynein binds to and crossbridges cytoplasmic microtubules. Proc Natl Acad Sci USA 76:5759–5763.

Hansson GK, Starkebaum GA, Benditt EP, Schwartz SM (1984): Fc-mediated binding of IgG to vimentin-type intermediate filaments in vascular endothelial cells. Proc Natl Acad Sci USA 81:3103–3107.

Harrison BC, Marchese-Ragona SP, Gilbert SP, Cheng N, Steven AC, Johnson KA (1993): Decoration of the microtubule surface by one kinesin head per tubulin heterodimer. Nature 362:73–75.

Hays TS, Salmon TE (1983): A non-microtubular component of sea urchin isolated spindles. J Cell Biol 97:44a.

Ip W, Hartzer MK, Pang Y-YS, Robson RM (1985): Assembly of vimentin *in vitro* and its implications concerning the structure of intermediate filaments. J Mol Biol 183:365–375.

Leslie RJ, Hird RB, Wilson L, McIntosh JR, Scholey JM (1987): Kinesin is associated with a nonmicrotubule component of sea urchin mitotic spindles. Proc Natl Acad Sci USA 84:2771–2775.

Lewis SA, Cowan NJ (1986): Anomalous placement of introns in a member of the intermediate filament multigene family: An evolutionary conundrum. Mol Cell Biol 6:1529–1534.

Lewis SA, Ivanov IE, Lee G-H, Cowan NJ (1989): Organization of microtubules in dendrites and axons is determined by a short hydrophobic zipper in microtubule-associated proteins MAP2 and tau. Nature 342:498–505.

Linck RW (1973): Chemical and structural differences between cilia and flagella from the lamellibranch mollusc *Aequipectens irradians*. J Cell Biol 12:951–981.

Linck RW (1976): Flagellar doublet microtubules: Fractionation of minor components and α-tubulin from specific regions of the A-tubule. J Cell Sci 20:405–439.

Linck RW (1990): Tektins and microtubules. In Miller KR (ed): Advances in Cell Biology, vol. 3. London: JAI Press Inc., pp 35–63.

Linck RW, Albertini DF, Kenney DM, Langevin GL (1982): Tektin filaments: Chemically unique filaments of sperm flagellar microtubules. Cell Motil Suppl 1:127–132.

Linck RW, Amos LA, Amos WB (1985): Localization of tektin filaments in microtubules of sea urchin sperm flagella by immunoelectron microscopy. J Cell Biol 100:126–135.

Linck RW, Goggin MJ, Norrander JM, Steffen W (1987): Characterization of antibodies as probes for structural and biochemical studies of tektins from ciliary and flagellar microtubules. J Cell Sci 88:453–466.

Linck RW, Langevin GL (1981): Reassembly of flagellar B (αβ) tubulin into singlet microtubules: Consequences for cytoplasmic microtubule structure and assembly. J Cell Biol 89:323–337.

Linck RW, Langevin GL (1982): Structure and chemical composition of insoluble filamentous components of sperm flagellar microtubules. J Cell Sci 58:1–22.

Linck RW, Olson GE, Langevin GL (1981): Arrangement of tubulin subunits and microtubule-associated proteins in the central-pair microtubule apparatus of squid *(Loligo pealei)* sperm flagella. J Cell Biol 89:309–322.

Linck RW, Stephens RE (1987): Biochemical characterization of tektins from sperm flagellar doublet microtubules. J Cell Biol 104:1069–1075.

Magin TM, Hatzfeld M, Franke WW (1987): Analysis of cytokeratin domains by cloning and expression of intact and deleted polypeptides in *Escherichia coli*. EMBO J 6:2607–2615.

Mandelkow EM, Thomas J, Cohen C (1977): Microtubule structure at low resolution by X-ray diffraction. Proc Natl Acad Sci USA 74:3370–3374.

Mandelkow E, Mandelkow EM (1989): Tubulin, microtubules, and oligomers: Molecular structure and implications for assembly. In Warner FD, McIntosh JR (eds): Cell Movement, vol. 2. New York: Alan Liss, Inc., pp 23–45.

Mandelkow EM, Schulthesis R, Rapp R, Muller M, Mandelkow E (1986): On the surface lattice of microtubules: Helix starts, protofilament number, seam, and handedness. J Cell Biol 102:1067–1073.

Mazia D (1968): Molecular units of the mitotic apparatus and the government of mitosis. In Letre H, Wagner G (eds): Aktuelle Probleme aus dem gebiet der cancerologie. New York: Springer-Verlag, pp 177–190.

McEwen B, Edelstein SJ (1980): Evidence for a mixed lattice in microtubules reassembled *in vitro*. J Mol Biol 139:123–145.

McKeon FD, Kirschner MW, Caput D (1986): Homologies in both primary and secondary structure between

nuclear envelope and intermediate filament proteins. Nature 319:463–468.
Meza I, Huang B, Bryan J (1972): Chemical heterogeneity of protofilaments forming the outer doublets from sea urchin flagella. Exp Cell Res 74:535–540.
Norrander JM, Amos LA, Linck RW (1992): Primary structure of tektin A1: Comparison with intermediate filament proteins and a model for its association with tubulin. Proc Natl Acad Sci USA 89: 8567–8571.
Norrander JM, Stephens RE, Linck RW (1988): Levels of ciliary 55kD tektin mRNA during sea urchin embryo development. J Cell Biol 107:20a.
Olmsted JB (1986): Analysis of cytoskeletal structures using blot-purified monospecific antibodies. Methods Enzymol 134:467–472.
Parry DAD, Steinert PM (1992): Intermediate filament structure. Curr Opin Cell Biol 4:94–98.
Pruss RM, Mirsley R, Raff MC, Thorpe R, Dowding AJ, Anderton BH (1981): All classes of intermediate filaments share a common antigenic determinant defined by a monoclonal antibody. Cell 27:419–428.
Quax-Jeuken YEFM, Quax WJ, Bloemendal H (1983): Primary and secondary structure of hamster vimentin predicted from the nucleotide sequence. Proc Natl Acad Sci USA 80:3548–3552.
Rebhun LI, Palazzo RE (1988): *In vitro* reactivation of anaphase B in isolated spindles of sea urchin egg. Cell Motil Cytoskel 10:197–209.
Robson B, Suzuki E (1976): Conformational properties of amino acid residues in globular proteins. J Mol Biol 107:327–356.
Schneider A, Hemphill A, Wyler T, Seebeck T (1988): Large microtubule-associated protein of *T. brucei* has tandemly repeated, near identical sequences. Science 241:459–462.
Smith EF, Sale WS (1992): Structural and functional reconstitution of inner dynein arms in *Chlamydomonas* flagellar axonemes. J Cell Biol 117:573–581.
Song Y-H, Mandelkow E (1993): Recombinant kinesin motor domain binds to β-tubulin and decorates microtubules with a B surface lattice. Proc Natl Sci USA 90:1671–1675.
Steffen W, Linck RW (1988): Evidence for tektins in centrioles and axonemal microtubules. Proc Natl Acad Sci USA 85:2643–2647.
Steffen W, Linck RW (1989a): Relationship between tektins and intermediate filament proteins: An immunological study. Cell Motil Cytoskel 14:359–371.
Steffen W, Linck RW (1989b): Tektins in ciliary and flagellar microtubules and their association with other cytoskeletal systems. In Warner FD, McIntosh JR (eds): Cell Movement, vol. 2. New York: Alan Liss, Inc., pp 67–81.
Steffen W, Linck RW (1992): Evidence for a non-tubulin spindle matrix and for spindle components immunologically related to tektin filaments. J Cell Sci 101:809–822.
Steinert PM, Steven AC, Roop DR (1985): The molecular biology of intermediate filaments. Cell 42:411–419.
Steinert PM, Roop DR (1988): Molecular and cellular biology of intermediate filaments. Annu Rev Biochem 57:593–625.
Stephens RE (1970): Thermal fractionation of outer doublet microtubules into A and B-subfiber components: α- and β-tubulin. J Mol Biol 47:353–363.
Stephens RE (1977): Differential protein synthesis and utilization during cilia formation in sea urchin embryos. Dev Biol 61:311–329.
Stephens RE (1989): Quantal tektin synthesis and ciliary length in sea-urchin embryos. J Cell Sci 92: 403–413.
Stephens RE, Oleszko-Szuts S, Linck RW (1989): Retention of ciliary ninefold structure after removal of microtubules. J Cell Sci 92:391–402.
Steven AC, Hainfeld JT, Trus BL, Wall JS, Steinert PM (1985): Conformity and diversity in the structures of intermediate filaments. Ann NY Acad Sci 455:371–380.
Tilney LG, Bryan J, Bush DJ, Fujiwara K, Mooseker MS, Murphy DR, Snyder DH (1973): Microtubules: Evidence for 13 protofilaments. J Cell Biol 59:267–275.
Wang E, Fischman D, Liem RKH, Sun TT (eds) (1985): Intermediate Filaments. NY Acad Sci 455:1–829.
Weber K, Plessmann U, Dodemont H, Kossmagk-Stephan K (1988): Amino acid sequences and homopolymer-forming ability of the intermediate filament proteins from an invertebrate epithelium. EMBO J 7:2995–3001.
Weber K, Plessmann U, Ulrich W (1989): Cytoplasmic intermediate filament proteins of invertebrates are closer to nuclear lamins than are vertebrate intermediate filament proteins; sequence characterization of two muscle proteins of a nematode. EMBO J 8: 3221–3227.
Weber K, Riemer D, Dodemont H (1991): Aspects of the evolution of the lamin/intermediate filament protein family: A current analysis of invertebrate intermediate filament proteins. Biochem Soc Trans 19:1021–1023.
Witman GB (1970): Fractionation and biochemical characterization of the flagella of *Chlamydomonas reinhardi*. J Cell Biol 47:229a.
Witman GB, Carlson K, Berliner J, Rosenbaum JL (1972a): *Chlamydomonas* flagella I. Isolation and electrophoretic analysis of microtubules, matrix, membranes and mastigonemes. J Cell Biol 54:507–539.
Witman GB, Carlson K, Rosenbaum JL (1972b): *Chlamydomonas* flagella II. The distribution of tubulins 1 and 2 in the outer doublet microtubules. J Cell Biol 54:540–555.
Wolin SL, Krohne G, Kirschner MW (1987): A new lamin in *Xenopus* somatic tissue displays strong homology to human lamin A. EMBO J 6:3809–3818.
Woodrum DT, Linck RW (1980): Structural basis of motility in the microtubular axostyle: Implications for cytoplasmic microtubule structure and function. J Cell Biol 87:404–414.

ABOUT THE AUTHORS

JAN M. NORRANDER is a postdoctoral research associate at the University of Minnesota in Minneapolis. She received her B.S. in biochemistry from the University of Minnesota in 1978. She received her Ph.D. in biochemistry from the University of Minnesota in 1984, after working in the laboratory of Joachim Messing on oligonucleotide-directed *in vito* mutagenesis and the expression in *E. coli* of the genes for the zein corn storage proteins. She is currently working in the laboratory of Richard Linck at the University of Minnesota. Her work focuses on the cloning, sequencing, and expression of the tektin proteins.

RICHARD W. LINCK is Professor of Cell Biology and Neuroanatomy at the University of Minnesota School of Medicine in Minneapolis, Minnesota, where he teaches graduate and medical courses in molecular cell biology. After receiving an A.B. from Stanford University in California, he pursued doctoral research in the laboratory of Raymond Stephens at Brandeis University, where he earned a Ph.D. in 1972 for work on the structure and biochemistry of ciliary and flagellar microtubules and dynein from marine invertebrates. This was followed by postdoctoral research at the Laboratory of Molecular Biology, Medical Research Council, Cambridge, England, where he worked with Aaron Klug and Linda Amos on the structural chemistry of microtubules. Dr. Linck's research has focused on the molecular biology and structure of certain microtubule proteins named *tektins,* and he is investigating their function in microtubules. Between 1975 and 1985 he held positions as Assistant and Associate Professor in the Department of Anatomy, Harvard Medical School, Boston, Massachusetts.

Microtubules: 221–228

Microtubule Stabilization

Robert L. Margolis and Didier Job

I. INTRODUCTION

Microtubules exhibit a full range of stability states in fulfilling their various cellular functions. They can exhibit rapid and stochastic growth and shrinkage cycles [Kirschner and Mitchison, 1986], or can treadmill while maintaining a relatively constant length [Margolis and Wilson, 1981], or can enter into states in which no subunit turnover is apparent over prolonged periods of time [Farrell, 1982]. These highly variant stability states may in principle be regulated by production of specific tubulin isotypes, by covalent modification of tubulin, or by association with proteins having a specific stabilizing function. The existence of several isotypes of both α- and β-tubulin is widespread among organisms. However, to date there has been no evidence that different tubulin isotypes generate differential polymer stability. In contrast, in *Drosophila* a single tubulin isotype was shown to be expressed for spermatocyte specific development but was used indiscriminately in this pathway to construct the dynamic microtubules of the meiotic spindle and the stable microtubules of the sperm axoneme [Kemphues et al., 1980].

There are multiple covalent modifications of tubulin. These include acetylation on lysine-40 of the β-subunit [Le Dizet and Piperno, 1987], tyrosination of the carboxyterminus of the α-subunit [Thompson, 1982], and the truncation of the C terminus of the α-subunit in mammalian brain by one tyrosyl residue or by a glutamyl-tyrosine dipeptide [Thompson, 1982; Paturle-Lafanechère et al., 1991]. The acetylation and tyrosination reactions are widespread among the phyla, and both apparently make use of enzymes specifically tailored to perform these modifications only on tubulin [Thompson, 1982]. However, the meaning of these modifications to cell physiology remains a mystery.

Specific antibodies have been raised that recognize acetylated subunits and tyrosinated, detyrosinated and nontyrosinatable tubulins. These antibodies show that tubulin is both highly acetylated [Schulze et al., 1987] and detyrosinated [Piperno and Fuller, 1985] in highly stable microtubules. These results could indicate that such covalent modifications are either directly responsible for the change in microtubule stability or could mark polymers for the binding of stabilizing proteins. However, it is evident from careful *in vivo* analysis that these changes occur as a result of microtubule stabilization and do not have a role in inducing microtubule stability [Khawaja et al., 1988]. The simple explanation is that stable microtubules make good substrates that permit sufficient time for these reactions to go to completion.

Exhaustive *in vitro* analyses of tubulin varying in its tyrosination state have confirmed that this covalent modification induces no measurable change in polymer stability [Khawaja et al., 1988; Paturle et al., 1989], nor does it alter the ability of microtubule associated proteins to stabilize polymers [Khawaja et al., 1988; Paturle et al., 1989]. Cilia and flagella have perhaps the most impressively stable microtubules. The outer doublet microtubules of

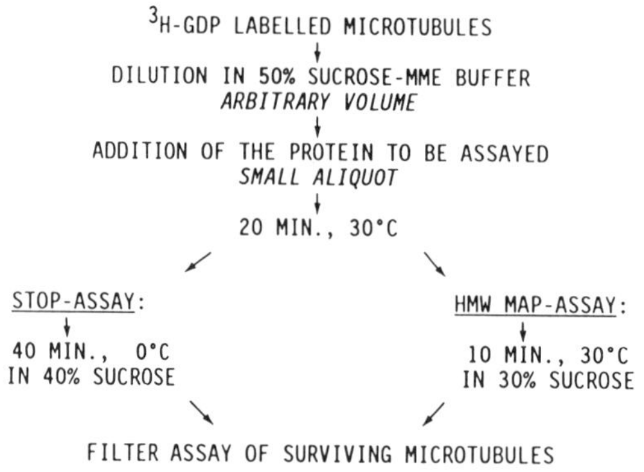

Fig. 1. *A general assay of microtubule stability effectors. Conditions for microtubule assembly and labeling are described by Pabion et al. [1984], Job et al. [1985], and Pirollet et al. [1988]. In the same articles, the STOP assay and the HMW MAP assay are extensively described.*

flagellar axonemes do not exchange subunits *in vivo*, and isolated axonemes retain their extreme stability state *in vitro* [Farrell, 1982]. Isolated axonemes may be disassembled by sonication. When their constituent tubulin is purified and assembled *in vitro*, it exhibits a temperature dependency and lability quite similar to that of mammalian brain tubulin [Farrell, 1982]. Thus, overwhelming evidence to date shows that pure tubulin from a variety of phyla assembles into relatively unstable polymers in the absence of associated proteins. It is therefore a substantial role of such proteins to stabilize polymers to fulfull their various specific functions.

A number of microtubule-associated proteins with such stabilizing functions has been identified. In general, these proteins have been shown to facilitate polymer assembly at the physiological temperature of mammalia [Timasheff and Grisham, 1980] and to dampen intrinsic polymer instability so that treadmilling dynamics prevail [Margolis and Wilson, 1981]. However, polymers assembled under these conditions remain highly labile to cold temperature and to dilution [Timasheff and Grisham, 1980; Job et al., 1985]. Clearly, these microtubule-associated protein associations do not account for the large number of microtubule stability states found in nature.

Brain microtubules are largely stable to cold temperature [Baas and Black, 1990] and to dilution, and mitotic spindle microtubule subpopulations can be highly stable to cold temperature [Brinkley and Cartwright, 1975; Margolis et al., 1990]. Various organelles composed of microtubules, including cilia and flagellae, contain highly stable polymers [Farrell, 1982]. The great majority of eukaryotic organisms exist and thrive at ambient temperatures well below the temperature at which pure tubulin assembly initiates *in vitro*. It has been our task to identify those associated proteins that induce states of microtubule stability concordant with those observed *in vivo* and to use these data to define the dynamics and regulation of microtubule polymers under real-life conditions.

To date, we have identified three proteins that are capable of inducing complete microtubule stability *in vitro*. They are STOP protein, myelin basic protein (MBP), and histone H1. Of these, there is compelling evidence that two, STOP and histone H1, have a physiological role in microtubule stabilization *in vivo*. It is presently unknown if MBP serves a similar *in vivo* function.

II. THE ASSAY

A sensitive assay for identifying microtubule stabilization *in vitro* has been essential to the isolation of specific stabilizing factors. The assay we use involves trapping radiolabeled microtubules on glass fiber filters after exposure to destabilizing conditions such as cold temperature or dilution. The survival of polymers during such challenges is proportional to the amount of stabilizing protein present.

The assay requires no covalent modification of tubulin. Polymers are assembled in the presence of [^{3}H]GTP. Each tubulin subunit incorporates a molecule of GTP, which is hydrolyzed to GDP during polymer assembly. The GDP is quantitatively retained in the polymer and serves as a marker for the assembled state, as well as for the relative position of the labeled subunit within the polymer [Margolis and Wilson, 1978].

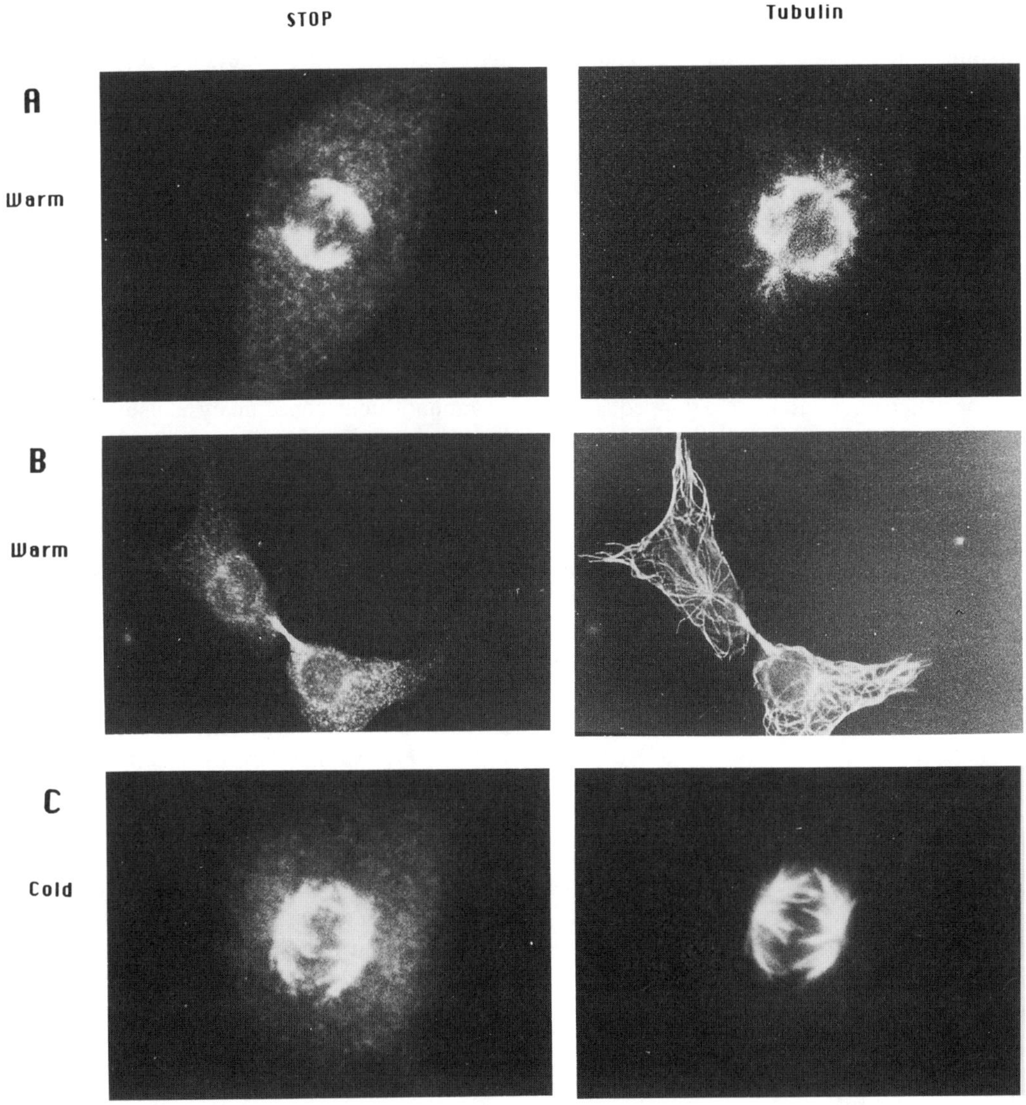

Fig. 2. *Double-label analysis of the relative distribution of STOP and tubulin in mitotic cells. Dorsal root ganglion cells, grown in primary culture, were prepared for immunofluorescence analysis with both rabbit anti-α- and β-tubulin [Chemicon, AB935, used at 1:60) and mouse anti-STOP antibodies. The distributions of the two antibodies were determined using secondary antibodies at 1:200 with different fluors (fluorescein, STOP; rhodamine, tubulin). The secondary antibodies (Chemicon) were purchased ''pre-cleared'' to avoid cross-species reactivity. Controls using both secondaries but only one primary were free of nonspecific fluorescence. The top two sets of images show cells fixed at physiological temperature; the bottom set has been fixed after 20 min exposure to 0°C.* **A:** *A metaphase cell is shown. At physiological temperature, only a subset of spindle microtubules contains STOP. Astral microtubules and interpolar microtubules, clearly apparent with antitubulin, do not appear to contain STOP antigen.* **B:** *Cells connected by a midbody bridge of microtubules show the presence of STOP in the midbody at physiological temperature, but show no organized presence of STOP on the interphase array of microtubules seen with tubulin antibody.* **C:** *At cold temperature, the kinetochore-to-pole microtubules resist disassembly (antitubulin image is at* **right**), *and this resistant population is coincident with the distribution of STOP antigen in these cells* **(left)**.

For an assay of microtubule stabilization to cold temperature, microtubule protein (either containing microtubule-associated proteins or composed of pure tubulin) is assembled in the presence of [^{3}H]GTP and a nucleotide regenerating enzyme system. After assembly at warm temperature, polymers are diluted into a sucrose buffer in which they are stable at warm temperature but unstable in the cold. Samples are then incubated briefly with protein fractions to be assayed and then exposed to 4°C for 40 minutes. After fixation with glutaraldehyde, samples are then applied to GF/C glass fiber filters on which microtubules are trapped. They are then assayed for radioactivity. Details of these assays, as well as controls, have been published elsewhere [Job et al., 1985; Pirollet et al., 1988]. As is evident from the fact that one can use microtubule-associated protein–containing or microtubule-associated protein–free polymers for this assay [Pirollet et al., 1988], microtubules are not stabilized measurably against cold temperature by the set of MAP proteins that recycle with tubulin from brain extracts.

III. STOP

The first microtubule cold stabilizing activity that we identified was a brain-specific protein designated *stable tubule only polypeptide* (STOP). We have used the assays described above to purify STOP to homogeneity from rat [Pirollet et al., 1988] and beef [Pirollet et al., 1992] brain. Curiously, it has an apparent mass of 145 kD in rat brain extracts and 220 kD in beef brain extracts.

Extensive analysis of STOP protein from rat brain has been performed. It is a Ca^{2+}-calmodulin–binding protein, and its activity is totally down regulated by Ca^{2+}-calmodulin [Pirollet et al., 1988, 1992]. It yields 50% stabilization of microtubules to cold temperature *in vitro* at a molar ratio of 1:100 tubulin dimers. STOP also has the property of protecting pure tubulin microtubules against dilution-induced disassembly at warm temperature [Margolis et al., 1990] and completely protects microtubule-associated protein–microtubules against disassembly induced by millimolar calcium [Pirollet et al., 1988]. Neuronal STOP at high concentration can make microtubules stable to a variety of physical assaults, generating what we have called *superstable microtubules* [Job and Margolis, 1984]. In contrast, at low molar concentrations, STOP is rarely and randomly dispersed on the polymer surface and does not protect microtubules against disassembly induced by shearing. The activity of STOP is regulated, directly or indirectly, by protein kinase activity [Job et al., 1983].

Immunofluorescence analysis using highly specific monoclonal antibodies raised against rat brain STOP has shown it to be present along stable microtubules of neuronal axons (unpublished data), and present on cold-stable microtubules of the mitotic spindle and midbody of neuronal cells in culture [Margolis et al., 1990]. It has been known for some time that kinetochore-associated microtubules of the mitotic spindle are especially stable to cold temperature [Brinkley and Cartwright, 1975; Margolis et al., 1990] and that they are specifically associated with Ca^{2+}-calmodulin [Welsh et al., 1979]. The correlation with the presence of STOP, which has these properties *in vitro*, suggests that STOP accounts for the special stability of kinetochore microtubules.

We now know that STOP activity is not confined to neuronal cells. Many nonneuronal tissues contain protein activities with similar column elution profiles and similar Ca^{2+}-calmodulin sensitivity [Pirollet et al., 1988]. Recently, a specific polyclonal antibody raised against neuronal STOP has shown a staining pattern on spindle microtubules in a variety of nonneuronal mammalian cells (unpublished data).

Sequence analysis of STOP cDNA, currently under way, will be used to identify definitively the nonneuronal STOPs and will also determine how they relate to neuronal STOP with respect to structure and function.

STOP at low molar ratio protects microtubules to cold temperature exposure but does not appear to interfere with treadmilling dynamics of polymers at elevated temperature. Microtubules exhibit the same rate of treadmilling, turning over their subunits in a unidirectional exchange mechanism, regardless of whether they contain STOP [Margolis et al., 1986]. The explanation for this behavior appears to be that STOP continuously slides on the surface of the polymer unless the polymer is challenged by cold temperature [Pabion et al., 1984; Margolis et al., 1986]. Sliding behavior thus permits microtubule subunits to transit from one end to the other of the polymer, unimpeded by the presence of STOP.

Sliding behavior of STOP was determined by a variant on the standard STOP assay. Microtubules were labeled by GTP only in their cold-stable region or in their cold-labile regions and then were assayed at time points for overall stability and for the stability of the region that was cold stable at the beginning of the incubation. The results [Pabion et al., 1984] clearly showed that polymers remained equally stable over time, but that the region of stability was constantly shifting. It was evident from control experiments that STOP must be freely sliding on the polymers and does not exhibit any measurable exchange between polymers at elevated temperature. The capacity of STOP to slide is not unique. It was subsequently shown that dynein also slides in a diffusional manner on microtubules in the absence of ATP [Vale et al., 1985].

The apparent ability of STOP to slide on microtubules and the unimpeded treadmilling of STOP-containing microtubules require a redefinition of microtubule stability. It is evident that STOP induces microtubule stability to a variety of physical and chemical assaults, but its stability does not eliminate the ordinary dynamic behavior of the polymer at elevated temperature.

The activity of neuronal STOP best corresponds to the following physical parameters: STOP slides freely on microtubules at elevated temperatures, but is not capable of exchanging from the polymer surface. If STOP is very abundant, the entire polymer is stable. When challenged with a destabilizing condition like cold temperature, the polymer disassembles to the position of a STOP protein that undergoes a conformational change and protects the new polymer end from further disassembly.

It is evident that microtubules associated with the kinetochore in the mitotic spindle are dynamic and exhibit a rapid poleward subunit flux or treadmilling behavior [Mitchison, 1989], as predicted [Margolis et al., 1978; Margolis and Wilson, 1981]. This behavior occurs despite the firm attachment of polymers at their two ends to the kinetochore and to the centrosome and also despite their apparently complete stability to cold temperature. All of these properties can be explained and permitted by the presence of STOP or a STOP-like protein on these microtubules. We have modeled anaphase motion of chromosomes poleward through a diffusional process that depends on the sliding of microtubules past a protein such as STOP that associates laterally with the polymer and is anchored at a site such as the centromere or centrosome [Garel et al., 1987]. We have called this mechanism *polymer guided diffusion*.

It has subsequently been shown that microtubules freely assemble or disassemble *in vitro* at their point of association with the kinetochore [Koshland et al., 1988], and a mechanism, equivalent to our proposed diffusion based process, was suggested on the basis of these observations for anaphase motility. Although a number of microtubule-based motor proteins have been identified in mitosis [Hyman et al., 1992; Nislow et al., 1992; Sawin et al., 1992; Yen et al., 1992], none thus far has shown activity moving chromosomes in anaphase, thus the diffusion-based mechanism is presently viable.

IV. MYELIN BASIC PROTEIN

In the process of exhaustively determining those proteins with microtubule cold-stabilizing activity in beef brain, we determined that there are only two proteins with detectable activity. One is the STOP protein described above, which immunofluorescence microscopy shows to be restricted in distribution to neuronal cells. The second is myelin basic protein [Pirollet et al., 1992]. Myelin basic protein (MBP) is present uniquely in oligodendroglial cells. These cells contain cold-stable microtubules (unpublished observations) but do not contain STOP, which is present only in neuronal cells (unpublished observations). It therefore appears that the two microtubule cold-stabilizing proteins in mammalian brain are distributed to mutually exclusive cell types. Like STOP, MBP binds Ca^{2+}-calmodulin, and its microtubule cold-stabilizing activity is abolished in the presence of Ca^{2+}-calmodulin [Pirollet et al., 1992]. MBP has previously been reported to interact with tubulin, and there is evidence for the association of MBP with microtubules in oligodendrocyte cell cultures [literature reviewed by Pirollet et al., 1992]. Our findings suggest that there may be a well-defined biological function for MBP that correlates with its physical association on microtubules in oligodendrocytes.

Microtubule interactions with the cell membrane may play an important role in cell morphogenesis. It is possible that the ability of microtubules to mold oligodendrocyte cell shape may be controlled by proteins such as MBP that can link between microtubules and the plasma membrane. Stabilization of microtubules within this framework can act to lock in changes in cell morphology during the outgrowth of oligodendrocyte projections. We are presently investigating microtubule stability and morphogenesis in shiverer mutant oligodendrocytes, mouse cells that are defective for MBP production [Popko et al., 1987], in order to define the specific *in vivo* stabilizing interactions between MBP and microtubules.

V. HISTONE H1

Cilia and flagella represent some of the most stable microtubule structures [Farrell, 1982]. The basis of the stability of sea urchin sperm flagellar microtubules has recently been elucidated by Multigner et al. [1992]. Microtubules were disassembled from axonemes with chaotropic agents, and the proteins solubilized in this fashion were assayed for their ability to stabilize microtubules to cold temperature using the glass fiber filter assay described above.

A single 34 kD protein had all the microtubule cold-stabilizing activity. Remarkably, sequence analysis showed it to be identical to sea urchin histone H1. Immunolocalization analysis showed histone H1 is indeed present along axonemal microtubules. This and other observations ruled out the possibility that histone H1 arose by extraction of contaminating sperm heads. These results, in summary, suggest an important new role for histone H1 and show there is a dual role for this histone in cell metabolism.

VI. SUMMARY

Thus far, three proteins have been identified that stabilize microtubules to extraordinary destabilizing conditions. Each is specific to a cell type or organelle, and each is known to codistribute with microtubules *in vivo*.

What is extraordinary is the diversity of these proteins, which have little sequence commonality and no apparent common origin. Although immunoanalysis with polyclonal antibodies indicates that neuronal STOP may be similar to proteins that stabilize microtubules in non-neuronal cells, it is nonetheless apparent that nature has solved the problem of stabilizing microtubule polymers for specific functions a number of times, and each time the solution has been unique. It is now the task to define what these proteins have in common that makes them so specifically active. Such points of commonality might suggest the mechanisms by which polymer stabilization occurs.

VII. REFERENCES

Baas PW, Black MM (1990): Individual microtubules in the axon consist of domains that differ in both composition and stability. J Cell Biol 111:495–509.

Brinkley BR, Cartwright J (1975): Cold labile and cold stable microtubules in the mitotic spindle of mammalian cells. Ann NY Acad Sci 253:428–439.

Farrell KW (1982): Purification and reassembly of tubulin from outer doublet microtubules. Methods Cell Biol 24:61–78.

Garel JR, Job D, Margolis RL (1987): Model of anaphase chromosome movement based on polymer-guided diffusion. Proc Natl Acad Sci USA 84:3599–3603.

Hyman AA, Middleton K, Centola M, Mitchison TJ, Carbon J (1992): Microtubule motor activity of a yeast centromere-binding protein complex. Nature 359: 533–535.

Job D, Margolis RL (1984): Isolation from bovine brain of a superstable microtubule subpopulation with microtubule seeding activity. Biochemistry 23:3025–3031.

Job D, Pabion M, Margolis RL (1985): Generation of microtubule stability subclasses by microtubule-associated proteins: Implications for the microtubule "dynamic instability" model. J Cell Biol 101: 1680–1689.

Job D, Rauch CT, Fishcer EH, Margolis RL (1983): Regulation of microtubule cold stability by calmodulin-dependent and independent phosphorylation. Proc Natl Acad Sci USA 80:3894–3898.

Kemphues KJ, Raff EC, Raff RA, Kauffman TC (1980): Mutation in a testes specific β tubulin in *Drosophila:* Analysis of its effects on meiosis and map location of the gene. Cell 21:445–451.

Khawaja S, Gundersen GG, Bulinski JC (1988): Enhanced stability of microtubules enriched in detyrosinated tubulin is not a direct function of detyrosination level. J Cell Biol 106:141–149.

Kirschner M, Mitchison T (1986): Beyond self-assembly: From microtubules to morphogenesis. Cell 45:329–342.

Koshland DE, Mitchison TJ, Kirschner MW (1988): Polewards chromosome movement driven by microtubule depolymerisation *in vitro*. Nature 331:499–504.

Le Dizet M, Piperno G (1987): Identification of an acetylation site of *Chlamydomonas* α tubulin. Proc Natl Acad Sci USA 84:5720–5724.

Margolis RL, Job D, Pabion M, Rauch CT (1986): Sliding of STOP proteins on microtubules: A model system for diffusion-dependent microtubule motility. Ann NY Acad Sci 466:306–321.

Margolis RL, Rauch CT, Pirollet F, Job D (1990): Specific association of STOP protein with microtubules *in vitro* and with stable microtubules in mitotic spindles of cultured cells. EMBO J 9:4095–4102.

Margolis RL, Wilson L (1978): Opposite end assembly and disassembly of microtubules at steady state *in vitro*. Cell 13:1–8.

Margolis RL, Wilson L (1981): Microtubule treadmills: Possible molecular machinery. Nature 293:705–711.

Margolis RL, Wilson L, Kiefer BI (1978): Mitotic mechanism based on intrinsic microtubule behavior. Nature 272:450–452.

Mitchison TJ (1989): Poleward microtubule flux in the mitotic spindle: Evidence from photoactivation of fluorescence. J Cell Biol 109:637–652.

Multigner L, Gagnon J, Van Dorsselaer A, Job D (1992): Stabilization of sea urchin flagellar microtubules by histone H1. Nature 360:33–39.

Nislow C, Lombillo VA, Kuriyama R, McIntosh JR (1992): A plus-end directed motor enzyme that moves antiparallel microtubules *in vitro* localizes to the interzone of mitotic spindles. Nature 539:543–547.

Pabion M, Job D, Margolis RL (1984): Sliding of STOP proteins on microtubules. Biochemistry 23:6642–6648.

Paturle L, Wehland J, Margolis RL, Job D (1989): Complete separation of tyrosinated detyrosinated and non tyrosinatable brain tubulin subpopulations using affinity chromatography. Biochemistry 28:2698–2704.

Paturle-Lafanechère L, Éddé B, Denoulet P, van Dorsselaer A, Mazarguil H, Le Caer JP, Wehland J, Job D (1991): Characterization of a major brain tubulin variant which cannot be tyrosinated. Biochemistry 30:10523–10528.

Piperno G, Fuller HT (1985): Monoclonal antibodies specific for an acetylated form of α-tubulin recognize the antigen in cilia and flagella from a variety of organisms. J Cell Biol 101:2085–2084.

Pirollet F, Derancourt J, Haiech J, Job D, Margolis RL (1992): Ca^{++}-calmodulin regulated effectors of microtubule stability in bovine brain. Biochemistry 31:8849–8855.

Pirollet F, Rauch CT, Job D, Margolis RL (1988): Monoclonal antibody to microtubule-associated STOP protein: Affinity purification of neuronal STOP activity and comparison of antigen with activity in neuronal and non neuronal cell extracts. Biochemistry 28:835–841.

Popko B, Puckett C, Lai E, Shine HD, Readhead C, Takahashi N, Hunt SW III, Sidman RL, Hood L (1987): Myelin deficient mice: expression of myelin basic protein and generation of mice with varying levels of myelin. Cell 48:713–721.

Sawin KE, Le Guellec K, Philippe M, Mitchison TJ (1992): Mitotic spindle organization by a plus-end directed microtubule motor. Nature 539:540–543.

Schulze E, Asai DJ, Bulinski JC, Kirschner M (1987): Posttranslational modifications and microtubule stability. J Cell Biol 105:2167–2177.

Thompson WC (1982): The cyclic tyrosination/detyrosination of alpha tubulin. Methods Cell Biol 24: 235–255.

Timasheff SN, Grisham LM (1980): *In vitro* assembly of cytoplasmic microtubules. Annu Rev Biochem 49: 565–591.

Vale RD, Soll DR, Gibbons IR (1985): One dimensional diffusion of microtubules bound to flagellar dynein. Cell 59:915–925.

Welsh MJ, Dedman JR, Brinkley BR, Means AR (1979): Tubulin and calmodulin. Effects of microtubules and microfilament inhibitors on localization in the mitotic apparatus. J Cell Biol 81:624–634.

Yen TJ, Li G, Shaar BT, Szilak I, Cleveland DW (1992): CENP-E is a putative kinetochore motor that accumulates just before mitosis. Nature 539:536–539.

ABOUT THE AUTHORS

ROBERT L. MARGOLIS is Director of Research, First Class (CNRS) at the Institut de Biologie Structurale in Grenoble, France, where he directs research on cytoskeletal proteins and centromeres, and on their cell cycle regulation. He received a B.A. degree in biology from Rutgers College in New Jersey in 1969, then completed a Ph.D. in molecular and cellular biology in 1975 at Wesleyan University in Connecticut in the laboratory of Dr. Barry Kiefer, working on the control of microtubule assembly. He then pursued postdoctoral research in the laboratory of Dr. Leslie Wilson, first at Stanford University, then at the University of California, Santa Barbara. In his postdoctoral work he demonstrated an intrinsic motor activity of microtubules, termed treadmilling or flux. From 1979 to 1991, Dr. Margolis was a member of the faculty at the Fred Hutchinson Cancer Research Center in Seattle, where he continued work on microtubule biochemistry, in particular stabilization by the protein STOP. He also demonstrated that the centromere protein, CENP-A, is a special histone. These projects continue to be pursued in Grenoble. In addition, his laboratory is presently concerned with the regulation of checkpoint controls in the mammalian cell cycle.

DIDIER JOB is Director of Research (INSERM) at the Département de Biologie Moléculaire et Structurale of the Centre d'Etudes Nucléaires in Grenoble, France. He received an M.D. degree in 1972 from the Université de Paris. He then spent five years as Assistant Professor (Chargé de Recherche) at the Institut Gustave Roussy in Villejuif, France, doing epidemiological research on diabetes. After turning to basic sciences, he received a Fogarty Fellowship in biochemical research at the University of Washington, Seattle, where he worked from 1979 to 1981. He then returned to Grenoble, where he established an independent biochemistry laboratory. He has since published extensively on microtubule dynamics and stabilization. Presently his laboratory is pursuing analysis of STOP protein, microtubule stabilization in flagellae, tubulin tyrosination, and centrosome function.

Microtubules: 229–249

The Biochemistry, Genetics, and Molecular Biology of Flagellar Dynein

George B. Witman, Curtis G. Wilkerson, and Stephen M. King

I. INTRODUCTION

The flagellar dyneins are large, multisubunit ATPases that make up the inner and outer arms of the axoneme and are responsible for generating the force that is the basis for flagellar movement. Our knowledge of the outer arm dynein is very extensive; this structure is currently the most well characterized of all dyneins and serves as a model for how these important mechanochemical enzymes are constructed and function. Our understanding of the inner arm dyneins is much less complete but is progressing at a rapid pace. For both the outer and inner arm dyneins, recent studies utilizing genetic and molecular biological approaches, as well as new methods for studying the force-generating properties of isolated dyneins *in vitro*, have been particularly productive. Our discussion emphasizes this recent work; reviews of other aspects of dynein biology can be found in the excellent volume edited by Warner et al. [1989].

II. OUTER ARM DYNEINS

A. Structure and Composition

Outer arm dyneins have been purified from the cilia and/or flagella of a diverse group of organisms (Table I). These enzymes appear in the electron microscope as two or three globular head domains (~12 nm in diameter) connected by stems to a common base [Johnson and Wall, 1983; Witman et al., 1983]. In addition, several smaller features emanating from both the heads and stems have been described for dyneins adsorbed to mica and prepared for electron microscopy by the quick freeze/deep etch procedure [Goodenough and Heuser, 1982, 1984]. One of these, the "B-link," which projects from the head domain and apparently interacts with the B-tubule of the adjacent outer doublet, also has been observed in highly contrasted ultrathin sections [Burgess et al., 1991]. However, it is not yet clear whether the other details represent true topographical features of the dynein or whether some or all derive from localized unfolding/denaturation of the particle following adsorption to the mica surface.

The number of head domains of the dynein particle is correlated with the number of heavy chains of mass >400 kD. Outer arm dyneins from protozoa and algae contain three heavy chains, while the remainder have only two; the functional significance (if any) of this dichotomy in outer arm design is not clear at present. Mass analysis performed by scanning transmission electron microscopy has demonstrated that outer arm dyneins containing two or three heavy chains have masses of ~1.25 and ~1.7 MD, respectively [Hastie et al., 1988; Johnson and Wall, 1983; Marchese-Ragona et al., 1987; Witman et al., 1983; reviewed in Marchese-Ragona and Johnson, 1988]. Further analysis of individual heads indicates that these

TABLE I. Species in Which Outer Arm Dyneins Have Been Examined

Species	Tissue/organelle	References
Protists		
Paramecium	Cilia	Larsen et al. [1991]
		Schroeder et al. [1990]
Tetrahymena	Cilia	Porter and Johnson [1983]
Algae		
Chlamydomonas	Flagella	Piperno and Luck [1979a]
		Pfister et al. [1982]
Invertebrates		
Aequipecten irradians	Sperm flagella	Linck [1973]
(scallop)	Gill cilia	
Oyster	Sperm flagella	Mohri et al. [1992]
Asterias amurensis	Sperm flagella	Mabuchi et al. [1976]
(starfish)		
Sea urchins		
Anthocidaris crassispina	Sperm flagella	Ogawa et al. [1990]
	Blastula cilia*	
Strongylocentrotus		
purpuratus	Sperm flagella	Sale et al. [1985]
Tripneustes gratilla	Sperm flagella	Bell et al. [1979]
Vertebrates		
Fish		
Salmo gairdneri (trout)	Sperm flagella	Ogawa et al. [1980]
		Gatti et al. [1989]
Notothenia coriiceps	Sperm flagella	Marchese-Ragona and Detrich [1990]
neglecta (Antarctic rockcod)		
Mammals		
Porcine	Tracheal cilia	Hastie et al. [1986)
Porcine	Sperm flagella	Gatti and Dacheux [1992]
Bovine	Sperm flagella	Pallini et al. [1982]
		Belles-Isles et al. [1986]
Bovine	Oviduct cilia	Pallini et al. [1982]

*This study involved an immunological, rather than a biochemical, analysis using monoclonal antibodies against sperm flagellar dynein components.

domains each have a mass of ~350–400 kD. Therefore, individual dynein heavy chains are too large to be totally included within a head domain and must contribute to both head and stem components of the complex.

The detailed polypeptide compositions of the four best-characterized dyneins are shown in Table II. In addition to two or three heavy chains, all outer arm dyneins examined to date contain at least two intermediate chains (50–150 kD) and up to 10 light chains (6–25 kD). The three-headed outer arm from *Chlamydomonas* dissociates upon high salt extraction from the axoneme to yield an 18S particle (the α/β-dimer) and a 12S particle (the γ-subunit). The former contains the α- and β-heavy chains, both intermediate chains, and eight light chains. The latter is composed of the γ-heavy chain and two light chains [Pfister et al., 1982; Piperno and Luck, 1979a]. The 12S and 18S components either remain as a three-headed structure or reassociate when the dynein is purified in the presence of Mg^{2+} [Takada et al., 1992]. This result implies that the two subunits associate *via* a Mg^{2+} salt bridge. Two-headed outer arms (and the α/β-dimer from *Chlamydomonas*) may be dissociated by low ionic strength dialysis, suggesting that they are held

TABLE II. Polypeptide Composition of Outer Arm Dyneins*

	Heavy chains**		Intermediate chains		Light Chains	
Species	Chain	Mass (kD)	Chain	Mass (kD)	Number	Mass (kD)
Chlamydomonas	α	480	IC78	76.5***	10†	8, 8, 11, 14, 14,
	β	440	IC69	63.5		16, 18, 19, 20, 22
	γ	415				
Sea urchin	α	475	IC1	122	4‡,§	24, 21, 16, 14
(*T. gratilla*)	β	475	IC2	90		
			IC3	76		
Tetrahymena	α	417	IC1	85¶	4?	Values not
	β	417	IC2	70		reported
	γ	403				
Trout	α	430	IC1	85	6‡	6, 7.5, 9, 11.5,
	β	415	IC2	73		19, 22
			IC3	65		
			IC4	63		
			IC5	57		

*The data quoted here for the composition of outer arm dyneins are from Pfister et al. [1982] (*Chlamydomonas*), Bell et al. [1979] (sea urchin), Porter and Johnson [1983] (*Tetrahymena*), and Gatti et al. [1989] (trout).
**Masses tabulated for the heavy chains are those determined by electrophoretic analysis. Molecular cloning indicates that the actual mass of the β-heavy chain from sea urchin dynein is 512, 804, suggesting that the values quoted here are underestimates by <10-20%.
***The masses of IC78 and IC69 were determined by molecular cloning by Wilkerson, King, and Witman (unpublished data) and Mitchell and Kang [1991], respectively.
†Determined by analysis of two-dimensional SDS-polyacrylamide gels.
‡Determined by analysis of one-dimensional SDS-polyacrylamide gels.
§Dynein from the sea urchin *S. purpuratus* has been reported to contain six light chains of 6, 8, 10, 13, 21, and 23 kD [Moss et al., 1992b].
¶In their original report, Porter and Johnson [1983] described an additional intermediate chain of M_r~100,000. More recently, this band has been found by Marchese-Ragona et al. [1989] to be a proteolytic fragment of the γ-heavy chain. Therefore, it is not included in this tabulation.

together by hydrophobic interactions [King et al., 1990; Pfister and Witman, 1984; Tang et al., 1982]. The solubilized heavy chains can be separated by sucrose density gradient centrifugation, presumably due to their associations with intermediate or light chains or, in the case of dyneins from fish and sea urchins, because the α-heavy chain aggregates to yield heterodisperse structures (a property that has made structural and functional analyses of this component problematic). Outer arm dynein from *Tetrahymena* has proved refractory to both ionic regimens; subfractionation of it has been achieved only by controlled proteolysis [Toyoshima, 1987].

Two of the dynein intermediate chains (IC78/IC69 in *Chlamydomonas,* IC2/IC3 in sea urchin and IC1/IC2 in trout) and at least four of the dynein light chains are tightly associated in an intermediate chain–light chain (IC–LC) complex [King et al., 1990, 1991b; Mitchell and Rosenbaum, 1986; Tang et al., 1982; reviewed by Witman et al., 1991]. The two intermediate chains interact with each other [King et al., 1991b; Mitchell and Rosenbaum, 1986], and both also bind to one or more of the light chains. Together, these components form a discrete subparticle that, at least in sea urchin and trout, can be removed from the remainder of the outer arm by low ionic strength dialysis and that sediments as a hydrodynamically compact structure at ~9S in sucrose density gradients. In

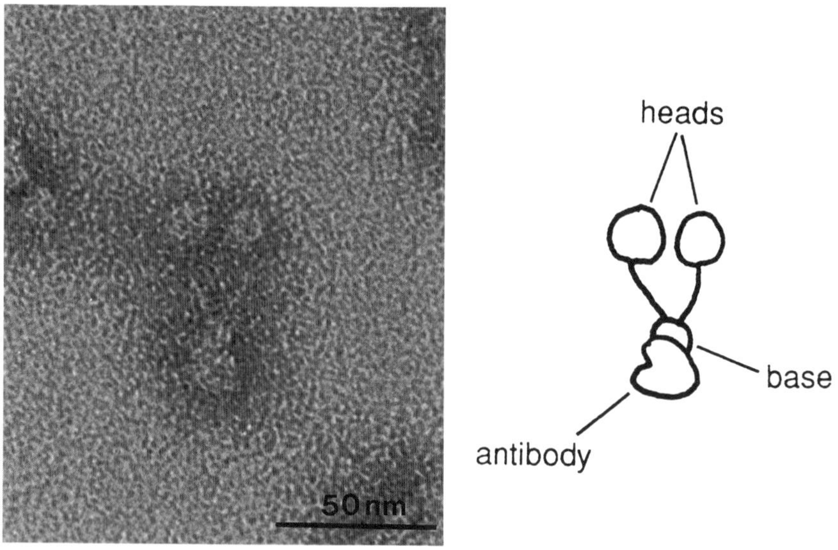

Fig. 1. *Electron micrograph and interpretive drawing of the α/β-dimer from* Chlamydomonas *outer arm dynein complexed with a monoclonal antibody against IC69 [see King and Witman, 1990]. The antibody is bound to the base of the dynein and indicates that the IC–LC complex is located within this domain. The sample was supported on a thin carbon film and negatively stained with uranyl acetate. ×440,000.*

Chlamydomonas, this complex tightly adheres to the β-heavy chain and is only removed under relatively denaturing conditions [Mitchell and Rosenbaum, 1986; see also King and Witman, 1990]. In *Chlamydomonas,* IC69 has been located at the base of the soluble dynein particle by immunoelectron microscopy using monoclonal antibodies (Fig. 1) [King and Witman, 1990]. As the other components of the IC–LC complex are tightly associated with IC69, they also are probably located at the base of the dynein. Although the additional intermediate chains found in dyneins from trout (ICs3–5) and sea urchin (IC1) associate most tightly with the β-heavy chain, their location relative to the head and stem domains has not yet been determined.

Many dynein heavy chains are tightly associated with a light chain that is not an integral part of the IC–LC complex. Unlike myosin, in which the essential and regulatory light chains associate with the head domain [for review, see Warrick and Spudich, 1987], recent biochemical and genetic evidence indicates that, at least for the β-heavy chain of *Chlamydomonas,* the tightly bound light chain is located in the stem or base domain [Sakakibara et al., 1993] (see below).

B. Differences Between Ciliary and Flagellar Dyneins

There have been few detailed studies of ciliary and flagellar dyneins from the same species. Pallini et al. [1982] reported minor differences in the heavy chains associated with bovine sperm flagella and oviductal cilia. More recently, Ogawa et al. [1990] described an immunological study of sperm flagella and blastula cilia from several species of sea urchin. A monoclonal antibody raised against the α-heavy chain of sperm outer arm dynein failed to recognize its ciliary counterpart, whereas four different anti-β-heavy chain antibodies reacted with both dyneins. This indicates that the ciliary α-heavy chain either is a modified version of the polypeptide found in flagella or is the product of a different gene. Interestingly, in *Chlamydomonas,* the α-heavy chain is the only outer arm component phosphorylated *in vivo* [Piperno and Luck, 1981]; it is modified at multiple sites [King and Witman, 1989]. If sea urchin dynein is similarly phosphorylated, the apparently different α-heavy chains detected in sea urchin dynein may simply reflect variation in the level of phosphorylation at a particular site.

In the same study, Ogawa et al. [1990] also examined the intermediate chains of dyneins from both sources and identified variations in the masses of immunologically related intermediate chains from a given species as well as between species. However, whether these differences reflect proteolysis during sample preparation or are indicative of posttranslational modifications, alternative splicing, or a multiplicity of intermediate chain genes remains to be determined.

C. Outer Arm Mutants

Analysis of mutant *Chlamydomonas* that lack outer dynein arms has implicated 12 different loci (*oda1–10, pf13A,* and *pf22A*) as essential for assembly of the outer arm into the axoneme (Table III) [reviewed by Luck and Piperno, 1989; Kamiya et al., 1989] and has provided much insight into the structure and function of this complex motor. The *oda* mutants all assemble essentially full-length flagella that lack outer arms and have a beat frequency approximately half that of wild-type. Restriction fragment length polymorphism (RFLP) mapping indicates that several of the *oda* mutations occur within structural genes for dynein polypeptides (Table III), suggesting that these polypeptides are essential for dynein assembly *per se*. The other *oda* gene products may be necessary for the expression of dynein genes, for assembly of dynein polypeptides into an outer arm, or for transport of a preformed dynein complex to the growing flagellum. The mutants *pf13A* and *pf22A* assemble only short flagella that are completely paralyzed. *pf22A* exerts some effect on the inner arms as well, suggesting either a defect of a more general nature or an altered interaction between the arms. However, the molecular basis for the differences between these two classes of mutant remains to be ascertained.

TABLE III. Outer Dynein Arm Mutants From *Chlamydomonas*

Mutant	Additional alleles	Polypeptide affected	References*
oda1		Not known	1
oda2	*pf28*	γ	1, 2, 3
oda3		Not known	1
oda4	*oda4-s7*, $sup_{pf}1$	β	1, 4, 5
oda5		Not known	1
oda6		IC69	1, 6
oda7		Not known	1
oda8		Not known	1
oda9		IC78	1, 9
oda10		Not known	1
oda11		α	7
pf13A		Not known	8
pf22A		Not known	8

*1, Kamiya [1988]; 2, Mitchell and Rosenbaum [1985]; 3, Wilkerson (unpublished data); 4, Sakakibara et al. [1992]; 5, Huang et al. [1982]; 6, Mitchell and Kang [1991]; 7, Sakakibara et al. [1991]; 8, Huang et al. [1979]; 9, King (unpublished data).

Mutants that assemble an outer dynein arm but lack specific outer arm components or contain truncated outer arm heavy chains also have been described. Mutation at the *oda11* locus results in an outer arm that is assembled into the flagellum but that does not contain either the α-heavy chain or the 16 kD light chain that is tightly associated with it [Sakakibara et al., 1991]. This mutant swims at a rate intermediate between those of outer-armless and wild-type strains. RFLP analysis using a probe for the α-heavy chain gene strongly suggests that *oda11* represents the structural gene for that chain. These results indicate that, unlike the other heavy chains, neither the α-heavy chain nor the 16 kD light chain is absolutely essential for dynein assembly or function.

The mutant $sup_{pf}1$, which is allelic with *oda4*, was originally isolated by Huang et al. [1982] as an extragenic suppressor of certain *pf* mutants. This mutation results in a small deletion within the β-heavy chain and restores motility to otherwise paralyzed radial spoke and central-pairless mutants. This result implicates the β-heavy chain as a target in the regulation of outer arm function mediated by other axonemal components.

Recently, Sakakibara et al. [1993] identified another *oda4* allele (*oda4-s7*) that also assembles an outer dynein arm. This mutant contains intact α- and γ-heavy chains but has a truncated β-heavy chain. The remaining portion of the β-heavy chain has a mass of ~160 kD and contains epitopes from the N terminus of the chain. Electron microscopic analysis indicates that this dynein lacks a head domain, supporting the idea that the N-terminal region of the heavy chain corresponds to the stem of the dynein complex. Interestingly, dynein from this mutant also contains the 19 kD light chain that is tightly associated with the β-heavy chain [Mitchell and Rosenbaum, 1986], suggesting that the light chain is attached to the stem or the base of the complex rather than to the head.

In cross section, the part of the outer arm that interacts with the B-tubule appears as a crescent-shaped structure formed from three globular domains. Outer arms from the mutant *oda11* lack the outermost unit [Sakakibara et al., 1991], whereas in *oda4-s7* the central lobe is missing [Sakakibara et al., 1993]. Thus these two globular units correspond to the α- and β-heavy chains, respectively. Presumably, the innermost lobe is formed from the γ-heavy chain. However, no alleles of *oda2* have yet been isolated that express a truncated γ-heavy chain and are competent to assemble an arm.

D. Molecular Structure of Dynein Heavy Chains

Several laboratories have employed proteolytic and chemical cleavage techniques to map the locations of various sites of structural and functional interest within the very large dynein heavy chains [King and Witman, 1987, 1988a,b; King et al., 1989, 1990; Lee-Eiford et al., 1986; Marchese-Ragona et al., 1989; Mocz et al., 1988, 1991; Tang and Gibbons, 1987]. These studies have now been augmented by the publication of the complete primary structure for the β-heavy chain from two species of sea urchin [Gibbons et al., 1991; Ogawa, 1991] and by partial sequences (reported in abstract form) for the three outer arm heavy chains from *Chlamydomonas*.

1. ATP-binding sites. Each dynein subunit exhibits an ATPase activity [Tang et al., 1982; Pfister and Witman, 1984; Piperno and Luck, 1979a], and photoaffinity labeling with 8-N_3ATP demonstrated that the hydrolytic sites are located within the heavy chains [Pfister et al., 1984, 1985]. Each heavy chain is cleaved at a single site (the VI site) by ultraviolet irradiation in the presence of ATP and vanadate [Lee-Eiford et al., 1986; King and Witman, 1987; King et al., 1990; Marchese-Ragona et al., 1989]. Cleavage at this site ablates the enzymatic activity of the polypeptide, suggesting that cleavage occurs at or near the hydrolytic site. The site maps to the location of the P-loop consensus sequence GPAGTGKT identified by molecular cloning (Fig. 2) [Gibbons et al., 1991; Ogawa, 1991]. P-loop motifs (A/GXXXXGKT/S) have been shown in other proteins to be involved in binding the phosphate moiety of nucleotides [Walker et al., 1982]. As vanadate acts as a phosphate analogue, this supports the hypothesis that photocleavage occurs at the site normally occupied by the γ-phosphate of ATP. Indeed, photolysis of the myosin heavy chain under V1 conditions occurs at S_{180} within the P-loop of that enzyme [Cremo et al., 1989].

Four other P-loop consensus sequences were found in the sea urchin β-chain (Fig. 2). One is located near the N terminus of the molecule in a region thought to correspond to the stem or base of the dynein. This region does not exhibit ATPase activity when separated from the head by proteolysis and purified [Ow et al., 1987], nor is it modified by photoaffinity analogues of ATP [Ow et al., 1987]. Thus available evidence does not support a functional nucleotide-binding site within this region.

The other sites are located C-terminal to the V1 site within the head domain. The third and fourth sites conform to the GXXXXGKS motif but are not particularly well conserved between sea urchin and *Chlamydomonas*. The fifth site, which is highly conserved, is unusual in having a Q rather than an S or T in the last position. This site apparently maps to the same location as a second site of vanadate-mediated photocleavage. Cleavage at this V2 site occurs

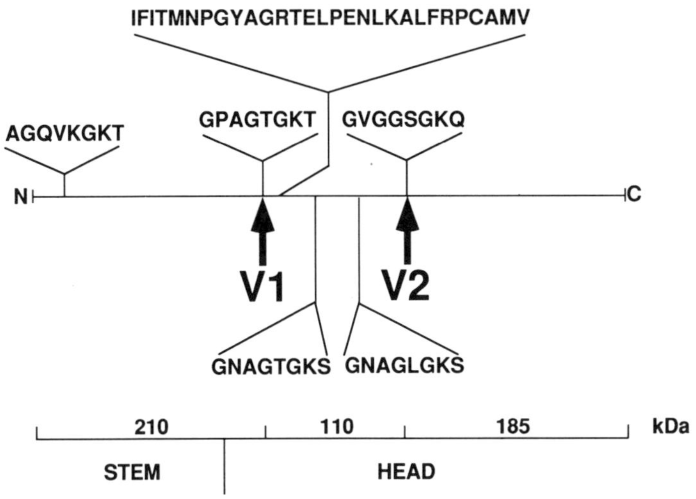

Fig. 2. *Diagram of the β-heavy chain from sea urchin sperm indicating the sites of vanadate-mediated photocleavage (V1 and V2) [Tang and Gibbons, 1987] and the locations of the five P-loop consensus sequences [Gibbons et al., 1991; Ogawa, 1991]. Also shown is the 30 residue sequence that is similar to the ATP-sensitive microtubule binding site of kinesin [Wilkerson et al., 1991; reviewed by Witman, 1992]. Masses for the various portions defined by the V1 and V2 sites are shown in kilodaltons, and the approximate boundary between the stem and head domains of the molecule also is indicated.*

in the presence of Mn^{2+} and oligomeric vanadate but is inhibited by the addition of ATP and ADP [Tang and Gibbons, 1987]. Interestingly, the α- and γ-heavy chains from *Chlamydomonas* contain additional sites cleaved under V2 conditions [King and Witman, 1987, 1988b]. The location of these other sites suggests that they may correspond to one or more of the remaining P-loops and that the susceptibility of the V2 sites within these polypeptides to photocleavage merely reflects minor variations in primary sequence.

Photoaffinity labeling of the β-heavy chains from *Chlamydomonas* and sea urchin revealed that $8\text{-}N_3ATP$ was incorporated near to the V2 site and over 100 kD from the V1 site [King et al., 1989; Mocz et al., 1988]. In *Chlamydomonas,* a second probe ($2\text{-}N_3ADP$) photoinserted close to the V1 site in both the α- and β-heavy chains. These observations were interpreted to indicate that parts of the hydrolytic site derived from widely separated regions of the polypeptide. However, in view of the sequence information now available, the same data would support the possibility that each heavy chain contains at least two nucleotide-binding sites. Further analysis of these P-loop motifs by site-directed mutagenesis will be required to answer this important point unambiguously.

2. ATP-sensitive binding to microtubules. Each heavy chain is thought to contain an ATP-sensitive microtubule-binding site located within the head domain of the molecule [Toyoshima, 1987]. Examination of the primary structures of the *Chlamydomonas* γ-heavy chain and sea urchin β-heavy chain has revealed a sequence near the V1 site (Fig. 2) [Wilkerson et al., 1991] that is similar to a region implicated in the ATP-sensitive microtubule binding of members of the kinesin superfamily [Yang et al., 1989; reviewed by Witman, 1992]. This 30 residue sequence is highly conserved between *Chlamydomonas* and sea urchin and is located in approximately the same position relative to the hydrolytic site in both dynein and kinesin. Alteration of this region in a dynein heavy chain by site-directed mutagenesis should clarify whether it has any role in the ATP-dependent microtubule binding that is essential for force transduction by dynein.

3. Phosphorylation. In *Chlamydomonas,* the α-heavy chain is the only outer arm component phosphorylated *in vivo* [Piperno and Luck, 1981]. Quantitative analysis indicates that the molecule contains 5–6 mol phosphate/mol enzyme [King and Witman, 1989], and sites have been mapped to both the N- and C-terminal regions of the molecule. However, it is not clear what (if any) role these modifications play *in vivo.* Indeed, as mentioned above, the mutant *oda11* is defective for the α-heavy chain yet assembles an outer arm that is at least partly functional [Sakakibara et al., 1991]. Finally, it is not known whether the heavy chains from other organisms are modified in a similar fashion.

E. Structure and Function of the Intermediate Chains

The intermediate chains from *Chlamydomonas,* sea urchin, and trout are immunologically related [King et al., 1985, 1990]. A monoclonal antibody (1869A) raised against IC69 from *Chlamydomonas* also reacts with IC3 from sea urchin and with IC2 from trout. Interestingly, an antibody against the other intermediate chain (IC78) from *Chlamydomonas* also recognizes trout IC2. This observation raised the possibility that the two intermediate chains within the IC–LC complex derived from a common ancestor. Both intermediate chains from *Chlamydomonas* now have been cloned and sequenced [IC69, Mitchell and Kang, 1991; IC78, Wilkerson, King and Witman, unpublished results]. The data indicate that indeed these molecules are related, being ~40% similar in their C-terminal halves; the N-terminal regions show little or no homology. A similar degree of conservation has been identified between IC69 and the 74 kD intermediate chain from cytoplasmic dynein [Paschal et al., 1992], indicating that both the flagellar and cytoplasmic dynein intermediate chains derive from a common ancestor.

In *Chlamydomonas,* genetic evidence indicates that both IC78 and IC69 are essential for the stable assembly of the outer arm [Mitchell and Kang, 1991; King, Wilkerson, and Witman, unpublished results]. Although the precise function of IC69 remains to be ascertained, some information on function is available for IC78. The latter has been shown to be in direct contact with α-tubulin *in situ* [King et al., 1991b] and to bind microtubules *in vitro* (King, Wilkerson and Witman, unpublished results). Combined with the localization of the IC–LC complex to the base of the dynein [King and Witman, 1990], these results suggest that IC78 is involved in the ATP-insensitive binding of the arm to the A-tubule of the outer doublets (see below).

Purified IC–LC complexes do not exhibit ATPase activity [Tang et al., 1982; King et al., 1990]. However, IC78 was labeled in an ultraviolet-dependent and ATP-sensitive manner by the photoactive ATP analogues 2-N_3ADP and 8-N_3ATP [King et al., 1989, 1991a; Pfister et al., 1985]. The labeling observed was highly specific in that IC69 (a homologue of IC78 present in equimolar amounts) was not modified. Incorporation of 8-N_3ATP into IC2 from sea urchin dynein also has been reported [Tang and Gibbons, 1987]. Although IC78 does not contain a classic P-loop consensus sequence, these results raise the tantalizing possibility that one intermediate chain within the outer arm contains an ATP-binding site perhaps involved in the regulation of arm activity.

F. Structural Binding to Microtubules

The outer dynein arm is permanently attached to the A-tubule by ATP-insensitive or structural bonds. Moss et al. [1992b] observed that the α-heavy chain of sea urchin dynein, but not the β/IC1 subunit (consisting of the β-heavy chain and IC1), was capable of forming ATP-insensitive bonds to microtubules, suggesting that the α-heavy chain has an important role in attaching the outer arm to the A-tubule in sea urchin sperm flagella. However, the α-heavy chain is not essential for assembly of the outer arm in *Chlamydomonas* [Sakakibara et al., 1991]. Moreover, binding of the outer arm to doublet microtubules in the latter species appears to require both the α/β-dimer and the γ-subunit [Fay and Witman, 1977]. As noted above, within the α/β dimer of *Chlamydomonas,* the 78 kD intermediate chain is a microtubule-binding protein. Further work will be necessary to understand this apparent difference between the dynein subunits involved in strucural binding to microtubules in these two species.

When *Tetrahymena, Chlamydomonas,* or sea urchin sperm outer arms are reconstituted onto outer doublet or brain microtubules at low stoichiometries *in vitro,* the arms bind in widely separated clusters, within which individual arms repeat at their normal 24 nm periodicity [Mitchell and Warner, 1980; Haimo and Rosenbaum, 1981; Moss et al., 1992b]. Single, separate arms are not observed. Therefore the binding of one outer arm to a microtubule increases the probability that another outer arm will bind to an adjacent site. Such cooperativity of binding may be due to lateral interactions between adjacent outer arms or to longitudinal perturbation of the tubulin lattice as a result of dynein-microtubule interaction.

G. Force Generation *In Vitro*

The development of an *in vitro* assay for monitoring force production by isolated axonemal dynein has made it possible to investigate the mechanochemical transducing activity of different dynein species and individual subunits of dynein [Paschal et al., 1987a; Vale and Toyoshima, 1988]. Paschal et al. [1987b] first showed that when isolated outer arm dynein from sea urchin sperm flagella was adsorbed to a glass chamber, and brain microtubules and ATP added, the microtubules attached to the glass-bound dynein and were translocated over the surface of the glass. The microtubules glided at rates of up to 7 μm/sec, which was much faster than that reported for other microtubule translocators (kinesin and cytoplasmic dynein) *in vitro*. The rate of gliding was independent of microtubule length, indicating that the velocity did not depend on the total number of arms participating in force production. This was in good agreement with earlier observations that, in disintegrating axonemes, the velocity of microtubule sliding did not depend on the extent of overlap between the sliding doublets [Takahashi et al., 1982]. Nevertheless, the *in vitro* result was surprising because, assuming that dynein produces force during only a small fraction of its mechanochemical cycle, velocity should increase as more asynchronous dyneins act on a microtubule. That this was not observed may indicate that the hydrolytic cycles of the glass-bound dyneins interacting with the microtubule were synchronized, presumably by the movement of the tubule. Such coupling of dynein activity to microtubule movement could be important in the coordination of flagellar bending *in vivo* [Brokaw, 1989]. Similar results were obtained by Vale and Toyoshima [1988] for the isolated outer arm dynein of *Tetrahymena;* these investigators also demonstrated that the isolated dynein moves toward the minus end of the microtubule, as expected from earlier studies on the polarity of outer doublet sliding in disintegrating axonemes [Sale and Satir, 1977].

Using the same assay, Sale and Fox [1988] found that the isolated β/IC1 subunit of sea urchin outer arm dynein was sufficient for microtubule translocation. Fractions containing the α-heavy chain or IC2 and IC3 did not support microtubule gliding. Interestingly, the β/IC1-subunit translocated microtubules at significantly faster rates than did the intact dynein, suggesting either that the α-heavy chain acted as a drag on the microtubules or that the activity of the β/IC1-subunit was down regulated by its interaction with the α-heavy chain. The latter possibility is consistent with the observation that the ATPase activity of the β-subunit of *Chlamydomonas* outer arm dynein is decreased when it is associated with the α-subunit [Pfister and Witman, 1984].

Moss et al. [1992a,b] also used the *in vitro* translocation assay, in conjunction with microtubule cosedimentation assays and observations of microtubule bundling *in vitro,* to investigate the force-producing and microtubule-binding properties of intact sea urchin sperm outer arm dynein and its individual subunits. The results showed that although the isolated β/IC1-subunit was capable of interacting with microtubules to produce force, it did not bind to microtubules in the absence of ATP. In contrast, the α-heavy chain, which did not translocate microtubules, was capable of forming ATP-sensitive (rigor) bonds to microtubules. Although these results do not rule out a role for the α-heavy chain in force generation, they raise the possibility that there is a segregation of function between the subunits of sea urchin outer arm dynein, and that the properties of the intact dynein reflect the sum of the properties of its individual subunits. The role of the β/IC1-subunit may be to generate microtubule sliding, whereas that of the α-heavy chain may be to bind to the opposing outer doublet in an ATP-dependent manner to resist or control that sliding. Isolated *Chlamydomonas* dyneins having defects in specific heavy chains should provide an independent means of testing this hypothesis.

The movement of short microtubules on outer arm dynein is not smooth and continuous, but exhibits frequent stops and starts, and occasional backward movements [Paschal et al., 1987b]. The backward movements are especially pronounced in the presence of vanadate and ATP or ADP [Vale et al., 1989], which causes the microtubules to cease unidirectional translocation and "shuttle" backward and forward. Quantitative analysis of this movement indicates that it is thermally driven; apparently under these conditions, which may mimic a step in dynein's mechanochemical cycle, the dynein is attached to the microtubule by a weak bond that permits it to diffuse longitudinally along the microtubule. These observations have led to the proposal that dynein may be a thermal ratchet that translocates along a microtubule by virtue of its Brownian movement and uses the energy of ATP hydrolysis to impose directionality on this movement [Vale and Oosawa, 1990].

The *in vitro* motility assay also is beginning to be used to determine how dynein is regulated biochemically. Hamasaki et al. [1991] have reported that the rate of microtubule translocation by *Paramecium* 22S dynein is doubled when the dynein is isolated from axonemes that have been pretreated with cAMP. This effect was correlated with the cAMP-dependent thiophosphorylation of a 29 kD polypeptide present in the 22S dynein fractions. However, these fractions contained only ~0.3 mol of the 29 kD polypeptide per mol of dynein, and only ~2% of the 29 kD polypeptides were thiophosphorylated by cAMP stimulation. Therefore it is not clear how the 29 kD protein is related to dynein, or how thiophosphorylation of such a small percentage of the molecules could have such a large effect on the microtubule translocation rate. A possibility suggested by Hamasaki and colleagues is that dyneins associated with the thiophosphorylated protein act as pacemakers that stimulate the activity of the other dyneins. Alternatively, it is possible that the 22S fraction, presumed to represent outer arm dynein, also contains substantial amounts of inner arm material, as is the case for the sucrose gradient-purified 18S dynein fraction from wild type *Chlamydomonas* [Goodenough et al., 1987]. If so, the 29 kD polypeptide may be part of an inner arm dynein that could be the dominant force-generating species in the preparation.

III. INNER ARM DYNEINS

A. Structure and Composition

The conventional image of the inner arm as seen in cross sections of axonemes is a superimposition of several structures differing in their morphology, composition, and functional roles. The first clear indication of this came from examination of rapidly frozen, deep-etched axonemes of invertebrate and protozoan species [Goodenough and Heuser, 1985]. Based on these images, it was proposed that there were two morphologically distinct types of inner arms: "triads," containing three heads, and "dyads," containing two heads. The triads and dyads were reported to alternate along the length of the axoneme, with one triad being followed by two dyads every 96 nm. Further analysis revealed that the triad was probably a two-headed structure adjoining a "bilobed unit" that lay close to the A-tubule [Goodenough and Heuser, 1989].

Subsequent studies have indicated that the inner arms are structurally more complex.

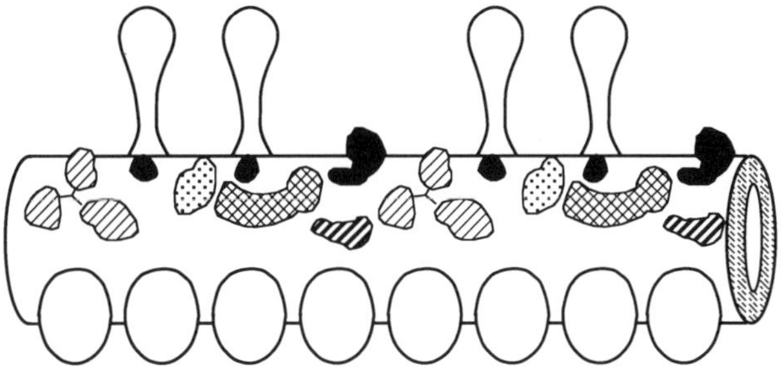

Structure	Name	Missing in	Polypeptide Composition
	I1	*pf9*	1α, 1β, 140, 110 kD
	I2	*ida4*	DHCs 2, 2' = 3a, 3b, 5 + ?
	Dynein Regulatory Complex	*pf2* + *pf3*	108, 83, 65, 60, 40, 29 kD
	—	*bop2*	152 kD + ?

Fig. 3. *Diagram of a portion of the A-tubule of a* Chlamydomonas *outer doublet as viewed from the B-tubule of the next outer doublet, illustrating the electron densities (filled with various patterns) that contribute to the inner arm image seen in flagellar cross sections. The unfilled oval structures along the lower edge of the A-tubule represent outer dynein arms; the unfilled elongated structures extending from the upper edge of the A-tubule represent the radial spokes.*

The table below the diagram indicates the names currently assigned to the structures, the Chlamydomonas *mutant lacking that structure, and the polypeptides currently correlated with the structure. For inner arm I2, the dynein heavy chain (DHC) nomenclature of both Piperno et al. [1990] and Kagami and Kamiya [1992] are given (see Table IV). The dynein regulatory complex is here presented as a single structure; one part of it is missing in* pf2 *and another part in* pf3. *The structure missing in* bop2 *has not yet been named. [Adapted from Mastronarde et al., 1992, with permission of the publisher.]*

Based on a comparison of freeze-etched and thin-sectioned images of chicken sperm flagella, Burgess et al. [1991a] concluded that the inner arms of at least this species were arranged in a triad–dyad–monad repeat. In an analysis of a tilt-series of thin longitudinal sections of *Chlamydomonas* axonemes, Muto et al. [1991] were able to resolve the inner arm region into 10 globular structures per repeating unit of 96 nm. These structures appeared to be arranged in two rows along the axoneme, suggesting that at least some of the densities represented structures not detected in the deep-etched images.

In a more recent study in which thin longitudinal sections of *Chlamydomonas* axonemes were analyzed by computer averaging, it was confirmed that the inner arm region can be resolved into 9 or 10 electron-dense lobes that repeat every 96 nm and that most (but not all) of these lobes are arranged in two parallel rows (Fig. 3) [Mastronarde et al., 1992]. Some of these lobes appear to represent different parts of a single arm. Their presence or absence in various mutants of *Chlamydomonas* allows them to be assigned to at least four different structures (Fig. 3).

1. Inner arm I1. The most well-characterized of the inner arm structures, termed *inner arm I1* [Piperno et al., 1990], is missing in the inner-arm defective mutant *pf9* [Mastronarde et al., 1992; Porter et al., 1992] and its allele *pf30* [Piperno et al., 1990] and in the idaA class of mutants, which includes another *pf9* allele, *ida1*, and two additional nonallelic strains, *ida2* and *ida3* [Kamiya et al., 1991]. This arm is located proximal to the first spoke (S1) of the radial spoke pair [Piperno et al., 1990] and appears to consist of two large, electron-dense lobes and a third, more distal, lobe that may be connected to the first two. Biochemical analyses of mutants lacking this arm indicate that inner arm I1 contains two dynein heavy chains, *1α* and *1β*, and at least one intermediate chain of 140 kD [Piperno et al., 1990] and possibly a second of ~100 kD [Smith and Sale, 1991; Porter et al., 1992]. Following extraction from the axoneme, these chains remain associated with one another and sediment as a discrete complex at 21S [Piperno et al., 1990; Smith and Sale, 1991; Porter et al., 1992]. Electron microscopy of the isolated particles indicates that they consist of two heads connected by thin stems to a prominent base [Smith and Sale, 1991]. Thus this arm may be structurally similar to the outer dynein arm, with the two heavy chains forming the two heads and extending to the base and the large globular base containing the intermediate chain(s). Based on its location relative to radial spoke S1, inner arm I1 corresponds to the "triad" of Goodenough and Heuser [1985].

2. Inner arm I2. The second type of structure, termed *inner arm I2* [Piperno et al., 1990], is defined in *Chlamydomonas* by the inner-arm defective mutant *ida4* [Kamiya et al., 1991]. Axonemes of *ida4* are missing a discrete lobe at the base of each radial spoke; a third, more centripetally directed lobe between the spoke pairs; and possibly a fourth, less distinct electron density located beneath the third lobe [Mastronarde et al., 1992].

There is still uncertainty concerning the number of heavy chains associated with these structures (see Section III. A.5.a., below). Piperno et al. [1990] originally reported that inner arm I2 was correlated with two dynein heavy chain bands, 2′ and 2, that could be resolved in SDS-polyacrylamide gels; these bands were present in mutant alleles of *pf9* lacking inner arm I1 but were missing in a mutant *pf23*, which appeared to be lacking both inner arms I1 and I2. More recently, Kagami and Kamiya [1992] reported that bands 2′ and 2 each contained two different dynein heavy chain

TABLE IV. Nomenclature of Inner Arm Dynein Heavy Chains

Piperno et al. [1990]*	Kagami and Kamiya [1992]
1α	1
1β	2b
2′	3a, 3b
2	2a, 5
3	4
3′	6

*In the nomenclature of Piperno et al. [1990], the primes indicate a minor satellite band on SDS-polyacrylamide gels.

species, which they termed *3a* and *3b*, and *2a* and *5*, respectively (Table IV). Of these, dynein heavy chains 3a, 3b, and 5 were missing in the mutant *ida4*. Therefore it is likely that these chains are components of the lobes missing in this mutant, but it is not clear how they are distributed between these structures. The lobes are widely separated from one another (>26 nm center to center) and do not appear to be physically connected. It is possible that each of the major lobes missing in the mutant is composed of a different dynein heavy chain. Alternatively, if one of the chains substitutes for another in different regions of the axoneme [Piperno and Ramanis, 1991] (see below), then all of the I2 lobes in any one part of the axoneme may be formed by a single type of chain.

When extracted from the axoneme and analyzed by sucrose density gradient centrifugation, the I2 polypeptides sediment at 11S [Piperno et al., 1990], and electron microscopy indicates that this 11S fraction contains dynein-like particles having a single head and stem [Smith and Sale, 1991]. Although it was assumed that these structures originated by dissociation of a two-headed dynein, it is possible that they actually came from dyneins that have only one head *in situ*. No one-headed dynein has yet been reported, but such a dynein would be in good agreement with the thin section data suggesting that inner arm I2 corresponds to single globular domains that are widely separated [Mastronarde et al., 1992].

It is intriguing that three dynein heavy chains are deleted together in *ida4*. This may indicate that they are posttranslational modifications of a single gene product, that they bind to a common receptor protein that is missing or defective in the mutant, or that the presence of one is necessary for the stability or assembly of the others.

3. The dynein regulatory complex. The third structure, termed the *spoke-specific suppressor system* [Huang et al., 1982] or the *dynein regulatory complex* [Piperno et al., 1992] because of its effect on mutants lacking the radial spokes (see Sale, this volume), consists of a crescent-shaped electron density lying just beneath and extending distal to the second spoke (S2) of the radial spoke pair (Fig. 3). The proximal portion of this crescent is absent in the *Chlamydomonas* mutant *pf2* [Mastronarde et al., 1992], which restores motility to radial spoke-deficient strains [Huang et al., 1982]; the distal portion is missing in mutant *pf3* (M. Porter and E. O'Toole,manuscript in preparation), which has the same phenotype [Huang et al., 1982]. Axonemes of mutant *pf2* are lacking polypeptides of 29, 40, 60, and 65 kD, whereas axonemes of mutant *pf3* are lacking the first two of these polypeptides together with polypeptides of 83 and 108 kD [Huang et al., 1982]. Therefore it seems likely that this crescent-shaped electron density consists of at least these six proteins. In a wild-type background, the *pf2* and *pf3* mutations cause a decrease in outer doublet sliding velocity but little change in beat frequency [Brokaw and Kamiya, 1987], which is similar to the phenotype observed for mutants deficient in inner arms I1 [Brokaw and Kamiya, 1987] and probably I2 [Kamiya et al., 1991]. No dynein heavy chains have been correlated with this structure, but mutants *pf2* and *pf3* have reduced amounts of dynein heavy chain bands 2′ and 2 [Piperno et al., 1992]. This has been interpreted to mean that there is a physical association between inner arm I2 and the dynein regulatory complex [Piperno et al., 1992]. However, an alternative possibility is that these mutants are completely missing a subset of the dynein heavy chains reported to run together in bands 2′ and 2 (Table IV).

4. The *bop2* defect. The fourth structure consists of a large, possibly bilobed, electron density situated between spokes S1 and S2 (Fig. 3) [Mastronarde et al., 1992] and missing in the *Chlamydomonas* mutant *bop2*, which has an abnormal waveform similar to that of other inner arm defective mutants [King et al., 1992]. Axonemes of this mutant are lacking a 152 kD phosphoprotein. Thus far no dynein heavy chains have been correlated with this structure, although analysis of the heavy chain content of the *bop2* mutant is not yet complete.

5. Outstanding questions. There are several fundamental questions yet to be answered definitively with regard to the structure and polypeptide composition of the inner arms.

a. How many inner arm dynein heavy chains are there? When axonemes of *Chlamydomonas* mutants lacking the outer dynein arms are analyzed by SDS-polyacrylamide gel electrophoresis, six bands are resolved in the high molecular weight region [Piperno et al., 1990; Kamiya et al., 1991]. The polypeptides represented by these bands all appear to be dynein heavy chains based on their extractability with 0.5 M NaCl and their susceptibility to cleavage by ultraviolet light in the presence of vanadate and $MgATP^{2-}$ [Piperno et al., 1990]. Recently, Kagami and Kamiya [1992] reported the separation, by FPLC, of seven inner arm dynein species. These seven species appear to contain eight distinct dynein heavy chains (see Table IV) based on the pattern of fragments obtained following ultraviolet-induced, vanadate-dependent photocleavage [Kagami and Kamiya, 1992]. However, at least two of these heavy chains are phosphorylated *in vivo* [Piperno and Luck, 1981], and it may be that some of the heavy chain species represent posttranslational variants of the same gene product that may or may not form different structures or have different locations in the axoneme. To begin to address this question, we have used the PCR and primers based on the highly conserved V1 P-loop and the potential ATP-dependent microtubule-binding domain of the outer arm γ-dynein heavy chain to isolate PCR clones and subsequently cDNA clones encoding *Chlamydomonas* dynein heavy chains. In an initial screen, clones were obtained for the α-, β-, and γ-dynein heavy chains of the outer arm, and for four new dynein heavy chains, which presumably represent inner arm polypeptides [Wilkerson et al., 1992]. Sequence information based on the new clones should make possible the production of specific antibodies that will be useful for probing the relationships between the different bands identified by gel electrophoresis and FPLC.

b. Where is the elusive inner arm I3? As discussed above, inner arm dynein heavy chains 1α and 1β form a heterodimer that constitutes inner arm I1. Three other heavy chains appear to form the structures collectively termed *I2*. The remaining two bands originally resolved by Piperno et al. [1990] and termed *3* and *3′* were postulated to constitute a third type of inner arm, I3 [Piperno et al., 1990]. However, the relationship of this postulated third type of arm to the electron densities seen in longitudinal sections of the axoneme is not clear, primarily because no mutant has yet been identified that is lacking both dynein heavy chains 3 and 3′. In addition, the composition of this third type of arm is clouded by the possibility that there are actually three dynein heavy chains (2a, 4, and 6 in Kagami and Kamiya's nomenclature; see Table IV) not yet accounted for. In the absence of a correlation between a missing structure and missing dynein heavy chains, the location of the extra dynein heavy chains in the axoneme must remain a matter of speculation. One possibility is that some or all of these chains will yet be shown to be components of the dynein regulatory complex or the structure missing in the mutant *bop2*.

c. Is there heterogeneity of inner arms along the axoneme? Although dynein heavy chains 1α and 1β are present in equimolar amounts, dynein heavy chains 2′ and 3′ are present in wild-type *Chlamydomonas* axonemes in lesser amounts than dynein heavy chains 2 and 3, respectively [Piperno et al., 1990]. Moreover, the relative intensities of 2′ and 2, and of 3 and 3′, vary depending on the length of the flagella [Piperno et al., 1990; Piperno and Ramanis, 1991]. Following up on these observations, Piperno and Ramanis [1991] examined the relative amounts of these dynein heavy chains in *Chlamydomonas* flagella isolated at various times during regeneration, in isolated axonemes in which the inner arms had been selectively extracted from the middle and distal regions of the outer doublets, and in mutants lacking various combinations of these polypeptides. The results led them to conclude 1) that dynein heavy chain 2 was located in the proximal and middle regions of the axoneme, whereas dynein heavy chain 2′ was located exclusively in the distal region; 2) that dynein heavy chain 3′ was located exclusively in the proximal portion of the axoneme, whereas dynein heavy chain 3 was contained in the middle and distal parts; and 3) that each of these heavy chains formed a distinct inner arm, with inner arm I2 being a homodimer composed of dynein heavy chains 2 or 2′ and inner arm I3 being a homodimer composed of dynein heavy chains 3 or 3′, depending on their location along the length of the axoneme. Such a differentiation of the inner arms along the length of the axoneme could have important implications for the functions of bend initiation and propagation (see Sale, this volume). However, at least the third conclusion seems difficult to reconcile with the most recent structural [Mastronarde et al., 1992] and biochemical data [Kagami and Kamiya, 1992]. Therefore confirmation of all or part of this model must await further research. Of particular importance will be a better understanding of the relationships between the different dynein heavy chains, and between the chains and the various electron densities observed by thin- section electron microscopy. Molecular genetic analysis and sequencing of inner arm dynein heavy chains (see above), together with the development of antibody probes capable of distinguishing between the major and minor dynein heavy chain species, should help to elucidate the relationships between the chains. Antibodies specific for each of the inner arm heavy chains also would permit an independent investigation of the distribution of the chains along the length of the axoneme.

d. What intermediate and light chains are associated with inner arms I2 and I3? More information is needed on the interactions between the heavy chain components of inner arms I2 and I3 and the lower molecular weight polypeptides reported to be associated with them. In sucrose density gradients, the polypeptides contained in dynein heavy chain bands 2′, 2, 3, and 3′ cosediment at 11S with polypeptides of 133, 42, 28, and 19 kD [Piperno et al., 1990]. The 42 kD polypeptide is similar or identical to actin [Piperno and Luck, 1979b], and the 19 kD polypeptide is identical or closely related to the Ca^{2+}-binding protein centrin (also called *caltractin* by Huang et al. [1988]) [Piperno et al., 1990]. Their distribution across the 11S peak is asymmetric, suggesting that they are not all uniquely associated with the same particle [Piperno et al., 1990]. The 28 kD polypeptide, termed *p28,* is missing in mutant *ida4,* and actin, centrin, and p28 are deficient in *pf2, pf3,* and other mutants with defects in the dynein regulatory complex [Piperno et al., 1992]. However, quantitative analysis of the amounts of these polypeptides in the various mutants does not permit a clear-cut correlation between these proteins and any one structure. The results suggest that these three lower molecular weight proteins may be part of a structural continuum between inner arm I2 (and perhaps I3) and the dynein regulatory complex, and that different parts of the actin–p28–centrin complex may be associated with different dynein arms and the dynein regulatory complex [Piperno et al., 1992].

6. Summary. Recent studies of mutants of *Chlamydomonas* that have defects in the inner dynein arms and related structures have yielded new information that reveals previously unsuspected structural complexity within the inner arm region. However, the different approaches utilized by the various laboratories working on this problem have led to different conclusions regarding inner arm structure and composition. The periodicity and heavy chain content of inner arm I1 is reasonably certain, but no consensus has yet been reached on how many additional inner arm dynein heavy chains there are, how these other dynein heavy chains are arranged along the axoneme, or how they interact with the smaller polypeptides that appear to be involved in the regulation of dynein arm activity. Further careful analysis of existing and new mutants that affect these structures will be necessary to reconcile all of the disparate observations into a coherent, consistent model.

B. *In Vitro* Motility

Vale and Toyoshima [1988] demonstrated that a 14S ATPase fraction from *Tetrahymena* cilia could translocate brain microtubules in an *in vitro* motility assay. This is a heterogeneous fraction [Marchese-Ragona and Johnson, 1989] which appears to contain at least some inner arm material [Warner et al., 1985]. An interesting property of this fraction was that it also generated torque, causing the microtubules to rotate in a clockwise manner as they moved forward during translocation. The rate of translocation by this fraction was about half of that observed for *Tetrahymena* outer arm dynein. However, because the fraction was not well characterized, these activities could not be ascribed with certainty to the inner arm dynein.

Kagami et al. [1990] reported that three *Chlamydomonas* inner-arm dynein subspecies, isolated by FPLC, could translocate microtubules in an *in vitro* system. More recently, Kagami and Kamiya [1992] found that six of the *Chlamydomonas* inner arm dynein heavy chains were capable of force production. These six species were, in their terminology, dynein heavy chains 2a, 3a, 3b, 4, 5, and 6 (Table IV). The maximum rates of translocation ranged from 2.7 to 11.5 μm/sec; in all cases the microtubules were translocated with plus ends leading, indicating that all of these dyneins are minus-end directed motors. All but dynein heavy chain 6 caused the microtubules to rotate during translocation. Interestingly, the ratio of translocation rate to rotation rate was not the same for all the dynein subspecies, suggesting that this ratio is determined by properties inherent in the dynein rather than in the microtubules.

Although Kagami and collaborators did not observe force production by inner arm I1, consisting of dynein heavy chains 1α and 1β, microtubule translocation by this last *Chlamydomonas* inner arm species was observed by Smith and Sale [1991]. In contrast to the rates observed for the other inner arm fractions, the isolated inner arm I1 produced very slow microtubule gliding velocities (0.76 μm/sec). The ability of inner arm I1 to produce force varied from preparation to preparation, which may explain why its force-generating properties were not observed by the other investigators. This arm also was capable of binding to and sedimenting with microtubules in an ATP-dependent manner, indicating that it can form a rigor bridge to microtubules. The other inner arm dynein heavy chains did not bind to the purified brain microtubules in the absence of ATP, suggesting that they have microtubule-binding properties similar to that of the β/IC1-subunit of sea urchin outer arm dynein [Moss et al., 1992a,b] (see above).

In summary, all of the *Chlamydomonas* inner arm dynein species are microtubule motors capable of generating force and, in most cases, torque. The motility-generating properties of these dyneins may be determined by their interactions with radial spokes [Smith and Sale, 1992a] (see also Sale, this volume), possibly via the dynein regulatory complex.

C. Inner Arm Assembly

Warner et al. [1985] demonstrated that components of the 14S ATPase fraction of *Tetrahymena* cilia could rebind to axonemes that had been extracted to remove their inner and outer arms. Electron microscopic examination of the reconstituted axonemes revealed that this material bound relatively specifically to the inner arm positions, providing the first unequivocal evidence that at least some of the 14S fraction from *Tetrahymena* cilia represented inner arm dynein. The binding was saturable, although some supernumerary arms were observed at other positions around the doublets when the reconstitution was carried out at high dynein/axoneme ratios. Rebinding was insensitive to ATP and vanadate, indicating that the inner arms were rebinding by their structural or "A-end" sites.

Recently, Smith and Sale [1992b] examined the rebinding of inner arm fractions from *Chlamydomonas* flagella. They found that the isolated inner arm I1 rebound to vacant I1 sites on doublet microtubules, but did not assemble onto outer arm sites, or sites normally occupied by the other inner arm species. Similarly, a mixture of components of inner arms I2 and I3 appeared to rebind specifically to the sites normally occupied by these arms. In both cases, binding was saturable. Particularly noteworthy was the fact that the arms rebound faithfully to sites on extracted doublets that were separated from the adjacent homologous sites by large gaps. Therefore the periodicity of the inner arms is not determined by cooperative interactions between these structures, as may be the case with the outer arms [Mitchell and Warner, 1980; Haimo and Rosenbaum, 1981; Moss et al., 1992b]. Rather, unique binding sites for each species of inner arm probably are determined by receptor proteins in or on the wall of the A-tubule.

This conclusion is supported by two other observations. First, inner arm subtypes bind specifically to their appropriate sites *in vivo* in *Chlamydomonas* mutants in which adjacent heterologous inner arm subtypes are missing [Piperno et al., 1990]. This would not occur if interactions between the arms were necessary to establish inner arm periodicity. Second, components of inner arm I2 and I3 fractions rebind to extracted outer doublets of *Chlamydomonas* [Smith and Sale, 1992b], but do not assemble onto microtubules assembled *in vitro* from purified brain tubulin [Smith and Sale, 1991]. This suggests that the outer doublets contain a protein or proteins, missing in brain microtubules, that are necessary for inner arm I2 and I3 binding. Perhaps assembly of brain tubulin in the presence of outer doublet fractions, followed by tests of the ability of the microtubules to bind inner arms, could lead to isolation and identification of the inner arm receptor proteins.

IV. INNER AND OUTER ARM FUNCTION IN VIVO

Gibbons and Gibbons [1973] found that, if the outer arms were selectively extracted from the axonemes of demembranated sea urchin sperm flagella by high-salt extraction, the axonemes continued to beat with normal waveform but reduced beat frequency. The reduction in beat frequency was proportional to the apparent number of arms removed; when ~40% of the total arms (i.e., greater than 80% of the outer arms, but few if any inner arms) were removed, beat frequency was reduced by ~50%. They concluded that the inner and outer arms were functionally equivalent in the generation of sliding between outer doublets and that components other than the dynein arms were responsible for determining flagellar waveform. Unfortunately, the complementary experiment—selective removal of the inner arms—could not be carried out in sea urchin. If this experiment had been done, a different conclusion might have been reached.

The equivalent experiment and its complement were made possible in *Chlamydomonas* by the discovery of mutations that selectively deleted inner or outer arms. Several mutants lacking the outer dynein arm have a beat frequency that is reduced to about one-half that of wild type; this is accompanied by subtle changes in waveform [Mitchell and Rosenbaum, 1985; Kamiya and Okamota, 1985; Brokaw and Kamiya, 1987]. Microtubule sliding velocity, which is proportional to (beat frequency) × (shear amplitude), was reduced to approximately one-third to one-fourth that of wild type [Brokaw and Kamiya, 1987]. This reduction is greater than the reduction in the total number of arms, indicating that the rate of sliding is not simply proportional to the number of arms remaining. In contrast, mutants lacking inner arm I1 exhibited a large decrease in shear amplitude, with little change in beat frequency [Brokaw and Kamiya, 1987]. A similar pattern was observed for mutants *pf2* and *pf3*, which have defects in the dynein regulatory complex (see above). More recently, it has been reported that mutant *ida4*, lacking inner arm I2, also has a beat frequency similar to that of wild type but probably has a reduced shear amplitude [Kamiya et al., 1991]. The different beat patterns observed in *Chlamydomonas* mutants lacking the inner and outer arms indicates that, although all of the arms probably contribute to microtubule sliding, they must have other functions, relating to the control of sliding and/or bending, that are different for the two types of arms. In particular, the inner arms appear to be intimately involved in the determination of flagellar waveform. They also may have an important role in the initiation of bending [Kamiya et al., 1991]. Elucidation of the specific functions of the outer arm dynein and the different inner arm dyneins, and the mechanisms by which they communicate with one another and with other components of the axoneme, will undoubtedly be a major goal of future research on these fascinating and ubiquitous structures.

ACKNOWLEDGMENTS

We are grateful to Dr. Susan Dutcher of the University of Colorado, Dr. Ritsu Kamiya of the University of Tokyo, and Dr. Mary Porter of the University of Minnesota for sharing information with us prior to publication. This work was supported by NIH grants GM30626, HD23858, CA12708 and by a grant from the Mellon Foundation.

V. REFERENCES

Bell CW, Fronk E, Gibbons IR (1979): Polypeptide subunits of dynein 1 from sea urchin sperm flagella. J Supramol Struct 11:311–317.

Belles-Isles M, Chapeau C, White D, Gagnon C (1986): Isolation and characterization of dynein ATPase from bull spermatozoa. Biochem J 240:863–869.

Brokaw CJ (1989): Operation and regulation of the flagellar oscillator. In Warner FD, Satir P, Gibbons IR (eds): Cell Movement, vol 1: The Dynein ATPases. New York: Alan R. Liss, Inc., pp 267–279.

Brokaw CJ, Kamiya R (1987): Bending patterns of *Chlamydomonas* flagella. IV. Mutants with defects in inner and outer dynein arms indicate differences in dynein arm function. Cell Motil Cytoskel 8:68–75.

Burgess SA, Carter DA, Dover SD, Wooley DM (1991a): The inner dynein arm complex: Compatible images from freeze etch and thin section methods of microscopy. J Cell Sci 100:319–328.

Burgess SA, Dover SD, Woolley DM (1991b): Architecture of the outer arm dynein ATPase in an avian sperm flagellum, with further evidence for the B-link. J Cell Sci 98:17–26.

Cremo CR, Grammer JC, Yount RG (1989): Direct chemical evidence that serine 180 in the glycine-rich loop of myosin binds to ATP. J Biol Chem 264: 6608–6611.

Fay RB, Witman GB (1977): The localization of flagellar ATPases in *Chlamydomonas reinhardii,* abstracted. J Cell Biol 75:266a.

Gatti J-L, Dacheux J-L (1992): Characterization of boar sperm dynein by immunological cross-reaction and photocleavage UV-vanadate dependent, abstracted. Second International Cell Motility Symposium, Nice—Sophia Antipolis, France.

Gatti J-L, King SM, Moss AG, Witman GB (1989): Outer arm dynein from trout spermatozoa: Purification, polypeptide composition and enzymatic properties. J Biol Chem 264:11450–11457.

Gibbons BH, Gibbons IR (1973): The effect of partial extraction of dynein arms on the movement of reactivated sea urchin sperm. J Cell Sci 13:337–357.

Gibbons IR, Gibbons BH, Mocz G, Asai DJ (1991): Multiple nucleotide-binding sites in the sequence of dynein β heavy chain. Nature 352:640–643.

Goodenough UW, Gebhart B, Mermall V, Mitchell DR, Heuser JE (1987): High pressure liquid chromatography fractionation of *Chlamydomonas* dynein extracts and characterization of inner arm dynein subunits. J Mol Biol 194:481–494.

Goodenough UW, Heuser JE (1982): Substructure of the outer dynein arm. J Cell Biol 95:798–815.

Goodenough UW, Heuser JE (1984): Structural comparison of purified dynein proteins with *in situ* dynein arms. J Mol Biol 180:1083–1118.

Goodenough UW, Heuser JE (1985): Substructure of inner dynein arms, radial spokes and the central pair/projection complex. J Cell Biol 100:2008–2018.

Goodenough UW, Heuser JE (1989): Structure of the soluble and in situ ciliary dyneins visualized by quick-freeze deep-etch microscopy. In Warner FD, Satir P, Gibbons IR (eds): Cell Movement, vol 1: The Dynein ATPases. New York: Alan R. Liss, Inc., pp 121–140.

Haimo LT, Rosenbaum JL (1981): Dynein binding to microtubules containing microtubule-associated protein. Cell Motil 1:499–516.

Hamasaki T, Barkalow K, Richmond J, Satir P (1991): cAMP-stimulated phosphorylation of an axonemal polypeptide that copurifies with the 22 S dynein arm regulates microtubule translocation velocity and swimming speed in *Paramecium.* Proc Natl Acad Sci USA 88:7918–7922.

Hastie AT, Dicker DT, Hingley ST, Kueppers F, Higgins ML, Weinbaum G (1986): Isolation of cilia from porcine tracheal epithelium and extraction of dynein arms. Cell Motil Cytoskel 6:25–34.

Hastie AT, Marchese-Ragona SP, Johnson KA, Wall JS (1988): Structure and mass of mammalian respiratory outer arm 19 S dynein. Cell Motil Cytoskel 11:157–166.

Huang B, Mengersen A, Lee VD (1988): Molecular cloning of cDNA for caltractin, a basal body-associated Ca^{2+}-binding protein: Homology in its protein sequence with calmodulin and the yeast CDC31 gene product. J Cell Biol 107:133–140.

Huang B, Piperno G, Luck DJL (1979): Paralyzed flagellar mutants of *Chlamydomonas reinhardtii* defective for axonemal doublet microtubule arms. J Biol Chem 254:3091–3099.

Huang B, Ramanis Z, Luck DJL (1982): Suppressor mutations in *Chlamydomonas* reveal a regulatory mechanism for flagellar function. Cell 28:115–124.

Johnson KA, Wall JS (1983): Structure and molecular weight of the dynein ATPase. J Cell Biol 96:669–678.

Kagami O, Kamiya R (1992): Translocation and rotation of microtubules caused by multiple species of *Chlamydomonas* inner-arm dynein. J Cell Sci 103:653–664.

Kagami O, Takada S, Kamiya R (1990): Microtubule translocation caused by three subspecies of inner arm dynein from *Chlamydomonas* flagella. FEBS Lett 264:179–182.

Kamiya R (1988): Mutations at twelve independent loci result in absence of outer dynein arms in *Chlamydomonas reinhardtii.* J Cell Biol 107:2253–2258.

Kamiya R, Okamoto M (1985): A mutant of *Chlamydomonas reinhardtii* that lacks the flagellar outer dynein arm but can swim. J Cell Sci 74:181–191.

Kamiya R, Kurimoto E, Muto E (1991): Two types of *Chlamydomonas* flagellar mutants missing different components of inner arm dynein. J Cell Biol 112:441–447.

Kamiya R, Kurimoto E, Sakakibara H, Okagaki T (1989): A genetic approach to the function of inner and outer arm dynein. In Warner FD, Satir P, Gibbons IR (eds): Cell Movement, vol 1: The Dynein ATPases. New York: Alan R. Liss, Inc., pp 209–218.

King SJ, Inwood WB, Power J, O'Toole ET, Dutcher SK (1992): Further heterogeneity of the inner dynein arm revealed by the *bop2* mutation (abstract). Mol Biol Cell 3:161a.

King SM, Witman GB (1987): Structure of the α and β heavy chains of the outer arm dynein from *Chlamydomonas* flagella. Masses of chains and sites of ultraviolet-induced vanadate-dependent cleavage. J Biol Chem 262:17596–17604.

King SM, Witman GB (1988a): Structure of the α and β heavy chains of the outer arm dynein from *Chlamydomonas* flagella. Location of epitopes and protease-sensitive sites. J Biol Chem 263:9244–9255.

King SM, Witman GB (1988b): Structure of the γ heavy chain of the outer arm dynein from *Chlamydomonas* flagella. J Cell Biol 107:1799–1808.

King SM, Witman GB (1989): Molecular structure of *Chlamydomonas* outer arm dynein. In Warner FD, Satir P, Gibbons IR (eds): Cell Movement, vol 1: The Dynein ATPases. New York: Alan R. Liss, Inc., pp 61–75.

King SM, Witman GB (1990): Localization of an intermediate chain of outer arm dynein by immunoelectron microscopy. J Biol Chem 265:19807–19811.

King SM, Gatti J-L, Moss AG, Witman GB (1990): Outer arm dynein from trout spermatozoa. Substructural oranization. Cell Motil Cytoskel 16:266–278.

King SM, Haley BE, Witman GB (1989): Structure of the α and β heavy chains of the outer arm dynein from *Chlamydomonas* flagella. Nucleotide binding sites. J Biol Chem 264:10210–10218.

King SM, Kim H, Haley BE (1991a): Strategies and reagents for photoaffinity labeling of mechanochemical proteins. Methods Enzymol 196:449–466.

King SM, Otter T, Witman GB (1985): Characterization of monoclonal antibodies against *Chlamydomonas* flagellar dyneins. Proc Natl Acad Sci USA 82:4717–4721.

King SM, Wilkerson CG, Witman GB (1991b): The M_r78,000 intermediate chain of *Chlamydomonas* outer arm dynein interacts with α-tubulin *in situ*. J Biol Chem 266:8401–8407.

Larsen J, Barkalow K, Hamasaki T, Satir P (1991): Structural and functional characterization of *Paramecium* dynein: Initial studies. J Protozool 38:55–61.

Lee-Eiford A, Ow RA, Gibbons IR (1996): Specific cleavage of dynein heavy chains by ultraviolet irradiation in the presence of ATP and vanadate. J Biol Chem 261:2337–2342.

Linck RW (1973): Chemical and structural differences between cilia and flagella from the lamellibranch mollusc, *Aequipecten irradians*. J Cell Sci 12:951–981.

Luck DJL, Piperno G (1989): Dynein arm mutants of *Chlamydomonas*. In Warner FD, Satir P, Gibbons IR (eds): Cell Movement, vol 1: The Dynein ATPases. New York: Alan R. Liss, Inc., pp 49–60.

Mabuchi I, Shimizu T, Mabuchi Y (1976): A biochemical study of flagellar dynein from starfish spermatozoa: Protein components of the arm structure. Arch Biochem Biophys 176:564–576.

Marchese-Ragona SP, Detrich HW III (1990): Purification and characterization of cytoplasmic and axonemal dyneins from antarctic fishes, abstracted. J Cell Biol 111:296a.

Marchese-Ragona SP, Johnson KA (1988): STEM analysis of the dynein ATPase. Electron Microsc Rev 1:141–153.

Marchese-Ragona SP, Johnson KA (1989): Structural and biochemical studies of the dynein ATPase. In Gagnon C (ed): Controls of Sperm Motility: Biological and Clinical Aspects. Boca Raton, FL: CRC Press, pp 203–217.

Marchese-Ragona SP, Facemyer KC, Johnson KA (1989): Structure of the α-, β-, and γ-heavy chains of 22 S outer arm dynein obtained from *Tetrahymena* cilia. J Biol Chem 264:21361–21368.

Marchese-Ragona SP, Gagnon C, White D, Belles-Isles M, Johnson KA (1987): Structure and mass analysis of 12 S and 19 S dynein obtained from bull sperm flagella. Cell Motil Cytoskel 8:368–374.

Mastronarde DN, O'Toole ET, McDonald KL, McIntosh JR, Porter ME (1992): Arrangement of inner dynein arms in wild-type and mutant flagella of *Chlamydomonas*. J Cell Biol 118:1145–1162.

Mitchell DR, Kang Y (1991): Identification of *oda6* as a *Chlamydomonas* dynein mutant by rescue with the wild-type gene. J Cell Biol 113:835–842.

Mitchell DR, Rosenbaum JL (1985): A motile *Chlamydomonas* flagellar mutant that lacks outer dynein arms. J Cell Biol 100:1228–1234.

Mitchell DR, Rosenbaum JL (1986): Protein–protein interactions in the 18 S ATPase of *Chlamydomonas* outer dynein arms. Cell Motil Cytoskel 6:510–520.

Mitchell DR, Warner FD (1980): Interactions of dynein arms with B subfibers of *Tetrahymena* cilia: Quantitation of the effects of magnesium and adenosine triphosphate. J Cell Biol 87:84–97.

Mocz G, Farias J, Gibbons IR (1991): Proteolytic analysis of domain structure in the β heavy chain of dynein from sea urchin sperm flagella. Biochemistry 30: 7225–7231.

Mocz G, Tang W-JY, Gibbons IR (1988): A map of photolytic and tryptic cleavage sites on the β heavy chain of dynein ATPase from sea urchin sperm flagella. J Cell Biol 106:1607–1614.

Mohri H, Wada S, Okuno M, Nakamura K (1992): Axonemal dynein of oyster sperm— A comparative study, abstracted. Second International Cell Motility Symposium, Nice—Sophia Antipolis, France.

Moss AG, Gatti J-L, Witman GB (1992a): The motile β/IC1 subunit of sea urchin sperm outer arm dynein does not form a rigor bond. J Cell Biol 118:1177–1188.

Moss AG, Sale WS, Fox LA, Witman GB (1992b): The α subunit of sea urchin outer arm dynein mediates structural and rigor binding to microtubules. J Cell Biol 118:1189–1200.

Muto E, Kamiya R, Tsukita S (1991): Double-rowed organization of inner dynein arms in *Chlamydomonas* flagella revealed by tilt-series thin-section electron microscopy. J Cell Sci 99:57–66.

Ogawa K (1991): Four ATP-binding sites in the midregion of the β heavy chain of dynein. Nature 352:643–645.

Ogawa K, Negishi S, Obika M (1980): Dynein 1 from rainbow trout spermatozoa: Immunological similarity between trout and sea urchin dynein 1. Arch Biochem Biophys 203:196–203.

Ogawa K, Yokota E, Hamada Y, Wada S, Okuno M, Nakajima Y (1990): The outer arm dynein α heavy chains of sea urchin sperm flagella and embryonic cilia are different. Cell Motil Cytoskel 16:58–67.

Ow RA, Tang W-JY, Mocz G, Gibbons IR (1987): Tryptic digestion of dynein 1 in low salt medium: Origin and properties of Fragment A. J Biol Chem 262: 3409–3414.

Pallini V, Bugnoli M, Mencarelli C, Scapagliati G (1982): Biochemical properties of ciliary, flagellar and cytoplasmic dyneins. Symp Soc Exp Biol 35:339–352.

Paschal BM, Collins CA, King SM, Moss AG, Vallee RB, Witman GB (1987a): *In vitro* gliding motility of axonemal dynein and MAP 1C, a dynein-like ATPase from calf brain, abstracted. J Cell Biol 105:96a.

Paschal BM, King SM, Moss AG, Collins CA, Vallee RB, Witman GB (1987b): Isolated flagellar outer arm dynein translocates brain microtubules *in vitro*. Nature 330:672–674.

Paschal BM, Mikami A, Pfister KK, Vallee RB (1992): Homology of the 74 kD cytoplasmic dynein subunit with a flagellar dynein polypeptide suggests an intracellular targeting function. J Cell Biol 118:1133–1143.

Pfister KK, Witman GB (1984): Subfractionation of *Chlamydomonas* 18 S dynein into two unique subunits containing ATPase activity. J Biol Chem 259: 12072–12080.

Pfister KK, Fay RB, Witman GB (1982): Purification and polypeptide composition of dynein ATPases from *Chlamydomonas reinhardtii*. Cell Motil 2:525–547.

Pfister KK, Haley BE, Witman GB (1984): The photoaffinity probe 8-azido-adenosine 5′-triphosphate selectively labels the heavy chain of *Chlamydomonas* 12 S dynein. J Biol Chem 259:8499–8504.

Pfister KK, Haley BE, Witman GB (1985): Labeling of *Chlamydomonas* 18 S dynein polypeptides by 8-azidoadenosine 5′-triphosphate, a photoaffinity analog of ATP. J Biol Chem 260:12844–12850.

Piperno G, Luck DJL (1979a): Axonemal adenosine triphosphatases from flagella of *Chlamydomonas reinhardtii*. Purification of two dyneins. J Biol Chem 254:3084–3090.

Piperno G, Luck DJL (1979b): An actin-like protein is a component of axonemes from *Chlamydomonas reinhardtii*. J Biol Chem 254:2187–2190.

Piperno G, Luck DJL (1981): Inner arm dyneins from flagella of *Chlamydomonas reinhardtii*. Cell 27: 331–340.

Piperno G, Mead K, Shestak W (1992): The inner dynein arms I2 interact with a "dynein regulatory complex" in *Chlamydomonas* flagella. J Cell Biol 118:1455–1463.

Piperno G, Ramanis Z (1991): The proximal portion of *Chlamydomonas* flagella contains a distinct set of inner dynein arms. J Cell Biol 112:701–709.

Piperno G, Ramanis Z, Smith EF, Sale WS (1990): Three distinct inner dynein arms in *Chlamydomonas* flagella: Molecular composition and location in the axoneme. J Cell Biol 110:379–389.

Porter ME, Johnson KA (1983): Characterization of the ATP-sensitive binding of *Tetrahymena* 30 S dynein to bovine brain microtubules. J Biol Chem 258:6575–6581.

Porter ME, Power J, Dutcher SK (1992): Extragenic suppressors of paralyzed flagellar mutations in *Chlamydomonas reinhardtii* identify loci that alter the inner dynein arms. J Cell Biol 118:1163–1176.

Sakakibara H, Mitchell DR, Kamiya R (1991): A *Chlamydomonas* outer arm dynein mutant missing the α heavy chain. J Cell Biol 113:615–622.

Sakakibara H, Takada S, King SM, Witman GB, Kamiya R (1993): A *Chlamydomonas* outer arm dynein mutant with a truncated β heavy chain. J Cell Biol 122: 653–661.

Sale WS, Fox LA (1988): Isolated β heavy chain subunit of dynein translocates microtubules *in vitro*. J Cell Biol 107:1793–1797.

Sale WS, Satir P (1977): Direction of active sliding of microtubules in *Tetrahymena* cilia. Proc Natl Acad Sci USA 74:2045–2049.

Sale WS, Goodenough UW, Heuser JE (1985): The substructure of isolated and *in situ* outer dynein arms of sea urchin sperm flagella. J Cell Biol 101:1400–1412.

Schroeder CC, Fok AK, Allen RD (1990): Vesicle transport along microtubular ribbons and isolation of cytoplasmic dynein from *Paramecium*. J Cell Biol 111:2553–2562.

Smith EF, Sale WS (1991): Microtubule binding and translocation by inner arm subtype I1. Cell Motil Cytoskel 18:258–268.

Smith EF, Sale WS (1992a): Regulation of dynein-driven microtubule sliding by the radial spokes in flagella. Science 257:1557–1559.

Smith EF, Sale WS (1992b): Structural and functional reconstitution of inner dynein arms in *Chlamydomonas* flagella axonemes. J Cell Biol 117:573–581.

Takada S, Sakakibara H, Kamiya R (1992): Three-headed outer arm dynein from *Chlamydomonas* that can functionally recombine with outer arm-missing axonemes. J Biochem 111:758–762.

Takahashi K, Shingyoji C, Kamimura S (1982): Microtubule sliding in reactivated flagella. Symp Soc Exp Biol 35:159–177.

Tang W-JY, Gibbons IR (1987): Photosensitized cleavage of dynein heavy chains: Cleavage at the V2 site by irradiation at 365 nm in the presence of oligovanadate. J Biol Chem 262:17728–17734.

Tang W-JY, Bell CW, Sale WS, Gibbons IR (1982): Structure of the dynein-1 outer arm in sea urchin sperm flagella. I. Analysis by separation of subunits. J Biol Chem 257:508–515.

Toyoshima YY (1987): Chymotryptic digestion of *Tetrahymena* 22 S dynein. I. Decomposition of three-headed 22 S dynein to one- and two-headed particles. J Cell Biol 105:887–895.

Vale RD, Oosawa F (1990): Protein motors and Maxwell's demons: does mechanochemical transduction involve a thermal ratchet. Adv Biophys 26:97–134.

Vale RD, Toyoshima YY (1988): Rotation and translocation of microtubules *in vitro* induced by dyneins from *Tetrahymena* cilia. Cell 52:459–469.

Vale RD, Soll DR, Gibbons IR (1989): One-dimensional diffusion of microtubules bound to flagellar dynein. Cell 59:915–925.

Walker JE, Saraste M, Runswick MJ, Gay NJ (1982): Distantly related sequences in the α- and β-subunits of ATP synthase, myosin, kinases and other ATP-requiring enzymes and a common nucleotide binding fold. EMBO J 1:945–951.

Warner FD, Perrault JG, McIlvain JH (1985): Rebinding of *Tetrahymena* 13 S and 21 S dynein ATPases to extracted doublet microtubules. The inner row and outer row dynein arms. J Cell Sci 77:263–287.

Warner FD, Satir P, Gibbons IR (1989): Cell Movement, vol 1: The Dynein ATPases. New York: Alan R. Liss, Inc.

Warrick HM, Spudich JA (1987): Myosin structure and function in cell motility. Annu Rev Cell Biol 3:379–421.

Wilkerson CG, King SM, Witman GB (1991): Sequence analysis of the γ heavy chain from the *Chlamydomonas* outer arm dynein, abstracted. J Cell Biol 115:167a.

Wilkerson CG, King SM, Witman GB (1992): Sequence analysis of multiple dynein heavy chain genes from *Chlamydomonas reinhardtii* (abstract). Mol Biol Cell 3:161a.

Witman GB (1992): Axonemal dyneins. Curr Opin Cell Biol 4:74–79.

Witman GB, Johnson KA, Pfister KK, Wall JS (1983): Fine structure and molecular weight of the outer arm dyneins of *Chlamydomonas*. J Submicrosc Cytol 15:193–197.

Witman GB, King SM, Moss AG, Wilkerson CG (1991): The intermediate chain/light chain complex: An important structural entity of outer arm dynein. In Baccetti B (ed): Comparative Spermatology, 20 Years After. New York: Raven Press, pp 439–443.

Yang JT, Laymon RA, Goldstein LSB (1989): A three-domain structure of kinesin heavy chain revealed by DNA sequence and microtubule binding analyses. Cell 56:879–889.

ABOUT THE AUTHORS

GEORGE B. WITMAN is Principal Scientist and Director of the Male Fertility Program at the Worcester Foundation for Experimental Biology in Shrewsbury, Massachusetts. Dr. Witman graduated with high honors in zoology from the University of California in 1967, and in 1972 earned his Ph.D. in cellular and developmental biology from Yale University, where he worked in Joel Rosenbaum's laboratory developing the first methods for fractionation and biochemical analysis of *Chlamydomonas* flagella. This was followed by postdoctoral research with Hewson Swift at the University of Chicago and with Gary Borisy at the University of Wisconsin. Dr. Witman then spent several years on the faculty at Princeton University. Since moving to the Worcester Foundation in 1981, he has been investigating the molecular basis for flagellar motility in several model systems. His most recent research is using molecular genetic and biochemical approaches in *Chlamydomonas* to elucidate the structure and mechanism of action of the dynein ATPases that generate the force for flagellar movement. He is the author of over 50 scientific publications and has been a member of the editorial board of *Cell Motility and the Cytoskeleton* since 1986.

CURTIS G. WILKERSON was awarded a B.Sc. in zoology by North Carolina State University in 1976, and in 1984 received his Ph.D. from the University of Georgia, where he investigated the mechanism of chloroplast biogenesis in *Euglena gracilis*. He then joined David Luck's laboratory, where he began his study of dynein heavy-chain function. In 1988 he became a member of George Witman's group at the Worcester Foundation for Experimental Biology, where he is currently studying the function of the *Chlamydomonas* γ dynein outer-arm heavy-chain subunit using molecular biological and genetic techniques.

STEPHEN M. KING was awarded a B.Sc. with honors in microbiology by the University of Kent, England, in 1979, and in 1982 received his Ph.D. from University College London, where he worked in Jerry Hyams' laboratory investigating the mechanism of mitosis in yeast. He then joined George Witman's group at the Worcester Foundation for Experimental Biology, where as a Senior Research Associate he studied the structure and biochemistry of dynein in *Chlamydomonas* and trout sperm flagella. In 1993 he traveled to Palmer Station, Antarctica, to investigate biochemical mechanisms of cold adaptation in the dyneins of fish that live at sub-zero temperatures. He was recently appointed Assistant Professor in the Department of Biochemistry at the University of Connecticut Health Center in Farmington, Connecticut.

Microtubules: 251–267

Molecular Characterization of Cytoplasmic Dynein

Erika L.F. Holzbaur, Atsushi Mikami, Bryce M. Paschal, and Richard B. Vallee

I. INTRODUCTION

Cytoplasmic dynein was initially identified as a minus end–directed microtubule motor [Paschal and Vallee,1987]. Scanning transmission electron microscopy revealed the protein to be indistinguishable in morphology and mass from two-headed axonemal forms of dynein [Vallee et al., 1988]. Enzymological, biochemical, and pharmacological evidence [Paschal et al., 1987; Lye et al., 1987; Euteneuer et al., 1988; Shpetner et al., 1988] added further support to the widespread existence of a true cytoplasmic form of dynein.

The force-producing and pharmacological characteristics of cytoplasmic dynein were consistent with a role for the enzyme in minus end–directed, or retrograde, organelle transport [Paschal and Vallee, 1987], a model that has been supported by subsequent work. Minus end–directed organelle movements were found to be inhibited by exposure to ultraviolet light in the presence of vanadate [Schnapp and Reese, 1989; Schroer et al., 1989], though bidirectional movements were inhibited by vanadate-mediated photocleavage in one system [Euteneuer et al., 1988]. Retrograde organelle transport was also found to be inhibited by immunodepletion of cytoplasmic dynein [Courtesy-Theulaz et al., 1992]. Immunocytochemistry has localized cytoplasmic dynein to minus end–directed organelles, including lysosomes and late endosomes [Lin and Collins, 1992].

Additional evidence has pointed to a potential role for cytoplasmic dynein in chromosome movement. Antibodies to subunits of cytoplasmic dynein have been reported to react with kinetochores between prometaphase and metaphase [Steuer et al., 1990; Pfarr et al., 1990]. Poleward prometaphase chromosome movement has been observed to occur *in vivo* at rates consistent with a role for cytoplasmic dynein [Rieder and Alexander, 1990], and dynein-like force production associated with kinetochores of isolated chromosomes has also been reported [Hyman and Mitchison, 1991]. Whether cytoplasmic dynein is actually involved in chromosome movement has yet to be proven conclusively, and whether its role is restricted to prometaphase remains unresolved.

A number of questions have arisen during the course of work on cytoplasmic dynein that are best addressed by molecular characterization. How closely related is the cytoplasmic enzyme to axonemal dynein? What aspects of dynein structure specify a cytoplasmic versus an axonemal role? What is the true range of functions of the cytoplasmic enzyme? How does the enzyme bind to subcellular structures? Many of these issues can be addressed by genetic and molecular genetic means, and the resolution of these and related issues represents the long-term goal of current research on the enzyme. We review the known structural components of cytoplasmic dynein and the substantial recent progress in the molecular cloning of most of these species.

II. STRUCTURE AND POLYPEPTIDE COMPOSITION OF CYTOPLASMIC DYNEIN

While both two-headed and three-headed forms of axonemal dynein have been identified, cytoplasmic dynein was found to be a two-headed structure [Vallee et al., 1988; Schnapp and Reese, 1989; Neely et al., 1990]. As was found for axonemal dyneins, the force-producing heads of the cytoplasmic enzyme are each connected by stalk regions to a common base.

The subunit compositions of cytoplasmic and axonemal dyneins are distinct. While all known forms of dynein contain catalytic heavy chains of similar size (>>400 kD, see below) corresponding in number to the number of force-producing heads, they differ in the content of lower molecular weight accessory subunits. Bovine brain cytoplasmic dynein contains two heavy chains, three 74 kD polypeptides, and approximately one copy each of four polypeptides of 59, 57, 55, and 53 kD (Fig. 1) [Paschal and Vallee, 1987; Vallee et al., 1988]. The overall composition of cytoplasmic dynein is distinct from that of axonemal dyneins, including those isolated from mammalian cilia and flagella [Hastie et al., 1986; Belles-Isles et al., 1986]. However, because axonemal dyneins differ markedly in subunit composition from each other, the importance of this observation is still uncertain. The subunit composition of cytoplasmic dynein does not appear to vary with species; a similar polypeptide composition has been noted in cytoplasmic dyneins from cow, rat, and chick [Paschal et al., 1987; Collins and Vallee, 1989; Steuer et al., 1990].

Cytoplasmic Dynein	Glued (Dynactin) Complex
532 kD	150 kD (135 kD)
74 kD	50 kD
59 kD	45 kD (centractin)
57 kD	
55 kD	
53 kD	

Fig. 1. *Polypeptide components of cytoplasmic dynein preparations. Components of the cytoplasmic dynein complex itself are listed at left [Paschal et al., 1987; Vallee et al., 1988]. Components of the* Glued *complex as determined by immunoprecipitation experiments (the 135 kD component is observed at variable levels) [Paschal et al., 1993], also referred to as the* dynactin complex *[Gill et al., 1991], are listed at right.*

The cytoplasmic dynein-associated polypeptides of 74, 59, 57, 55, and 53 kD appear to bind tenaciously to the complex, copurifying with the high molecular weight species at a constant and reproducible stoichiometry through a variety of fractionation procedures [Paschal et al., 1987]. This evidence along with sequence comparisons between cytoplasmic and axonemal dynein accessory polypeptides (see below) indicate that some, and probably all, of these polypeptides are true subunits of the cytoplasmic dynein holoenzyme.

In contrast, a number of additional polypeptides have been identified in cytoplasmic dynein preparations [Collins and Vallee, 1989], which can be separated from the enzyme by chromatography or immunoselection [Schroer and Sheetz, 1991; Paschal et al., 1993]. This group of polypeptides, of which the most prominent members are species of 150, 50, and 45 kD, cofractionate and are thought to comprise a complex distinct from cytoplasmic dynein (Fig. 1). The functional relationship between this complex, referred to as the *Glued* or *dynactin* complex after the 150 kD component [Holzbaur et al., 1991; Gill et al., 1991; Paschal et al., 1993], and cytoplasmic dynein is not fully clear, though it has been reported that the complex is required for cytoplasmic dynein-mediated vesicular transport *in vitro* [Gill et al., 1991]. As discussed below, the molecular analysis of the components of this complex has provided insight into its composition and structure and has suggested an interaction between the actin- and microtubule-based networks within the cell [Holzbaur et al., 1991; Lees-Miller et al., 1992a; Clark and Meyer, 1992; Paschal et al., 1993].

III. MOLECULAR ANALYSIS OF THE CYTOPLASMIC DYNEIN HEAVY CHAIN

The cloning and sequencing of overlapping cDNAs encoding the cytoplasmic dynein heavy chain of two species—*Dictyostelium discoideum* [Koonce et al., 1992] and rat [Mikami et al., 1993]—has been completed. Comparisons between the sequences of the cytoplasmic dynein heavy chain from two widely divergent species, and comparisons of these sequences with that determined for the β-heavy chain of sea urchin sperm flagellar outer arm dynein [Gibbons et al., 1991; Ogawa, 1991], have provided important new information on the structural organization of the dynein molecule.

A partial cDNA clone encoding the cytoplasmic dynein heavy chain from *Dictyostelium* was selected by antibody screening of an expression library. The full-length sequence was then isolated by using PCR to walk along the *Dictyostelium* genomic DNA [Koonce et al., 1992]. Mikami et al. [1993] selected a partial cDNA clone encoding the mammalian cytoplasmic dynein heavy chain using an oligonucleotide probe based on amino acid sequence obtained from a tryptic fragment of the polypeptide. A complete set of overlapping cDNAs encoding the full-length sequence from rat was obtained by hybridization screening and the polymerase chain reaction (PCR). The deduced molecular weights for the *Dictyostelium* and rat polypeptides were found to be 538 and 532 kD, respectively.

Figure 2 shows a dot matrix comparison of the deduced amino acid sequence of the rat cytoplasmic dynein heavy chain with the sequence of the sea urchin flagellar dynein β-heavy chain. Extensive homology may be seen, with the highest degree of conservation existing within the central region of the polypeptides. The striking exception to this pattern is the N-terminal third of the sequences where no evidence of conservation is seen. This region is, however, clearly conserved between the sequences for the rat and *Dictyostelium* cytoplasmic dyneins [Mikami et al., 1993]. The difference between the N-terminal regions of the axonemal and cytoplasmic dynein heavy chains suggests that this region may be involved in specifying functions unique to each of the dynein forms. This region may, itself, specify binding to organelles, to the kinetochore, or to the microtubules of the axoneme. Alternatively, this divergent sequence may bind to accessory subunits specific to each of the dynein forms.

A. Multiple ATP-Binding Consensus Sequences Within the Heavy Chain Sequence

Five phosphate-binding loop (P-loop) consensus sequence elements (GXXXXGKT/S), which are characteristic of nucleotide-binding proteins [Walker et al., 1982; Fry et al., 1986], were identified within the sea urchin axonemal dynein heavy chain sequence [Gibbons, 1991; Ogawa, 1991]. The cytoplasmic dynein heavy chains lack the first of these elements, which is located within the nonconserved N-terminal region of the dynein sequence. However, the cytoplasmic polypeptide encodes the other four consensus elements, which are located within the central third of the polypeptide (Fig. 2). The first of these elements, GPAGTGKT, beginning at amino acid 1,852 in the sea urchin sequence, was tentatively identified as the principle site of ATP hydrolysis for flagellar dynein, based on its position within the heavy chain and on the size of the polypeptides produced by the vanadate photocleavage reaction. Consistent with an important role for this P-loop consensus element in dynein function, it is completely conserved in both the *Dictyostelium* and rat sequences, and the region immediately surrounding this consensus element is the most highly conserved part of the molecule. Over 100 amino acid residues spanning this region, the rat and sea urchin sequences are 59% identical. The other elements are less well conserved, but still show significant relatedness; sequence identities of 33, 25, and 36% were determined for similar-sized regions surrounding the second, third, and fourth P-loop motifs, respectively [Mikami, 1993].

The functional significance of the multiple P-loop elements is unknown. Potentially, all four

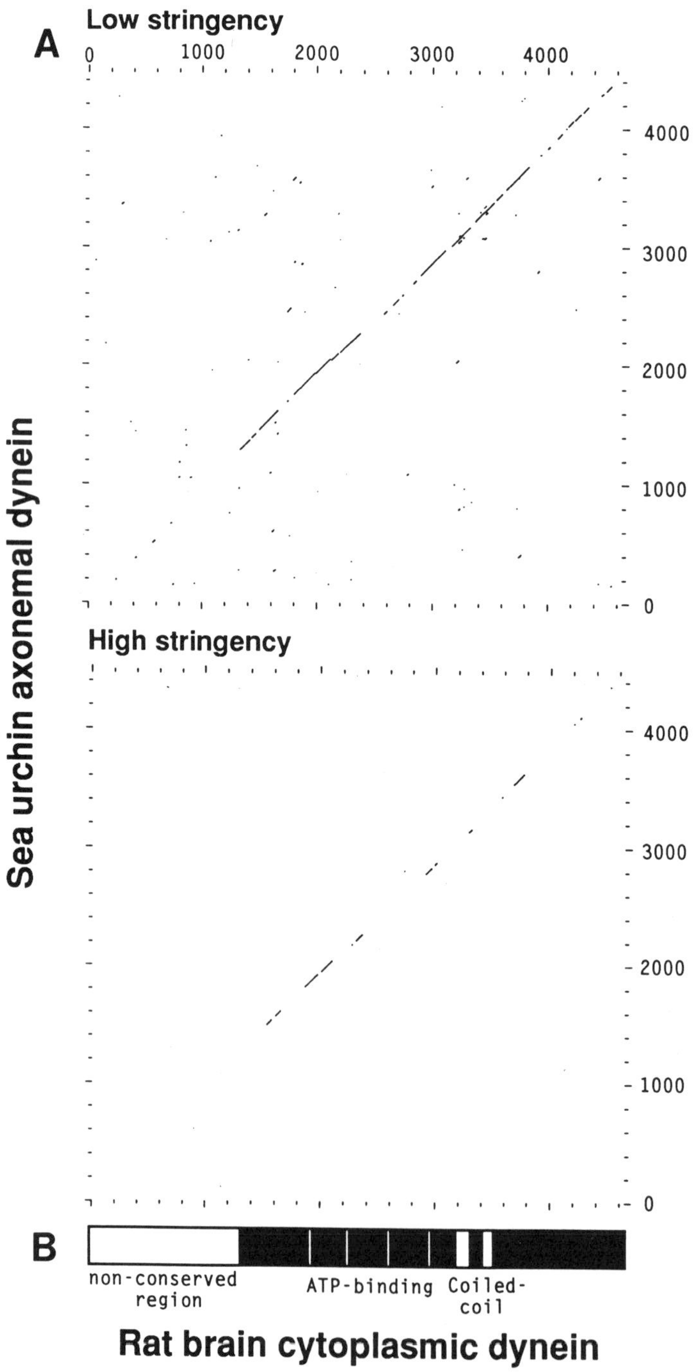
A
Low stringency
0
1000
2000
3000
4000
High stringency
Sea urchin axonemal dynein
B
non-conserved
region
ATP-binding
Coiled-
coil
Rat brain cytoplasmic dynein

of the sequences conserved between flagellar and cytoplasmic dynein heavy chain polypeptides are capable of binding nucleotide. However, presteady-state kinetic analyses of axonemal dynein have yielded a value of one ATP hydrolysis site per 750,000 D, approximately corresponding to one site per heavy chain within the dynein holoenzyme [Johnson, 1983; Shimizu et al., 1983]. Also, the rate of ATP hydrolysis was observed to decrease in parallel with vanadate-mediated photocleavage at the V1 site [Lee-Eiford et al., 1986; Gibbons et al., 1987], again suggesting a single ATP hydrolysis site per heavy chain. Furthermore, no evidence for cooperative kinetic interactions between multiple ATP-binding sites has been observed for the axonemal dynein ATPase [Shimizu et al., 1983; Holzbaur and Johnson, 1989a,b]. Nonetheless, clear sequence conservation among the four centrally located P-loop motifs of both cytoplasmic and axonemal dyneins suggests that they could all be functionally significant. Further studies, including the detailed kinetic characterization of the cytoplasmic dynein mechanochemical cycle and targeted mutagenesis of the dynein heavy chain, may provide insight into this question.

Most of the mass of the cytoplasmic dynein heavy chain is thought to form the globular head domain of 327 ± 50 kD [Vallee et al., 1988; Johnson and Wall, 1983]. The remainder of the polypeptide may extend from the head to form the stalks linking the heads to the common base of the molecule. Within the cytoplasmic dynein heavy chain sequences two extended runs of predicted α-helix were observed of about 100 amino acids each that fit a heptad repeat pattern [Koonce et al., 1992; Mikami et al., 1993]. These regions have the potential to form a coiled-coil structure and may therefore mediate the twining of two separate heavy chain stalks to form a nucleus at the base of the molecule to which the intermediate and light chains may bind.

Fig. 2. *Primary and secondary structural analyses of the rat cytoplasmic dynein heavy chain gene.* **A:** *Alignment of the predicted amino acid sequence of the cytoplasmic dynein heavy chain from rat [Mikami et al., 1993] with the sequence of the sea urchin axonemal dynein β-heavy chain [Gibbons et al., 1991; Ogawa, 1991] at low and high stringencies.* **B:** *Schematic diagram of the cytoplasmic dynein heavy chain sequence indicating the positions of the P-loop elements that are conserved between the cytoplasmic and axonemal dynein sequences, as well as the predicted regions of coiled-coil secondary structure within the cytoplasmic dynein sequence. The N-terminal 30% of the cytoplasmic and axonemal dynein heavy chains are not conserved. [Reproduced from Mikami et al., 1993, with permission of the publisher.]*

B. Comparison of the Cytoplasmic Dynein Heavy Chain Sequences to Other Motor Enzymes

As described above, the region of highest similarity between the cytoplasmic and flagellar dynein heavy chain sequences is localized around the putative principal site for ATP hydrolysis. However, this region was not found to be conserved in the myosin and kinesin superfamilies, including the minus end–directed kinesin-like motor NCD [Endow et al., 1990; McDonald et al., 1990]. Therefore, although there is good evidence for a common ancestry of axonemal and cytoplasmic dyneins, there is little to suggest that kinesin and dynein share a common evolutionary predecessor.

C. How Many Cytoplasmic Dynein Genes?

Structural analysis of cytoplasmic dynein has revealed a two-headed enzyme, but only a single cytoplasmic dynein heavy chain gene has been detected by Southern blotting in both *Dictyostelium* and rat. Furthermore, RFLP analysis of mouse genomic DNA as well as the localization of the mouse gene to chromosome 12 [Mikami et al., 1993] are consistent with a single gene.

For mammalian dyneins, the cytoplasmic heavy chains can be differentiated into two bands under some conditions of electrophoresis [Vallee et al., 1988]. Multiple forms of the heavy chain may be generated by alternative splicing or differential posttranslational processing. However, no evidence for multiple transcripts has yet been obtained in either rat or *Dictyostelium*. Some posttranslational processing of the cytoplasmic dynein heavy chain has already been observed; as isolated, the heavy chains of the enzyme are phosphorylated (Holzbaur and Vallee, unpublished data). It remains to be determined if posttranslational processing accounts for the multiple isoforms observed by electrophoresis or if these forms vary significantly in either their catalytic or functional properties.

The apparent size and diversity of the kinesin superfamily, whose members share a putative motor domain homologous to kinesin, has suggested that each of these homologues may be specialized for a particular cellular role, with significant degeneracy in overlapping functions also observed. The observation of a single cytoplasmic dynein heavy chain gene suggests the targeting of the motor enzyme to a particular cellular binding site or function may be specified by either its accessory subunits or other cellular factors, such as the *Glued* complex (described below).

IV. MOLECULAR ANALYSIS OF THE 74 kD CYTOPLASMIC DYNEIN SUBUNIT

The complete sequence for a cDNA encoding the 74 kD intermediate chain of cytoplasmic dynein has been determined [Paschal et al., 1992]. A full-length clone was isolated from a rat brain cDNA library by screening with an oligonucleotide ''guessmer'' based on protein sequence data. The open reading frame encodes a polypeptide of 643 amino acids with a molecular weight of 72,753 D and a calculated pI of 4.9. Hybridization of the cDNA probe to Northern ern blots revealed a single band of 2.9 kb (Fig. 3).

Electrophoretic analysis of mammalian cytoplasmic dynein consistently reveals at least two, and sometimes three, bands in the 74 kD region [Paschal et al., 1992]. All of these species reacted with a monoclonal anti-74 kD antibody region [Paschal et al., 1992; Dillman and Pfister, 1992]. Further evidence for heterogeneity came from N-terminal microsequencing of proteolytic fragments of the polypeptide. Although eight of nine peptide sequences were found to match completely with the protein sequence predicted from the characterized cDNA clone, the ninth sequence was found to be partially divergent. This polypeptide matched the cDNA sequence over only the first 12 residues, but then diverged completely for the last seven residues.

To test the possibility that multiple forms of the 74 kD polypeptide are generated by alternative splicing from a single gene, Paschal et al. [1992] used PCR of first-strand cDNA to generate products spanning different regions of the molecule. A short isoform was identified in this manner, which contained a 21 amino acid deletion, resulting in a polypeptide with a predicted molecular weight of 70,383 (Fig. 3). The deletion mapped close to the site of divergence between the peptide and the deduced amino acid sequences, although the two sequences differed by seven amino acids. Taken together, these data may account for at least three 74 kD isoforms, generated by apparent alternative splicing. However, the full-length sequence for only a single one of these isoforms is currently known.

Southern blots probed with the cDNA encoding the 74 kD polypeptide revealed a complex pattern (Paschal and Vallee, unpublished data). Furthermore, a monoclonal antibody against the 74 kD cytoplasmic dynein polypeptide also recognized a similar-sized, but distinct, polypeptide in mammalian cilia and flagella [Paschal et al., 1992]. These observations suggest the existence of a family of dynein accessory polypeptides, perhaps representing cytoplasmic and axonemal forms (see below).

A. Structural Analysis of the 74 kD Cytoplasmic Dynein Subunit

A striking aspect of the predicted polypeptide sequence of the 74 kD cytoplasmic dynein subunit is a cluster of charged amino acids within the first 220 residues of the sequence. This region was also determined to have a high probability of α-helical secondary struc-

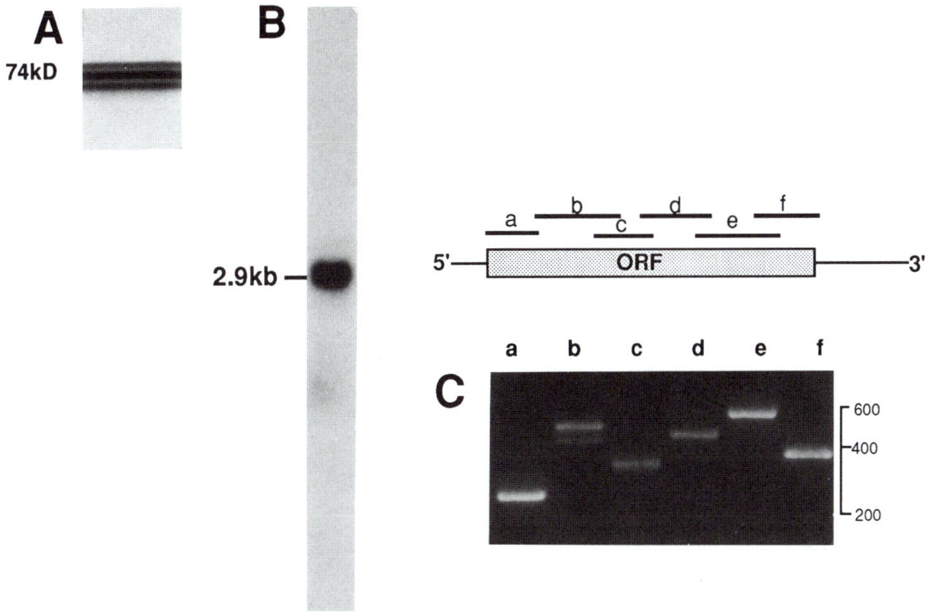

D

```
              102
PCR           SVSTPSEAGSQDDLGPLTR—————————————————RLNKLGVSPEPT...
p74        ...SVSTPSEAGSQDDLGPLTRTLQWDTDPSVLQLQSDSELGRRLNKLGVSPEPT...
peptide 4     SVSTPSEAGSQDSGDGAVG
```

Fig. 3. *Relationship among the multiple isoforms of the 74 kD cytoplasmic dynein intermediate chain.* **A:** *Multiple isoforms of the 74 kD cytoplasmic dynein subunit are resolved by SDS-PAGE analysis. The polypeptides were resolved on a 5–10% gradient acrylamide gel and visualized by silver staining.* **B:** *Northern blot of adult rat brain poly(A)*$^+$ *RNA probed with an* EcoRI *fragment of the cDNA clone encoding the 74 kD intermediate chain, showing a single broad band of 2.9 kb.* **C:** *PCR analysis of first strand brain cDNA to identify regions of sequence diversity. Primers were designed to yield products diagrammed above. Primers spanning region b yielded two products.* **D:** *Comparison of the divergent sequences identified by PCR, cDNA cloning, and N-terminal microsequencing. Three apparent isoforms of the 74 kD polypeptide were identified. All methods are described by Paschal et al. [1992]. [Reproduced from Paschal et al., 1992, with permission of the publisher.]*

ture. The paired motifs $RK(X)_{10}$ KKRK and $KK(X)_{10}$RKRR were also detected within this region, beginning at amino acids 13 and 42, respectively. These motifs are similar to the nuclear targeting signal of *Xenopus* nucleoplasmin [Robbins et al., 1991]. A functional role for this motif within the 74 kD sequence has not been identified.

B. Homology of the 74 kD Cytoplasmic Dynein Sequence to the Intermediate Chains of Axonemal Dyneins

A general search of the databases with the predicted amino acid sequence of the cytoplasmic dynein 74 kD polypeptide did not reveal any significant homology with other known sequences. However, direct comparisons of the predicted amino acid sequence of the 74 kD polypeptide with those reported for the 70 kD intermediate chain of flagellar dynein from *Chlamydomonas* [Mitchell and Kang, 1991] did reveal a distant relationship between the cytoplasmic dynein polypeptide and each of the two axonemal species. A comparison of the aligned sequences revealed that the C-terminal halves of 74 and 70 kD polypeptides are 26% identical (Fig. 4) [Paschal et al., 1992]. The N-terminal halves of the two polypeptides were found to

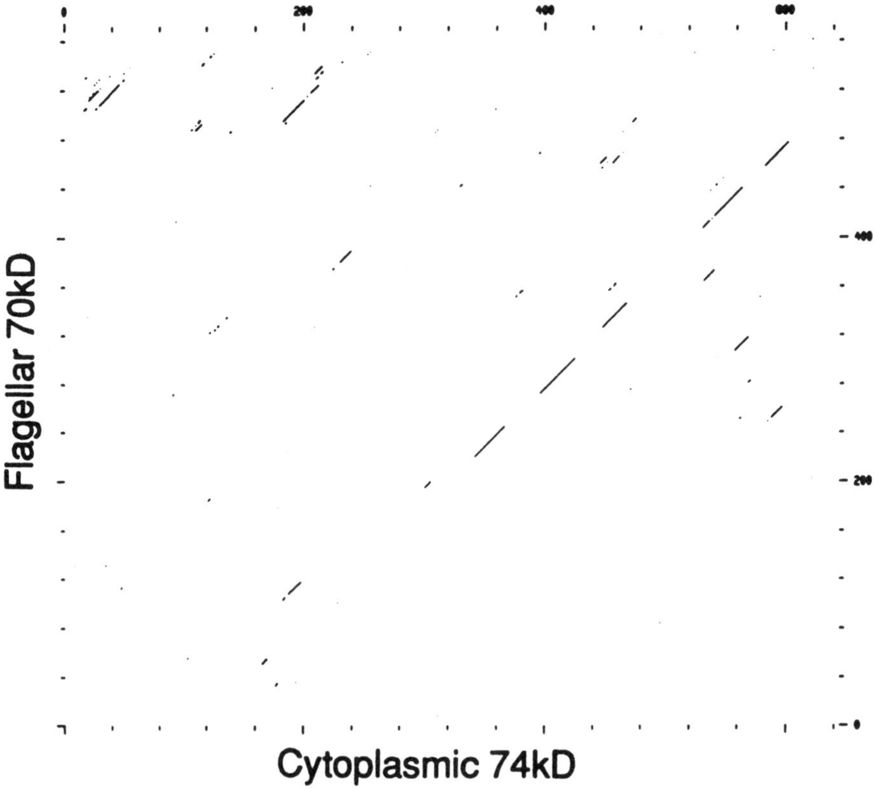

Fig. 4. *Alignment of the deduced polypeptide sequences of the 74 kD subunit of cytoplasmic dynein and the 70 kD subunit of* Chlamydomonas *flagellar dynein. The C-terminal halves of the two sequences are most closely related; the sequence identity in this region was determined to be 26.4%. Comparisons are from Paschal et al. [1992]. [Reproduced from Paschal et al., 1992, with permission of the publisher.]*

be unrelated. A similar relationship was also found between the 74 kD polypeptide and the 78 kD intermediate chain from *Chlamydomonas* [King et al., 1992]. Again, the C-terminal halves of the polypeptides were found to be related, but the N-terminal halves were divergent.

The relatedness observed between the carboxyl halves of the cytoplasmic and axonemal dynein polypeptides suggests that this region may be involved in a common function, such as binding to the dynein heavy chain [Paschal et al., 1992]. The divergence observed within the N-terminal sequences of the cytoplasmic and axonemal intermediate chains suggests that these regions are specific for either the cytoplasmic or axonemal environments of the motor enzymes. These sequences may encode the targeting signals that direct cytoplasmic dynein to bind to either organelle or kinetochore or that mediate the structural binding of axonemal dynein to the A-tubule of the axoneme. This hypothesis is consistent with the results of studies on the axonemal enzyme, which have mapped the 78 and 70 kD intermediate chains of *Chlamydomonas* to the base of the molecule, where they have been reported to interact directly with tubulin to form a structural attachment [King and Witman, 1990; King et al., 1991, 1992].

According to this hypothesis, the differences among the apparent isoforms of the 74 kD polypeptide identified by Paschal et al. [1992], which map near the N terminus of the polypeptide, may play a significant role in the spe-

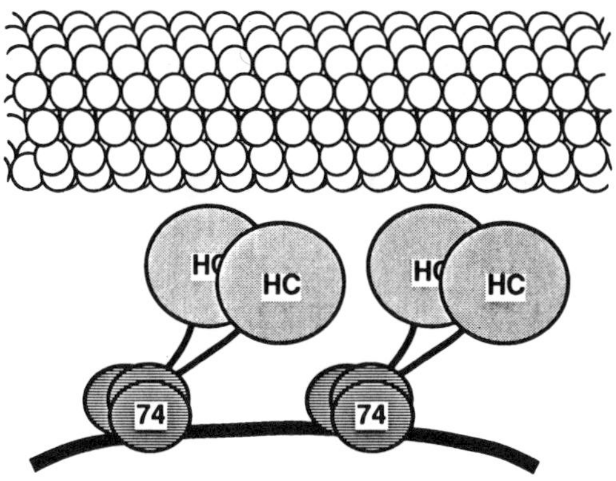

Fig. 5. *Schematic diagram of the proposed functional role of the 74 kD intermediate chains. The relationship observed between the 70 and 78 kD subunits of* Chlamydomonas *flagellar dynein, which are implicated in the binding of axonemal dynein to the A subfiber microtubule of the outer doublet, and the 74 kD cytoplasmic dynein subunit suggests an analogous role for this polypeptide. The 74 kD polypeptide may mediate the binding of cytoplasmic dynein to the surface of membranous organelles and kinetochores.*

cific targeting of cytoplasmic dynein within the cell. For example, the binding of cytoplasmic dynein to either the surface of a vesicle or to the kinetochore may be mediated by different isoforms of the 74 kD polypeptide, thus allowing a single heavy chain to be directed to distinct cellular sites (Fig. 5).

V. OTHER SUBUNITS OF CYTOPLASMIC DYNEIN

Efforts are underway to obtain primary sequence information on the 53–59 kD subunits of cytoplasmic dynein. The four electrophoretic species are reduced to two by phosphatase treatment, which are closely related in primary sequence as judged by N-terminal microsequencing of tryptic and Endo-glu fragments (Hughes et al., 1993). cDNA clones have been selected from chicken [Gill et al., 1992] and rat [Hughes et al., 1993] libraries, and the complete primary sequences should be available soon. So far, no clear relationship with other polypeptides has been detected.

VI. CHARACTERIZATION OF THE *GLUED* (DYNACTIN) COMPLEX

A number of polypeptides have been observed to copurify partially with cytoplasmic dynein [Collins and Vallee, 1989; Steuer et al., 1990; Holzbaur et al., 1991]. Further work has shown that polypeptides of 150, 62, 50, and 45 kD form a distinct complex within the cytosol that is microtubule associated and that sediments at 20 S, yet is distinct from cytoplasmic dynein (Fig. 1) [Gill et al., 1991; Paschal et al., 1993].

The 150 kD polypeptide was originally observed as a substoichiometric component of cytoplasmic dynein preparations, cosedimenting with the enzyme at 20 S [Collins and Vallee, 1989; Holzbaur et al., 1991]. Further fractionation of cytoplasmic dynein by ion exchange chromatography led to the separation of a group of polypeptides from the main cytoplasmic dynein peak [Gill et al., 1991]. This fraction, which was enriched in polypeptides of 150, 50, and 45 kD, was reported to be a required activator for cytoplasmic dynein-mediated vesicular motility in an *in vitro* assay [Gill et al., 1991]. The peak fraction of cytoplasmic dynein alone was found to mediate microtubule gliding but not to promote the motility of salt-washed vesicles along microtubules. Readdition of the fracion enriched in the 150 kD polypeptide restored vesicle motility.

The pre-existence of the complex in cytosol was demonstrated by immunoprecipitation with a monoclonal antibody raised against the 50 kD polypeptide [Paschal et al., 1993]. Immunoprecipitation from rat brain cytosol resulted in the coprecipitation of the 150, 62, and 45 kD components in a ratio very similar to that seen in cytoplasmic dynein preparations. Surprisingly, cytoplasmic dynein heavy and intermediate chains were undetectable in the 50 kD immunoprecipitates, and the components of the new complex were undetectable in parallel immunoprecipitates with an antibody directed against the 74 kD cytoplasmic dynein subunit.

As a further test for the existence of a preformed complex, total rat brain cytosol was fractionated by sucrose density gradient centrifugation. Analysis of the resulting fractions by Western blotting indicated that the 150, 50, and 45 kD components of the complex were present in the cytosol in a single complex that sedimented at 20 S.

The cDNAs encoding the 150 and 45 kD polypeptides of the complex have been cloned and characterized in detail, as described below, and in each case the identification of homology with a previously characterized sequence has been informative. We refer to the complex by the name *Glued* in reference to the homology detected between the 150 kD polypeptide and the product of the *Glued* gene in *Drosophilia* [Holzbaur et al., 1991; Paschal et al., 1993]; the complex has also been referred to as dynactin in reference to its role in activating dynein-based vesicular transport [Gill et al., 1991].

VII. MOLECULAR CHARACTERIZATION OF p150^{Glued}

A polyclonal antibody was used to isolate cDNA clones encoding the 150 kD polypeptide from a rat brain cDNA expression library [Holzbaur et al., 1991]. The full-length 4.4 kb cDNA was isolated by hybridization screening, and verified by N-terminal microsequencing of peptide fragments of the purified polypeptide. Northern blot analysis indicated the existence of a single-sized mRNA within adult rat brain, although the polypeptide characteristically appears as a doublet of 150 and 135 kD on SDS-polyacrylamide gels [Collins et al., 1989; Steuer et al., 1990; Holzbaur et al., 1991]. Southern blot analysis is consistent with the existence of a single gene in both rat and mouse [Holzbaur et al., 1991].

The sequence of a cDNA clone encoding the equivalent polypeptide in chick, referred to as *dynactin,* has subsequently been reported [Gill et al., 1991]. The predicted amino acid sequence is about 80% identical to that of the rat form, though the chick cDNA clone encodes a polypeptide of only 117 kD. Northern blot analysis using the cDNA to probe various tissues from chick indicated some evidence of heterogeneity, though whether the 117 kD dynactin polypeptide represents a short isoform or an incomplete sequence remains to be determined.

A. Homology of the 150 kD Cytoplasmic Dynein-Associated Polypeptide to the *Drosophila* Gene *Glued*

Comparison of the predicted amino acid sequence for the 150 kD polypeptide to the protein sequence databases revealed homology to a sequence from *Drosophila* (Fig. 6) [Holzbaur et al., 1991]. Homology was observed throughout the length of the rat sequence to the 148 kD product of the *Drosophila Glued* locus [Swaroop et al., 1987]. The predicted amino acid sequences of the two polypeptides were found to be 31% identical and 53% similar. The two polypeptides are most closely related in their N-terminal halves, which are 36% identical. Based on the observed extent of homology between the predicted amino acid sequence of the rat polypeptide and that of *Drosophila,* we now refer to the rat polypeptide as p150^{Glued}.

The original mutation at the *Glued* locus, Gl^1, is dominant, exhibiting pleiotropic developmental defects in heterozygotes, which are particularly apparent in the aberrant eye and optic lobe formation of the mutant flies [Plough and Ives, 1935; Meyerowitz and Kankel, 1978; Garen and Kankel, 1983]. Significantly, however, both the homozygous Gl^1 mutation and the null mutation are lethal. The flies die early in development. Genetic mosaic analysis has shown a failure to recover cells homozygous for Gl^1. This "cell lethal" phenotype was interpreted as suggesting a role for the *Glued* gene in the viability and/or reproduction of the individual cell [Harte and Kankel, 1982].

The 150 kD polypeptide may represent the mammalian homologue of this essential gene. This identification is consistent with the known characteristics of both polypeptides. If the 150 kD polypeptide is a required cofactor for organelle and vesicle transport driven by cyto-

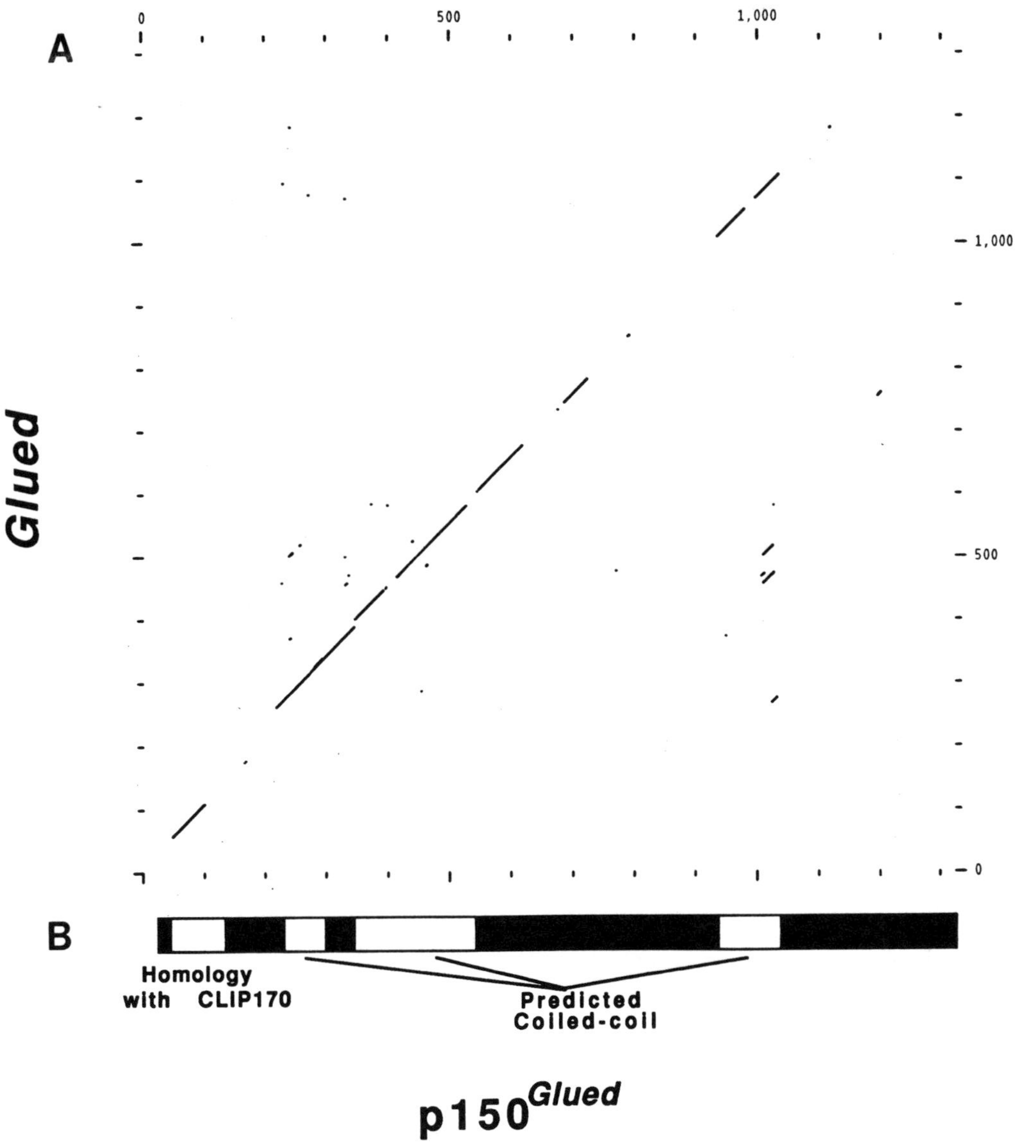

Fig. 6. *Primary and secondary structural analysis of p150*[Glued]. **A:** *Dot plot of the alignment of the predicted protein sequences of p150*[Glued] *and the* Glued *gene product from* Drosophila. *The two sequences are 36% identical over the first 750 amino acids; a comparison of the entire two sequences shows 31% identity and 53% similarity [Holzbaur et al., 1991].* **B:** *Schematic diagram of the deduced polypeptide sequence for p150*[Glued]. *A putative microtubule-binding domain, identified by homology with a similar element in the microtubule-endosome linker CLIP170 [Pierre et al., 1992], and the predicted regions of coiled-coil secondary structure are indicated.*

plasmic dynein *in vivo,* as has been suggested [Gill et al., 1991], then it is likely that loss of this species would result in a lethal disruption of cellular function, as was observed for *Glued.* Alternatively, *Glued* and the 150 kD polypeptide from rat may represent related species, perhaps members of a multigene family. However, no evidence for additional related polypeptides has yet been found. Further analysis of the biochemistry of the *Glued* polypeptide should clarify the relationship and may then allow the application of *Drosophila* genetics to the investigation of the biological role of this complex.

B. Structural Analysis of p150^{Glued}

Within the N-terminal of the *Glued* and p150^{Glued} polypeptides is a potential microtubule binding motif. Pierre et al. [1992] recently reported the sequence of CLIP170, which is thought to link endocytic vesicles to microtubules *in vivo*. The cDNA sequence for CLIP170 was found to encode two repeats of a motif that is present in a single copy in p150^{Glued} (Fig. 6), in the *Drosophila Glued* sequence [Swaroop et al., 1987], and in BIK1, an apparent microtubule-associated protein from *Saccharomyces cerevisiae* [Trueheart et al., 1987]. CLIP170 has been shown to bind to microtubules in an *in vitro* assay. Specific site mutations within either repeat of this conserved motif were determined to reduce significantly the extent of microtubule binding observed [Pierre et al., 1992]. However, the sequence is not related to previously characterized microtubule-binding motifs.

By analogy to CLIP170, the conserved motif may mediate the binding of the *Glued* complex to microtubules. Apart from this sequence p150^{Glued} and CLIP170 are not related in primary sequence, but do contain similar extensive regions of coiled-coil secondary structure (Fig. 6). Secondary structure predictions of the p150^{Glued} sequence indicate extended α-helical potential, which also fits the heptad repeat motif. Two stretches of amino acids, an N-terminal stretch of ~380 residues and a C-terminal stretch of ~130 residues, have the potential to form a coiled-coil interaction either intra- or intermolecularly [Holzbaur et al., 1991; Gill et al., 1991]. These regions of the polypeptide may mediate an intramolecular association or may bind the polypeptide to the other components of the complex.

The higher-order structures of the two polypeptides seem to be distinct: CLIP170 sediments at 5.7S [Pierre et al., 1992], whereas p150^{Glued} is a component of a 20S complex [Paschal et al., 1993; Gill et al., 1991]. Despite these differences, the polypeptides may have a similar function *in vivo*. Both CLIP170 and p150^{Glued} have been implicated in the association of vesicles with microtubules. CLIP170 appears to mediate this interaction directly [Pierre et al., 1992]. A potential interaction of p150^{Glued} with vesicles is suggested by the results of Gill et al. [1991]. Evidence for a direct interaction with microtubules has come from expression studies which revealed co-localization of the N-terminal region of p150^{Glued} with microtubules in transfected cells, and co-sedimentation with microtubules of *in vitro* translated p150^{Glued} [Waterman-Storer et al., 1993].

Expression of p150^{Glued} has been observed in a range of tissues and cell types. Although expression may be required in all cells, as is suggested by the analysis of genetic mosaics in *Drosophila* [Harte and Kankel, 1982], the 150 kD polypeptide has been found to be highly enriched in neuronal cells at all stages of development (Melloni and Holzbaur, unpublished data). The distribution observed is consistent with a role for the species in the retrograde transport of vesicles along the axon.

VIII. CLONING AND CHARACTERIZATION OF CENTRACTIN

As described above, the major polypeptide of the *Glued* complex is a species of 45 kD. N-terminal microsequencing of proteolytic fragments of the 45 kD polypeptide led to the unexpected observation that it is a novel form of actin [Lees-Miller et al., 1992a; Paschal et al., 1993]. The observed sequences were found to be ~50% identical to conventional actin species, which is far below the expected level of sequence conservation observed between actin family members. However, the sequence was no more closely related to the previously characterized nonconventional actins from yeast [Schwob and Martin, 1992; Lees-Miller et al., 1992b].

The complete cDNA sequence for the 45 kD polypeptide was determined for both canine [''centractin,'' Clark and Meyer, 1992] and human [''actin-RPV,'' Lees-Miller et al., 1992a] forms. The predicted sequences are identical. Evidence has also been obtained for a second closely related isoform of centractin, termed β-centractin. The α- and β-isoforms are 90% identical [Clark and Meyer, 1993]. Sufficient polypeptide sequence data have been obtained to indicate that the calf brain *Glued* complex contains α-centractin exclusively [Paschal et al., 1993]. Thus, in contrast to conventional actins, it appears that the sole supramolecular state of α-centractin is as a component of a complex. In view of the sequence similarity between the two centractin isoforms, we speculate that β-centractin may be a component of a similar complex. Whether α- and β-isoforms have distinct functions or reflect differences in tissue-specific or developmental expression remains to be determined.

It is uncertain whether centractin may form polymers under physiological conditions. This hypothesis is supported by the observed stoichiometry of the immunoprecipitated complex. The high ratio of centractin to the other polypeptides of the complex suggests multiple monomers of centractin that may self-associate to form short filaments. Electron microscopy of the complex has revealed a short filament with the diameter of an actin filament, with associated amorphous material [Schroer et al., 1992]. These short filaments may then be capped at each end by one of the other components of the complex to form a structure that sediments as a single peak at 20S.

IX. DISTRIBUTION OF THE *GLUED* COMPLEX

The cellular role of the *Glued* complex is not yet clear, nor is the nature of its relationship with cytoplasmic dynein. The observation that a mutation in the *Drosophila* gene *Glued* causes a lethal defect early in development and the complete sequence conservation between canine contractin and human actin-RPV strongly suggest a critical role for the *Glued* complex within the cell. A working hypothesis is that the *Glued* complex is required for the intracellular targeting of cytoplasmic dynein (Fig. 7). The complex may function as a vesicle-bound receptor or, alternatively, it may form a transient link between the microtubule and the vesicle or organelle, which then allows cytoplasmic dynein to bind to both of these species. This model (Model III in Fig. 7), although slightly more complex, is supported by preliminary data that implicate p150^{Glued} in direct microtubule binding [Waterman-Storer et al., 1993].

Immunocytochemistry to probe the intracellular distribution of the complex has provided some interesting but potentially misleading indications of its cellular function. Gill et al. [1991], using a monoclonal antibody to dynactin, described a punctate cytoplasmic pattern. Detergent extraction of the cells prior to fixation revealed only centrosomal staining. Meyer and Clark [1992] reported a strong centrosomal distribution for centractin, using a polyclonal antibody. A punctate cytoplasmic distribution was also observed. Using polyclonal antibodies to p150^{Glued}, we have also observed strong centrosomal staining along with a punctate distribution throughout the cytoplasm [Paschal et al., 1993]. The association of the complex with microtubules was probed by exposing cultured cells to nocodazole for 3 hours to depolymerize the microtubules. Under these conditions, centrosomal staining by anti-p150^{Glued} was gradually lost, and bright punctate cytoplasmic staining was observed. The loss of centrosomal staining in response to nocodazole suggests that the *Glued* complex is only peripherally associated with the centrosome, presumably via the minus ends of microtubules. This behavior is consistent with an interaction between the complex and cytroplasmic dynein, though a direct physical interaction between the two supramolecular structures has not been observed *in vitro* [Paschal et al., 1993]. Further studies will be necessary to determine the nature of the possible interaction of the *Glued* complex with microtubules and/or cytoplasmic dynein (Fig. 7).

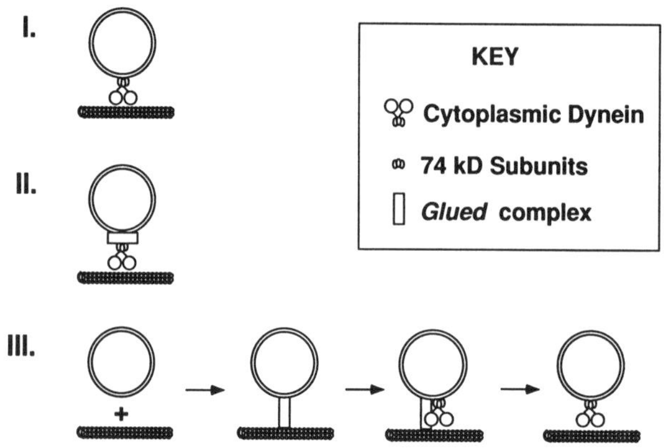

Fig. 7. *Models for the role of the 74 kD cytoplasmic dynein subunits and the* Glued *(dynactin) complex in organelle binding. Model I depicts a direct interaction between cytoplasmic dynein and the surface of an organelle mediated by the 74 kD intermediate chains. Model II depicts an interaction between cytoplasmic dynein and an organelle involving the* Glued *complex in the role of an adaptor. Model III depicts a sequential reaction in which the* Glued *complex first mediates a static attachment of an organelle to the microtubule perhaps via the CLIP-170 related microtubule-binding site within p150*^{Glued}. *The complex is then envisioned to facilitate binding of cytoplasmic dynein to the organelle. Finally, motility is achieved upon dissociation of the* Glued *complex from the organelle, releasing the organelle from its static attachment to the microtubule.*

X. SUMMARY

The molecular analysis of cytoplasmic dynein is well under way, and sequence information on all of its subunits and most of the subunits of the *Glued* complex should soon be available. The existing primary sequence data have already provided important new insights into the relationship between cytoplasmic and axonemal dyneins, and have dramatically advanced our understanding of how these large and complex structures are organized. The molecular probes that are becoming available should now allow for genetic tests of cytoplasmic dynein function. In addition, expression of recombinant polypeptides should be of great value in defining the force-producing domain of the molecule, determining how its activity is regulated, and learning how it is targeted to subcellular organelles.

NOTE ADDED IN PROOF

While this review was in preparation, cloning of cytoplasmic dynein heavy chain from a number of species reached or was nearing completion. These include *S. cerevisiae* (laboratories of K. Bloom and I. Gibbons), *C. Elegans* (J. Lye and co-workers), *N. crassa* (laboratory of M. Plamann), *D. melanogaster* (laboratory of T. Hayes), and rat (laboratory of N. Hirokawa).

XI. REFERENCES

Belles-Isles M, Chapeau C, White D, Gagnon C (1986): Isolation and characterization of dynein ATPase from bull spermatozoa. Biochem J 240:863–869.

Clark SW, Meyer DI (1992): Centractin is an actin homologue associated with the centrosome. Nature 359: 246–250.

Clark SW, Meyer DI (1993): Long lost cousins of actin. Curr Biol 3:54–55.

Collins CA, Vallee RB (1989): Preparation of microtubules from rat liver and testis: cytoplasmic dynein is a major microtubule associated protein. Cell Motil Cytoskel 14:491–500.

Corthesy-Theulaz I, Pauloin A, Pfeffer SR (1992): Cytoplasmic dynein participates in the centrosomal localization of the Golgi complex. J Cell Biol 118:1333–1345.

Dillman JF, Pfister KK (1992): The 74 kD polypeptide of cytoplasmic dynein is phosphorylated *in vitro* and *in vivo*. Mol Biol Cell 3:162a.

Endow SA, Henikoff S, Soler-Niedziela L (1990): Mediation of meiotic and early mitotic chromosome segregation in *Drosophila* by a protein related to kinesin. Nature 345:81–83.

Euteneuer U, Koonce MP, Pfister KK, Schliwa M (1988): An ATPase with properties expected for the organelle motor of the giant amoeba, *Reticulomyxa*. Nature 332:175–178.

Fry DC, Kuby SA, Mildvan AS (1986): ATP binding site of adenylate kinase: Mechanistic implications of its homology with *ras*-encoded p21, F_1 ATPase, and other nucleotide-binding proteins. Proc Natl Acad Sci USA 83:907–911.

Garen SH, Kankel DR (1983): Golgi and genetic mosaic analyses of visual system mutants in *Drosophila melanogaster*. Dev Biol 96:445–466.

Gibbons IR, Gibbons BH, Mocz G, Asai DJ (1991): Multiple nucleotide-binding sites in the sequence of dynein β heavy chain. Nature 352:640–643.

Gibbons IR, Lee-Eiford A, Mocz G, Phillipson CA, Tang W-JY, Gibbons BH (1987): Photosensitized cleavage of dynein heavy chains. J Biol Chem 262:2780–2786.

Gill SR, Salcedo AF, Cleveland DW, Schroer TA (1992): Cloning of a cytoplasmic dynein 50 kD subunit (abstract). Mol Biol Cell 3:161.

Gill SR, Schroer TA, Szilak I, Steuer ER, Sheetz MP, Cleveland DW (1991): Dynactin, a conserved, ubiquitously expressed component of an activator of vesicle motility mediated by cytoplasmic dynein. J Cell Biol 115:1639–1650.

Harte PJ, Kankel DR (1982): Genetic analyses of mutations at the *Glued* locus and interacting loci in *Drosophila melanogaster*. Genetics 101:477–501.

Hastie AT, Dicker DT, Hingley ST, Kueppers F, Higgins ML, Weinbaum G (1986): Isolation of cilia from porcine tracheal epithelium and extraction of dynein arms. Cell Motil Cytoskel 6:25–34.

Holzbaur ELF, Hammarback JA, Paschal BM, Kravit NG, Pfister KK, Vallee RB (1991): Homology of a 150K cytoplasmic dynein-associated polypeptide with the *Drosophila* gene *Glued*. Nature 351:579–583.

Holzbaur ELF, Johnson KA (1989): ADP release is rate limiting in steady-state turnover by the dynein adenosinetriphosphatase. Biochemistry 28:5577–5585.

Holzbaur ELF, Johnson KA (1989): Microtubules accelerate ADP release by dynein. Biochemistry 28:7010–7016.

Hughes SM, Herskovits JA, Vaughan KA, Vallee RB (1993): Molecular characterization of 53/55 and 57/59 kD cytoplasmic dynein subunits. (abstract) Mol Biol Cell (in press).

Hyman AA, Mitchison TJ (1991): Two different microtubule-based motor activities with opposite polarities in kinetochores. Nature 351:206–211.

Johnson KA (1983): The pathway of ATP hydrolysis by dynein. J Biol Chem 258:13825–13832.

Johnson KA, Wall JS (1983): Structure and molecular weight of the dynein ATPase. J Cell Biol 96:669–678.

King SM, Wilkerson CG, Witman GB (1991): The Mr 78,000 intermediate chain of *Chlamydomonas* outer arm dynein interacts with α-tubulin *in situ*. J Biol Chem 266:8401–8407.

King SM, Wilkerson CG, Witman GB (1992): Molecular cloning and domain analysis of the Mr 78,000 intermediate chain of *Chlamydomonas* outer arm dynein. Mol Biol Cell 3:2a.

King SM, Witman GB (1990): Localization of an intermediate chain of outer arm dynein by immunoelectron microscopy. J Biol Chem 265:19807–19811.

Koonce MP, Grissom PM, McIntosh JR (1992): Dynein from *Dictyostelium:* Primary structure comparisons between a cytoplasmic motor enzyme and flagellar dynein. J Cell Biol 119:1597–1604.

Lee-Eiford A, Ow RA, Gibbons IR (1986): Specific cleavage of dynein heavy chains by ultraviolet irradiation in the presence of ATP and vanadate. J Biol Chem 261:2337–2342.

Lees-Miller JP, Helfman DM, Schroer TA (1992a): A vertebrate actin-related protein is a component of a multisubunit complex involved in microtubule-based vesicle motility. Nature 359:244–246.

Lees-Miller JP, Henry G, Helfman DM (1992b): Identification of *act2*, an essential gene in the fission yeast *Schizosaccharomyces pombe* that encodes a protein related to actin. Proc Natl Acad Sci USA 89:80–83.

Lin SXH, Collins CA (1992): Immunolocalization of cytoplasmic dynein to lysosomes in cultured cells. J Cell Sci 101:125–137.

Lye RJ, Porter ME, Schloey JM, McIntosh JR (1987): Identification of a microtubule-based cytoplasmic motor in the nematode *C. elegans*. Cell 51:309–318.

McDonald HB, Goldstein LSB (1990): Identification and characterization of a gene encoding a kinesin-like protein in *Drosophila*. Cell 61:991–1000.

Meyerowitz EM, Kankel DR (1978): A genetic analysis of visual system development in *Drosophila melanogaster*. Dev Biol 62:112–142.

Mikami A, Paschal BM, Mazumdar M, Vallee RB (1993): Molecular cloning of the retrograde transport motor cytoplasmic dynein (MAP1C). Neuron 10:787–796.

Mitchell DR, Kang K (1991): Identification of oda6 as a *Chlamydomonas* dynein mutant by rescue with the wild-type gene. J Cell Biol 113:835–842.

Neely MD, Erickson HP, Boekelheide K (1990): HMW-2, the Sertoli cell cytoplasmic dynein from rat testis, is a dimer composed of nearly identical subunits. J Biol Chem 265:8691–8698.

Ogawa K (1991): Four ATP binding sites in the midregion of the β heavy chain of dynein. Nature 352:643–645.

Paschal BM, Holzbaur ELF, Pfister KK, Clark S, Meyer D, Vallee RB (1993): Characterization of a 50 kD polypeptide in cytoplasmic dynein preparations reveals a complex with $p150^{Glued}$ and a novel actin. J Biol Chem 268:15,318–15,323.

Paschal BM, Mikami A, Pfister KK, Vallee RB (1992): Homology of the 74-kD cytoplasmic synein subunit with a flagella dynein polypeptide suggests an intracellular targeting function. J Cell Biol 118:1133–1143.
Paschal BM, Shpetner HS, Vallee RB (1987): MAP 1C is a microtubule-activated ATPase which translocates microtubules *in vitro* and has dynein-like properties. J Cell Biol 105:1273–1282.
Paschal BM, Vallee RB (1987): Retrograde transport by the microtubule-associated protein MAP 1C. Nature 330:181–183.
Pfarr CM, Coue M, Grisson PM, Hays TS, Porter ME, McIntosh JR (1990): Cytoplasmic dynein is localized to kinetochores during mitosis. Nature 345:263–265.
Pierre P, Scheel J, Rickard JE, Kreis TE (1992): CLIP-170 links endocytic vesicles to microtubules. Cell 70:887–900.
Plough HH, Ives PT (1935): Induction of mutations by high temperature in *Drosophila*. Genetics 20:42–69.
Rieder CL, Alexander SP (1990): Kinetochores are transported poleward along a single astral microtubule during chromosome attachment to the spindle in next lung cells. J Cell Biol 110:81–95.
Robbins J, Dilworth SM, Laskey RA, Dingwall C (1991): Two interdependent basic domains in nucleoplasmin nuclear targeting sequence: Identification of a class of bipartite nuclear targeting sequence. Cell 64:615–623.
Schnapp BJ, Reese TS (1989): Dynein is the motor for retrograde axonal transport of organelles. Proc Natl Acad Sci USA 86:1548–1552.
Schroer TA, Heuser JE, Helfman DM, Lees-Miller JP (1992): Actin-RPV: The major component of the dynactin complex, activator of cytoplasmic dynein. Mol Biol Cell 3:2a.
Schroer TA, Sheetz MP (1991): Two activators of microtubule-based vesicle transport. J Cell Biol 115:1309–1318.
Schroer TA, Steuer ER, Sheetz MP (1989): Cytoplasmic dynein is a minus-end directed motor for membranous organelles. Cell 56:937–946.
Schwob E, Martin RP (1992): New yeast actin-like gene required late in the cell cycle. Nature 355: 179–182.
Shimizu T, Johnson KA, Holzbaur ELF (1983): Kinetic evidence for multiple dynein ATPase sites. J Biol Chem 258:13841–13846.
Shpetner HS, Paschal BM, Vallee RB (1988): Characterization of the microtubule-activated ATPase of brain cytoplasmic dynein (MAP1C). J Cell Biol 107:1001–1009.
Steuer E, Wordeman L, Schroer TA, Sheetz MP (1990): Localization of cytoplasmic dynein to mitotic spindles and kinetochores. Nature 345:266–268.
Swaroop A, Swaroop M, Garen A (1987): Sequence analysis of the complete cDNA and encoded polypeptide for the *Glued* gene of *Drosophila melanogaster*. Proc Natl Acad Sci USA 84:6501–6505.
Trueheart J, Boeke JD, Fink G (1987): Two genes required for cell fusion during yeast conjugation: evidence for a phenomone-induced surface protein. Mol Cell Biol 7:2316–2328.
Vallee RB, Wall JS, Paschal BM, Shpetner HS (1988): Microtubule-associated protein 1C from brain is a two-headed cystolic dynein. Nature 332:561–563.
Walker JE, Saraste M, Runswick MJ, Gay NJ (1982): Distantly related sequences in the α- and β-subunits of ATP synthase, myosin, kinases and other ATP-requiring enzymes and a common nucleotide binding fold. EMBO J 1:945–951.
Waterman-Storer C, Karki S, Holzbaur ELF (1993): Analysis of the p150^{Glued}-centractin complex reveals a microtubule-binding function *in vivo* and *in vitro*. (abstract) Mol Biol Cell (in press).

ABOUT THE AUTHORS

ERIKA L. F. HOLZBAUR is Assistant Professor of Biochemistry at the University of Pennsylvania School of Veterinary Medicine in Philadelphia, where she teaches graduate courses in biochemistry and cell biology. After receiving a B.S. degree with high honors from the College of William and Mary in Virginia, she attended the Pennsylvania State University to study the kinetics and thermodynamics of the microtubule-dynein ATPase cycle with Dr. Kenneth A. Johnson, and she received a Ph.D. in biochemistry in 1987. This was followed by a postdoctoral fellowship in the laboratory of Dr. Richard Vallee at the Worcester Foundation for Experimental Biology, where she focused on the molecular biology and regulation of the microtubule-based motor cytoplasmic dynein. Since 1992 Dr. Holzbaur has been an assistant professor at the University of Pennsylvania, where her research interests continue to focus on cytoplasmic dynein and on the microtubule-associated *Glued* complex (see Holzbaur et al., *Nature* 351:579–583). Her awards include election into Phi Beta Kappa, an NIH postdoctoral fellowship, and an award from the Research Foundation of the University of Pennsylvania.

ATSUSHI MIKAMI is a postdoctoral fellow at the Worcester Foundation for Experimental Biology in Worcester, Massachusetts. After receiving an M.D. from Kyoto University, Japan, he worked for two years as a physician at Kyoto University Hospital and at another hospital. He pursued doctoral research at Kyoto University, mostly in the laboratory of late Shosaku Numa, earning a Ph.D. in 1990 for work on the molecular biology of the voltage-dependent calcium channel. After two more years at Kyoto University, he started post-doctoral research in the laboratory of Richard B. Vallee, his current position. Since then he has been working on the molecular biology of the cytoplasmic dynein.

BRYCE M. PASCHAL is currently a postdoctoral research associate in the laboratory of Larry Gerace at the Scripps Research Institute in La Jolla, California. His present research, which is supported by the Damon Runyon-Walter Winchell Cancer Research Fund, is focused on understanding the cell biological and molecular basis of protein transport through the nuclear pore using reconstituted *in vitro* assays. After receiving his B.A. in biology from the University of North Carolina at Chapel Hill, he pursued doctoral studies at the University of Massachusetts Medical School. His thesis research was carried out at the adjacent Worcester Foundation for Experimental Biology (Shrewsbury, Massachusetts) in the laboratory of Richard B. Vallee. Dr. Paschal's work in the Vallee laboratory resulted in the discovery of the retrograde factor, also known as cytoplasmic dynein, the enzyme which powers the movement of membrane-bound organelles along microtubules in the direction opposite to that of kinesin (Paschal and Vallee, *Nature* 330:181–183, 1987; Paschal, Shpetner, and Vallee, *J. Cell Biol.* 105:1273–1282, 1987).

RICHARD B. VALLEE is Principal Scientist at the Worcester Foundation for Experimental Biology in Shrewsbury, Massachusetts. He is also Professor of Cell Biology and of Biochemistry at the University of Massachusetts Medical School, where he teaches graduate cell biology. After receiving a B.A. with honors at Swarthmore College in Pennsylvania, he pursued doctoral research in the laboratory of Robley Williams, Jr., at Yale University, where he earned an M.Phil in 1970 and a Ph.D. in 1974 for work on reversible protein denaturation, using lactate dehydrogenase as a model system. This was followed by postdoctoral work at the University of Wisconsin, Madison, where he studied microtubule assembly and the structure and function of microtubule-associated proteins, or MAPs. Since 1978, he has continued his investigation of these proteins. Among his accomplishments are the discovery of the retrograde transport motor protein MAP 1C (cytoplasmic dynein) and of the microtubule-associated endocytic GTPase dynamin. He is a fellow of the AAAS, a member of the editorial board of the *Journal of Biological Chemistry,* an associate editor of *Cell Motility and the Cytoskeleton*, and an editor of *Methods in Enzymology.*

Microtubules: 269–284

The Kinesin Superfamily

Andrea Pereira and Lawrence S.B. Goldstein

I. INTRODUCTION

Microtubule-based motility is critical for the differentiation, function, and proliferation of eukaryotic cells. Two groups of motors are responsible for various aspects of this motility; each consists of related members with specialized functions. One group is the cytoplasmic dyneins [reviewed by Vallee and Shpetner, 1990]. The second group, and the topic of this review, is the kinesin superfamily, which is comprised of the kinesin heavy chain (KHC) and related proteins (kinesin-like proteins [KLPs]).

The kinesin superfamily is defined by two features. First, all members share substantial (35%–40%) sequence homology to the KHC motor domain, and each KLP motor domain contains short segments of extremely high conservation within the microtubule-binding region, the ATP-binding site, and other areas of unknown function of the KHC motor. Second, at the sequence level each KLP has a unique nonmotor domain. Although many of these molecules share some predicted structural features (e.g., α-helical-coiled coil) with other members of this superfamily, the general lack of sequence conservation led to the hypothesis that the nonmotor-domain specifies cargo attachment or other specialized functions [reviewed by Goldstein, 1991]. This hypothesis may be true in part, but interestingly, there exist two KLPs with strikingly similar function (and presumably cargo) that share no sequence similarity within their nonmotor domains [Saunders and Hoyt, 1992].

The study of the kinesin superfamily is of tremendous importance for several reasons. At a biochemical level, kinesin and KLPs are being used to answer questions about the mechanisms of force transduction and control of direction of movement. At a cellular and organismal level, questions about the biological function of different motors and their integration into cellular architecture are being studied. Thus studies of the kinesin superfamily are giving us important insights into diverse biological problems such as the mechanisms of formation and function of the endoplasmic reticulum, the mitotic and meiotic spindles, and the axoplasmic transport system.

We begin with a review of three specific families within the kinesin superfamily, the kinesin, bimC, and KAR3 families. The kinesin family is discussed first since it is the archetype of this superfamily, about which a great deal is known. We then discuss other members of the superfamily for which no grouping into families is yet possible. Finally, we discuss and compare several different approaches that have been used for identifying new KLPs.

II. SPECIFIC FAMILIES

In this section we will review what is known about three specific families within the kinesin superfamily. When possible we will focus our attention on genetic analyses of mutants that have led to conclusions about cellular functions. Intriguingly, not only are cellular functions similar within a family, but pairwise sequence comparisons between the different motor domains also identify the same three groups (Fig. 1) [Hoyt and Saunders, 1992; McDonald and Goldstein, 1990; Meluh and Rose, 1990; Navone et al., 1992].

A. The Kinesin Family

Although a general idea of kinesin function has emerged from *in vitro* assays, little is known about kinesin's specific biological role. The genetic analyses of the *Drosophila* KHC and the *Caenhorhabditis elegans* KHC as well as antisense inhibition studies in cultured cell systems have given us considerable insight into this problem. From these studies, it can be concluded that the members of this family are involved in the transport of membranous organelles toward the cell periphery as well as to the synapse. To put these studies into context, we will briefly review what is known about the structure and function of kinesin.

1. Kinesin structure. Kinesin is the only member of the kinesin superfamily that has been isolated in its native form and comparatively well characterized biochemically. Kinesin was first discovered during studies on the mechanism of squid axonal transport. It was found that axonal transport could be reconstituted using purified organelles, microtubules, and ATP. Since earlier pharmacological experiments had shown that axoplasmic vesicles stably bind to microtubules in the presence of the nonhydrolyzable ATP analog AMP-PNP [Lasek and Brady, 1985], this property was used to purify a soluble factor that could increase the rate of organelle movement. This soluble factor, or kinesin, also moves latex beads toward the plus end of microtubules *in vitro* [Vale et al., 1985].

Kinesin is highly conserved throughout phylogeny. Since the initial purification from squid, kinesin has been purified from a number of diverse sources and shown in each case to be a heterotetramer consisting of two heavy chains (KHCs) of 110–135 kD and two light chains of 58–70 kD [Bloom et al., 1988; Kuznetsov et al., 1988; Saxton et al., 1988; Scholey et al., 1985]. The deduced protein sequences of the cloned and sequenced KHC from *Drosoph-*

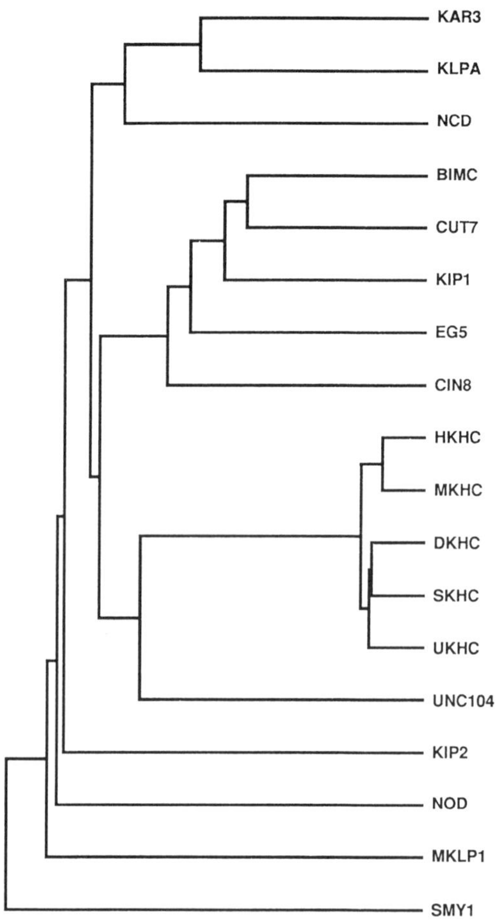

Fig. 1. *Dendrogram showing the similarity among relationships in a mutually conserved region within the putative motor domains of various kinesin superfamily members. Note the clustering within the KHC, BIMC, and KAR3 families. The dendrogram was created using the program Pileup from the UWGCG suite of sequence analysis programs [Devereux et al., 1984]. The analysis was done on sequences within the motor domain from just upstream of the consensus nucleotide-binding sequence (starting with GYN in* Drosophila *KHC) to a sequence near the end of the obvious region of similarity among all the superfamily members (LGG in* Drosophila *KHC); this corresponds to residues 83–300 in* Drosophila *kinesin heavy chain [Yang et al., 1989]. Sequence names and references are as cited in the text with the exception of HKHC, MKHC, DKHC, SKHC, and UKHC, which refer to human, mouse,* Drosophila*, squid, and sea urchin kinesin heavy chain, respectively [Yang et al., 1989; Kosik et al., 1990; Kato, 1991; Wright et al., 1991; Navone et al., 1992].*

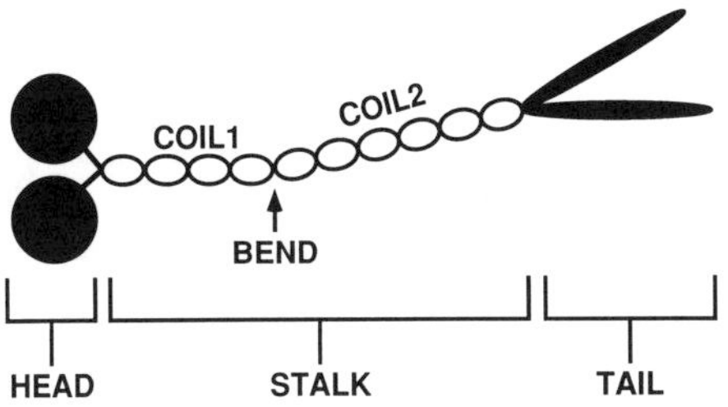

Fig. 2. *Diagram of the kinesin heavy chain domain structure. The solid circles at the left (N-terminal end) represent the head or motor domain. The chain-like structure in the middle represents the α-helical coiled-coil stalk region with a bend in the middle. The solid ellipses at the right end indicate the globular tail domain. This diagram is based on the data described in the text.*

ila, sea urchin, mouse, human, and squid are highly similar throughout their length [Kato, 1991; Kosik et al., 1990; Navone et al., 1992; Wright et al., 1991; Yang et al., 1988]. For example, the amino acid sequences of *Drosophila* and human KHC are 63% identical. The motor or head region (defined below) is the most conserved part of the protein, since it is 76% identical between human and *Drosophila* [Navone et al., 1992].

Much progress has been made on understanding the KHC domain structure (Fig. 2). Based on sequence analysis, the KHC is predicted to have three structural domains: an N-terminal globular domain or head, an extended α-helical coiled-coil stalk domain, and a smaller globular domain at the C-terminus. This domain structure can also be seen by electron microscopy [Amos, 1987; Hirokawa et al., 1989; Scholey et al., 1989]. Two different kinds of experiments have mapped the motor function of kinesin to the head region. In one study electron microscopy demonstrated that an antibody that inhibits motility *in vitro* decorates the N-terminal globular heads [Scholey et al., 1989]. The second study used truncated forms of the KHC to demonstrate that the first 450 amino acids of the globular N-terminal domain can move microtubules *in vitro* [Yang et al., 1990]. This motor domain has recently been narrowed down to the first 339 amino acids (R. Stewart and L.S.B. Goldstein, unpublished observations).

While little is known about the function of the nonmotor domains of the KHC, the structure of this region is better understood. Biochemical and electron microscopic studies verified that the stalk domain of *Drosophila* KHC is an α-helical coiled coil as predicted from the sequence [de Cuevas et al., 1992]. The end of the stalk and/or tail domain is the site of light chain interaction based on electron microscopy analysis of bovine brain kinesin [Hirokawa et al., 1989] and an *in vitro* analysis of the *Drosophila* light chain interaction with this region of the KHC (A. Gauger and L.S.B. Goldstein, in preparation).

2. Kinesin function. Several functions for kinesin have been proposed as a result of immunolocalization studies. For example, antikinesin antibodies stain vesicles but not spindles in mammalian interphase cells, suggesting a role in vesicle transport [Hollenbeck, 1989; Pfister et al., 1989]. In other immunolocalization analyses kinesin is found in the spindle poles of mammalian and amphibian cells [Neighbors et al., 1988] and in a nonmicrotubule component of the sea urchin spindle [Leslie et al., 1987; Wright et al., 1991]. Although spindle-associated kinesin immunoreactivity in certain cell types suggests a possible mitotic function, further work is still needed to address this issue more directly.

Several *in vitro* studies implicate kinesin in the formation of the endoplasmic reticulum (ER). First, the vanadate sensitivity of formation of an ER-like membrane network *in vitro* is similar to that of kinesin, and the rate of tubule extension is similar to that of kinesin-induced organelle movement [Dabora and Sheetz, 1988]. Second, an ER-like tubule network was observed to form upon addition of ATP and microtubules to a membrane-contaminated kinesin preparation [Vale and Hotani, 1988]. These observations, coupled with findings showing colocalization and motility of ER elements along microtubules, suggest that a kinesin motor could be involved in ER formation [Lee et al., 1989; Terasaki et al., 1986].

Antibody inhibition studies have also yielded *in vivo* evidence for the function of kinesin in nonneural cells. In one study two different antikinesin antibodies caused the centripetal collapse of tubular lysosomes when these antibodies were scrape-loaded into macrophages [Hollenbeck and Swanson, 1990]. Similarly, antikinesin antibody injections suppressed the centrifugal translocation of pigment granules toward the plus ends of microtubules in melanophores [Radionov et al., 1991]. A surprising result is the collapse of the intermediate filament network in fibroblasts injected with antikinesin antibodies [Gyoeva and Gelfand, 1991]. One explanation for this latter finding is that there are interactions of a membranous organelle system with both microtubules and intermediate filaments. Such an interaction may be maintaining the intermediate filament array. Hence, when the force between kinesin and the microtubules is released, collapse is observed. As we discuss below, kinesin appears to also have a nonneuronal function in *C. elegans,* but, curiously *Drosophila* kinesin appears to have a more specific neural function [Gho et al., 1992; Saxton et al., 1991].

In *C. elegans,* the *unc-116* locus encodes the KHC [de Feo et al., 1991]. At the organismal level, *unc-116* animals have a paralysis phenotype indicative of a defect in neuromuscular function. At the cellular level, a thorough electron microscopic analysis of animals homozygous for a severe *unc-116* allele [Hall et al., 1991] found defects in secretory and excretory cells that strongly suggest a role for kinesin in membrane translocation. These defects include an absence of rough ER, an accumulation of late stage lysosomes in the somata of secretory cells, a build up of dense-core secretory granules, and misshapen excretory cells with abnormally positioned canaliculi. A problem with axon outgrowth, which also requires membrane transport, was also observed. There were no abnormalities in the distribution of synaptic vesicles. Unless KHC functions differently in *C. elegans,* this latter observation is not consistent with the view that the KHC transports synaptic vesicles *in vivo* [Hirokawa et al., 1989].

It is also difficult to explain the phenotype of *Drosophila* KHC mutants by a defect in synaptic vesicle transport, although the mutant phenotype does indicate a problem with axonal transport. These animals die as severely undersized larvae or pupae with malformed posterior ends that do not respond to touch [Saxton et al., 1991] and also exhibit progressive loss of mobility. A temperature-sensitive allele was used to observe the behavior of adults raised at permissive temperature. Over the course of several days at nonpermissive temperature these adults cease all general activity. Together the larval and adult phenotype suggest a neuronal or neuromuscular abnormality, which could result from a defect in secretion or in the transport of any one of a number of different organelles at the neuromuscular junction or synapse.

A physiological study of *Drosophila* KHC mutants lends further support to the hypothesis that kinesin functions in axonal transport *in vivo* [Gho et al., 1992]. Both weakened action potentials and reduced neurotransmitter release were observed in larval motor neurons. This defect is graded in severity from the anterior to posterior end of the abdomen, presumably because the motor neurons in the more posterior segments are longer and therefore more vulnerable to defects in axonal transport. Since an electron microscopic analysis of mutant larval motor neurons does not reveal any morphological differences (even in synaptic vesicle distribution), the problem of determining kinesin's cargo, vesicular or other, still remains.

Until further work is done, the phenotype of kinesin-deficient *Drosophila* cannot be related to any one specific microtubule-dependent transport process. However, the results of the *unc-116* study demand a similar morphological analysis of the *Drosophila* mutations. It is possible that *Drosophila* KHC has a similar role in membrane-bound organelle translocation in nonneural cells but that such defects are not so severe as to cause lethality. One important finding from the *Drosophila* KHC phenotypic analysis was that this gene is probably inessential for embryonic mitoses. Since a role for the KHC in mitosis was suggested by some of the immunolocalization data, this discrepancy again raises the possibility that kinesin function varies between species.

At this point the *Drosophila* KHC and *unc-116* mutant phenotypes do not reveal a defect in fast axonal transport of known cargo. However, since axonal transport has only been studied in large neurons such as the squid giant fiber and the vertebrate visual system, we know virtually nothing about the various components of axonal transport in *Drosophila* and the phenotypes that would result from its disruption. It may, however, be possible to identify the kinds of vesicles that are being affected by the KHC mutation. The organization of the invertebrate CNS should greatly facilitate this type of analysis. Since the CNS neurons have their soma in the cortex and their synapses in the neuropil, a defect in the transport of any number of different neural antigens and histochemical markers should be apparent.

The last KHC mutant phenotype we discuss results from the suppression of KHC expression in cultured hippocampal neurons [Ferreira et al., 1992]. These particular cells were used because they send out elaborate arborizations that are obviously dependent on membrane translocation during their formation. The inhibition of KHC expression in these cells still allows neurite extension, but the overall neurite length is significantly reduced. This KHC mutant phenotype further suggests a role for kinesin in membrane translocation. This phenotype may also include an effect on axonal transport *in vivo* since the distribution of GAP43 and synapsin I was found to be dramatically affected in the kinesin null cells. These two proteins remained in the cell body and were not transported to the synapse. The significance of this latter finding is unclear, since there is conflicting evidence regarding the type of transport (fast or slow) used for synapsin I [Baitinger and Willard, 1987; Petrucci et al., 1991]. Furthermore, the synthesis and transport of GAP43 is a response to axonal growth [Kalil and Skene, 1986]; hence a perturbation of neurite expression rather than an axonal transport defect could be inhibiting GAP43 transport.

The functional analyses discussed above suggest that the KHC may function in the transport of membrane-delimited organelles including the ER, lysosomes, and secretory granules. However, either the specific function of the KHC varies in different organisms and cell types, or certain aspects of the mutant defect may have gone undetected in some of the mutant phenotype analyses. A caveat to this interpretation is that some proposed KHC functions may instead be carried out by other superfamily members. Since some KLPs would be expected to have similar biochemical and antigenic properties to kinesin, many of the studies using antikinesin antibodies and pharmacological inhibitors may have actually examined the effect of these reagents on cross-reacting KLP(s) rather than on kinesin itself. For example, although the antikinesin antibodies that inhibit pigment granule dispersion [Rodionov et al., 1991] and cause intermediate filament collapse [Gyoeva and Gelfand, 1991] were raised and affinity purified against the kinesin head, it is possible that these antibodies can recognize and inhibit kinesin-like proteins expressed in melanophores and fibroblasts. Furthermore, the mutational analyses of other KLPs provide strong evidence for their involvement in processes as diverse as vesicle transport and the separation of spindle pole bodies.

B. The bimC Family

The bimC family consists of representatives from *Aspergillus nidulans* (bimC), *Schizosaccharomyces pombe* (cut7), *Saccharomyces cerevisae* (CIN8 and KIP1), *Drosophila* (KLP2 or urchin), and *Xenopus* (Eg5). Considerable information is available about the cellular function of this family, because in all but one case (Eg5) a mutant phenotype is known and these phenotypes are remarkably similar across the phylogenetic scale. Each of these proteins appear to play a role in moving the microtubule organizing centers (MTOCs) apart during spindle formation. A further role of the CIN8 and KIP1 proteins is to keep the spindle poles apart until the time of anaphase [Saunders and Hoyt, 1992].

The first identified member of this family was *bimC*, which was discovered in a screen for temperature-sensitive lethals that are blocked in mitosis [Morris, 1976]. The cloning and sequence determination of *bimC* revealed that it has sequence similarity only to the motor domain of KHC but not to other domains [Enos and Morris, 1990]. However, the overall structure of bimC does share some general features with the KHC. The bimC C terminus is also predicted to form a small globular domain, which may be separated from the N-terminal globular head domain by a long α-helical coiled-coil region.

A closer examination of the *bimC* mutant phenotype revealed a specific defect in the initiation of spindle assembly. This block was found to be due to a failure of spindle pole body (the fungal equivalent of the higher eukaryotic MTOC or centrosome) separation after their replication [Enos and Morris, 1990]. Other microtubule-based processes such as nuclear migration were unimpaired in these mutants. Since polar growth continued at the restrictive temperature in *bimC* strains, the authors deduced that microtubule-dependent vesicle transport was also normal. These results must be interpreted with the caveat that *bimC* may not be a null mutation.

cut 7 was isolated in a screen similar to that which uncovered *bimC*. The *cut7* phenotype is almost identical to that of *bimC* [Hagan and Yanagida, 1990]. Antitubulin immunofluorescent staining of *cut7* cells blocked at the start of mitosis showed an abnormal V-shaped staining pattern. This was interpreted to result from the inability of the microtubules extending from the spindle pole bodies (SPBs) to slide apart. Cell viability during interphase was found to be unaffected by this mutation. Subsequent cloning and sequence analysis revealed homology to the kinesin motor domain.

Further information about the function of this family comes from immunolocalization analyses using cut7 antibodies [Hagan and Yanagida, 1992]. These antibodies were raised against the *E. coli* expressed cut7 nonmotor domain and thus are likely to be specific for this particular KLP. Also, cut7 reactivity is absent in cells that are *cut7* deficient. As expected from the mutant phenotype, anti-cut7 antibodies label the spindle poles of mitotic cells during prometaphase. The spindle midzone is also labeled, which may indicate a further function of cut7 after spindle assembly. Indeed, such a function for this family is suggested by phenotypic analysis of the *Sac. cerevisiae* members of this family (discussed below). cut7 reactivity is also present with a similar distribution in the spindles of meiotic cells.

The *Sac. cerevisiae* members of this family, *CIN8* and *KIP1*, were discovered to encode KLPs through two independent approaches. In the first, *CIN8* was identified in a screen for mutations that cause chromosome loss [Hoyt et al., 1990]. *KIP1* was recovered as a second gene that could complement the defects in *CIN8* mutants. Cloning and sequence analysis then revealed homology to the kinesin superfamily [Saunders and Hoyt, 1992]. In the second approach, *KIP1* was isolated in a polymerase chain reaction (PCR) screen that searched for new KLPs based on conserved sequence motifs [Roof et al., 1992]. Since Roof et al. did not observe a phenotype caused by loss of *KIP1* expression, they reasoned that there may exist a separate gene that encodes a function redundant to that of *KIP1*. A synthetic lethal screen then resulted in the recovery of *CIN8* mutants.

The phenotype of cells doubly mutant for *CIN8* and *KIP1* is strikingly similar to that of *bimC* and *cut7* mutants [Hoyt et al., 1992; Roof et al., 1992; Saunders and Hoyt, 1992]. As mentioned above, either of these genes can be deleted with no effect on viability at physiological temperature; however, double deletions are inviable due to a block in mitosis. Subsequently, an ultrastructural analysis of cells bearing a deletion of the *KIP1* gene and a temperature-sensitive mutation of *CIN8* shifted to the nonpermissive temperature was conducted. These cells showed not only a failure of SPB separation but also the presence of a bridge between the SPB (the bridge is a structure that is observed between SPBs prior to migration [Byers, 1981]).

An elegant experiment by Saunders and Hoyt [1992] demonstrated a further requirement for either *CIN8* or *KIP1* following spindle assembly. Since in yeast spindle assembly, but not elongation, continues in the absence of chromosome replication, the DNA synthesis inhibitor hydroxyurea was used to arrest cells at preanaphase. When arrested cells doubly mutant for both *KIP1* and a temperature-sensitive allele of CIN8 were released into fresh medium at high temperature, they remained blocked in mitosis even though spindle morphologies were normal prior to the temperature shift. When these blocked cells were examined by electron microscopy, they were found to contain duplicated SPBs that were adjacent to one another rather than segregated to opposite ends of the spindle. Unlike the double mutants that were blocked prior to spindle assembly (described above), the SPBs of these cells contained long microtubules extending into the nucleus. These observations suggest a further requirement for CIN8 or KIP1 function in maintaining pole separation prior to anaphase.

The *Drosophila* representative of this group is called *urchin*. This mutation was found to map to the same cytogenetic location as *KLP2*, a gene encoding a KLP identified in a PCR screen [Stewart et al., 1991]. There are two lines of evidence that *KLP2* and *urchin* are the same gene (M. Heck et al., in preparation). First, one *urchin* allele is the result of a P element insertion within the 5′-untranslated region of *KLP2*. Second, a P element containing a KLP2 cDNA rescues the lethality of homozygous *urchin* animals.

The *urchin* phenotype has been well characterized cytologically (P. Wilson et al., in preparation). This phenotype is exactly what one would predict for the *Drosophila* homologue of the yeast and fungal mutations described above. A careful analysis of a mitotic population in homozygous *urchin* larvae shows that these cells are tremendously polyploid and oversized. Since in *Drosophila* chromosome replication will continue in the absence of spindle formation, this phenotype could result from a block in spindle formation. Wilson (in preparation) has determined that centrosome duplication but not segregation has occurred. Together these observations suggest that the wild-type *urchin* function may be like its yeast and fungi relatives in segregating the newly replicated MTOC.

The last known member of this family is Eg5. Eg5 was discovered during a general analysis of gene expression during *Xenopus* early development [Le Guellec et al., 1991]. The amino acid sequence of Eg5, like bimC described above, also predicts a three domain structure; an N-terminal globular region, an extended central region, and a C-terminal globular domain. Within this central region, Eg5 appears to also share some similarity with bimC at the sequence level. Both of these KLPs contain conserved leucines at every sixth position throughout about 200 amino acids of this region which may be functionally significant or may suggest that both proteins have an α-helical coiled coil. Since there is no classical genetic system available for this, the mutant phenotype for this KLP is unknown. To circumvent this problem, it may be possible to observe the effect of anti-Eg5 antibody injections on mitosis.

To summarize, the studies described above suggest that the different members of the bimC family share a similar function during spindle assembly. Since the motor domain sequence similarity among members of this family is higher than that found with other superfamily members, the motor domains of these proteins might be specialized, e.g., perhaps for exerting force between the microtubule fibers emanating from the SPBs. If indeed the tail functions as a cargo link, it is intriguing that although most of the nonmotor domains of the individual bimC members share no significant sequence similarity, a small conserved region of 40 amino acid residues has been found within the tail region of KLP2, Eg5, cut7, and bimC (M. Heck et al., in preparation). While this sequence may be required for specific cargo binding, it is difficult to explain why CIN8 and KIP1 do not contain this sequence. Perhaps, if the bimC family forms part of a higher order molecular structure such as a bridge between antiparallel spindle fibers, then this sequence could be part of another protein within the bridge of *Sac. cerevisiae*.

C. The KAR3 Family

Three members of the KAR3 family have been identified; KAR3 (*Sac. cerevisiae*), ncd (*D. melanogaster*), and KLPA (*A. nidulans*) [McDonald and Goldstein, 1990; Meluh and Rose, 1990] (O'Connell et al., in preparation). They all share two predicted structural features. First, the structural organization of these three KLPs is reversed relative to the other members of this superfamily such that the motor domains of the KAR3 family are located at the C rather than the N terminus. Second, even though there is no detectable sequence similarity outside of the motor domain, the predicted structure is the same. Thus the motor domains are separated from a smaller globular domain at the N-terminus by a region predicted to be an α-helical coiled coil. Third, the N-terminal globular domain is basic and proline rich, similar to the microtubule-binding region of MAP2, tau, and *Drosophila* 205K MAP [Irminger-Finger et al., 1990; Lee et al., 1989; Lewis et al., 1988]. Indeed, there is evidence that a second microtubule-binding region is contained within the nonmotor domains of KAR3. This site has not yet been mapped relative to the globular domain [Meluh and Rose, 1990].

The KAR3 family may have the added distinction of being minus-end directed motors with respect to microtubule polarity. This has been demonstrated *in vitro* for ncd and is inferred for KAR3 and KLPA owing to their structural similarity [McDonald et al., 1990; Walker et al., 1990]. Interestingly, genetic analyses also raise the possibility that KAR3 might be minus-end directed (discussed below).

KAR3 was initially identified as a mutant of *Sac. cerevisiae* defective in the process of karyogamy, in which two haploid nuclei fuse during conjugation [Conde and Fink, 1976; Polaina and Conde, 1982]. This process is dependent on cytoplasmic microtubules that extend between the two nuclei. A morphological comparison of wild-type and *KAR3* mutant zygotes revealed two differences [Meluh and Rose, 1990]. The first difference was in nuclear migration. Shortly after cell fusion, wild-type zygotes contained either a single fused nucleus or two closely associated nuclei while the mutant zygotes always contained two separated nuclei. The second difference was in microtubule associations. Immunofluorescent staining with an antitubulin antibody revealed a short, densely stained bar of fluorescence between two wild-type nuclei. Similar staining of mutant zygotes showed long cytoplasmic microtubules that were not interconnected between the two nuclei. This observation led to the hypothesis that KAR3 mediates the association of antiparallel microtubules between nuclei that are destined to fuse.

The subcellular localization of KAR3 further supports this hypothesis [Meluh and Rose, 1990]. Anti-β-galactosidase immunofluorescence of KAR3–lacZ hybrid proteins revealed localization of the protein on long cytoplasmic microtubules that extend into the cytoplasmic extensions of cells about to conjugate. This staining was dependent on the presence of cytoplasmic microtubules since cells treated simultaneously with nocodazole and mating pheromone were not stained. Interestingly, if the α-factor treatment preceded the nocodazole treatment, the KAR3–lacZ hybrid protein was found to prevent the same microtubules from depolymerizing, suggesting that KAR3 can stabilize these fibers. Unsuccessful attempts were also made to localize the KAR3 protein with a specific anti-KAR3 antibody. However, anti-KAR3 staining of cytoplasmic microtubules was exhibited by cells bearing the *kar3-1* mutation, which is a semidominant allele predicted to encode a protein that binds more strongly than the wild-type protein to microtubules. These data, together with the identification of KAR3 as a KLP, strongly suggest a role for this protein in the association, and perhaps intermicrotubule sliding, of cytoplasmic microtubules during nuclear fusion.

KAR3 also plays an important albeit nonessential role during mitosis. Mutations of *KAR3* cause slowed mitotic growth and accumulation of cells with a single nucleus and a short mitotic spindle. Further insight into the role of KAR3 in spindle function resulted from a screen for suppressors of the *CIN8 KIP1* double mutant [Saunders and Hoyt, 1992]. As described above, double mutant *cin8–3 kip1–*Δ are temperature sensitive for growth at higher temperature owing to a defect in spindle assembly. One of these two gene products continues to be needed to maintain SPB separation until anaphase. However, at higher temperature *cin8–3 kip1–*Δ*kar3–*Δ are viable and when arrested at metaphase are less likely to have collapsed spindles. KAR3 may therefore generate the force that draws the two SPBs back together. This force might be generated by a minus-end directed motor with respect to the hypothesized polarity of spindle microtubules. The idea that KAR3 functions as a minus-end directed motor that slides antiparallel microtubule tor that slides antiparallel microtubules together explains both the defect in nuclear fusion during karyogamy and the suppression of *cin8–3 kip1–*Δ.

The ncd protein, like KAR3 also may function to mediate the interactions between microtubules. In this case, the microtubule array is found in the *Drosophila* female meiotic spindle. From Northern blot analysis, ncd may be present in other cells but the *ncd* mutation only affects chromosome behavior during female meiosis and the first few mitotic divisions of the embryo. Meiotic spindles from *ncd* mutant females show various abnormalities that could possibly be attributed to a problem in spindle assembly [Hatsumi and Endow, 1992; Kimble and Church, 1983]. These include monopolar spindles, a misshapen and disorganized appearance, and splayed half spindles.

Further cytological analyses with a specific anti-ncd antiserum has shown ncd to be distributed in two different patterns within the wild-type spindle. In one class of spindles, ncd is localized to a region that extends from the border of the midzone toward the pole region, but not in the region where the microtubules converge at the pole (H. McDonald and L.S.B. Goldstein, in preparation). In the other spindles, ncd is more sharply excluded from a narrower region thought to be the midzone. Since two classes of spindles (12 μm and 25 μm) are also observed with respect to length, these staining differences might be related to stage. These data do not appear to support the hypothesis that ncd interacts with antiparallel microtubules because staining is not always seen in the region thought to contain overlapping microtubules of opposite polarity. At this time, however, the interpretation of the ncd localization analysis suffers from our poor understanding of the wild-type *Drosophila* spindle architecture.

D. Single Member Families

Several KLPs that do not fall into one of the three families discussed above have been identified. These KLPs share less sequence similarity with all other KLPs individually than is found among the members within a single family. In addition, for those in which there exist corresponding mutations or other functional tests, their implied biological roles are very diverse. The existence of these families may be explained in part by the fact that the identification of KLPs in any one organism is not complete [Stewart et al., 1991]. In some cases, however, a one-member family may be the actual situation in nature. These would be either the more recently evolved members and/or those that perform highly specialized or species-specific functions. We now briefly summarize the existing information about some of these individual KLPs.

1. No distributive disjunction (nod). nod is an example of a KLP with a highly specialized primary function. The absence of *nod* function during meiosis was observed to cause nondisjunction and loss of nonrecombinant chromosomes during *Drosophila* female meiosis [Carpenter, 1973]. More recently, a further understanding of *nod* function came about quite suddenly when a structural analysis of the *nod* locus revealed homology to the motor domain of kinesin as well as to other KLPs. Although the tail domain of nod shares no homology with any known protein the similarity to the KHC motor domain led to the hypothesis that nod generates an antipoleward force on achiasmate chromosomes [Rasooly et al.,

1991]. This force is believed to substitute for the antipoleward force provided by chiasma at the metaphase plate.

To understand the *nod* mutant phenotype, it is necessary first to review recent work on spindle assembly in wild-type *Drosophila* [Theurkauf and Hawley, 1992]. In female *Drosophila,* meiosis I is arrested at metaphase. In wild-type oocytes the nonexchange chromosomes are observed to be positioned between the metaphase plate and the spindle pole. The exact position of these chromosomes may be size dependent. During the course of this study it was also observed that meiotic spindle assembly is quite unusual in wild-type *Drosophila* females. Spindle assembly appears to be initiated from prepolymerized cytoplasmic microtubules of unknown polarity by the centrally located chromosomes. Once a short bipolar spindle is established the achiasmate fourth chromosome moves poleward away from the main chromosome mass. If the X chromosomes are achiasmate, they are only observed apart from the mass in longer spindles.

Great insight into the precise function of the *nod* gene comes from cytological analysis that indicates a requirement for *nod* during the time of spindle assembly to keep certain chromosomes centrally located [Theurkauf and Hawley, 1992]. In *nod* oocytes, spindle assembly appears to be initiated in the same manner as in wild-type oocytes. The first departure from normality is seen once the early short spindle is visible. At this time the fourth chromosomes are dissociated and found free in the cytoplasm. Achiasmate X chromosomes show similar misbehavior, although one X chromosome was usually observed to be associated with a pole. Together these observations led the authors to suggest that *nod*-induced chromosome loss results from a failure of the achiasmate chromosomes to reassociate with the spindle after slipping off into the cytoplasm. Similarly, what appears to be nondisjunction is a result of a chromosome's reassociation to the wrong pole rather than movement to the wrong pole during anaphase. These observations also indicate that the force exerted by nod is acting to keep achiasmate chromosomes centrally located during the process of spindle assembly. Since it is unclear if there actually is a poleward force at this time, the hypothesized role of nod in the generation of the "anti-poleward force" may or may not be correct.

Two observations also suggest a role for nod during mitosis, albeit one that may be redundant. First, nod mRNA is expressed in the mitotic population of the ovariole and throughout the time of cell division during the *Drosophila* life cycle [Zhang et al., 1990]. Second, a cold-sensitive mutation of *nod*, nod^{DTW}, has been isolated and exhibits an embryonic lethal phenotype [Rasooly et al., 1991; Wright, 1974]. This phenotype can be concluded to be due to a mitotic defect from two observations. First, nod^{DTW} embryos grown at nonpermissive temperature exhibit "abnormal mitotic figures and multipolar spindles" [Wright, 1974]. Second, an analysis of mitotic chromosome behavior in nod^{DTW} animals reveals chromosome loss as well as morphological abnormalities indicative of cell death. These effects are more severe at cold temperature [Baker et al., 1978]. Since the *nod* null mutation is viable and only affects female meiosis, it may be possible to identify and genetically delete a locus whose function is redundant to that of *nod*.

2. SMY1. SMY1 is a *Sac. cerevisiae* KLP discovered in a screen for suppressors of a myosin mutation, *myo2* [Lillie and Brown, 1992]. *myo2* cells are defective in the process of secretion, which in yeast is dependent on microfilaments. It is therefore perplexing as to how overexpression of SMY1 can suppress such a defect. Even though genetic nulls of *SMY1* have no observable phenotype, the double mutant *myo2 smy1* combination is lethal. This synthetic lethal interaction has been suggested to reflect a requirement for a secondary microtubule-based system that becomes essential in the absence of actin-based transport. A precedent for this interpretation is a recent report of squid axoplasmic vesicles that can move on both of these filamentous systems [Kuznetsov et al., 1992]. The explanation for this fascinating *SMY1/MYO2* genetic interaction awaits the further characterization of SMY1.

3. *unc-104*. The mutant phenotype analysis of the *C. elegans* KLP unc-104 strongly indicates that it specifically transports synaptic vesicles to the synapse. At a behavioral level, different mutations effect various kinds of complex behaviors. Depending on the strength of the particular allele, these effects range from slowed locomotion to cessation of feeding. Most likely as a consequence of feeding defects, some strains are slow to grow and reproduce [Hall and Hedgecock, 1991]. Although mutant animals exhibit no histological abnormalities, a morphological basis for the behavioral phenotype is ultrastructurally evident. In loss of function *unc-104* mutations, the chemical synapse does not form and an accumulation of synaptic vesicles occurs in the soma. Very few vesicles are found in the axon, reflecting defective anterograde transport. There is also a problem with neuromuscular junction development, which may be a consequence of the defective vesicle transport. Furthermore, *unc-104* is not required for axonal growth or tubulovesicular localization. The latter observation is not surprising considering that *unc-116* (the *C. elegans* KHC) may be responsible for these processes as discussed above. Hence within the same cell type different members of the kinesin superfamily are transporting different classes of organelles.

4. Mitotic KLP1 (MKLP1). MKLP1 is a human KLP that has been implicated in both spindle organization [Nislow et al., 1990] and movement during anaphase B [Nislow et al., 1992]. Previous to the sequence analysis of the MKLP1 gene, this protein was identified as the antigen recognized by the monoclonal antibody CHO1 [Nislow et al., 1990]. In CHO and PtK1 cells, the antibody CHO1 stains spindles with increasing intensity from prophase to anaphase [Nislow et al., 1990; Sellito and Kuriyama, 1988]. During anaphase the antigen is found in the midzone region of microtubule overlap. Injection of the CHO1 antibody into CHO1 cells results in metaphase-arrested cells with disrupted spindle poles and microtubule organization [Nislow et al., 1990]. Another indication that MKLP1 could function during anaphase is the ability of bacterially expressed MKLP1 to both bundle and slide antiparallel microtubules *in vitro* [Nislow et al., 1992].

III. SCREENS FOR NEW KLPS

In this section we describe and compare the results of screens that have been used to identify new KLPs. Attempts have also been made to use these screens for estimating the size of the kinesin superfamily in a single organism [Endow and Hatsumi, 1991]. Some of the KLP genes that were detected in these screens have been discovered previously through the cloning of genetically identified loci as described above (i.e., *ncd, KIP1,* and so forth). Since little is known about the function of the new KLPs that have been identified in these screens, they cannot yet be divided into families.

A. KLPs Identified by the PCR

The first attempt to search directly for new KLPs used the PCR [McDonald and Goldstein, 1990]. Degenerate primers corresponding to two regions of the KHC motor that are conserved between KAR3 and kinesin were used to amplify *Drosophila* genomic DNA. These primers surround the region proposed to be the microtubule-binding domain [Yang et al., 1990]. Sequence analysis of 25 cloned amplification products revealed only one novel sequence. Comparison of this new sequence with that of the newly isolated *ncd* gene [Endow et al., 1990) revealed that the novel product was amplified from the *ncd* gene.

Two other attempts were made to identify new KLP genes in *Drosophila*. In the first of these, degenerate primers that correspond to the ATP-binding region and a second motor region of high conservation were used to amplify genomic DNA as well as several cDNA libraries [Endow and Hatsumi, 1991]. The resulting amplification products were then directly used for *in situ* hybridization probes to salivary gland chromosomes. The authors concluded from the number of hybridization sites that as many as 30 KLP genes exist in *Drosophila*. Given that these products were not cloned and sequenced and that the PCR can result in artifacts [McDonald and Goldstein, 1990; Stewart et al., 1991], this method may be inaccurate for estimating superfamily size. In the second attempt, Stewart et al. [1991] isolated six new KLP genes. In this case the amplification products were cloned and sequenced and proven to correspond to bona-fide KLPs. Furthermore, an *in vitro* motility assay has recently been used to demonstrate that at least three of these new KLP genes encode motor proteins (R. Stewart, N. Barton, P. Pesavento, A. Pereira, and L.S.B. Goldstein, unpublished observations). The Stewart et al. report makes no attempt to estimate the total number of KLP genes that may exist because certain genes that should have been amplified were not. For example, under the conditions used the *nod* gene was not detected even though it should have been isolated with the primers used. Alternatively, *nod* may have been amplified but selected against during the cloning procedure for some unknown reason.

A PCR screen has also been used to isolate KLP genes from *Sac. cerevisae* [Roof et al., 1992]. Using similar primers to those used in the PCR screens described above for *Drosophila,* two new KLP genes were identified. One of these, *KIP1* was found to correspond to the same locus as described above (see The bimC Family, above) [Hoyt et al., 1990]. The second new KLP gene, *KIP2,* has no observable mutant phenotype. It is unclear why only two KLP genes were isolated in this screen, since, as discussed earlier, *Sac. cerevisiae* contains at least three other KLPs: KAR3, SMY1, and CIN8.

B. KLPs Identified in Antibody Screens

The approach of first identifying the protein and then working back to the gene was used in an attempt to isolate KLPs from two different systems. In one study, Sawin et al. [1992] used *Xenopus* egg extracts because they provide a powerful system for analyzing spindle function *in vitro*. The hope was to identify KLPs that are spindle and/or kinetochore components. In another study Cole et al. [1992] used sea urchin eggs to address the more general question of how many KLPs exist in a single mitotic cell type. In both cases antipeptide antibodies to conserved regions of the motor domain of kinesin were assayed for reactivity against microtubule-binding proteins by immunoblotting. Although this work resulted in the identification of novel microtubule-binding proteins that may or may not be KLPs, it did not answer the question of how many KLPs actually exist in these cell types. Since some KLPs are known to contain amino acid variations of the antigens used for these antibodies, they may not be recognized by the antipeptide antibodies. Furthermore, the antibodies were only assayed on a high speed supernatant and it is not yet evident that other KLPs will be as soluble as kinesin.

The immunological search in sea urchins resulted in the discovery of a complex of three proteins of 85, 95, and 115 kD. The 85 and 95 kD are both detected by the antipeptide antibodies [Cole et al., 1992]. It is unknown how they are related to one another. The 115 kD protein was found to copurify with these two proteins during the fractionation of ATP-eluted sea urchin microtubule-binding proteins. Together this complex of three proteins was observed to bundle microtubules *in vitro,* and the further addition of ATP induced microtubule sliding. If this complex does indeed contain KLPs, this finding is significant in that it is the first time KLP(s) other than kinesin have been isolated in its native form. Furthermore, it indicates that these kinesin heavy chain relative(s) can form an oligomeric structure different from that of native kinesin.

The *Xenopus* study resulted in the identification of five candidate proteins [Sawin et al., 1992]. These potential KLPs have similar microtubule-binding properties to kinesin and are not recognized by specific antikinesin antibodies. Although these proteins must also share at least two antigenic determinants with kinesin, the extent of their amino acid sequence similarity to kinesin is unknown. In the same study these antibodies were used to label kinetochores and spindles in mitotic PtK2 cells.

It is unknown how this staining pattern relates to the proteins detected in *Xenopus* by immunoblotting. More work is also needed to determine whether these are *Xenopus* KLPs and whether they are spindle or kinetochore associated.

The results of the two antibody screens prove the value of this approach. They have resulted in the discovery of new potential KLPs in those organisms that do not allow classic genetic studies. Furthermore, this type of screen may have resulted in the first isolation of a KLP in its native form as opposed to the other KLPs that have only been analyzed as expressed proteins from *E. coli*. At another level, it is problematic to use this approach to answer questions about superfamily size. This would necessitate that the potential KLPs be cloned by expression vector screening and then sequenced. Another drawback is that it is not feasible to sample every or even most cell types of *Xenopus* or sea urchin. The PCR approach is potentially the most valuable and allows one to survey the whole genome.

IV. CONCLUSIONS

Our knowledge of the kinesin superfamily reveals much about how a single class of motor performs very diverse microtubule-based functions. Genetic analyses revealed that these functions are carried out by multiple related proteins within the same organism and sometimes within the same cell. A major problem that now remains to be addressed is how variations in motor sequence are related to structure and function. This will become approachable once we better understand the tertiary structure of the motor domain and how it produces mechanical energy.

The study of the kinesin superfamily has not just furthered our understanding of motor function *per se* but has also unveiled deeper levels of complexity within certain biological processes. For example, the analyses of *nod* and *ncd* have led us to a better understanding of the highly unusual assembly of the female meiotic spindle in *Drosophila* (and perhaps in other organisms). Similarly, studies of SMY1 have revealed the possible interchange in the use of microfilaments and microtubules during yeast secretion. Finally, the curious need for redundant motors during SPB separation in some organisms but not in others is evident from studies on the bimC family. Our insight into each of these processes as well as numerous others will continue to grow as the functional analysis of this superfamily continues.

ACKNOWLEDGMENTS

We thank Russell Stewart, Nelson Barton, Paul Dobner, and especially Anna Pereira for their help and valuable comments on the manuscript.

V. REFERENCES

Amos LA (1987): Kinesin from pig brain studied by electron microscopy. J Cell Sci 87:105–111.

Baitinger C, Willard M (1987): Axonal transport of synapsin I–like proteins in rabbit retinal ganglion cells. J Neurosci 7:3723–3735.

Baker BS, Carpenter ATC, Ripoll P (1978): The utilization during mitotic cell division of loci controlling meiotic recombination and disjunction in *Drosophila melanogaster.* Genetics 90:531–578.

Bloom GS, Wagner MC, Pfister KK, Brady ST (1988): Native structure and physical properties of bovine brain kinesin and identification of the ATP-binding subunit polypeptide. Biochemistry 27:3409–3416.

Byers B (1981): Cytology of the yeast life cycle. In Strathern JN, Jones EW, Broach JR (eds): The Molecular Biology of the Yeast *Saccharomyces:* Life Cycle and Inheritance. Cold Spring Harbor, NY: Cold Spring Harbor Laboratory, pp 59–96.

Carpenter ATC (1973): A meiotic mutant defective in distributive disjunction in *Drosophila melanogaster.* Genetics 73:393–428.

Cole DG, Cande WZ, Baskin RJ, Skoufias DA, Hogan CJ, Scholey JM (1992): Isolation of a sea urchin kinesin-related protein using peptide antibodies. J Cell Sci 101:291–301.

Conde J, Fink GR (1976): A mutant of *Saccharomyces cerevisiae* defective for nuclear fusion. Proc Natl Acad Sci USA 73:3651–3655.

Dabora SL, Sheetz MP (1988): The microtubule-dependent formation of a tubulovesicular network with characteristics of the ER from cultured cell extracts. Cell 54:27–35.

de Cuevas M, Tao T, Goldstein LSB (1992): Evidence that the stalk of *Drosophila* kinesin heavy chain is an alpha-helical coiled coil. J Cell Biol 116:957–965.

de Feo G, Patel N, and Mancillas JR (1991): Soc Neurosci Abstr 17:58.

Devereux J, Haeberli P, Smithies O (1984): A comprehensive set of sequence analysis programs for the VAX. Nucleic Acids Res 12:387–395.

Endow SA, Henikoff S, Soler-Niedzeila L (1990): Mediation of meiotic and early mitotic chromosome segregation in *Drosophila* by a protein related to kinesin. Nature 345:81–83.

Endow SA, Hatsumi M (1991): A multimember kinesin gene family in *Drosophila*. Proc Natl Acad Sci USA 88:4424–4427.

Enos AP, Morris NR (1990): Mutation of a gene encoding a kinesin-like protein blocks nuclear division in *A. nidulans*. Cell 60:1019–1027.

Ferreira A, Niclas J, Vale RD, Banker G, Kosik KS (1992): Suppression of kinesin expression in cultured hippocampal neurons using antisense oligonucleotides. J Cell Biol 117:595–606.

Gho M, McDonald K, Ganetsky B, Saxton WM (1992): Effects of kinesin mutations on neuronal function. Science 258:313–316.

Goldstein LSB (1991): The kinesin superfamily: Tails of functional redundancy. Trends Cell Biol 1:93–98.

Gyoeva FK, Gelfand VI (1991): Coalignment of vimentin intermediate filaments with microtubules depends on kinesin. Nature 353:445–448.

Hagan I, Yanagida M (1990): Novel potential mitotic motor protein encoded by the fission yeast *cut7* gene. Nature 347:563–566.

Hagan I, Yanagida M (1992): Kinesin-related cut7 protein associates with mitotic and meiotic spindles in fission yeast. Nature 356:74–76.

Hall DH, Hedgecock EM (1991): Kinesin-related gene *unc-104* is required for axonal transport of synaptic vesicles in *C. elegans*. Cell 65:837–847.

Hall DH, Plenefisch J, Hedgecock EM (1991): Ultrastructural abnormalities of kinesin mutant *unc-116*. J Cell Biol 115:389a.

Hatsumi M, Endow SA (1992): Mutants of the microtubule motor, nonclaret disjunctional, affect spindle structure and chromosome movement in meiosis and mitosis. J Cell Sci 101:547–559.

Hirokawa N, Pfister KK, Yorifuji H, Wagner MC, Brady ST, Bloom GS (1989): Submolecular domains of bovine brain kinesin identified by electron microscopy and monoclonal antibody decoration. Cell 56:867–878.

Hollenbeck PJ (1989): The distribution, abundance, and subcellular localization of kinesin. J Cell Biol 108: 2335–2342.

Hollenbeck PJ, Swanson JA (1990): Radial extension of macrophage tubular lysosomes supported by kinesin. Nature 346:864–866.

Hoyt MA, Stearns T, Botstein D (1990): Chromosome instability mutants of *Saccharomyces cerevisae* that are defective in microtubule-mediated processes. Mol Cell Biol 10:223–234.

Hoyt MA, Saunders WS (1992): Two *S. cerevisiae* kinesin-related gene-products required for mitotic spindle assembly. J Cell Biol 118:109–120.

Irminger-Finger I, Laymon RA, Goldstein LSB (1990): Analysis of the primary sequence and microtubule-binding region of the *Drosophila* 205K MAP. J Cell Biol 111:2563–2572.

Kalil K, Skene JHP (1986): Elevated synthesis of an axonally transported protein correlates with axon outgrowth in normal and injected pyramidal tracts. J Neurosci 6:2563–2570.

Kato K (1991): Sequence analysis of twenty mouse brain cDNA clones selected by specific expression patterns. J Neurosci 2:704–711.

Kimble M, Church K (1983): Meiosis and early cleavage in *Drosophila melanogaster* eggs: Effects of the Claret non-disjunctional mutation. J Cell Sci 62: 301–318.

Kosik KS, Orecchio LD, Schnapp B, Inouye H, Neve RL (1990): The primary structure and analysis of the squid kinesin heavy chain. J Biol Chem 265:3278–3283.

Kuznetsov SA, Vaisberg YA, Shanina NA, Magretova NN, Chernyak VY, Gelfand VI (1988): The quarternary structure of bovine brain kinesin. EMBO J 7:353–356.

Kuznetsov SA, Langford GM, Weiss DG (1992): Actin dependent organelle movement in squid axoplasm. Nature 356:722–725.

Lasek RJ, Brady ST (1985): Attachment of transported vesicles to microtubules is facilitated by AMPPNP. Nature 316:645–647.

Lee C, Ferguson M, Chen LB (1989): Construction of the endoplasmic reticulum. J Cell Biol 109:2045–2055.

Lee G, Neve RL, Kosik KS (1989): The microtubule binding domain of tau. Neuron 2:1615–1624.

Le Guellec R, Paris J, Couturier A, Roghi C, Philippe M (1991): Cloning by differential screening of a *Xenopus* cDNA that encodes a kinesin-related protein. Mol Cell Biol 11:3395–3398.

Leslie RJ, Hird RB, Wilson L, McIntosh JR, Scholey JM (1987): Kinesin is associated with the non-microtubule component of sea urchin mitotic spindles. Proc Natl Acad Sci USA 2771–2775.

Lewis SA, Wang D, Cowan NJ (1988): Microtubule-associated protein MAP2 shares a microtubule-binding motif with tau protein. Science 242:936–939.

Lillie SH, Brown SS (1992): Suppression of a myosin defect by a kinesin-related gene. Nature 356:358–361.

McDonald HB, Goldstein LSB (1990): Identification and characterization of a gene encoding a kinesin-like protein in *Drosophila*. Cell 61:991–1000.

McDonald HB, Stewart RJ, Goldstein LSB (1990): The kinesin-like *ncd* protein of *Drosophila* is a minus end-directed microtubule motor. Cell 63:1159–1165.

Meluh PB, Rose MD (1990): *KAR3*, a kinesin-related gene required for yeast nuclear fusion. Cell 60:1029–1041.

Morris NR (1976): Mitotic mutants of *Aspergillus nidulans*. Genet Res 26:237–254.

Navone F, Niclas J, Hom-Booher N, Sparks L, Bernstein HD, McCaffrey G, Vale RD (1992): Cloning and expression of a human kinesin heavy chain gene: Interaction of the COOH-terminal domain with cytoplasmic microtubules in transfected CV-1 cells. J Cell Biol 117:1263–1275.

Neighbors BW, Williams RC, McIntosh JR (1988): Localization of kinesin in cultured cells. J Cell Biol 106:1193–1204.

Nislow C, Sellito C, Kuriyama R, McIntosh JR (1990): A monoclonal antibody to a mitotic microtubule-associated protein blocks mitotic progression. J Cell Biol 111:511–522.

Nislow C, Lombillo VA, Kuriyama R, McIntosh JR (1992): A plus-end directed motor that cross-links and moves antiparallel microtubules in vitro localizes to the interzone of mitotic spindles. Nature 359:543–547.

Petrucci TC, Macioce P, Paggi P (1991): Axonal transport kinetics and posttranslational modification of synapsin I in mouse retinal ganglion cells. J Neurosci 11:2938–2946.

Pfister KK, Wagner MC, Stenoien DL, Brady ST, and Bloom GS (1989): Monoclonal antibodies to kinesin heavy chains stain vesicle-like structures, but not microtubules in cultured cells. J Cell Biol 1989:1453–1463.

Polaina J, Conde J (1982): Genes involved in the control of nuclear fusion during the sexual cycle of *Saccaromyces cerevisiae*. Mol Gen Genet 186:253–258.

Rasooly RS, New CN, Zhang P, Hawley RS, Baker BS (1991): The *lethal(1)TW-6^{cs}* mutation of *Drosophila melanogaster* is a dominant antimorphic allele of *nod* and is associated with a single base change in the putative ATP-binding domain. Genetics 129:409–422.

Rodionov VI, Gyoeva FK, Gelfand VI (1991): Kinesin is responsible for centrifugal movement of pigment granules in melanophores. Proc Natl Acad Sci USA 88:4956–4960.

Roof DM, Meluh PB, Rose MD (1992): Kinesin-related proteins required for assembly of the mitotic spindle. J Cell Biol 118:95–108.

Saunders WS, Hoyt MA (1992): Kinesin-related proteins required for structural integrity of the mitotic spindle. Cell 70:451–458.

Sawin KE, Mitchison TJ, Wordeman LG (1992): Evidence for kinesin-related proteins in the mitotic aparatus using peptide antibodies. J Cell Sci 101:303–313.

Saxton WM, Porter ME, Cohn SA, Scholey JM, Raff EC, Macintosh JR (1988): *Drosophila* kinesin: Characterization of microtubule motility and ATPase. Proc Natl Acad Sci USA 85:1109–1113.

Saxton WM, Hicks J, Goldstein LSB, Raff EC (1991): Kinesin heavy chain is essential for viability and neuromuscular functions in *Drosophila,* but mutants show no defects in mitosis. Cell 64:1093–1102.

Scholey JM, Heuser J, Yang JT, Goldstein LSB (1989): Identification of globular mechanochemical heads of kinesin. Nature 338:355–357.

Scholey JM, Porter ME, Grissom PM, McIntosh JR (1985): Identification of kinesin in sea urchin eggs, and evidence for its localization in the mitotic spindle. Nature 318:483–486.

Sellito C, Kuriyama R (1988): Distribution of a matrix component of the midbody during the cell cycle in chinese hamster ovary cells. J Cell Biol 106:431–439.

Stewart RJ, Pesavento PA, Woerpel DN, Goldstein LSB (1991): Identification and partial characterization of six members of the kinesin superfamily in *Drosophila*. Proc Natl Acad Sci USA 88:8470–8474.

Terasaki M, Chen LB, Fujiwara K (1986): Microtubules and the endoplasmic reticulum are highly interdependent structures. J Cell Biol 103:1557–1568.

Theurkauf WE, Hawley RS (1992): Meiotic spindle assembly in *Drosophila* females: Behavior of non-exchange chromosomes and the effects of mutations in the *nod* kinesin-like protein. J Cell Biol 116: 1167–1180.

Vale RD, Goldstein LSB (1990): One motor, many tails, an expanding repertoire of force-generating enzymes. Cell 60:883–885.

Vale RD, Hotani H (1988): Formation of kinesin networks on vitro by kinesin-driven microtubule movement. J Cell Biol 107:2233–2241.

Vale RD, Schnapp BJ, Mitchison T, Steuer E, Reese TS, Sheetz MP (1985): Different axoplasmic proteins generate movement in opposite directions along microtubules *in vitro*. Cell 43:623–632.

Vallee RB, Shpetner HS (1990): Motor proteins of cytoplasmic microtubules. Ann Rev Biochem 59:909–932.

Walker RA, Salmon ED, Endow SA (1990): The *Drosophila claret segregation* protein is a minus end-directed motor molecule. Nature 347:780–782.

Wright BD, Henson JH, Wedaman KP, Willy PJ, Morand JN, Scholey JM (1991): Subcellular localization and sequence of sea urchin kinesin heavy chain: Evidence for its association with membranes in the mitotic apparatus and interphase cytoplasm. J Cell Biol 113: 511–536.

Wright TRF (1974): A cold-sensitive zygotic lethal causing high frequencies of nondisjunction during meiosis I in *Drosophila melanogaster* females. Genetics 76:511–536.

Yang JT, Saxton WM, Goldstein LSB (1988): Isolation and characterization of the gene encoding the heavy chain of *Drosophila* kinesin. Proc Natl Acad Sci USA 85:1864–1868.

Yang JT, Laymon RA, Goldstein LSB (1989): A three domain structure of kinesin heavy chain revealed by DNA sequence and microtubule binding analyses. Cell 56:879–889.

Yang JT, Saxton WM, Stewart RJ, Raff EC, Goldstein LSB (1990): Evidence that the head of kinesin is sufficient for force generation and motility in vitro. Science 249:42–47.

Zhang PAKB, Goldstein LSB, Hawley RS (1990): A kinesin-like protein required for distributive chromosome segregation in *Drosophila*. Cell 62:1053–1062.

ABOUT THE AUTHORS

ANDREA PEREIRA is a research associate in the lab of Lawrence S.B. Goldstein at Harvard University. After receiving a B.A. from Regis College in Weston Massachusetts, she began graduate study at Brandeis University, where she worked on *Drosophila* nervous system development with Kalpana White and L.E. Cannon. She received a Ph.D. in biology from Brandeis in 1987. Since that time she has been using a classical genetic approach to understanding microtubule-associated protein function in *Drosophila*.

LAWRENCE S.B. GOLDSTEIN is Professor of Cellular and Developmental Biology at Harvard University in Cambridge, Massachusetts, where he teaches undergraduate and graduate courses in cell biology. After receiving a B.A. in biology magna cum laude from the University of California, San Diego, he pursued doctoral research in the laboratory of Larry Sandler in the Department of Genetics at the University of Washington in Seattle, where he earned a Ph.D. in 1980 for work on the genetics and cytology of meiosis in *Drosophila*. This was followed by postdoctoral research at the University of Colorado in Boulder, Colorado, where he worked with J. Richard McIntosh on the genetics and biochemistry of microtubule proteins in *Drosophila*. After a short stint at MIT in the laboratory of Barbara Meyer, Dr. Goldstein's research has been pursued at Harvard University since 1984. His work has focused on the genetics, biochemistry, and physiology of cytoskeletal protein function in *Drosophila*, in particular on kinesin microtubule motors. He is a member of the editorial board of the *Journal of Cell Biology*.

MICROTUBULE ORGANIZATION AND FUNCTION IN CELLS

Microtubules: 287–301

Dynamics of Microtubule Assembly *In Vivo*

Linda Wordeman and Timothy J. Mitchison

I. INTRODUCTION

Microtubules are stiff, and they resist compressive and bending forces [Bailly et al., 1991; Hotani and Miyamoto, 1990; Winckler and Solomon, 1991]. This characteristic is compatible with the conception that microtubules serve a variety of architectural functions in living cells. Assembled microtubule arrays serve as flexible, yet resistive substrates for the operation of mechanochemical enzymes in ciliary and flagellar movement. Microtubules serve as rails for the movement of intracellular vesicles and other components. Microtubules are resistant to the compressive effects of the actomyosin system [Bailly et al., 1991], and assembled microtubule arrays are required for the proper positioning of the mitotic spindle and the nucleus in a large number of cell types.

Paradoxically, it was discovered within the last decade that microtubules in living cells exist along with tubulin subunits in an unusual dynamic equilibrium termed *dynamic instability*. Individual microtubules alternate stochastically between periods of polymerization and depolymerization, leading to rapid exchange between tubulin subunits and polymer. Rapid turnover of microtubule polymer via dynamic instability is a highly nonequilibrium behavior and requires the energy of GTP hydrolysis. The cell expends a certain amount of energy to support the constant turnover of microtubule polymer. Furthermore, poisons that disrupt or suppress microtubule dynamics also inhibit microtubule-dependent cellular functions.

The aim of this chapter is to describe the dynamic behavior of microtubules in cells and to examine how microtubule dynamics in cells differs from that observed for pure tubulin assembled *in vitro*. We discuss the mechanisms and incentives for modulating microtubule dynamics *in vivo* in an attempt to gain some insight into the functional advantages of dynamic microtubules. Other comprehensive reviews on microtubule dynamics [Gelfand and Bershadsky, 1991], molecular mechanisms of dynamic instability [Kirschner and Mitchison, 1986; Cassimeris et al., 1987], and mitotic microtubule dynamics [for further information, see McIntosh and Koonce, 1989; McIntosh and Hering, 1991; Mitchison, 1988; Mitchison and Sawin, 1990] (see also McIntosh, this volume) have been published.

II. DYNAMIC INSTABILITY OF MICROTUBULES *IN VITRO*

Polymerized microtubules exhibit structural polarity. This statement is supported by several lines of evidence, including X-ray fiber diffraction [Amos and Baker, 1979; Mandelkow, 1986], the uniform angle of laterally bound dynein [Allen and Borisy, 1974], hook decoration [Mandelkow and Mandelkow, 1979], and the kinetics of microtubule assembly [Allen and Borisy, 1974]. Furthermore, it has been shown that arrays of microtubules in cells exhibit a uniform polarity relative to each other [Heidemann and McIntosh, 1980; Euteneuer and

McIntosh, 1981]. This strongly implies that the polarity of microtubules is functionally significant.

The property of microtubules described as "dynamic instability" [Mitchison and Kirschner, 1984] also reflects microtubule polarity. Microtubules in solution, in the absence of accessory proteins, exhibit alternating transitions between phases of elongation and rapid shortening. The transitions between these phases are abrupt and stochastic. This behavior, summarized by the expression dynamic instability, occurs at both ends of the microtubule. Yet the respective ends of an individual microtubule are different in their dynamic properties. Before examining microtubule dynamics in cells, it is worth having a qualitative understanding of the dynamic properties of microtubules in solution.

The most thorough study of the dynamic properties of microtubules *in vitro* is that of Walker et al. [1988]. (See also Bailey, this volume). In this study the authors determined the rate constants for both elongation and rapid shortening. The transition frequencies between phases of elongation and rapid shortening were also measured for individual microtubules. In order to facilitate discussion, the abrupt transition between elongation and rapid shortening was described as *catastrophe* and the transition between rapid shortening and elongation was termed *rescue*. That nomenclature will also be used in this review.

Microtubules assembled axonemes and visualized by videomicroscopy exhibit the following properties: 1) The rate of elongation at both ends of the microtubule is proportional to the free tubulin concentration, and the elongation rate is twice as fast at the plus end of the microtubule. There is a substantial dissociation rate contributing to the kinetics of microtubule assembly. 2) The rate of rapid shortening is independent of tubulin concentration for both ends, was approximately 10-fold faster than the variable elongation rate, and is similar for the two ends of the microtubule. 3) The plus end of the microtubule exhibits a higher catastrophe frequency and a lower rescue frequency than the minus end. Hence, the overall length changes, during both depolymerization and polymerization, are greater for the plus end. Higher concentrations of tubulin reduce the frequency of catastrophe at both ends of the microtubule and increase the rescue frequency at the minus end.

From this study and others [reviewed by Gelfand and Bershadsky, 1991], we can make the following generalization regarding the dynamic behavior of microtubules in solution: Although both ends of a microtubule in solution exhibit the property of dynamic instability, the plus end is more dynamic and the minus end is less dynamic or more "stable." Therefore, although the minus end of a microtubule is more sluggish, with respect to assembly, and is often shorter than the plus end when assembled from the same seed, the minus end is also more stable to disassembly than is the plus end.

This property is supported by microtubule-severing experiments. In these studies, microtubules mechanically severed at the plus end of the microtubule rapidly depolymerized. The severed minus end of such microtubules, however, remained static or continued to elongate. These results imply that the plus end of the microtubule is capped, perhaps by a layer of subunits with bound GTP (GTP cap) [Hill and Carlier, 1983], whereas the mechanism by which the minus end is stabilized is unknown. Interestingly, similar results were obtained whether the experiments were performed on pure microtubules assembled onto axonemes *in vitro* [Walker et al., 1989] or with grasshopper spermatocyte mitotic spindles *in situ*, which presumably contain many accessory and regulatory-associated proteins [Nicklas et al., 1989].

III. FACTORS THAT MODULATE THE DYNAMIC PROPERTIES OF MICROTUBULES

Dynamic instability was originally discovered using pure tubulin assembled *in vitro* [Mitchison and Kirschner, 1984]. Nevertheless, several early studies showed that similar behavior occurs *in vivo* [Cassimeris et al., 1988; Sammak and Borisy, 1987; Schulze and

TABLE I. Microtubule-Associated Proteins

Protein	M_r (kD)	Localization	References
MAP 1A	350	Neural tissue* and nonneural cells	Bloom et al.[1984]
MAP 1B (MAP 5)	325	Neural tissue	Garner et al. [1990] Hammarback et al. [1991]
MAP 1C (dynein)	300	All cells	Vallee et al. [1988]
Vesikin	295	Neural tissue	Cell Motil. Cyt. 10:246
MAP 2A, 2B	270	Neural tissue	Burgoyne and Cumming [1984]
MAP 4	200	Many cell types	Parysek et al. [1984]
MAP 3	180	All tissues tested	Huber et al. [1985]
Kinesin	120	All cells	Vale et al. [1985]
Dynamin	100	Neural tissue	Obar et al. [1990]
Chartins	70–80	Neural tissue	Peng et al. [1985]
MAP 2C	70	Neural tissue	Garner et al. [1988]
Tau	50–65	Neural tissue	Cleveland et al. [1972]

*Many of these MAPs were first identified in brain, however, homologs may yet be identified in non-neural cells.
[Adapted from Avila, 1990, with permission of the publisher.]

Kirschner, 1987]. How accurately does the behavior of microtubules assembled *in vitro* from pure tubulin reflect the behavior of microtubules in cells? Several classes of microtubule-associated proteins (MAPs) bind to microtubules both *in vivo* and *in vitro* [Vallee et al., 1986; Avila, 1992]. These MAPs (see Table I), especially in the case of neuronal tissue, are developmentally and spatially regulated in their distribution [Olmsted et al., 1986; Matus, 1988; Vera et al., 1988]. Furthermore, these MAPs have been shown *in vitro* to bind to and stabilize microtubules [Cleveland et al., 1977; Mandelkow et al., 1988] at substoichiometric quantities [Job et al., 1985]. Finally, it has been observed *in vitro* that phosphorylation inhibits the assembly-promoting activity of brain MAPs [Hoshi et al., 1992] possibly by decreasing the affinity of stabilizing MAPs for microtubules [Raffaelli et al., 1992]. These results suggest that the binding of relatively dephosphorylated MAPs to microtubules promotes microtubule stability by inhibiting the dynamic instability that is intrinsic to such microtubules. Microtubule dynamics could then be increased, as is observed when cells enter mitosis, by simply phosphorylating stabilizing MAPs, causing them to become dissociated from microtubules. While this interpretation is supported by studies in which active MAP-2 kinase added to *Xenopus* egg extracts increased microtubule dynamics from the interphase to the mitotic level, presumably by phosphorylating MAP-2 [Gotoh et al., 1991], other researchers believe that this view is oversimplified and that active destabilizing factors are also involved [Walker et al., 1988; Belmont et al., 1990; Faruki et al., 1992].

Posttranslational modifications of tubulin subunits, such as acetylation and detyrosination of α-tubulin and the covalent phosphorylation of β-tubulin, have been correlated with differentially stable subsets of microtubules [reviewed by Greer and Rosenbaum, 1989]. Both detyrosinated and acetylated tubulin subunits have been shown to be enriched in stable populations of microtubules [Gunderson et al., 1987; Piperno et al., 1987; Schulze and Kirschner, 1987]. Yet experimentally detyrosinating α-tubulin has no effect on microtubule stability or on microtubule assembly properties [Maruta et al., 1986; Khawaja et al., 1988; Idriss et al., 1991]. This apparent contradiction has led researchers to speculate that posttranslational modifications of tubulin subunits may promote the binding of MAPs and, in this manner, lead to the stabilization of microtubules in cells. This conclusion is, however, based primarily on the circumstantial evidence that the C terminus of tubulin serves both as the site of posttranslational modification and as the site for the binding of many MAPs [Edde et al. 1990; Cross et al., 1991]. There is little or no evidence showing that posttranslational modifications of tubulin directly increase MAP binding. Also, no MAP has been shown to exhibit the same distribution as is seen for stable microtubules in tissue culture cells. This lack of a functional correlation between microtubule stability and posttranslational modifications have led many researchers to suggest that microtubule stabilization may be signalled by an accumulation of posttranslational modifications such as acetylation or detyrosination but may not directly result from such modifications. Thus tubulin acetylation and detyrosination remain enigmatic. Their evolutionary conservation implies that they have an important role, perhaps in modulating microtubule dynamics. This role, however, has yet to be discovered.

IV. MICROTUBULE DYNAMICS IN INTERPHASE CELLS

Given the plethora of mechanisms available for modulating microtubule dynamics, what exactly is the behavior of microtubules in cells? With a few exceptions [Baas et al., 1988, 1989; Troutt and Burnside, 1988a,b] microtubules in living cells have their minus ends capped, meaning that they are associated with some structure such as a spindle pole, spindle pole body, or centrosome. Furthermore, the interior of cells is difficult to probe with either DIC or fluorescence microscopy. Hence, studies of microtubule dynamics in living cells have primarily involved the plus ends of the microtubules located at the periphery of the cell [Sammak and Borisy, 1987]. The dynamics of individual microtubules have been observed in living cultures of newt lung epithelial cells [Cassimeris et al., 1988]. The individual microtubules observable at the thin leading edge of such cells exhibited dynamic instability behavior that was similar to that observed in purified microtubules *in vitro*. Microtubules elongated at a constant velocity and rapidly shortened at a rate at least twofold faster than the elongation rate. Furthermore, the transitions between elongation and shortening occurred abruptly and stochastically. The rate of rapid shortening *in vivo* (17.3 μm/min) was statistically equivalent to that observed *in vitro* (22 μm/min) for a tubulin concentration of 16 μM (a concentration chosen because it corresponds to the estimate of the free tubulin concentration cells [Hiller and Weber, 1978]). Curiously, however, the elongation rate was fivefold faster *in vivo* (7.2 μm/min) than that observed *in vitro* (1.4 μm/min). The frequency of rescue (transition from rapid shortening to elongation) was similar in cells to that measured *in vitro*. However, the catastrophe frequency was approximately 10-fold higher *in vivo*. Finally, a small stable population of microtubules was observed that did not either elongate or shorten during a 15 minute observation time. Thus microtubules in interphase cells exhibit, at least qualitatively, the same behavior as is observed *in vitro* with the exceptions of a faster elongation rate and a greater frequency of catastrophe.

A microtubule undergoing rapid shortening has a certain probability of suddenly transiting to the elongating phase. This has been called the *rescue frequency.* The rescue frequency (0.05/sec *in vitro*) is high enough that a microtubule of a length of at least 10 μm can expect to be rescued before shortening completely away to nothing. The average microtubule length in an epithelial cell, such as a BSC1 (African Green monkey kidney) cell, is 20 μm. Therefore it is unlikely that a microtubule in such a cell would shrink more than halfway to the centrosome before elongating again. This property has been illustrated by Sammak et al. [1987], in which they found that the half-time for microtubule replacement is greatest at the leading edge of cells in culture. They have referred to this property of microtubules in cells as "tempered dynamic instability." This property is illustrated, for a fibroblast-type cell, in Figure 1a. In this diagram, dynamic and sta-

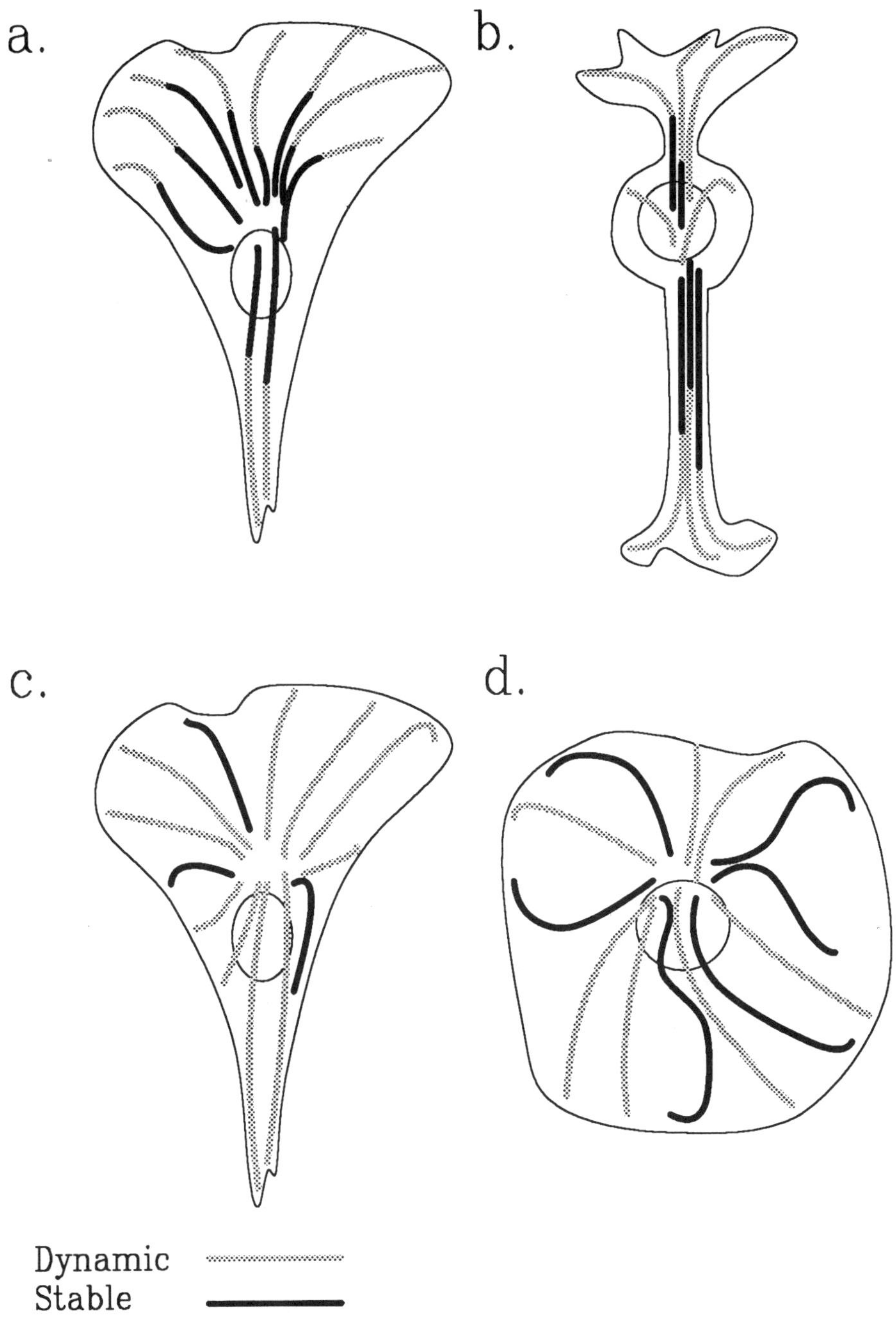

Fig. 1. *Diagram of stable and dynamic microtubules in cells. Microtubules may be partially stabilized along their length* **(a,b)** *or exist as a stable subpopulation within an array of dynamic microtubules* **(c,d)**. *Dynamic distal microtubule ends may reflect tempered dynamic instability [Sammak et al., 1987] in a. Actual gradients of posttranslational modifications may accompany stabilization of proximal microtubule ends [Baas and Black, 1990] in axons and dendrites (b). Many tissue culture cells have a small subpopulation of metastable microtubules [Schulze and Kirshner, 1987] that exist in close proximity (c). Stable microtubules have been observed to increase in number in confluent cultures of epithelial cells (d) but not fibroblasts [Pepperkok et al., 1990].*

ble microtubules are a reflection of the half-time of microtubule turnover for the entire population of microtubules in different regions of the cell. Walker et al. [1988] argue that shorter half-times for tubulin turnover at the plus end of microtubules are related to the rescue frequency per unit length. In this case, the decreased half-time for microtubule turnover further from the edge of the cell would not result from the polymer in that region being inherently more stable, but rather simply because of its distance from the plus end of the microtubule.

Does the cell take advantage of this intrinsic property of individual microtubules to produce stabilized or partially stabilized arrays of microtubules? There are reasons to suspect that older microtubule polymer, such as that which is closer to the centrosome, is modified by the cell cytoplasm in order to promote stability. Gradients of detyrosination are seen on the very slowly assembling pellicular microtubules of the trypanosome [Sherwin and Gull, 1989]. The experimenters suggest that the microtubules of tissue culture cells, in contrast, are so highly dynamic that they appear generally tyrosinated. In addition, Baas and Black [1990] found that axonal microtubules proximal to the cell body were more resistant to nocodazole, suggesting that the axonal microtubules contain domains that differ in stability. Such a differentiated neuron is diagrammed in Figure 1b. This phenomenon has also been observed in *Chlamydomonas* flagella [Dentler and Adams, 1992]. In axons, immunoelectron microscopy revealed microtubules with a sharp demarcation between tyrosinated (dynamic) tubulin at the plus end of the microtubule and detyrosinated/acetylated (stable) regions at the minus end. As microtubules in axons can be quite long, this would seem to fit the model that the minus ends of long microtubules are more persistent, due to a relatively high probability of rescue, and are therefore more likely to undergo posttranslational modification and stabilization. However, the time-course of posttranslational modification suggests that stabilization occurs prior to posttranslational modification [Baas et al., 1991]. In these cells, posttranslational modification is likely to be a consequence of increased stability rather than its cause, as discussed earlier. One might imagine that the proximal microtubule segments in an undifferentiated cell, such as the one diagrammed in Figure 1a, might, by virtue of their inherently lower turnover time, be subsequently stabilized during neuronal differentiation (Fig. 1b), with proximal microtubule segments stabilized by MAP-binding or posttranslational modifications. It should be noted, however, that Baas et al. [1991] found that the microtubules in dendrites versus axons exhibited similar lability and turnover dynamics, despite the fact that dendrites contain microtubules of different polarities [Baas et al., 1988] and different MAP compositions [Matus, 1988].

Epithelial and fibroblastic cells that exhibit tempered dynamic instability generally do not have gradients of modified tubulin or MAP-binding along a single microtubule [Schulze and Kirshner, 1987]. These cells tend to show a subpopulation of microtubules that are stable to exchange of labeled tubulin over long time periods (approx. 1 hour) [Schulze and Kirshner, 1987]. This is shown in Figure 1c. These microtubules, unlike the axonal microtubules observed by Baas et al. [1991], are stable and often completely detyrosinated all along their length. They have a characteristic wavy appearance. They may correspond to the individual microtubules observed *in vivo* by Cassimeris et al. [1988] that neither shortened nor elongated for extended periods of observation.

Incubating cells in varying levels of microtubule-depolymerizing drugs has been a useful technique for revealing not only different stable domains of a single microtubule but also subclasses of microtubules of differential stability within interphase cells. This technique has been used in conjunction with injections of labeled tubulin in different cell types and different culture conditions to show that interphase microtubule dynamics are cell type specific. Wadsworth and McGrail [1990] showed that fibroblast (CHO) cell microtubules disassembled more rapidly than those of epithelial

(PtK1) cells. Furthermore, Pepperkok et al. [1990] have shown that a subpopulation of stable microtubules (turnover half-time, 25 minutes) appears in MDCK (epithelial cells) that have reached confluence. Just such an epithelial cell is diagrammed in Figure 1d. Both subconfluent MDCK and Vero (fibroblast) cells had an average half-time for microtubule turnover of 10 minutes. However, the Vero cells did not show a subpopulation of metastable microtubules. These studies indicate that interphase microtubule dynamics vary for different cell types and can change with different environmental conditions. This is illustrated in Figure 2, in which two adjacent PtK2 cells exhibit strikingly different extents of biotin–tubulin incorporation despite the fact that they were injected with biotin–tubulin almost simultaneously.

Microtubule turnover, which is a reflection of microtubule dynamics, is the sum of four components of individual microtubule behavior: 1) polymerization rate, 2) depolymerization rate, 3) frequency of catastrophe, and 4) frequency of rescue. Unless each of these parameters is measured for individual microtubules it is impossible to determine what aspect of microtubule dynamics is altered in a cell-type-specific manner and why.

Shelden and Wadsworth [1992] have quantified dynamic instability parameters for fibroblasts and epithelial cells and have determined that both the polymerization and depolymerization rate is significantly greater for fibroblasts as for epithelial cells. Furthermore, while the frequency of catastrophe is approximately the same for both cell types, the frequency of rescue is greater for epithelial cells. Hence, microtubules in epithelial cells are more stable than those in fibroblasts. Polymerization rate, frequency of catastrophe, and frequency of rescue were all significantly higher in both cell types than was observed *in vitro* and these parameters were variable in different cell types. Therefore, one would be tempted look for cytoplasmic factors that increase or influence these aspects of microtubule behavior and might be regulated in a cell-type-specific manner.

Microtubule turnover in cells is a reflection of the dynamic instability inherent to the microtubules plus the superimposition of regulatory factors. The contribution that each of these makes to microtubule turnover in cells is easily confused unless single microtubules are observed. Stabilization may be limited to a portion of a single microtubule, as Baas and Black [1990] have seen in axons; or stabilization may involve an entire subpopulation of complete microtubules (Fig. 1). It is tempting to speculate that stabilization of entire microtubules involves some sort of capping mechanism, whereas partial stabilization may result from lateral interactions along the microtubule. In *Chlamydomonas* lateral interactions with the actin-based cytoskeleton has been implicated in promoting stability of flagellar microtubules [Dentler and Adams, 1992].

V. MICROTUBULE DYNAMICS IN MITOTIC CELLS

The search for factors that influence microtubule dynamics is particularly enticing with regard to the transition from interphase to mitosis. As first shown by fluorescence photobleaching recovery, microtubule turnover is approximately 20-fold faster in mitotic versus interphase cells [Saxton et al., 1984]. This increase in microtubule turnover accompanies a dramatic reorganization of the long interphase microtubule arrays to form the generally shorter microtubules of the mitotic spindle [McIntosh et al., 1975]. It is more difficult to visualize single microtubules in the mitotic spindle, due to the high density of close-packed microtubules, than it is in interphase cells. Therefore information regarding the dynamic behavior of individual microtubules in interphase *versus* mitosis has been obtained from studies of microtubules assembled in mitotic or interphase cytoplasmic extracts from *Xenopus* eggs. These studies have shown that mitotic microtubules are generally shorter and have a higher frequency of catastrophe [Belmont et al., 1990], and this effect can be triggered by the addition of bacterially synthesized cyclin B to micro-

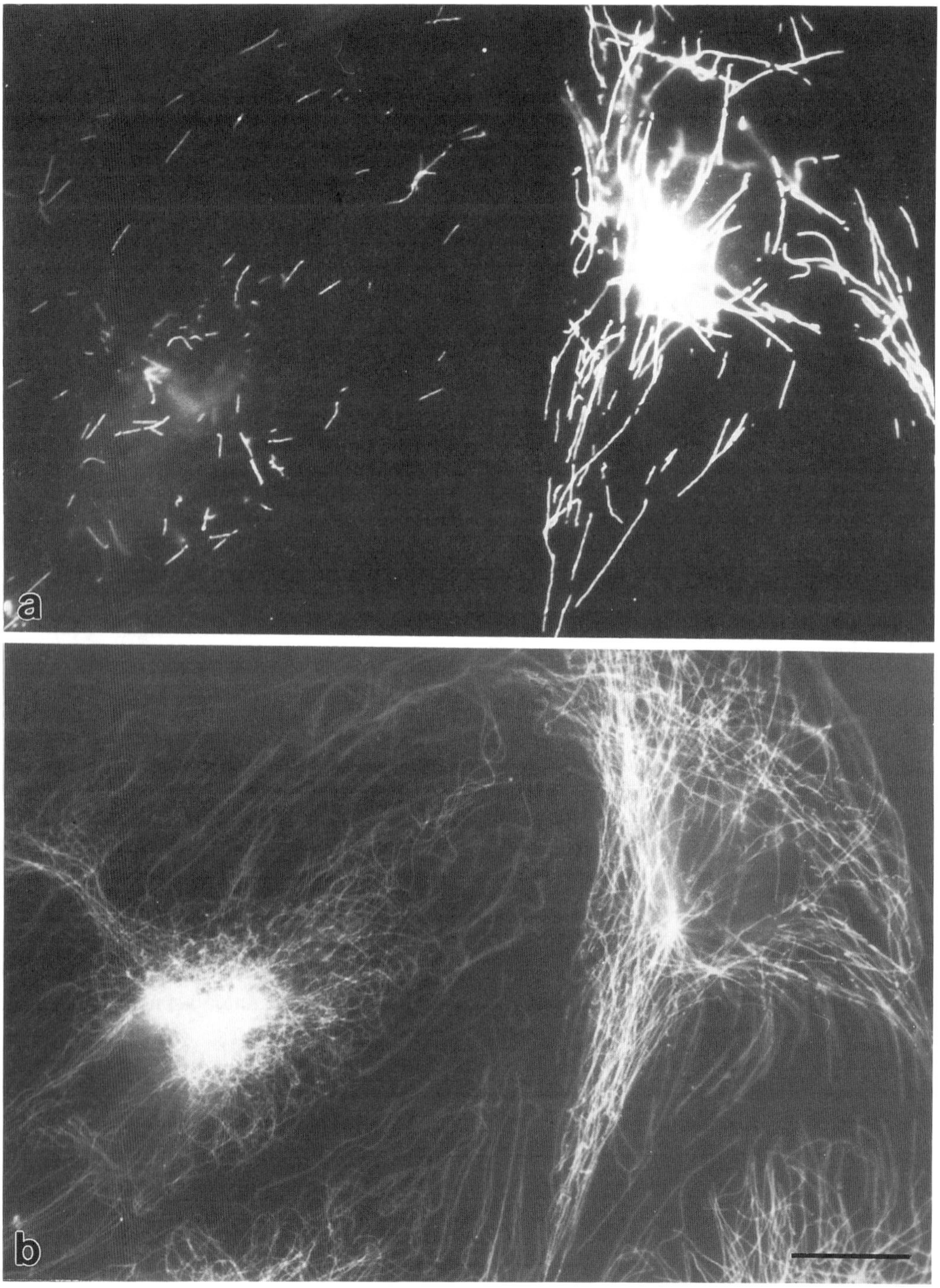

Fig. 2. *Microtubule turnover in PtK cells. Biotin–tubulin was injected into two adjacent PtK2 cells at, respectively, 4 minutes 51 seconds (***left***) and 4 minutes 42 seconds (***right***) prior to fixation at 30°C and processed for indirect immunofluorescence. Biotin–tubulin incorporation* **(a)** *and total tubulin* **(b)** *immunofluorescence are shown. Note the difference in incorporation between the two cells in culture. Bar = 10 μm.*

tubule-containing extracts [Verde et al., 1992]. A factor that induces increased frequencies of catastrophe in response to cell cycle–dependent phosphorylation events would be a fortuitous discovery, indeed. Additionally, a family of protein kinases known as ERKs has been identified. ERKs become activated in response to mitogenic signals and phosphorylate, among other things, microtubule-associated proteins, suggesting that they might have a role in microtubule reorganization [Crews et al., 1992].

In addition to microtubule-associated proteins, phosphorylation, and putative catastrophe-induced factors, there is another factor that influences microtubule dynamics in the mitotic spindle, and that is the capping of the plus end of spindle microtubules. Fluorescence redistribution after photobleaching, originally used to demonstrate high rates of microtubule turnover in the mitotic spindle, also revealed that a subset of microtubules, those likely to be comprising the kinetochore fiber, turn over more slowly [Wadsworth and Salmon, 1986]. Subsequent studies showed that microtubules whose plus ends are capped by kinetochores (chromosome associated) incorporate new tubulin at their plus ends at a slower rate than nearby uncapped microtubules. This tubulin incorporation can be detected up until midanaphase [Wadsworth et al., 1989]. The newly incorporated polymer fluxes slowly, relative to the rate of microtubule depolymerization, toward the spindle pole [Geuens et al., 1989; Mitchison, 1989]. In addition, microtubule flux in the spindle may not result from capping by kinetochores *per se* but by association with chromatin or, perhaps, any capping structure [Sawin and Mitchison, 1991]. Furthermore, the incorporation of tubulin subunits at the kinetochore in response to flux may have superimposed upon it dynamic activity of kinetochore microtubules (Fig. 3). The incorporation of labeled tubulin onto the plus ends of microtubules in the spindle midzone occurs in a manner consistent with microtubules exhibiting dynamic instability in that region [Shelden and Wadsworth, 1990].

The incorporation of tubulin subunits into kinetochore fibers of PtK1 cells declines by late anaphase [Wadsworth et al., 1989]. At this time, the incorporation of tubulin subunits onto the plus ends of midzone microtubules is maximal in PtK1 cells, eventually decreasing dur-

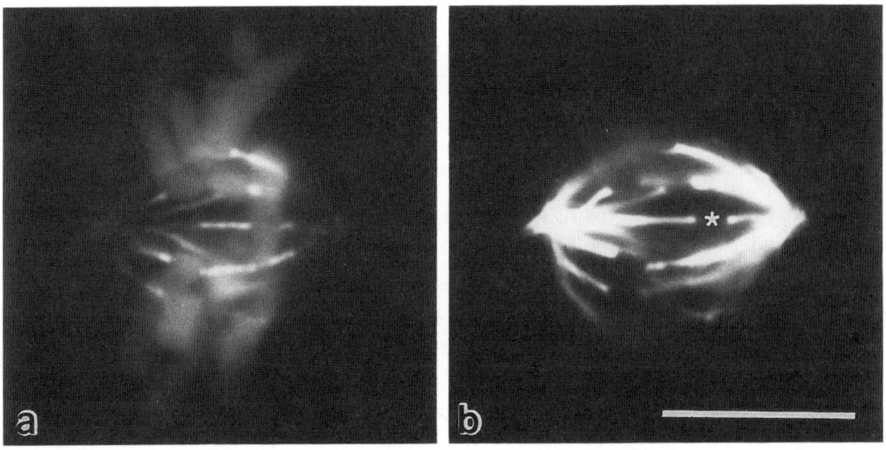

Fig. 3. *Incorporation of biotin–tubulin subunits into the proximal ends of kinetochore microtubules. A prometaphase PtK2 cell was injected with biotin–tubulin 5 minutes prior to fixation at 30°C. Biotin–tubulin incorporation and bleed-through chromosome fluorescence are shown in* **a**. *Total tubulin is shown in* **b**. *Microtubule attachment for two sister chromatids can be seen in b and is marked (**). *The incorporation of biotin–tubulin reflects the dynamic movement of the chromosomes superimposed on kinetochore microtubule flux. Bar = 10 μm.*

ing telophase [Shelden and Wadsworth, 1990]. The timing and extent of tubulin subunit incorporation into the mitotic spindle involves not just the dynamics of individual microtubules but also the reorganization of the entire mitotic spindle. For example, tubulin incorporation in the spindle midzone is intimately associated with spindle elongation (anaphase B) [Masuda and Cande, 1988; Saxton and McIntosh, 1987; Shelden and Wadsworth, 1990], and tubulin incorporation in the kinetochore fiber is obstructed, presumably, by the encroaching kinetochore during anaphase [Gorbsky and Borisy, 1989; Mitchison, 1989; Wadsworth et al., 1989].

VI. THE FUNCTION OF DYNAMIC INSTABILITY *IN VIVO*

Paradoxically, many of the functions attributed to microtubules require microtubules as rigid rails to support the directional movement of cell components, the maintenance of cell shape, and as an integral component of cilia and flagella. Superficially, it would seem that a rigid, less dynamic polymer would be as serviceable as a dynamic one. However, maintaining dynamic microtubules is essential for the proper functioning of certain microtubule-dependent processes.

Low levels of microtubule drugs have been used to subtly alter the balance between microtubule polymer and monomer (see Wilson, this volume). The mitotic spindle appears to be the most sensitive to alterations in microtubule dynamics and therefore to have the greatest requirement for dynamic microtubules. Compounds such as colchicine, colcemid, nocodazole and the vinca alkaloids shift the dynamic equilibrium of microtubules toward depolymerization, in the case of colchicine by forming a tubulin monomer:colchicine complex that binds to the end of the microtubule, inhibiting further polymerization [Margolis and Wilson, 1977; Sternlicht and Ringel, 1979; Jordan and Wilson, 1990]. These drugs resulted in aberrant metaphase chromosome orientation even when the mitotic spindle morphology was normal [Jordan et al., 1992]. Hence, these antimitotic drugs can block mitosis just by inhibiting the dynamics of spindle microtubules rather than completely depolymerizing the spindle microtubules or disrupting the spindle morphology. The antimitotic drug taxol has the opposite effect from the previously described antimitotic drugs, that is, it promotes the assembly of tubulin monomer into polymer [Minotti et al., 1991]. Although mitotic spindles can be taxol stabilized during isolation *in vitro* and spindle elongation subsequently reactivated [Masuda and Cande, 1988], spindle elongation in cells is inhibited by taxol. Despite the fact that bulk polymerization is promoted by taxol, the individual microtubules become equalized in length resulting in a loss of microtubule–microtubule overlap in the spindle midzone. The asters become large but lose their elliptical or asymmetric structure [Amin-Hanjani and Wadsworth, 1991; Hamaguchi et al., 1987].

Genetic lesions may alter microtubule dynamics in yeast [Stearns et al., 1990] and in tissue culture cells [Cabral et al., 1986]. These mutations result in either hypersensitivity (often rescued by taxol) or resistance (often taxol hypersensitive) to microtubule depolymerizing drugs. In *Saccharomyces cerevisiae,* most of these mutations map to tubulin genes. However, three other genes have been identified (CIN1, CIN2, CIN4), mutations that can suppress tubulin mutations and also can result in increased chromosome loss.

Therefore it appears that dynamic microtubules are essential to mitotic spindle function. It is likely that the dynamic nature of mitotic microtubules is important in allowing the rapid restructuring of microtubule arrays in response to cytosolic signals. Furthermore, Ault et al. [1991] have proposed that dynamic microtubules actually position chromosomes within the mitotic spindle during prometaphase. Both assembly-promoting and assembly-inhibiting drugs stopped oscillatory chromosome movement. The final position of static chromosomes, however, depends on the microtubule length distribution imposed by the specific drug used.

The effect of drugs that alter microtubule dynamics is more subtle in interphase cells. Generally, a loss of labile subpopulations of microtubules seen in low concentrations of microtubule depolymerizing drugs while stable microtubule arrays may remain unaffected [Binet et al., 1990]. Loss of subpopulations of labile microtubules results in the inhibition of some microtubule-dependent cell functions but not all [Oliver and Specian, 1991]. Hence it is likely that there are some microtubule dependent processes in which stable, as opposed to dynamic, microtubules are preferred. These processes may involve the establishment of fixed shapes in stationary, differentiated cells and stable polarized arrays in epithelial sheets.

There appears to be a correlation between motility and dynamic microtubules. As previously discussed, more motile cells such as fibroblasts tend to have more dynamic microtubules than less motile epithelial cells. Furthermore, the highly motile tip of the growing axon shows the greatest sensitivity to microtubule drugs [Bamburg et al., 1986]. Therefore dynamic microtubules may be required in order to accommodate the alterations in cell shape involved when cells probe the environment during migration. Preferential stabilization of subsets of microtubules may occur in order to promote (or as a result of) stable changes in cell shape such as occurs during axonal outgrowth [Kirschner and Mitchison, 1986]. It may be that one function of microtubule dynamics is passive space filling of cells undergoing rapid cell shape changes. Microtubules are capable of rapidly polymerizing or depolymerizing to accommodate these alterations. Microtubules are somewhat rigid, and they are capable of deforming membranes upon polymerization [Hotani and Miyamoto, 1990]. The tendency of microtubules to depolymerize spontaneously and rapidly keeps them from opposing membrane retractions until such time as the cell makes a decision to maintain a particular asymmetry. This "decision" may involve stabilizing microtubules by capping, by lateral interactions with membrane components, or by microtubule bundling.

VII. CONCLUSION

Microtubules *in vitro* exhibit dynamic instability in which individual microtubules exist in two states: 1) polymerization and 2) rapid depolymerization; and the transitions between these states, 3) catastrophe and 4) rescue. Dynamic instability behavior has been observed for individual microtubules both interphase and mitotic cells. Evidence suggests that the dynamics of individual microtubules is modulated by putative factors that alter the frequencies of catastrophe and rescue, resulting in, for example, higher levels of microtubule turnover in mitotic cells versus interphase. Different levels of dynamic activity have been described both in subpopulations of microtubules in close proximity and in different regions of the same microtubule. These two situations represent additional circumstances in which cytosolic factors may be involved in modulating microtubule dynamics by, in these cases, stabilizing microtubules. In the mitotic spindle the astral and interzone microtubules are dynamic, whereas the kinetochore microtubules are capped and hence are less dynamic. New polymer incorporates at the kinetochore and fluxes toward the spindle pole. Mutations or drugs that alter the dynamic properties of microtubules can severely inhibit microtubule-dependent processes; however, the exact mechanism by which this occurs is unknown. This is because the functional contribution that microtubule dynamics makes to microtubule-dependent processes is unclear. One can only state that microtubule-ssociated proteins, microtubule motor proteins, intracellular signalling, and the dynamic behavior of individual microtubules appear to operate in concert in order to perform microtubule-dependent cell functions.

VIII. REFERENCES

Allen C, Borisy GG (1974): Structural polarity and directional growth of microtubules of *Chlamydomonas* flagella. J Mol Biol 72:642–654.

Amin-Hanjani S, Wadsworth P (1991): Inhibition of spindle elongation by taxol. Cell Motil Cytoskeleton 20:136–144.

Amos LA, Baker TS (1979): The three-dimensional structure of tubulin protofilaments. Nature 279: 607–612.

Ault JG, DeMarco AJ, Salmon ED, Rieder CL (1991): Studies on the ejection properties of asters: Astral microtubule turnover influences the oscillatory behavior and positioning of mono-oriented chromosomes. J Cell Sci 99:701–710.

Avila J (1990): Microtubule dynamics. FASEB J 4: 3284–3290.

Avila J (1992): Microtubule functions. Life Sci 50:327–334.

Baas PW, Black MM (1990): Individual microtubules in the axon consist of domains that differ in both composition and stability. J Cell Biol 111:495–509.

Baas PW, Slaughter T, Brown A, Black MM (1991): Microtubule dynamics in axons and dendrites. J Neurosci Res 30:134–153.

Baas PW, Deitch JS, Black MM, Banker GA (1988): Polarity orientation of microtubules in hippocampal neurons: Uniformity in the axon and nonuniformity in the dendrite. Proc Natl Acad Sci USA 85:8335–8339.

Baas PW, Black MM, Banker GA (1989): Changes in microtubule polarity orientation during the development of hippocampal neurons in culture. J Cell Biol 109: 3085–3094.

Bailly E, Celati C, Bornens M (1991): The cortical actomyosin system of cytochalasin D-treated lymphoblasts. Exp Cell Res 196:287–293.

Bamburg JR, Bray D, Chapman K (1986): Assembly of microtubules at the tip of growing axons. Nature 321:788–790.

Belmont LD, Hyman AA, Sawin KE, Mitchison TJ (1990): Real-time visualization of cell cycle-dependent changes in microtubule dynamics in cytoplasmic extracts. Cell 62:579–589.

Binet S, Chaineau E, Fellous A, Lataste H, Krikorian A, Couzinier JP, Meininger V (1990): Immunofluorescence study of the action of navelbine, vincristine, and vinblastine on mitotic and axonal microtubules. Int J Cancer 46:262–266.

Bloom GS, Luca FC, Vallee RB (1984): Widespread cellular distribution of MAP-1A (microtubule-associated protein 1A) in the mitotic spindle and on interphase microtubules. J Cell Biol 98:331–340.

Burgoyne RD, Cumming R (1984): Ontogeny of microtubule-associated protein 2 in rat cerebellum: Differential expression of the doublet polypeptides. Neuroscience 11:156–167.

Cabral FR, Brady RC, Shibler MJ (1986): A mechanism of cellular resistance to drugs that interfere with microtubule assembly. Ann NY Acad Sci 466:745–756.

Cassimeris LU, Walker RA, Pryer NK, Salmon ED (1987): Dynamic instability of microtubules. Bioessays 7: 149–154.

Cassimeris L, Pryer NK, Salmon ED (1988): Real-time observations of microtubule dynamic instability in living cells. J Cell Biol 107:2223–2231.

Cleveland DW, Hwo SY, Kirschner MW (1977): Purification of tau, a microtubule-associated protein that induces assembly of microtubules from purified tubulin. J Mol Biol 116:207–225.

Crews CM, Alessandrini A, Erickson RL (1992): Erks: their fifteen minutes has arrived. Cell Growth Differ. 3:135–142.

Cross D, Dominguez J, Maccioni RB, Avila J (1991): MAP-1 and MAP-2 binding sites at the C-terminus of beta-tubulin. Studies with synthetic tubulin peptides. Biochemistry 30:4362–4366.

Dentler WL, Adams C (1992): Flagellar microtubule dynamics in *Chlamydomonas:* Cytochalasin D induces periods of microtubule shortening and elongation. And colchicine induces disassembly of the distal, but not proximal, half of the flagellum. J Cell Biol 117: 1289–1298.

Do CV, Sears EB, Gilbert SP, Sloboda RD (1988): Vesikin, a vesicle associated ATPase from squid axoplasm and optic lobe, has characteristics in common with vertebrate brain MAP 1 and MAP 2. Cell Motil Cytoskeleton 10:246–254.

Edde B, Rossier J, Le Caer JP, Desbruyeres E, Gros F, Denoulet P (1990): Posttranslational glutamylation of alpha-tubulin. Science 247:83–85.

Euteneuer U, McIntosh JR (1981): Polarity of some motility-related microtubules. Proc Natl Acad Sci USA 78:372–376.

Faruki S, Doree M, Karsenti E (1992): cdc2 kinase-induced destabilization of MAP2-coated microtubules in *Xenopus* egg extracts. J Cell Sci 101:69–78.

Garner CC, Brugg B, Matus A (1988): A 70-kilodalton microtubule-associated protein (MAP2c), related to MAP2. J Neurochem 50:609–615.

Garner CC, Garner A, Huber G, Kozak C, Matus A (1990): Molecular cloning of microtubule-associated protein 1 (MAP1A) and microtubule-associated protein 5 (MAP1B): Identification of distinct genes and their differential expression in developing brain. J Neurochem 55:146–154.

Gelfand VI, Bershadsky AD (1991): Microtubule dynamics: Mechanism, regulation, and function. Annu Rev Cell Biol 7:93–116.

Geuens G, Hill AM, Levilliers N, Adoutte A, DeBrabander M (1989): Microtubule dynamics investigated by microinjection of *Paramecium* axonemal tubulin: Lack of nucleation but proximal assembly of microtubules at the kinetochore during prometaphase. J Cell Biol 108:939–953.

Gorbsky GJ, Borisy GG (1989): Microtubules in the kinetochore fiber turn over in metaphase but not in anaphase. J Cell Biol 106:653–662.

Gotoh Y, Nishida E, Matsuda S, Shiina N, Kosako H, Shiokawa K, Akiyama T, Ohta K, Sakai H (1991): *In vitro* effects on microtubule dynamics of purified *Xenopus* M phase-activated MAP kinase. Nature 349: 251–254.

Greer K, Rosenbaum JL (1989): Post-translational modifications of tubulin. In Warner FD, McIntosh JR (eds): Cell Movement, vol 2. New York: Alan R. Liss, pp 47–66.

Gundersen GG, Khawaja S, Bulinski JC (1987): Post-polymerization detyrosination of alpha-tubulin: A mechanism for subcellular differentiation of microtubules. J Cell Biol 105:251–264.

Hammarback JA, Obar RA, Hughes SM, Vallee RB (1991): MAP1B is encoded as a polyprotein that is processed to form a complex N-terminal microtubule-binding domain. Neuron 7:129–139.

Heidemann SR, McIntosh JR (1980): Visualization of the structural polarity of microtubules. Nature 286:517–519.

Hill T, Carlier MF (1983): Steady-state theory of the interference of GTP hydrolysis in microtubule assembly. Proc Natl Acad Sci USA 80:7234–7238.

Hiller G, Weber K (1978): Radioimmune assay for tubulin: A quantitative comparison of the tubulin content of different established tissue culture cells and tissues. Cell 14:795–804.

Hoshi M, Ohta K, Gotoh Y, Mori A, Murofushi H, Sakai H, Nishida E (1992): Mitogen-activated-protein-kinase-catalyzed phosphorylation of microtubule-associated proteins, microtubule-associated protein 2 and microtubule-associated protein 4, induces an alteration in their function. Eur J Cell Biochem 203:43–52.

Hotani H, Miyamoto H (1990): Dynamic features of microtubules as visualized by dark-field microscopy. Adv Biophys 26:135–156.

Huber G, Alaimo-Beuret D, Matus A (1985): MAP3: characterization of a novel microtubule-associated protein. J Cell Biol 100:496–507.

Idriss H, Stammers DK, Ross CK, Burns RG (1991): The dynamic instability of microtubules is not modulated by alpha-tubulin tyrosinylation. Cell Motil Cytol 20:30–37.

Job D, Pabion M, Margolis RL (1985): Generation of microtubule stability subclasses by microtubule-associated proteins: implications for the microtubule "dynamic instability" model. J Cell Biol 101:1680–1689.

Jordan MA, Wilson L (1990): Kinetic analysis of tubulin exchange at microtubule ends at low vinblastine concentrations. Biochemistry 29:2730–2739.

Jordan MA, Thrower D, Wilson L (1992): Effects of vinblastine, podophyllotoxin and nocodozole on mitotic spindles. Implications for the role of microtubule dynamics in mitosis. J Cell Sci 102:401–416.

Khawaja S, Gundersen GG, Bulinski JC (1988): Enhanced stability of microtubules enriched in detyrosinated tubulin is not a direct function of the detyrosination level. J Cell Biol 93:576–582.

Kirschner M, Mitchison T (1986): Beyond self-assembly: From microtubules to morphogenesis. Cell 45:329–342.

Mandelkow E (1986): X-ray fiber diffraction of cytoskeletal fibers. Methods Enzymol 134:633–657.

Mandelkow EM, Lange G, Jagla A, Spann U, Mandelkow E (1988): Dynamics of the microtubule oscillator: Role of nucleotides and tubulin-MAP interactions. EMBO J 7:357–365.

Mandelkow EM, Mandelkow E (1979): Junctions between microtubule walls. J Mol Biol 129:135–148.

Margolis RL, Wilson L (1977): Addition of colchicine-tubulin complex to microtubule ends: The mechanism of substoichiometric colchicine poisoning. Proc Natl Acad Sci USA 74:3466–3470.

Maruta H, Greer K, Rosenbaum JL (1986): The acetylation of alpha-tubulin and its relationship to the assembly and disassembly of microtubules. J Cell Biol 103:571–579.

Masuda H, Cande WZ (1988): The mechanism of anaphase spindle elongation: Uncoupling of tubulin incorporation and microtubule sliding during *in vitro* spindle reactivation. J Cell Biol 107:623–633.

Matus A (1988): Microtubule-associated proteins: Their potential role in determining neuronal morphology. Annu Rev Neurosci 11:29–44.

McIntosh JR, Hering GE (1991): Spindle fiber action and chromosome movement. Annu Rev Cell Biol 7:403–426.

McIntosh JR, Koonce MP (1989): Mitosis. Science 246:622–628.

McIntosh JR, Cande WZ, Snyder JA (1975): Structure and physiology of the mammalian mitotic spindle. In Inoue S, Stephens RE (eds): Molecules and Cell Movement. New York: Raven Press, pp 31–76.

Minotti AM, Barlow SB, Cabral F (1991): Resistance to anti-mitotic drugs in Chinese hamster ovary cells correlates with changes in the level of polymerized tubulin. J Biol Chem 266:3987–3994.

Mitchison TJ (1988): Microtubule dynamics and kinetochore function in mitosis. Annu Rev Cell Biol 4: 527–549.

Mitchison TJ (1989): Polewards microtubule flux in the mitotic spindle: Evidence from photoactivation of fluorescence. J Cell Biol 109:637–652.

Mitchison TJ, Kirschner MW (1984): Dynamic instability of microtubule growth. Nature 312:237–242.

Mitchison TJ, Sawin KE (1990): Tubulin flux in the mitotic spindle: where does it come from and where is it going? Cell Motil Cytol 16:93–98.

Nicklas RB, Lee GM, Rieder CL, Rupp G (1989): Mechanically cut mitotic spindles: Clean cuts and stable microtubules. J Cell Sci 94:415–423.

Obar RA, Collins CA, Hammarback JA, Shpetner HS, Vallee RB (1990): Molecular cloning of the microtubule-associated mechanochemical enzyme dynamin reveals homology with a new family of GTP-binding proteins. Nature 347:256–261.

Oliver MG, Specian RD (1991): Cytoskeleton of intestinal goblet cells: Role of microtubules in baseline secretion. Am J Physiol 260:G850–G857.

Olmsted JB, Asnes CF, Parysek LM, Lyon HD, Kidder GM (1986): Distribution of MAP-4 in cells and in adult and developing mouse tissues. Ann NY Acad Sci 466:292–305.

Parysek LM, Wolosewick JJ, Olmsted JB (1984): MAP 4: A microtubule-associated protein specific for a subset of tissue microtubules. J Cell Biol 99:2287–2296.

Peng I, Binder LI, Black MM (1985): Cultured neurons contain a variety of microtubule-associated proteins. Brain Res 361:200–211.

Pepperkok R, Bre MH, Davoust J, Kreis TE (1990): Microtubules are stabilized in confluent epithelial cells but not in fibroblasts. J Cell Biol 111:3003–3012.

Piperno G, LeDizet M, Chang X-J (1987): Microtubules containing acetylated alpha-tubulin in mammalian cells in culture. J Cell Biol 104:289–302.

Raffaelli N, Yamauchi PS, Purich DL (1992): Microtubule-associated protein autophosphorylation alters in vitro microtubule dynamic instability. FEBS Lett 296:21–24.

Riederer B, Cohen R, Matus A (1986): MAP5: A novel brain microtubule-associated protein under strong developmental regulation. J Neurocytol 15:763–775.

Sammak PJ, Borisy GG (1987): Direct observation of microtubule dynamics in living cells. Nature 332:724–726.

Sammak PJ, Gorbsky GJ, Borisy GG (1987): Microtubule dynamics in vivo: A test of mechanisms of turnover. J Cell Biol 104:395–405.

Saxton WM, McIntosh JR (1987): Interzone microtubule behavior in late anaphase and telophase spindles. J Cell Biol 105:875–886.

Sato-Yoshitake R, Shiomura Y, Miyasaka H, Hirokawa N (1989): Microtubule-associated protein 1B: Molecular structure, localization, and phosphorylation-dependent expression in developing neurons. Neuron 3:229–238.

Sawin KE, Mitchison TJ (1991): Poleward microtubule flux mitotic spindles assembled *in vitro*. J Cell Biol 112:941–954.

Saxton WM, Stemple DL, Leslie RJ, Salmon ED, McIntosh JR (1984): Tubulin dynamics in cultured mammalian cells. J Cell Biol 99:2175–2186.

Schulze E, Kirschner M (1987): Dynamic and stable populations of microtubules in cells. J Cell Biol 104: 277–288.

Shelden E, Wadsworth P (1990): Interzonal microtubules are dynamic during spindle elongation. J Cell Sci 97:273–281.

Shelden E, Wadsworth P (1992): Observation and quantification of individual microtubule behavior *in vivo:* Microtubule dynamics are cell type specific. J Cell Biol (in press).

Sherwin T, Gull K (1989): Visualization of detyrosination along single microtubules reveals novel mechanisms of assembly during cytoskeletal duplication in *Trypanosomes*. Cell 57:211–221.

Stearns T, Hoyt MA, Botstein D (1990): Yeast mutants sensitive to antimicrotubule drugs that define three genes that affect microtubule function. Genetics 124:251–262.

Sternlicht H, Ringel I (1979): Colchicine inhibition of microtubule assembly via copolymer formation. J Biol Chem 254:10540–10550.

Troutt LL, Burnside B (1988a): Microtubule polarity and distribution in teleost photoreceptors. J Neurosci 8:2371–2380.

Troutt LL, Burnside B (1988b): The unusual microtubule polarity in teleost retinal pigment epithelial cells. J Cell Biol 107:1461–1464.

Vale RD, Reese TS, Sheetz MP (1985): Identification of a novel force-generating protein, kinesin, involved in microtubule-based motility. Cell 42:39–50.

Vallee RB, Bloom GS, Luca FC (1986): Different structure and distribution of the high molecular weight microtubule associated proteins, MAP1 and MAP2. Ann NY Acad Sci 466:134–144.

Vallee RB, Wall JS, Paschal BM, Shpetner HS (1988): Microtubule-associated protein 1C from brain is a two-headed cytosolic dynein. Nature 332:561–563.

Vera JC, Rivas CI, Maccioni RB (1988): Heat-stable microtubule protein MAP-1 binds to microtubules and induces microtubule assembly. FEBS Lett 232:159–162.

Verde F, Dogterom M, Stelzer E, Karsenti E, Leibler S (1992): Control of microtubule dynamics and length by cyclin A- and cyclin B-dependent kinases in *Xenopus* egg extracts. J Cell Biol 118:1097–1108.

Wadsworth P, McGrail M (1990): Interphase microtubule dynamics are cell type-specific. J Cell Sci 95:23–32.

Wadsworth P, Salmon ED (1986): Analysis of the treadmilling model during metaphase of mitosis using fluorescence redistribution after photobleaching. J Cell Biol 102:1032–1038.

Wadsworth P, Shelsen E, Rupp G, Rieder CL (1989): Biotin–tubulin incorporates into kinetochore fiber microtubules during early but not late anaphase. J Cell Biol 109:2257–2265.

Walker RA, Inoue S, Salmon ED (1989): Assymmetric behavior of severed microtubule ends after ultraviolet-microbeam irradiation of individual microtubules *in vitro*. J Cell Biol 108:931–937.

Walker RA, O'Brien ET, Pryer NK, Soboeiro MF, Voter WA, Erickson HP, Salmon ED (1988): Dynamic instability of individual microtubules: Rate constants and transition frequencies. J Cell Biol 107:1437–1448.

Winckler B, Solomon F (1991): A role for microtubule bundles in the morphogenesis of chicken erythrocytes. Proc Natl Acad Sci USA 88:6033–6037.

ABOUT THE AUTHORS

LINDA WORDEMAN is a research associate in the Department of Pharmacology at the University of California in San Francisco. Dr. Wordeman received her B.A. with honors in molecular biology at the University of California, Berkeley, where she studied cell division in the early leech embryo in the laboratory of Dr. Gunther Stent. She received her Ph.D. from the Zoology Department at Berkeley in 1987 for work with Dr. W. Zac Cande on the regulation of mitotic spindle reactivation *in vitro*. Since that time she has been identifying and characterizing microtubule-dependent motor activities associated with mitotic chromosomes in the laboratory of Dr. Tim Mitchison at the University of California, San Francisco. Dr. Wordeman is a member of Phi Beta Kappa, and has held National Science Foundation, Helen Hay Whitney, and Giannini fellowships.

TIMOTHY J. MITCHISON is Assistant Professor in the Department of Pharmacology at the University of California, San Francisco. After receiving his B.A. in biochemistry from Merton College, Oxford (U.K.), in 1980, he pursued graduate studies under Marc Kirschner in the Department of Biochemistry and Biophysics at the University of California, San Francisco, focusing on the isolation and properties of microtubule-organizing centers and the dynamics of microtubule arrays. Upon obtaining his Ph.D. in biochemistry in 1984, he remained in Dr. Kirschner's laboratory as a postdoctoral research fellow, concentrating on microtubule dynamics *in vivo* and on kinetochore structure and function. Dr. Mitchison's research papers have appeared in such journals as *Nature*, the *Journal of Cell Biology*, and *Cell*. His current research is in the role of microtubule dynamics and motors in mitotic chromosome separation. He is also doing research in actin dynamics in cell locomotion.

Microtubules: 303–324

Cell Cycle Regulation of Centrosome Function

Frédéric Tournier and Michel Bornens

I. INTRODUCTION

Living cells are able to divide through a timed series of events, the order of which is critical for cell survival. Our understanding of how the cell cycle progression is controlled in eukaryotic cells has increased a great deal in recent years with the unraveling of the central elements of the mitotic clock, which acts as a general control in the induction of mitosis. A most striking feature of the biochemical M-phase control mechanism is certainly its universal character and the fact, established by genetic analysis in both budding and fission yeasts, that the same CDC28/cdc2 gene function is required twice during the cell cycle, at two critical phases, just before the G_1/S transition at the so-called START point and later at the G_2 to M-phase transition. The role of $p34^{cdc2}$ for the initiation of mitosis is much better understood than its biochemical function at START [for reviews, see Maller, 1990; Nurse, 1990]. One main reason for that difference comes from the existence of elementary segmentation cycles during the early development of amphibian and marine eggs, which have long been model systems for a biochemical approach to the cell cycle. In these systems, the division machinery is uncoupled from growth requirements, which have been met during oogenesis, and this makes the cell cycle autonomous and perfectly timed upon fertilization or activation. These systems have been quite important for understanding the role of $p34^{cdc2}$ and the regulation of its activity by mitotic cyclins [Kirschner, 1992].

Untangling the cell division cycle in somatic cells will involve the understanding of the coupling between cell growth and cell division. Cell cycle decisions must integrate cell growth progression. Growth rate must therefore be monitored by appropriate mechanisms. This is likely to be achieved by numerous interconnected feedback loops and checkpoint controls of several types, which have been most clearly defined by Hartwell and Weinert [1989].

The centrosome–microtubule system is the major structural basis of the mitotic spindle, and its involvement in the mechanics of chromosome segregation is one of the most anciently recognized features of the cell division cycle. As often emphasized by Mazia [1978], the essential bipolarity of the division process rests on the centrosome duplication cycle. There are further reasons to believe that, beyond this participation in the mitotic spindle, the centrosome–microtubule system could participate in the coupling between cell growth and cell division. The major reasons are threefold:

1. There is growing appreciation of the role of microtubules in numerous aspects of cell activity. This role can be observed at the level of individual microtubules and at that of the entire centrosome–microtubule system. a) Microtubules determine the cellular distribution and to some extent the dynamics of most membrane compartments [see Bornens, 1991]. Therefore assembly of new membrane compartments during cell growth would be expected to require

more microtubules. b) At the cell level, the centrosome acting as a unique nucleating center tightly associated with the nucleus provides the microtubular network with an asymmetrical organization and a unique polarity, the individual microtubules being anchored to the centrosome by their minus end [Euteneuer and McIntosh, 1981]. This feature, together with the abundance of cytoplasmic motors capable of anterograde or retrograde movements, represents a powerful means for organizing cytoplasmic membranes. Therefore, the behavior and dynamics of microtubules could therefore in some way sum up, or integrate, information on the assembly of new membranes belonging to all cell compartments, since the pool of assembled tubulin should increase in size accordingly. Increase in the number of microtubules should also be reflected by the increase in the nucleating capacity of the centrosome. This nucleating capacity could be in itself a way, for example, to obtain a global estimation of the tubulin polymer within the cell. All these features make microtubules good candidates for participating in the monitoring of mass increase.

2. The duplication cycle of the centrosome encompasses all four phases of the cell cycle. In most cellular systems, the major events that characterize centrosome duplication anticipate the major cell cycle decisions: Initiation of centriole budding in animal cells, or of spindle pole body duplication in yeasts, precedes the initiation of S-phase; separation of the duplicated centrosomes, or spindle pole bodies, precedes or marks the onset of mitosis. This temporal sequence would suggest that controls on the cell division cycle could have entry points connected with the progression of the centrosome duplication cycle. Furthermore, the doubling of centrosomes represents a discontinuous process that could be used as a counting mechanism in monitoring time and mass increase (by titering some cytoplasmic factor, for example). It parallels the doubling of chromosomes and accordingly could be involved in the maintenance of the correlation between cell mass and ploidy. In conclusion, the centrosome–microtubule system has properties well designed to monitor time and mass increase.

3. Cellular reproduction not only entails replication of genetic material and segregation of duplicated sets of chromosomes but also reproduction and distribution of all cytoplasmic compartments. It is the whole cell that must divide: nucleus and cytoplasm. Cell strategies to reproduce and divide the different cytoplasmic compartments are not yet fully documented and might vary from one compartment to the other. For most of the cytoplasmic compartments, the accuracy or fidelity of the distribution might not be as critical as it is for genetic material but most probably must be high over the long term. The centrosome–microtubule system is apparently the only cytoplasmic compartment that deserves a precise duplication mechanism in the cytoplasm. This is indeed necessary for the establishment of a functional bipolar spindle to segregate chromosomes but could also control the distribution of all the other cytoplasmic compartments. This would appear as a logical consequence of the first point discussed above and would seem more economic than a specific duplication mechanism for each individual compartment, which would further require coordination with the others. This was actually the prevalent view during the early days of cell biology when Van Beneden and Boveri saw the centrosome as "the division organ or "the division center," capable of coordinating cytoplasmic and nuclear division [Wilson, 1925]. Recent experimental data do not conflict with this view: Terasaki and Jaffe [1991], using a living dye, observed the redistribution of the endoplasmic reticulum of sea urchin eggs after fertilization. Results, which are cartooned in Figure 1, show that the cell-wide reticulum of the egg is first concentrated by the aster of microtubules on the sperm centrosome while the latter is moving toward the female pronucleus. It is then separated into two domains along with the two separating asters born of centrosome duplication, when they are on their way to establish the mitotic spindle. In this particular example the reticulum is distributed, through numerous interactions with microtubules, into two probably quite equivalent parts *before* chromosome

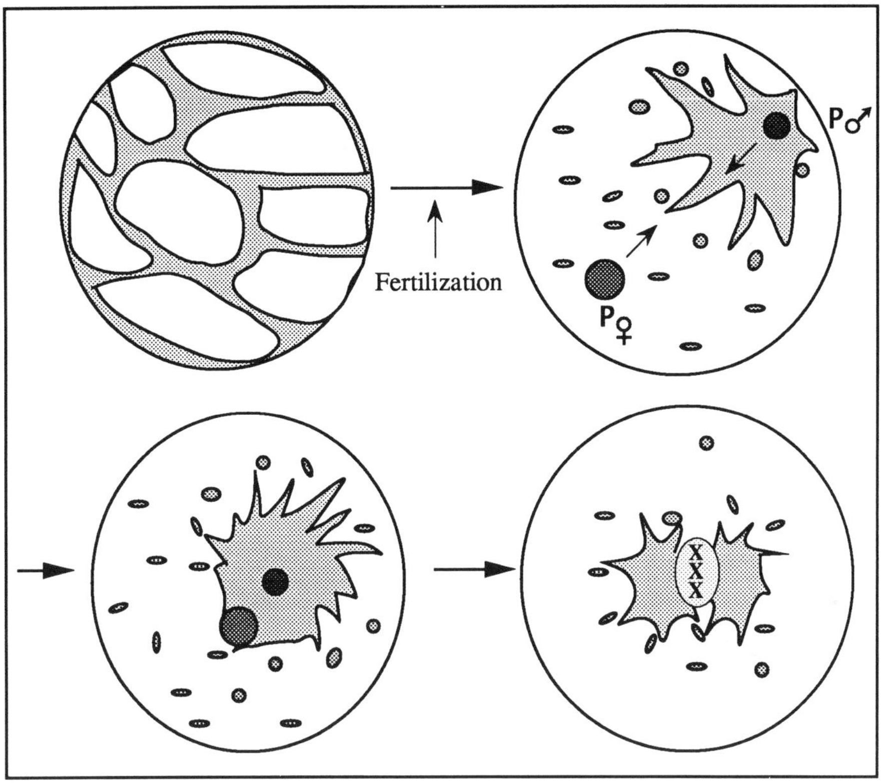

Fig. 1. *Reorganization of the cell-wide endoplasmic reticulum of the sea urchin egg during fertilization and pronuclear migration. [Scheme adapted from the report of Terasaki and Jaffe, 1991.]*

separation is initiated at the onset of anaphase. If this is the general rule for all membrane cytoplasmic compartments, the prophase to metaphase part of mitosis would ensure a general bipartition of cytoplasmic compartments in the vicinity of the spindle poles where they would sit from metaphase to anaphase, when asters are shortening, before interacting again with lengthening microtubules from telophasic asters to be redistributed in each future daughter cell.

Paradoxically, an argument in support of this view could be found in asymmetrical divisions during development. Asymmetrical division is one of the few basic mechanisms that allow sister cells to differentiate from each other [Horvitz and Herskovitz, 1992]. It is controlled by the position of the mitotic spindle within the cell. In some species, like *Caenorhabditis elegans,* there exists in addition a rotation by 90° of the mitotic spindle before each mitosis in one of the two original cell lineages [Hyman, 1989]. This is achieved through a physical interaction between the centrosome and the cell cortex. Therefore one could see from such examples the adaptative value of using the centrosome–microtubule duplication–separation cycle to segregate cellular organelles: Not only does it respect a principle of parsimony but it also allows flexibility in controlling the nuclear to cytoplasmic ratio during development. Polar bodies are indeed extreme examples of this principle.

To grow and to divide, a cell must face spatial and topological constraints. The centrosome–microtubule system seems the most obvious candidate for resolving them. In both the interphasic and the mitotic cell, the microtubular network would act as a determinant for the organization of the cytoplasm and for the fidelity of cell division, the separation of duplicated centrosomes defining the plane of division. What about plant cells that have microtubules but no centrosome? The cell wall is one of the major structural and topological features of these cells and the overall system cell wall plus microtubules seem to be the answer. The assembly of the marginal band of microtubules appears as the major element for anticipating the plane of division [Flanders et al., 1990].

If one accepts the general view developed above as a working hypothesis, one has to address two main questions: What are the mechanisms underlying centrosome duplication or assembly? How is the centrosome duplication cycle controlled and coordinated with the general cell cycle control? We review below the available data on these questions from our and others' laboratories.

II. CENTRIOLE ASSEMBLY AND CENTROSOME DUPLICATION

The continuity of the centrosome has been a conundrum in cell biology since its discovery more than a century ago. A most general feature is the semiconservative process of centrosome duplication that holds for all kinds of centrosomes, from the simple spindle pole bodies in fungi to the complex centriole-containing centrosomes in animal cells. A further twist, however, apparently linked to the capacity of the centriole structure to act as a template for the assembly of "9 + 2" cilia or flagella in animal and sometimes in vegetal cells, has been the numerous examples of centriole basal body assembly by number rather than by the two to four usual mode. The principles underlying these alternative modes of assembly are no better known than are those underlying the semiconservative mode. There are some recent indications that they could, however, have features in common (see below).

The great range of differences in structure and complexity of microtubule nucleating centers among evolutionary divergent eukaryotes does not result apparently from an evolutionary change from the simple to the complex but rather from the other way around: Recent data on the molecular phylogeny of large ribosomal RNAs indicate that the centriolar structure and the associated "9 + 2" flagellum are very ancient inventions, present at the apparition of the early eukaryotic cells [Baroin et al., 1988]. The spinde pole bodies (SPBs) of yeasts [Winey and Byers, 1992], the nucleus-associated bodies (NABs) of social amoebae (Roos, 1975; Omura and Fukui, 1985], and the lack of recognizable centrosome in the evolutionary recent plants are derived features.

There are two types of experimental approaches that have been developed to address the question of centrosome duplication. One approach uses genetic analysis, which is particularly efficient in unicellular systems in which centrosome functions are accessible by mutation. Such is the case in yeasts, in which the SPB is also more simple and amenable to cytological description at the ultrastructural level in a rather unambiguous way. This approach has been recently reviewed in detail [Winey and Byers, 1992; Kimble and Kuriyama, 1992].

The structural complexity of the centrosome in animal cells is considerable [Bornens et al., 1987; Paintrand et al., 1992] (see Fig. 2) and is not amenable to genetic analysis, except perhaps during early development in *Drosophila*, in which mitotic mutants have been isolated [for review, see Glover, 1992]. The semiconservative mode of reproduction, a daughter centriole forming and segregating with its mother, has been recently reinvestigated by an original approach and satisfactorily demonstrated [Kochanski and Borisy, 1990]. The reproductive capacity of centrosomes clearly depends on the number of centrioles, as demonstrated in marine eggs by Sluder and Rieder [1985]. An alternative approach to decipher the mechanism of centrosome duplication has been to use the *Xenopus* egg in which isolated centrosomes are powerful and exclusive parthenogenetic agents. This is the approach we have taken in our laboratory and that is reviewed here.

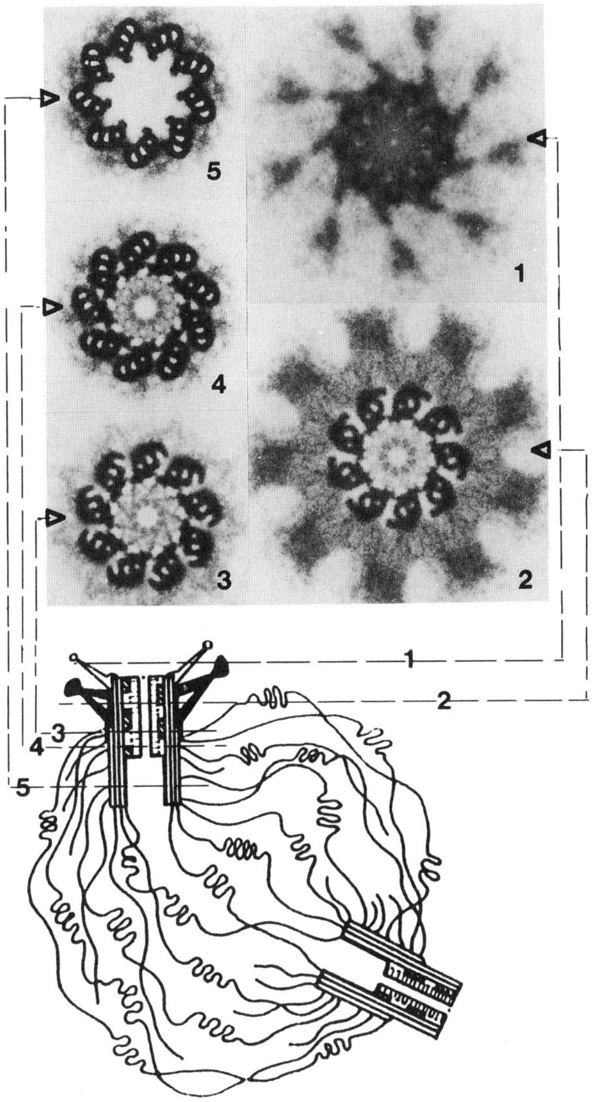

Fig. 2. *Transverse sectioning (shown here as the average images of an ideal 40° angular sector, recalculated from the complete sections; × 90,000) through the mother centriole from a centrosome isolated from a human lymphoblast (KE37 cell line). A rough estimate of the level of each section (numbered 1 to 5 from proximal to distal ends) is indicated on the drawing of the centrosome (bottom). A proximodistal polarity is obvious and includes the typical nine triplets of microtubules in the proximal part with a tangential A–C link between adjacent triplets (levels 4 and 5) and nine unlinked doublets in the distal part (levels 2 and 3) and a set of nine distal appendages (section 1) and of nine subdistal appendages (section 2) that are not observed on the daughter centriole. Therefore, the polarity of the centrosome is manifest within the whole centrosome organization, which possesses the mother–daughter polarity, and within the centriole architecture, which possesses the proximodistal polarity.*

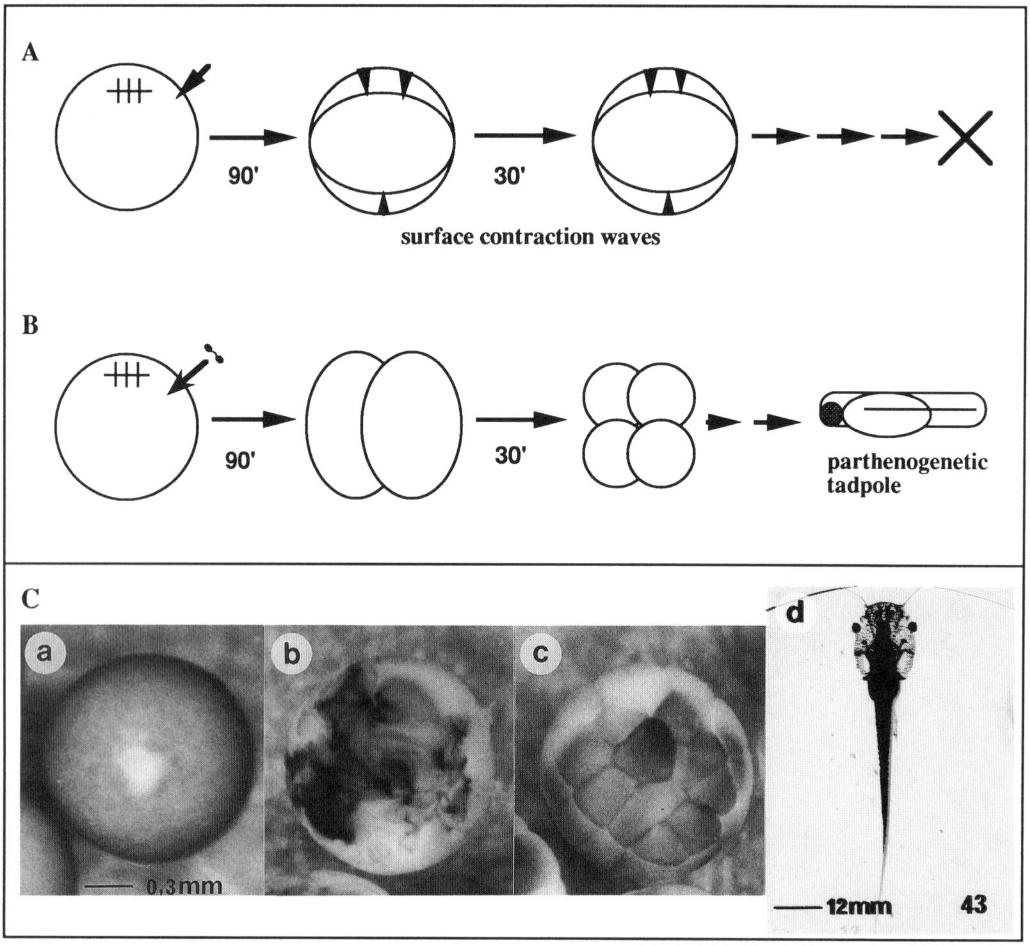

Fig. 3. *The parthenogenetic assay in* Xenopus laevis. **A:** *Pricking the* Xenopus *egg triggers cell cycle events and activates the mitotic clock.* **B:** *Injection of an heterologous centrosome at the time of pricking leads to cellularisation and parthenogenetic development.* **C:** **a,** *mature oocyte;* **b,** *oocyte 4 hours after pricking (note the absence of cleavage and the reorganization of the cortical pigment);* **c,** *a blastula 4 hours after the injection of a heterologous centrosome in a mature oocyte;* **d,** *parthenogenetic tadpole 43 days after the injection of a heterologous centrosome in a mature oocyte.*

A. Parthenogenetic Development in *Xenopus* Requires an Exogenous Centrosome

In marine eggs, from sea urchins or surf clams, for example, centrosome assembly can be induced by various treatments [Kuriyama and Borisy, 1983]. KCl-activated oocytes from surf clam have been recently shown to allow, within a time as short as 4 minutes, the assembly of individual centrioles that are already duplicated by 15 minutes [Palazzo et al., 1992].

In contrast, frog eggs are unable to assemble a centrosome from its elements in the absence of a preexisting centrosome. Isolated centrosomes are exclusive parthenogenetic agents in amphibian eggs [Maller et al., 1976], which makes this system favorable to study the initiation of centrosome duplication (Fig. 3).

B. Centrosomes From Divergent Species Are Active

The most striking feature of the centrosome-induced parthenogenesis in *Xenopus* is that it does not require species specificity. Heterologous centrosomes from sea urchin [Maller et al., 1976], mouse (Karsenti et al., 1984], man [Bornens et al, 1987; Klotz et al., 1990] and several other mammal species including rat and calf [Tournier et al., 1991b] have been proven active in triggering early development. Evolutionary conservation of centrosome function can be demonstrated in experimental systems in which heterologous centrosomes are able to promote the assembly of microtubules from the endogenous tubulin. But in order to induce parthenogenesis, the injected centrosome must be able not only to assemble microtubules from the egg tubulin but also to duplicate and to interact properly with the nuclear compartment of the egg while participating in the assembly of a functional mitotic spindle. In other words, centrioles from a foreign centrosome must be able to direct the assembly of procentrioles from endogenous components within the oocyte, leading to two chimeric centrosomes that then will trigger the first cleavage (see below). This interspecific assembly strongly argues for a high conservation of essential components of centrosomes or centrioles, which would parallel the exceptional stability of the centriolar edifice during evolution.

C. Centrosome Integrity Is Required for Its Activity

As an approach to identify the active components in the heterologous centrosome capable of complementing the egg, centrosomes were dissociated in several ways [Klotz et al., 1990]. All attempts to obtain the activity in a soluble form have failed.

Centrosomes isolated from human KE37 cells and sequentially extracted with salts, urea, or detergents,as well as after treatment with nucleases or proteases, were monitored by protein analysis, Western blotting with specific markers for centrioles and pericentriolar material, and transmission electron microscopy. Structural modifications were correlated with the loss of microtubule nucleating activity *in vitro* and with the loss of parthenogenetic activity after injection into eggs. Neither the parthenogenetic activity nor the nucleating activity were modified by nuclease treatments alone or used together with a dissociating agent. These latter experiments were designed to address the recurrent idea that a basal body/centriolar nucleic acid would govern the mode of inheritance of the centrosome [see Hall et al., 1989; Johnson and Rosenbaum, 1990; Kuroiwa et al., 1990; Johnson and Dutcher, 1991]. Clearly the parthenogenetic activity does not require the presence of nucleic acids in the injected centrosomes. This result does not eliminate the possibility that centrosomes are complemented by some nucleic acid in the egg. The possibility that renaturation or complementation of the centrosome could take place in the egg cytoplasm was supported by the observation that the parthenogenetic activity was preserved under conditions that abolished the microtubule nucleating activity *in vitro*. The recovery of the nucleating activity of urea-treated centrosomes when incubated in the egg cytoplasm has been further demonstrated by Buendia et al. [1992] and shown even to reach a higher level than that of centrosomes before treatment with urea. However, a role of a nucleic acid in this complementation appears unlikely. By contrast, both the nucleating activity and the parthenogenetic activity were very sensitive to proteases.

The parthenogenetic activity was therefore associated with an insoluble proteinacious structure that was not significantly simpler than the native centrosome. A correlation between the loss of the parthenogenetic activity and the disorganization of the centriolar triplets was observed during urea treatment. This is not sufficient to prove that centrioles are the parthenogenetic agent, however, as we were unable to prepare centrioles free of the surrounding material. The orthogonal budding of a procentriole from the wall of a parental centriole is a rather uncanny process and does not suggest obvious mechanisms, contrary to the distal growth of the axoneme on the centriole/basal body. Is the centriolar architecture required in the injected centrosome to support the assembly of a procentriole in the egg? Other results indicate that the ninefold radially symmetrical structural motif of the centriole, if necessary, is not sufficient. These results are, however, compatible with the general idea of the centrosome acting as a structural template in the initiation of centrosome assembly.

Some initial results concern the discovery that centrosomes isolated from calf thymocytes display an inactive configuration [Tournier et al., 1991b] (see below). A second set of results includes our recent observation that the lack of species barrier in the parthenogenetic assay is not absolute. Although divergent invertebrate species such as sea urchins [Maller et al., 1976] and probably others [Rostand, 1968] do contain centrosomes that can complement *Xenopus* eggs, insects do not: Centrosomes isolated from *Drosophila* syncitial embryos were shown to be quite active for nucleating microtubules from the egg tubulin but unable to trigger parthenogenetic development (Tournier, Santamaria, and Bornens, unpublished observations). Centrosomes from *Drosophila* contain centrioles that display the typical nine-fold radial symmetry but that are very short, looking more like a disc than a cylinder [Szöllözi et al., 1986]. One possibility would be that centrioles from *Drosophila* and those from *Xenopus* are too different to be compatible: Orthogonal budding of a procentriole from *Xenopus* components would be precluded in the vicinity of a centriole from *Drosophila*. In contrast, centriole/basal bodies from sea urchin sperm and centrioles from vertebrates are very much alike.

D. The Duplication of Heterologous Centrosomes Can Be Initiated in *Xenopus* Eggs

We addressed the possibility that the heterologous centrosome would have to be at a particular stage in its own duplicative cycle to induce cleavage. It was important to decide if centrosome duplication could really be initiated in response to the egg cytoplasm, or if it only could be terminated, the new centrioles elongating from preformed buds? Centrosomes isolated from human lymphoblastic cells synchronized in G_1 or G_2 and from peripheral human lymphocytes that are arrested in G_0 were shown to possess similar activities [Tournier et al., 1989], indicating that duplication of heterologous centrosomes can be triggered in *Xenopus* eggs.

E. Centrosomes Can Be in an Inactive Configuration

Centrosomes isolated from calf thymocytes were incapable of triggering egg cleavage [Tournier et al., 1991b]. This could not be related simply to the quiescent state of the thymocytes, as centrosomes from quiescent human peripheral lymphocytes were previously shown to be active [Tournier et al., 1989]. On the other hand, centrosomes isolated from other bovine cells or from thymocytes of other species have also been tested and shown to be active. These centrosomes have a peculiar structure in that the two centrioles are tightly associated with each other by their proximal ends, giving a linear configuration to the centrosome [Komesli et al., 1989]. These results are still puzzling, as they do not allow a simple interpretation for the origin of the linear structure of centrosomes in calf thymocytes (see below). However, they allow a simple working hypothesis for the lack of activity of such centrosomes: The initiation of procentriole budding could not occur as long as the two centrioles are closely linked to each other [Tournier et al., 1991b]. Accordingly, evidence has been obtained that they do not duplicate in the eggs [Tournier et al., 1991a] (see below).

F. Centriole Duplication in a Cell-Free System

The duplication of the injected centrosome is supposed to be one of the essential steps in the formation of the first mitotic spindle within the *Xenopus* egg, but was not demonstrated until recently due to the fact that, contrary to ma-

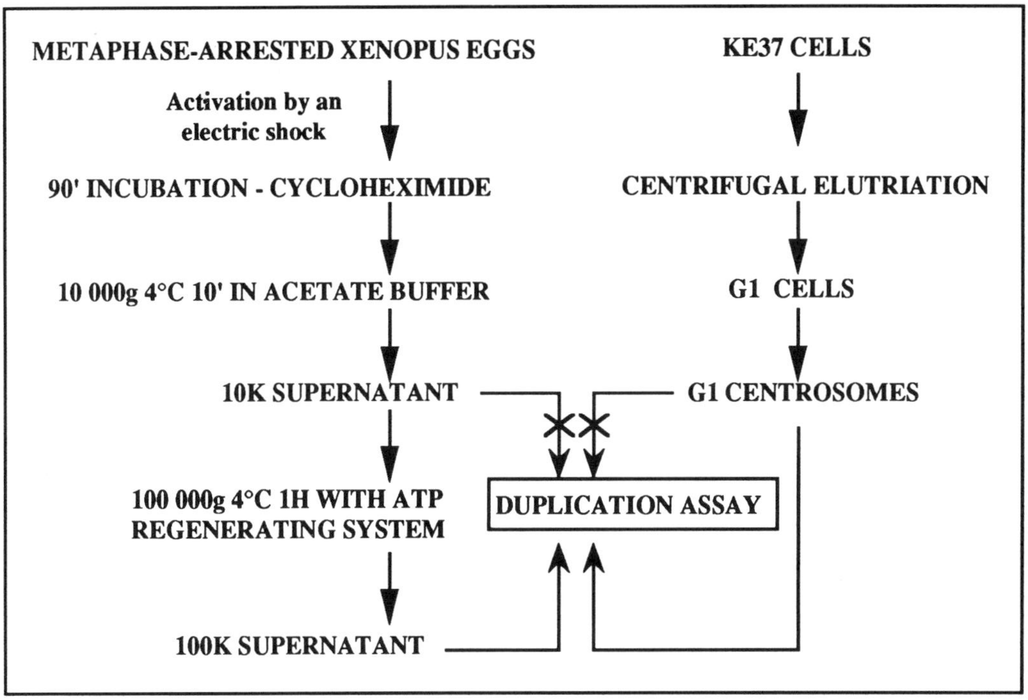

Fig. 4. *Centrosome duplication assay in cell-free extract from* Xenopus *eggs.*

rine eggs, amphibian eggs are not transparent to visible light. An alternative approach was to design an *in vitro* assay for obtaining the initiation of centriole duplication using *Xenopus* egg extracts [Tournier et al., 1991a]. The protocol is shown in Figure 4. This method should allow us to identify the factors controlling the initiation of procentriole budding. We observed, for example, that centrioles did not support procentriole budding when incubated in low-speed extracts (10,000*g*; see Fig. 4). This assay also showed that initiation of procentriole budding can occur in cycloheximide-treated extracts, a result that would be expected in the description of the embryonic cycle as oscillat ing between two stable states (low and high maturation promoting factor [MPF] activity) that suffice to trigger cell cycle events [Hara et al., 1980; Murray and Kirschner, 1989]. We did not succeed, however, in obtaining several rounds of duplication, suggesting either that the duplication process is not completed under our *in vitro* conditions or that centrosome replicates only once when the cell cycle clock is arrested. The latter possibility is the situation for DNA that once replicated fails to reinitiate unless the nuclear envelope is disassembled during mitosis to give access to a putative "licensing factor" [Laskey et al., 1989]. Results obtained *in vivo* in *Xenopus* embryos and in sea urchin eggs dismiss this possibility: Several rounds of centrosome duplication were observed in the absence of any protein synthesis [Gard et al., 1990; Sluder et al., 1990]. The likely explanation to our results is that the components necessary for assembling full-length centrioles are limiting in comparison to the number of centrosomes incubated in the extract. The minimum number required to make the experiment feasible is 500–1,500 centrosomes per microliter of extract. This figure is close to the number of centrosomes produced per embryo (which has a volume of about 1 μl) at midblastula transition [Gard et al., 1990]. From such a figure, we may conclude that a single centrosome duplication at most could take place under our conditions.

This assay, however, allowed us to observe that centrosomes from calf thymocytes, which do not trigger egg cleavage [Tournier et al., 1991a] (see above), were unable to support procentriole budding in extracts in which KE37 centrosomes did form buds. This result strongly suggests, but does not directly demonstrate, that in parthenogenetic *Xenopus* eggs the first cleavage requires the duplication of the injected centrosome. Specific egg proteins must be able to assemble on the two centrioles of the heterologous centrosome that once duplicated will form the two poles of the first mitotic spindle, each pole containing a chimeric centrosome. The linear configuration of the calf thymus centrosomes apparently represents a locked configuration (see below).

An alternative approach to obtain a cell-free system that executes centriole duplication has been recently reported [Palazzo et al., 1992]. In this case, one makes use of marine oocytes in which centrosome assembly can be triggered by various treatments. A mixture of lysates from potassium chloride–activated oocytes from surf clam, which assemble single centrioles within 4 minutes, and from unactivated oocytes, which do not spontaneously assemble centrioles, constitutes a system in which centrioles assemble and duplicate. Heterologous centrosomes have not been tested in this system. Precursor material for centriole assembly has not yet been characterized in unactivated marine oocytes. One possibility is that this material is in some preassembled form that does not exist in *Xenopus* oocytes.

These cell-free systems should open the way toward an understanding of the major features of centriole assembly and centrosome duplication. As a matter of fact, they illustrate the two possible ways of assembling centrioles, the so-called centriolar pathway in the *Xenopus* system, which requires a structural continuity since a preexisting centriole is necessary, and the acentriolar pathway in the artificially activated surf clam oocyte, which can overcome this requirement. In the latter case, as in other marine eggs [Kuriyama and Borisy, 1983], the acentriolar pathway is triggered upon artificial activation for the first round of centriole assembly only. Further centriole assembly is then achieved by duplication according to the centriolar pathway. The artificially induced acentriolar pathway of assembly appears therefore as an alternative to the normal centriolar pathway set up by the sperm centrosome upon fertilization. This suggests that the two pathways have much in common and that they both may help us to understand the mechanisms underlying centriole assembly. Centriole assembly, however, cannot proceed *de novo* in somatic cells as recently demonstrated in a rather definitive manner by Maniotis and Schliwa [1991].

G. Centriolar and Acentriolar Pathways for Centriole Assembly

The presence of centrioles within the centrosome of most animal cells has long been a matter of debate, since it is not a general feature throughout the evolutionary range of eukaryotes [see Bornens and Karsenti, 1984; Bornens, 1992]. By contrast, the participation of basal bodies as templates in the growth of cilia and flagella is well established. There are examples of cell differentiation in metazoa that involve the growth of numerous cilia or flagella and therefore needs as a prerequisite the assembly of numerous centriole/basal bodies [for review, see Dirksen, 1991]. In this case, assembly takes place according to the acentriolar pathway rather than the centriolar pathway: New centriole/basal bodies assemble at once from a material that accumulates in the vicinity of the centrosome–Golgi apparatus. The latter feature suggests that the structural discontinuity may not be absolute. There are, however, examples of differentiation in which simultaneous assembly of numerous centriole/basal bodies occurs with complete structural discontinuity. One case has been particularly well documented in the fern *Marsilea* where hundreds of centriole-basal bodies assemble from an enigmatic blepharoplast during the germination of the microspores [Mizukami and Gall, 1966; Hepler, 1976]. There are also examples of differentiation in unicellular eukaryotes in which structural discontinuity is complete: The protozoan *Naegleria gruberi,* for example, grows as an amoeba that is deprived of centrioles, and, in response to a suitable environmental signal, such as suspension in nonnutrient buffer, differentiates into a flagellate with a fixed body shape and two to four flagella within an hour after changing the environment [Fulton, 1971].

Since the end product in all these pathways of assembly is the highly conserved centriole/basal body structure, one would expect the various pathways to have steps in common and that essential components for controlling the assembly are conserved in all of them. As mentioned above, the switch from one pathway to the other demonstrated by marine eggs upon artificial activation is a further argument for this view. Finally, the fact that the centriolar pathway can proceed across species, as demonstrated in the *Xenopus* system, is the strongest indication of a general conservation of some components essential for centriole structure and/or assembly.

There is indeed growing evidence from immunological, biochemical, and genetic approaches that some centrosomal components could be conserved among divergent species. A prominent protein from the centrosome-like structure in unicellular *Chlamydomonas,* the so-called centrin or caltractin, a 20 kD Ca^{2+}-binding protein, has important sequence identity with both *Chlamydomonas* calmodulin and yeast CDC31 gene product required for SPB duplication in the yeast *Saccharomyces cerevisiae* [Huang et al., 1988a,b]. Antibodies against centrin have demonstrated immunological cross-reactivity with centrosomes in many eukaryotes [Baron and Salisbury, 1988] and with a 62/64 kDa protein in human centrosomes [Moudjou et al., 1991]. A specific tubulin gene, first discovered in *Aspergillus nidulans* [Oakley and Oakley, 1989; Oakley et al., 1990], has been shown to code for a γ-tubulin specifically located at the SPB of *A. nidulans* and now identified in almost all species tested [Zheng et al., 1991; Stearns et al., 1991; Oakley, 1992]. On the other hand, several autoimmune sera and several monoclonal antibodies raised against centrosome structures or mitotic spindles have revealed many different cross-reactions that suggest a general conservation of the centrosome components [for review, see Kimble and Kuriyama, 1992; Fuller et al., 1992] or of components participating in the assembly of centriole/basal bodies. Among a library of monoclonal antibodies raised by immunizing mice with centrosomes isolated from human KE37 lymphoblasts, the monoclonal antibody CTR210 was particularly interesting in this last respect: Besides its strong reaction with human and other mammal centrosomes, CTR210 was found to react with centrosomes, or microtubule organizing areas, in most of very divergent species tested, including the primitive eukaryotic dinoflagellate *Crypthecodinium cohnii* [Perret et al., 1991], the ciliates *Paramecium tetraurelia* [Keryer et al., 1990], and *Paraurostyla weissei* [Fleury et al., 1993]. In several cases, including mammal cells [Perret et al., 1992] and hypotrich ciliates [Fleury et al., 1993], the decoration by CTR210 was associated with the centrosome itself and with structures that were apparently upstream in the process of assembling centrosomes or centriole/basal bodies. This is particularly striking in the hypotrich *Paraurostyla weissei* in which a new ciliature is apparently able to reform after the disappearance of the former one. CTR210 has revealed a conspicuous but transient system of ''scaffolding tracks'' making a continuity between the old and the new ciliature [Fleury et al., 1993]. From such a result, one may infer that centriolar and acentriolar pathways may not be that different, the main divergence being due to the lengthening of an upstream step in the process of assembly. In both cases, there would be a necessary continuity. Interestingly, where identified, CTR210 antigen corresponds to an apparently conserved protein of MW ~70 kD. Independent evidence demonstrates in one of the species studied that the antigen belongs to the HSP70 family [Perret et al., 1992]. A concentration of the constitutive form of HSP70 in the mitotic poles of animal cells has also been reported from an immunofluorescent study [Rattner, 1990]. These results suggest that specific chaperone proteins are associated with the centrosome in eukaryotic cells. Chaperonins are highly conserved proteins required in the folding and transport of proteins as well as in the assembly and disassembly of protein complexes [Ellis, 1987; Rothman, 1989]. Therefore the data gathered in divergent species with CTR210 could open a promising way for rationalizing the still enigmatic process of centrosome assembly.

III. THE CENTROSOME DUPLICATION CYCLE AND CELL CYCLE CONTROL

The important issue here is to identify the mechanism that coordinates the centrosome duplication cycle and the nuclear cycle. Centrosome duplication, like DNA duplication, must happen only once in each cell cycle, and both duplication cycles must be coupled in order to lead to two viable daughter cells. Either the two pathways are intrinsically dependent on each other, some events being common to both pathways, or they are independent but coordinated through extrinsic control mechanisms that ensure a correct temporal order of events [Hartwell and Weinert, 1989]. There are several experimental arguments in favor of the second possibility.

A. The Centrosome Cycle is Independent of DNA Synthesis

In eggs or early embryos whose cells are not growth limited, centrosome duplication can be completely dissociated from the nucleus division cycle. This was demonstrated in sea urchins by Sluder et al. [1986], who removed the nucleus from the eggs at prophase of the first mitosis, and in starfish by Picard et al. [1988], who further investigated whether the centrosome could duplicate in the absence of any material from the oocyte germinal vesicle since this material becomes largely cytoplasmic in the egg. Enucleation was carried out before or after hormonal induction of meiotic maturation and fertilization. In both cases, the sperm centrosome underwent periodic duplications so as to reach the number of centrosomes corresponding to the blastomere number (2^9) at the midblastula transition. This led to uncleaved eggs with 2^9 centrosomes in the first case and to 2^9 blastomeres lacking nuclei in the second one. The latter result was previously obtained in this species by aphidicolin treatment of the eggs to inhibit nuclear replication [Nagano et al., 1981]. Dissociation between centrosomal and nuclear division has also been demonstrated in eggs of *Drosophila* carrying the recessive, maternal-effect mutation *gnu* [Freeman et al., 1986]. Phenotypic copies of the *gnu* mutation have been produced by aphidicolin treatment of the eggs [Raff and Glover, 1988].

In growth-limited cells, DNA synthesis is not required for centrosome duplication [Rattner and Phillips, 1973; Kuriyama and Borisy, 1981a]. There is, however, some requirement for the presence of the nucleus. This could simply reflect a requirement for some gene products at the initiation step. Alternatively, the physical association between centrosome and nucleus is necessary for procentriole formation [Kuriyama and Borisy, 1981a].

B. Centrosome Duplication Is Independent of the Mitotic Clock

Several rounds of centrosome duplication were observed in the absence of any protein synthesis in *Xenopus* embryos and in sea urchin eggs [Gard et al., 1990; Sluder et al., 1990]. This is indeed a strong argument for viewing the centrosome duplication cycle governed independently of the mitotic clock. Another argument could be taken from examples of somatic cells in which there is uncoupling between the mechanisms of centriole assembly and those that control cell cycle progression: Postmitotic cells that are differentiating in ciliated cells (see above) are able to produce massive centriole assembly, although according to the acentriolar pathway rather than the two to four centriolar pathway observed in proliferative cells. These examples suggest that the basic steps of centrosome assembly do not need the context of the cell cycle to be completed, but that the pattern of this assembly may change depending on the proliferative status of the cell. For example, centrosome duplication was shown to be asynchronous in *Xenopus* embryos and in sea urchin eggs, in the absence of cyclin synthesis [Gard et al., 1990; Sluder et al., 1990], suggesting that under normal conditions it is likely kept in phase with the cytoplasmic clock.

There is, however, the possibility that mechanisms extrinsic to both the centrosome duplication pathway and the mitotic kinase activation pathway are responsible for coordinating both pathways in growing cells. This is suggested by the existence of the gene *ESP1* (*extra spindle pole body*) in *S. cerevisiae*, which exerts a negative control on SPB duplication [Baum et al., 1988]: Its gene product would participate in the mechanism that allows the SPB to duplicate only once in each cell cycle. Interestingly, extra SPBs duplicate asynchronously.

There are two important events during the centrosome cycle in somatic cells, which corresponds to major cell cycle decisions: The initiation of centriole duplication during G_1 or at the G_1–S border, after the early separation of the two orthogonal centrioles inherited by each daughter cell [Kuriyama and Borisy, 1981b] and the separation of the duplicated centrosomes at the G_2–M border. The mechanisms that maintain the proper order of events along the centrosome cycle up to its completion are not known, but could resemble those that govern the duplication cycle of the DNA. We know, for example, that the dependence of M-phase on the completion of S-phase is not intrinsic but must be maintained by regulatory mechanisms that can be disrupted by mutation or chemical treatments or that can be absent during early development [Enoch and Nurse, 1991]. The same holds apparently for the centrosome cycle, since separation of centrosomes at the onset of mitosis can be unconnected from centrosome duplication. This has been produced artificially in the classic experiment of Mazia et al. [1960] in which sea urchin eggs were treated with β-mercaptoethanol: spindle poles were apparently normal but had only half of the reproductive capacity. This effect has been fully analyzed by Sluder and Rieder [1985], who showed that the splitting apart of mother and daughter centrioles took place without a previous duplication of each of them. It is also what happens during meiosis when there is no centrosome duplication between meiosis I and II in sea urchin oocytes [Sluder et al., 1989], just like there is no DNA synthesis.

The possibility that the mitotic kinase itself participates in the phasing mechanism, by transiently interrupting the process of centrosome duplication during mitosis, cannot be ruled out, since the centrosome undertakes profound structural modifications at this stage that could be incompatible with the progression of the duplication process. And indeed the duplication cycle overlaps on segments of two successive cell cycles: It ends in the early postmitotic part of the cell cycle next to the one it has initiated (see Fig. 6 below). This would simply ensure that centrosome duplication happens only once during each cell cycle.

C. The Mitotic Kinase Interacts With the Centrosome

Several studies have attempted to localize the components of the mitotic kinase in cells. They most often relied only on an immunofluorescent approach. More rarely were they corroborated by cell fractionation.

An immunocytochemical study of two human cultured cells (HeLa and KE37 cell lines), making use of two different antibodies, has demonstrated that in interphase $p34^{cdc2}$ is distributed in an apparently random manner in the cytoplasm and occupies DNA-excluding areas in an irregular manner in the nucleus [Bailly et al., 1989]. Some of the $p34^{cdc2}$ becomes associated with the centrosome during G_2-phase and at the beginning of M-phase. Cell fractionation corroborates the conclusions from immunochemistry and further demonstrates that there is an enrichment of the slow migrating forms of $p34^{cdc2}$ in isolated centrosomes and other Triton-insoluble fractions at this time. The protein homologue of the $p13^{suc1}$ from

Schizosaccharomyces pombe was found to be distributed in HeLa cells in a manner strikingly similar to that observed for p34^{cdc2}, with one exception. Although the presence of both proteins in the centrosomes of cells late in the cell cycle was obvious, even more so for the p13^{suc1} homologue than for the p34^{cdc2} homologue, the centrosomal staining with the anti-p13^{suc1} antibody was definitely seen throughout the cell cycle. These results have been confirmed by more recent studies [Rattner et al., 1990; Krek and Nigg, 1991]. Divergent results were reported in other mammal cells, indicating either a strict nuclear [Riabowol et al., 1989] or cytoplasmic [Akhurst et al., 1989] localization of p34^{cdc2}. These contradictory results were established by immunofluorescence with a single antibody. p34^{cdc2} has been reported to be cytoplasmic in *Sac. cerevisiae* [Wittenberg et al., 1987] and nuclear in *Sch. pombe* [Booher et al., 1989]. Only cell fractionation would allow assessment of the validity of these claims.

The subcellular distribution of mitotic cyclins in mammal cells has demonstrated striking differences for cyclin A and cyclins B1 or B2 [Pines and Hunter, 1991; Bailly et al., 1992]. Cyclin A accumulation is mainly realized in the nucleus. A minor and very transient association of cyclin A, from prophase to metaphase, to the separating centrosomes has also been detected [Bailly et al., 1992]. Cyclin B1 accumulation in human cells is entirely cytoplasmic, part of it being associated with the centrosome. Shortly before nuclear membrane breakdown, cyclin B1 becomes almost entirely nuclear at the onset of mitosis, during the separation of the duplicated centrosomes, which maintain an association with cyclin B1. In metaphase, cyclin B1 is mainly concentrated in the mitotic spindle area and at the poles before disappearing at the metaphase–anaphase transition. In agreement with the cytological data, cell fractionation indicates that cyclin B1 accumulates mainly in a Triton X100–resistant form, the amount of which is not modified by disrupting microtubules.

Almost identical results have been recently reported for the distribution of cyclin B2 in chicken cells [Gallant and Nigg, 1992]. In *Drosophila,* the distribution of mitotic cyclins between the nucleus and the cytoplasm is roughly similar to the distribution in vertebrates cells [Maldonado-Codina and Glover, 1992]. Starfish oocytes have also been shown to possess sticking similarities with somatic cells [Ookata et al., 1992]: Cyclin B is entirely cytoplasmic in immature oocytes, arrested in G_2, and redistributed into the germinal vesicle after reactivation of the oocyte with 1-methyladenine. In the yeast *Sch. pombe*, in which cyclin A is absent, the yeast homologue of cyclin B would be concentrated into the nucleus [Booher et al., 1989] and at the SPB [Alfa et al., 1990]. If confirmed by cell fractionation, the nuclear localization of most of the *cdc13* product could be linked to the process of endomitosis.

The interaction of B-type cyclins from yeast [Alfa et al., 1990] to *Drosophila* [Debec and Montmory, 1992; Maldonado-Codina and Glover, 1992], chick [Gallant and Nigg, 1992] and man [Bailly et al., 1992] with the centrosome is therefore general. There are two features of this interaction that should be noted in human cells: It is an early event that is observed as soon as the accumulation of cyclin B1 is sufficient to be immunodetected in cells. The centrosomal staining is very small and corresponds to a single spot on each of the two separating centrosomes at the onset of mitosis. To be understood, this aspect needs ultrastructural study.

D. Is the Centrosome a Substrate for Mitotic Kinase?

There are indications that the mitotic kinase may trigger entry into mitosis by direct phosphorylation of abundant structural proteins [for review, see Nigg, 1991] and could also participate in the arrest of membrane traffic [Bailly et al., 1991]. Numerous observations point to the possibility that phosphorylation could regulate centrosome structure and activity during mitosis. Phosphoproteins are known to be associated with the mitotic poles at the onset of mitosis [Vandre et al., 1986; Thibodeau and Vincent, 1991; Sellitto and Kuriyama, 1988;

Kimble and Kuriyama, 1992]. Centrosome-specific immunostaining often reveals an increase in size during G_2, whereas a dramatic decrease takes place at the onset of anaphase [Gosti-Testu et al., 1986; Bailly et al., 1989], which is likely to correspond to structural modifications correlated with a rise and fall in the microtubule nucleating activity [Robbins et al., 1968; Rieder and Borisy, 1982]. All this might be controlled by phosphorylation. The interaction of $p34^{cdc2}$ in G_2–M, when the activity is maximum, suggests that it is directly responsible for the centrosomal modifications that go along with the G_2–M transition.

Use of *Xenopus* cell-free systems suggests also a direct role of $p34^{cdc2}$ in both the regulation of the nucleating activity of centrosomes. Cyclin A $p34^{cdc2}$ and cyclin B $p34^{cdc2}$ would both act on the dynamics of microtubules but with a significant difference in the net shortening effect on microtubule length [Belmont et al., 1990; Verde et al., 1990, 1991, 1992], whereas cyclin A $p34^{cdc2}$ would act in addition on the nucleating capacity of the centrosome [Buendia et al., 1992]. CSF-arrested *Xenopus* egg extracts (mitotic extracts) have been shown to activate the microtubule-nucleating activity of SPBs from the fission yeast *Sch. pombe* that are otherwise unable to nucleate in interphase extracts [Masuda et al., 1992]. Moreover, this ATP-dependent conversion is sensitive to protein kinase inhibitors and to alkaline phosphatase and render the SPBs recognizable by MPM2, a monoclonal antibody reacting with phosphorylated antigens [Vandre et al., 1986]. Whether these effects are produced by $p34^{cdc2}$ itself or by other activities downstream $p34^{cdc2}$ is not yet known. Masuda et al. [1992] were able to obtain the appearance of MPM2-reacting antigens on SPBs with an active purified $p34^{cdc2}$ complex, without converting them into active microtubule-nucleating centers. They conclude that the ability of SPBs to nucleate microtubules after exposure to CSF-arrested extracts is activated through a downstream pathway regulated by cdc2 kinase. Another possibility is, however, that it is the cyclin A–dependent cdc2 kinase that is active, as suggested by the results of Buendia et al. [1992].

However, in order to identify centrosomal substrates of $p34^{cdc2}$, it is necessary to undertake a more general study of centrosome-associated kinases, since both regulatory and catalytic subunits of the cAMP-dependent protein kinase II have been localized in the centrosome and in the Golgi apparatus [Nigg et al., 1985; De Camilli et al., 1986]. Thus a 350 kD centrosomal protein has been identified as an RII-binding protein in human centrosomes [Keryer et al., 1993a]. Moreover, the possibility that cross-regulation between protein kinase A and $p34^{cdc2}$ protein kinase could occur is suggested by two recent results: RIIα and $p34^{cdc2}$ have been shown to coimmunoprecipitate in a human cellular system [Tournier et al., 1992]; both RIIα and RIIβ have been shown to be phosphorylated during mitosis *in vivo* or by $p34^{cdc2}$ *in vitro* and to be functionally modified by this phosphorylation [Keryer et al., 1991, 1993b].

E. Does the Centrosome Cycle Exert Feedback Control Over the Cell Cycle?

The transition between embryonic and somatic cell cycles is the setting along the MPF-activating pathway of entry points for feedback mechanisms capable of affecting the progression of the cell cycle in order to couple growth and division. One of the major features of the somatic cell cycle is the existence of a postmitotic phase during which DNA and centrosome duplication are prevented, a feature that allows decisions for growth and entry into the stepwise events of the cell cycle to be made according to external signals [Murray and Kirschner, 1989].

The involvement of the centrosome, centrioles, or SPBs in the control of the G_0 to G_1 transition has been proposed in the past [Byers and Goetsh, 1975; Brooks et al., 1980]. Some experimental support for this proposal was reported from studies showing a temporal correlation between structural modifications of the centrosome and cell commitment to divide [Tucker et al., 1979; Sherline and Mascardo, 1982]. No causal link, however, was ever established. The structure–function relationship

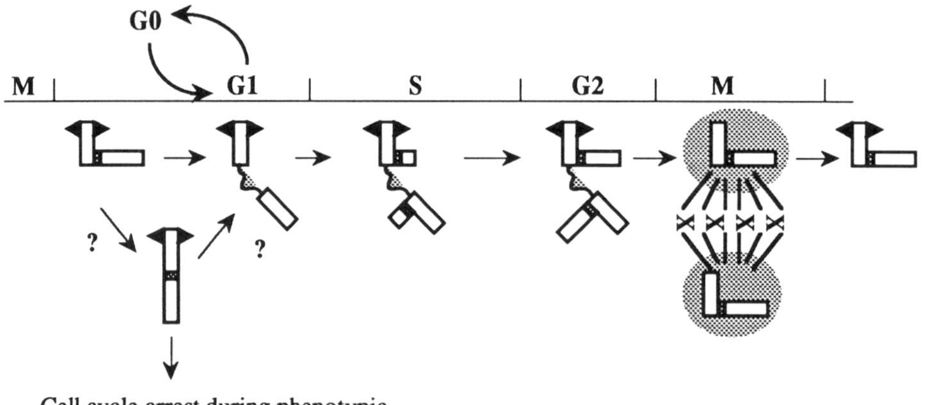

Fig. 5. *Scheme for centrosome duplication during the cell cycle. Postmitotic cells inherit a centrosome that is still in a replicative form (in which the two centrioles are in a strict orthogonal configuration). The dissociation of the centrioles is necessary but not sufficient for the reinitiation of the budding of new centrioles. The linear organization of centrosomes from calf thymus would represent a locked configuration at an intermediary step in the dissociation process due either to thymic differentiation or to some other cause [see Tournier et al., 1991b].*

observed in *Xenopus* eggs with centrosomes isolated from calf thymocytes represents new information on this issue. In a proliferating cell, daughter centrioles do not dissociate from the parent centrioles before the beginning of the next cell cycle, in early G_1 when the so-called disorientation of centrioles occurs [Kuriyama and Borisy, 1981b]. As long as the centrioles remain orthogonally associated, they can be considered as being in a replicative form: Their duplication is not completed, and a new duplicative round apparently cannot reinitiate (procentrioles budding on two orthogonally associated centrioles have never been reported). The linearly associated configuration of the centrosome from calf thymocytes could be an intermediate step toward dissociation/disorientation (see Fig. 5), normally very transient as it is rarely observed in proliferating cells but which, for some reason, would be permanent in calf thymocytes. Whatever the origin of the block in these cells [Tournier et al., 1991b], it precludes duplication, strongly suggesting the possibility that among the various requirements to be met for a cell to enter the proliferative cycle at the G_0–G_1 transition, one of them concerns the initiation of the centrosome cycle.

Inhibition of mitosis by antimitotic drugs is a well established phenomenon and has long been interpreted as the direct consequence of microtubule disassembly [De Brabander et al., 1976]. The possibility that mitotic block is in fact indirect, due to a checkpoint mechanism that controls the state of microtubules before completion of mitosis, has been suggested by genetic approaches [Hoyt et al., 1991; Li and Murray, 1991; see also Nurse, 1991]. It is apparently the dynamics of the microtubules rather than their mass that is the important parameter [Jordan et al., 1992]. Sluder [1988] has also demonstrated that spindle microtubules participate in the timing of the mitosis portion of the cell cycle in echinoderm eggs: Disrupting microtubules lengthen but do not block mitosis. When most of the interphasic microtubules and all of the mitotic microtubules are lacking in *Aspergillus,* due to the disruption of the gene coding for γ-tubulin, chromosomes do not segregate equally as expected, but the cell cycle apparently is not blocked [Oakley et al., 1990].

Neither the content of DNA nor the mitotic index are modified. The nucleus enlarges as there is no nuclear division before entering G_1. If the microtubule-dependent mitotic block observed with drug treatment is really linked to the dynamics of the microtubules, one may expect γ-tubulin, normally concentrated at the SPB, and involved in the nucleation of microtubules, to participate in this mechanism.

A report by Maniotis and Schliwa [1991] supports the idea that centrosomes play a critical part in control of the cell cycle, at the G_2–M transition [Bailly and Bornens, 1992]. Maniotis and Schliwa [1991] have shown that African green monkey BSC1 cells do not enteı mitosis following removal of the centrosome. Centriole-free karyoplasts failed to initiate nuclear membrane disassembly, chromosome condensation, or spindle formation even though they enter and presumably complete S phase. These results point to the existence of a checkpoint mechanism that makes the activation of the M-phase controlling protein kinase tightly dependent on the successful duplication of the centrosome. It is important to note here that is is apparently not through its microtubule nucleating activity *per se* that centrosome would affect the onset of mitosis, for two major reasons: 1) Centriole-free karyoplasts can reestablish a unique astral microtubule array focused on a perinuclear region where the centriole location would be expected. This aster is apparently functional as it concentrates the Golgi apparatus in its center. 2) Cultured cells enter mitosis in the complete absence of microtubules or, on the contrary, when the whole cellular tubulin is artificially assembled. It is the centrosome that is likely to play the important role in this checkpoint mechanism. The experiments of Maniotis and Schliwa [1991] did not look for the presence of γ-tubulin at the center of the acentriolar asters. From the strict localization of γ-tubulin, one may infer that centrosomal amputation totally depletes γ-tubulin from the karyoplasts or suppresses a specific cellular addressing pathway for γ-tubulin.

How the coordination of centrosome duplication and kinase activation might be achieved is unknown but could require direct interaction of the mitotic kinase itself, an interaction that has been substantiated in many cell systems (see above). It is noteworthy that from the differential localization of cyclin A, mainly nuclear, and cyclin B, totally cytoplasmic [Pines and Hunter, 1991; Bailly et al., 1992], one may infer that centrosomal amputation preferentially depletes cyclin B-associated cdc2 kinase. The interaction of cyclin B with cytoplasmic structures demonstrated in human cells [Bailly et al., 1992] might be a constitutive feature of the mechanism whereby a subset of the mitotic kinase is recruited and subsequently targeted toward mitotic substrates. The centrosomal association of both A and B types of cyclins at the onset of mitosis could provide the cell with feedback control on $p34^{cdc2}$ activation that would ensure phasing of the centrosome cycle with the mitotic clock. For instance, one can imagine that a late centrosomal process, such as the completion or the separation of the two duplicated centrosomes, which represents one of the earliest mitotic events, might constitute the last checkpoint before the large-scale activation of the cyclinB-$p34^{cdc2}$ complex at the onset of mitosis.

IV. CONCLUSION

Cell division is a temporal process exhibiting an essential oscillatory character, the intention of most current research on cell cycle control. Likewise, it is a spatial process exhibiting an essential bipolar character, present in any example of cell division. If, as emphasized in the Introduction, the essential bipolarity of the division process rests on the centrosome duplication cycle, the initiation of centrosome duplication early in the cell cycle would have significance: Cell division has to be anticipated, or prepared, from the very start of the cell cycle. And the discrete events of the slow and irreversible process of centrosome duplication, which anticipate the major cell-cycle decisions, would represent landmarks for the coupling between cell growth and cell division.

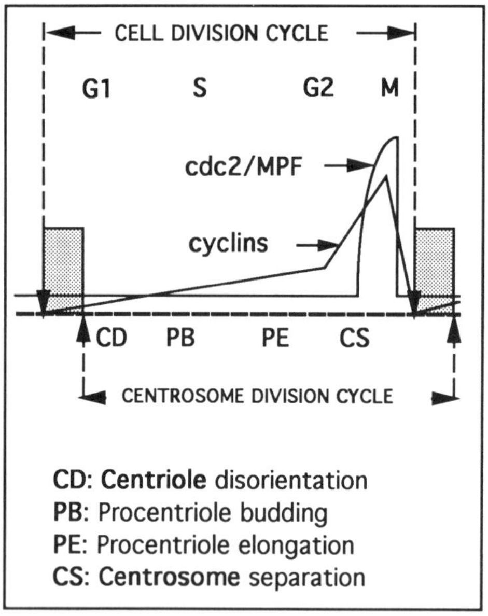

Fig. 6. *The centrosome cycle overlaps on two successive cell division cycles.*

The evidence that the centrosome duplication cycle is independent of p34^{cdc2} and cyclins raises many questions on the mechanisms that control centrosome assembly and on those that couple centrosome duplication to the cyclin-p34^{cdc2} cycle. The overlap between the centrosome cycle and two successive cell cycles (Fig. 6) could allow flexibility in the coupling during G_1 phase, before reentry into the proliferative cycle. One of the central questions in the cell cycle regulation is to understand "how an all-or-none, off-or-on response is generated biochemically from the continuously graded stimulus that can be provided by growth factors" [Pardee, 1989]. As a single copy organelle, the centrosome might be part of the answer. Furthermore, the centrosome in animal cells contains two centriole copies that are not equivalent: They have been assembled during different cell division cycles, and only one possesses appendages (see Fig. 2). This supposes that the complete centrosome cycle encompasses at least two complete successive cell cycles. This polarity within the centrosome organization may have consequences for the polarity of cell division and the temporal order of one cycle with respect to the next [see Holmes and Dutcher, 1989; Beech et al., 1991]. One example has been recently illustrated in the simpler haploid division of the budding yeast *Sac. cerevisiae* [Vallen et al., 1992] where the product of the gene *KAR1,* which is required for SPB duplication and nuclear fusion, is associated with only one of the two SPBs after duplication, usually the one that enters the bud. Thus the real biological significance of the centrosome duplication cycle is still to be determined.

V. REFERENCES

Akhurst RJ, Flavin NB, Worden J, Lee MG (1989): Intracellular localization and expression of mammalian CDC2 protein during myogenic differentiation. Differentiation 40:36–41.

Alfa CE, Ducommun B, Beach D, Hyams JS (1990): Distinct nuclear and spindle pole body populations of cyclin-cdc2 in fission yeast. Nature. 347:680–682.

Bailly E, Bornens M (1992): Centrosome et cell cycle. Nature 355:300–301.

Bailly E, Dorée M, Nurse P, Bornens M (1989): p34^{cdc2} is associated with the centrosome at G_2/M transition and enters vesicles at anaphase in human cells. EMBO J 8:3985–3995.

Bailly E, McCaffrey M, Touchot N, Zahraoui A, Goud B, Bornens M (1991): Phosphorylation of two small GTP-binding proteins of the rab family by p34cdc2. Nature. 350:715–718.

Bailly E, Pines J, Hunter T, Bornens M (1992): Cytoplasmic accumulation of cyclin B1 in human cells: Association with a detergent-resistant compartment and with the centrosome. J Cell Sci 101:529–545.

Baroin A, Perasso R, Qu LH, Brugerolle G, Bachellerie JP, Adoutte A (1988): Partial phylogeny of the unicellular eukaryotes based on rapid sequencing of a portion of 28S ribosomal RNA. Proc Natl Acad Sci USA 85:3474–3478.

Baron AT, Salisbury JL (1988): Identification and localization of a novel, cytoskeletal, centrosome-associated protein in PtK2 cells. J Cell Biol 107:2669–2678.

Baum P, Yin C, Goetsch L, Byers B (1988): A yeast gene essential for regulation of spindle pole duplication. Mol Cell Biol 8:5386–5397.

Beech PL, Heimann K, Melkonian M (1991): Development of the flagellar apparatus during the cell cycle in unicellular algae. Protoplasma 164:23–47.

Belmont LD, Hyman AA, Sawin KE, Mitchison TJ (1990): Real-time visualization of the cell cycle-dependent changes in microtubule dynamics in cytoplasmic extracts. Cell 62:579–589.

Booher RN, Alfa CE, Hyams JS, Beach DH (1989): The fission yeast cdc2/cdc13/suc1 protein kinase: Regulation of catalytic activity and nuclear localization. Cell 58:485–497.

Bornens M (1991): Cell polarity: Intrinsic or externally imposed? New Biol 3:627–636.

Bornens M (1992): Structure and functions of isolated centrosomes. In Kalnins VI (ed): The Centrosome. New York: Academic Press, pp 1–43.

Bornens M, Karsenti E (1984): The centrosome. In Bittar EE (ed): Membrane: Structure and Functions. New York: John Wiley and Sons, vol 6, pp 100–171.

Bornens M, Paintrand M, Berges J, Marty MC, Karsenti E (1987): Structural and chemical characterization of isolated centrosomes. Cell Motil Cytoskel 8:238–249.

Brooks RF, Bennett DC, Smith JA (1980): Mammalian cell cycles need two random transitions. Cell 19: 493–504.

Buendia B, Draetta G, Karsenti E (1992): Regulation of the microtubule nucleating activity of centrosomes in *Xenopus* egg extracts: Role of cyclin A–associated protein kinase. J Cell Biol 116:1431–1442.

Byers B, Goetsch L (1975): Behavior of spindles and spindle plaques in the cell cycle and conjugation of *Saccharomyces cerevisiae*. J Bacteriol 124:511–523.

De Brabander M, Van De Veire J, Aerts RML, Borgers FEM, Janssen PAJ (1976): The effect of methyl-5-[2-thienylcarbonyl]-1H-benzimidazol-2yl) carbamate, a new synthetic antitumoral drug interfering with microtubules, on mammalian cells cultured *in vitro*. Cancer Res 36:905–916.

De Camilli P, Moretti M, Denis Donini S, Lohman SM (1986): Heterogeneous distribution of the c-AMP receptor protein kinase R-II in the nervous system: Evidence for its intracellular accumulation on microtubules, microtubule organizing centers and in the area of the golgi complex. J Cell Biol 103:189–203.

Debec A, Montmory C (1992): Cyclin B is associated with centrosomes in *Drosophila* mitotic cells. Biol Cell 75:121–126.

Dirksen ER (1991): Centriole and basal body formation during ciliogenesis revisited. Biol Cell 77:31–38.

Ellis RJ (1987): Proteins as molecular chaperones. Nature 328:378–379.

Enoch T, Nurse P (1991): Coupling M phase and S phase: Controls maintaining the dependence of mitosis on chromosome replication. Cell 65:921–923.

Euteneuer U, McIntosh JR (1981): Polarity of some motility-related microtubules. Proc Natl Acad Sci USA 78:372–376.

Flanders DJ, Rawlins DJ, Shaw PJ, Lloyd CW (1990): Nucleus-associated microtubules help determine the division plane of plant epidermal cells: Avoidance of four-way junctions and the role of cell geometry. J Cell Biol 110:1111–1122.

Fleury A, LeGuyader H, Iftode F, Laurent M, Bornens M (1993): A scaffold for basal body patterning revealed by a monoclonal antibody in the hyptotrich ciliate *Paraurostyla weissei*. Dev Biol 157:285–302.

Freeman M, Nüsslein-Volhard C, Glover DM (1986): The dissociation of nuclear and centrosomal division in *gnu*, a mutation causing giant nuclei in *Drosophila*. Cell 46:457–468.

Fuller SD, Kenney JM, Karsenti E (1992): Centrosomes. Curr Opin Struct Biol 2:264–274.

Fulton C (1971): Centrioles. In Beerman W, Reinert J, Ursprung H (eds): New York: Springer-Verlag, pp 170–221.

Gallant P, Nigg E (1992): Cyclin B2 undergoes cell cycle-dependent nuclear translocation and, when expressed as a non-destructible mutant, causes mitotic arrest in HeLa cells. J Cell Biol 117:213–224.

Gard DL, Hafezi S, Zhang T, Doxsey ST (1990): Centrosome duplication continues in cycloheximide-treated *Xenopus* blastulae in the absence of a detectable cell cycle. J Cell Biol 110:2033–2042.

Glover D (1992): The centrosome in cell division and development of *Drosophila*. In Kalnins VI (ed): The centrosome. New York: Academic Press, pp 219–234.

Gosti-Testu F, Marty MC, Berges J, Maunoury R, Bornens M (1986): Identification of centrosomal proteins in a human lymphoblastic cell line. EMBO J 5:2545–2550.

Hall JL, Ramanis Z, Luck DJL (1989): Basal body/centriolar DNA: Molecular genetic studies in *Chlamydomonas*. Cell 59:121–132.

Hara K, Tydeman P, Kirschner MW (1980): A cytoplasmic clock with the same period as division cycle of *Xenopus* eggs. Proc Natl Acad Sci USA 77:462–466.

Hartwell LH, Weinert TA (1989): Checkpoints: Controls that ensure the order of cell cycle events. Science 246:629–634.

Hepler PK (1976): The blepharoplast of Marsilea: Its *de novo* formation and spindle association. J Cell Sci 21:361–390.

Holmes JA, Dutcher SK (1989): Cellular asymmetry in *Chlamydomonas reinhardtii*. J Cell Sci 94:273–285.

Horvitz HR, Herskovitz I (1992): Mechanisms of asymetric cell division: Two Bs or, not two Bs, that is the question. Cell 68:237–255.

Hoyt MA, Totis L, Roberts BT (1991): *S. cerevisie* genes required for cell cycle arrest in response to loss of microtubule function. Cell 66:507–517.

Huang B, Mangersen A, Lee VD (1988a): Molecular cloning of cDNA for caltractin, a basal body-associated Ca^{2+}-binding protein: Homology in its protein sequence with Calmodulin and the yeast cdc31 gene product. J Cell Biol 107:133–140.

Huang B, Watterson DM, Lee VD, Schibler MJ (1988b): Purification and characterization of a basal body-associated Ca^{2+}-binding protein. J Cell Biol 107: 121–131.

Hyman AA (1989): Centrosome movement in the early divisions of *C. elegans:* A cortical site determining centrosome position. J Cell Biol 109:1185–1193.

Johnson DE, Dutcher SK (1991): Molecular studies of the linkage group XIX of *Chlamydomonas rienhardtii:* Evidence against a basal body location. J Cell Biol 113:339–346.

Johnson KA, Rosenbaum JL (1990): The basal bodies of *Chlamydomonas reinhardtii* do not contain immunologically detectable DNA. Cell 62:615–619.

Jordan MA, Thrower D, Wilson L (1992): Effects of vinblastine, podophyllotoxin and nocodazole on mitotic spindles. J Cell Sci 102:401–416.

Karsenti E, Newport J, Hubble R, Kirschner MW (1984) Interconversion of metaphase and interphase microtubule arrays as studied by the injection of centrosomes and nuclei into *Xenopus* eggs. J Cell Biol 98:1730–1745.

Keryer G, Cavadore JC, Bornens M (1991): Mitotic phosphorylation of the regulatory subunit of cAMP-dependent protein kinase II by cdc2 kinase. J Cell Biol 115:284a, 1644.

Keryer G, Iftode F, Bornens M (1990): Identification of proteins associated with microtubule-organizing centres in the oral apparatus of the ciliate *Paramecium tetraurelia*. J Cell Sci 97:553–563.

Keryer G, Rios RM, Landmark BF, Skalhegg B, Lohmann SM, Bornens M (1993a): A high-affinity binding protein for the regulatory subunit of cAMP-dependent protein kinase II in the centrosome of human cells. Exp Cell Res 204:230–240.

Keryer G, Luo Z, Cavadore JC, Erlichman J, Bornens M (1993b): Phosphorylation of the regulatory subunit of type IIβ cAMP-dependent protein kinase by cyclin B/p34^{cdc2} kinase impairs its binding to microtubule-associated protein 2. Proc Natl Acad Sci USA 90:5418–5422.

Kimble M, Kuriyama R (1992): Functional components of microtubule-organizing centers. Int Rev Cyto 136:1–50.

Kirschner MW (1992): The cell cycle then and now. TIBS 17:281–285.

Klotz C, Dabauvalle MC, Paintrand M, Weber H, Bornens M, Karsenti E (1990): Parthenogenesis in *Xenopus* eggs requires centrosomal integrity. J Cell Biol 110:405–415.

Kochanski RS, Borisy GG (1990): Mode of centriole duplication and distribution. J Cell Biol 109:1599–1605.

Komesli S, Tournier F, Paintrand M, Margolis R, Job D, Bornens M (1989): Mass isolation of calf thymus centrosomes: The identification of a specific configuration. J Cell Biol 109:2869–2878.

Krek W, Nigg E (1991): Mutations of p34cdc2 phosphorylation sites induce premature mitotic events in HeLa cells: Evidence for a double block to p34cdc2 kinase activation in vertebrates. EMBO J 10:3331–3341.

Kuriyama R, Borisy GG (1981a): Microtubule-nucleating activity of centrosomes in Chinese hamster ovary cells is independent of the centriole cycle but coupled to the mitotic cycle. J Cell Biol 91:822–826.

Kuriyama R, Borisy GG (1981b): Centriole cycle in Chinese hamster ovary cells as determined by whole mount electron microscopy. J Cell Biol 91:814–821.

Kuriyama R, Borisy GG (1983): Cytasters induced within unfertilized sea urchin eggs. J Cell Sci 61:175–189.

Kuroiwa T, Yorihuzi T, Yabe N, Ohta T, Uchida H (1990): Absence of DNA in the basal body of *Chlamydomonas reinhardtii* by fluorimetry using a video-intensified microscope photon-counting system. Protoplasma 158:155–164.

Laskey RA, Fairman MP, Blow JJ (1989): S phase of the cell cycle. Science 246:609–613.

Li R, Murray AW (1991): Feedback control of mitosis in budding yeast. Cell 66:519–531.

Maldonado-Codina G, Glover DM (1992): Cyclins A and B associate with chromatin and the polar regions of spindles, respectively, and do not undergo complete degradation at anaphase in syncitial *Drosophila* embryos. J Cell Biol 116:967–976.

Maller J (1990): *Xenopus* oocyte and the biochemistry of cell division. Biochemistry 29:3157–3166.

Maller J, Poccia D, Nishioka D, Kido P, Gerhart J, Hartman H (1976): Spindle formation and cleavage in *Xenopus* eggs injected with centriole containing fractions from sperm. Exp Cell Res 99:285–294.

Maniotis A, Schliwa M (1991): Microsurgical removal of centrosomes blocks cell reproduction and centriole generation in BSC-1 cells. Cell 67:495–504.

Masuda H, Sevic M, Cande WZ (1992): *In vitro* microtubule-nucleating activity of spindle pole bodies in fission yeast *S. pombe:* Cell cycle-dependent activation in *Xenopus* cell-free extracts. J Cell Biol 117:1055–1066.

Mazia D (1978): Origin of twoness in cell reproduction. ICN–UCLA Symp Mol Cell Biol 12:1–14.

Mazia D, Harris PJ, Bibring T (1960): The multiplicity of mitotic centers and the time-course of their duplication and separation. J Biophys Biochem Cytol 7:1–10.

Mizukami I, Gall J (1966): Centriole replication: II. Sperm formation in the fern Marsilea, and the cycad, Zamia. J Cell Biol 29:97–111.

Moudjou M, Paintrand M, Vigues B, Bornens M (1991): A human centrosomal protein is immunologically related to basal body-associated proteins from lower eucaryotes and is involved in the nucleation of microtubules. J Cell Biol 115:129–140.

Murray AW, Kirschner MW (1989): Dominoes and clocks: The union of two views of the cell cycle. Science 246:614–621.

Nagano H, Hirai S, Okana K, Ikegami S (1981): A chromosomal cleavage of fertilized starfish eggs in the presence of aphidicolin. Dev Biol 85:409–415.

Nigg E (1991): The substrates of the cdc2 kinase. Semin Cell Biol 2:261–270.

Nigg EA, Schafer G, Hiltz H, Eppenberger H (1985): Cyclic-AMP–dependent protein kinase type II is associated with the Golgi complex and with centrosomes. Cell 41:1039–1051.

Nurse P (1990): Universal control mechanism regulating onset of M-phase. Nature 344:508.

Nurse P (1991): Checkpoints and spindles. Nature 354:356–358.

Oakley BR (1992): γ-Tubulin: The microtubule organizer? Trends Cell Biol 2:1–5.

Oakley CE, Oakley BR (1989): Identification of γ-tubulin, a new member of the tubulin superfamily encoded by mipA gene of *Aspergillus nidulans*. Nature 338: 662–664.

Oakley BR, Oakley CE, Zoon Y, Yung MK (1990): γ-Tubulin is a component of the spindle pole body that is essential for microtubule function in *Aspergillus nidulans*. Cell 61:1289–1301.

Omura F, Fukui Y (1985): Dictyostelium MTOC: Structure and linkage to the nucleus. Protoplasma 127: 212–221.

Ookata K, Hisanaga S, Okano T, Tachibana K, Kischimoto T (1992): Relocation and distinct subcellular localization of p34cdc2-cyclin B complex at meiosis reinitiation in starfish oocytes. EMBO J 11:1763–1772.

Paintrand M, Moudjou M, Delacroix H, Bornens M (1992): Centrosome organization and centriole architecture: Their sensitivity to divalent cations. J Struct Biol 108:107–128.

Palazzo RE, Vaisberg E, Cole RW, Rieder CL (1992): Centriole duplication in lysates of *Spisula solidissima* oocytes. Science 256:219–221.

Pardee AB (1989): G_1 events and regulation of cell proliferation. Science 246:603–608.

Perret E, Albert M, Bordes N, Bornens M, Soyer-Gobillard MO (1991): Microtubular spindle and centrosome structures during the cell cycle in a dinoflagellate *Crypthecodinium cohnii:* An immunocytochemical study. BioSystems 24:53–65.

Perret E, Moudjou M, Geraud ML, Derancourt J, Dubois P, Soyer-Gobillard M-O, Bornens M (1992): A 72 kDa molecular chaperone is constitutively associated with centrosomes, from dinoflagellates to human cells. Mol Biol Cell 3 (Suppl): 346a.

Picard A, Harricane MC, Labbe JC, Doree M (1988): Germinal vesicle components are not required for the cell cycle oscillator of the early starfish embryo. Dev Biol 128:121–128.

Pines J, Hunter T (1991): Human cyclin A and B1 are differentially located in the cell and undergo cell cycle dependent nuclear transport. J Cell Biol 115:1–17.

Raff JW, Glover DM (1988): Nuclear and cytoplasmic mitotic cycles continue in *Drosophila* embryos in which DNA synthesis is inhibited with Aphidicolin. J Cell Biol 107:2009–2019.

Rattner JB (1990): HSP70 is constitutively associated with the centrosome in HeLa cells. Exp Cell Res 195: 110–113.

Rattner JB, Lew J, Wang JH (1990): p34cdc2 kinase is localized to distinct domains within the mitotic apparatus. Cell Motil Cytoskel 17:227–235.

Rattner JB, Phillips SG (1973): Independence of centriole formation and DNA synthesis. J Cell Biol 57: 359–367.

Riabowol K, Draetta G, Brizuela L, Vandre D, Beach D (1989) The cdc2 kinase is a nuclear protein that is essential for mitosis in mammalian cells. Cell 57: 393–401.

Rieder CR, Borisy GG (1982): The centrosome cycle in PtK2 cells: Asymmetric distribution and structural changes in the pericentriolar material. Biol Cell 44:117–132.

Robbins E, Jentzsch G, Micali A (1968): The centriole cycle in synchronized HeLa cells. J Cell Biol 36: 329–339.

Roos UP (1975): Fine structure of an organelle associated with the nucleus and cytoplasmic microtubules in the cellular slime mould *Polyshondylium violaceum*. J Cell Sci 18:315–326.

Rostand J (1968): La parthénogénèse (ou reproduction virginale) des vertébrés. Conférence du 30 novembre 1968 (Palais de la Découverte, A345, 1969).

Rothman JE (1989): Polypeptide chain binding proteins: Catalysis of protein folding and elated processes in cells. Cell 59:591–601.

Sellitto C, Kuriyama R (1988): Distribution of pericentriolar material in multipolar spindles induced by colcemid treatment in chinese hamster ovary cells. J Cell Sci 89:57–65.

Sherline P, Mascardo R (1982): EGF induced centrosomal separation: Mechanism and relationship to mitogenesis. J Cell Biol 95:316–322.

Sluder G (1988): Control mechanism of mitosis: the role of spindle microtubules in the timing of mitotic events. Zool Sci 5:653–665.

Sluder G, Miller FJ, Cole R, Rieder CL (1990): Protein synthesis and the cell cycle: Centrosome reproduction in sea-urchin eggs is not under translational control. J Cell Biol 110:2025–2032.

Sluder G, Miller FJ, Lewis K, Davison ED, Rieder CL (1989): Centrosome inheritance in starfish zygotes: Selective loss of the maternal centrosome after fertilization. Dev Biol 131:567–579.

Sluder G, Miller FJ, Rieder CL (1986): The reproduction of centrosomes. Nuclear versus cytoplasmic controls. J Cell Biol 103:1873–1881.

Sluder G, Rieder CL (1985): Centriole number and the reproductive capacity of spindle poles. J Cell Biol 100:887–896.

Stearns T, Evans L, Kirschner MW (1991): γ-Tubulin is a highly conserved component of the centrosome. Cell 65:825–836.

Szöllösi A, Ris H, Szöllösi D, Debec A (1986): A centriole-free *Drosophila* cell line. A high voltage EM study. Eur J Cell Biol 40:100–104.

Terasaki M, Jaffe LA (1991): Organization of the sea urchin egg endoplasmic reticulum and its reorganization at fertilization. J Cell Biol 114:929–940.

Thibodeau A, Vincent V (1991): Monoclonal antibody CC-3 recognizes phosphoproteins in interphase and mitotic cells. Exp Cell Res 195:145–153.

Tournier F, Cyrklaff M, Karsenti E, Bornens M (1991a): Centrosomes competent for parthenogenesis in *Xeno-*

pus eggs support budding in cell-free extracts. Proc Natl Acad Sci USA 88:9929–9923.

Tournier F, Karsenti E, Bornens M (1989): Parthenogenesis in *Xenopus* eggs injected with centrosomes from synchronized human lymphoid cells. Dev Biol 136: 321–329.

Tournier F, Komesli S, Paintrand M, Job D, Bornens M (1991b): Centrosome-induced parthenogenetic development of *Xenopus* eggs: Structure–activity relationship. J Cell Biol 113:1361–1369.

Tournier S, Raynaud F, Gerbaud P, Lohman SM, Dorée M, Evain-Brion D (1992): Association of type II cAMP-dependent protein kinase with p34cdc2 protein kinase in human fibroblasts. J Biol Chem 266:19018–19022.

Tucker R, Pardee A, Fujiwara K (1979): Centriole ciliation is related to quiescence and DNA synthesis in 3T3 cells. Cell 17:527–535.

Vallen EA, Scherson TY, Roberts T, Van Zee K, Rose MD (1992): Asymmetric mitotic segregation of the yeast spindle pole body. Cell 69:505–515.

Vandre D, Davis F, Rao P, Borisy GG (1986): Distribution of cytoskeletal proteins sharing a conserved phosphorylated epitope. Eur J Cell Biol 41:72–81.

Verde F, Berrez JM, Antony C, Karsenti E (1991): Taxol induced microtubule asters in mitotic extracts of *Xenopus* eggs: Requirement for phosphorylated factors and cytoplasmic dynein. J Cell Biol 112: 1177–1187.

Verde F, Dogterom M, Stelzer E, Karsenti K, Leibler S (1992): Control of microtubule dynamics and length by cyclin A and cyclin B-dependent kinases in *Xenopus* egg extracts. J Cell Biol 118:1097–1108.

Verde F, Labbé JC, Dorée M, Karsenti E (1990): Regulation of microtubule dynamics by cdc2 protein kinase in cell-free extracts of *Xenopus* eggs. Nature 343:233–238.

Wilson EB (1925): The Cell in Development and Heredity. New York: MacMillan.

Winey M, Byers B (1992): Spindle pole body of *S. cerevisiae:* A model for genetic analysis of the centrosome cycle. In Kalnins VI (ed): The Centrosome. New York: Academic Press, pp 197–218.

Wittenberg C, Richardson SL, Reed SI (1987): Subcellular localization of a protein kinase required for cell cycle initiation in *Saccharomyces cerevisiae:* Evidence for an association between the CDC28 gene product and the insoluble cytoplasmic matrix. J Cell Biol 105:1527–1538.

Zheng Y, Jung MK, Oakley BR (1991): γ-Tubulin is present in *Drosophila melanogaster* and *Homo sapiens* and is associated with the centrosome. Cell 65: 817–823.

ABOUT THE AUTHORS

FRÉDÉRIC TOURNIER is Attaché Temporaire d'Enseignement et de Recherche at the University Paris 7, where he teaches undergraduate courses dealing with topics in general and developmental biology. He did his doctoral research in Michel Bornens's laboratory, completing his thesis in 1992 on duplication of centrosomes of animal cells.

MICHEL BORNENS is Directeur de Recherche at CNRS in the Centre de Génétique Moléculaire in Gif-sur-Yvette, France, where he heads the Biology of Cell Cycle and Cell Motility Department. After receiving his bachelor's degree at l'Institut National Agronomique in Paris, he pursued doctoral research at the Institute for Cancer Research of the CNRS in Villejuif, where he earned a Ph.D in 1969 for work on the isolation and biochemical analysis of the lipid composition of the nuclear envelope from rat hepatocytes. This was followed by postdoctoral research at the McArdle Laboratory for Cancer Research in Madison, Wisconsin, where he worked with Charles B. Kasper on the protein composition of the nuclear envelope. From 1972 to 1982 he worked in the Molecular Biology Department of the Pasteur Institute, Paris, on the heparin-induced solubilization of chromatin from rat nuclei and on the mitogenesis induced by lectins in rat thymocytes. Since 1977 Dr. Bornens's research has been focusing on the cytoskeleton and is principally devoted to the study of the structure and functions of the centrosome from animal cells.

Microtubules: 325–341

The Higher Plant Microtubule Cycle

Anne Marie Lambert and Clive W. Lloyd

I. INTRODUCTION

The microtubule (MT) assemblies formed in the higher plant cell are different from those of animal tissue cells in several distinct ways: The interphase MTs are tightly associated with the plasma membrane but can change their alignment in response to environmental and hormonal factors; they somehow give way to a ring of MTs (the preprophase band) that anticipates the cortical division site, there is a prophase spindle that forms around the nuclear envelope and is succeeded by an acentriolar mitotic spindle, and there is a characteristic cytokinetic apparatus (the phragmoplast) that grows out centrifugally until it contacts the mother wall at the site vacated by the preprophase band. All of these must be studied in plant cells, but there are physical difficulties in microinjecting, and even immunostaining, walled and usually vacuolated cells. In addition, the absence of identifiable and easily located MT nucleating sites has made it difficult to understand the MT cycle in detail.

The special features of the microtubule cycle in higher plants were identified by Ledbetter and Porter as early as 1964, and the mechanisms that control these events and particularly relations with other cytoskeletal components have been debated ever since [for reviews, see Seagull, 1989; Baskin and Cande, 1990; Lambert et al., 1991; Lloyd, 1989, 1991a,b; Palevitz, 1991; Staiger and Lloyd, 1991; Gunning, 1992; Smirnova and Bajer, 1992]. Rather than recover this field, we shall concentrate on several significant areas in which advances have recently been made. For instance, an increasing knowledge of plant tubulin diversity [for reviews, see Fosket, 1989; Hussey et al., 1991; Fosket and Morejohn, 1992] and of plant microtubule proteins [Cyr, 1991; Schellenbaum et al., 1992] and actin-associated proteins [McCurdy and Williamson, 1991] has given new insights into potential regulation of the microtubule cycle. In addition, although there are still problems with identifying higher plant microtubule nucleation sites, recent fluorescent analogue cytochemistry studies have shown more about the relationships between different MT arrays [Vantard et al., 1990; Zhang et al., 1990; Asada et al., 1991; Lambert, 1993].

Comparison of microtubule formation and dynamics in normal somatic cells and within cells that do not form a cellulosic wall, such as meiotic or endosperm cells, sets the scene for the questions concerning the origin of distinct microtubule populations issued from a common tubulin cytosolic pool. The main questions concern: 1) the origin, turnover, and reorganization of cortical, preprophase band, and nucleus-associated microtubules and the possible role of actin in these events; 2) the identification of microtubule assembly and/or nucleation sites during the organization of the plant mitotic spindle; and 3) how and where phragmoplast microtubules originate during cytokinesis.

II. CELL CYCLE-DEPENDENT ASSEMBLY/DISASSEMBLY AND/OR REARRANGEMENT OF MICROTUBULE ARRAYS

The distribution of plant MTs throughout the division cycle has been thoroughly documented [Gunning and Wick, 1985; Seagull, 1989; Baskin and Cande, 1990; Lambert et al., 1991; Palevitz, 1991; Staiger and Lloyd, 1991; Gunning, 1992; Wick, 1991a,b; Williamson, 1991]. Because the information has been drawn from studies on fixed cells, little is known about the dynamic aspects, but functional assays based on the incorporation of labeled tubulin are beginning to change that [Zhang et al., 1990; Vantard et al., 1990; Asada et al., 1991].

A. Cortical and Preprophase Band Microtubule Arrays

A fundamental specialization of the interphase plant cell is the presence of cortical MTs linked to the plasma membrane. Additionally, depending on the stage of the cell cycle, MTs may emerge from the nuclear surface and radiate toward the cell periphery. These nucleus-associated MTs (NAMTs) are prominent at two stages of the cell cycle: postcytokinesis, when the new interphase array is being formed; and during preprophase, when the cell displays its commitment to divide again [Lloyd, 1989; Palevitz, 1991; Lambert et al., 1991; Staiger and Lloyd, 1991; Wick, 1991a,b]. These NAMTs, not visible in quiescent epidermal cells, become conspicuous around the nucleus as it begins to migrate away from the cortex in preparation for a division induced by tissue wounding [Lloyd et al., 1991].

The first signs of cortical MT aggregation are difficult to time relative to DNA replication but seem to occur at the end of S phase. The progressive formation of the preprophase band is therefore achieved in G_2 (although earlier activity cannot be ruled out) and may take as long as 120 minutes in meristematic cells [Mineyuki et al., 1988; Gunning and Sammut, 1990]. In vacuolated cells, where the nucleus clearly moves to the center of the cell to divide, migration seems to start just before DNA replication [Venverloo and Libbenga, 1987]. The appearance of NAMTs in such cells also coincides with the movement of the nucleus away from the cortex and so these would appear to form at G_1–S. Although the exact timing of their initiation is imprecise, these NAMTs can at least be taken as a marker of the G_2 state.

In nonwalled cells, which do not possess cortical or preprophase band microtubules, as in *Haemanthus* endosperm, for example, NAMTs are present during the whole cell cycle, although their density is higher in the premitotic stage [Bajer and Mole-Bajer, 1982; De Mey et al., 1982; Schmit et al., 1983]. In addition to MTs, actin is also present in the preprophase band [Kakimoto and Shibaoka, 1987; Palevitz, 1987; Traas et al., 1987], as well as in the strands that connect the nucleus to the cortex [Kakimoto and Shibaoka, 1987; Traas et al., 1987; Goodbody et al., 1991]. In *Haemanthus* cells, actin is also present during the whole cell cycle [Schmit and Lambert, 1987]. The distribution and probably therefore the dynamics of actin and MTs are codependent, and hypotheses have been proposed for their participation in division plane alignment [Katsuta and Shibaoka, 1988; Flanders et al., 1990a; Lloyd, 1991a]. The premitotic stages are characterized by the migration of the nucleus toward the cell center, where it is anchored by transvacuolar strands that gradually reorganize to form a transverse sheet, the phragmosome, as illustrated in Figure 1 [Flanders et al., 1990a; Lloyd, 1991a; Staiger and Lloyd, 1991]. F-actin and microtubules in these transvacuolar strands are involved in this process and bridge the nucleus to the cortical preprophase band. The cytoplasmic microtubules are disassembled during spindle formation, but F-actin remains in the cytoplasm to connect the outgrowing phragmoplast to the cortex [Lloyd and Traas, 1988]. The interdependent dynamics and the stuctural continuity between the NAMTs and the cortical microtubules suggest that perinuclear microtubules and actin filaments are not only actively involved in the realignment of preexisting cortical microtubules into the preprophase band [Lloyd, 1989] but integrate into, or may

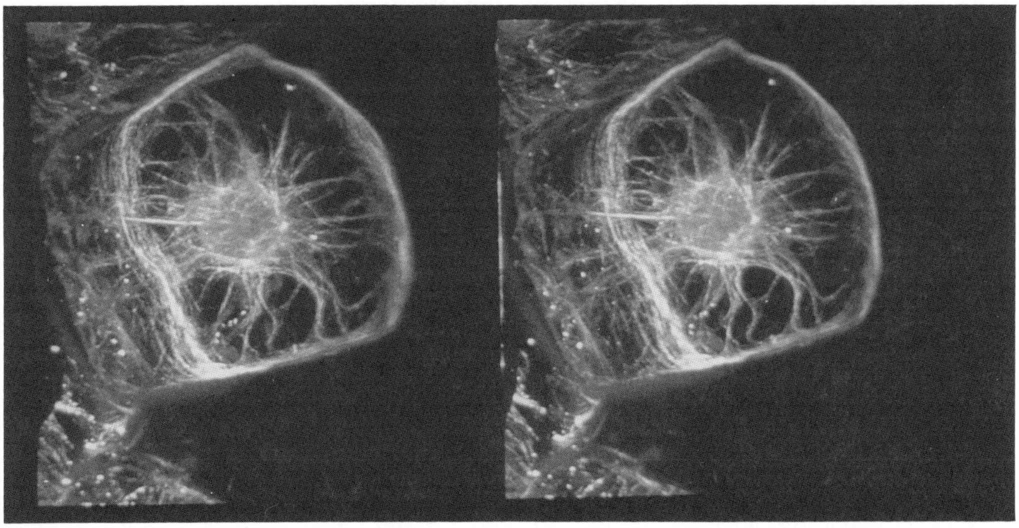

Fig. 1. *Stereopair of a 3-dimensional reconstruction of an antitubulin-labeled* Nautilocalyx lynchii *epidermal cell, imaged by confocal laser scanning microscopy. This shows the nucleus-associated MTs that are beginning to gather in the division plane. This also shows the connection between the nucleus and the preprophase band. [Reproduced from Goodbody et al., 1991, with permission of the publisher.]*

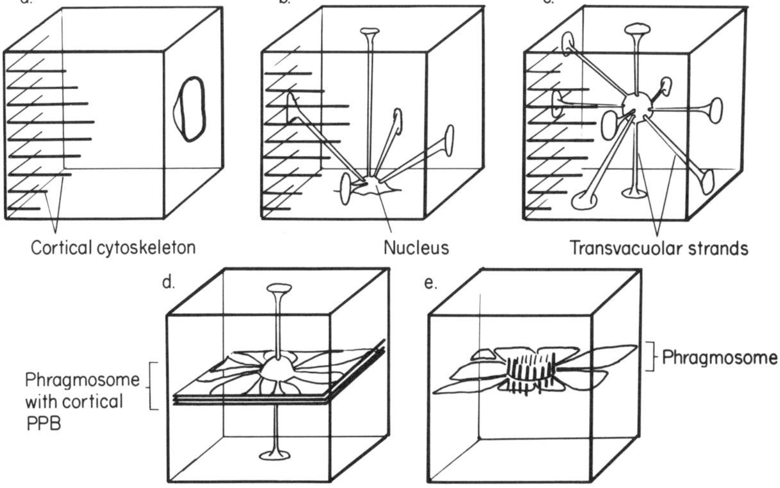

Fig. 2. *Model for the involvement of the cytoskeleton in division plane formation in vacuolated, walled cells.* **a:** *The interphase cytoskeleton is chiefly cortical and consists of actin filaments (which also run in various directions, deeper in the cytoplasm) and MTs.* **b:** *In response to wounding or to a mitogenic signal, the nucleus begins to migrate and is markedly associated with actin filaments.* **c:** *At the center of the cell the premitotic nucleus is the focus for both actin filaments and MTs.* **d:** *Gradually, these radiating strands become reorganized into the phragmosome: a transvacuolar baffle that marks the division plane and anchors the central nucleus. Both MTs and actin filaments are in the perinuclear strands and the cortical preprophase band.* **e:** *However, during cytokinesis, while there are MTs and actin filaments in the phragmoplast, there is only apparently actin, and not MTs, to connect the cytokinetic apparatus to the cortex. [Reproduced from Lloyd, 1991a, with permission of the publisher.]*

even form, the preprophase band [Flanders et al., 1990a,b; reviewed by Lloyd, 1991a; Staiger and Lloyd, 1991], as illustrated in Figure 2. Drug studies show that MTs are required for the transverse alignment of cortical F-actin [McCurdy and Gunning, 1990] and that preprophase band MTs do not form a tight band in the presence of cytochalasin D [Mineyuki and Palevitz, 1990]. Such studies demonstrate the codependency of the two systems for proper cortical alignment.

Relations between preexisting and "new" polymers that might be of different stability, within the preprophase band, are not yet defined. The elegant centrifugation experiments of Murata and Wada [1989] [see also Wada and Murata, 1991], although on fern protonema, may provide clues to the mechanisms/signals controlling division plane determination in higher plant cells. Displacement of the nucleus before preprophase band formation generally caused preprophase bands to form at the new location of the nucleus. Centrifugation during or after preprophase band formation resulted in a new preprophase band forming at the new site of the nucleus, yielding cells with two preprophase bands. In this case the old preprophase band did not disassemble until the sedimented nucleus entered telophase or progressed as far as early interphase; however, the new preprophase band disappeared on cue, before metaphase. This suggests that the position of preprophase band MTs depends on the location of the nucleus during G_2 and that some factor in the nuclear region is normally responsible for depolymerization of cortical MTs before metaphase. The presence of transmembrane proteins in association with cortical microtubules has been described in tobacco BY2 cells [Akashi and Shibaoka, 1991]. In the same cell type, Sonobe [1990a] has obtained a crude protein fraction that depolymerizes cortical MTs on protoplast ghosts. This requires ATP and is not effective on the MTs of isolated spindles or phragmoplasts. Using an antibody raised against the PSTAIR motif, Mineyuki et al. [1991] have shown by immunofluorescence that the antigen is concentrated in the preprophase band of onion meristematic cells. This raises the possibility that a plant homologue of $p34^{cdc2}$ kinase (which has been shown to affect the behavior of MTs associated with animal centrosomes [Verde et al., 1990]) may be involved in the stage-specific behavior of cortical MTs. However, because the PSTAIR motif is found in other kinases, further evidence is required to confirm the presence of $p34^{cdc2}$. Protein phosphorylation is an obvious candidate for the cyclic control of MT organization, but the evidence is thus far tantalizing rather than definitive. For instance, Katsuta and Shibaoka [1992] have shown that cortical MTs attached to plasma membrane disks torn out of protoplasts can be eluted with ATP and that this can be inhibited by kinase inhibitors. However, substrate-attached preprophase and interphase MTs show no difference in ATP or kinase inhibitor sensitivity, suggesting that if phosphorylation does play a part in the cyclic stability of MTs it is likely to depend on the presence of other factors in the cytoplasm.

B. Plant Mitotic Microtubules

The induction of mitosis in fungi and yeasts, as in animal cells, is characterized by the duplication and further separation of centrosomes or spindle bodies and by drastic changes in their nucleation activity and microtubule turnover. This suggests a cell cycle regulation of centrosomal function that is, at least in part, associated with $p34^{cdc2}$ kinase [Rattner et al., 1990; Verde et al., 1990]. However, in higher plants that are free of defined centrosomes such a correlation is not possible thus far [Baskin and Cande, 1990; Lambert et al., 1991].

In higher plant cells that possess a cellulosic wall, the preprophase band forms among the interphase cortical array; later, the preprophase band coexists with the bipolar prophase spindle that forms around the nucleus [Gunning and Sammut, 1990; Gunning, 1992; Wick, 1991a]. Because the preprophase band and the interphase array physically and temporally overlap, it is possible that the preprophase band is made by a reutilization of the existing cortical MTs. However, the NAMTs seen in G_2 must arise

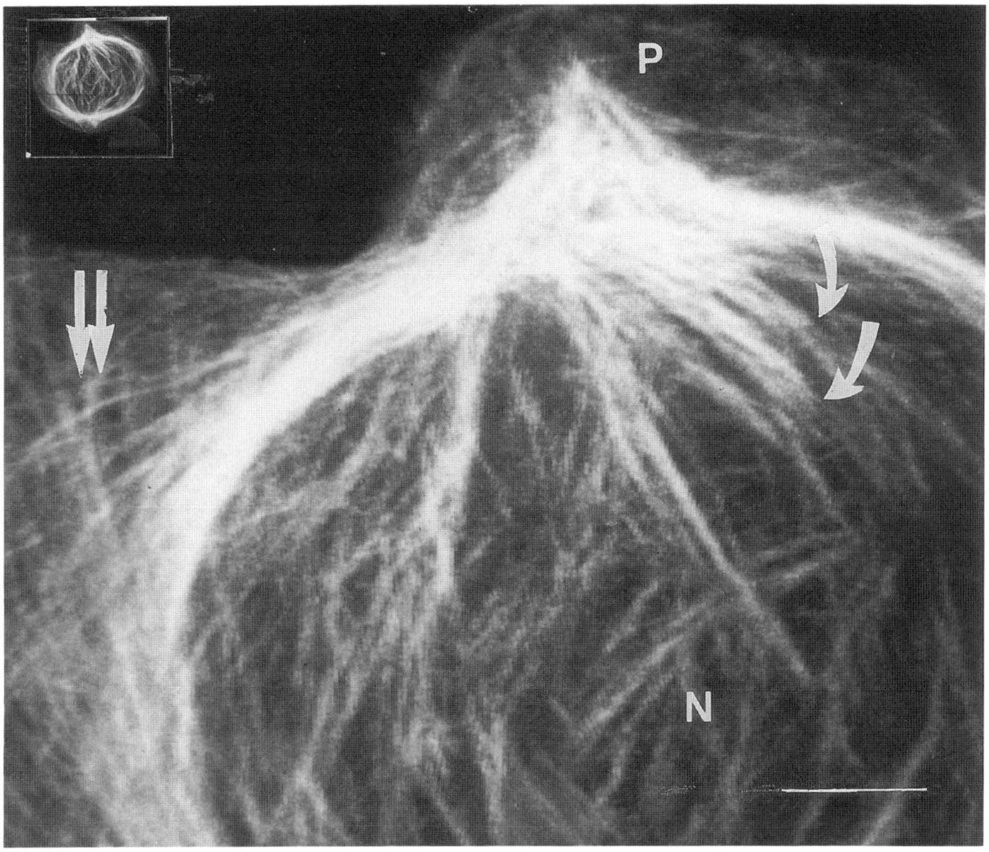

Fig. 3. *Confocal microscopy of* Haemanthus *endosperm cell in prophase at the time of the nuclear envelope breakdown. Antitubulin labeling of microtubules. Detail of one half-spindle where a defined pole is seen without the presence of any centrosomal structure. Microtubules emerging from the poles (P) penetrate into the nucleus (N) and are captured by kinetochores (arrows) as shown on serial optical sections. These data suggest that microtubule-organizing center-like components that were on the nuclear surface during interphase are redistributed into the spindle poles, where they nucleate highly dynamic mitotic microtubules. Criss-crossing of microtubules assembled at the two opposite poles is detected in the cytoplasm (double arrow). Bar = 10 μm.*

de novo from the nuclear surface, since, with rare exception, none are reported for G_1 cells. But during the next transition, does the spindle utilize tubulin from the decaying preprophase band? In cells, such as *Haemanthus* endosperm that have no cellulosic wall, there are no preprophase bands or cortical MTs since only NAMTs are present (Fig. 3). In both cases, therefore, the situation is entirely different from what occurs in centrosome equipped cells, as in animals.

The preprophase band is not disassembled prior to formation of the prophase spindle [e.g., Wick and Duniec, 1983] around the nucleus in walled cells, and a continuous transition is observed between cytoplasmic microtubules and spindle organization in nonwalled cells [Schmit et al., 1983]. The so-called prophase spindle that is formed in both cell types before nuclear envelope breakdown would appear to be an expression of the enhanced activity of MT nucleating sites at the nuclear surface, as seen in Figure 3 [for reviews, see Lambert et al., 1991; Smirnova and Bajer, 1992].

How these bipolar, oriented microtubules are formed without centrosomal assistance is obscure, but multiple and transitory sites of microtubule convergence are often observed on the nuclear surface of prophase *Haemanthus* cells. They progressively form two major aster-like arrays, while chromatin is progressively condensed inside the nucleus in late G_2 [Schmit et al., 1983]. This indicates that the chromosomal and microtubule cycles are concomitantly regulated. These first signs of microtubule reorganization and/or assembly on the nuclear surface predict the future formation of spindle poles and have been shown to be associated with calmodulin [Vantard et al., 1985; Wick et al., 1985], which is later distributed to the polar regions during mitosis of most animal or plant cell types. Striking microtubule convergence at the spindle poles, which later seems to be moved away from the nucleus, as observed by confocal imaging (Fig. 3), suggests the presence of centrosomal-like material at these polar regions.

In meristematic cells the preprophase band has generally disassembled by metaphase at the latest [Wick, 1991a; Gunning, 1992]. This results in an increase in the cytosolic tubulin pool upon nuclear envelope breakdown and can be correlated with the sudden microtubule assembly that contributes to the development of spindle fibers. Polar nucleation of these fibers might therefore be identical to the process found in wall-free *Haemanthus* cells, as in both cell types the individualization of spindle poles has already been achieved by this stage. Despite the absence of a centrosome, these prophase mitotic microtubules in plants demonstrate similar dynamic properties to centrosomally originated astral arrays in animal cells. This is supported by the rapid incorporation of microinjected reporter tubulin into living prophase cells of *Tradescantia* stamen hairs [Zhang et al., 1990]. The fact that exogenous tubulin is incorporated into preprophase band, spindle and then phragmoplast demonstrates that these three assemblies are formed from a common tubulin pool.

Common epitopes between plant and animal centromere/kinetochore proteins have been demonstrated with human autoantibodies from CREST patients [Mole-Bajer et al., 1990; Liu and Palevitz, 1991]. This indicates a degree of conservation of centrosome/kinetochore antigens between the animal and plant kingdoms. Further investigations will certainly prove the role of motor proteins in chromosome movements on the spindle as is now established in animals.

C. Phragmoplast Microtubules and Cytokinesis

During late anaphase, the plant phragmoplast develops in the interzone and is formed of two sets of microtubules of opposite polarity [Euteneuer et al., 1982] that intermingle as a ring at the equatorial region. The phragmoplast develops equally well in normal-walled cells, *Haemanthus* endosperm cells, and even in cytoplasmic fragments [Bajer and Mole-Bajer, 1982, 1986]. This indicates that, although the structure usually forms at a particular stage of the cell cycle, it can be uncoupled from nuclear divisions, as in the delayed cytokineses that cellularize the endosperm. Furthermore, it is well established that phragmoplasts can form between nonsister nuclei, or even by self-reorganization of polymers in nondividing cytoplasts [Smirnova and Bajer, 1992]. This weakens the concept that the phragmoplast forms as a spindle remnant. New functional assays *in vivo* and *in vitro* provide additional evidence on the origin of the phragmoplast and are discussed now in connection with the related topic of plant MT nucleation sites.

III. HIGHER PLANT MICROTUBULE NUCLEATION SITES: IDENTIFICATION AND FUNCTIONAL PROBES

In the majority of animal cells, as well as in yeasts or fungi, most if not all interphase or mitotic microtubules emerge from the centrosome(s), there being one centrosome in interphase, and two at the opposite poles during mitosis, with the minus end of microtubules being anchored at the centrosome. The cytoskeletal configuration of higher plant cells suggests various possibilities. There could be one immobile set of nucleation sites from which all the MT arrays emerge, or there could be one set of nucleation sites that moves around the cell, polymerizing different arrays according to the cell cycle (cf. Mazia's concept [1987] of a flexible centrosome), or there could be multiple, independent, immobile nucleation sites, each of which is dedicated to a single MT array and is only active at the appropriate time of the cell cycle. The term *microtubule nucleation site* (MTNS) is preferable to microtubule-organizing center in plants since MTs are not polymerized upon an organized template and since there are clear cases in which MT organization is a secondary process (e.g., the cortical array).

A. The Plant Nuclear Surface as an MTNS

The plant nuclear surface was first proposed as a putative MTNS on the basis of structural data indicating that the anchorage of cytoplasmic microtubules to the nuclear envelope was often associated with the presence of a dense perinuclear material [Lambert, 1980]. This idea was supported by antitubulin staining showing cytoplasmic MTs to be associated with the nucleus at particular stages of the cell cycle [De Mey et al., 1982; Bajer and Mole-Bajer, 1982; Schmit et al., 1983; Wick and Duniec, 1983; Falconer et al., 1988; Lambert et al., 1991; Staiger and Lloyd, 1991; Gunning, 1992].

Microinjected fluorescent tubulin in living cells of *Tradescantia* stamenal hairs [Zhang et al., 1990] or *Paramecium* axonemal tubulin incorporated into lysed *Haemanthus* endosperm cells [Vantard et al., 1990] provides functional evidence that the nuclear surface of the higher plant cell functions as a microtubule nucleation site. This is supported by recent studies with monoclonal antibodies to pericentriolar material of isolated animal centrosomes [Chevrier et al., 1992]. These antibodies stain the nuclear surface of different plant cell types such as *Haemanthus* endosperm or maize root tip cells [Chevrier et al., 1992]. However, no staining could be detected at the cell cortex. Confocal imaging of these labeled cells indicates the presence of multiple reactive sites regularly distributed on interphase or telophase nuclei that correspond to previously detected anchoring sites of microtubules at the nuclear envelope [Lambert et al., 1991]. If the nuclear surface is "the" ("major") plant MTNS, the distal plus end of cytoplasmic microtubules must therefore be directed toward the cell membrane. This is in agreement with the incorporation of *Paramecium* tubulin indicating an elongation of intracellular plant microtubules at their distal end, while new, short, and entirely labeled microtubules are seen at the nuclear envelope, suggesting that that is where nucleation occurs [Vantard et al., 1990].

Previously, human autoimmune antibodies (5051 serum) were also shown to decorate the periphery of the nucleus and the spindle poles of higher plant cells [Clayton et al., 1985; Wick, 1985]. The significance of this cross-reactivity by scleroderma sera has been questioned [Harper et al., 1989], but higher plant polypeptides recognized by anticentrosomal antibodies are now being studied and could help to resolve this problem (Stoppin, personal communication).

B. Cortical and Nucleus-Associated Microtubules: A Common Origin?

In cells such as *Haemanthus* endosperm that have no cortical MT array, nucleus-associated microtubules are present from telophase to the next prophase [Bajer and Mole-Bajer, 1982; De Mey et al., 1982; Schmit et al., 1983; review by Lambert et al., 1991]. In walled cells, however, it is far less clear that NAMTs are always present beyond a brief phase following cytokinesis. Wick [1985] and Clayton et al. [1985] showed that cytoplasmic MTs radiated from the reforming nuclear surfaces even while traces of the phragmoplast remained. Once the cortical array becomes ordered, such NAMTs are generally not reported. Hence, the idea has emerged that cortical MTs are nucleated at the nuclear surface in early G_1, migrating out to the cortex where they become increasingly ordered, but that NAMTs disappear by the time this ordering phase is under way.

In large, vacuolated epidermal cells, Flanders et al. [1990b] were able to show by confocal microscopy and image reconstruction that cortical MTs, which were absent in metaphase, returned during early cytokinesis via MT cables radiating through the vacuole from nucleus to cortex. The earliest cortical MTs were in branched V shapes that did not form parallel arrays while the NAMT bundles were present. This view has been confirmed by Gunning [1992] for several meristematic cells in a careful confocal study. However, Gunning did observe that there were some (relatively scarce) MTs between nucleus and cortex in cells with well-ordered cortical MTs. Baluska et al. [1992] also reported NAMTs in G_1 phase cells of the slowly dividing quiescent center of maize. However, in nondividing leaf epidermal cells no such MTs could be detected by confocal microscopy until the nucleus began to migrate away from the cortex toward the center for a wound-induced mitosis [Lloyd et al., 1991]. At the center of the cell, the nucleus becomes the focus for a starburst of NAMTs from which it appears that the preprophase band begins to develop. γ-Tubulin is a member of the tubulin superfamily and appears to be localized to microtubule organizing centers. Recently Liu et al. [1993] have used antibodies to detect this antigen in higher plants. It is associated in a puncture manner with cortical microtubules and accumulates around the nucleus as the preprophase band forms. In summary, the evidence points toward the nuclear surface as the source of MTNSs during early interphase and PPB and spindle formation. It is not certain where additional MTs are nucleated in mature cortical arrays: the observations of Gunning [1992] and Baluska et al. [1992] hint that even cortical MTs can have a nuclear origin, but with the advent of γ-tubulin the idea of cortical nucleation cannot be ruled out.

C. Polar Microtubule Assembly in Mitosis

Concerning the assembly of polar mitotic microtubules, it is highly probable that spindle microtubules are progressively assembled and/or nucleated from diffuse and reorganized MTNSs that originate from the nuclear surface. They are then redistributed at first on the nuclear periphery leading to transitory microtubule converging or assembly sites and then migrate progressively away from the nucleus as seen during formation of aster-like arrays [Schmit et al., 1983]. Using monoclonal antibodies directed against pericentriolar material of animal centrosomes, it has been possible to detect MTNS-like material at the nuclear surface of various plant cells during interphase and telophase, but it has not been possible to detect such material during mitosis [Chevrier et al., 1992].

Polar microtubule assembly varies during the mitotic cycle, suggesting that regulatory processes operate as in cells equipped with centrosomes. In cells with or without a preprophase band, the mitotic microtubules are highly unstable and turn over rapidly, within a few minutes as observed from microinjection assays of reporter tubulins in mitotic *Tradescantia* stamen hairs [Zhang et al., 1991] or *Haemanthus* endosperm cells (Schmit, personal communication).

This situation indicates that the plant pernuclear MTNS components are able to reorganize and to be redistributed at the onset of mitosis under cell cycle regulating processes. Such processes are likely to involve protein kinase activity. Cell cycle-dependent protein phosphorylation in higher plants has been recently identified with MPM-2 antibody, and the maximal reactivity is observed in G_2 [Traas et al., 1992]. The presence of calmodulin raises the question of a putative calcium/calmodulin regulation of microtubule stability during plant mitosis [Vantard et al., 1985; Wick et al., 1985; Lambert and Vantard, 1986].

D. Does the Phragmoplast Contain MT Nucleating Activity?

The phragmoplast arises out of the spindle interzone. Hook decoration studies showed that overlapping ends of *Haemanthus* phragmoplast MTs interdigitated at their plus ends, sharing the same polarity therefore as that of the half-spindles [Euteneuer et al., 1982]. This suggested that the phragmoplast could be a remnant of the spindle, whose MTs had originally been nucleated at the spindle poles. However, phragmoplasts expand beyond the interpolar corridor, forming an ever-widening circle. MTs could depolymerize at the center and repolymerize at the edges, although it is possible that stable MTs are merely passively displaced outward [Bajer and Mole-Bajer, 1972]. Recent tubulin incorporation studies provide new evidence on the origin of phragmoplast MTs. Both the microinjection of exogenous fluorescent tubulin into living cells [Zhang et al., 1990] and the incorporation of a reporter tubulin into lysed cells [Vantard et al., 1990] indicate intense MT polymerization in the interzone with an increased fluorescence during the development and lateral extension of the phragmoplast, as illustrated in Figure 4. These data suggest that mitotic interzonal microtubules and phragmoplast microtubules utilize the same soluble pool of tubulin marked by the exogenous tubulin. This also indicates an important and unanticipated turnover of mitotic and phragmoplast microtubules under steady-state conditions. However, it cannot be decided from the intense equatorial fluorescence whether the existing interdigitating interzonal microtubules are turning over (treadmilling) or whether there is new nucleated assembly. Another elegant investigation based on successive pulses of exogenous fluorescent and nonfluorescent tubulins into isolated phragmoplasts from tobacco BY-2 cultured cells by Asada et al. [1991] confirmed the intense tubulin incorporation at the equator, at the putative plus end of microtubules. After a second incorporation of nonfluorescent tubulin the fluorescent zone split into two bands that moved away from each other and from the equator. This behavior appeared to be GTP dependent since it was not inhibited by the nonhydrolyzable nucleotide analogues GMP-PNP and AMP-PNP but was preferentially induced by GTP. The authors suggest that tubulin, added to the overlapping plus ends, is moved away from the equator by a form of MT sliding. However, the results are also consistent with a flux of tubulin subunits through the phragmoplast MTs after addition at the plus end. In a recent paper, Yasuhara et al. [1992] isolated a 100 kD protein from phragmoplasts by elution with ATP, but it does not appear to be a candidate for the GTPase since its binding is not sensitive to GTP.

The equatorial region of the phragmoplast contains dense material that resembles the amorphous material around centrioles. In the cell plate some antigens are recognized by antibodies raised against pericentriolar material (Lambert et al., unpublished data), suggesting putative MTNS activity at least in later stages of phragmoplast development. Such material may also be associated with stabilized microtubule fragments in the equatorial interdigitating region that could act as seeds for a new microtubule assembly, as occurs in axons.

In phragmoplasts that form between nonsister nuclei, as observed in *Haemanthus* syncitia, there is no spindle remnant. However, there is still structural continuity between microtubules forming the dense phragmoplast ring and microtubules emerging from the nuclear surface [De Mey et al., 1982; Bajer and Mole-Bajer, 1982; Schmit et al., 1983]. These observations support the hypothesis that microtubules nucleated by the nuclear surface may take active part in formation of the phragmoplast between nonsister nuclei. This may contribute as well to phragmoplast expansion that occurs when daughter nuclei are being formed in telophase and nucleate new microtubules. In these cases the distal plus ends of these microtubules would interdigitate at the equator, as expected from polarity data [Euteneuer et al., 1982]. In this region they may become more stable due to the interdigitation and bundling processes.

In conclusion, the origin of the phragmoplast may well be dependent on microtubules that were originally assembled at the poles or at the nuclear surface, depending on how and when the phragmoplast is formed in the cell cycle. The new sites of MT assembly, revealed by incorporation studies to be at the equator, may be based on preexisting microtubule fragments (originating from previous interdigitating microtubules) that are centrifugally displaced by the growth of the cell plate. It cannot yet be ruled out that the phragmoplast contains its own nucleating material. Whichever hypothesis is correct, phragmoplast MTs turn over more rapidly than previously thought.

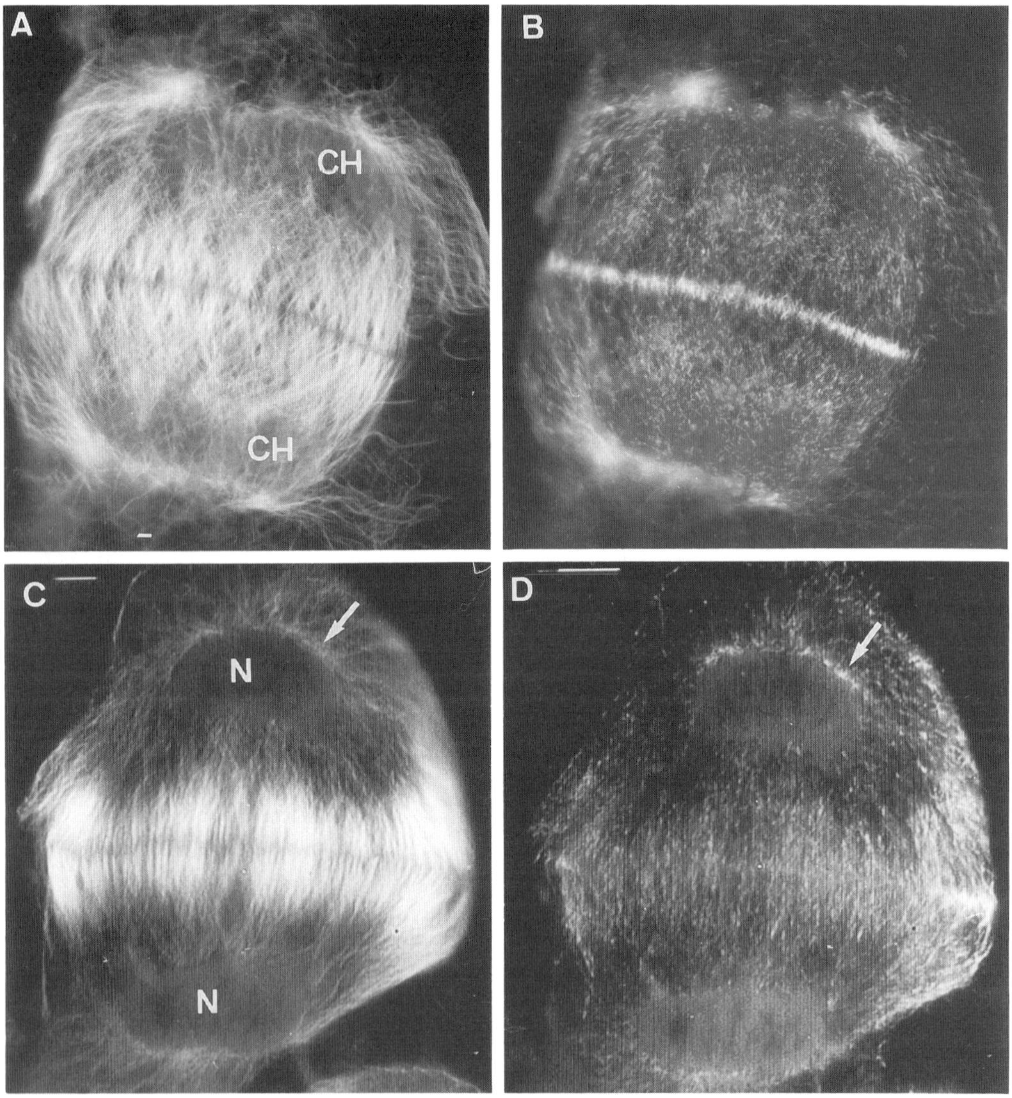

Fig. 4. *Incorporation of* Paramecium *axonemal tubulin into saponin-lysed cells of* Haemanthus *at two successive stages of telophase.* **A,B:** *The same early stage.* **C,D:** *Another cell at a later stage. Double label immunofluorescence. A/C: The whole intracellular plant microtubule arrays detected by rhodamine/antibody directed against brain tubulin. B/C: The same cell labeled with FITC-specific antibody directed against* Paramecium *axonemal tubulin that recognizes only microtubule segments that incorporated the exogenous* Paramecium *tubulin. An intense incorporation of* Paramecium *tubulin is detected at the equatorial region (B) at the start of telophase while tubulin-incorporated fragments are seen within the whole phragmoplast in later stage. The data, complemented with electron microscopic studies using gold-labeled* Paramecium *antibodies, indicate an elongation process of microtubules distal from the poles in the interzone and an intense assembly at the equator, implying that the equatorial region may also function as a MTNS, particularly during phragmoplast extension. [Reproduced from Vantard et al., 1990, with permission of the publisher.]*

IV. ACTIN CYCLE/MICROTUBULE CYCLE

A large body of evidence obtained during the last few years has revealed new features of the plant actin cytoskeleton that seem to be mostly specific to the higher plant cells [Lloyd, 1989; Lambert et al., 1991; McCurdy and Williamson, 1991; Pierson and Cresti, 1992]. They suggest that in most plant cells studied thus far, independent of the presence of the cell wall, actin microfilaments intermingle with microtubules and are often found parallel to most microtubule arrays. The codistribution of plant actin polymers and microtubules in cortical and preprophase band arrays, within and around the mitotic spindle, as well as in the phragmoplast during cytokinesis, raises the fundamental question of the role of F-actin in the reorganization of the cytoskeleton and particularly microtubules. The nucleation sites of actin filaments and their turnover are difficult to assess. It would be important to know if an F-actin cycle is superimposed onto the microtubule cycle in higher plant cells and to know more of the mechanisms that would control the codynamics of these different polymers. Microinjection of fluorescent phalloidin in living mitotic cells of *Haemanthus,* at minimal concentrations that avoid stabilization, revealed the active stretching of actin-labeled filaments on the surface of the microtubular spindle during chromosome movement [Schmit and Lambert, 1990]. This indicates microfilament–microtubule interactions and also the flexibility of F-actin bundles in living plant cells. These data suggest that under these experimental conditions, the cortical F-actin population remains as an elastic cage around the mitotic spindle while microtubules turn over more actively.

The role of actin in mitosis remains debatable as cytochalasins do not inhibit chromosome movement. During phragmoplast formation, however, the actin ring that appears at the equatorial region seems to correspond to an active assembly of new short microfilaments as seen in phalloidin-microinjected endosperm cells [Schmit and Lambert, 1990]. This suggests the presence of coregulated processes of microtubule and F-actin polymerization and/or nucleation during plant cytokinesis as also suggested in cultured cells [Lloyd and Traas, 1988] or miniprotoplasts [Sonobe, 1990b]. A better knowledge of microtubule- and actin-associated proteins is now one important goal in plant cytoskeletal research. Few associated proteins are known. The heavy chain of myosin [Parke et al., 1986] and a calcium-dependent protein kinase (CDPK) [Putnam-Evans et al., 1989; McCurdy and Williamson, 1991] have been shown to colocalize with F-actin in dividing higher plant cells, suggesting a possible actomyosin-based regulation of vesicle transport. The G-actin– and phosphoinositide-binding protein profilin has recently been detected in birch pollen [Valenta et al., 1991] and may have a role in transducing extracellular signals into dramatic changes of actin organization.

V. THE DIVERSITY OF HIGHER PLANT TUBULINS: ITS POSSIBLE ROLE IN MICROTUBULE PROPERTIES AND FUNCTIONS

The existence of multiple tubulin genes allows the possibility that different isoforms may possess different roles *in vivo,* potentially by interacting with different microtubule-associated proteins (MAPs). However, tubulin isoforms do not seem to be specific to particular microtubule assemblies. Hussey et al. [1987], for example, showed that all tubulin isotypes identifiable in carrot cells were utilized by the interphase MTs of detergent-extracted cytoskeletons. Multiple tubulin isoforms have been reported in plants [Fosket, 1989; Hussey et al., 1991; Fosket and Morejohn, 1992], and some of them are developmentally regulated. Six α- and nine β-tubulin genes have been described recently in *Arabidopsis* and the transcripts of several of these isoforms accumulate preferentially in roots, leaves, or flowers [Kopczak et al., 1992; Snustad et al., 1992; Carpenter et al., 1992]. In pollen development, for example, microtubules possess particular dynamics and activity that could be TUA1 functionally specific as suggested in *Arabidopsis* [Carpenter

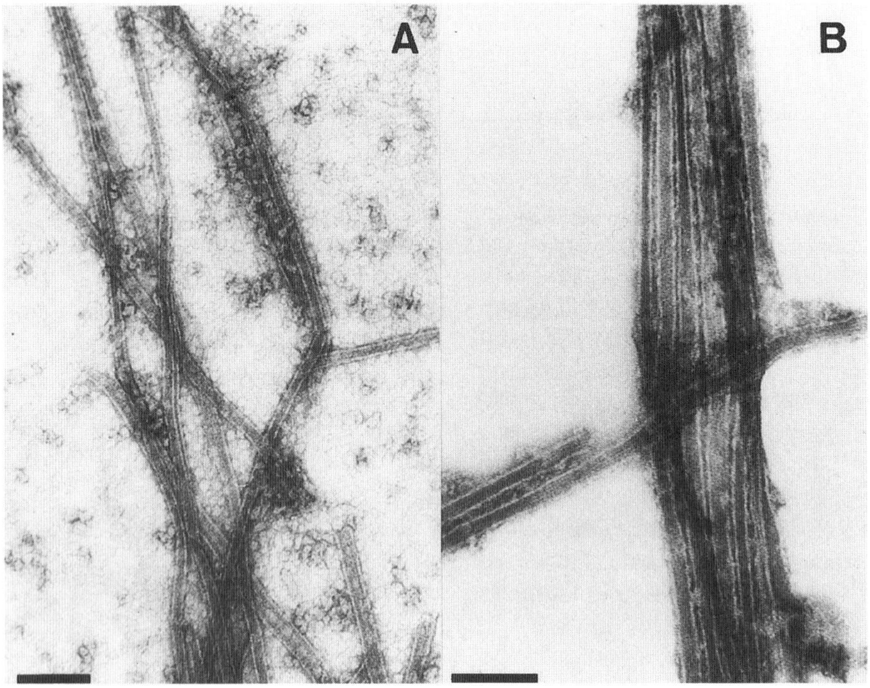

Fig. 5. In vitro *microtubule bundling induced by the presence of higher plant microtubule-associated proteins.* **A:** *Control, purified brain microtubule assembly; no bundling is observed.* **B:** *The same preparation in the presence of a MAP fraction purified from maize-cultured cells. Bar = 10 μm. [Reproduced from Vantard et al., 1991, with permission of the publisher.]*

et al., 1992]. The TUA1 gene encodes a divergent α tubulin where the divergence takes place near the C terminus and within the N-terminal 70 amino acids [Kopczack et al., 1992], regions implicated in binding to MAPs and calcium. These data suggest the hypothesis that tubulin isoforms could play a role in binding specific MAPs that regulate the rate of microtubule assembly and stability and consequent microtubule function. However, it is not yet established that a specific microtubule-based process is present in pollen, which requires a particular tubulin and which cannot be substituted by another tubulin.

VI. HIGHER PLANT MAPS AND THE MT CYCLE

MAPs can account for the assembly and stabilization of microtubules at low tubulin concentration, for their ability to interact with other microtubules and/or other cytoskeletal polymers or organelles, and for regulating their anchoring and bundling properties (Fig. 5). The properties of MAPs are extensively developed in other chapters of this book, and there are indications that plant MAPs will be generically similar. Higher plant MAPs have been characterized by their ability to bind to taxol-stabilized MTs from brain [Cyr and Palevitz, 1989; Cyr, 1991b]. Vantard et al. [1991] [see also, Schellenbaum et al., 1993] have used taxol-mediated assembly of plant microtubules to identify MAPs from maize suspension cells. In these studies the molecular weights ranged between 39 and 114 kD. These plant MAPs initiate and promote purified tubulin assembly from both brain or plant origins under conditions of nonefficient self-assembly and induce microtubule bundling *in vitro*. One of these proteins of about 83 kD in maize is immunologically related to neural tau [Vantard et al., 1991]. As plant MAPs coassemble with brain and plant tubulins it is suggested that their microtubule-binding domain is highly conserved through evolution. Although algae are beyond the scope of this review, it is worth mentioning that Maekawa et al. [1990] have identified in *Dichotomosiphon tuberosus* a protein of 90 kD that is recognized by an antibody raised against a synthetic peptide to the tubulin-binding domain of the bovine adrenal medullary gland 190 kD MAP. Several such tubulin-binding repeats occur in tau, MAP2, and nonbrain MAP4 [references in Schellenbaum et al., 1992b], uniting them as a family of structural MAPs. These also share the property of heat stability, and it is interesting that the algal MAP is found in the heat stable fraction.

As plant microtubules exhibit important *in situ* bundling activity (as in cortical or preprophase band arrays, or during the drastic reorganization of the cytoskeleton at the onset of mitosis, and during phragmoplast formation), different MAPs may regulate differential stability and function of these distinct microtubule populations. The tobacco cell line BY-2 is proving extremely useful since it can yield high levels of mitotic synchrony. Recently, Yasuhara et al. [1992] isolated phragmoplasts from these cells and showed that a 100 kD protein, eluted with ATP, could bind to taxol-stabilized brain MTs. The protein could also be isolated from spindles, although there is no indication as to whether it associates with interphase or preprophase band MTs.

Immunocytochemical investigations are rare [Cyr and Palevitz, 1987], and the activities of particular MAPs in the living cell are mostly unknown in higher plant cells. It will be important to know if the functions of higher plant MAPs are developmentally regulated as in neural systems. As higher plants lack defined centrosomes, specific MAPs could be involved in the control of the critical concentration of intracellular plant tubulin and the formation and stabilization of microtubular seeds [Cyr, 1991b]. STOP proteins (Margolis, this volume) have not yet been demonstrated unambiguously in higher plants, although cold-stable microtubules are observed in kinetochore microtubule bundles that resist low temperature [Lambert and Bajer, 1977] and alkaloids. The selective distribution of calmodulin along kinetochore fibers [Vantard et al., 1985; Wick et al., 1985] suggests that kinetochore–microtubule dynamics and disassembly in anaphase might be controlled by a calcium/calmodulin-dependent kinase that could involve STOP-like associated proteins activity in plants [Lambert and Vantard, 1986]. Microtubule stability in lysed protoplasts has been also shown to be calcium/calmodulin dependent via MAPs [Cyr, 1991a,b].

Motor MAPs, such as dynein and kinesin, remain poorly documented in plants and will not be discussed here. Nevertheless, their activity in plant kinetochore microtubule functions and in intracellular microtubule-mediated transport is highly likely as has been well documented in animal systems. An immunoreactive homologue of mammalian kinesin has recently been detected in pollen tubes [Tiezzi et al., 1992], announcing rapid progress in this field in the near future.

VII. CYTOSKELETAL MUTANTS IN HIGHER PLANTS: AN IMPORTANT TOOL IN THE INVESTIGATION OF THE MICROTUBULE CYCLE

Mutants that affect cell division, growth and development, as well as organelle transport represent valuable systems in the evaluation of cytoskeletal function. Higher plant cytoskeletal mutants are now being investigated in several systems. For example, mutations that cause male sterility in plants are often related to defects in meiosis and/or pollen abnormalities. The phenotypes of some of these mutants in maize suggest that they may be caused by cytoskeletal abnormalities [Staiger and Cande, 1993]. By examining the microtubule cycle in maize plants homozygous for the recessive allele *dv (dv/dv)*, Staiger and Cande [1990] found that a defect in spindle pole organization during prophase to metaphase transition may involve microtubule-organizing center organization and induce abnormal formation of multinucleated daughter cells. If results from fungal genetics are a good indicator, transposon-mediated mutagenesis in plants should reveal exciting information concerning the role of the different cytoskeletal components during the cell cycle and in plant development.

VIII. CONCLUDING REMARKS

From recent tubulin incorporation studies, we are beginning to understand a little more about the sites and the dynamics of MT assembly. A wide variety of immunocytochemical and fluorescent analogue cytochemical studies point to the concept of a major, diffuse plant MTNS whose components are distributed on the nuclear surface of all higher plant cells. This diffuse higher plant MTNS seems to be reorganized and redistributed at the polar regions, during mitosis, independent of the presence or absence of the cell wall. During early interphase and during the G_2 events leading to preprophase band formation, MTs emerge from the nucleus. This perinuclear and during mitosis polar, origin of plant microtubules has been assessed by the incorporation of exogenous tubulins into plant cells, and the sites where microtubules elongate have been identified at most stages of the cell cycle. The nucleus therefore seems to be the primary site of MT initiation, but the phragmoplast may possess independent MT nucleating capacity. If there are MTNSs in the phragmoplast, then it is important to know what controls their activity independently of nucleation at perinuclear sites, and the advent of fluorescent analogue cytochemistry should help here. It is still not clear whether MTs interpolated into the mature cortical array derive from the nucleus or from cortical nucleation sites.

It is now evident that the codistribution of actin filaments and MTs survives some drastic reorganizations (i.e., preprophase band formation) but not others (i.e., mitosis). Further progress is likely to be made in identifying actin- and MT-associated proteins and then by understanding how they are subject to external signals and cell cycle controls.

IX. REFERENCES

Akashi T, Shibaoka H (1991): Involvement of transmembrane proteins in association of cortical microtubules with the plasma membrane in tobacco BY2 cells. J Cell Sci 98:169–174.

Asada T, Sonobe S, Shibaoka H (1991): Microtubule translocation in the cytokinetic apparatus of cultured tobacco cells. Nature 350:238–241.

Bajer AS, Mole-Bajer J (1972): Spindle dynamics and chromosome movement. Int Rev Cytol Suppl 3:1–263.

Bajer AS, Mole-Bajer J (1982): Asters, poles, and transport properties within spindle-like microtubule arrays. Cold Spring Harbor Symp Quant Biol 46:263–283.

Bajer AS, Mole-Bajer J (1986): Reorganization of microtubules in endosperm cells and cell fragments of the higher plant *Haemanthus*. J Cell Biol 102:263–281.

Baluska F, Parker JS, Barlow PW (1992): Specific patterns of cortical and endoplasmic microtubules associated with cell growth and tissue differentiation in roots of maize. J Cell Sci 103:191–200.

Baskin TI, Cande WZ (1990): The structure and function of the mitotic spindle in flowering plants. Annu Rev Plant Physiol Plant Mol Biol 41:277–315.

Carpenter JL, Picense SE, Snustad DP, Silflow CD (1992): Preferential expression of an alpha tubulin gene of *Arabidopsis* in pollen. Plant Cells 4:557–571.

Chevrier V, Komesli S, Schmit AC, Vantard M, Lambert AM, Job D (1992): A monoclonal antibody, raised against mammalian centrosomes and screened by recognition of plant microtubule organizing centers, identifies a pericentriolar component in different cell types. J Cell Sci 101:823–835.

Clayton L, Black MC, Lloyd CW (1985): Microtubule nucleating sites in higher plant cells, identified by an autoantibody against pericentriolar material. J Cell Biol 101:319–324.

Cyr RJ (1991a): Calcium-calmodulin affects microtubule stability in lysed protoplasts. J Cell Sci 100:311–318.

Cyr RJ (1991b): Microtubule-associated proteins in higher plants. In Lloyd CW (ed): The Cytoskeletal Basis of Plant Form and Growth. New York: Academic Press, pp 57–67.

Cyr RJ, Palevitz BA (1989): Microtubule-binding proteins from carrot. I. Initial characterization and microtubule bundling. Planta 177:245–260.

De Mey J, Lambert AM, Bajer AS, Moeremans M, De Brabander M (1982): Visualization of microtubules in interphase and mitotic plant cells of *Haemanthus* endosperm with the immunogold staining method. Proc Natl Acad Sci USA 79:1898–1902.

Euteneuer U, Jackson WT, McIntosh JT (1982): Polarity of spindle microtubules in *Haemanthus* endosperm. J Cell Biol 94:644–653.

Falconer MM, Donalson G, Seagull RW (1988): MTOCs in higher plant cells: An immunofluorescent study of microtubule assembly sites following depolymerization by APM. Protoplasma 144:46–55.

Flanders DJ, Rawlins DJ, Shaw PJ, Lloyd CW (1990a): Nucleus-associated microtubules help determine the division plane of plant epidermal cells: avoidance of four-way junctions and the role of cell geometry. J Cell Biol 110:1111–1122.

Flanders DJ, Rawlins DJ, Shaw PJ, Lloyd CW (1990b): Re-establishment of the interphase microtubule array in vacuolated plant cells studied by confocal microscopy and 3-D imaging. Development 110:897–904.

Fosket DE (1989): Cytoskeletal proteins and their genes in higher plants. In Stumpf PK (ed): The Biochemistry of Plants. New York: Academic Press, pp 392–454.

Fosket DE, Morejohn LC (1992): Structural and functional organization of tubulin. Annu Rev Plant Physiol Plant Mol Biol 100:49–92.

Goodbody KC, Venverloo CJ, Lloyd CW (1991): Laser microsurgery demonstrates that cytoplasmic strands anchoring the nucleus across the vacuole of premitotic plant cells are under tension. Development 113: 931–939.

Gunning BES (1992): Use of confocal microscopy to examine transitions between successive microtubule arrays in the plant cell division cycle. In Shibaoka H (ed): Cellular Basis of Plant Growth and Development. Proc VII Int Japan Prize Biol Symp. Osaka: Osaka University, pp 145–155.

Gunning BES, Sammut M (1990): Rearrangements of microtubules involved in establishing cell division planes start immediately after DNA synthesis and are completed just before mitosis. Plant Cell 2:1273–1282.

Gunning BES, Wick SM (1985): Preprophase band, phragmoplast and spatial control of cytokinesis. J Cell Sci Suppl 2:157–179.

Harper JDI, Mitchison JM, Williamson RE (1989): Does the autoimmune serum 5051 specifically recognize microtubule-organizing centres in plant cells? Cell Biol Int Rep 13:471–483.

Hussey PJ, Traas JA, Gull K, Lloyd CW (1987): Isolation of cytoskeletons from synchronized plant cells: The interphase microtubule array utilises multiple tubulin isotypes. J Cell Sci 88:225–230.

Hussey PJ, Snustad DP, Silflow CD (1991): Tubulin gene expression in higher plants. In CW Lloyd (ed): The Cytoskeletal Basis of Plant Growth and Form. New York: Academic Press, pp 15–27.

Kakimoto T, Shibaoka H (1987): Actin filaments and microtubules in the preprophase band and phragmoplast of tobacco cells. Protoplasma 140:151–156.

Katsuta J, Shibaoka H (1988): The roles of the cytoskeleton and the cell wall in nuclear positioning in tobacco BY2 cells. Plant Cell Physiol 29:403–413.

Katsuta J, Shibaoka H (1992): Inhibition by kinase inhibitors of the development and the disappearance of the preprophase band of microtubules in tobacco BY-2 cells. J Cell Sci 103:397–405.

Kopczak SD, Haas NA, Hussey PJ, Silflow CD, Snustad DP (1992): The small genome of *Arabidopsis* contains at least six expressed a tubulin genes. Plant Cell 4:539–547.

Lambert AM (1980): The role of chromosomes in anaphase trigger and nuclear envelope activity in spindle formation. Chromosoma 76:295–308.

Lambert AM (1993): Microtubule-organizing centers in higher plants. Curr Opin Cell Biol 5:116–122.

Lambert AM, Bajer AS (1977): Microtubule distribution and reversible arrest of chromosome movement induced by low temperature. Cytobiologie 15:1–23.

Lambert AM, Vantard M (1986): Calcium and calmodulin as regulators of chromosome movement during mitosis in higher plants. In Trewavas (ed): Molecular and Cellular Aspects of Calcium in Plant Development. NATO ASI series. New York: Plenum, pp 175–183.

Lambert AM, Vantard M, Schmit AC, Stoeckel H (1991): Mitosis in plants. In Lloyd CW (ed): The Cytoskeletal Basis of Plant Growth and Form. New York: Academic Press. pp 199–208.

Ledbetter MC, Porter KR (1964): Morphology of microtubules of plant cells. Science 144:872–874.

Liu B, Mare J, Joshi HC, Palevitz BA (1993): A γ-tubulin-related protein associated with the microtubule arrays of higher plants in a cell cycle-dependent manner. J Cell Sci 104:1217–1228.

Liu B, Palevitz BA (1991): Kinetochore fiber formation in dividing generative cells of *Tradescantia*. J Cell Sci 98:475–482.

Lloyd CW (1989): The plant cytoskeleton. Curr Opin Cell Biol 1:30–35.

Lloyd CW (1991a): Cytoskeletal elements of the phragmosome establish the division plane in vacuolated higher plant cells. In Lloyd CW (ed): The Cytoskeletal Basis of Plant Growth and Form. New York: Academic Press, pp 245–257.

Lloyd CW (1991b): Probing the plant cytoskeleton. Nature 350:189–190.

Lloyd CW, Traas JA (1988): The role of F-actin in determining the division plane of carrot suspension cells: Drug studies. Development 102:211–222.

Lloyd CW, Venverloo CJ, Goodbody KC, Shaw PJ (1991): Confocal laser microscopy and three-dimensional reconstruction of nucleus-associated microtubules in the division plane of vacuolated plant cells. J Microsc 166:99–109.

Maekawa T, Ogihara S, Murofushi H, Nagai R (1990): Green algal microtubule associated protein with a molecular weight of 90kDa which bundles microtubules. Protoplasma 158:10–18.

Mazia D (1987): The chromosome cycle and the centrosome cycle in the mitotic cycle. Int Rev Cytol 100: 49–92.

McCurdy DW, Williamson RE (1991): Actin and actin associated proteins. In Lloyd CW (ed): The Cytoskeletal Basis of Plant Growth and Form. New York: Academic Press, pp 3–14.

Mineyuki Y, Wick SM, Gunning BES (1988): Preprophase bands of microtubules and the cell cycle: Kinetics and experimental uncoupling of their formation from the nuclear cycle in onion root tip cells. Planta 174:518–526.

Mineyuki Y, Palevitz BA (1990): Relationship between preprophase band organization, F-actin and the division site in *Allium*. Fluorescence and morphogenetic studies on cytochalasin-treated cells. J Cell Sci 97:283–295.

Mineyuki Y, Yamashita M, Nagahama Y (1991): p34cdc2 kinase homologue in the preprophase band. Protoplasma 162:182–186.

Mole-Bajer J, Bajer AS, Zinkowski RP, Balczon RD, Brinkley BR (1990): Autoantibodies from a patient with scleroderma CREST recognized kinetochores of the higher plant *Haemanthus*. Proc Natl Acad Sci USA 87:3599–3603.

Murata T, Wada M (1989): Reorganization of microtubules during preprophase band development in *Adiantum protonema*. Protoplasma 151:73–80.

Palevitz BA (1987): Actin in the preprophase band of *Allium cepa*. J Cell Biol 104:1515–1519.

Palevitz BA (1991): Microtubule rearrangement, translocation and reutilization. In Lloyd CW (ed): The Cytoskeletal Basis of Plant Growth and Form. New York: Academic Press, pp 45–55.

Parke J, Miller C, Anderson BH (1986): Higher plant myosin heavy chain identified using monoclonal antibody. Eur J Cell Biol 41:9–13.

Pierson ES, Cresti M (1992): Cytoskeleton and cytoplasmic organization of pollen and pollen tubes. Int Rev Cytol 140:73–125.

Putnam-Evans C, Harmon AC, Palevitz BA, Fechheimer M, Cormier MJ (1989): Calcium-dependent protein kinase is localized with F-actin in plant cells. Cell Motil Cytoskel 12:12–22.

Rattner JB, Lew J, Wang JH (1990): p34 cdc2 kinase is located to distinct domains with the mitotic apparatus. Cell Motil Cytoskel 17:227–235.

Schellenbaum P, Vantard M, Lambert AM (1992): Higher plant microtubule-associated proteins: A survey. Biol Cell 76:359–364.

Schellenbaum P, Vantard M, Peter C, Fellous A, Lambert AM (1993): Coassembly properties of higher plant microtubule-associated proteins (MAPs) with purified brain and plant tubulins. Plant J 3:253–260.

Schmit AC, Vantard M, Lambert AM (1983): Aster-like microtubule centers establish spindle polarity during interphase-mitosis transition in higher plant cells. Plant Cell Rep 2:285–288.

Schmit AC, Lambert AM (1987): Characterization and dynamics of cytoplasmic F-actin in higher plant endosperm cells during interphase, mitosis and cytokinesis. J Cell Biol 105:2157–2166.

Schmit AC, Lambert AM (1990): Microinjected fluorescent phalloidin *in vivo* reveals F-actin dynamics and assembly in higher plant cells. Plant Cell 2:129–138.

Seagull RW (1989): The plant cytoskeleton. CRC Crit Rev Plant Sci 8:131–167.

Smirnova EA, Bajer AS (1992): Spindle poles in higher plant mitosis. Cell Motil Cytoskel 23:1–17.

Snustad DP, Haas NA, Kopczak SD, Silflow CD (1992): The small genome of *Arabidopsis* contains at least nine expressed β tubulin genes. Plant Cell 4:549–556.

Sonobe S (1990a): ATP-dependent depolymerization of cortical microtubules by an extract in tobacco BY-2 cells. Plant Cell Physiol 31:1147–1153.

Sonobe S (1990b): Cytochalasin B enhances cytoplasmic cleavage in miniprotoplasts isolated from cultured tobacco cells. Protoplasma 155:239–242.

Staiger CJ, Cande WZ (1990): Microtubule distribution in dv, a maize mutant defective in the prophase to metaphase transition. Dev Biol 139:231–242.

Staiger CJ, Cande WZ (1993): Cytoskeletal analysis of maize meiotic mutants. In Omrod JC, Francis D (eds): Molecular and Cell Biology of the Plant Cell Cycle. Amsterdam: Kluwer Academic Publishers, pp 157–171.

Staiger CJ, Lloyd CW (1991): The plant cytoskeleton. Curr Opin Cell Biol 3:33–42.

Tiezzi A, Moscatelli A, Cai G, Bartalesi A, Cresti M (1992): An immunoreactive homolog of mammalian kinesin in *Nicotiana tabacum* pollen tubes. Cell Motil Cytoskel 21:132–137.

Traas JA, Beven AF, Doonan JD, Cordwener J, Shaw PJ (1992): Cell-cycle-dependent changes in labelling of specific phosphoproteins by the monoclonal antibody MPM-2 in plant cells. Plant J 2:723–732.

Traas JA, Doonan JD, Rawlins DJ, Shaw PJ, Watts J, Lloyd CW (1987): An actin network is present in the cytoplasm throughout the cell cycle of carrot cells and associates with the dividing nucleus. J Cell Biol 105:387–395.

Valenta R, Duchene M, Pettenburger K, Sillaber C, Valent P, Bettelheim P, Breitenbach M, Rumpold H, Kraft D, Scheiner O (1991): Identification of profilin as a novel pollen allergen: IgE autoreactivity in sensitized individuals. Science 253:557–560.

Vantard M, Lambert AM, De Mey J, Picquot P, Van Eldik LJ (1985): Characterization and immunocytochemical distribution of calmodulin in higher plant endosperm cell: Localization in the mitotic apparatus. J Cell Biol 101:488–499.

Vantard M, Levilliers N, Hill AM, Adoutte A, Lambert AM (1990): Incorporation of *Paramecium* tubulin into higher plant cells reveals functional sites of microtubule assembly. Proc Natl Acad Sci USA 87:8825–8829.

Vantard M, Schellenbaum P, Fellous A, Lambert AM (1991): Characterization of maize microtubule-associated proteins, one of which is immunologically related to tau. Biochemistry 30:9334–9340.

Venverloo CJ, Libbenga KR (1987): Regulation of the plane of cell division in vacuolated cells. 1, The function of nuclear positioning and phragmosome formation. J Plant Physiol 131:267–284.

Verde F, Labbe J, Doree M, Karsenti E (1990): Regulation of microtubule dynamics by cdc2 protein kinase in cell free extracts of *Xenopus* eggs. Nature 343: 233–238.

Wada M, Murata T (1991): The cytoskeleton in fern protonemata growth in relation to photomorphogenesis. In Lloyd CW (ed): The Cytoskeletal Basis of Plant Form and Growth. New York: Academic Press, pp 277–288.

Wick SM (1985): Immunofluorescence microscopy of tubulin and microtubule arrays in plant cells III: Transition between mitotic/cytokinetic and interphase microtubule arrays. Cell Biol Int Rep 9:357–371.

Wick S (1991a): The preprophase band. In Lloyd CW (ed): The Cytoskeletal Basis of Plant Growth and Form. New York: Academic Press, pp 231–244.

Wick SM (1991b): Spatial aspects of cytokinesis in plant cells. Curr Opin Cell Biol 3:253–260.

Wick SM, Duniec J (1983): Immunofluorescence microscopy of tubulin and microtubule arrays in plant cells. I. Preprophase band development and concomitant appearance of nuclear envelope associated tubulin. J Cell Biol 97:235–243.

Wick SM, Muto S, Duniec J (1985): Double immunofluorescence labelling of calmodulin and tubulin in dividing plant cells. Protoplasma 126: 198–206.

Williamson RE (1991): Orientation of cortical microtubules in interphase plant cells. Int Rev Cytol 129: 135–206.

Yasuhara H, Sonobe S, Shibaoka H (1992): ATP-sensitive binding to microtubules of polypeptides extracted from isolated phragmoplasts of tobacco BY-2 cells. Plant Cell Physiol 33:601–608.

Zhang D, Wadsworth P, Hepler PK (1990): Microtubule dynamics in living dividing plant cells: Confocal imaging of microinjected fluorescent brain tubulin. Proc Natl Acad Sci USA 87:8820–8824.

ABOUT THE AUTHORS

ANNE MARIE LAMBERT is Professor of Cell Biology at the University Louis Pasteur in Strasbourg, France, where she teaches graduate and postgraduate courses on topics in plant developmental biology. She is a group leader at the Centre National de la Recherche Scientifique (C.N.R.S.) in the Institute of Plant Molecular Biology in Strasbourg. She pursued doctoral reserach in the laboratory of Professor Alice Gagnieu in Strasbourg, where she earned her Ph.D in 1971 for work on the dynamics of plant microtubules. As a Fullbright fellow in the group of Drs. A. and J. Bajer at the University of Oregon, she did research in plant mitosis. Back in France, she strated up her laboratory, where the theme of research is at present focused on the characterization of higher plant microtubule-nucleating and/or organizing centers (MTOCs) and microtubule-associated proteins (MAPs). She is a member of the editorial boards of the *Plant Journal* and *Biology of the Cell.*

CLIVE W. LLOYD is Senior Principal Scientific Officer in the Department of Cell Biology at the John Innes Institute, Norwich, England. He is engaged in full-time research into the role of the cytoskeleton in plant morphogenesis. After receiving first class honors in zoology at the University College of Wales, Aberystwyth, he continued in that department to investigate the involvement of cell surface glycoproteins in the adhesion of animal cells. After being awarded his Ph.D. in 1971, he went, as Beit Memorial Research Fellow, to Dr. G.M.W. Cook's laboratory at the Strangeways Laboratory, Cambridge, England, to continue his research into cell surface carbohydrates in animal cell adhesion. Next he joined Dr. D.A. Rees at the Unilever Research Sharnbrook Laboratory, where ideas on the importance of the underlying cytoskeleton were beginning to emerge. In 1979 Dr. Lloyd turned his attention to plants and the part that the different cytoskeletal arrays play in regulating cell shape and, in particular, the plane of cell division. This theme was maintained when he moved, as Royal Society Senior Research Fellow in the Biological Sciences, to the John Innes Institute, where he is now a staff member. He is a member of the editorial boards of the *Journal of Cell Science, Protoplasma,* and *Cell Motility and the Cytoskeleton.*

Microtubules: 343–365

Kinesin-Based Organelle Transport

Jonathan M. Scholey and Ronald D. Vale

I. INTRODUCTION

A characteristic property of eukaryotic cells is their ability to use energy liberated from ATP hydrolysis to maintain a dynamic yet highly organized state. To a large extent this is accomplished through the use of intracellular transport systems that move and position subcellular components. At the molecular level, such transport systems (Fig. 1) consist of four basic components: 1) a fibrillar protein (actin or tubulin) that self-assembles into a cytoskeletal fiber (microfilament or microtubule), serving as a track along which intracellular components are moved; 2) a motor protein that splits ATP and transports particles along the fiber (myosins along actin; kinesins and dyneins along tubulin; 3) the cargo (e.g., membranous organelles, chromosomes, or macromolecules) that is attached to the motor protein and is thus transported along the fiber; and 4) the fuel, usually MgATP, that provides the energy required to move the cargo.

The transport system consisting of tubulin, kinesin, membranes, and MgATP was discovered less than a decade ago [Vale et al., 1985a], but much has already been learned about the kinesin molecule and its role in using MgATP as a fuel to drive the movement of membranous cargo toward the fast polymerizing–depolymerizing (plus) ends of polar microtubule tracks within cells. The aim of this discussion is to provide a brief description of current ideas concerning the biological functions and molecular mechanism of action of the kinesin-based intracellular transport system. We focus on kinesin itself, with the expectation that descriptions of the transport of membrane-bound organelles (and other cargoes) by kinesin-related proteins and dyneins will be covered elsewhere in this volume (see Pereira and Goldstein; Collins and Holtzbauer et al., this volume). For additional information concerning microtubule-based motor proteins and intracellular transport, the reader is directed to several reviews that were recently published [Bloom, 1992; Brady, 1991; Gelfand, 1989; Huitorel, 1988; McIntosh and Porter, 1989; Sawin and Scholey, 1991; Schroer and Sheetz, 1991; Sheetz, 1987; Vale, 1987, 1990, 1992; Vallee and Shpetner, 1990].

II. ROLE OF MICROTUBULES IN INTRACELLULAR MEMBRANE ORGANIZATION AND VESICLE TRAFFICKING

The membrane-bound organelles of eukaryotic cells at interphase form a system of specialized compartments that occupy characteristic positions in the cytoplasm; microtubules (MTs) play important roles in maintaining the composition, organization, and position of many of these organelles, as well as in moving materials packaged into vesicles from one compartment to another [Schroer and Sheetz, 1991]. For example, MT-based motors are likely to be involved in the organization of endoplasmic reticulum (ER) membranes that often are dispersed throughout the eukaryotic cell cytoplasm (Fig. 2), whereas lysosomal and

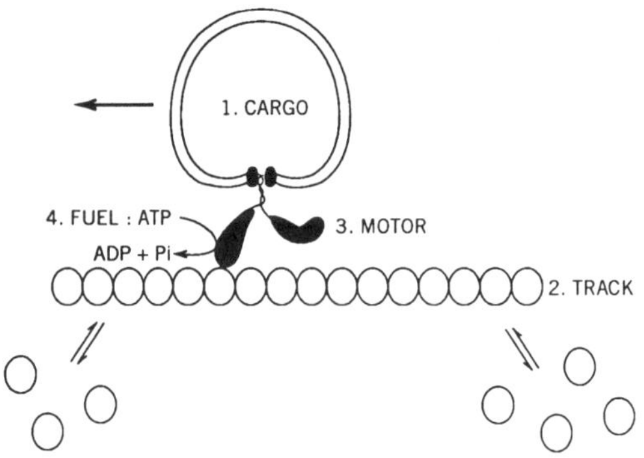

Fig. 1. *Molecular design of intracellular transport systems* **(1)**. *The cargo, e.g., a membrane-bound organelle, is moved along* **(2)** *a fibrous track, e.g., a microtubule, by* **(3)** *a motor protein, e.g., kinesin using energy released by burning* **(4)** *a fuel, e.g., MgATP.*

Golgi membranes are usually found clustered around the nucleus, a position that they achieve by moving toward the minus ends of MTs [Kelly, 1990]. Material newly synthesized on the ER/Golgi system is thought to be delivered to surface locations by vesicle trafficking along MT tracks, whereas endocytosed material is packaged into endocytic transport vesicles that are moved by an MT-dependent process via the prelysosomal compartment to lysosomes at the cell center. In addition, MT-based transport is thought to move material from the salvage compartment back to the ER [Kelly, 1990].

Within many cells, the arrays of MTs that serve as tracks for intracellular membrane transport radiate from "organizing centers" (MTOCs) such as centrosomes, with their plus ends pointing away from the MTOC. Thus membrane-bound organelles that track toward the plus ends of MTs move away from the MTOC, whereas those moving toward the MTOC display minus end–directed transport. The MTOC is located adjacent to the nucleus in most cells, thereby providing an effective transport system between the cell center and the peripheral plasma membrane.

Certain cells reorganize their MT networks to provide specialized transport needs. During differentiation to a columnar epithelium, MDCK cells lose their perinuclear MTOC and nucleate microtubule growth (again from the minus end) from their apical membrane [Bacallao et al., 1989]. This organization presumably allows these polarized cells to direct membrane traffic between the apical and basolateral domains of the cells using microtubule polarity as a guide for vectoral delivery. Another unusual type of microtubule organization is found in polarized neurons. Here the microtubule polarity of the cell body and axons conforms to the "conventional" model, even though axonal microtubules are not connected to the centrosome; plus ends are oriented toward the nerve

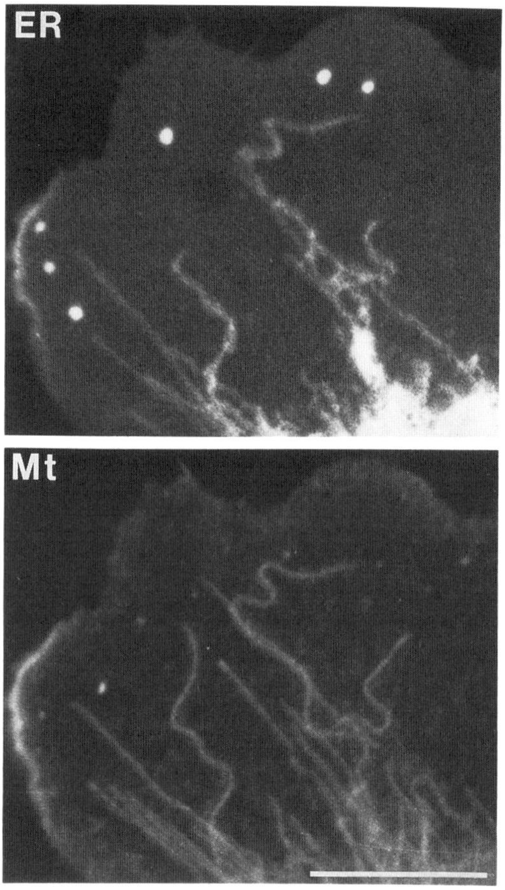

Fig. 2. *Double-label immunofluorescence micrograph showing that endoplasmic reticulum tubules (ER) and microtubules (Mt) are coincident in cultured cells. Microtubules and ER in an A6* Xenopus *kidney cell line are stained, respectively, with antitubulin and DiOC6(3). [Micrograph courtesy of Dr. Mark Terasaki.]*

terminal and minus ends point toward the cell body. In dendrites, however, microtubule polarity appears to be random [Bass et al., 1989]. Since microtubule minus ends here can be directed away from the cell body, this may provide a mechanism of transporting organelles that typically accumulate at microtubule minus ends (e.g., Golgi membranes) into the dendritic process, where they can be used for local protein synthesis.

Much of the membrane traffic in cells occurs in the form of small vesicles (40–300 nm diameter) that serve as intermediate carriers of components in the secretory and endocytotic transport pathways. What appear to be discrete vesicles can be easily seen moving rapidly through the cytoplasm of cells by video-enhanced differential interference contrast microscopy. However, a variety of membrane compartments, including the Golgi, ER, lysosomes, and endosomes also can extend elongated tubules through the cytoplasm. Tubular extensions of membranes become particularly evident shortly after adding brefeldin A to cells and depend on microtubules for their formation. Extension of membranes into a tubular configuration is energetically unfavorable, but can be driven by membrane-bound microtubule motors that pull the membrane as they move along a microtubule track [Dabora and Sheetz, 1988; Vale and Hotani, 1988]. The function of microtubule motor-driven formation of membrane tubules is not known; it may be an intermediate stage in vesicle budding or could create membrane subcompartments for sorting of internal contents. Attachment of motors probably occurs at only certain sites on these large membrane compartments, but this specificity may be disrupted by brefeldin A treatment.

The organization of membrane compartments can also change dramatically during the cell cycle and during development. Dramatic changes in the organization of both microtubules and membranes accompany the transition between interphase and mitosis. At the interphase–mitotic transition, the microtubule array is rapidly disassembled, and microtubules are reorganized in association with the MTOCs at the spindle poles, to form the bipolar spindle that segregates sister chromatids [Sawin and Scholey, 1991]. In many cells, membrane-bound organelles (such as the Golgi apparatus) are fragmented, excluded from the spindle, and randomly dispersed through the cytoplasm so that they can then be partitioned stochastically to two daughter cells [Warren, 1989]. Interestingly though, different cell types appear to vary in the extent to which membranes associate with spindle microtubules. In cells of the early sea urchin embryo, for example, dynamic ER membranes are associated with MT arrays of the mitotic apparatus, where microtubule-based ER membrane transport may have an important organizational function in concentrating the ER around the spindle [Henson et al., 1989; Terasaki and Jaffe, 1991]. By the blastula stage, however, the ER becomes dispersed during mitosis and no longer is concentrated in the spindle [Henson et al., 1989; Wright et al., 1991].

While it is clear that microtubule-based membrane transport is significant for a variety of eukaryotic cell types, certain cell types are of particular interest because they transport particles in an obvious, exaggerated manner [Bray, 1992]; these include the transport of pigment granules in chromatophores, the movement of ribosomes along MTs in insect nutritive tubes, the bidirectional transport of membrane-bound organelles along MTs in feeding tentacles of *Reticulomyxa* and bidirectional vesicle transport along MTs in axons. Indeed, the study of fast axonal vesicle transport in squid giant axons led to the discovery of kinesin-based organelle transport [reviewed by Vale et al., 1986].

III. DISCOVERY AND PURIFICATION OF KINESIN

The term *fast axonal transport* is used to describe the transport of membrane-bound organelles along nerve axons between the cell body and the synapse at rates as high as 5 μm/sec. Using video-enhanced light microscopy, it was observed that this organelle transport persists in axoplasm extruded from the squid giant axon, allowing a biochemical analysis of the molecular mechanism of fast axonal transport; this vesicle transport was soon demonstrated to be microtubule based, ATP dependent, and driven by a presumptive motor protein that exists primarily in a soluble form. However, the important discovery made by Lasek and Brady [1985] that AMPPNP causes organelles to bind tightly to microtubules in an immotile state suggested that this analogue might induce a "rigor-like" attachment of the organelle motors to MTs, further suggesting that MTs prepared in AMPPNP might serve as affinity matrices for identifying and isolating the motors.

Accordingly, experiments conducted with cytosol from axons [Vale et al., 1985a; Brady, 1985] and independently from sea urchin eggs [Scholey et al., 1985] demonstrated that a polypeptide of 110–135 kD copurified with MTs prepared in the presence but not in the absence of AMPPNP (Fig. 3). ATP eluates of squid axonal MTs contained the 110 kD polypeptide and induced the gliding of microtubules over glass coverslips in a video-enhanced light microscopic motility assay, which was used to assay the purification of the motor protein by column chromatography. The purified squid axon protein is a complex of 110 kD "heavy chains" and 60/75 kD "light chains" that uses ATP to move MTs at approximately 0.5 μm/sec and was named *kinesin* [Vale et al., 1985a; reviewed by Vale et al., 1986]. The sea urchin egg protein was subsequently shown to display similar MT gliding, whereas Brady [1985] observed that chick brain MTs prepared in AMPPNP were enriched in ATPase activity as well as in kinesin polypeptides, in support of the hypothesis that kinesin is a mechanochemical ATPase. Indeed, Kuznetsov and Gelfand [1986] convincingly demonstrated that purified kinesin displays MT-activated MgATPase activity.

Kinesin has now been isolated from a wide variety of sources, including squid axons and vertebrate neuronal tissues [Vale et al., 1985a; Brady, 1985], adrenal medulla [Murofushi et al., 1988], sea urchin eggs [Scholey et al., 1985], and *Drosophila* embryos [Saxton et al., 1988], and the protein has been detected immunologically in a variety of animal cell types plus at least one higher plant cell [Tiezzi et al., 1992]. It should be noted, however, that because a large number of proteins are related to kinesin by sequence similarity in their motor domains (see Goldstein, this volume), the definitive identification of a protein as a true kinesin requires the demonstration that its en-

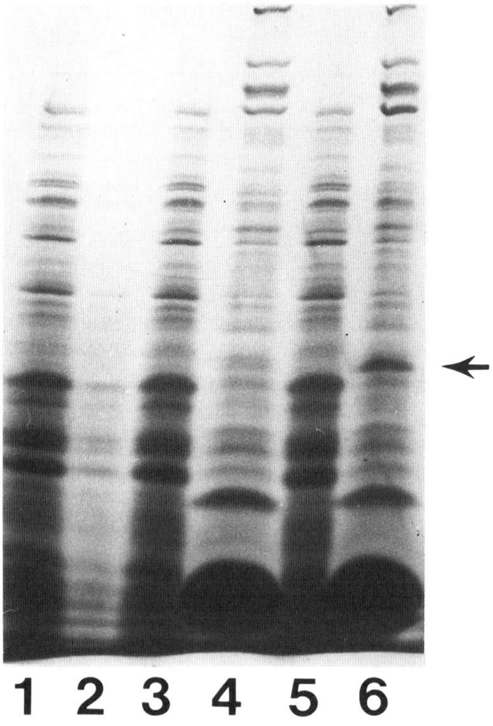

Fig. 3. *Identification of kinesin based on its AMPPNP-induced cosedimentation with microtubules. This Coomassie-stained gel shows MT pellets* **(lanes 2, 4, 6)** *and corresponding supernatants* **(lanes 1, 3, 5)** *prepared from sea urchin egg cytosol treated with colchicine (1,2) (to inhibit tubulin polymerization), taxol alone (3,4) (to enhance tubulin polymerization), and taxol plus AMPPNP (5,6). The "blobs" at the bottoms of lanes 4 and 6 are tubulin in the MT pellets. Note that the 130 kD heavy chain of sea urchin egg kinesin is present only in the MTs prepared from AMPPNP-treated cytosol (lane 6, arrow) [see also earlier reports from Vale et al., 1985; Brady, 1985]. [Adapted from Scholey et al., 1985.]*

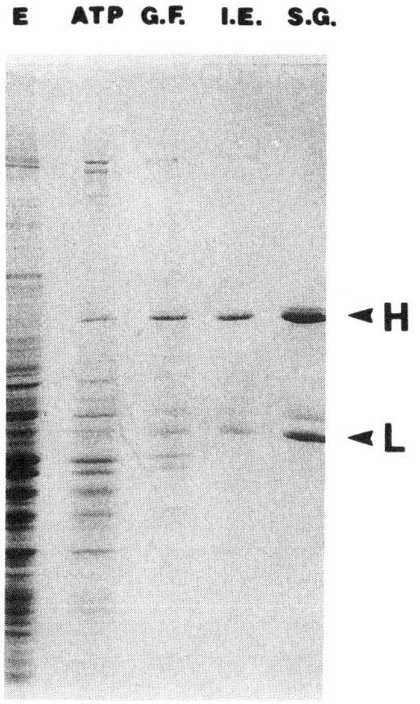

Fig. 4. *Purification of bovine brain kinesin. AMPPNP MTs were sedimented from cytosolic extracts (E), and MAPs were eluted from the MTs using ATP. Kinesin was purified from the eluate by gel filtration (G.F.), ion exchange chromatography (I.E.), and sucrose density gradient centrifugation (S.G.). The 124 kD heavy (H) and light (L) chains are indicated. [Adapted from Wagner et al., 1989.]*

tire amino acid sequence is homologous to the conventional kinesin [Yang et al., 1989; Kosik et al., 1990; Wright et al., 1991; Navone et al., 1992].

A variety of procedures for purifying kinesin have been published, most of which depend on the copurification of kinesin with MTs in the absence of ATP or the presence of a nucleotide analogue, usually AMPPNP or PPPi, followed by subsequent elution of kinesin from MTs by MgATP and column chromatography or density gradient centrifugation [e.g., Vale et al., 1985; Buster and Scholey, 1991; Hackney, 1991; Murofushi et al., 1988; Wagner et al., 1991] (see Fig. 4, for example). Biochemical analysis has illuminated the basic subunit composition, ATPase, and motor activities of purified kinesins; the enzyme is generally thought to be a heterotetramer comprising a pair of heavy chains of 110–135 kD together with equimolar quantities of 60–85 kD light chains

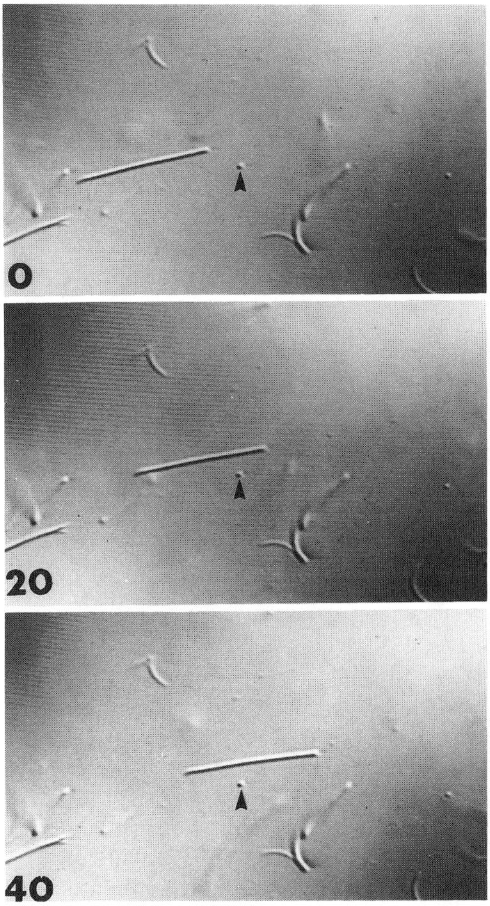

Fig. 5. *Purified kinesin adsorbed onto a glass coverslip of a video-enhanced light microscope induces the gliding of axonemes (or MTs) over the glass surface in the presence of MgATP. Shown here are three frames taken at 20 second intervals [for details, see Porter et al., 1987].*

[Bloom et al., 1988; Kuznetsov et al., 1988; Johnson et al., 1990] (Fig. 4) but there is evidence for heterogeneity in subunit composition, and kinesin fractions depleted of light chains can be obtained [Hackney et al., 1991; Sadhu and Taylor, 1992]. Kinesin displays MT activated MgATPase activity [Brady, 1985; Hackney, 1988; Kuznetsov and Gelfand, 1986; Wagner et al., 1989] that is used to drive MT-based movement [Cohn et al., 1987], although MT-activated MgATPase activity can apparently be uncoupled from motility by some monoclonal

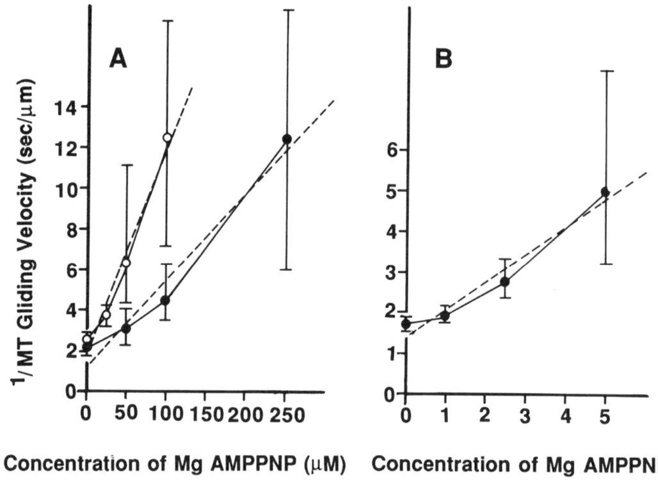

Fig. 6. *AMPPNP inhibits kinesin-driven MT gliding in a motility assay. These Dixon plots show the effects of increasing concentrations of MgAMPPNP on MT gliding velocity in the presence of* **A:** *250 μM MgATP (open circles) or 500 μm MgATP (closed circles) or* **B:** *10 mM MgATP. The dotted line shows the best linear regression for each set of data. AMPPNP is a potent inhibitor of kinesin-driven transport, but there is a clear deviation from linearity that reflects the exertion of drag forces by coverslip–kinesin–AMPPNP–MT "rigor" cross-links at high motor density as discussed by Cohn et al. [1987, 1989] and by Howard et al. [1989]. [Adapted from Cohn et al., 1989.]*

antikinesins [Ingold et al., 1988] or by limited proteolysis of kinesin [Kuznetsov et al., 1989]. The basic substrate and inhibitor sensitivity of kinesin-driven movement of MTs over glass surfaces in a videomicroscopic motility assay (e.g., Fig. 5) has been characterized in detail [Cohn et al., 1989; Shimizu et al., 1991]. In accordance with its ability to induce strong binding of kinesin to MTs, AMPPNP is found to be a potent inhibitor of kinesin-driven motility (Fig. 6). However, arguably the most important functional property of purified kinesin is its ability to use MgATP as a substrate (K_m = 60 μM; V_{max} = 0.6 μM/sec for sea urchin egg kinesin [Cohn et al., 1989]) to move particles toward the plus ends of MTs in a motility assay [Vale et al., 1985b; Porter et al., 1987] (see Fig. 7), leading to the hypothesis that kinesin functions as a plus end–directed microtubule-based motor (within most, if not all, cells).

IV. EVIDENCE THAT KINESIN IS A PLUS END MEMBRANE MOTOR

Videomicroscopic motility assays (below) reveal that purified kinesin moves particles toward the plus ends of MTs [Vale et al., 1985b; Porter et al., 1987], leading to proposals that it transports secretory vesicles to the plasma membrane and drives the radial dispersion of membranes such as the ER along MT tracks [Vale, 1987; Kelly, 1990]. Evidence in support of the idea that kinesin functions as a motor that moves membranes toward the plus ends of MTs within cells has been obtained with three experimental approaches: the localization of kinesin to membrane-bounded organelles within cells, the inhibition of kinesin activity *in vivo,* and the analysis of membrane transport using motility assays performed in cell-free systems or reconstituted from isolated components.

Immunocytochemical experiments performed on a variety of cell types support the idea that

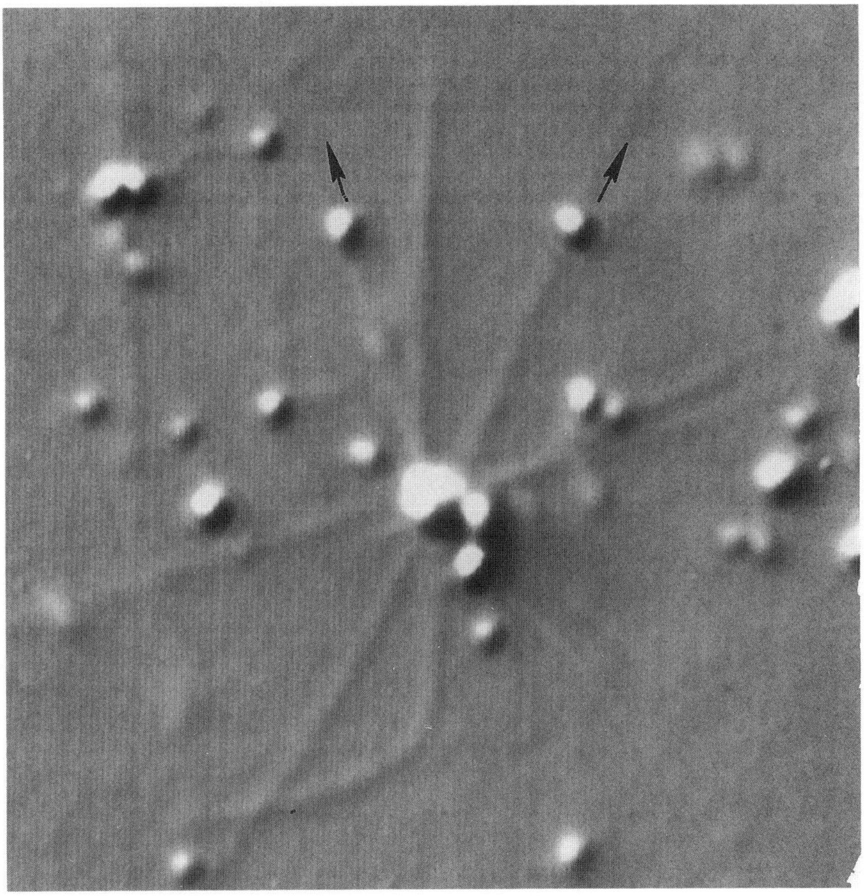

Fig. 7. *Purified kinesin moves carboxylated latex beads toward the plus ends of MTs nucleated from isolated centrosomes. Shown here are microtubules nucleated from purified centrosomes as visualized by differential interference contrast microscopy. Arrows show beads that have attached to and are moving along the microtubules [for details, see Vale et al., 1985b].*

kinesin is associated with intracellular membranes. For example, kinesin antibodies stained punctate vesicle-like structures in extruded squid axoplasm [Brady et al., 1990] (Fig. 8) and labeled anterogradely transported membrane-bound organelles in ligated mouse nerves [Hirokawa et al., 1991], supporting the hypothesis that kinesin functions as a motor for anterograde fast axonal transport of organelles. Antikinesins also stained detergent-extractable membrane vesicles, organelles, and tubules in cultured vertebrate cells [Pfister et al., 1989; Hollenbeck, 1989; Brady and Pfister, 1991], frog egg cortex [Houliston and Elinson, 1991], the mitotic apparatuses of early embryonic sea urchin cells, and the cytoplasm of terminally differentiated sea urchin coelomocytes [Wright et al., 1991] (Figs. 9, 10). In several of these studies, it was observed that membrane-bound kinesin codistributes with MTs, supporting the hypothesis that kinesin transports membranes along MT tracks [Pfister et al., 1989; Hollenbeck, 1989; Brady and Pfister, 1991; Wright et al., 1991].

In several experiments, kinesin activity has been disrupted in cells or cell extracts either genetically or using antibodies. For example, kinesin antibodies interfere with organelle trans-

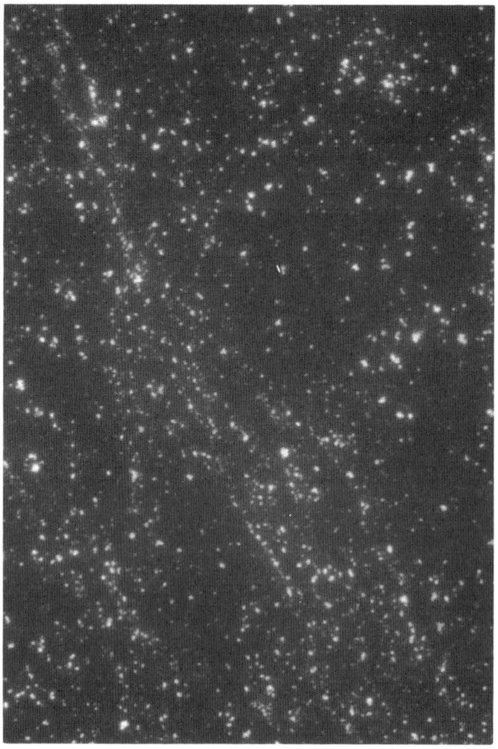

Fig. 8. *Immunolocalization of kinesin to punctate vesicle-like structures in squid (*Loligo pealei*) axoplasm using a monoclonal antibody to the kinesin heavy chain (H2). TRITC-labeled goat antimouse IgG was used as a secondary antibody [for details, see Brady et al., 1990]. (Micrograph courtesy of Dr. George Bloom.)*

port in squid axoplasm [Brady et al., 1990], and the introduction of antibodies into cells inhibits tubular lysosome extension in rabbit macrophages [Hollenbeck and Swanson, 1990] (Fig. 11) and pigment granule dispersion in melanophores [Rodionov et al., 1990]. These results provide compelling evidence for the idea that kinesin drives plus end–directed membrane transport in cells. This role for kinesin is also supported by experiments with *Drosophila* containing kinesin heavy chain mutations that give rise to impaired neuromuscular function [Saxton et al., 1991]. Concordant results were obtained with kinesin antisense oligonucleotides that interfere with kinesin expression in rat hippocampal neurons. These kinesin-depleted neurons have impaired neurite outgrowth and do not transport vesicles containing synapsin I and GAP-43 to nerve terminals [Ferreira et al., 1992] (Fig. 12).

Evidence for an involvement of kinesin in membrane transport has been sought by studying the pharmacological properties of organelle transport and tubular membrane network formation in extracts of various cell types. Most striking is the observation that AMPPNP, which stabilizes the attachment of kinesin to MTs, also induces attachment of organelles to MTs in squid axoplasm [Brady et al., 1985; Lasek and Brady, 1985]. However, in other respects the pharmacological properties of organelle transport in axoplasm differ from those of kinesin [Leopold et al., 1990], possibly reflecting the presence of multiple organelle motors in the crude extracts. It is likely that kinesin contributes to vesicle transport and membrane network formation in a variety of cell extracts, but its precise role relative to other motors has rarely been determined [Allan and Vale, 1991; Dabora and Sheetz, 1988; Pryer et al., 1986] (see Fig. 13).

The reconstitution of organelle transport using purified kinesin has provided evidence in support of the notion that kinesin can power organelle transport. For example, Urrutia et al. [1991] reported that purified kinesin will transport adrenal medullary chromaffin granules along MTs. However, Schroer et al. [1988] reported that kinesin alone was unable to transport vesicles along MTs, and invoked a requirement for soluble accessory cofactors for organelle transport. Further work will be required to resolve this interesting discrepancy.

What is the actual identity of kinesin's membranous cargo in cells? Kinesin and the ER have been shown to codistribute in vertebrate cultured cells [Hollenbeck, 1989], in the frog egg cortex [Houliston and Elinson, 1991], in the mitotic apparatus of early embryonic sea urchin cells [Wright et al., 1991], and in the cytoplasm of sea urchin coelomocytes [Henson et al., 1992], consistent with a role for kinesin in the radial dispersion of ER tubules. In a subset of coelomocytes, antikinesins also stained endosomes, consistent with a role for kinesin

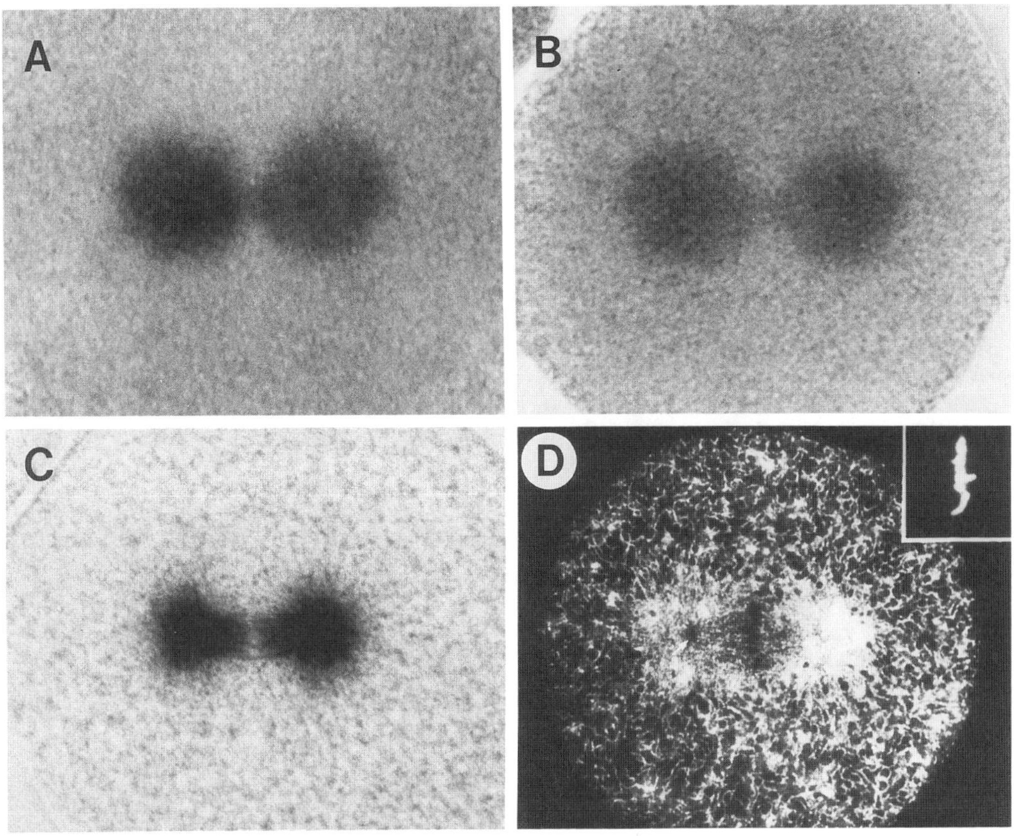

Fig. 9. *Immunolocalization of kinesin in mitotic cells of the early sea urchin embryo. First mitotic sea urchin eggs were stained with monoclonal antibodies specific for kinesin* (**A, C**) *and the luminal ER protein calsequestrin* (**B, D**). *A and B are fixed whole-mount cells showing the codistribution of kinesin and ER membranes. Following detergent lysis, kinesin associates (by ATP-sensitive bonds) with MTs (C), consistent with its role in transporting ER membranes along MTs in the MA. The distribution of membranes is clearer in frozen thin sections (D) [for details, see Henson et al., 1990; Wright et al., 1991].*

in recycling endosomal membranes out to the cell surface, but no codistribution with mitochondria, the Golgi apparatus, or lysosomes was observed [Henson et al., 1992]. Brady and Pfister [1991], on the other hand, found kinesin associated with a variety of classes of membrane-bound organelles, including mitochondria in cultured cells. These results indicate that kinesin may associate with different subcellular membranous organelles in different tissues or in the same cell under different physiological circumstances.

Does kinesin have functions other than membrane transport? Kinesin has been localized to stress fibers [Okuhara et al., 1989] as well as spindle poles and primary cilia [Neighbors et al., 1988] in cultured cells; however, it is unclear whether these results reflect the association of kinesin with nonmembranous intracellular structures, whether the antikinesins are labeling sites of membrane-bound kinesin accumulation, or if the antibodies are cross-reactive with a kinesin-related protein (see Goldstein, this volume). In the case of sea urchins, it was possible to distinguish between these possibilities; the first kinesin immunolocalization experiments suggested that kinesin is associated with the mitotic apparatus of early sea urchin

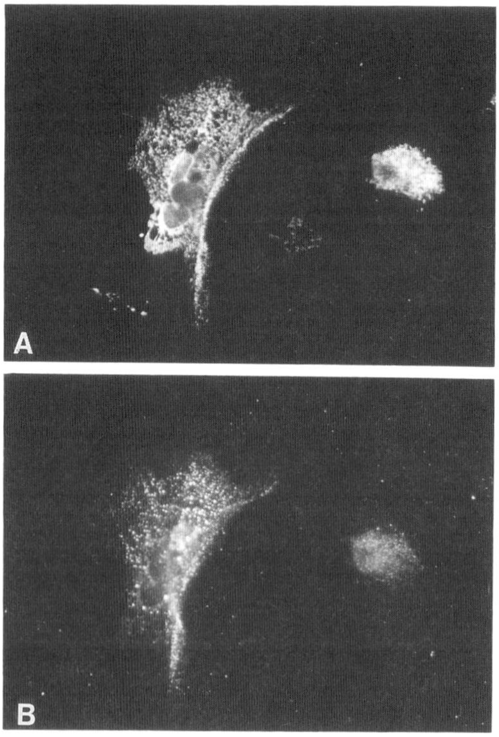

Fig. 10. *Colocalization of kinesin and ER membranes in motile and stationary sea urchin coelomocytes. Cells were double labeled with anti-sea urchin calsequestrin* (**A**)*, a resident ER protein, and antikinesin* (**B**). *The staining pattern for these two proteins appear to be regionally codistributed in the cytoplasm. Note that the motile cell on the left has a widespread distribution of kinesin and the ER, while in the stationary cell on the right the patterns are more perinuclear [for details, see Henson et al., 1992]. (Micrograph courtesy of Dr. John Henson.)*

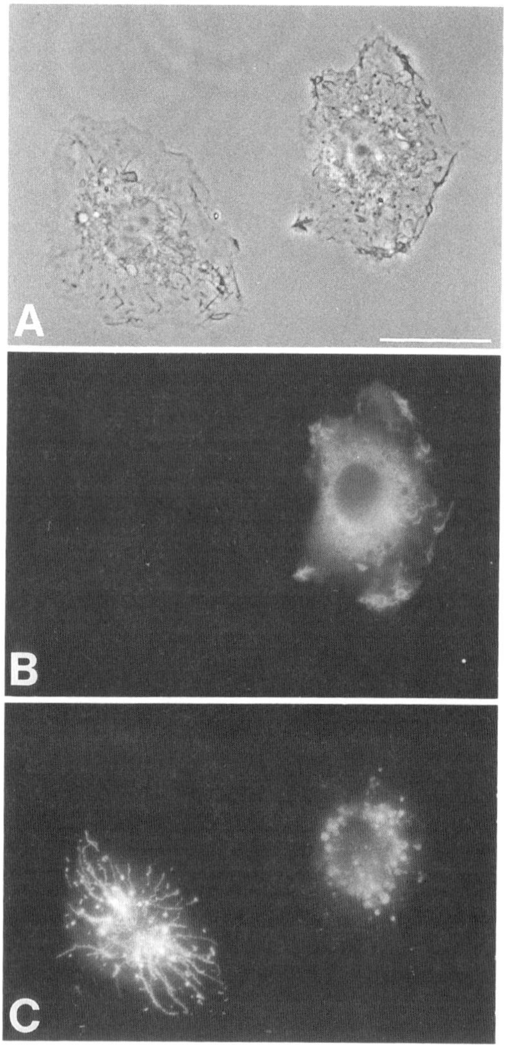

Fig. 11. *Inhibition of macrophage tubular lysosome extension by antikinesin.* **A:** *Two macrophages from a population that had its lysosomal compartment fluorescently labeled by pinocytosis of Texas Red–ovalbumin and was then scrape loaded with SUK4 monoclonal antikinesin [Ingold et al., 1988], fixed, and stained with an FITC-conjugated second antibody against the antikinesin.* **B:** *The cell at right received antikinesin, and the one at left did not.* **C:** *The Texas Red channel shows that the MT-dependent radial transport of tubular lysosomes normally seen in macrophages* (**left**) *is inhibited by antikinesin* (**right**) *[for details, see Hollenbeck and Swanson, 1990]. Bar = 20 μm. (Micrograph courtesy of Dr. Peter Hollenbeck.)*

embryos [Scholey et al., 1985; Leslie et al., 1987], and subsequent work with monoclonal antibodies that distinguish kinesin from kinesin-related proteins demonstrated an association with membranes that accumulate in the spindle region [Wright et al., 1991] (Fig. 9). These results suggest that kinesin may be involved in membrane organization in the spindle, but not in movements of chromosomes or assembly of the mitotic spindle. Microinjection of anti-kinesin antibodies causes collapse of intermediate filaments in tissue culture cells

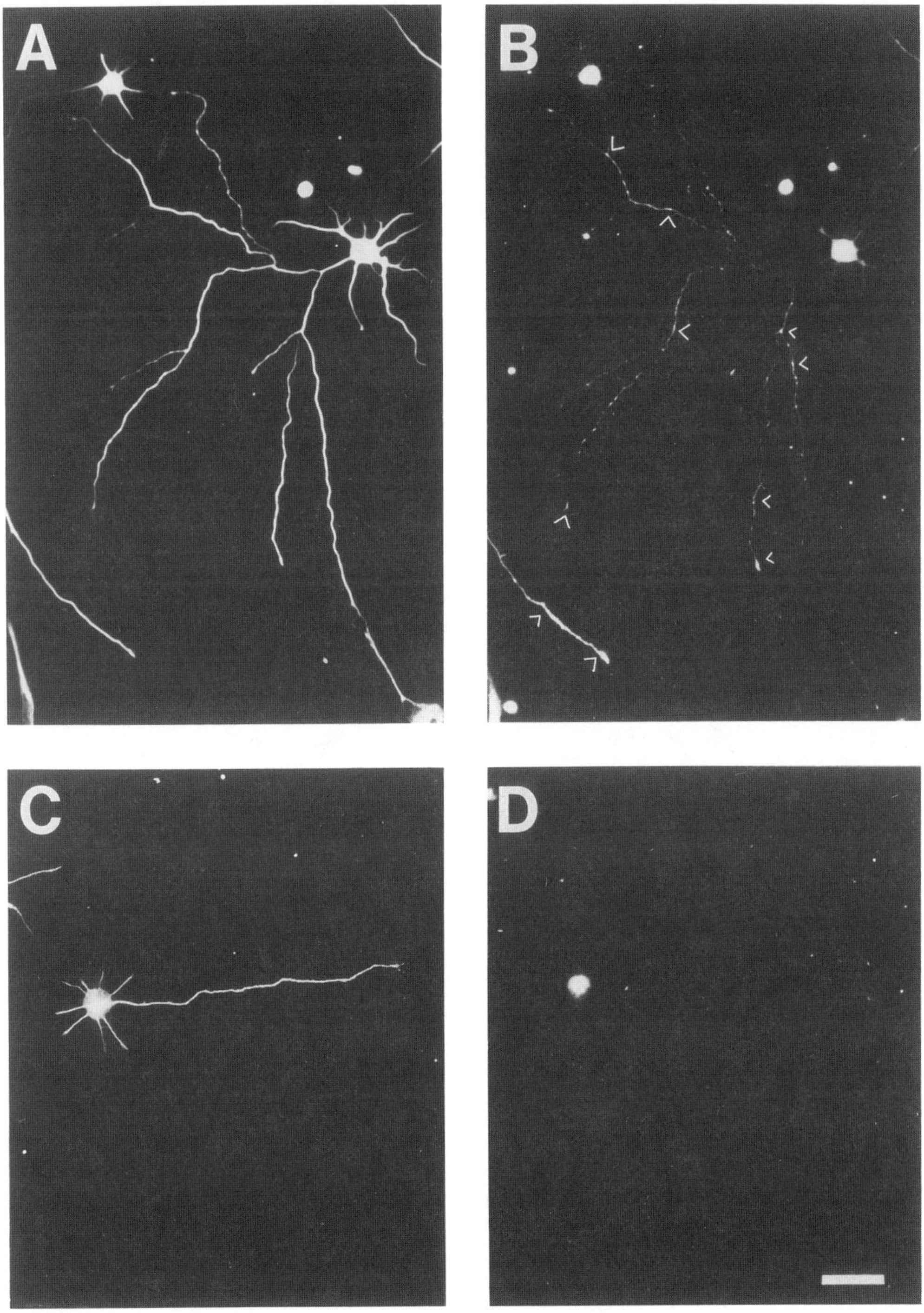

Fig. 12. *Effect of kinesin antisense oligonucleotides on neurite outgrowth and synapsin I transport in primary cultures of rat hippocampal neurons.* **A,B:** *Typical untreated neuron.* **C,D:** *Typical kinesin-depleted cell that was treated for 2 days with antisense oligonucleotides. A and C show immunofluorescence staining with an antitubulin antibody, and B and D show the distribution of synapsin I. Kinesin "null" cells show an inhibition of neurite outgrowth (cf. A and C) and fail to transport synapsin I out of the cell body (cf. B and D). These results are consistent with a role for kinesin in anterograde axonal transport [for details, see Ferreira et al., 1992].*

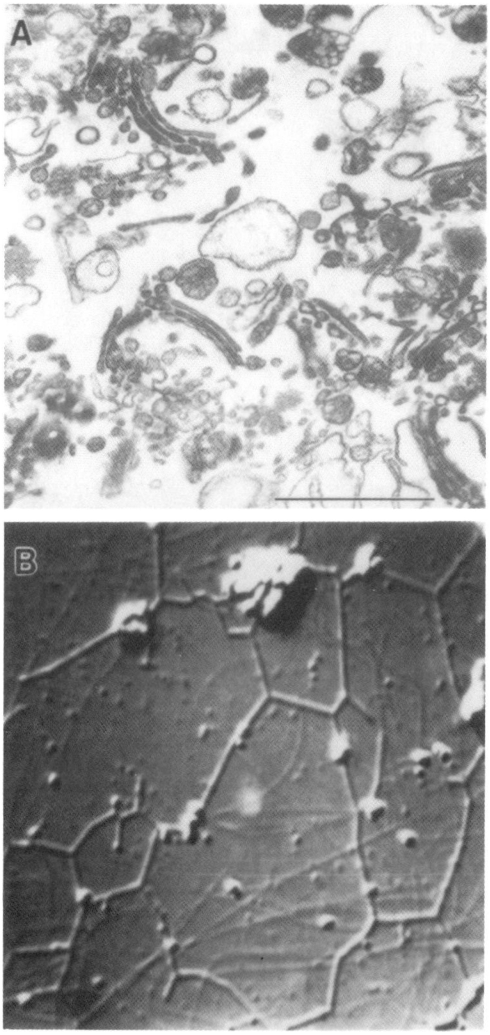

Fig. 13. *Formation of membrane tubules in a rat liver Golgi preparation by plus end–directed microtubule-based movement. The rat liver Golgi preparation (shown by thin-section electron microscopy in* **A**) *is combined with an interphase high speed supernatant from* Xenopus *eggs. Membranes bind to microtubules that polymerize from tubulin in the supernatant and extend tubules along the microtubules by plus end–directed transport.* **B:** *Light micrograph of membrane network.*

[Gyoeva and Gelfand, 1991]. This result indicates that kinesin may cause the peripheral dispersion of intermediate filaments, but it is unclear whether kinesin attaches directly to the filaments or whether the filaments get pulled to the periphery by membrane vesicles moving toward the plus-ends of microtubules.

In light of all these data, it would be difficult to escape the conclusion that kinesin functions as a motor protein that drives plus end–directed membrane transport, but the possibility that it also moves other intracellular components along MTs in cells cannot be excluded. In addition, the relationship of kinesin to other plus end–directed membrane motors that probably exist in eukaryotic cells has yet to be resolved.

V. THE KINESIN MOLECULE

Tremendous progress has been made in understanding the structure and the domain organization of kinesin. Electron microscopy, biochemical approaches, molecular cloning, and bacterial expression have all contributed in unique ways to this effort.

As visualized by electron microscopy after rotary shadowing, kinesin appears as an elongated molecule with two small globular heads at one end, a very long rod domain, and a fan-shaped tail domain at the opposite end of the protein [Amos, 1987; Hirokawa et al., 1989; Scholey et al., 1989; Hisanaga et al., 1989; Hackney et al., 1992] (Fig. 14). This overall shape is somewhat reminiscent of myosin II. Kinesin is ~80 nm in length from head to tail, but the molecule can bend at a distinct hinge in the middle of the rod domain. Decoration of kinesin with monospecific antibodies coupled with electron microscopy has provided information on the positions of the heavy and light chains in this structure [Hirokawa et al., 1989; Scholey et al., 1989]. Antibodies directed against the heavy chain bind to sites found in the head, rod, and tail domains, while light chain antibodies localize exclusively to the tail domain [Hirokawa et al., 1989]. These results indicate that a dimer of two heavy chains gives rise to the heads and stalk of the kinesin molecule and that both the heavy and light chains comprise the fan-shaped tail domain [Hirokawa et al., 1989].

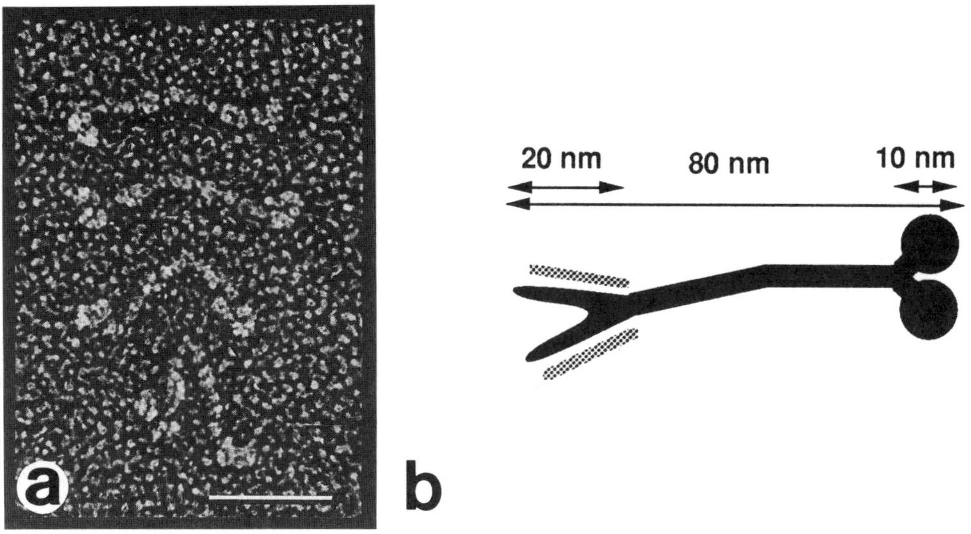

Fig. 14. *Kinesin structure as determined by electron microscopy.* **a:** *Electron micrograph illustrating the flexible hinge in the middle of the elongated stalk.* **b:** *Model of the structure depicting the location of heavy (black) and light (stippled) chains and dimensions of the different domains. The force-generating heads are located at the right.*

The cloning and sequencing of cDNAs encoding the *Drosophila* heavy chain provided complementary information on the organization of the kinesin molecule [Yang et al., 1988, 1989, 1990]. In general, secondary structure predictions based on the amino acid sequence were consistent with the shape of the protein as seen by electron microscopy. The N-terminal ~400 amino acids were predicted to form α-helices, β-sheets, and turns, which would be expected to fold into a globular configuration, such as the small head domain seen at the end of kinesin. The subsequent 485 amino acids were predicted to form an α-helical coiled-coil domain, whose calculated length is similar to the rod domain seen by electron microscopy. In direct experimental confirmation of this prediction, this segment of the heavy chain has been expressed in *Escherichia coli* and shown to self-assemble into an α-helical coiled coil that leads to dimer formation [de Cuevas et al., 1992]. The presence of several glycines and prolines in the rod domain between amino acids 580 and 610 signifies a disruption of the α-helix. The position of this predicted hinge corresponds well to the kink or bend in the rod domain visualized by electron microscopy. The final ~100 amino acids of the heavy chain is predicted to form a second globular domain that, together with the light chains, is thought to give rise to the fan-shaped tail.

The three structurally distinct domains of the kinesin molecule are thought to perform different specialized functions. Of the three, the best understood is the globular domain at the N-terminus, which appears to function as an autonomous force-generating unit. First, an isolated N-terminal domain, which contains the only consensus site for ATP binding, can be produced by expression in a reticulocyte lysate system [Yang et al., 1989] or by proteolytic digestion of the native molecule [Ingold et al., 1988; Kuznetsov et al., 1989], and it exhibits ATP-dependent binding to microtubules and microtubule-activated ATPase activity. A monoclonal antibody, SUK 4, that inhibits kinesin-induced motility *in vitro* also reacts against the N-terminal region of the kinesin heavy chain and decorates the two small heads visible in the electron microscope, leading to the hypoth-

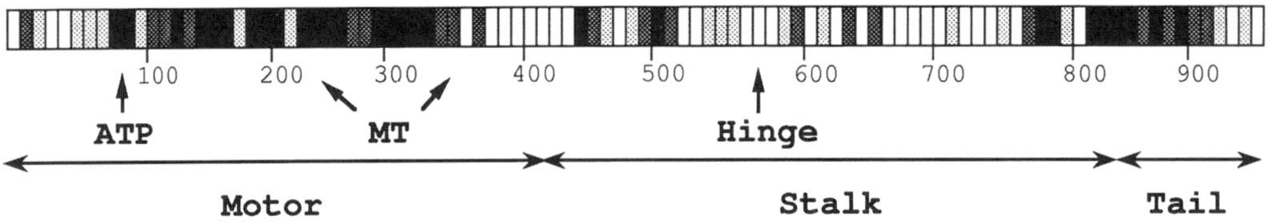

Fig. 15. *Domain structure and conservation of the kinesin heavy chain. The upper diagram illustrates the amino acid identity between the human and* Drosophila *kinesin heavy chains; black box, 9 or 10 amino acid residues out of 10 are identical; dark stipple, 7 or 8 identical residues out of 10; light stipple, 5 or 6 identical residues out of 10. Amino acid residues numbered from the N-terminal are indicated on the bottom. The approximate boundaries of the motor, stalk, and tail domains as defined by secondary structure predictions and functional studies are indicated, as are the sites for ATP and microtubule binding. A discontinuity in the α-helical coiled-coil domain of the stalk (Hinge) is noted.*

esis that the amino terminal "heads" function as motor domains [Ingold et al., 1988; Scholey et al., 1989]. Convincing support for this hypothesis was obtained by Yang et al. [1990], who provided a direct demonstration of force-generating activity by the N-terminal domain; these workers expressed the N-terminal 447 amino acids of kinesin joined to α-spectrin in bacteria and showed that this fusion protein induced microtubule movement in an *in vitro* assay. The sequence of the N-terminal motor domain is highly conserved among kinesin heavy chains that have been cloned from *Drosophila,* squid, sea urchin, and humans [Yang et al., 1989; Kosik et al., 1990; Wright et al., 1991; Navone et al., 1992] (Fig. 15), and the most conserved elements in this domain probably play important roles in nucleotide binding and hydrolysis, microtubule binding, and force generation. Many of these elements are also conserved in members of the kinesin-like motor protein superfamily (see Goldstein, this volume).

The α-helical coiled-coil domain is generally thought to be important for allowing two kinesin heavy chains to dimerize and perhaps for providing a rigid spacer between the chemomechanical heads and an attachment site on the organelle. However, in addition to this purely structural role, this domain may have regulatory functions that have yet to be elucidated. For example, portions of the coiled-coil domain are far more highly conserved between species (Fig. 15) than would be predicted if their only purpose was to maintain a coiled-coil configuration [Yang et al., 1989; Kosik et al., 1990; Wright et al., 1991; Navone et al., 1992]. Some of these regions may be involved in binding interactions with other polypeptides (e.g., light chains or a receptor protein) or lipids. Furthermore, although not conserved in sequence, the position of the hinge domain is maintained in conventional kinesins from all species examined thus far. This hinge domain allows the kinesin molecule to fold back onto itself so that the motor and tail domains come in close proximity. This folded state may modulate the activity of the motor, as is discussed in a subsequent section. The melting temperatures of the α-helical coiled-coil segments on either side of the hinge also differ from one another, implying some different function of the proximal and distal rod segments. Transitions between α-helix and random coil could modulate the force-generating process of kinesin, as has been suggested for myosin. Thus the α-helical coiled coil of kinesin may be more interesting than once suspected, although its precise roles remain to be elucidated.

Even more mysterious is the globular tail domain of kinesin, which sequence analysis predicts to be highly basic (pI > 11) and thus positively charged at physiological pH. The amino acid sequence of this region is as highly conserved as is the motor domain (Fig. 15), implying some important function or binding interaction. Since this domain is thought to anchor the kinesin onto specific organelles, it has been hypothesized that this region may interact with a specific membrane receptor or with lipid head groups; indeed, given the likely function of kinesin as a membrane motor, it will be important to determine the location of the membrane binding site in the kinesin molecule and the relative roles of the heavy and light chains in membrane binding. Unpublished work by Skoufias et al. suggests that the highly basic C-terminal end of the heavy chain binds saturably and with high affinity to membranes *in vitro,* with the light chains being nonessential. If this result reflects the situation within cells, it may mean that the light chains located at the tail end of the kinesin molecule play a regulatory or targeting role. Little is known at present about the binding component in the membrane; however, very recently a 160 kD integral ER membrane protein that binds to kinesin immobilized on an immunoaffinity resin was nominated as a candidate ER receptor protein and named *kinectin* [Toyoshima et al., 1992].

The kinesin tail may bind to structures other than membranes. For example, the bacterially expressed kinesin tail domain binds in an ATP-insensitive manner to microtubules, single kinesins can cross-bridge microtubules *in vitro* via interactions at both the motor and tail domains, and a microtubule distribution was observed when the human kinesin tail domain was expressed in African green monkey epithelial cells (CV-1) [Navone et al., 1992]. Furthermore, kinesin can apparently induce MT–MT sliding in a motility assay, leading to the formation of astral MT arrays [Urrutia et al., 1991]. These results indicate that the tail domain may contain an MT binding site, which would enable the intact molecule to cross-bridge MTs. Whether kinesin mediates MT–MT interactions under any physiological circumstances, however, remains to be determined. There is also evidence that kinesin can radially disperse vimentin filaments [Gyoeva and Gelfand, 1991] and can associate with actin-based stress fibers under certain conditions [Okuhara et al., 1989] in cultured cells, but the physiological significance of these findings and the mechanism of attachment of kinesin to vimentin and actin filaments are not yet clearly understood.

The functions of the kinesin light chains are also poorly understood. The reported heavy to light chain stoichiometries have been variable (ranging from 2:1 to 1:1), and recently it was shown that a subset of bovine brain kinesin can be isolated without any associated light chais [Hackney et al., 1991]. Thus the light chains may not be an obligatory subunit of the motor, as was once thought. Light chains have been postulated to participate in kinesin attachment to organelles, although relatively little direct evidence supports this claim. Hackney and co-workers [1991] have found that the light chains may repress kinesin ATPase activity. Thus, even though the light chains are located at the tail domain, they may modulate the enzymatic activity of the motor domain located at the opposite end of the molecule. The light chains have recently been cloned, and the sequence predicts that they are largely α-helical [Cyr et al., 1991]. Alternative splicing, occurs in the 3′-end of the translated region, which probably contributes, at least in part, to the heterogeneity that has been observed in the light chains. It is tempting to speculate that light chain diversity is somehow involved in attaching kinesin to different intracellular cargoes, but, here again, there is currently no evidence to support this speculation.

VI. BIOPHYSICS OF KINESIN-DRIVEN MOTILITY

How many motors are necessary to move organelles? Does transport occur via discrete jumps or through a more continuous movement? How is the activity of the motor domain regulated? Answers to these questions require an understanding of the mechanism by which chemical energy is transduced into work for any motor protein. Although this problem has been studied for several decades with muscle myosin, a detailed mechanism of the chemomechanical transduction process is still not in hand. Furthermore, it is becoming clear that kinesin motility differs from myosin-induced movement in a number of respects. Below, we will highlight various characteristics of kinesin-induced motility, compare them with myosin, and also indicate how these features may allow kinesin to serve efficiently as an organelle motor.

One aspect of understanding the force-generating process is to define the enzymatic pathway by which kinesin hydrolyzes ATP. The goal of this endeavor is to define the rate constants that govern transitions between intermediate states in the cycle. From this information, which requires transient state kinetic analysis, it is possible to ascertain which steps are rate-limiting and which are associated with large free energy changes that might give rise to force generation. Such studies [Hackney, 1988; Hackney et al., 1989; Sadhu and Taylor, 1992] have provided a tentative model of the kinesin ATP hydrolysis pathway in the absence of microtubules. As observed for myosin, ATP binds to kinesin's active site in a two-step reaction; an initial rapid equilibrium is followed by an isomerization to a tighter binding state. A burst of phosphate formation occurs (stoichiometry of one phosphate per active site) before steady-state turnover, indicating that product release is rate-limiting in the kinesin ATPase cycle, as it is for myosin. However, whereas phosphate release is the slow step for myosin, phosphate appears to be very rapidly released from kinesin. Instead, ADP release is the rate-limiting step in the kinesin ATPase reaction [Hackney, 1988], as it is for dynein [Johnson, 1985]. The reaction pathway in the presence of MTs is less well defined at the present time, although it appears as though polymer binding substantially increases the rate at which ADP is released from the kinesin active site. Hydrolysis may be the rate-limiting step in the MT-activated ATPase cycle.

Force-generating proteins must undergo a cyclic interaction with MTs, so that they remain tightly bound while producing force and then weakly bound when searching for a new subunit further along the filament. A characteristic property of kinesin is that this two-headed motor seems to be tightly attached to MTs for the majority of the time. This deduction arose from experiments aimed at determining the number of kinesin molecules required to move an MT. As the density of kinesin on a surface is progressively decreased, the dependence of MT attachment and movement as a function of kinesin density revealed a Hill coefficient of 1, indicating that single kinesin molecules can support movement [Howard et al., 1989] (Fig. 16). Similar conclusions were reached for the number of kinesins required to move a latex bead along an MT [Block et al., 1990]. In contrast, similar types of experiments conducted with muscle myosin indicate that 5 to 20 motors may be needed to transport actin filaments along a surface [Uyeda et al., 1991].

One might imagine that this difference in the minimal motor number between kinesin and myosin would reflect a difference in force generated by the two motors, but this may not be the case. The amount of force needed to overcome the viscous drag on a moving MT or actin filament is remarkably small (0.01 pN) compared with the force produced by single kinesin or myosin motors (probably in the range of a pN), and in fact a single motor can generate sufficient force to move a very large bead or organelle in an aqueous environment. What is important instead is the "duty ratio" of the motor domain, i.e., the ratio of the time a motor domain and its filament remain attached, to the time taken to complete the mechanochemical cycle. Each kinesin head is thought to have a high duty ratio (>0.5), so it attaches to tubulin for a large fraction of its ATPase cycle. Consequently, there is a high probability that one of its two heads is attached at any time, thereby enabling the two-headed kinesin molecule to hang onto its filamentous track for long time intervals. In contrast, the duty ratio of the myosin-II motor domain is low (0.05), so a large number of heads are required to hang onto the actin filament [Uyeda et al., 1991]. The high duty ratio and long attachment time of kinesin is important, because once the motor is detached from its filamentous track, random

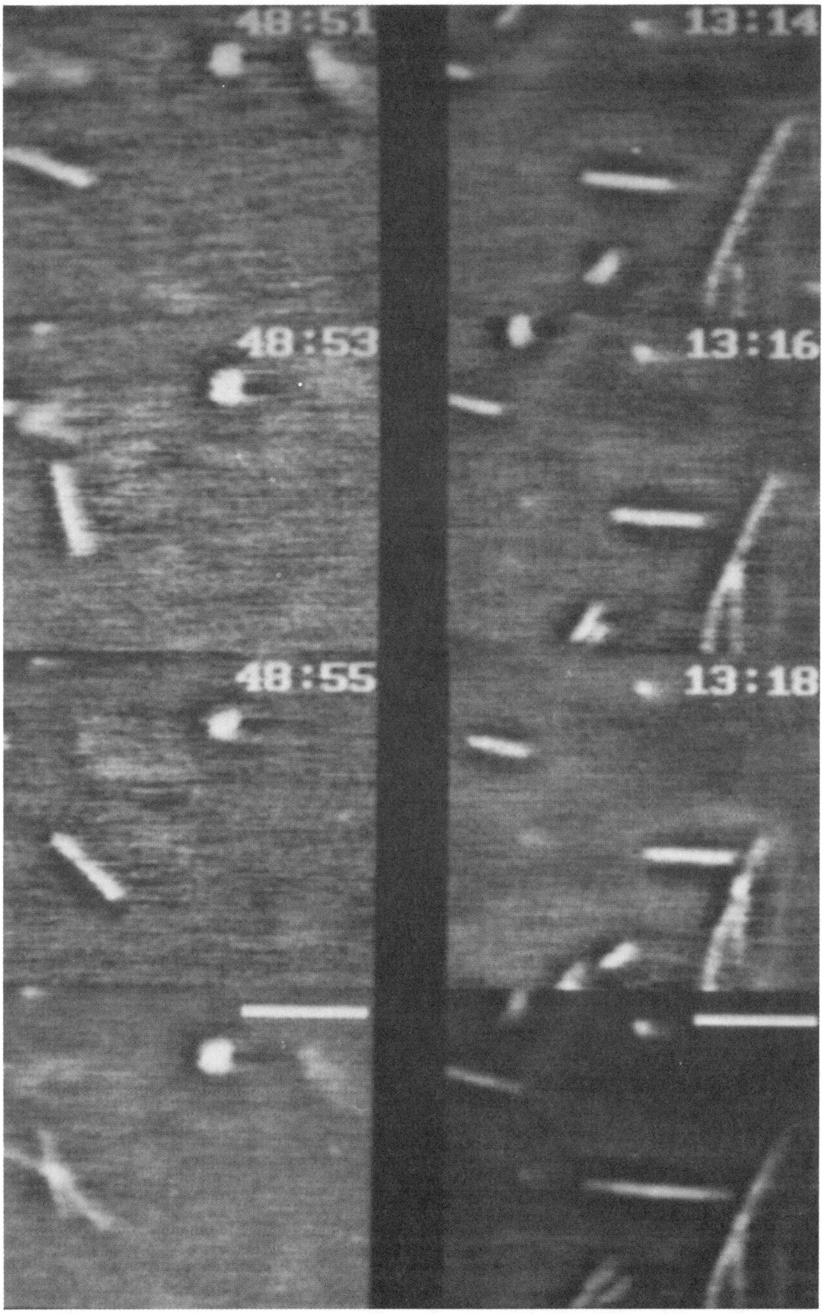

Fig. 16. *Movement of microtubules by single kinesin molecules* in vitro *(left-hand panels). Microtubules (seen by dark field microscopy) are attached to and transported by single kinesins attached to a glass surface. Numbers designate time (mins:secs). As they move forward, microtubules pivot about a nodal point of attachment (seen by overlapping the three time points in the bottom panel) where the kinesin molecule is presumably located. When the end of the microtubule reaches this nodal point, it subsequently dissociates from the surface and diffuses into the solution. For comparison, the right-hand panels show a microtubule moving in a relatively straight path over a surface coated with a high density of kinesin; here the MT does not pivot about a nodal point (bottom).*

Brownian motion acts rapidly (<1 ms) to separate the object being transported (such as a bead or organelle) from the MT. Thus the nearly continuous attachment of kinesin to MTs would allow single kinesin molecules to function very efficiently in transporting organelles for long intervals (hundreds or thousands of ATPase cycles) without releasing the organelle from the MT. Electron microscopic observations support the idea that very few motors (one to five) may be involved in moving organelles during axoplasmic transport. Myosin, on the other hand, works in large ensembles in muscle (millions of motors), and hence individual motors do not need to maintain prolonged contact with actin in order to achieve efficient contraction.

Another intriguing property of kinesin-driven movement is that the velocity of transport is completely independent of the number of motors driving the movement. The manner in which kinesin "balances" the velocity when additional motors participate in transport is unclear. However, this may have practical consequences for organelle transport since all membranes, irrespective of the number of attached motors, would move at the same rate through the cytoplasm. Myosin again appears to behave differently from kinesin, as the velocity of transport increases saturably with motor number, reaching a plateau when up to 10–20 motors interact per actin filament [Uyeda et al., 1991].

Kinesin must undergo some sort of conformational change during the cycle that produces a mechanical displacement or "step." The quest to detect these steps is challenging, since they are presumably small in magnitude (<10 nm) and since they may occur quite abruptly. The first step in this direction was made by Gelles et al. [1988], who used a computational centroid tracking method to measure small (1–2 nm) displacements of kinesin-coated beads moving along MTs from video records. Because of the limited temporal resolution of video, these analyses were performed at very low ATP concentrations at which the movement occurred very slowly. Their analysis, however, showed episodes of 4 nm jumps with intervening pauses; the magnitude of the displacements corresponds to the tubulin monomer spacing in the microtubule. Since the movement occurred at very low ATP concentration, it is not absolutely certain whether these steps correspond to discrete power strokes or to episodes when the rigor (no nucleotide) bond releases and then reforms. Thus higher temporal resolution techniques than video are being developed to examine movement occurring at maximal velocities and at low motor number. These efforts may detect any elemental force-generating episodes that power motility.

VII. REGULATION OF KINESIN-BASED MOTILITY

Organelle transport can be modulated by external signals. The best known example is the chromatophore cells of fish scales, which respond to epinephrine by transporting their large melanophore granules out along MTs toward the cell periphery. The movement can be induced in detergent-permeabilized cell models by cAMP, which mediates its effects in this system through cAMP-dependent protein kinase [Rozdzial and Haimo, 1986]. This plus end–directed transport appears to be kinesin driven, since microinjection of affinity purified, anti-kinesin polyclonal antibodies blocks the anterograde movement of pigment granules [Rodionov et al., 1991]. However, it is still unclear whether kinesin is a direct target of the kinase. Reaggregation of the pigment granules, which appears to be mediated by cytoplasmic dynein, requires the activation of the phosphatase calcineurin [Thaler and Haimo, 1990].

Organelle transport is also controlled by posttranslational modifications that occur during the cell cycle [Allan and Vale, 1991]. Interphase extracts from frog eggs support robust organelle transport along MTs; however, little organelle transport is observed in mitotic extracts. The levels of soluble motor activity, as measured by bead movement along MTs, is similar in interphase and mitotic extracts. Thus the difference in the levels of organelle transport in these two cell cycle states may be due to a regulation in the association of the motors with the organelles.

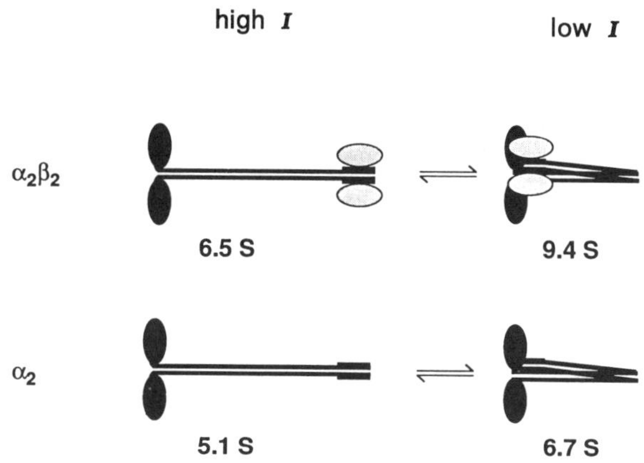

Fig. 17. *Diagrams illustrating the folded and unfolded states of kinesin. The heavy chain dimer (α_2) and the heavy and light chain heterotetramer ($\alpha_2\beta_2$) both undergo an ionic strength-dependent conformational change between an extended form (***left***, high ionic strength) and a folded conformation (***right***, low ionic strength). Heavy chains (or α-subunits) are black, and light chains (β-subunits) are pale ovals [see Hackney et al., 1992, for details]. (Figure generously provided by David Hackney.)*

Axonal transport may also require modulation of motor activity. Hirokawa et al. [1991] have shown that vesicles being transported to the nerve terminal contain both kinesin and cytoplasmic dynein. Since these vesicles are moving in the opposite direction to that produced by dynein, these results would suggest that dynein is being carried to the terminal in an inactive form. Modifications could then occur at the terminal that would activate dynein (and perhaps inactivate kinesin) to carry vesicles back to the cell body by retrograde transport.

At the present time, very little is known about molecular mechanisms that regulate kinesin (or other microtubule motor) activity. A variety of investigators have found that kinesin (in particular the kinesin light chain) is phosphorylated; however, the physiological consequences of these posttranslational modifications are unknown. It has also been found that kinesin exists in a folded conformation, analogous to smooth muscle myosin, where its tail bends around to the head domain [Hackney et al., 1992] (Fig. 17). Perhaps, analogous to the situation with smooth muscle myosin, this folded form of kinesin, which constitutes the majority of kinesin isolated from cell extracts, may represent a less active "storage" form of the enzyme. Indeed, soluble kinesin may need to be kept in an enzymatically repressed conformation, so that it does not needlessly hydrolyze ATP. By this scenario, a posttranslational modification, or binding to an activating protein, may be required to convert folded kinesin into an active, extended form. In this context, it is interesting that Schroer et al. [1988] proposed that soluble factors are needed for kinesin-driven organelle transport, but whether they are involved in unfolding kinesin to an active state is unknown. The regulation of kinesin activity may also be relevant to the finding that kinesin appears to be concentrated on intracellular membranes according to immunocytochemical observations, yet it is usually extracted into the cytosolic rather than the organellar fraction upon cell lysis [Vale et al., 1985; Brady, 1985; Scholey et al., 1985; Hollenbeck, 1989].

VIII. RELATIONSHIP OF KINESIN-BASED ORGANELLE TRANSPORT TO OTHER INTRACELLULAR TRANSPORT SYSTEMS

While most studies of the biological functions of kinesin have focused on its role as a motor protein for membrane transport, there are indications that kinesin can also transport other intracellular components along MTs; additional potential cargoes include MTs [Urrutia et al., 1991; Navone et al., 1992], spindle poles [Neighbors et al., 1988], intermediate filaments [Gyoeva and Gelfand, 1991], and microfilaments [Okuhara et al., 1989]. In addition, it is clear that kinesin-based membrane transport works in concert with other intracellular membrane transport systems. For example, the finding that cytoplasmic dynein can generate force for moving MTs [Lye et al., 1987] in a minus end–directed [Paschal and Vallee, 1987] or bidirectional [Euteneur et al., 1988] manner suggests that the concerted action of kinesin and dynein could drive movement of membranes in both directions within cells. Other members of the kinesin superfamily (see Goldstein, this volume) have been shown to be minus end–directed MT motors (i.e., ncd protein) and membrane motors (i.e., unc 104 protein); thus multiple kinesins, possibly displaying overlapping functions, are likely to participate in the organization and transport of intracellular membranes. Finally, there are actin-based membrane motors, and it has recently become apparent that actin and tubulin-based membrane transport systems are closely interrelated. For example, a kinesin-related protein can functionally replace an actin-based vesicle motor in yeast [Lille and Brown, 1992], and squid axons contain vesicles that move on both MTs and microfilaments [Kuznetsov et al., 1992]. Thus, in addition to defining the detailed mechanism of kinesin-driven membrane transport and its regulation, future work should better define the range of physiological functions mediated by kinesin and how kinesin-driven transport is integrated into the complex network of intracellular transport systems that appears to operate in eukaryotic cells.

IX. REFERENCES

Allan VJ, Vale RD (1991): Cell cycle control of microtubule-based membrane transport and tubule formation *in vitro*. J Cell Biol 113:347–359.

Amos LA (1987): Kinesin from pig brain studied by electron microscopy. J Cell Sci 87:105–111.

Bacallao R, Antony C, Dotti C, Karsenti E, Stelzer EH, Simons K (1989): The subcellular organization of Mardin-Darby canine kidney cells during the formation of a polarized epithelium. J Cell Biol 109: 2817–2832.

Bass PW, Black MM, Banker GA (1989): Changes in microtubule polarity orientation during the development of hippocampal neurons in culture. J Cell Biol 109: 3085–3094.

Block SM, Goldstein LS, Schnapp BJ (1990): Bead movement by single kinesin molecules studied with optical tweezers. Nature 348:348–352.

Bloom GS (1992): Motor proteins for cytoplasmic microtubules. Curr Opin Cell Biol 4:66–73.

Bloom GS, Wagner MC, Pfister KK, Brady ST (1988): Native structure and physical properties of bovine brain kinesin and identification of the ATP-binding subunit polypeptide. Biochemistry 27:3409–3416.

Brady ST (1985): A novel brain ATPase with properties expected of the fast axonal transport motor. Nature 317:73–75.

Brady ST (1991): Molecular motors in the nervous system. Neuron 7:521–533.

Brady ST, Lasek RJ, Allen RD (1985): Video microscopy of fast axonal transport in extruded axoplasm. Cell Motil 5:81–101.

Brady ST, Pfister KK (1991): Kinesin interactions with membrane bounded organelles *in vivo* and *in vitro*. J Cell Sci Suppl 14:103–108.

Brady ST, Pfister KK, Bloom GS (1990): A monoclonal antibody against kinesin inhibits both anterograde and retrograde fast axonal transport in squid axoplasm. Proc Natl Acad Sci USA 87:1061–1065.

Bray D (1992): Cell Movements. New York: Garland Publishing.

Buster D, Scholey JM (1991): Purification and assay of kinesin from sea urchin eggs and early embryos. J Cell Sci [Suppl] 14:109–115.

Cohn SA, Ingold AL, Scholey JM (1987): Correlation between the ATPase and microtubule translocating activities of sea urchin egg kinesin. Nature 328:160–163.

Cohn SA, Ingold AL, Scholey JM (1989): Quantitative analysis of sea urchin egg kinesin-driven microtubule motility. J Biol Chem 264:4290–4297.

Cyr JL, Pfister KK, Bloom GS, Slaughter CA, Brady ST (1991): Molecular genetics of kinesin light chains: Generation of isoforms by alternative splicing. Proc Natl Acad Sci USA 88:10114–10118.

Dabora SL, Sheetz MP (1988): Cultured cell extracts support organelle movement on microtubules *in vitro*. Cell Motil Cytoskel 10:482–495.

de Cuevas M, Tao T, Goldstein LS (1992): Evidence that the stalk of *Drosophila* kinesin heavy chain is an alpha-helical coiled coil. J Cell Biol 116:957–965.

Euteneur U, Koonce MP, Pfister KK, Schliwa M (1988): An ATPase with properties expected for the organelle motor of the giant amoeba, *Retinculomyxa*. Nature 332:176–178.

Ferreira A, Niclas J, Vale RD, Banker G, Kosik KS (1992): Suppression of kinesin expression in cultured hippocampal neurons using antisense oligonucleotides. J Cell Biol 117:595–606.
Gelfand VI (1989): Cytoplasmic microtubular motors. Curr Opin Cell Biol 1:63–66.
Gelles J, Schnapp BJ, Sheetz MP (1988): Tracking kinesin-driven movements with nanometer scale precision. Nature 331:450–453.
Gyoeva FK, Gelfand VI (1991): Coalignment of vimentin intermediate filaments with microtubules depends on kinesin. Nature 353:445–448.
Hackney DD (1988): Kinesin ATPase: Rate-limiting ADP release. Proc Natl Acad Sci USA 85:6314–6318.
Hackney DD (1991): Isolation of kinesin using initial batch ion exchange. Methods Enzymol 196:175–181.
Hackney DD, Levitt JD, Suhan J (1992): Kinesin undergoes a 9 S to 6 S conformational transition. J Biol Chem 267:8696–8701.
Hackney DD, Levitt JD, Wagner DD (1991): Characterization of alpha 2 beta 2 and alpha 2 forms of kinesin. Biochem Biophys Res Commun 174:810–815.
Hackney DD, Malik AS, Wright KW (1989): Nucleotide-free kinesin hydrolyzes ATP with burst kinetics. J Biol Chem 27:15943–15948.
Henson JH, Begg DA, Beaulieu SM, Fishkind DJ, Bonder EM, Terasaki M, Lebeche D, Kaminer B (1989): A calsequestrin-like protein in the endoplasmic reticulum of the sea urchin: Localization and dynamics in the egg and first cell cycle embryo. J Cell Biol 109:149–161.
Henson JH, Nesbitt D, Wright BD, Scholey JM (1992): Immunolocalization of kinesin in sea urchin coelomocytes: Associations of kinesin with intracellular organelles. J Cell Sci 103:309–320.
Hirokawa N, Pfister KK, Yorifuji H, Wagner MC, Brady ST, Bloom GS (1989): Submolecular domains of bovine brain kinesin identified by electron microscopy and monoclonal antibody decoration. Cell 56:867–878.
Hirokawa N, Sato-Yoshitake R, Kobayashi N, Pfister KK, Bloom GS, Brady ST (1991): Kinesin associates with anterogradely transported membranous organelles *in vivo*. J Cell Biol 114:295–302.
Hisanaga S, Murofushi H, Okuhara K, Sato R, Masuda Y, Sakai H, Hirokawa N (1989): The molecular structure of adrenal medulla kinesin. Cell Motil Cytoskel 12:264–272.
Hollenbeck PJ (1989): The distribution, abundance and subcellular localization of kinesin. J Cell Biol 108: 2335–2342.
Hollenbeck PJ, Swanson JA (1990): Radial extension of macrophage tubular lysosomes supported by kinesin. Nature 346:864–866.
Houliston E, Elinson RP (1991): Evidence for the involvement of microtubules, ER, and kinesin in the cortical rotation of fertilized frog eggs. J Cell Biol 114: 1017–1028.
Howard J, Hudspeth AJ, Vale RD (1989): Movement of microtubules by single kinesin molecules. Nature 342:154–158.
Huitorel P (1988): From cilia and flagella to intracellular motility and back again: A review of a few aspects of microtubule-based motility. Biol Cell 63:249–258.
Ingold AL, Cohn SA, Scholey JM (1988): Inhibition of kinesin-driven microtubule motility by monoclonal antibodies to kinesin heavy chains. J Cell Biol 107: 2657–2667.
Johnson CS, Buster D, Scholey JM (1990): Light chains of sea urchin kinesin identified by immunoadsorption. Cell Motil Cytoskeleton 16:204–213.
Johnson KA (1985): Pathway of the microtubule-dynein ATPase and structure of dynein: A comparison with actomyosin. Annu Rev Biophys Chem 14:161–188.
Kelly RB (1990): Microtubules, membrane traffic, and cell organization. Cell 61:5–7.
Kosik KS, Orecchio LD, Schnapp B, Inouye H, Neve RL (1990): The primary structure and analysis of the squid kinesin heavy chain. J Biol Chem 265: 3278–3283.
Kuznetsov SA, Gelfand VI (1986): Bovine brain kinesin is a microtubule-activated ATPase. Proc Natl Acad Sci USA 83:8530–8534.
Kuznetsov S, Langford GM, Weiss DG (1992): Actin dependent organelle movement in squid axoplasm. Nature 356:722–725.
Kuznetsov SA, Vaisberg YA, Rothwell SW, Murphy DB, Gelfand VI (1989): Isolation of a 45-kDa fragment from the kinesin heavy chain with enhanced ATPase and microtubule-binding activities. J Biol Chem 264:589–595.
Kuznetsov SA, Vaisberg EA, Shanina NA, Magretova NN, Chernyak VY, Gelfand VI (1988): The quarternary structure of bovine brain kinesin. EMBO J 7:353–356.
Lasek RJ, Brady ST (1985): Attachment of transported vesicles to microtubules in axoplasm is facilitated by AMPPNP. Nature 316:645–647.
Leslie RJ, Hird RB, Wilson L, McIntosh JR, Scholey JM (1987): Kinesin is associated with a nonmicrotubule component of sea urchin mitotic spindles. Proc Natl Acad Sci USA 84:2771–2775.
Leopold PL, Snyder R, Bloom GS, Brady ST (1990): Nucleotide specificity for the bidirectional transport of membrane-bounded organelles in extruded axoplasm. Cell Motil Cytoskel 15:210–219.
Lillie SH, Brown SS (1992): Suppression of a myosin defect by a kinesin-related gene. Nature 356:358–361.
Lye RJ, Porter ME, Scholey JM, McIntosh JR (1987): Identification of a microtubule-based cytoplasmic motor in the nematode *C. elegans*. Cell 51:309–318.
McIntosh JR, Porter ME (1989): Enzymes for microtubule-dependent motility. J Biol Chem 264:6001–6004.
Murofushi H, Ikai A, Okuhara K, Kotani S, Aizawa H, Kumakura K, Sakai H (1988): Purification and characterization of kinesin from bovine adrenal medulla. J Biol Chem 263:12744–12750.

Navone F, Niclas J, Hom-Booher N, Sparks L, Bernstein HD, McCaffrey G, Vale RD (1992): Cloning and expression domain of a human kinesin heavy chain gene: Interaction of the COOH terminal with cytoplasmic microtubules in transfected CV-1 cells. J Cell Biol 117:1263–1275.
Neighbors BW, Williams RC Jr, McIntosh JR (1988): Localization of kinesin in cultured cells. J Cell Biol 106:1193–1204.
Okuhara K, Murofushi H, Sakai H (1989): Binding of kinesin to stress fibers in fibroblasts under condition of microtubule depolymerization. Cell Motil Cytoskel 12:71–77.
Paschal BM, Vallee RB (1987): Retrograde transport by the microtubule-associated protein MAP 1C. Nature 330:181–183.
Pfister KK, Wagner MC, Stenoien DL, Brady ST, Bloom GS (1989): Monoclonal antibodies to kinesin heavy and light chains stain vesicle-like structures, but not microtubules, in cultured cells. J Cell Biol 108:1453–1463.
Porter ME, Scholey JM, Stemple DL, Vigers GP, Vale RD, Sheetz MP, McIntosh JR (1987): Characterization of the microtubule movement produced by sea urchin egg kinesin. J Biol Chem 262:2794–2802.
Pryer NK, Wadsworth P, Salmon ED (1986): Polarized microtubule gliding and particle saltations produced by soluble factors from sea urchin eggs and embryos. Cell Motil Cytoskel 6:537–548.
Rodionov VI, Gyoeva FK, Gelfand VI (1991): Kinesin is responsible for centrifugal movement of pigment granules in melanophores. Proc Natl Acad Sci USA 88:4956–4960.
Rozdzial MM, Haimo LT (1986): Bidirectional pigment granule movements of melanophores are regulated by protein phosphorylation and dephosphorylation. Cell 47:1061–1070.
Sawin KE, Scholey JM (1991): Motor proteins in cell division. Trends Cell Biol 1:122–129.
Sadhu A, Taylor EW (1992): A kinetic study of the kinesin ATPase. J Biol Chem 267:11352–11359.
Saxton WM, Hicks J, Goldstein LS, Raff EC (1991): Kinesin heavy chain is essential for viability and neuromuscular functions in *Drosophila,* but mutants show no defects in mitosis. Cell 64:1093–1102.
Saxton WM, Porter ME, Cohn SA, Scholey JM, Raff EC, McIntosh JR (1988): *Drosophila* kinesin: Characterization of microtubule motility and ATPase. Proc Natl Acad Sci USA 85:1109–1113.
Scholey JM, Heuser J, Yang JT, Goldstein LS (1989): Identification of globular mechanochemical heads of kinesin. Nature 338:355–357.
Scholey JM, Porter ME, Grissom PM, McIntosh JR (1985): Identification of kinesin in sea urchin eggs, and evidence for its localization in the mitotic spindle. Nature 318:483–486.
Schroer TA, Schnapp BJ, Reese TS, Sheetz MP (1988): The role of kinesin and other soluble factors in organelle movement along microtubules. J Cell Biol 107:1785–1792.
Schroer TA, Sheetz MP (1991): Functions of microtubule-based motors. Annu Rev Physiol 53:629–652.
Sheetz MP (1987): What are the functions of kinesin? BioEssays 7:165–168.
Shimizu T, Furusawa K, Ohashi S, Toyoshima YY, Okuno M, Malik F, Vale RD (1991): Nucleotide specificity of the enzymatic and motile activities of dynein, kinesin, and heavy meromyosin. J Cell Biol 112: 1189–1197.
Terasaki M, Jaffe LA (1991): Organization of the sea urchin egg endoplasmic reticulum and its reorganization at fertilization. J Cell Biol 114:929–940.
Thaler CD, Haimo LT (1990): Regulation of organelle transport in melanophores by calcineurin. J Cell Biol 111:1939–1948.
Tiezzi A, Moscatelli A, Cai G, Bartalesi A, Cresti M (1992): Immunoreactive homolog of mammalian kinesin in *Nicotiana tabacum* pollen tubes. Cell Motil Cytoskel 21:132–137.
Toyoshima I, Yu H, Sheetz MP (1992): Kinectin, a major kinesin-binding protein on ER. J Cell Biol 118:1121–1131.
Urrutia R, McNiven MA, Albanesi JP, Murphy DB, Kachar B (1991): Purified kinesin promotes vesicle motility and induces active sliding between microtubules *in vitro.* Proc Natl Acad Sci USA 88:6701–6715.
Uyeda TQP, Warrick HM, Kron SJ, Spudich JA (1991): Quantized velocities are low myosin densities in an *in vitro* motility assay. Nature 352:307–311.
Vale RD (1987): Intracellular transport using microtubule-based motors. Annu Rev Cell Biol 3:347–378.
Vale RD (1990): Microtubule-based motor proteins. Curr Opin Cell Biol 2:15–22.
Vale RD (1992): Microtubule motors: Many new models off the assembly line. Trends Biochem Sci 200:300–304.
Vale RD, Hotani H (1988): Formation of membrane networks in vitro by kinesin-driven microtubule movement. J Cell Biol 107:2233–2241.
Vale RD, Reese TS, Sheetz MP (1985a): Identification of a novel force-generating protein, kinesin, involved in microtubule-based motility. Cell 42:39–50.
Vale RD, Schapp BJ, Mitchison T, Steuer E, Reese TS, Sheetz MP (1985b): Different axoplasmic proteins generate movement in opposite directions along microtubules *in vitro.* Cell 43:623–632.
Vale RD, Scholey JM, Sheetz MP (1986): Kinesin: Possible biological roles of a new microtubule motor. Trends Biochem Sci 11:464–468.
Vallee RB, Shpetner HS (1990): Motor proteins of cytoplasmic microtubules. Annu Rev Biochem 59:909–932.
Wagner MC, Pfister KK, Bloom GS, Brady ST (1989): Copurification of kinesin polypeptides with microtubule-stimulated Mg-ATPase activity and kinetic analysis of enzymatic properties. Cell Motil Cytoskel 12: 195–215.
Wagner MC, Pfister KK, Brady ST, Bloom GS (1991): Purification of kinesin from bovine brain and assay of microtubule-stimulated ATPase activity. Methods Enzymol 196:157–175.

Warren G (1989): Mitosis and membranes. Nature 342:857–858.

Wright BD, Henson JH, Wedaman KP, Willy PJ, Morand JN, Scholey JM (1991): Subcellular localization and sequence of sea urchin kinesin heavy chain: Evidence for its association with membranes in the mitotic apparatus and interphase cytoplasm. J Cell Biol 113: 817–833.

Yang JT, Laymon RA, Goldstein LS (1989): A three-domain structure of kinesin heavy chain revealed by DNA sequence and microtubule binding analyses. Cell 56:879–889.

Yang JT, Saxton WM, Goldstein LS (1988): Isolation and characterization of the gene encoding the heavy chain of *Drosophila* kinesin. Proc Natl Acad Sci USA 85:1864–1868.

Yang JT, Saxton WM, Stewart RJ, Raff EC, Goldstein LS (1990): Evidence that the head of kinesin is sufficient for force generation and motility *in vitro*. Science 249:42–47.

ABOUT THE AUTHORS

JONATHAN M. SCHOLEY is Associate Professor in the Section of Molecular and Cellular Biology, Division of Biological Sciences, at the University of California, Davis. His teaching duties include a graduate course in cellular biochemistry and an undergraduate course in the structure and function of biomolecules. Dr. Scholey received a B.Sc. with first class honors in cell and molecular biology from the MRC Cell Biophysics Unit/Department of Biophysics at Kings College, London. He received a Ph.D. in molecular biology from Trinity College, Cambridge, where he carried out research in the MRC Laboratory of Molecular Biology, focusing on the regulation of cytoplasmic myosin polymerization and ATPase activity by phosphorylation. Dr. Scholey then carried out postdoctoral research on the identification of microtubule-based motors that may participate in mitosis, working in the laboratory of Dr. J.R. McIntosh at the University of Colorado in Boulder. His recent research has involved the purification and analysis of kinesins.

RONALD D. VALE is Associate Professor in Pharmacology and Biochemistry at the University of California, San Francisco, where he teaches pharmacology to medical students and cell biology to graduate students. After receiving a B.A. from the University of California, Santa Barbara, he pursued graduate studies with Dr. Eric Shooter on nerve growth factor receptors in the Department of Neurobiology at Stanford University and received a Ph.D. in 1985 in neurosciences. Postdoctoral work was performed with Dr. Mike Sheetz and Tom Reese at the Marine Biological Laboratory at Woods Hole, Massachusetts; during this time, he examined the molecular mechanism of axonal transport. Since 1986, Dr. Vale has pursued studies on kinesin and dynein motors at the University of California, San Francisco. He received the Pfizer Award in Enzyme Chemistry in 1991 and the Young Investigator Award from the Biophysical Society in 1993. He is presently a monitoring editor for the *Journal of Cell Biology.*

Microtubules: 367–380

Dynein-Based Organelle Movement

Christine A. Collins

I. INTRODUCTION

The participation of microtubules in intracellular motility has been well documented over the past two decades [reviewed by Vale, 1987; Schroer and Sheetz, 1991a; Vallee and Bloom, 1991]. Organelle transport has been shown to be dependent on the presence of an intact microtubule array and requires ATP, suggesting the involvement of microtubule-dependent motor molecules. Two microtubule-associated ATPases, cytoplasmic dynein and kinesin, exhibit properties consistent with a role in cytoplasmic motility. The evidence for kinesin as an anterograde organelle translocator has been greatly strengthened by recent molecular and genetic experiments (see Scholey and Vale, this volume). While structural and biochemical properties of purified cytoplasmic dynein have been described by investigators working with a variety of cell and tissue systems (listed in Table I), there are fewer studies that address the role of dynein in organelle motility, the topic of this discussion.

II. EVIDENCE THAT DYNEIN IS A RETROGRADE ORGANELLE TRANSLOCATOR

The evidence that cytoplasmic dynein is an organelle translocator has come from three lines of inquiry. First, are the characteristics of dynein enzymatic activity consistent with what has been observed for retrograde organelle motility *in vivo?* Second, is dynein present in the appropriate cellular and subcellular locations to function as a retrograde organelle motor? Finally, can dynein-dependent retrograde organelle translocations be reconstituted *in vitro?* As described below, the answers to these questions have supported the argument for dynein as a retrograde translocator, but the interpretation of the experimental results has not been straightforward in all cases.

A. Biochemical Characteristics of Retrograde Transport

1. Inhibitors of organelle motility. A role for a dynein-like enzyme in cytoplasmic motility was suspected early in investigations of organelle transport with permeabilized cell models, based on the effects of known inhibitors of axonemal dynein. Treatment with vanadate or erythro-9-[3-(2-hydroxynonyl)]adenine (EHNA) has been shown to inhibit vesicle motility in several systems [Beckerle and Porter, 1982; Clark and Rosenbaum, 1982; Forman, 1982; Stearns and Ochs, 1982; Forman et al., 1983a]. Vanadate is an inhibitor of many ATP-utilizing enzymes and is a particularly potent inhibitor of ion pump ATPases [Cantley et al., 1978] and dyneins from both axonemal and cytoplasmic sources [Gibbons et al., 1978; Kobayashi et al., 1978; Lye et al., 1987; Shpetner et al., 1988]. Its use is somewhat limited, however, in that it does not readily enter intact cells and thus must be used in lysed cell models or for *in vitro* assays [Penningroth, 1986]. EHNA has been shown to inhibit selectively retrograde but not anterograde transport of a subset of organelles in axons [Forman et al., 1983b]. This adenosine analogue inhibits axonemal [Penn-

TABLE I. Structural and Biochemical Characteristics of Cytoplasmic Dynein

Structural properties			
1.	Size and macromolecular structure[a–f]		
	Mass = 1.2×10^6		
	Two globular head domains joined by flexible stalks to a globular base		
2.	Subunit composition and stoichiometry[f–h]		
	2 subunits of > 400 kD		
	3 subunits of 74 kD		
	1 subunit each of 59, 57, 55, 53 kD		
3.	UV/vanadate cleavage products[c–g,i,j]		
	Polypeptides generated from the high molecular weight subunit of approximately 185 and 225 kD		
Enzymatic characteristics			
1.	Nucleotide specificity[g,h,j,l]		
	Nucleotidase activity	CTP = TTP >> UTP >> GTP >> ITP = ATP	
	Microtubule gliding	ATP only	
2.	Inhibitors[g,j,l,m]		
	ATPase activity*	Vanadate	10–50 μM
		NEM	1 mM
		EHNA	1 mM
		AMPPNP	5 mM
	Microtubule gliding**	Vanadate	5–50 μM
		NEM	0.1 mM
		EHNA	1 mM
		AMPPNP	2 mM
3.	Kinetic constants[f–h,j–l]		
	Nucleotidase activity	K_m (ATP)	15 μM
		K_m (CTP)	1 mM
		K_m (MT)	0.16 mg/ml
		V_{max}	186–500 nmol/min/mg microtubule activated
	Microtubule gliding rates		1.25–2.0 μm/sec

*Concentrations of inhibitor that decrease ATPase by >50%.
**Concentrations of inhibitor that block motility.
[a]Vallee et al. [1988].
[b]Amos [1989].
[c]Gilbert and Sloboda [1989].
[d]Schroeder et al. [1990].
[e]Schnapp and Reese [1989].
[f]Neely et al. [1990].
[g]Paschal et al. [1987].
[h]Collins and Vallee [1989].
[i]Schroer et al. [1989].
[j]Lye et al. [1987].
[k]Paschal and Vallee [1987].
[l]Shpetner et al. [1988].
[m]Neely and Boekelheide [1988].

ingroth, 1986] and cytoplasmic [Shpetner et al., 1988, Neely and Boekelheide, 1988] dyneins, but has effects on protein methylation, DNA synthesis, and actin assembly as well [Penningroth, 1986].

The sensitivity of retrograde axonal transport to *N*-ethylmaleimide (NEM) has also been taken as support for a role for cytoplasmic dynein in motility, as this enzyme in purified form is more sensitive to inhibition by NEM than is kinesin [Lye et al., 1987; Paschal and Vallee, 1987; Shpetner et al., 1988]. NEM is not a specific inhibitor of dynein, as many enzymes would be affected by this sulfhydryl alkylating agent. In several studies investigating vesicle motility in axoplasm, vanadate and NEM were found to affect anterograde and retrograde transports similarly [Forman et al., 1983a; Pfister et al., 1989; Leopold et al., 1990; Weiss et al., 1991]. Thus the lack of specific inhibitors of dynein makes it difficult to assess the contribution of dynein to transport activities in whole cell systems. In addition, the pharmacological properties of vesicle transport in whole or permeabilized cells may not be reproduced when using purified components.

2. Polarity of force production. Purified cytoplasmic dynein has been found to promote gliding of microtubules toward their plus ends, a property that is consistent with force generation of a dynein-bound organelle on a fixed microtubule in the opposite direction, toward the minus end [Paschal and Vallee, 1987]. Both inner and outer arm axonemal dyneins appear to generate force similarly [Sale and Satir, 1977; Fox and Sale, 1987], and *in vitro* motility assays with purified axonemal dyneins confirm this polarity [Vale and Toyoshima, 1988]. This suggests that, at least for axonemal dyneins, the polarity observed *in vitro* is the same as that demonstrated *in vivo*. However, we do not know the structural or enzymatic characteristics that lead to unidirectional force production. For example, ncd, a protein containing a motor domain with sequence similarity to kinesin, promotes retrograde rather than anterograde motility *in vitro* [McDonald et al., 1990; Walker et al., 1990]. Preparations of cytoplasmic dynein from *Reticulomyxa* exhibit both retrograde and anterograde motility [Euteneuer et al., 1988], raising the possibility that one or both microtubule motor proteins may function bidirectionally *in vivo*.

3. Nucleotide specificity. Cytoplasmic dynein hydrolyzes both purine and pyrimidine nucleotides, many at greater rates than found with ATP as substrate [Shpetner et al., 1988; Collins and Vallee, 1989]. The nucleotidase activity and motility characteristics of several motor molecules have been examined, and the resulting nucleotide "fingerprint" has been purported to be useful in the identification of motility enzymes in complex mixtures [Shimizu et al., 1991]. Based on such an argument, it has been concluded that both anterograde and retrograde motility in *Reticulomyxa* are powered by the same (or very similar) enzyme(s), in this case, cytoplasmic dynein [Schliwa et al., 1991]. Dynein-mediated microtubule gliding has been shown to be ATP specific [Lye et al., 1987; Paschal and Vallee, 1987]. However, in squid axoplasm both retrograde and anterograde motility were supported to similar extents by nucleotides other than ATP, a result more in keeping with the substrate specificity of kinesin [Leopold et al., 1990].

Taken together with the pharmacological data described above, the biochemical characteristics of vesicle transport in permeabilized cell models do not clearly distinguish between dynein and kinesin, and in some cases they are not consistent with those found for either motor molecule as assayed *in vitro* [Leopold et al., 1990; Weiss et al., 1991]. However, similar inconsistencies have been found in studies of axonemal motility, where ciliary reactivation exhibits different nucleotide requirements than does microtubule translocation powered by axonemal dynein [Shimizu et al., 1991]. Additionally, the motile activity of force-producing enzymes has been shown to be more sensitive to inhibitors than nucleotide hydrolysis [Lye et al., 1987; Paschal and Vallee, 1987; Shpetner et al., 1988; Cohn et al., 1989].

The disparate pharmacology and nucleotide specificity data obtained in cell models compared with those using purified mixtures do not rule out the participation of dynein in organelle motility, but suggest at least three possible explanations for these results. First, retrograde and anterograde motors act synergistically in the cell, either in a complex or coordinately regulated, such that inhibition of either of the pair results in similar effects on both directions of movement. Second, retrograde and anterograde motors (dynein and kinesin, or perhaps other unidentified motility factors) share many biochemical properties and cannot be distinguished from one another in crude systems on the basis of current pharmacological studies as compared with data obtained using purified proteins. Finally, in some cases (e.g., squid axoplasm, *Reticulomyxa*), there may be one or more bidirectional motors.

B. Cellular and Subcellular Dynein Distribution

Cytoplasmic dynein is an abundant protein in brain tissue, where it is expected to function in retrograde axonal transport. Consistent with this, antibody to cytoplasmic dynein stained axons of central and peripheral nerves [Yoshida et al., 1992]. However, cytoplasmic dynein is abundant in nonneuronal cells as well [Collins and Vallee, 1989], and its cellular content correlates to some extent with the activity of organelle traffic within each cell type [Yoshida et al., 1992].

Cytoplasmic dynein has been immunolocalized at the subcellular level to membrane-bound structures, including lysosomes, which would be expected to undergo retrograde transport [Gilbert and Sloboda, 1989; Koonce and McIntosh, 1990; Hirokawa et al., 1990; Lin and Collins, 1992]. However, electron microscopic analysis of peripheral axons shows dynein localization on both anterograde and retrograde vesicles [Hirokawa et al., 1990]. In addition, a study of dynein association with vesicle membranes *in vitro* shows that dynein is present on synaptic vesicles, an organelle that would be expected to undergo primarily anterograde movement [Lacey and Haimo, 1992a]. These data can be taken as evidence for an association of dynein in an inhibited or inactive form with anterograde vesicles. Thus localization or binding studies alone may not provide information concerning the function of dynein in specific motile events. Cytoplasmic dynein has also been localized to kinetochores, spindle poles, and spindle microtubules [Pfarr et al., 1990; Steuer et al., 1990; Lin and Collins, 1992]. This leaves open the question of whether dynein is multifunctional, acting as both a retrograde organelle translocator and a mitotic motor, or whether multiple species of dynein-related proteins exist that cannot be distinguished from one another using the immunochemical reagents currently available.

C. Dynein-Mediated Organelle Transport in Reconstituted Systems

The primary evidence for cytoplasmic dynein function in organelle transport has come from studies utilizing ultraviolet/vanadate-mediated photocleavage [Euteneuer et al., 1989; Schnapp and Reese, 1989; Schroer et al., 1989]. Cleavage of the polypeptide backbone of dynein by vanadate in the presence of ultraviolet light and ATP has been used as a diagnostic tool for the identification of dynein and its associated functions [Gibbons and Mocz, 1991].

Retrograde organelle transport in cytosols prepared from squid axoplasm [Schnapp and Reese, 1989] and from chick embryo fibroblasts [Schroer et al., 1989] has been observed to be selectively inhibited by ultraviolet/vanadate treatment. Addition of exogenous purified dynein partially restored retrograde motility in one case [Schroer et al., 1989], but had little or no effect in others [Euteneuer et al., 1989; Schnapp and Reese, 1989]. However, it has been shown that ultraviolet/vanadate-cleaved dynein can bind to organelles and thus would inhibit the ability of added dynein to promote motility [Schroer et al., 1989; Yu et al., 1992].

Ultraviolet/vanadate treatment can induce cleavage of other ATPases besides dynein. The most well-studied example is myosin II [Cremo et al., 1991]. Though myosin inhibition and cleavage by ultraviolet/vanadate require somewhat different conditions than those found for dynein, the effects of this treatment on other myosins, some of which may also be responsible for organelle motility [Pollard et al., 1991; Kuznetsov et al., 1992] have not been examined. Ionic strength also affects vanadate inhibition of dynein, but not myosin [Kobayashi et al., 1978]; thus higher concentrations of vanadate are required for inhibition of dynein in lysed cells or in cell extracts, and these higher concentrations would be less specific for dynein. It is difficult to control for irreversible inhibitory effects of vanadate on other components of the motility complex, as vanadate is easily reduced in crude protein mixtures, binds tightly to some proteins, and can be reoxidized to varying degrees by atmospheric oxygen [Penningroth, 1986]. Vesicle motility assays involving dynein and cytosolic factors are difficult to carry out, with very few numbers of moving particles and rates of movement that are lower than that in axoplasm or live cells [Schroer et al., 1988, 1989; Schnapp and Reese, 1989; Weiss et al., 1991]. There has thus far been no effort to select for populations of vesicles that might show active retrograde movement, and the salt-stripped vesicles [Schroer et al., 1988] used for motility assays may, in addition, have lost components essential for appropriate regulation of motility.

In spite of these technical difficulties, the available evidence suggests that cytoplasmic dynein can function as a retrograde motor *in vitro*. It remains to be determined whether dynein plays a role in specific retrograde translocation events *in vivo*. Further evidence for a role for dynein in organelle transport will likely come from the development of more specific inhibitors (either drugs or antibodies), through biochemical reconstitution of functional organelle motility complexes and through molecular and genetic approaches.

III. FUNCTIONS OF A RETROGRADE TRANSLOCATOR

Assuming that cytoplasmic dynein does, in fact, mediate retrograde organelle motility *in vivo*, the species of organelles transported by this motor may be dictated in large part by the organization of microtubules within the cytoplasm. In axonal processes, microtubules are longitudinally arrayed and of uniform polarity, with the plus ends distal [Burton and Paige, 1981; Heidemann et al., 1981]. In cultured fibroblasts, the majority of the microtubules are nucleated at the centrosome, with plus ends distal [Karsenti et al., 1984]. Microtubules in polarized epithelial cells are organized in a parallel fashion with their minus ends at the apical surface [Tucker et al., 1986; Achler et al., 1989; Bacallao et al., 1989]. A retrograde translocator may therefore be involved in the minus end–directed movement of different classes of organelles, depending on the cell type. In addition, the reorganization of cellular compartments following mitosis may rely on the function of a retrograde translocator.

A. Retrograde Fast Axonal Transport

The movement of organelles over long distances in the axons of nerve cells is a manifestation of the same motility mechanisms operative in other cell types [reviewed by Vallee and Bloom, 1991]. Vesicles translocated by dynein in the axon would include lysosomes and multivesicular degradative bodies [Tsukita and Ishikawa, 1980] and endocytic vesicles carrying trophic factors from target organs to the nerve cell body [Schwab et al., 1981]. One complicating factor in axonal transport is the large distance that must be covered by translocating organelles. This makes it unlikely that newly synthesized or reutilized dynein would simply diffuse down the length of a neuronal process in order to associate with vesicles for the return trip back to the cell body. There is likely to be a mechanism for anterograde transport of dynein in an inactive form, perhaps in association with translocating vesicles, to the nerve terminal for subsequent activation and association with its target organelles.

As described above, evidence to support this hypothesis has come from dynein localization to both anterograde and retrograde vesicles in studies examining peripheral nerves [Hirokawa et al., 1990]. In nonneuronal cells, this process may be unnecessary or take a different form because of smaller distances involved in the distribution of dynein throughout the cytoplasm.

B. Retrograde Transport in Cultured Fibroblasts

In fibroblasts, a retrograde transporter is likely involved in the movement of endocytic organelles formed at the plasma membrane toward the lysosomes and Golgi apparatus clustered around the nucleus. Early stages of endocytosis are not apparently dependent on microtubules [reviewed by Gruenberg and Howell, 1989], but transfer of carrier vesicle contents to late endosomes [Gruenberg et al., 1989) and late endosome transfer to lysosomes [Brown et al., 1980; Oka and Weigel, 1983] have been shown to require an intact microtubule network and would be expected to utilize a minus end–directed microtubule motor.

Dynein translocation activity may also be required for the restoration of perinuclear distributions of organelles following mitosis or during recovery from treatment with microtubule-disruptive drugs. Peripherally distributed lysosomes and endosomes follow microtubule paths back toward the nucleus during recovery from nocodozole treatment [Herman and Albertini, 1984; Matteoni and Kreis, 1987; Lin and Collins, 1992]. As these redistributed organelles appear to retain dynein, they are primed for retrograde translocation upon restoration of the microtubule array [Lin and Collins, 1992]. Similarly, the accumulation of fragmented Golgi vesicles near the nucleus following drug treatment requires ATP and microtubules [Turner and Tartakoff, 1989; Ho et al., 1989]. Cytoplasmic dynein has been shown to be required for capture and perinuclear accumulation of Golgi-derived membranes within semiintact cells [Corthesy-Theulaz et al., 1992]. It is not clear, however, whether the maintenance of organelle distribution within nondividing cells requires the continued presence and activity of a microtubule motor or whether other stable interactions with the cytoskeleton are involved [Bloom and Brashear, 1989; Mithieux and Rousset, 1989; Scheel and Kreis, 1991; Goltz et al., 1992; Karecla and Kreis, 1992].

C. Membrane Traffic in Polarized Epithelial Cells

Because of the parallel arrangement of microtubules in polarized epithelial cells, a retrograde organelle motor might be expected to function in endocytosis at the basal, but not the apical, surface. Conversely, dynein would be expected to mediate transport to and secretion at the apical surface. Microtubules have been shown to be required for polarized secretion at the apical (but not basolateral) domain of vesicles derived from the trans-Golgi network as well as transcytotic vesicles carrying material from the basolateral surface, consistent with a role for dynein in these processes [Rindler et al., 1987; Eilers et al., 1989; reviewed by Mostov et al., 1992]. Dynein may also be involved in secretion from Sertoli cells, since the microtubules in these cells are of uniform polarity with their minus ends toward the lumen of seminiferous tubules [Redenbach and Vogl, 1991]. Dynein is present in high concentration in rat testis and in Sertoli cells in particular [Neely and Boekelheide, 1990; Collins and Vallee, 1989; Hall et al., 1992; Yoshida et al., 1992]. Dynein has also has been proposed to function in spermatid maturation [Hall et al., 1992].

Several endocytic fusion and transfer events have been reconstituted *in vitro* [reviewed by Gruenberg and Howell, 1989; Schmid, 1992]. The efficiency of fusion between apical and basolateral endocytic vesicles was increased in the presence of microtubules as well as dynein and kinesin [Bomsel et al., 1990]. The ability of dynein and kinesin to bind and translocate organelles along microtubules likely increases the probability of an encounter between vesicles that will undergo fusion events [Bomsel et al., 1990]. Cytoplasmic dynein has also been proposed to function in the segregation of endosomal vesicles by allowing the transport of ligand-containing vesicles away from those carrying receptors [Goltz et al., 1992]. The function of microtubule motors in endocytosis and secretion is likely to be one of increasing efficiency and specificity of the process, since the small distances involved might not otherwise require an active process.

IV. BIOCHEMICAL AND IMMUNOCYTOCHEMICAL ASPECTS OF DYNEIN–VESICLE INTERACTIONS

Cytoplasmic dynein has been isolated from many sources by coassembly with microtubules from cytosolic extracts, implying that a considerable amount of the cellular content of the protein is soluble [Lye et al., 1987; Paschal et al., 1987; Collins and Vallee, 1989]. A decrease in diffuse staining with antidynein antibody was found following detergent extractions in immunolocalization studies, also suggesting the presence of a soluble pool of dynein in cultured cells [Steuer et al., 1990; Lin and Collins, 1992]. However, the appearance of vesicular staining by immunofluorescence microscopy [Gilbert and Sloboda, 1989; Koonce and McIntosh, 1990; Pfarr et al., 1990; Lin and Collins, 1992], as well as immunogold labeling of peripheral nerves [Hirokawa et al., 1990], indicates that cellular dynein is membrane associated. Experiments attempting to reconcile these conflicting results have shown that the cell or tissue type and homogenization conditions used for subcellular fractionations can influence the extent to which dynein is found associated with membranes (Delgado-Partin, Lin, Tuma, and Collins, unpublished data).

A. Requirement for Other Factors in Promoting Specificity and Function

Membrane-bound and/or cytosolic factors appear to be required in addition to motor molecules for reconstitution of organelle transport [Schroer et al., 1988]. A complex of polypeptides has been identified in embryonic chick brain dynein preparations, termed *dynactin complex* [Schroer and Sheetz, 1991b]. The major polypeptide constituents of this complex are a 150 kD doublet and 45 and 50 kD species [Gill et al., 1991]. Several polypeptides of this complex were first identified as potential subunits or dynein-associated proteins in cytoplasmic dynein preparations from brain and nonneuronal sources [Collins and Vallee, 1989; Holzbaur et al., 1991]. It has since been shown that the dynactin complex can be fractionated away from dynein in the presence of high salt, and the dynactin complex alone will sediment at 20S in sucrose gradients [Gill et al., 1991]. Therefore it is unclear whether the copurification of these polypeptides with dynein is functionally important or is a consequence of similar molecular size. The 150 and 45 kD components of this complex, also referred to as the *Glued* complex after the 150 kD component [Holzbaur et al., 1991], have been analyzed at the molecular level [Gill et al., 1991; Holzbaur et al., 1991; Lees-Miller et al., 1992] (see also Holzbaur et al., this volume). Immunolocalization using antibodies raised against the 150 kD component gives similar results to that found using monoclonal antibodies to the intermediate chain of dynein [Gill et al., 1991], but discrete organelle populations containing this polypeptide have not been identified. The means by which this complex activates dynein is unclear. It apparently affects the efficiency of dynein-mediated motility rather than the rate of movement, perhaps indicating that it is involved in organelle recognition or binding interactions [Schroer and Sheetz, 1991b].

B. Mode of Interaction Between Dynein and Membranes

For dynein to function as a vesicle motor, it must physically interact with the cytoplasmic surface of the membranous organelle. It would be logical, on the basis of specificity considerations, that membrane proteins (receptors) be involved in the recognition, binding, and, perhaps, activation of dynein. An integral membrane protein in endoplasmic reticulum that binds kinesin has recently been identified [Toyoshima et al., 1992]. The ability of dynein to recognize and bind with high affinity to organelles may not depend to the same extent on peripheral or integral membrane proteins. Dynein can bind saturably to synaptic vesicles that have been salt extracted or protease treated and can bind as well to pure phospholipid vesicles [Lacey and Haimo, 1992a,b], though proteolysis of microsomal vesicles eliminates dynein binding to this class of organelle [Yu et al., 1992]. Like binding interactions that have been described for myosins I [Adams and Pollard, 1989; Hayden et al., 1990], dynein may be able to bind to cell membranes through an interaction with lipids, and further specificity or activation may be mediated through integral and peripheral accessory proteins.

Organelles stripped of peripheral proteins with salt are capable of retrograde motility in the presence of cytoplasm, which presumably contributes soluble dynein and other factors. Motility of these vesicles suggests that peripherally associated membrane proteins are not required for dynein function, that they are present as well in the cytoplasm, or that this assay does not reflect a requirement for such proteins [Schroer et al., 1988, 1989; Gilbert and Sloboda, 1989; Schnapp and Reese, 1989].

V. REGULATION OF DYNEIN-MEDIATED VESICLE TRANSPORT

To examine regulation of motility, it is helpful to investigate both naturally occurring and artificially induced disruptions in processes attributable to cytoplasmic dynein. Mitosis provides a natural block in vesicle transport activity in the cell. A difference in the ability of mitotic and interphase cell extracts to promote organelle motility has been observed, yet microtubule gliding activity of partially purified motor molecules from these extracts was unchanged [Allan and Vale, 1991]. It was suggested that the regulation of transport may therefore occur at the level of association of the motor molecule with organelle membranes. However, cytoplasmic dynein remains associated with at least some organelles throughout mitosis, indicating that other means of regulation may also be operative [Lin and Collins, 1992]. Cell cycle control perhaps implies a regulatory phosphorylation cascade, but the role of phosphorylation in regulation of dynein binding to membranes or vesicle motility has yet to be addressed. Phosphorylated factors have, however, been shown to be important for the dynein-dependent formation of microtubule asters in egg extracts [Verde et al., 1991]. Phosphorylation pathways also appear to be involved in the highly regulated movement of pigment granules in fish scale chromatophores [Lynch et al., 1986; Rozdial and Haimo, 1986].

Other studies examining regulation of motor activity involve treatments of cells, such as by using microtubule drugs or by decreasing cytoplasmic pH through treatment with acetate. Acidification of the cytoplasm in fibroblasts causes peripheral dispersion of endosomes and lysosomes, organelles that generally accumulate around the nucleus [Heuser, 1989]. In MDCK cells, low cytoplasmic pH induced a fragmentation of the endosomal compartment, and a microtubule-dependent shift of both endosomes and lysosomes to the basal domain [Parton et al., 1991]. In both cases, normally perinuclear (microtubule minus end–associated) organelles became distributed toward regions of the cell containing plus ends of microtubules. It has been suggested that low cytoplasmic pH inhibits retrograde transport and/or stimulates anterograde transport [Heuser, 1989; Parton et al., 1991]. Dynein binding to lysosomes has been shown to be stable under low cytosolic pH conditions [Lin and Collins, 1992]. This implies that dynein remains associated with organelles not undergoing active retrograde transport and that regulation of dynein activity under these conditions takes place at other than the level of binding interactions.

VI. UNRESOLVED QUESTIONS AND AREAS FOR FURTHER STUDY

A. Regulation of Vesicle Transport

Clearly, an important area for future work is the consideration of mechanisms of regulation of motile activity in the cell. Regulation of cytoplasmic dynein is possible at several levels: 1) organelle and microtubule affinity and binding specificity; 2) alterations in the rate of enzyme turnover; 3) coupling between ATP hydrolysis and force production; and 4) coordination with other components of the motility machinery, such as kinesin and accessory factors.

The binding interactions as measured through immunolocalization, or via biochemical experiments examining binding exogenous dynein to purified vesicle preparations, may vary with cell type, the presence or absence of accessory proteins, the number of potential receptor sites on the organelle, and the activity of the purified dynein. It has been suggested that as few as one motor molecule is sufficient to move an organelle [Howard et al., 1989; Ashkin et al., 1990; Block et al., 1990]. Observed differences in dynein binding affinity or in concentration of available binding sites on different classes of organelles may therefore not be interpretable in terms of motile function.

One of the hallmarks of the cytoskeletal motor proteins is their ability to be activated by their particular filament system. Thus myosin ATP turnover is stimulated by actin; and kinesin and dynein ATPases (axonemal and cytoplasmic) are stimulated by microtubules. However, it is not clear that microtubule activation is a requirement for function of a cytoplasmic motor. Dynein ATPase is stimulated only two-to fivefold by microtubules [Paschal et al., 1987; Shpetner et al., 1988; Collins and Vallee, 1989]. Calculations of energy requirements suggest that very few molecules of a motor such as dynein would be required to maintain the rates of motion observed *in vivo* [Ashkin et al., 1990]. This low degree of activation of dynein by microtubules is quite striking, however, when compared with kinesin. Under some conditions, kinesin can be activated 1,000-fold by microtubules [Kuznetsov and Gelfand, 1986]. The regulation of nucleotide hydrolysis by these two microtubule motor proteins therefore appears to be quite different. Additional, as yet unidentified, factors may be involved in the modulation of nucleotidase activity once tertiary complexes between organelle, motors, and microtubules are formed.

It is likely that there is regulation of force production distinct from effects of microtubules on ATPase activity. For axonemal dyneins, ATP hydrolysis can be uncoupled from force production under some circumstances. It is likely that this is the case for cytoplasmic dynein as well. For example, cytoplasmic dynein in association with organelles and microtubules is unable to promote efficient vesicle motility in the absence of the dynactin complex [Schroer and Sheetz, 1991b], though ATP hydrolysis is active in preparations lacking these polypeptides [Shpetner et al., 1988]. The function of the dynactin complex or other accessory proteins may be to promote efficient utilization of energy obtained from ATP hydrolysis for force production.

One attractive model for coordination of intracellular transport is that both dynein and kinesin are part of a multicomponent complex that functions in vesicle motility [Schroer and Sheetz, 1991a; Yu et al., 1992]. Such an arrangement could allow for the activation of one motor and the simultaneous inactivation of the other. In such a complex, motility function might be sensitive to inhibition of only one of its components. Pharmacological data in permeabilized cells, as well as observations that organelles *in vivo* rarely undergo uniformly unidirectional movement, but often undergo frequent reversals of direction support this idea. That soluble regulatory or accessory factors may be shared between kinesin and dynein comes from experiments in which antibody affinity columns were used to deplete cytosol of kinesin [Schroer et al., 1988]. The resulting cytosol was unable to promote either anterograde or retrograde transport efficiently, suggesting that required factors for cytoplasmic dynein mediated motility had been depleted as well. Experiments examining kinesin and dynein binding to microsomes indicate that there may be competition for the same membrane site under some circumstances [Yu et al., 1992].

However, coordinated activity of the two motors does not require them to be present in a complex. The distributions of kinesin and dynein in the cell do not strictly overlap, at least in terms of where the proteins are most concentrated [Lin et al., 1991]. While an integral membrane "receptor" has been identified for kinesin [Toyoshima et al., 1992], such a binding protein has not yet been described for dynein, and each motor can function independently in motility and binding assays. Further work is required in order to understand the relationship between kinesin and dynein localization and activity. In addition, in light of recent evidence that actin and actin associated motor molecules may participate in vesicle motility [Kuznetsov et al., 1992], it will be interesting to determine whether there is coordinated activity of the two cytoskeletal motility systems.

B. Functional and Isoform Diversity

There is evidence to suggest that cytoplasmic dynein functions as a bidirectional motor in the freshwater alga *Reticulomyxa* [Euteneuer et al., 1988, 1989]. Is the same molecular species responsible for both retrograde and anterograde movements in this organism, or are closely related but not identical dynein-like enzymes responsible for the observed results? If the same molecule is involved, what form of regulation is involved in the switching mechanism, what does this do to the physical properties of the dynein molecule and its binding characteristics, and does this occur in other organisms as well?

An additional function suggested for cytoplasmic dynein is that of a mitotic motor. Assuming a multifunctional base structure that could interact with multiple binding sites [Paschal et al., 1992], the retrograde force-producing characteristics of cytoplasmic dynein could function in mitosis as well as in organelle transport. Cytoplasmic dynein has a functional ATP-insensitive microtubule-binding site, as implicated from microtubule cross-linking studies [Amos, 1989]. This binding site may be operative in the observed colocalization of cytoplasmic dynein with mitotic microtubules [Pfarr et al., 1990; Steuer et al., 1990; Lin and Collins, 1992]. However, the question remains as to why dynein binds to mitotic microtubules but not to the interphase cytoskeleton. The difference may be due to protein modifications or to microtubule-associated protein content in the mitotic apparatus, or perhaps to the presence of a mitosis-specific form of dynein.

The antibodies currently used for dynein localization do not apparently distinguish between different dynein populations, for example, between vesicular and spindle-associated dynein [Pfarr et al., 1990; Lin and Collins, 1992]. Molecular cloning studies have thus far indicated only one gene for the dynein heavy chain (see Holzbaur et al., this volume). Thus, not only are the two heads of the intact cytoplasmic dynein molecule identical, but all dynein isoforms, if they exist, must also share this common subunit. Recently, the dynein intermediate chain sequence has been determined [Paschal et al., 1992]. At least three forms of this polypeptide exist, and it will be of interest to determine whether all dyneins contain one of each of these polypeptides or whether subunit shuffling may lead to isoform and functional diversity. Additional dynein subunits are currently under investigation at the molecular level. Posttranslational modifications may also be involved in regulation and production of dyneins with different properties, such as switching from retrograde to anterograde, and mitotic versus interphase binding.

VII. SUMMARY

Accumulating evidence points to cytoplasmic dynein as a retrograde organelle translocator. There is a need to define further the requirement for microtubule-based retrograde motility during endocytosis and secretion and in the establishment and maintenance of polarity in cells. Additional biochemical characterization of cytoplasmic dynein and receptor and accessory proteins will be necessary in order to develop *in vitro* correlates of retrograde motility. Systems amenable to molecular and genetic manipulation will no doubt provide considerable information regarding dynein function and the regulatory pathways involved in the control of intracellular organelle traffic in the near future.

VIII. REFERENCES

Achler C, Filmer D, Merte C, Drenckhahn D (1989): Role of microtubules in polarized delivery of apical membrane proteins to the brush border of the epithelium. J Cell Biol 109:179–189.

Adams RJ, Pollard TD (1989): Binding of myosin-I to membrane lipids. Nature 340:565–568.

Allan VJ, Vale RD (1991): Cell cycle control of microtubule-based membrane transport and tubule formation *in vitro*. J Cell Biol 113:347–359.

Amos LA (1989): Brain dynein crossbridges microtubules into bundles. J Cell Sci 93:19–28.

Ashkin A, Schutze K, Dziedzic JM, Euteneuer U, Schliwa M (1990): Force generation of organelle transport measured *in vivo* by an infrared laser trap. Nature 348: 346–348.

Bacallao R, Antony C, Dotti C, Karsenti E, Stelzer K, Simons K (1989): The subcellular organization of Madin-Darby canine kidney cells during the formation of a polarized epithelium. J Cell Biol 109: 2817–2832.

Beckerle MC, Porter KR (1982): Inhibitors of dynein activity block intracellular transport in erythrophores. Nature 295:701–703.

Block SM, Goldstein LSB, Schnapp BJ (1990): Bead movement by single kinesin molecules studied with optical tweezers. Nature 348:348–352.

Bloom GS, Brashear TA (1989): A novel 58-kDa protein associates with the Golgi apparatus and microtubules. J Biol Chem 264:16083–16092.

Bomsel M, Parton R, Kuznetsov SA, Schroer TA, Gruenberg J (1990): Microtubule- and motor-dependent fusion *in vitro* between apical and basolateral endocytic vesicles from MDCK cells. Cell 62:719–731.

Brown KD, Friedkin M, Rozengurt E (1980): Colchicine inhibits epidermal growth factor degradation in 3T3 cells. Proc Natl Acad Sci USA 77:480–484.

Burton PR, Paige JL (1981): Polarity of axoplasmic microtubules in the olfactory nerve of the frog. Proc Natl Acad Sci USA 78:3269–3273.

Cantley LC, Cantley LG, Josephson L (1978): A characterization of vanadate interactions with the (Na/K)-ATPase. J Biol Chem 253:7361–7368.

Clark TG, Rosenbaum JL (1982): Pigment particle translocation in detergent permeabilized melanophores of *Fundulus heteroclitus*. Proc Natl Acad Sci USA 79:4655–4659.

Cohn SA, Ingold AL, Scholey JM (1989): Quantitative analysis of sea urchin egg kinesin-driven microtubule motility. J Biol Chem 264:4290–4297.

Collins CA, Vallee RB (1989): Preparation of microtubules from rat liver and testis: Cytoplasmic dynein is a major microtubule associated protein. Cell Motil Cytoskel 14:491–500.

Corthesy-Theulaz I, Pauloin A, Pfeffer AR (1992): Cytoplasmic dynein participates in the centrosomal localization of the Golgi complex. J Cell Biol 118: 1333–1345.

Cremo CR, Grammer JL, Yount RG (1991): Vanadate-mediated photocleavage of myosin. Methods Enzymol 196:442–449.

Eilers U, Klumperman J, Hauri HP (1989): Nocodazole, a microtubule-active drug, interferes with apical protein delivery in cultured intestinal epithelial cells (Caco-2). J Cell Biol 108:13–22.

Euteneuer U, Johnson KB, Schliwa M (1989): Photolytic cleavage of cytoplasmic dynein inhibits organelle transport in *Reticulomyxa*. Eur J Cell Biol 50:34–40.

Euteneuer U, Koonce MP, Pfister KK, Schliwa M (1988): An ATPase with properties expected for the organelle motor of the giant amoeba, *Reticulomyxa*. Nature 332:176–178.

Forman DS (1982): Vanadate inhibits saltatory organelle movement in a permeabilized cell model. Exp Cell Res 141:139–147.

Forman DS, Brown KJ, Livengood DR (1983a): Fast axonal transport in permeabilized axons is inhibited by vanadate. J Neurosci 3:1279–1288.

Forman DS, Brown KJ, Promersberger ME (1983b): Selective inhibition of retrograde transport by erythro-9-[3-(2-hydroxynonyl)]adenine. Brain Res 272:194–197.

Fox LA, Sale WS (1987): Direction of force generated by the inner row of dynein arms on flagellar microtubules. J Cell Biol 105:1781–1787.

Gibbons IR, Cosson MP, Evans JA, Gibbons BH, Houck B, Martinson KH, Sale WS, Tang W-JY (1978): Potent inhibition of dynein adenosine triphosphatase and of the motility of cilia and sperm flagella by vanadate. Proc Natl Acad Sci USA 75:2220–2224.

Gibbons IR, Mocz G (1991): Photocatalytic cleavage of proteins with vanadate and other transition metal complexes. Methods Enzymol 196:428–442.

Gilbert SP, Sloboda RD (1989): A squid dynein isoform promotes axoplasmic vesicle translocation. J Cell Biol 109:2379–2394.

Gill SR, Schroer TA, Szilak I, Steuer ER, Sheetz MP, Cleveland DW (1991): Dynactin, a conserved, ubiquitously expressed component of an activator of vesicle motility mediated by cytoplasmic dynein. J Cell Biol 115:1639–1650.

Goltz JS, Wolkoff AW, Novikoff PM, Stockert RJ, Satir P (1992): A role for microtubules in sorting endocytic vesicles in rat hepatocytes. Proc Natl Acad Sci USA 89:7026–7030.

Gruenberg J, Griffiths G, Howell KE (1989): Characterization of the early endosome and putative endocytic carrier vesicles *in vivo* and with an assay of vesicle function *in vitro*. J Cell Biol 108:1301–1316.

Gruenberg J, Howell KE (1989): Membrane traffic in endocytosis: Insights from cell-free assays. Annu Rev Cell Biol 5:453–481.

Hall ES, Eveleth J, Jiang C, Redenback DM, Boekelheide K (1992): Distribution of the microtubule-dependent motors cytoplasmic dynein and kinesin in rat testis. Biol Reprod 46:817–828.

Hayden SM, Wolenski JS, Mooseker MS (1990): Binding of brush border myosin I to phospholipid vesicles. J Cell Biol 111:443–451.

Heidemann S, Landers JM, Hamborg MA (1981): Polarity orientation of axonal microtubules. J Cell Biol 91:661–665.

Herman B, Albertini DF (1984): A time-lapse video image intensification analysis of cytoplasmic organelle movements during endosome translocation. J Cell Biol 98:565–576.

Heuser J (1989): Changes in lysosome shape and distribution correlated with changes in cytoplasmic pH. J Cell Biol 108:855–864.

Hirokawa N, Sato-Yoshitake R, Yoshida T, Kawashima T (1990): Brain dynein (MAP 1C) localizes on both anterogradely and retrogradely transported membranous organelles *in vivo*. J Cell Biol 111:1027–1037.

Ho WC, Allan VJ, van Meer G, Berger EG, Kreis TE (1989): Reclustering of scattered Golgi elements occurs along microtubules. Eur J Cell Biol 48:250–263.

Holzbaur ELF, Hammarback JA, Paschal BM, Kravit NG, Pfister KK, Vallee RB (1991): Homology of a 150K cytoplasmic dynein-associated polypeptide with the Drosophila gene glued. Nature 351:579–583.

Howard J, Hudspeth AJ, Vale RD (1989): Movement of microtubules by single kinesin molecules. Nature 342:154–158.

Karecla PI, Kreis TE (1992): Interaction of membranes of the Golgi complex with microtubules *in vitro*. Eur J Cell Biol 57:139–146.

Karsenti E, Kobayashi S, Mitchison T, Kirschner M (1984): Role of the centrosome in organizing the interphase microtubule array: Properties of cytoplasts containing or lacking centrosomes. J Cell Biol 98:1763–1776.

Kobayashi T, Martenson T, Nath J, Flavin M (1978): Inhibition of dynein ATPase by vanadate, and its possible use as a probe for the role of dynein in cytoplasmic motility. Biochem Biophys Res Commun 81:1313–1318.

Koonce MP, McIntosh JR (1990): Identification and immunolocalization of cytoplasmic dynein in *Dictyostelium*. Cell Motil Cytoskel 15:51–62.

Kuznetsov SA, Gelfand VI (1986): Bovine brain kinesin is a microtubule-activated ATPase. Proc Natl Acad Sci USA 83:8530–8534.

Kuznetsov SA, Langford GM, Weiss DG (1992): Actin-dependent organelle movement in squid axoplasm. Nature 356:722–725.

Lacey ML, Haimo LT (1992a): Cytoplasmic dynein is a vesicle protein. J Biol Chem 267:4793–4798.

Lacey ML, Haimo LT (1992b): Cytoplasmic dynein-membrane binding is mediated by a direct interaction with phospholipids. Mol Biol Cell 3:163a.

Lees-Miller JP, Helfman DM, Schroer TA (1992): A vertebrate actin-related protein is a component of a multisubunit complex involved in microtubule-based vesicle motility. Nature 359:244–246.

Leopold PL, Snyder R, Bloom GS, Brady ST (1990): Nucleotide specificity for the bidirectional transport of membrane-bounded organelles in isolated axoplasm. Cell Motil Cytoskel 15:210–219.

Lin SXH, Collins CA (1992): Immunolocalization of cytoplasmic dynein to lysosomes in cultured cells. J Cell Sci 101:125–137.

Lin SXH, Pfister KK, Collins CA (1991): Distribution of cytoplasmic dynein-associated organelles along microtubules and co-localization with kinesin. J Cell Biol 115:168a.

Lye RJ, Porter ME, Scholey JM, McIntosh JR (1987): Identification of a microtubule-based cytoplasmic motor in the nematode *C. elegans*. Cell 51:309–318.

Lynch TJ, Wu B-Y, Taylor JD, Tchen TT (1986): Regulation of pigment organelle translocation. II. Participation of a cAMP-dependent protein kinase. J Biol Chem 216:4212–4216.

Matteoni R, Kreis TE (1987): Translocation and clustering of endosomes and lysosomes depends on microtubules. J Cell Biol 105:1253–1265.

McDonald HB, Stewart RJ, Goldstein LSB (1990): The kinesin-like ncd protein of *Drosophila* is a minus end–directed microtubule motor. Cell 63:1159–1165.

Mithieux G, Rousset B (1989): Identification of a lysosome membrane protein which could mediate ATP-dependent stable association of lysosomes to microtubules. J Biol Chem 264:4664–4668.

Mostov K, Apodaca G, Aroeti B, Okamoto C (1992): Plasma membrane protein sorting in polarized epithelial cells. J Cell Biol 116:577–583.

Neely MD, Boekelheide K (1988): Sertoli cell processes have axoplasmic features: An ordered microtubule distribution and an abundant high molecular weight microtubule-associated protein (cytoplasmic dynein). J Cell Biol 107:1767–1776.

Neely MD, Erickson HP, Boekelheide K (1990): HMW-2, the sertoli cell cytoplasmic dynein from rat testis, is a dimer composed of nearly identical subunits. J Biol Chem 265:8691–8698.

Oka JA, Weigel PH (1983): Microtubule-depolymerizing agents inhibit asialo-orosomucoid delivery to lysosomes but not its endocytosis in isolated rat hepatocytes. Biochim Biophys Acta 763:368–376.

Parton RG, Dotti CG, Bacallao R, Kurtz I, Simons K, Prydz K (1991): pH-induced microtubule-dependent redistribution of late endosomes in neuronal and epithelial cells. J Cell Biol 113:261–274.

Paschal BM, Mikami A, Pfister KK, Vallee RB (1992): Homology of the 74 kD cytoplasmic dynein subunit with a flagellar dynein polypeptide suggests an intracellular targeting function. J Cell Biol 118:1133–1143.

Paschal BM, Shpetner HS, Vallee RB (1987): MAP 1C is a microtubule-activated ATPase which translocates microtubules *in vitro* and has dynein-like properties. J Cell Biol 105:1273–1282.

Paschal BM, Vallee RB (1987): Retrograde transport by the microtubule-associated protein MAP 1C. Nature 330:181–183.

Penningroth SM (1986): Erythro-9-[3-(2-hydroxynonyl)]-adenine and vanadate as probes for microtubule-based cytoskeletal mechanochemistry. Methods Enzymol 134:477–487.

Pfarr CM, Coue M, Grissom PM, Hays TS, Porter ME, McIntosh JR (1990): Cytoplasmic dynein is localized to kinetochores during mitosis. Nature 345:263–265.

Pfister KK, Wagner MC, Bloom GS, Brady ST (1989): Modification of the microtubule binding and ATPase activities of kinesin by *N*-ethylmaleimide (NEM) suggests a role for sulfhydryls in fast axonal transport. Biochemistry 28:9006–9012.

Pollard TD, Doberstein SK, Zot HG (1991): Myosin I. Annu Rev Physiol 53:653–681.

Redenbach DM, Vogl AW (1991): Microtubule polarity in Sertoli cells: A model for microtubule-based spermatid transport. Eur J Cell Biol 54:277–290.

Rindler MJ, Ivanov IE, Sabatini DD (1987): Microtubule-acting drugs lead to the nonpolarized delivery of the influenza hemagglutinin to the cell surface of polarized Madin-Darby canine kidney cells. J Cell Biol 104:231–241.

Rozdial MM, Haimo LT (1986): Bidirectional pigment granule movements of melanophores are regulated by protein phosphorylation and dephosphorylation. Cell 47:1061–1070.

Sale WS, Satir P (1977): Direction of active sliding of microtubules in Tetrahymena cilia. Proc Natl Acad Sci USA 74:2045–2059.

Sawin KE, Scholey JM (1991): Motor proteins in cell division. Trends Cell Biol 1:122–129.

Scheel J, Kreis TE (1991): Motor protein independent binding of endocytic carrier vesicles to microtubules in vitro. J Biol Chem 266:18141–18148.

Schliwa M, Shimizu T, Vale RD, Euteneuer U (1991): Nucleotide specificities of anterograde and retrograde organelle transport in *Reticulomyxa* are indistinguishable. J Cell Biol 112:1199–1203.

Schmid SL (1992): The mechanism of receptor-mediated endocytosis: More questions than answers. BioEssays 14:589–596.

Schnapp BJ, Reese TS (1989): Dynein is the motor for retrograde axonal transport of organelles. Proc Natl Acad Sci USA 86:1548–1552.

Schroeder CC, Fok AK, Allen RD (1990): Vesicle transport along microtubular ribbons and isolation of cytoplasmic dynein from *Paramecium*. J Cell Biol 111:2553–2562.

Schroer TA, Schnapp BJ, Reese TS, Sheetz MP (1988): The role of kinesin and other soluble factors in organelle movement along microtubules. J Cell Biol 107:1785–1792.

Schroer TA, Sheetz MP (1991a): Functions of microtubule-based motors. Annu Rev Physiol 53:629–652.

Schroer TA, Sheetz MP (1991b): Two activators of microtubule-based vesicle transport. J Cell Biol 115:1309–1318.

Schroer TA, Steuer ER, Sheetz MP (1989): Cytoplasmic dynein is a minus end-directed motor for membranous organelles. Cell 56:937–946.

Schwab ME, Heumann R, Thoenen H (1982): Communication between target organs and nerve cells: Retrograde axonal transport and site of action of nerve growth factor. Cold Spring Harbor Symp Quant Biol 1:125–134.

Shimizu T, Furusawa K, Ohashi S, Toyoshima YY, Okuno M, Malik F, Vale RD (1991): Nucleotide specificity of the enzymatic and motile activities of dynein, kinesin and heavy meromyosin. J Cell Biol 112:1189–1197.

Shpetner HS, Paschal BM, Vallee RB (1988): Characterization of the microtubule-activated ATPase of brain cytoplasmic dynein (MAP 1C). J Cell Biol 107:1001–1009.

Stearns M, Ochs RL (1982): A functional in vitro model for studies of intracellular motility in digitonin permeabilized erythrophores. J Cell Biol 94:727–739.

Steuer ER, Wordeman L, Schroer TA, Sheetz MP (1990): Localization of cytoplasmic dynein to mitotic spindles and kinetochores. Nature 345:266–268.

Toyoshima I, Yu H, Steuer ER, Sheetz MP (1992): Kinectin, a major kinesin-binding protein on ER. J Cell Biol 118:1121–1131.

Tsukita S, Ishikawa H (1980): The movement of organelles in axons: Electron microscopic identification of anterogradely and retrogradely transported organelles. J Cell Biol 84:513–530.

Tucker JB, Milner MJ, Currie DA, Muir JA, Forrest DA, Spencer MJ (1986): Centrosomal microtubule organizing centers and a switch in the control of protofilament number for cell surface associated microtubules during *Drosophila* wing morphogenesis. Eur J Cell Biol 41:279–289.

Turner JR, Tartakoff AM (1989): The response of the Golgi complex to microtubule alterations: The roles of metabolic energy and membrane traffic in Golgi complex organization. J Cell Biol 109:2081–2088.

Vale RD (1987): Intracellular transport using microtubule-based motors. Annu Rev Cell Biol 3:347–378.

Vale RD, Toyoshima YY (1988): Rotation and translocation of microtubules *in vitro* induced by dyneins from *Tetrahymena* cilia. Cell 52:459–469.

Vallee RB, Bloom GS (1991): Mechanisms of fast and slow axonal transport. Annu Rev Neurosci 14:59–92.

Vallee RB, Shpetner HS (1990): Motor proteins of cytoplasmic microtubules. Annu Rev Biochem 59:909–932.

Vallee RB, Wall JS, Paschal BM, Shpetner HS (1988): Microtubule-associated protein 1C from brain is a two-headed cytosolic dynein. Nature 332:561–563.

Verde F, Berrez J-M, Antony C, Karsenti E (1991): Taxol-induced microtubule asters in mitotic extracts of *Xenopus* eggs: Requirement for phosphorylated factors and cytoplasmic dynein. J Cell Biol 112:1177–1187.

Walker RA, Salmon ED, Endow SA (1990): The *Drosophila* claret segregation protein is a minus-end directed motor molecule. Nature 347:780–782.

Weiss DG, Seitz-Tutter D, Langford GM (1991): Characteristics of the motor responsible for the gliding of native microtubules from squid axoplasm. J Cell Sci Suppl 14:157–161.

Yoshida T, Takanari H, Izutsu K (1992): Distribution of cytoplasmic and axonemal dyneins in rat tissues. J Cell Sci 101:579–587.

Yu H, Toyoshima I, Steuer ER, Sheetz MP (1992): Kinesin and cytoplasmic dynein binding to brain microsomes. J Biol Chem 267:20457–20464.

ABOUT THE AUTHOR

CHRISTINE A. COLLINS is Assistant Professor of Cell, Molecular, and Structural Biology at Northwestern University Medical School in Chicago, Illinois. Her teaching responsibilities include dental and medical courses in biochemistry, and an advanced topics graduate course in cellular and molecular aspects of the cytoskeleton. After receiving a B.S., magna cum laude, in chemistry and biology from the University of Minnesota, Duluth, she went on to graduate work at Michigan State University. She earned a Ph.D. in biochemistry in 1983 for her analysis of enzyme and lipid components of lysosomal membranes in the laboratory of Dr. William Wells. She carried out postdoctoral research on microtubules and microtubule-associated proteins in the laboratory of Dr. Richard Vallee at the Worcester Foundation for Experimental Biology in Shrewsbury, Massachusetts. Since 1985 Dr. Collins' research has focused on the biochemical characterization and cellular function of microtubule-associated nucleotidases, including cytoplasmic dynein.

Microtubules: 381–392

Mechanisms of Flagellar Movement: Functional Interactions Between Dynein Arms and the Radial Spoke–Central Apparatus Complex

Elizabeth F. Smith and Winfield S. Sale

I. INTRODUCTION

The eukaryotic cilium or flagellum is a motile organelle found on cells as diverse as protozoa, such as *Paramecia* and *Chlamydomonas,* spermatozoa of vertebrates and some invertebrates, and mammalian epithelia. Despite the diversity of cells that utilize flagella or cilia for motile functions, the basic structural and functional components of this organelle persist and are widely conserved. The basis of such persistent conservation remains a mystery [for review, see Sleigh, 1974; Warren et al., 1989].

Recent experimentation has focused on the mechanism and regulation of movement [Satir 1985], morphogenesis [Lefebvre and Rosenbaum, 1986], and membrane functions [Dentler, 1990; Bloodgood, 1990]. These topics are reviewed in several excellent articles and will not be covered here. In this discussion we include only studies of functional interactions between dynein arms and the radial spoke– central apparatus complex. Based on structural, physiological, and genetic evidence described below, the radial spoke–central apparatus complex appears to play a regulatory role in normal motility.

II. STRUCTURAL ORGANIZATION OF THE AXONEME

Cilia and flagella are composed of a highly ordered and complex core of proteins, the axoneme, which is organized into the well known "9 + 2" arrangement of microtubules and microtubule-associated proteins (Fig. 1). The major structural components relevant to this discussion are best illustrated in transverse section and include the nine peripheral doublet microtubules composed of A- and B-tubules and a central pair of singlet microtubules with their associated projections. Attached to the A-tubule of each doublet microtubule are the inner and outer rows of dynein arms that project toward the next adjacent doublet and the radial spokes that project toward the central pair. In longitudinal images the rows of outer and inner dynein arms are seen as regularly repeating structures (Fig. 1, bottom). The spokes are also regularly repeating structures (96 nm period) but may repeat in groups of two or three, depending on the organism from which the axonemes were obtained [Warner and Satir, 1974; Goodenough and Heuser, 1985]. More detail of axonemal macromolecular structure has been achieved using the technique of quick-freeze, rotary shadow electron microscopy, and the results were reviewed earlier [Goodenough and Heuser, 1989].

Axoneme complexity is also reflected in composition. Two-dimensional gel electrophoresis has shown that isolated axonemes are composed of more than 200 polypeptides [Luck, 1984]. Furthermore, gel electrophoresis of mutant axo-

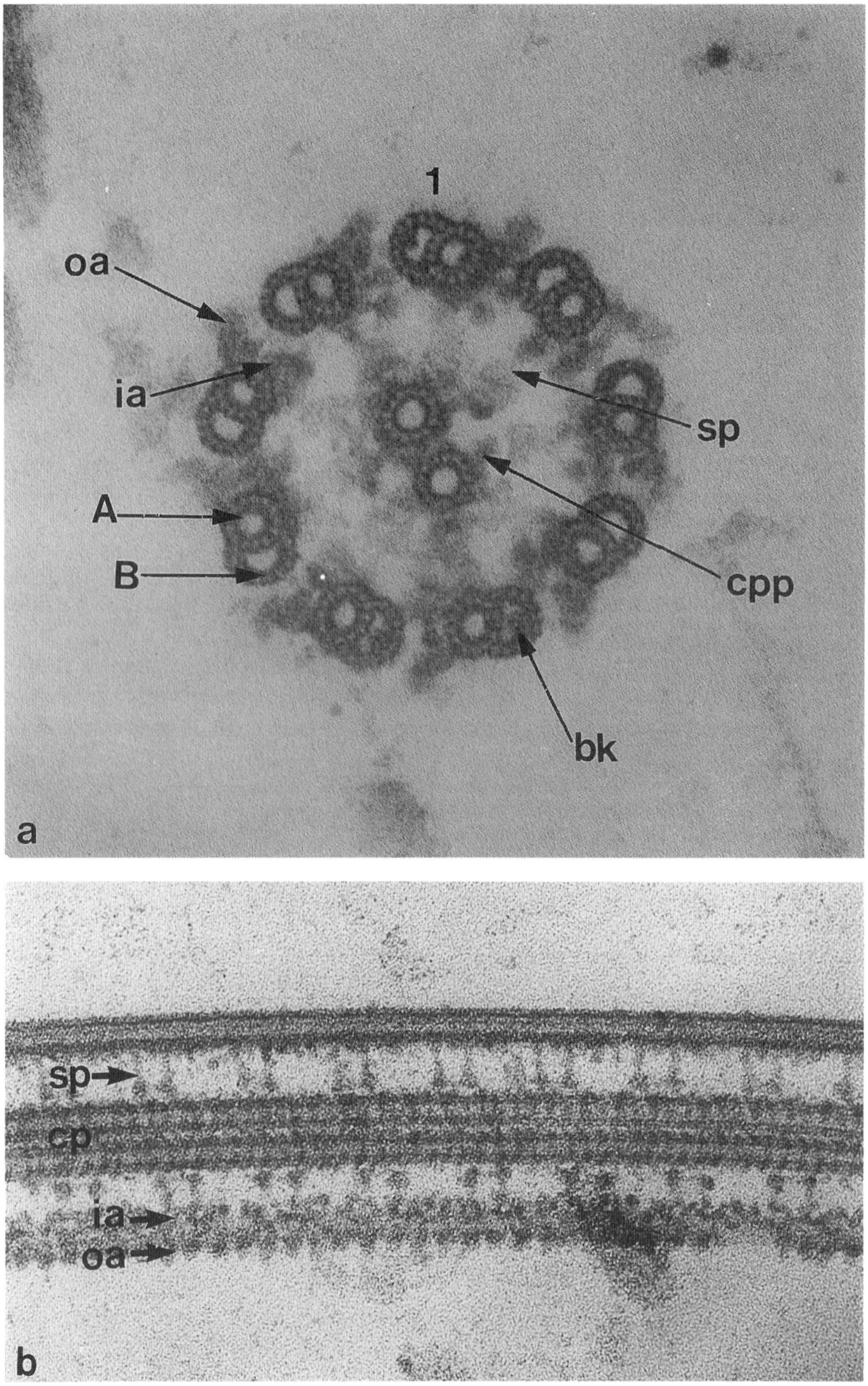

Fig. 1. *Thin section electron micrographs of transverse* **(a)** *and longitudinal* **(b)** *sections of wild-type* Chlamydomonas *axonemes. In transverse section. A, B, A- and B-subtubules of microtubule doublets; oa, ia, outer and inner dynein arms, respectively; sp, spokes; cpp, central pair projections; bk, beaks; 1, doublet number 1 as defined by Hoops and Witman [1983]; other doublets numbered in the direction of the dynein arms. In the longitudinal image the rows of spokes and dynein arms are indicated.*

nemes missing specific structures [i.e, spokes, dynein arms; Luck et al., 1977; Luck, 1984; Huang et al., 1981; Huang, 1986], or those selectively extracted by biochemical procedures, has led to the discovery that many axonemal components are multipolypeptide complexes. For example, based on studies of *Chlamydomonas* and several other organisms, the radial spokes are composed of 17 different polypeptides [Piperno et al., 1981; Curry and Rosenbaum, 1993], the central apparatus is composed of 23 polypeptides excluding tubulin [Dutcher et al., 1984], and the outer dynein arms are composed of at least 11 polypeptides (see Witman, this volume) [Huang et al., 1979; Piperno and Luck, 1979; Witman, 1989, 1992].

Even more astonishingly complex is the molecular organization of the inner row of dynein arms. In contrast to the outer row of dynein arms, which have a 24 nm repeat period along each microtubule and are homogeneous in structure and composition, the inner row is composed of several different subforms of dynein arm, each specialized in composition and location in the axoneme (Witman, this volume) [Goodenough and Heuser, 1985, 1989; Luck and Piperno, 1989; Sale et al., 1989; Piperno et al., 1990; Piperno and Ramanis, 1991; Kamiya et al., 1991; Burgess et al., 1991; Smith and Sale, 1991; Muto et al., 1992; Mastronarde et al., 1992]. The inner arms are attached to the A-tubule of each doublet near the base of the spokes and repeat in complex groups in exact register with the radial spokes (see Witman, Fig. 3, this volume). As discussed below, this close organization provides for potential structural or functional interactions of inner arms and radial spokes.

III. A SLIDING MICROTUBULE MODEL OF AXONEMAL MOTILITY

The normal oscillatory beating of cilia and flagella is believed to result from coordinated sliding movements between doublet microtubules, leading to formation and propagation of bends [Satir, 1985; Sugino and Naitoh, 1982; Brokaw, 1972]. Early structural studies determined that ciliary microtubules are inextensible [Satir, 1965, 1968] and led to the formal description of a sliding microtubule model of ciliary and flagellar bending [Satir, 1968; Brokaw, 1972]. More recent evidence has clearly indicated that microtubule sliding accompanies axoneme bending. For example, Shingyogi et al. [1977] iontophoretically applied ATP to specific regions of flagellar axonemes suspended in a reactivation buffer. As predicted by the sliding microtubule model, application of ATP to the midportion of the axoneme resulted in formation of a pair of bends equal in dimension but opposite in direction. Furthermore, Brokaw [1989, 1991] has directly demonstrated that microtubules slide during axonemal beating (Fig. 2). Using colloidal gold beads attached to individual microtubule doublets of demembranated sperm axonemes, Brokaw was able to record relative displacements of the gold particles during reactivated beating and thus confirm that doublet microtubule sliding accompanies axonemal bending.

A variety of evidence indicates that relative sliding of doublet microtubules results from shear produced by the dynein arms through a mechanochemical cycle. This cycle involves changes in angular orientation of the arms concomitant with the hydrolysis of ATP and results in their cyclic detachment and reattachment to the B-tubule [Gibbons, 1988; Omoto, 1991; Johnson, 1985; Porter and Johnson, 1989; Vale et al., 1989; Moss et al., 1992a,b]. This conclusion is founded in data derived from a variety of experimental approaches, including direct observations of ATP-induced microtubule sliding in axonemes [Summers and Gibbons, 1971; Kurimoto and Kamiya, 1991], identification and characterization of a "rigor" state of the axoneme [Gibbons and Gibbons, 1974; Okuno, 1980; Sale and Gibbons, 1979], functional reconstitution of dynein with dynein-depleted axonemes [Gibbons and Gibbons, 1976, 1979; Sakakibara and Kamiya, 1989; Smith and Sale, 1992a,b], kinetic analysis of dynein ATPase activity and microtubule binding [Porter and Johnson, 1982, 1989; Johnson, 1985], and reconstitution of motility *in vitro* using purified

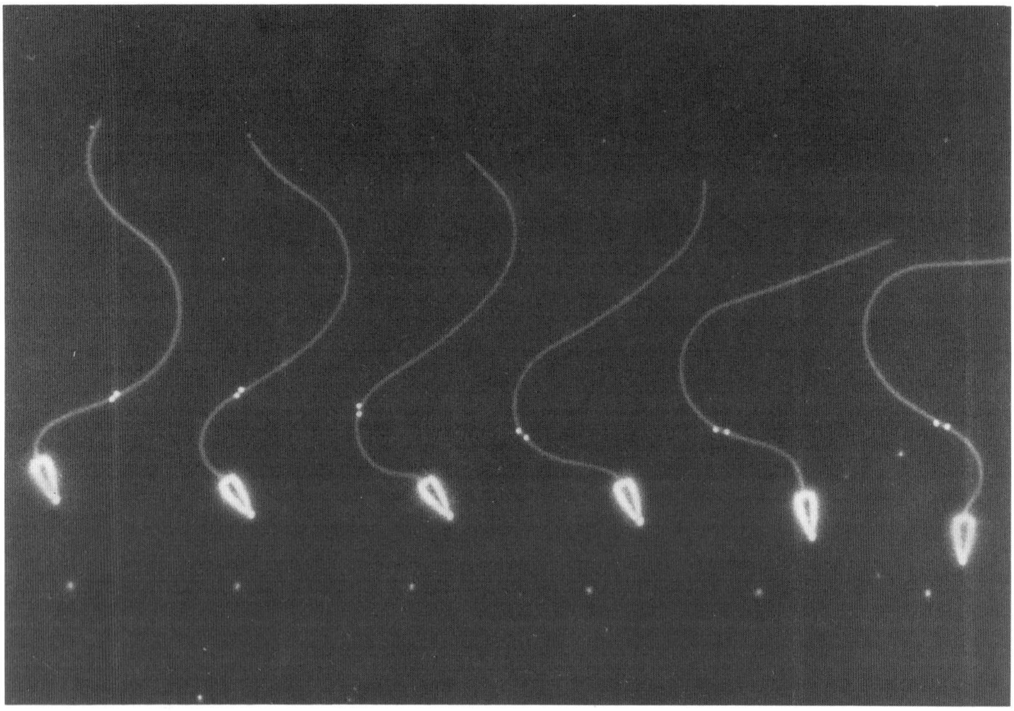

Fig. 2. *Multiple exposure, moving film, photograph of a demembranated sea urchin spermatozoon reactivated with ATP. Gold beads attached to the exposed outer doublet microtubules of the sperm flagellum allow direct measurement of sliding between microtubules as the flagella bends. [Reproduced from Brokaw, 1991, Journal of Cell Biology, 1991, 114:1201–1216, by copyright permission of the Rockefeller University Press.]*

dynein and microtubules [Vale and Toyoshima, 1988; Paschal et al., 1987; Sale and Fox, 1988; Vale et al., 1989; Moss et al., 1992a,b].

Although much is known about the composition, structure, and mechanism of axonemal dyneins, little is known about how dynein is regulated or how microtubule sliding is converted into the bends that characterize ciliary and flagellar motility. Based on analysis of waveforms and calculation or measurement of shear, bending appears to result from variations in microtubule sliding along the axoneme [Brokaw, 1989, 1991; Gibbons, 1981, 1982]. Furthermore, because axonemal dyneins generate force in a single direction relative to microtubules [Sale and Satir, 1977; Fox and Sale, 1987], every dynein on all doublets cannot be active simultaneously. Such unregulated activity would presumably result in a rigid flagellum in which no bending would occur. Thus, to generate a bend in one direction, dynein on one side of the axoneme must be active and dynein on the opposite side inactive. To bend in the opposite direction, the active and inactive sides must be "switched" [for discussion, see Satir and Sale, 1977; Satir, 1985; Sale, 1986]. Therefore, the question is, how is dynein activity regulated around the axis and along the length of the axoneme [Sale, 1986; Omoto and Kung, 1980]?

Mathematical models of flagellar movement include theoretical mechanical feedback mechanisms such as degree of curvature regulating the timing and location of dynein activity [Brokaw, 1982; Sugino and Naitoh, 1982]. Experimental evidence is lacking that describes an explicit mechanism of dynein regulation and coordination of microtubule sliding. However, analysis of mutant axonemes in *Chlamyodomas* have given clues and suggest a functional interaction between dynein arms and the radial spoke–central apparatus complexes.

IV. EVIDENCE FOR RADIAL SPOKE–CENTRAL APPARATUS FUNCTIONAL ROLES

A. Structural Interaction Between Radial Spokes and Central Apparatus

Based on their locations in the axoneme and evidence described below, the spokes and central apparatus appear to play a key role in dynein driven microtubule sliding. As illustrated in Figure 1, the radial spokes are attached at their base to the A-microtubule near the inner dynein arms and terminate in globular spoke heads where they appear to interact with central pair projections. In structural studies of *Elliptio* gill cilia fixed in various bending positions, Warner and Satir [1974] observed that the radial spokes maintain transient contacts with the central pair projections. After analyzing the tilt of the spokes in bent and straight regions, they concluded that spokes must detach and reattach to the central pair projections during microtubule sliding. Thus, based on this structural interaction, the central pair projections could conceivably provide a regional signal that is relayed through the spokes to affect active microtubule sliding locally.

Possible functional interaction between the central apparatus and the spokes is further illustrated by the finding that the central pair complexes of some cilia and flagella rotate with respect to the peripheral doublet microtubules during bending. For example, studies of cilia from *Paramecium* demonstrate that the central pair rotates once per beat cycle and is accompanied by a propagated twist that has a period close to that predicted for the propagating ciliary bend [Omoto and Kung, 1979, 1980]. Rotation of the central pair has also been observed for flagella of other species [e.g., Omoto and Witman, 1981; Hosokawa and Miki-Noumura, 1987]. Therefore, one hypothesis is that the central pair projections periodically sweep the radial spoke heads, much like a distributor, and induce a local signal that is transmitted through the spokes, ultimately regulating dynein-induced microtubule sliding [Omoto and Kung, 1980]. This mechanism is appealing since we know that regulation must occur in the axoneme in a proximal to distal direction as well as circumferentially. On the other hand, the central pair apparatus does not appear to rotate in all cilia and flagella [e.g., Tamm and Tamm, 1981]. However, the absence of central complex rotation does not rule out central pair interaction with the radial spokes. Rather, the rotation observed in some organisms may represent a refinement of preexisting regulatory mechanisms, providing such axonemes a wide array of bending patterns.

B. Genetic Evidence for Functional Interactions Between Radial Spoke–Central Apparatus Complexes and Dynein Arms

The roles of the radial spokes and central apparatus have been studied by structural and functional analyses of wild-type and mutant cells. *Chlamydomonas* mutants that are missing the central apparatus or radial spokes have paralyzed flagella [Huang, 1986], yet retain the ability to undergo sliding between doublet microtubules in a sliding disintegration assay [Witman et al., 1978; Smith and Sale, 1992b]. In this assay, microtubule sliding is uncoupled from flagellar bending by adding a protease and ATP to isolated axonemes [Summers and Gibbons, 1971; Okagaki and Kamiya, 1986; Kurimoto and Kamiya, 1991]. The protease evidently releases structural or functional restraints and results in ATP-induced sliding disruption of the axoneme. Since central pair and spoke-deficient mutants retain the ability to slide, their motility defect does not appear to be in the force-generating mechanism but in a control system that regulates force generation. Thus these data indicated that the radial spokes and central apparatus are necessary for normal ciliary and flagellar motility and serve to convert dynein-driven microtubule sliding into bending [cf. Summers and Gibbons, 1971; Witman et al., 1978; Warner and Satir, 1974]. How this might be accomplished is unknown.

The discovery of suppressor mutations in *Chlamydomonas* that restore oscillatory beating to paralyzed mutants lacking radial spokes or the central apparatus without restoring these structures has provided some clues [Huang et al., 1982]. These unusual and extremely informative mutant cells were discovered in a reversion analysis of paralyzed *Chlamydomonas* strains defective for either radial spokes or the central apparatus. Extragenic mutations were found that suppressed flagellar paralysis (i.e., restored motility) in the original mutant strains without restoring the missing structures (i.e., spokes or central pair components) [Huang et al., 1982; Huang, 1986].

Based on detailed genetic and phenotypic analysis, Huang et al. [1982] described two classes of suppressor mutants. One class restored motility when combined with either central pair defective (paralyzed) or spoke defective (paralyzed) mutations, and one class only restored motility when combined with spoke defective mutants. One mutant in the latter class of suppressor mutations, sup-pf 1, contains a molecular defect in the β-heavy chain of the outer dynein arms [Huang et al., 1982], and, as discussed below, new evidence indicates that additional suppressor mutants may contain defects in the inner arm dyneins [Piperno et al., 1992; Porter et al., 1992]. Predictably, none of the suppressor mutants restored motility when combined with additional dynein-defective mutants.

The primary conclusion from study of these suppressor mutants was that paralysis associated with radial spoke and central pair defects reveals a control system that inhibits flagellar movement when these components are missing. Thus the suppressor mutations restore motility by bypassing the inhibited state (i.e., bypassing the inhibitory control system) resulting from missing spokes or central apparatus components. A prediction of this conclusion is that in wild-type cells the spokes and central apparatus serve to activate or signal the control system providing active dynein-driven microtubule sliding and flagellar motility. As described below this prediction is consistent with new data that indicate that the radial spokes and central complex activate dynein-driven microtubule sliding [Smith and Sale, 1992b].

The suppressor mutants also made it possible to analyze flagellar motility in the absence of radial spokes or central pair structures [Brokaw et al., 1982; Brokaw and Luck, 1985] and potentially determine their role in wild-type cells. Studying flagellar motility of recombinant radial spoke defective/suppressor cells by computer-assisted motion analysis revealed that, in contrast to wild-type flagella with asymmetric bending patterns, the flagella of the recombinant cells could only produce large-amplitude, symmetric waveforms. Evidently the spokes are not required for bend initiation or propagation, but may be required to convert simple symmetric bends into the asymmetric bends required for efficient forward motility in *Chlamydomonas* [Brokaw et al., 1982]. However, this interpretation is complicated by the presence of modified dyneins in the suppressed recombinant cells. In every case studied thus far suppressors of paralysis resulting from spoke–central pair defects have defects in one or more dynein components [Porter et al., 1992; Piperno et al., 1992]. In some cases, a series of proteins originally referred to as the *spoke-specific suppressor system* are also missing in the suppressor mutants and have more recently been referred to as the *dynein regulatory complex,* ''drc'' [Piperno et al., 1992]. The dynein regulatory complex is discussed below.

C. Experimental Evidence for Radial Spoke Regulation of Dynein-Driven Microtubule Sliding

The data described above suggest that paralysis resulting from mutations that disrupt radial spoke or central pair structures is the consequence of inactivation of dynein throughout the axoneme. The suppressor mutations, which restore motility, revealed a control system that, in the absence of the spoke–central complex, inhibits the activity of the dynein [reviewed by Huang, 1986]. Thus one model is that the central pair–spoke interaction sends a signal to the putative control system (the "drc") turning the control system off and resulting in local activation of dynein arms.

To test the hypothesis that radial spokes affect dynein function, we analyzed microtubule sliding velocities in isolated axonemes from mutant strains of *Chlamydomonas* that are missing radial spoke components [Smith and Sale, 1992b]. In the sliding disintegration assay where sliding is uncoupled from flagellar bending, we predicted that a regulatory interaction between spokes and dynein would be reflected in altered sliding velocities. We found that a spokeless mutant (*pf14,* which has paralyzed flagella) had sliding velocities that were significantly slower than those of wild type. Therefore, in the absence of the radial spokes, dynein (either inner or outer arms) is apparently not as efficient in translocating the doublet microtubules.

For further study of spoke–central apparatus–dynein interaction we focused on the inner dynein arms because of their location (see Witman, this volume) and special features that suggest they might be targets of regulation. First, mutants that are missing the outer row of dynein arms are motile, whereas mutants containing all outer dynein arms but missing the inner row of dynein arms are paralyzed [Brokaw and Kamiya, 1987; Kamiya et al., 1989]. Therefore the inner dynein arms are both necessary and sufficient for motility. Second, unlike the outer dynein arms, the inner dynein arms are heterogeneous in composition. As noted above, there are multiple subforms of inner arm dynein each composed of unique heavy chains and an assortment of intermediate and light chains (see Witman, this volume) [Piperno et al., 1990]. Additionally, the inner dynein arm subforms have unique locations along the length of the axoneme; they repeat in structural groups every 96 nm in specific and precise positions relative to the radial spokes [Piperno et al., 1990, 1991; Muto et al., 1991; Mastronarde et al., 1992]. This heterogeneity could be indicative of functional specialization. Finally, a subset of inner arm components are phosphoproteins and may possess actin and the calcium-binding protein caltractin/centrin as a subunit [Piperno and Luck, 1976, 1981; Piperno et al., 1992]. Furthermore, we have previously shown that the inner dynein arms could be structurally and functionally reconstituted *in vitro* providing a new experimental approach for the exchange of dynein arms on axonemes of distinct composition [Smith and Sale, 1992a]. Thus, for further study we focused on double mutants missing both the spokes (*pf14*) and the outer dynein arms (*pf28*), and referred to as *pf14pf28*.

In comparing sliding velocities of a mutant that lacks outer dynein arms, *pf28* (motile flagella), with the mutant that lacks both radial spokes and outer dynein arms, *pf14pf28* (immotile flagella), we found that axonemes from *pf14pf28* had significantly slower sliding velocities than *pf28* axonemes, indicating that the microtubule translocating efficiency of the inner dynein arms is affected by the presence or absence of spokes. We hypothesized that in the presence of spokes the inner dynein arms, or a subset of inner arms, are in an activated state (symbolized DA*), and in the absence of spokes the arms are in a state of reduced activity (symbolized DA). To determine whether dynein was activated through a modification induced by the spokes, we used an *in vitro* reconstitution system to switch the inner dynein arms of *pf28* (DA*) axonemes with those of *pf14pf28* axonemes (DA) [Smith and Sale, 1992a,b].

We had previously shown that inner dynein arms can be isolated and precisely reconstituted onto both mutant and extracted axonemes missing subsets or all of the inner dynein arms [Smith and Sale, 1992a]. The reconstituted axonemes are both structurally and functionally restored based on quantitative analysis of composition, structure, and function. Each inner arm subtype binds exclusively to the appropriate position, and the reconstituted axonemes have restored microtubule sliding velocities. Thus two possible outcomes for the dynein switching experiment were possible, depending on whether the activated state of the inner dynein arms

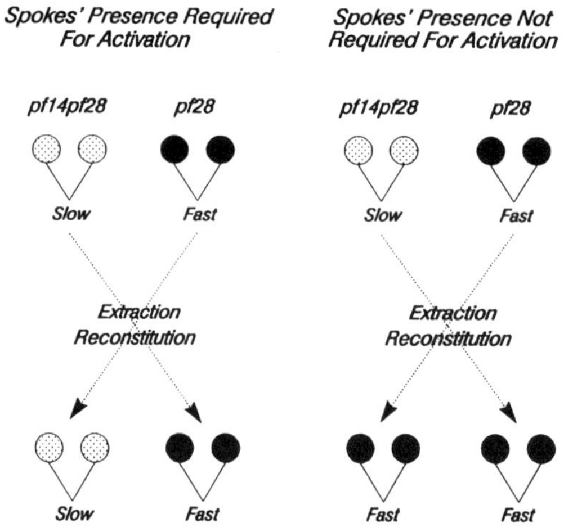

Fig. 3. *Experimental design and predictions for a dynein switching experiment. The reconstitution experiment was used to test whether the radial spokes are required for activation of rapid dynein-induced microtubule sliding. Closed circles indicate an activated dynein state (DA*), capable of inducing rapid microtubule sliding, and the stippled circles represent a latent or inactivated dynein state (DA). For reconstitution, extracted axonemes and appropriate dynein fraction were combined at ratios where excess dynein was added, and samples were incubated at room temperature for 15 minutes prior to performing sliding disintegration. The sliding disintegration assay was performed as described previously [Smith and Sale, 1992a].*

(DA*) is contingent on the presence of the spokes or whether the dynein maintains the activated state following reconstitution. The experiment is illustrated in Figure 3. If the activated state is dependent on the presence of the spokes, then inner arms from *pf28* reconstituted onto *pf14pf28* axonemes should now induce slow sliding velocities (DA* → DA), and the *pf28* axonemes reconstituted with dynein derived from *pf14pf28* axonemes should still slide at rapid velocities (Fig. 3, left). However, if the spoke-induced modification of inner dynein arms (DA*) survives reconstitution and is maintained in the absence of spokes, the *pf14pf28* axonemes reconstituted with *pf28* inner arms should now have rapid sliding velocities characteristic of *pf28* (Fig. 3, right).

The results of these experiments are summarized in Table I in Smith and Sale [1992b]. Reconstitution experiments with inner dynein arms that had been exposed to spokes (in *pf28*) or had not been exposed to spokes (from *pf14pf28*) indicated that the radial spokes induced a modification of the inner dynein arms and that the modification was maintained in the subsequent absence of radial spokes (Fig. 3, right). Both *pf28* and *pf14pf28* axonemes reconstituted with their respective extracts slide at their original velocities, indicating that the extraction and reconstitution procedure does not by itself induce any changes in dynein activity. Extracted axonemes from *pf28* reconstituted with inner arms derived from *pf14pf28* (DA) slide at the velocity of intact *pf28* (DA*), and extracted axonemes from *pf14pf28* reconstituted with inner arms derived from *pf28* (DA*) also slide at the velocity of intact *pf28* axonemes (DA*). Therefore the inner dynein arms from *pf28*, once exposed to radial spokes, remain in an activated state following extraction and reconstitution without the radial spokes. We could not rule out the possibility that certain inner arm subforms may be differentially affected, but we could eliminate other possible explanations.

The increased sliding velocity for *pf14pf28* axonemes reconstituted with inner arms from *pf28* was not due to dynein arms binding to extraneous positions on the extracted *pf14pf28* axonemes, was not due to transfer of the "activating factor" (spoke components, enzymes) in extracts from the *pf28* axonemes onto the *pf14pf28* axonemes, and was not a consequence of differences in inner arm composition associated with flagella of varying length [see Smith and Sale, 1992b]. Our interpretation was that the spokes activate the inner dynein arms by a modification that survives high salt extraction and dialysis required for reconstitution. This interpretation was further supported by comparing sliding velocities of other axonemes defective for radial spokes (*pf14*, *pf1*, *pf17*, and *pf27*) with wild-type axonemes (137c). Axonemal microtubules from all of these mutants slide at one-half to one-third the velocity of wild type, indicating that reduced microtubule sliding velocities are not unique to the *pf14* mutation and that the outer arms cannot overcome the inactivation caused by the lack of radial spokes.

Outer dynein arm activity might also be affected by the spokes in wild-type axonemes as suggested by analysis of the above-mentioned suppressor mutants in which the requirement of radial spokes for movement is bypassed by a defect in the beta heavy chain of the outer dynein arms (*sup-pf1*). Calculations of sliding velocities [Brokaw et al., 1982] as well as those we measured from sliding disintegration (*pf17* compared to *sup-pf1* and the double mutant *sup-pf1pf17*) indicate that the suppressor mutation increases microtubule sliding velocities in the absence of the spokes. Our results indicate activation of dynein-induced microtubule sliding activity by the radial spokes. We are currently trying to understand what this modification may be, including the possibility of differences in phosphorylation of dynein arm components. Perhaps relative differences in phosphorylation result in activation or inactivation of dynein arms.

An additional question generated by these experiments is, if the radial spokes regulate dynein, what regulates the radial spokes? We have tried similar experiments with mutants missing the central apparatus or subsets of central apparatus components in order to begin investigating the possibility of the central pair regulating the spokes and in turn regulate dynein activity. With the exception of a mutant that is missing a subset of central pair projections, all mutants examined thus far with defects in the central pair have slow microtubule sliding velocities in the sliding disintegration assay. These data, although preliminary, also support the model of the central apparatus ultimately regulating dynein activity. To date, no direct functional data exist to indicate that the central pair actually interacts with the radial spokes to regulate dynein.

D. A Dynein Regulatory Complex in Flagella

An additional question is, do the spokes affect dynein activity directly or through other axonemal components? Recently, Piperno et al. [1992] discovered a set of six polypeptides they refer to as the *dynein regulatory complex* (drc), which appears to be associated with or in close proximity to inner dynein arms, referred to as *I2* and *I3*. Several suppressor mutants that restore motility to radial spokeless mutants are missing subsets of proteins from the drc and often contain a reduced amount of inner arm components. The exact location of the dynein regulatory complex has not been determined but since missing drc components often result in a reduced amount of I2 and I3, it is deduced that the drc is in close proximity to and may be responsible for the assembly of inner arm components. This hypothesis is supported by the structural studies of flagella from *Chlamydomonas* mutant *pf2* by Mastronarde et al. [1992]. Therefore, this complex might serve as attachment sites for inner dynein arms and possibly represent a structural and functional/regulatory link between inner dynein arms and radial spokes.

V. CONCLUSIONS

Although it is clear that dynein provides the force for flagellar motility, we are only beginning to understand how various axonemal components interact to regulate dynein and produce complex waveforms that characterize normal motility. One focus is on the role of the radial spokes and central apparatus in regulating dynein activity. This understanding is derived from functional, structural, biochemical, and genetic approaches, which suggest that the radial spokes, central apparatus, and the drc regulate dynein-induced microtubule sliding.

In this discussion we have provided an abridged review. Now, the challenge remains to determine how these complexes direct the timing and location of active sliding, and to determine the exact chemical mechanism that mediates changes in dynein activity.

VI. REFERENCES

Bloodgood RA (1990): Ciliary and Flagellar Membranes. New York: Plenum Press.

Brokaw CJ (1972): Flagellar movement: A sliding filament model. Science 178:455–462.

Brokaw CJ (1982): Models for oscillation and bend propagation by flagella. Symp Soc Exp Biol 35:313–338.

Brokaw CJ (1989): Direct measurements of sliding between outer doublet microtubules in swimming sperm flagella. Science. 243:1593–1596.

Brokaw CJ (1991): Microtubule sliding in swimming sperm flagella: Direct and indirect measurements on sea urchin and tunicate spermatozoa. J Cell Biol 114:1201–1216.

Brokaw CJ, Kamiya R (1987): Bending patterns of Chlamydomonas flagella IV: Mutants with defects in inner and outer dynein arms indicate differences in arm function. Cell Motil Cytoskel 8:68–75.

Brokaw CJ, Luck DJL (1985): Bending patterns of *Chlamydomonas* flagella III. A radial spoke head deficient mutant and a central pair deficient mutant. Cell Motil 5:195–208.

Brokaw CJ, Luck DJL, Huang B (1982): Analysis of movement of Chlamydomonas flagella: The function of the radial-spoke system is revealed by comparison of wild-type and mutant flagella. J Cell Biol 100: 1875–1883.

Burgess S, Carter D, Dover S, Wooley D (1991): The inner dynein arm complex: Compatible images from freeze etch and thin sections. J Cell Sci 100:319–320.

Curry AM, Rosenbaum JL (1993): Flagellar radial spoke: A model molecular genetic system for studying organelle assembly. Cell Motil Cytoskel 24:224–232.

Dentler WL (1990): Linkages between microtubules and membranes. In Bloodgood RA (ed): Ciliary and Flagellar Membranes. New York: Plenum Press, pp 31–64.

Dutcher SK, Huang B, Luck DJL (1984): Genetic dissection of the central pair microtubules of the flagella of *Chlamydomonas reinhardtii*. J Cell Biol 98:229–236.

Fox LA, Sale WS (1987): Direction of force generated by the inner row of dynein arms on flagellar microtubules. J Cell Biol 105:1781–1787.

Gibbons IR (1981): Transient flagellar waveforms during intermittent swimming in sea urchin sperm. II. Analysis of tubule sliding. J Muscle Res Cell Motil 2:83–130.

Gibbons IR (1982): Sliding and bending in sea urchin sperm flagella. Symp Soc Exp Biol 35:225–287.

Gibbons IR (1988): Dynein ATPases as microtubule motors. J Biol Chem 263:15837–15840.

Gibbons BH, Gibbons IR (1974): Properties of flagellar "rigor waves" produced by abrupt removal of ATP from actively swimming sea urchin sperm. J Cell Biol 63:970–985.

Gibbons BH, Gibbons IR (1976): Functional recombination of dynein 1 with demembranated sea urchin sperm partially extracted with KCl. Biochem Biophys Res Commun 73:1–6.

Gibbons BH, Gibbons IR (1979): Relationship between the latent adenosine triphosphatase state of dynein 1 and its ability to recombine functionally with KCl-extracted sea urchin sperm flagella. J Biol Chem 254:197–201.

Goodenough UW, Heuser JE (1985): Substructure of inner dynein arms, radial spokes, and the central pair projection complex of cilia and flagella. J Cell Biol 100:2008–2018.

Goodenough UW, Heuser JE (1989): Structure of the soluble and *in situ* ciliary dyneins visualized by quick-freeze deep-etch microscopy. In Cell Movement, vol 1. New York: Alan R. Liss, pp 121–140.

Hoops HJ, Witman GB (1983): Outer doublet heterogeneity reveals structural polarity related to beat direction in *Chlamydomonas* flagella. J Cell Biol 97:902–908.

Hosokawa Y, Miki-Noumura T (1987): Bending motion of *Chlamydomonas* axonemes after extrusions of the central pair. J Cell Biol 105:1297–1301.

Huang B (1986): *Chlamydomonas reinhardtii,* a model system for the genetic analysis of flagella structure and motility. Int Rev Cytol 99:181–215.

Huang B, Piperno P, Luck DJL (1979): Paralyzed flagellar mutants of *Chlamydomonas reinhardtii* defective for axonemal doublet microtubule arms. J Biol Chem 254:3091–3099.

Huang B, Piperno G, Ramanis Z, Luck DJL (1981): Radial spokes of *Chlamydomonas* flagella: Genetic analysis of assembly and function. J Cell Biol 88:80–88.

Huang B, Ramanis Z, Luck DJL (1982): Suppressor mutations in *Chlamyodomonas* reveal a regulatory mechanism for flagellar function. Cell 28:115–124.

Johnson KA (1985): Pathway of the microtubule–dynein ATPase and the structure of dynein: A comparison with actomyosin. Annu Rev Biophys Chem 14:161–188.

Kamiya R, Kurimoto E, Muto E (1991): Two types of *Chlamydomonas* flagellar mutants missing different components of inner-arm dynein. J Cell Biol 112: 441–447.

Kamiya R, Kurimoto E, Sakakibara H, Okagaki T (1989): A genetic approach to the function of inner and outer arm dynein. In Warner FD, Satir P, Gibbons IR (eds): Cell Movement, vol 1. New York: Alan R. Liss, pp 209–218.

Kurimoto E, Kamiya R (1991): Microtubule sliding in flagellar axonemes of *Chlamydomonas* mutants missing inner and outer arm dynein: Velocity measurements on new types of mutants with an improved method. Cell Motil Cytoskel 19:275–281.

Lefebvre PA, Rosenbaum JL (1986): Regulation and assembly of ciliary and flagellar proteins during regeneration. Annu Rev Cell Biol 2:517–546.

Luck DJL (1984): Genetic and biochemical dissection of the eukaryotic flagellum. J Cell Biol 98:789–794.

Luck DJL, Piperno G (1989): Dynein arm mutants of *Chlamydomonas*. In Warner FD, Satir P, Gibbons IR (eds): Cell Movement, vol 1. New York: Alan R. Liss, pp 49–60.

Luck DJL, Piperno G, Ramanis Z, Huang B (1977): Flagellar mutants of *Chlamydomonas*: Studies of radial spoke-defective strains by dikaryon and revertant analysis. Proc Natl Acad Sci USA 74:3456–3460.

Mastronarde D, O'Toole E, McDonald K, McIntosh JR, Porter M (1992): Arrangement of inner dynein arms in wild type and mutant flagella of *Chlamydomonas*. J Cell Biol 118:1145–1162.

Moss A, Gatti J, Witman GB (1992a): The motile beta/IC subunit of sea urchin sperm outer arm dynein does not form a rigor bond. J Cell Biol 118:1177–1188.

Moss A, Sale W, Fox L, Witman GB (1992b): The alpha-subunit of sea urchin outer arm dynein mediates structural and rigor binding to microtubules. J Cell Biol 118:1189–1200.

Muto E, Kamiya R, Tsukita S (1991): Double-rowed organization of inner dynein arms in *Chlamydomonas* flagella revealed by tilt series thin-section electron microscopy. J Cell Sci 99:57–66.

Okagaki T, Kamiya R (1986): Microtubule sliding in mutant *Chlamydomonas* axonemes devoid of outer or inner dynein arms. J Cell Biol 103:1895–1902.

Okuno M (1980): Inhibition and relaxation of sea urchin sperm flagella by vandate. J Cell Biol 85:712–725.

Omoto C (1991): Mechanochemical coupling in cilia. Int Rev Cytol 131:255–292.

Omoto C, Kung C (1979): The pair of central tubules rotates during ciliary beat in *Paramecium*. Nature 279:532–534.

Omoto C, Kung C (1980): Rotation and twist of the central-pair microtubules in the cilia of *Paramecium*. J Cell Biol 87:33–46.

Omoto C, Witman GB (1981): Functionally significant central pair rotation in a primitive eukaryote. Nature 290:708–710.

Paschal BM, King S, Moss A, Collins C, Vallee R, Witman G (1987): Isolated flagellar outer arm dynein translocates brain microtubules *in vitro*. Nature (Lond) 330:672–674.

Piperno G, Huang B, Ramanis Z, Luck DJL (1981): Radial spokes of *Chlamydomonas* flagella: Polypeptide composition and phosphorylation of stalk components. J Cell Biol 88:73–79.

Piperno G, Luck DJL (1976): Phosphorylation of axonemal proteins in *Chlamydomonas reinhardtii*. J Biol Chem 251:2161–2167.

Piperno G, Luck DJL (1979): Axonemal adenosine triphosphatases from flagella of *Chlamydomonas reinhardtii*. J Biol Chem 254:3084–3090.

Piperno G, Luck DJL (1981): Inner arm dyneins from flagella of *Chlamydomonas reinhardtii*. Cell 27: 331–340.

Piperno G, Mead K, Shestak W (1992): The inner dynein arms I2 interact with a "dynein regulatory complex" in *Chlamydomonas* flagella. J Cell Biol 118:1455–1464.

Piperno G, Ramanis Z (1991): The proximal portion of *Chlamydomonas* flagella contains a distinct set of inner dynein arms. J Cell Biol 112:701–709.

Piperno G, Ramanis Z, Smith EF, Sale WS (1990): Three distinct inner dynein arms in *Chlamydomonas* flagella: Molecular composition and location in the axoneme. J Cell Biol 110:379–389.

Porter M, Johnson K (1982): Characterization of ATP sensitive binding of 30s dynein to brain MTs. J Biol Chem 258:6575–6581.

Porter ME, Johnson KA (1989): Dynein structure and function. Annu Rev Cell Biol 5:119–151.

Porter M, Power J, Dutcher S (1992): Extragenic suppressors of paralyzed flagellar mutations in *Chlamydomonas* identify loci that alter the inner dynein arms. J Cell Biol 118:1163–1176.

Sakakibara H, Kamiya R (1989): Functional recombination of outer dynein arms with outer arm-missing flagellar axonemes of a *Chlamydomonas* mutant. J Cell Sci 92:77–83.

Sale WS (1986): The axonemal axis and Ca-induced asymmetry of active microtubule sliding in sea urchin sperm tails. J Cell Biol 102:2042–2052.

Sale WS, Fox LA (1988): Isolated β-heavy chain subunit of dynein translocates microtubules *in vitro*. J Cell Biol 107:1793–1798.

Sale WS, Fox LA, Milgram SA (1989): Composition and organization of the inner row of dynein arms. In Warner FD, Satir P, Gibbons IR (eds): Cell Movement, vol 1. New York: Alan R. Liss, pp 89–102.

Sale WS, Gibbons IR (1979): Study of the mechanism of vanadate inhibition of the dynein cross-bridge cycle in sperm flagella. J Cell Biol 82:291–298.

Sale WS, Satir P (1977): Direction of active sliding of microtubules in *Tetrahymena* cilia. Proc Natl Acad Sci USA 74:2045–2059.

Satir P (1965): Studies on cilia II. Examination of the distal region of the ciliary shaft and the role of the filaments in motility. J Cell Biol 26:805–834.

Satir P (1968): Studies on cilia III. Further studies on the cilia tip and a "sliding filament" model of ciliary motility. J Cell Biol 39:77–94.

Satir P (1985): Switching mechanisms in the control of ciliary motility. In Satir B (ed): Modern Cell Biology. New York: Alan R. Liss, pp 1–46.

Satir P, Sale WS (1977): Tails of *Tetrahymena*. J Protozool 24:498–501.

Shingyoji C, Murakami A, Takahashi K (1977): Local reactivation of Triton-extracted flagella by iontophoretic application of ATP. Nature 265:269–270.

Sleigh MA (1974): Patterns of movement of cilia and flagella. In Sleigh MA (ed): Cilia and Flagella. New York: Academic Press.

Smith EF, Sale WS (1991): Microtubule binding and translocation by inner dynein arm subtype II. Cell Motil Cytoskel 18:258–268.

Smith EF, Sale WS (1992a): Structural and functional reconstitution of inner dynein arms in *Chlamydomonas* flagellar axonemes. J Cell Biol 117:573–581.

Smith EF, Sale WS (1992b): Regulation of dynein-driven microtubule sliding by the radial spokes in flagella. Science 257:1557–1559.

Sugino K, Naitoh Y (1982): Simulated cross bridge patterns corresponding to ciliary beating in *Paramecium*. Nature 295:609–611.

Summers K, Gibbons IR (1971): Adenosine triphosphate-induced sliding of tubules in trypsin treated flagella of sea urchin sperm. Proc Natl Acad Sci USA 68: 3092–3096.

Tamm S, Tamm S (1981): Ciliary reversal without rotation of axonemal structures in ctenophore comb plates. J Cell Biol 89:495–509.

Vale RD, Soll DR, Gibbons IR (1989): One dimensional diffusion of microtubules bound to flagellar dynein. Cell 59:915–925.

Vale RD, Toyoshima YY (1988): Rotation and translocation of microtubules *in vitro* induced by dyneins from *Tetrahymena* cilia. Cell 52:459–469.

Warner FD, Satir P (1974): The structural basis of ciliary bend formation. Radial spoke positional changes accompanying microtubule sliding. J Cell Biol 63: 35–63.

Warner FD, Satir P, Gibbons IR, McIntosh JR (eds) (1989): Cell Movement, vols 1, 2. New York: Alan R. Liss.

Witman GB (1989): Composition and molecular organization of the dyneins. In Cell Movement, vol 1. New York: Alan R. Liss, pp 25–35.

Witman GB (1992): Axonemal dyneins. Curr Opin Cell Biol 4:74–79.

Witman GB, Plummer J, Sander G (1978): *Chlamydomonas* flagellar mutants lacking radial spokes and central tubules. J Cell Biol 76:729–747.

ABOUT THE AUTHORS

ELIZABETH F. SMITH is a postdoctoral fellow in Genetics and Cell Biology at the University of Minnesota (Twin Cities). After receiving a B.A. from Agnes Scott College in Decatur, Georgia, she pursued doctoral research in the laboratory of Winfield S. Sale at Emory University in Atlanta, Georgia, where she earned a Ph.D. in 1992 for work on dynein subforms in the flagella of the green alga *Chlamydomonas reinhardtii* (Piperno et al., *J. Cell Biol.* 110:379 1990; Smith and Sale, *J. Cell Biol.* 117:573 1992; Smith and Sale, *Science* 257:1557, 1992). This work was followed by her current postdoctoral research at the University of Minnesota with Paul A. Lefebvre. Dr. Smith's postdoctoral work focuses on understanding the mechanism of flagellar motility by investigating axonemal structures that may interact with dynein to regulate flagellar beating. She is currently supported by an NSF-RTG grant for cytoskeleton research at the University of Minnesota, and has recently received a postdoctoral fellowship from the American Cancer Society. Upon completion of her postdoctoral research, she plans to pursue an academic research position.

WINFIELD S. SALE is Professor of Anatomy and Cell Biology at Emory University in Atlanta, where he teaches cell biology and is Director of the Graduate Program in Cell and Developmental Biology. After receiving his A.B. in Physiology at the University of California, Berkeley, he pursued doctoral research at Berkeley in the laboratory of Peter Satir, where he earned a Ph.D. in 1977 for work on the mechanism of ciliary motility and on analysis of patterns of microtubule sliding in cilia. This was followed by postdoctoral research, supported by a fellowship from the Helen Hay Whitney Foundation, at the University of Hawaii in Ian Gibbons' laboratory, on the biochemistry of dynein ATPases. Dr. Sale's research has continued with focus on understanding the structural and functional domains in dynein. Most recently, and while working with Elizabeth F. Smith, he has used mutant strains of *Chlamydomonas* to reveal structural and functional heterogeneity in dyneins in flagella and to reveal functional interactions of dynein arms with other axonemal structures. He has held an NIH Research Career Development Award and is a member of the editorial board of *Cell Motility and the Cytoskeleton*.

Microtubules: 393–412

Structure and Molecular Biology of the Kinetochore

W.C. Earnshaw

I. INTRODUCTION

The terms *kinetochore* and *centromere* were both originally proposed to describe the point at which the mitotic spindle attaches to chromosomes [Rieger et al., 1976]. I here adopt the definition suggested by Reider [1982] in his comprehensive review of the kinetochore and use the term *centromere* to describe the chromatin structure in this region of the chromosome. I will use the term *kinetochore* to refer to the substructure at the surface of the centromere into which the microtubules insert (see Fig. 2, below).

Studies of the centromeres of the budding yeast *Saccharomyces cerevisiae* and the fission yeast *Schizosaccharomyces pombe* reveal that the heart of the centromere/kinetochore region of the chromosome is composed of specific DNA sequences. These sequences provide binding sites for proteins that carry out the specialized functions of this region of the chromosome. To date the centromeric DNAs and their binding proteins have best been characterized in *Sac. cerevisiae,* and I will therefore begin the discussion with this organism.

II. STRUCTURE OF THE *SAC. CEREVISIAE* CENTROMERE/KINETOCHORE

A. DNA

The budding yeast has 16 chromosomes ranging from 0.26 to 3 Mb in length. The position of centromeres on these chromosomes is specified by cis-acting DNA sequences, called *CEN* sequences. CEN sequences can function autonomously and if inserted into self-replicating plasmid vectors will transform those vectors into functional minichromosomes that segregate with high efficiency during mitosis [Clarke and Carbon, 1980]. In such experimental systems the test chromosome is lost at a rate of about once in 10^5 cell divisions [Clarke, 1990; Schulman and Bloom, 1991]. This ability to endow a vector with mitotic stability has been used as an assay to clone functional centromere sequences from genomic DNA libraries.

To date functional centromeres have been cloned from 13 of the budding yeast chromosomes [Schulman and Bloom, 1991]. Both DNA sequence analysis and functional experiments have revealed that these centromeres share a common tripartite organization, with two regions of highly conserved sequence flanking a central more variable region. The three centromeric subdomains are designated (in the 5′ to 3′ direction) CDE I (Centromere DNA Element I, 8 bp), CDE II (78–86 bp) and CDE III (25 bp) (Fig. 1). These three elements are all that it takes to make a functional centromere in budding yeast: A 125 bp fragment spanning CDE I to CDE III retains full centromere function in the mitotic stability assay mentioned above [Cottarel et al., 1989]. CEN DNA functions autonomously in chromosome segregation: The CEN region from one chromosome will direct the segregation of other chromosomes perfectly well, and it will do so

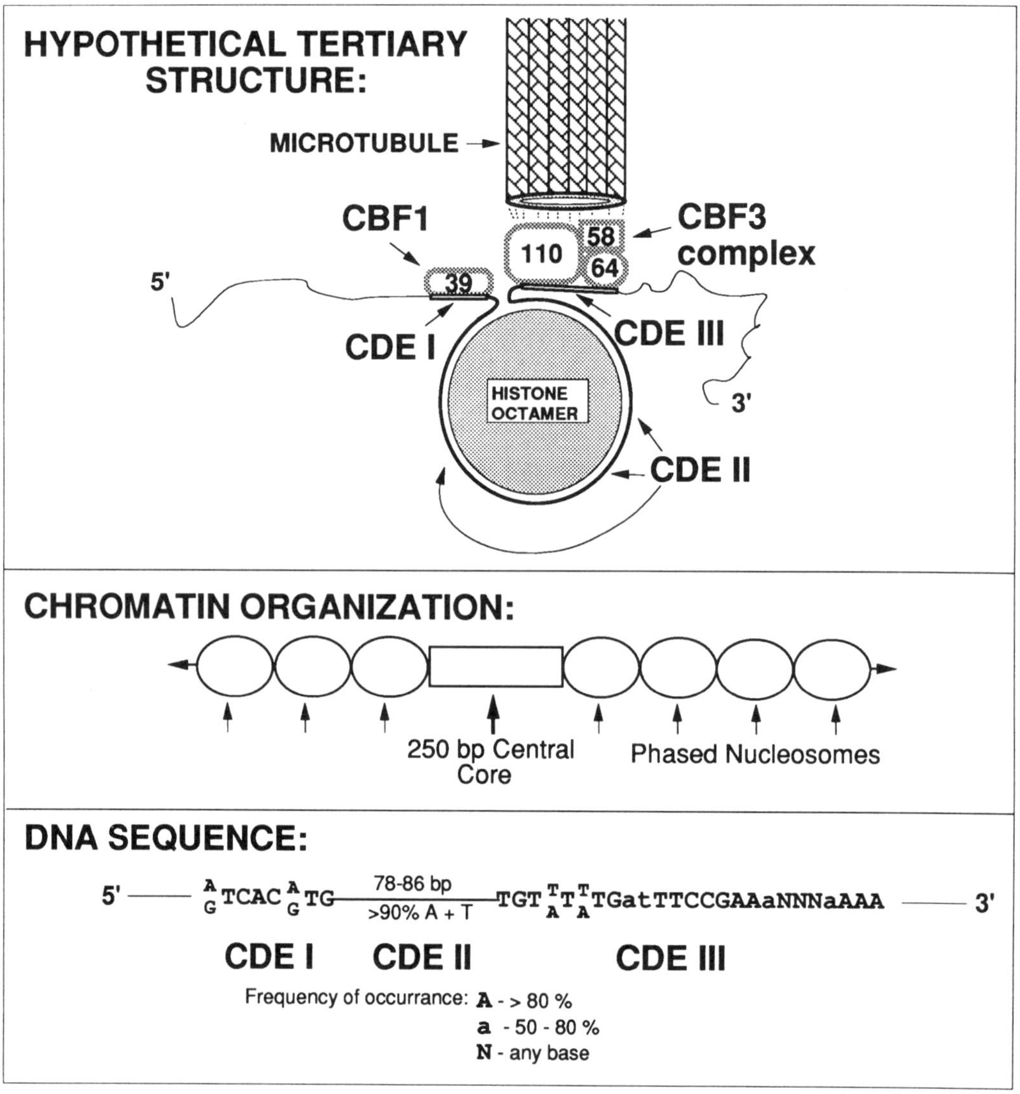

Fig. 1. *The kinetochore of the budding yeast,* Saccharomyces cerevisiae.

when inserted into the chromosome in either orientation [Clarke and Carbon, 1983]. As is discussed below, this DNA appears to specify a microtubule-binding site on the yeast chromosome.

The 8 bp CDE I sequence is short enough that it might be expected to occur numerous times in the *Sac. cerevisiae* genome. Indeed, this is the case, and this sequence is found upstream from a number of genes. Because CDE I is found at numerous locations in the genome that are not centromeres, this sequence is unlikely to specify centromere function on its own (since this would mean that some yeast chromosomes would contain several centromeres). In fact, CDE I appears to have only an accessory role in centromere function. If this element is deleted from a centromere on a plasmid minichromosome, the fidelity of minichromosome segregation goes down only about

10–60-fold [Gaudet and Fitzgerald-Hayes, 1989; Murphy and Fitzgerald-Hayes, 1990].

The second of the centromere elements, CDE II, does not have a single consensus sequence. Instead, two attributes of this region appear to be important: its length (78–86 bp) and overall base composition. CDE II is extremely A + T-rich (typically >90%), with the As and Ts often organized in short homopolymeric runs. If CDE II is replaced by random sequence DNA of the correct length, CEN function decreases by about 100-fold. If, on the other hand, it is replaced by randomly ordered A + T-rich DNA of the correct length, the fidelity of chromosome transmission drops only about 10-fold [Gaudet and Fitzgerald-Hayes, 1989; Murphy and Fitzgerald-Hayes, 1990].

It is worth noting that the length of CDE II corresponds almost exactly to the length of a single turn of DNA about the nucleosome core. Thus, if the yeast centromere is located in a nucleosomal particle (see below), CDE I and CDE III will be adjacent in space (Fig. 1). Consistent with this view, alterations in the length of CDE II decrease CEN function. Partial deletions of CDE II have been observed to lower CEN function by 1,000–4,000-fold. Similarly, increasing the length of CDE II approximately twofold decreases the fidelity of chromosome transmission by about 200-fold [Gaudet and Fitzgerald-Hayes, 1989; Murphy and Fitzgerald-Hayes, 1990].

The 25 bp CDE III domain occurs only at centromeres. CDE III has pseudodyad symmetry, centered around a G/C bp. If this base pair is mutated to A/T, CEN function is essentially abolished and the fidelity of chromosome transmission drops about 100,000-fold [Gaudet and Fitzgerald-Hayes, 1989; Jehn et al., 1991]. A similar loss of function occurs if CDE III is inverted relative to the other two CEN DNA elements, an unexpected result given the pseudodyad symmetry of CDE III. Certain base changes elsewhere in this element also have strongly disruptive effects, although modification of some of the highly conserved base pairs has surprisingly little effect on CEN function [Gaudet and Fitzgerald-Hayes, 1989; Jehn et al., 1991].

B. Protein

The centromeres of *Sac. cerevisiae* chromosomes are packaged into nucleoprotein structures that are distinct from that of the surrounding nucleosomal chromatin. This was first seen when yeast chromatin was digested with micrococcal nuclease or DNase 1. Such experiments revealed that the CEN DNA is located in a region of 220–250 bp that is relatively inaccessible to these nucleases [Bloom and Carbon, 1982]. This protected region is flanked by a region of highly ordered nucleosomes. The presence of the specialized chromatin structures at yeast centromeres requires normal histone expression. If the expression of either histone H4 or H2B is repressed, the organization of the centromeric chromatin is abnormal [Saunders et al., 1990]. This suggests that the 250 bp protected region could be organized around a modified nucleosome [Bloom et al., 1989]. This specialized chromatin structure is correlated with centromere function. A yeast centromere containing a nonfunctional CDE III sequence does not adopt the nuclease-resistant core structure [Schulman and Bloom, 1991].

The idea that yeast centromeres would have a specialized chromatin structure led to a prediction for a simple way to make a conditional centromere. Bloom and coworkers predicted that if a strong inducible promoter were placed adjacent to a centromere so that the promoter would direct transcription toward the centromere, activation of transcription might bombard the centromere with polymerases and disrupt the centromeric chromatin complex [Hill and Bloom, 1987]. This proved to be correct. Insertion of such a promoter next to a centromere results in inactivation of that centromere when cells are grown under inducing conditions (although the chromatin structure of the centromere does not appear to be disrupted). It is tempting to speculate that the large transcriptionally inactive heterochromatin regions that flank centromeres in higher eukaryotes may serve as a buffer against transcriptional encroachment.

The specialized chromatin structure at yeast centromeres endows chromosomes with the ability to bind to microtubules. This has been demonstrated directly through the use of an assay for minichromosome binding to microtubules *in vitro* [Kingsbury and Koshland, 1991]. This assay uses cosedimentation to detect the binding of microtubules to plasmid minichromosomes. In order for binding to be observed, the minichromosomal DNA must be present as chromatin isolated from cells. Binding requires the presence of functional CDE II and CDE III sequences on the minichromosome: No requirement for CDE I has been detected [Kingsbury and Koshland, 1991]. It now appears highly likely that this binding involves components of the yeast CBF3 complex (described below).

The CDE I sequence defines a binding site for a protein that has been designated variously CBF1, CP1, and CPF1 [reviewed by Schulman and Bloom, 1991]. The CDE I sequence is present both at centromeres and 5′ of a number of yeast genes, and the role of CBF1 *in vivo* may be complex. For example, cells in which the CBF1 gene is disrupted are auxotrophic for methionine [Cai and Davis, 1990]. Thus CBF1 may function both as a centromere component and as a transcription factor [but see Mellor et al., 1991]. This provides an interesting analogy between centromeres and replication origins (ARS elements), which also appear to require the binding of transcription factors for full function *in vivo* [Marahrens and Stillman, 1992]. The exact identity of the transcription factor does not appear to matter much in the case of ARS function. I am not aware of any experiments in which CDE I has been replaced by a binding site for another transcription factor, but it is possible that such an operation might also yield a functional centromere.

When the gene for CBF1 was disrupted, cells remained viable, but the rate of chromosome loss increased only about 10-fold [Cai and Davis, 1990]. Thus the interaction of CBF1 and CDE1 is not essential for chromosome partitioning in mitosis, but does increase the fidelity of the process.

At present no proteins have been shown to bind specifically to CDE II. Demonstration of such binding may be difficult due to the unusual DNA sequence composition of this region (>90% A/T rich). A number of abundant proteins have been shown to bind preferentially to DNA containing runs of A and T, and these would be expected to make direct purification of CDE II–specific proteins difficult. However, in a recent screen for temperature-sensitive mutants of a *Sac. cerevisiae* strain carrying an elongated CDE II element (130 bp), one mutant was obtained that upon cloning and sequencing shows a marked resemblance to a known mammalian centromere protein—CENP-A (Stoller and Fitzgerald-Hayes, personal communication). It will be interesting to determine whether this protein shows a preferential association with CDE II and to determine its role in CEN function.

Mutational analyses point to CDE III as being critical for centromere function. There has consequently been considerable interest in proteins that may bind to this region, and a number of strategies have been used to find them. A complex of proteins with CDE III–binding activity (termed *CBF3*) has been purified 160,000-fold [Lechner and Carbon, 1991]. CBF3 consists of three major subunits of 110, 64 and 58 kD, although several minor components are also present in variable amounts in the final fractions. CBF3 has several unusual and very interesting properties.

First and foremost, assembly of CBF3 onto plasmid minichromosomes endows those chromosomes with the abilities to bind to microtubules and to move along them in an ATP-dependent manner [Hyman et al., 1992]. This suggests strongly that the CBF3 fraction contains a motor component, the identity of which remains to be determined.

Second, the assembly of CBF3 onto CEN DNA is complex [Lechner and Carbon, 1991]. CBF3 must be phosphorylated in order to bind to CDE III: Phosphatase treatment of the purified fraction completely abolishes binding. In addition, CBF3 requires the presence of an assembly factor in order to bind to CDE II *in vitro*. Purified CBF3 alone is inactive in DNA binding; however, full activity is observed when either column flow-through or purified casein is added. Although neither the exact role of phosphorylation in DNA binding nor the identity of the true biological assembly factor is known, both clearly offer possibilities for the regulation of centromere structure, and presumably function. In particular, given the emerging paradigm that cell cycle transitions are regulated through posttranslational modification of proteins, this offers a potential mechanism for regulating centromere function across the cell cycle. Consistent with this idea, minichromosomes isolated from mitotic cells were found to bind microtubules *in vitro* better than their counterparts isolated from cultures in exponential growth [Kingsbury and Koshland, 1991].

A second attempt to identify CDE III–interacting proteins used a genetic screen for yeast genes that, when expressed in high copy number, could suppress weak mutations in CDE III (Shero and Hieter, 1991]. One such gene, termed *MCK1* (meiosis and centromere regulatory kinase), was unable to suppress mutations in CDE I or CDE II, but did suppress two independent CDE III mutations. The *MCK1* gene encodes a putative serine: threonine kinase. It is tempting to speculate that *MCK1* may be responsible for the phosphorylation of CBF3.

The major protein components of CBF3 are now being characterized in molecular detail. The gene for the 110 kD component has been cloned and appears unrelated to other known proteins (Jiang et al., 1993]. Interestingly, a temperature-sensitive mutant for this protein had previously been obtained. The phenotype of this mutant, known as *ndc10*, is that the chromosomes dissociate from the mitotic spindle at nonpermissive temperature [Goh and Kilmartin, 1993]. Localization of the *ndc10* protein using specific antibodies indicates that it is positioned close to the spindle pole body in mitotic cells, as would be expected for a component of the kinetochore. Thus the combination of the activity of CBF3 *in vitro* and the phenotype of the *ndc10* mutant provide strong support for the notion that CBF3 is important for centromere function in *Sac. cerevisiae*.

The gene encoding a second CBF3 component has also been isolated [Doheny et al., 1993]. This protein, which has a predicted molecular weight of 58 kD, appears to be required for interaction of minichromosomes with microtubules *in vitro*. Although definitive proof is not yet available, currently available observations are consistent with this protein being one of the two lower molecular weight components of CBF3.

We presume that *in vivo* CDE I, II, and III, combine with their associated proteins to form the structure of the budding yeast kinetochore. At present, nothing is known about the ultrastructure of these kinetochores (see Fig. 1 for a speculative model). Yeast chromosomes do not condense into morphologically distinct units during mitosis. Instead, serial section electron microscopy reveals that one or two microtubules typically connect yeast chromosomes to the spindle pole body. These microtubules appear to end directly in chromatin fibers [Peterson and Ris, 1976].

III. STRUCTURE OF THE *SCH. POMBE* CENTROMERE/KINETOCHORE

The study of centromere organization in the fission yeast *Sch. pombe* has provided both conceptual and technological challenges. The conceptual challenges arise as a result of the observation that *Sch. pombe* centromeres apparently have a completely different organization from their counterparts in the budding yeast. This raises the spectre that general principles of CEN DNA organization may be difficult to formulate.

Sch. pombe has three chromosomes ranging in size from 3.5 to 5.5 Mb. This is only 2–10-fold larger than the largest chromosomes of *Sac. cerevisiae,* and yet the centromeres in *Sch. pombe* are at least 300–600-fold longer than their counterparts in budding yeast (spanning 35–110 kb [Clarke, 1990; Murakami et al., 1991]). To date, the smallest regions of chromosomes 1 and 2 required to construct functional centromeres in *Sch. pombe* are 40 and 80 kb in length, respectively [Hahnenberger et al., 1989, 1991]. A different approach yielded a 120 kb stable derivative of chromosome 3 [Niwa et al., 1989].

The centromeres of the two yeasts differ in more than size. *Sac. cerevisiae* centromeres are not only short, they are also flanked relatively closely by active genes. In contrast, the centromeres of *Sch. pombe* contain a central unique sequence core 4–7 kb in length flanked by large regions of nontranscribed DNA. This nontranscribed DNA is organized into complex repeating units that have been variously designated K-L-B and dg and dh [reviewed by Schulman and Bloom, 1991]. It has been argued that the repeated DNA in *Sch. pombe* centromeres forms a region of inactive chromatin analogous to the heterochromatin found in the centromeres of higher eukaryotes. However, this idea has yet to be supported by direct evidence. In fact, it has recently been observed that the *Sch. pombe* centromere is peppered with tRNA genes [Takahashi et al., 1991; Kuhn et al., 1991]. While technical difficulties have thus far precluded the determination of whether these genes are actively transcribed, this certainly raises the possibility that the entire *Sch. pombe* centromere may not be entirely transcriptionally inert.

The study of *Sch. pombe* centromeres has proven to be technically difficult for two principal reasons. First, the regions of DNA involved are large and contain multiple repeated sequence elements. Second, the repeated DNA segments flanking the central core region have an inverted orientation. This has made cloning in *Escherichia coli* extremely difficult, since the presence of large inverted repeats in that organism typically leads to rearrangement of the DNA. Nonetheless, it now appears that many of these difficulties have been solved, and the functional analysis of *Sch. pombe* centromeres is now under way.

The identification of protein components that interact with *Sch. pombe* centromeres is likely to be difficult because of the large size of the DNA regions involved. A preliminary analysis of the chromatin structure of *Sch. pombe* centromeres of chromosomes 1 and 2 has been reported, and the results indicate that the organization of this region is complex [Polizzi and Clarke, 1991]. This study revealed several regions of altered chromatin structure throughout the centromere, suggesting that numerous specific binding factors may be involved.

IV. ULTRASTRUCTURE OF THE MAMMALIAN KINETOCHORE

Centromeres of mammalian chromosomes are constricted regions where the two sister chromatids are most tightly paired. The development of modern fixatives and techniques for thin-section electron microscopy led to the discovery of a button-like structure about half a micron in diameter at the surface of the centromeric chromatin [Luykx, 1965; Brinkley and Stubblefield, 1966; Jokelainen, 1967]. This structure is the kinetochore (Fig. 2).

Mammalian kinetochores have four morphologically distinct layers, although in any one preparation only three of these can be detected (one layer is observed only in the absence of microtubules; another only in their presence). The innermost layer is a disk of densely packed heterochromatin called the *inner plate* [Rieder, 1982]. The inner plate is continuous with the surface of the centromeric chromatin. This feature of the kinetochore is somewhat variable in appearance: Often it is indistinguishable from the surrounding chromatin. The inner plate is not observed on the chromosomes of cells that have been held in mitosis as a consequence of exposure to microtubule-depolymerizing drugs such as colcemid [Rieder, 1982]. Such treatments result in the hypercondensation of the chromosomes. This suggests that the inner plate may normally be a region of chromatin that is distinct simply because it is more condensed than the surrounding heterochromatin.

The function of the inner plate remains mysterious. This region may be responsible for the assembly of a robust kinetochore that is capable of withstanding the stresses and strains of mitotic movements [Bernat et al., 1991]. In

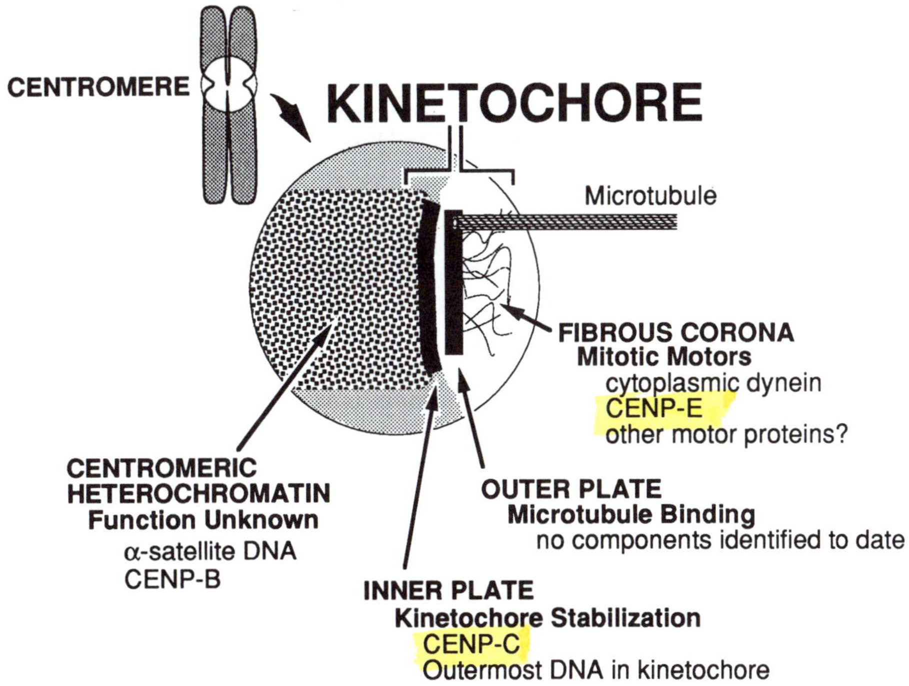

Fig. 2. *The human kinetochore.*

addition, recent evidence suggests that the inner plate might be where the mammalian equivalent of CEN DNA is located [Cooke et al., 1993].

The second layer of the kinetochore is more appropriately considered as an *interstitial space* between the two plates. This space is a region 30–40 nm across that is not stained under standard protocols for electron microscopy. Although it is possible that the interstitial space contains some component that is not stained by the various heavy metal reagents used, the most likely interpretation is that this is simply a space that separates the inner and outer kinetochore plates. Recent high-resolution tomographic analysis has revealed that this space is traversed by numerous fibers that appear to connect the inner and outer regions of the kinetochore [McEwen et al., 1993].

The most striking and consistent feature of the kinetochore is the *outer plate,* a dense structure that is typically 0.5 μm in diameter and 30–40 nm thick. This is the predominant site of microtubule attachment to the chromosome. When seen in ultrathin sections, this outer plate has a bipartite structure resembling either two thin laminae separated by a narrow space or, more commonly, two "braided" fibers having the appearance of a line of figure eights laid on their sides [Rattner and Bazett-Jones, 1989; Bernat et al., 1991; McEwen et al., 1993]. In whole mounts, the outer plate has a fibrous appearance [Rattner, 1986]. The detailed architecture of the outer plate is not yet established: It may be composed of interwoven fibers, or it could have another, more complex organization.

The outer plate forms the primary point of attachment for spindle microtubules. This is not the only point of microtubule attachment to the centromere, however. Microtubules are also observed to penetrate the chromatin immediately surrounding the kinetochore, and some microtubules appear to pass through the outer plate and end in the inner plate [Rieder, 1990].

The outer surface of the kinetochore is associated with a poorly defined filamentous meshwork known as the *fibrous corona*. The fibrous corona is seen only in kinetochores that do not have large numbers of associated microtubules. Immunoelectron microscopy reveals that the fibrous corona contains at least one mechanochemical motor protein [cytoplasmic dynein; Wordeman et al., 1991]. This is consistent with the observation that the initial interaction between chromosomes and spindle microtubules in newt pneumocytes appears to involve the lateral association of the fibrous corona region with microtubules that extend well beyond the chromosome [Rieder et al., 1989; Merdes and De Mey, 1990]. These early associations are characterized by a rapid minus end–directed movement of the chromosome along the microtubule at rates resembling those characteristic of the action of cytoplasmic dynein.

The multilaminar kinetochore structure is observed only during mitosis. During interphase, the centromere persists as a condensed ball of heterochromatin that is intermingled with, and morphologically indistinguishable from, the other condensed chromatin within the nucleus [Cooke et al., 1990]. The kinetochore plates appear on the surface of the centromere during late prophase. The mature multilaminar structure is typically visible by the time the chromosome comes in contact with microtubules at the onset of mitotic prometaphase following nuclear envelope breakdown [Roos, 1973].

V. STRUCTURE OF THE MAMMALIAN CENTROMERE/KINETOCHORE: DNA

The functional centromere (CEN) DNA from higher eukaryotes has not yet been identified. This is at least partly due to the structural complexity of the centromere of most higher eukaryotes. For example, the centromere of chromosome 21 (the smallest human chromosome at ~40 Mb) has been estimated to span ≥5 Mb.

Typically, the centromere contains enormous expanses of repetitive (satellite) DNA. For example, the centromeres of human chromosomes contain large amounts (0.3–5 Mb) of a repeated DNA family termed α-*satellite* DNA. This makes up approximately 5% of the human genome [see Willard, 1990, for a recent review]. The sequence organization of α-satellite DNA is complex, being based on a highly polymorphic monomer 171 bp in length [Willard and Waye, 1987]. Monomers are organized into characteristic higher order repeats that differ from chromosome to chromosome. How, then, is one to find the functional CEN DNA in this mass of repeated DNA? Several strategies have been attempted, none with convincing success to date.

It is possible that α-satellite DNA itself comprises human CEN DNA. There is experimental evidence both for and against this hypothesis. First, digestion of centromeric satellite DNAs with restriction enzymes and analysis of Southern blots of pulsed field gels reveal the enormous extent of the centromeric α-satellite DNA arrays: Regions of simple sequence DNA spanning up to 3–5 Mb (megabases) have been detected [Wevrick and Willard, 1989; Jabs et al., 1989]. One might argue that there is simply no room in the centromere for other CEN DNA sequences, thereby awarding the CEN DNA role to α-satellite DNA more or less by default.

More direct evidence for a role of α-satellite DNA in centromere function was obtained by transfecting a green monkey cell line with the human α-satellite DNA repeat from chromosome 17 [Haaf et al., 1992]. In certain cell lines, the integrated DNA was found to become highly amplified in culture. This DNA bound to centromeric autoantigens detected with anticentromere antibodies from patients with rheumatic disease (see below), suggesting that it adopts at least some aspects of the structure of normal centromeres. In addition, in at least one of the cell lines, the chromosomes containing human α-satellite DNA were found to be unstable in mitosis, as though the transfected DNA had acquired at least partial centromere function. However, the authors also observed cell lines containing transfected α-satellite DNA in chromosomes that appeared to exhibit normal mitotic stability [Haaf et al., 1992]. This experiment provides a strong suggestion that α-satellite DNA might have CEN function under some circumstances, but the authors could not rule out the possibility that the regions undergoing DNA amplification had acquired sequences from the green monkey centromere during the process of selection in culture.

Two observations argue against α-satellite DNAs being the human equivalents of yeast CEN DNA. First, mitotically stable dicentric marker chromosomes have α-satellite at both their active and inactive centromeres, and yet do not show anaphase bridges [see references of Earnshaw and Migeon, 1985]. Second, a mitotically stable human marker chromosome observed during prenatal diagnosis has been found that does not have α-satellite DNA detectable by *in situ* hybridization with any of the probes available to date (J. Zackowsi, S. Schwartz, and W. Earnshaw, unpublished data; H. Willard, personal communication). This chromosome could either contain only very low levels of α-satellite DNA, or have an α-satellite DNA array composed of divergent monomers that have yet to be cloned. Interestingly, this chromosome does not exhibit a prominent centromeric primary constriction, suggesting that α-satellite DNA may have a role in establishment of this aspect of centromere structure.

Further evidence against a role for satellite DNAs as CEN DNA comes from the interspecific variation shown in the distribution of these DNAs. For example, mouse centromeres contain large amounts of two satellites, termed *major* and *minor* [Singer, 1982; Pietras et al., 1983]. These show quite different distributions in *Mus musculus* and *Mus spretus*. In *Mus musculus,* the minor satellite is apparently clustered in the vicinity of the kinetochore [Wong et al., 1989; Joseph et al., 1989; Kipling et al., 1991]. In *Mus spretus,* the minor satellite is actually the major centromeric repetitive DNA and is distributed throughout the entire primary constriction and not localized to the vicinity of the kinetochore [Wong et al., 1990; Narayanswami et al., 1992].

How, then, is one to find CEN DNA in the presence of those overwhelming amounts of satellite DNA? I will touch briefly on four strategies that have been adopted. These are 1) screening of DNA libraries enriched in middle repetitive DNA for sequences that are located at centromeres; 2) use of antibody probes for known centromeric proteins to enrich for centromeric DNA; 3) the search for functional CEN DNA sequences using transfection assays analogous to the minichromosome stability assays developed in the yeasts; 4) examination of stable minichromosomes that are highly enriched in centromeric sequences. A fifth method, direct cloning of centromeric DNAs in yAC vectors, is also being undertaken in several laboratories, and anecdotal information available at the time of writing indicates that the entire centromere of the human Y chromosome may have been cloned in such a series of yAC vectors.

The construction of a DNA library containing middle-repetitive DNA sequences originally led to the cloning of human telomere DNA [Moyzis et al., 1988]. This library has now been screened for sequences that are enriched at centromeres. Several such sequences were found, and their characterization has yielded a surprising result: They correspond to previously known families of satellite DNA [Grady et al., 1992].

The new clones were enriched in members of the satellite II and III families that together comprise about 5% of the human genome. They consist of a highly polymorphic sequence array based on a 5 bp consensus sequence $(GGAAT)_n$, together with other embedded repeats [Singer, 1982]. The embedded repeats contain a stretch of eight nucleotides also found in *Sac. cerevisiae* CDE III. Satellites II and III have classically been described as being confined to pericentric regions on chromosomes 1, 9, and 16 as well as the long arm of the Y chromosome [Singer, 1982]. However, the new study revealed that when more sensitive methods for fluorescent *in situ* hybridization were employed, these sequences could be detected at all centromeres of human chromosomes

[Grady et al., 1992]. They were also detected in the genomes of chicken, *Drosophila*, sea urchin, and maize. The role of this DNA in centromere function is not currently known; however, the authors propose that it may carry out some or all of the functions of CEN DNA.

When human anticentromere antibodies were used to obtain DNA clones enriched in centromeric DNA, most of the clones contained α-satellite DNA [Masumoto et al., 1989b]. This led to the discovery of the specific binding of CENP-B protein to this DNA [Masumoto et al., 1989a], but has not as yet yielded other candidates for functional centromere DNA.

In one recent study, high molecular weight DNA remaining after nuclease digestion of sonicated human chromosomes was isolated and cloned [Hadlaczky et al., 1991]. A novel 13.8 kb clone, reportedly located at centromeres and present in 16–32 copies per genome, was cotransfected into mouse cells with a selectable marker. Several stable cell lines carrying the selectable marker were found to contain a mitotically stable minichromosome that had sequences from both the 13.8 kb clone and the marker. This report suggested that the 13.8 kb clone might contain functional human centromere sequences. The results should be regarded with caution for two reasons, however. 1) When *in situ* hybridization was performed using the 13.8 kb clone, a signal was found solely near the telomere of chromosome 9 (one to two copies per genome) and not at centromeres [McGill et al., 1992; Cooper and Tyler-Smith, 1992]. 2) As in the study of α-satellite transfection discussed above, the authors could not exclude that the minichromosome had acquired functional centromere sequences from the host cell [Hadlaczky et al., 1991].

The strategy of using naturally occurring stable minichromosomes as a source of potential centromere DNA has been tried in at least three systems. First, when human–hamster somatic cell hybrids expressing a human gene capable of rescuing a hamster mutant in the succinate dehydrogenase gene (centromere linked on chromosome 1 in the human) were obtained, several of these were found to contain the human DNA on a stable minichromosome [Carine et al., 1989]. This minichromosome was small: about 2 Mb in size. Initial studies indicated that portions of the α-satellite and satellite III arrays had been deleted, suggesting that this chromosome might possess a minimal centromere. However, subsequent studies suggested that some more complex rearrangements might have occurred during the genesis of the minichromosome (I. Scheffler, personal communication).

A second study was performed in *Drosophila melanogaster,* where a tiny derivative of the X chromosome was obtained in 1953 subsequent to irradiation of fly stock [described by Lindsley and Grell, 1968]. This chromosome, known as *Dp(1;f)1187,* is largely (~80%)l composed of centromeric heterochromatin. DNA from the *Dp(1;f)1187* chromosome was recently excised from pulsed field gels and used to screen a library of *Drosophila* DNA [Abad et al., 1992]. One clone that was obtained has a new *Drosophila* satellite DNA with a 12 base pair consensus sequence (CCCGTACTCGGT). This is termed *dodeca* satellite. Dodeca satellite probes hybridize to the centromere of one *Drosophila* autosome, and also the centromeres of human chromosomes 9 and 15 as well as to the long arm of the Y chromosome. Dodeca satellite appears to adopt a non-B-form conformation, as has been found for human satellite II and satellite III DNAs [Abad et al., 1992; Grady et al., 1992]. Its role in centromere function, if any, remains unknown.

In an ongoing collaboration, our laboratory is exploiting a naturally occurring human minichromosome that we estimate to contain 2–5 Mb of DNA (G. Stetten and WCE, unpublished). This chromosome has been obtained by microdissection from chromosome spreads, and several PCR-based libraries have been obtained from it (D. Johnson, G. Stetten and WCE, unpublished). We are currently in the process of screening these libraries for sequences that bind to the CENP-C protein (see below).

In summary, although numerous attempts have been made to find them, the identity of the CEN DNA sequences from the chromosomes of higher eukaryotes remains a mystery.

VI. IS DNA AN INTEGRAL STRUCTURAL COMPONENT OF THE KINETOCHORE?

Classic cytological evidence suggests that kinetochores are assembled from multiple functional subunits. It has recently been proposed that α-satellite DNA is an integral component of each of these subunits, and that a folded chromatin fiber containing this DNA is the principal structural component of the outer kinetochore plate [Zinkowski et al., 1991].

The subunit organization of mammalian kinetochores was best demonstrated by careful analysis of a model system for the forced entry of cells into mitosis. In this system, cell cycle controls become perturbed when cells that have been arrested in S phase by treatment with hydroxyurea are subsequently exposed to caffeine [Schlegel and Pardee, 1986]. These cells enter mitosis without undergoing detectable DNA replication.

In a surprising and elegant series of experiments it was found that this experimental protocol appears to cause the detachment of kinetochores from the chromosomes [Brinkley et al., 1988]. These detached kinetochores, together with associated DNA, undergo a normal series of mitotic movements while the bulk of the chromosomal mass remains static to one side of the cell. When the number of detached kinetochores was scored, in some cases it appeared to be higher than the number of kinetochores that would be expected in the cell type in question. This was interpreted as support for the idea that the treatment causes kinetochores to dissociate into functional subunits [Zinkowski et al., 1991].

Further evidence for the existence of such subunits was obtained when chromosomes on glass slides were exposed to vigorous shear forces and the positions of α-satellite DNA and centromeric antigens subsequently examined [Zinkowski et al., 1991]. We had originally obtained evidence that the centromere regions of human chromosomes could become split into subdomains when the chromosomes were subjected to shear forces [Earnshaw and Rothfield, 1985]. The more recent results were interpreted in terms of a model in which the kinetochore is constructed from functional subunits that are organized around the repeating units of α-satellite DNA [Zinkowski et al., 1991]. The outer plate of the kinetochore was proposed to be a specialized chromatin structure, like that at yeast centromeres.

We have recently obtained experimental evidence that is inconsistent with this conclusion [Cooke et al., 1993]. Noting that existing evidence for the involvement of DNA in kinetochore plate structure was indirect [reviewed by Rieder, 1982], we decided to use two independent methods to map the distribution of DNA in and around the mammalian kinetochore. In this study, both immunoelectron microscopy with an antibody to DNA and osmium ammine-B staining failed to reveal DNA within the outer layer of the mammalian kinetochore while clearly showing that the inner plate of the kinetochore is DNA rich. We did not interpret our results as conclusive proof that there is no DNA in the outer kinetochore plate or the outer regions of the kinetochore but rather as a strong indication that DNA, if present, is a very minor component [Cooke et al., 1993].

This led us to reinterpret earlier experiments that used electron spectroscopic imaging to analyzing thin sections of mitotic chromosomes. A strong phosphorous signal in the outer kinetochore plate found in those studies was suggested to arise due to the presence of DNA [Rattner and Bazett-Jones, 1989]. We now suspect that this signal may arise from the presence of an abundance of phosphoproteins in the outer plate [Cooke et al., 1993]. This is in agreement with the observation that the monoclonal antibody MPM-2 (specific for mitotic phosphoproteins) reacts with the kinetochore region [Vandre et al., 1984].

These results do not rule out the involvement of DNA in organization of the kinetochore structure, nor do they argue against a subunit mode for kinetochore structure. Rather, they imply that mammalian CEN DNA is likely to be concentrated elsewhere in the kinetochore, perhaps in the vicinity of the inner kinetochore plate. The outer plate is most likely a proteinaceous structure that is assembled raft-like on the structural template provided by the chromatin of the inner plate.

VII. PROTEINS OF THE MAMMALIAN CENTROMERE/KINETOCHORE

A. Autoantigens

The identification of mammalian centromere proteins first became possible with the discovery of antibodies that target the centromere in sera from autoimmune patients with scleroderma spectrum disease. The use of these anticentromere antibodies (ACA) led to the identification of four major centromere proteins. These are designated CENP-A (17 kD), CENP-B (80 kD), CENP-C (140 kD) [Earnshaw and Rothfield, 1985], and CENP-D (50 kD) [Kingwell and Rattner, 1987]. These proteins all bind DNA, although they do so with differing specificities and are located in distinct regions of the centromere/kinetochore.

1. CENP-A. CENP-A is a highly divergent variant of histone H3 that is found only in centromeres. This was originally predicted based on its size and charge and on the observation that it cosediments on sucrose gradients with nucleosome cores following digestion of chromosomes with nucleases [Palmer and Margolis, 1987]. The prediction was subsequently confirmed when CENP-A from bull sperm was purified to apparent homogeneity and characterized by microsequencing of proteolytic fragments [Palmer et al., 1991].

CENP-A is probably involved in the specific packaging of the centromeric chromatin. Like H3, CENP-A is probably a DNA-binding protein, but one that does not recognize specific DNA sequences. Thus the protein probably targets to the centromere through interactions with other centromere components. At present, nothing is known about the distribution of CENP-A in centromeres or about the identities of other proteins with which it might interact. The identification of a potential homologue of CENP-A in *Sac. cerevisiae* (Stoller and Fitzgerald-Hayes, personal communication) should be a great boon in determining the role of this protein in centromere function.

2. CENP-B. CENP-B (M_r in SDS-PAGE, 80 kD; true MW, 65,000) is an acidic protein that is distributed throughout the centromeric heterochromatin beneath the kinetochore [Earnshaw et al., 1987; Cooke et al., 1990]. This distribution is explained by the observation that CENP-B binds to a 17 bp sequence (the CENP-B box) in α-satellite DNA [Masumoto et al., 1989a]. Interestingly, the CENP-B box motif is also present in mouse minor satellite even though this satellite is otherwise unrelated to human α- satellite DNA [Masumoto et al., 1989a].

CENP-B is a multidomain protein, with two highly acidic hinge regions in its C-terminal one-third and a proline-rich hinge motif in the N-terminal region. The N-terminal domain (roughly 130 amino acids) is responsible for the localization of the protein to centromeres *in vivo*, apparently as a result of recognition of the CENP-B box sequence [Pluta et al., 1992; Sugimoto et al., 1992; Muro et al., 1992]. This region of the protein is absolutely conserved between human and murine CENP-Bs, and it shows some similarity to the family of helix-loop-helix DNA binding proteins [Sullivan and Glass, 1991]. However, the protein has a number of significant differences from helix-loop-helix proteins. For example, the CENP-B box is not palindromic, and although the protein apparently does interact with DNA as a dimer [Muro et al., 1992], it does not appear that the putative helix-loop-helix region is involved in dimerization. It thus appears likely that CENP-B exhibits a novel mechanism for binding to the DNA.

The function of CENP-B is not known. It has been argued that the protein is involved in interacting directly with microtubules. This was originally suggested when CENP-B was found to cross-link to exogenously added tubulin [Balczon and Brinkley, 1987; Balczon et al., 1990] and in recent experiments in which CENP-B was shown to bind to microtubules

[Armas-Portela et al., 1992]. However, two observations argue against CENP-B playing a major role in chromosome–microtubule interactions *in vivo*. First, the distribution of CENP-B in chromosomes (beneath the kinetochore and undetectable in the outer kinetochore plate [Cooke et al., 1990] does not mirror the distribution of microtubule-binding sites, which are primarily in the outer kinetochore plate. Second, CENP-B is found at the inactive centromeres of two stable dicentric chromosomes [Earnshaw et al., 1989; Earnshaw and Jabs, unpublished data]. These chromosomes contain two centromeres as defined by C-banding and the ability to bind anti-centromere antibodies, but are stable in mitosis. True dicentric chromosomes undergo cycles of breakage and reannealing in anaphase as a consequence of the attachment of the two centromeres on a single sister chromatid to opposite spindle poles. The fact that this does not occur in the stable dicentric chromosomes argues strongly that the inactive centromeres of these chromosomes do not make stable attachments to the mitotic spindle. Thus it is unlikely that direct CENP-B binding to tubulin is relevant *in vivo*.

Examination of the CENP-B sequence also led to the suggestion that the protein might interact with MAPs (microtubule-associated proteins), since the protein contains several sequences that resemble a putative MAP-binding sequence found on β-tubulin [Brinkley, 1990]. The cross-linking of CENP-B to exogenously added tubulin could be explained if the tubulin were recruited to centromeres through binding to CENP-B–bound MAPs. The significance *in vivo* of this potential for association between CENP-B and MAPs remains to be explored.

Analysis of the distribution of CENP-B boxes in α-satellite DNA suggests that CENP-B might be involved in the assembly of the specialized heterochromatin that occurs at the primary constriction. The CENP-B box is located at positions 127–143 of the consensus sequence of human α-satellite DNA [Pluta et al., 1992]. Earlier studies of nucleosome positioning in African green monkey α-satellite DNA indicated that a significant percentage of the nucleosomes occupy a predominant phase with respect to the underlying DNA sequence, such that the internucleosomal linker is located near positions corresponding to 118–126 of the human consensus array [reviewed by Simpson, 1990]. According to a widely accepted form of the solenoid model of chromatin higher order packing [Widom and Klug, 1985], the linker region would be expected to be located primarily in the interior of the 30 nm chromatin fiber. Binding of the highly acidic CENP-B molecule to this region could have profound effects on the solenoid structure, possibly introducing disruptions to the lattice [Pluta et al., 1992]. Although disruption of the solenoid in heterochromatin might *a priori* appear to be unlikely, given that centromeres typically appear more condensed than do the chromosomal arms, it is possible that the dimerization capability of CENP-B endows the protein with the ability to cross-link adjacent chromatin fibers [Pluta et al., 1992; Muro et al., 1992]. Thus the higher order organization of chromatin containing CENP-B might be profoundly different from that of bulk chromatin.

3. CENP-C. CENP-C (M_r in SDS-PAGE, 140 kD; true MW, 107,000) is a basic protein that is highly concentrated in the region of the inner kinetochore plate. This protein has recently been cloned and shows no significant similarity to sequences or motifs currently present in the database [Saitoh et al., 1992].

Analysis of the distribution of CENP-C on dicentric chromosomes suggested that this protein might have some role in kinetochore function. CENP-C was completely undetectable at the inactive centromere on a dicentric human chromosome 13 [Earnshaw et al., 1989]. This was the same centromere in which CENP-B was readily detected.

CENP-C is a possible candidate for one of the proteins involved in the interaction with human CEN DNA. In unpublished experiments, we obtained evidence that CENP-C binds DNA (H. Saitoh and W.C. Earnshaw, unpublished data). Although we do not yet know if sequence-specific interactions are involved in this binding, the highly localized distribution of CENP-C on chromosomes leads us to suspect this to be the case. As stated above, CENP-C is highly concentrated in the region of the inner kinetochore plate, apparently extending outward to the inner surface of the outer plate [Saitoh et al., 1992]. This is also the outermost location of DNA in the kinetochore [Cooke et al., 1993]. Future experiments should elucidate the nature of the CENP-C–DNA interaction.

4. CENP-D. CENP-D (M_r in SDS-PAGE, 50 kD; true MW, 44,847) is recognized by a subset of sera from patients with anticentromere autoantibodies [Kingwell and Rattner, 1987]. Surprisingly, this protein appears to correspond to a known cell cycle regulatory protein, RCC1 (regulator of chromosome condensation). RCC1 is a component of the normal cellular control circuitry that prevents cells from entering mitosis until DNA replication is complete [Nishimoto et al., 1992].

When a ~ 47 kD chromosomal protein reactive with ACAs was selectively solubilized from chromosomes by exposure to the intercalating agent chloroquine, purified, and subjected to microsequencing, the sequence was found to correspond exactly to that of the RCC1 protein [Bischoff et al., 1990]. Thus, although the authors of the study did not draw this conclusion, it appears highly likely that CENP-D corresponds to RCC1.

These results raise some uncertainty as to the distribution of RCC1 protein in mitotic cells. Antibodies raised against cloned RCC1 protein were originally found to react strongly with interphase nuclei, but not with mitotic chromosomes [Ohtsubo et al., 1989]. More recent studies using alternative fixation procedures reveal that the RCC1 protein is indeed present on chromosomes, though not concentrated at centromeres (T. Sekiguchi, T. Nishimoto, and W.C. Earnshaw, unpublished data). This is surprising, since ACAs affinity purified from the 47 kD chloroquine-soluble protein decorate centromeres specifically. The best resolution to the paradox appears to be that the autoantibodies recognize a posttranslational modification that is specific for a subset of the RCC1 protein that is concentrated at centromeres. Evidence reported previously suggested that ACA recognize posttranslational modifications shared by CENP-A, -B, and -C [Earnshaw et al., 1987].

The specific role (if any) of RCC1 protein at centromeres is unknown, as is the mechanism by which it delays the entry of cells into mitosis. Both genetic and biochemical studies have shown that the protein interacts with a nuclear *ras*-like GTP-binding protein, termed *ran* [Bischoff and Ponstingl, 1991; Matsumoto and Beach, 1991].

B. Other Proteins

The use of antibodies to known proteins involved in intracellular motility to stain chromosomes and mitotic cells revealed the presence of both cytoplasmic dynein [Pfarr et al., 1990; Steuer et al., 1990] and kinesin-related polypeptides at centromeres [Sawin et al., 1992]. The localization of cytoplasmic dynein in the fibrous corona of the kinetochore has been confirmed by immunoelectron microscopy [Wordeman et al., 1991]: The distribution of the other components is not well established, although preliminary evidence suggests that some of them might be found in the fibrous corona, while others may be internal to this (L. Wordeman, personal communication).

The search for novel protein components of the centromere/kinetochore has principally involved the development of monoclonal antibodies to chromosome scaffolds. (CENP-B and CENP-C are components of the mitotic chromosome scaffold fraction [Earnshaw et al., 1984]). Of the several monoclonal antibodies that have thus far been obtained most are known only from immunoblotting and/or immunofluorescence [Pankov et al., 1990; Compton et al., 1991]. However, two of these antigens have been cloned and characterized in greater detail.

CENP-E is a large polypeptide (M_r = 250–300 kD, MW = 312,000) that is seen at centromeres starting in prometaphase [Yen et al., 1991]. The protein is concentrated at centromeres through metaphase but appears to be quantitatively released from the chromosomes at the onset of anaphase. During the final stages of mitosis, CENP-E is apparently associated with the overlapping polar microtubules of the central spindle. This distribution is remarkably similar to the distribution of several other antigens that are chromosomal for a portion of mitosis and spindle associated from anaphas onward [reviewed by Earnshaw, 1991]. CEN is not detected at other times during th cycle, and recent evidence reveals tha graded as cells exit mitosis [Yen et

The cDNA sequence of CENP-E yie extremely surprising result. CENP-E is a m ber of the kinesin superfamily [Yen et al., 1992]. This is by far the largest of this superfamily of motor proteins yet to be identified. (For example, its α-helical rod domain is four times longer than that of conventional kinesin.) The strong implication that CENP-E will have motor functions suggests that this protein, which is chromosomal through metaphase, may play a role both in congression of the chromosomes to the metaphase plate and in anaphase B spindle elongation.

The second of the centromere-associated antigens that has been characterized by molecular cloning is designated NuMA (nuclear mitotic apparatus–associated protein [Lydersen and Pettijohn, 1980]). This large (MW = 237,000 [Yang et al., 1992; Compton et al., 1992]) protein turns out to be associated with centromeres only in cells in which the microtubules have been dissociated by colcemid [Compton et al., 1991]. In normal mitotic cells, the protein is concentrated at the spindle poles and shows little association with the chromosome. NuMA may be representative of a class of proteins that show an adventitious association with the kinetochore when microtubules are disassembled [see Earnshaw, 1991].

VIII. ASSEMBLY OF THE MAMMALIAN KINETOCHORE

The CENP antigens are required for normal mitosis *in vivo:* injection of anticentromere autoantibodies disrupts mitotic events [Simerly et al., 1990; Bernat et al., 1990]. Interestingly, antibodie njected into mitotic cells produce no phe Instead, they exert their effect whe o cells during interphase, w ast two distinct aspects ably.

on purified from sera of pa s is introduced into the cell nu ime prior to about 3 hours before e assembly of a trilaminar kineto inhibited when the cells subsequently mitosis [Bernat et al., 1990, 1991]. Chro osomes in such injected cells are capable of binding to microtubules, but are apparently unable to move along them either toward the spindle poles or toward the metaphase plate. These chromosomes have a reduced level of cytoplasmic dynein at the kinetochores (R. Bernat, W.C. Earnshaw, and L. Wordeman, in preparation).

Two important conclusions may be drawn from this experiment. First, microtubule binding by the kinetochore and chromosome movement at mitosis are distinct events. This is fundamentally different from vesicle movement, where binding and movement are inter-linked. Second, the results suggest that chromosome movement requires the presence of the trilaminar kinetochore structure.

Introduction of purified ACAs into the nucleus of HeLa cells during the 2 hours immediately preceding mitosis results in a block of the metaphase–anaphase transition. Chromosomes in these cells assemble trilaminar kinetochores, but these are apparently unstable. Kinetochores are only observed in these injected cells in the absence of microtubules [Bernat et al., 1990; 1991]. The chromatin in the region of the inner kinetochore plate appears less condensed than normal in these cells, possibly due to antibody interference with the binding of CENP-C. In fact, microinjection of rabbit antibodies to CENP-C at any point during interphase causes a similar phenotype (J. Tomkiel and W.C. Earnshaw, unpublished data). These chromosomes complete prometaphase congression events and achieve a metaphase alignment, but are unable to make the transition into anaphase. This phenotype suggests that the kinetochore may be directly involved in the metaphase checkpoint.

Cells monitor the alignment of chromosomes at the metaphase plate, prior to initiating anaphase. For example, if even one maligned chromosome remains in a newt pneumocyte, the cell will delay for a considerable period before entering anaphase [Zirkle, 1970b; Rieder and Alexander, 1989]. This monitoring mechanism has recently been termed the *metaphase checkpoint* [Li and Murray, 1991; Hoyt et al., 1991]. The underlying biochemical mechanisms are unknown, although an early study suggested that the kinetochore might be involved, presumably by transmitting an inhibitory signal until its chromosome achieves a successful bipolar alignment at the metaphase plate [Zirkle, 1970a].

Two genetic screens in *Sac. cerevisiae* have recently identified mutants that appear to be defective in the metaphase checkpoint control. These mutants, known as *mad* [metaphase arrest defective, Li and Murray, 1991] or *bub* [budding uninhibited by benzimadazole, Hoyt et al., 1991] have identified a number of complementation groups. The role of the MAD and BUB gene products in the timing of normal mitotic events is not known, though several of the gene products have been cloned and are in the process of being characterized further. It is not yet known if any of the *mad* or *bub* mutants affect genes involved in kinetochore structure.

One explanation for the metaphase arrest induced by injection of anti-CENP-C antibodies is that these altered kinetochores are unable to switch from the inhibitory to the permissive state, even though their chromosomes have achieved a bipolar alignment. Whatever the mechanism, the antibody injection results suggest that the kinetochore is somehow involved in the operations of the metaphase checkpoint. The *mad* and *bub* mutants may ultimately identify other components of this signaling network, or they may conceivably lead to the identification of other kinetochore components.

IX. CONCLUSIONS

It now appears that the first kinetochore that will be understood in molecular detail is that of the budding yeast, whose DNA sequences are well known and presumptive polypeptide components have been identified; development of functional assays is well under way. A major challenge will be to relate the conclusions from this system to the much more elaborate kinetochores of the higher eukaryotes. These kinetochores are much more amenable to ultrastructural study, and the availability of autoantibodies against centromeric proteins provided a fortuitous way to identify certain of the relevant polypeptides. Unfortunately, identification of the CEN DNA from higher eukaryotes has not yet been accomplished. Finally, the striking differences in organization between the centromeres of budding and fission yeasts raise the possibility that the mammalian kinetochore will have its own unique organization.

X. REFERENCES

Abad JP, Carmena M, Baars S, Saunders RDC, Glover DM, Ludena P, Sentis C, Tyler-Smith C, Villasante A (1992): Dodeca satellite: A conserved G+C-rich satellite from the centromeric heterochromatin of *Drosophila melanogaster*. Proc Natl Acad Sci USA. 89:4663–4667.

Armas-Portela R, Kremer L, Avila J (1992): The centromere protein CENP-B behaves as a microtubule-associated protein. Acta Histochem [Suppl] 41S:37–43.

Balczon R, Accavitti MA, Brinkley BR (1990): Identification of a 40 × 103 Mr centromere-associated protein in cultured mammalian cells. J Cell Sci 97:705–713.

Balczon RD, Brinkley BR (1987): Tubulin interaction with kinetochore proteins: Analysis by *in vitro* assembly and chemical cross-linking. J Cell Biol 105:855–862.

Bernat RL, Borisy GG, Rothfield NF, Earnshaw WC (1990): Injection of anticentromere antibodies in interphase disrupts events required for chromosome movement at mitosis. J Cell Biol 111:1519–1533.

Bernat RL, Delannoy MR, Rothfield NF, Earnshaw WC (1991): Disruption of centromere assembly during interphase inhibits kinetochore morphogenesis and function in mitosis. Cell 66:1229–1238.

Bischoff FR, Maier G, Tilz G, Ponstingl H (1990): A 47-kDa human nuclear protein recognized by antikinetochore autoimmune sera is homologous with the protein encoded by RCC1, a gene implicated in onset of chromosome condensation. Proc Natl Acad Sci USA 87:8617–8621.

Bischoff FR, Ponstingl H (1991): Mitotic regulator protein RCC1 is complexed with a nuclear ras-related polypeptide. Proc Natl Acad Sci USA 88:10830–10834.
Bloom K, Hill A, Kenna M, Saunders M (1989): The structure of a primitive kinetochore. TIBS 14:223–227.
Bloom KS, Carbon J (1982): Yeast centromere DNA is in a unique and highly ordered structure in chromosomes and small circular minichromosomes. Cell 29:305–317.
Brinkley BR (1990): Centromeres and kinetochores: Integrated domains on eukaryotic chromosomes. Curr Opin Cell Biol 2:446–452.
Brinkley BR, Stubblefield E (1966): The fine structure of the kinetochore of a mammalian cell *in vitro*. Chromosoma 19:28–43.
Brinkley BR, Zinkowski RP, Mollon WL, Davis FM, Pisegna MA, Pershouse M, Rao PN (1988): Movement and segregation of kinetochores experimentally detached from mammalian chromosomes. Nature 336:251–254.
Cai M, Davis RW (1990): Yeast centromere binding protein CBF1, of the helix-loop-helix protein family, is required for chromosome stability and methionine protrophy. Cell 61:437–446.
Carine K, Jacquemin-Sablon A, Waltzer W, Mascarello J, Scheffler IE (1989): Molecular characterization of human minichromosomes with centromere from chromosome 1 in human-hamster hybrid cells. Somat Cell Mol Genet 15:445–460.
Clarke L (1990): Centromeres of budding and fission yeasts. Trends Genet 6:150–154.
Clarke L, Carbon J (1980): Isolation of a yeast centromere and construction of functional small circular chromosomes. Nature 287:504–509.
Clarke L, Carbon J (1983): Genomic substitution of centromeres in *Saccharomyces cerevisiae*. Nature 305: 23–28.
Compton DA, Szilak I, Cleveland DW (1992): Primary structure of NuMA, an intranuclear protein that defines a novel pathway for segregation of proteins at mitosis. J Cell Biol 116:1395–1408.
Compton DA, Yen TJ, Cleveland DW (1991): Identification of novel centromere/kinetochore associated proteins using monoclonal antibodies generated against human mitotic chromosome scaffolds. J Cell Biol 112:1083–1097.
Cooke CA, Bernat RL, Earnshaw WC (1990): CENP-B: A major human centromere protein located beneath the kinetochore. J Cell Biol 110:1475–1488.
Cooke CA, Bazett-Jones DP, Earnshaw WC, Rattner JB (1993): Mapping DNA within the mammalian kinetochore. J Cell Biol 120:1083–1091.
Cooper KI, Tyler-Smith C (1992): The putative centromere-forming sequence λCM8 is a single copy sequence and is not a component of most human centromeres. Hum Mol Genet 1:753–754.
Cottarel G, Shero J, Hieter P, Hegemann JH (1989): A 125-base pair CEN6 DNA fragment is sufficient for complete meiotic and mitotic centromere functions in *Saccharomyces cervisiae*. Mol Cell Biol 9: 3342–3349.
Doheny KF, Sorger PK, Hyman AA, Tugendreich S, Spencer F, Hieter P (1993): Identification of essential components of the *S. cerevisiae* kinetochore. Cell 73:761–774.
Earnshaw WC (1991): When is a centromere not a kinetochore? J Cell Sci 99:1–4.
Earnshaw WC, Halligan N, Cooke C, Rothfield N (1984): The kinetochore is part of the chromosome scaffold. J Cell Biol 98:352–357.
Earnshaw WC, Machlin PS, Bordwell B, Rothfield NF, Cleveland DW (1987): Analysis of anti-centromere autoantibodies using cloned autoantigen CENP-B. Proc Natl Acad Sci USA 84:4979–4983.
Earnshaw WC, Migeon B (1985): A family of centromere proteins is absent from the latent centromere of a stable isodicentric chromosome. Chromosoma 92:290–296.
Earnshaw WC, Ratrie H, Stetten G (1989): Visualization of centromere proteins CENP-B and CENP-C on a stable dicentric chromosome in cytological spreads. Chromosoma 98:1–12.
Earnshaw WC, Rothfield N (1985): Identification of a family of human centromere proteins using autoimmune sera from patients from scleroderma. Chromosoma 91:313–321.
Gaudet AM, Fitzgerald-Hayes M (1989): The function of centromeres in chromosome segregation. In Strauss P, Wilson S (eds): The Eukaryotic Nucleus: Molecular Structure and Macromolecular Assemblies. Caldwell, NJ: The Telford Press.
Goh P-Y, Kilmartin JV (1993): NDC10: A gene involved in chromosome segregation in *Saccharomyces cerevisiae*. J Cell Biol 121:503–512.
Grady DL, Ratliff RL, Robinson DL, McCanlies EC, Meyne J, Moyzis RK (1992): Highly conserved repetitive DNA sequences are present at human centromeres. Proc Natl Acad Sci USA 89:1695–1699.
Haaf T, Warburton PE, Willard HF (1992): Integration of human α-satellite DNA into simian chromosomes: Centromere protein binding and disruption of normal chromosome segregation. Cell 70:1–20.
Hadlaczky G, Praznovszky T, Cserpán I, Keresø J, Péterfy M, Kelemen I, Atalay E, Szeles A, Szelei J, Tubak V, Burg K (1991): Centromere formation in mouse cells cotransformed with human DNA and a dominant marker gene. Proc Natl Acad Sci USA 88:8106–8110.
Hahnenberger KM, Baum MP, Plizzi CM, Carbon J, Clarke L (1989): Construction of functional artificial minichromosomes in the fission yeast *Schizosaccharomyces pombe*. Proc Natl Acad Sci USA: 86:577–581.
Hahnenberger KM, Carbon J, Clarke L (1991): Identification of DNA regions required for mitotic and meiotic functions within the centromere of *Schizosaccharomyces pombe* chromosome I. Mol Cell Biol 11:2206–2215.

Hill A, Bloom K (1987): Genetic manipulation of centromere function. Mol Cell Biol 7:2397–2405.

Hoyt MA, Totis L, Roberts BT (1991): *S. cerevisiae* genes required for cell cycle arrest in response to loss of microtubule function. Cell 66:507–517.

Hyman AA, Middleton K, Centola M, Mitchison TJ, Carbon J (1992): Microtubule-motor activity of a yeast centromere-binding protein complex. Nature 359:533–536.

Jabs EW, Goble CA, Cutting GR (1989): Macromolecular organization of human centromeric regions reveals high-frequency, polymorphic macro DNA repeats. Proc Natl Acad Sci USA 86:202–206.

Jehn B, Niedenthal R, Hegemann JH (1991): *In vivo* analysis of the *Saccharomyces cerevisiae* centromere CDEIII sequence: Requirements for mitotic chromosome segregation. Mol Cell Biol 11:5212–5221.

Jiang W, Lechner J, Carbon J (1993): Isolation and characterization of a gene (CBF2) specifying a protein component of the budding yeast kinetochore. J Cell Biol 121:513–519.

Jokelainen PT (1967): The ultrastructure and spatial organization of the metaphase kinetochore in mitotic rat cells. J Ultrastruct Res 19:19–44.

Joseph A, Mitchell AR, Miller OJ (1989): The organization of mouse satellite DNA at centromeres. Exp Cell Res 183:494–500.

Kingsbury J, Koshland D (1991): Centromere-dependent binding of yeast minichromosomes to microtubules *in vitro*. Cell 66:483–495.

Kingwell B, Rattner JB (1987): Mammalian kinetochore/centromere composition: A 50 kDa antigen is present in the mammalian kinetochore/centromere. Chromosoma 95:403–407.

Kipling D, Ackford HE, Taylor BA, Cooke HJ (1991): Mouse minor satellite DNA genetically maps to the centromere and is physically linked to the proximal telomere. Genomics 11:235–241.

Kuhn RM, Clarke L, Carbon J (1991): Clustered tRNA genes in *Schizosaccharomyces* pombe centromeric DNA sequence repeats. Proc Natl Acad Sci USA 88:1306–1310.

Lechner J, Carbon J (1991): A 240 kd multisubunit protein complex, CBF3, is a major component of the budding yeast centromere. Cell 64:717–725.

Li R, Murray AW (1991): Feedback control of mitosis in budding yeast. Cell 66:519–531.

Lindsley D, Grell EH (1968): Genetic Variations of *Drosophila melanogaster.* Washington: Carnegie Institution.

Luykx P (1965): The structure of the kinetochore in meiosis and mitosis in *Urechis* eggs. Exp Cell Res 39:643–657.

Lydersen BK, Pettijohn DE (1980): Human specific nuclear protein that associates with the polar region of the mitotic apparatus: Distribution in a human/hamster hybrid cell. Cell 22:489–499.

Marahrens Y, Stillman B (1992): A yeast chromosomal origin of DNA replication defined by multiple functional elements. Science 255:817–823.

Masumoto H, Masukata H, Muro Y, Nozaki N, Okazaki T (1989a): A human centromere antigen (CENP-B) interacts with a short specific sequence in alphoid DNA, a human centromeric satellite. J Cell Biol 109:1963–1973.

Masumoto H, Sigimoto K, Okazaki T (1989b): Alphoid satellite DNA is tightly associated with centromere antigens in human chromosomes throughout the cell cycle. Exp Cell Res 181:181–196.

Matsumoto T, Beach D (1991): Premature initiation of mitosis in yeast lacking RCC1 or an interacting GTPase. Cell 66:347–360.

McEwen BF, Arena JT, Frank J, Rieder CL (1993): Structure of the colcemid treated PtK1 kinetochore outer plate as determined by high voltage electron microscopic tomography. J Cell Biol 120:301–312.

McGill NJ, Fantes J, Cooke H (1992): λCM8, a human sequence with putative centromeric function, does not map to the centromere but is present in one to two copies at 9qter. Hum Mol Genet 1:749–751.

Mellor J, Rathjen J, Jiang W, Dowell SJ (1991): DNA binding of CPF1 is required for optimal centromere function but not for maintaining methionine prototrophy in yeast. Nucleic Acids Res 19:2961–2969.

Merdes A, De Mey J (1990): The mechanism of kinetochore-spindle attachment and polewards movement analyzed in PtK2 cells at the prophase-prometaphase transition. Eur J Cell Biol 53:313–325.

Moyzis RK, Buckingham JM, Cram LS, Dani M, Deaven LL, Jones MD, Meyne J, Ratliff RC, Wu JR (1988): A highly conserved repetitive DNA sequence $(TTAGGG)_n$, present at the telomeres of human chromosomes. Proc Natl Acad Sci USA 85:6622–6626.

Murakami S, Matsumoto T, Niwa O, Yanagida M (1991): Structure of the fission yeast centromere cen3: Direct analysis of the reiterated inverted region. Chromosoma 101:214–221.

Muro Y, Masumoto H, Yoda K, Nozaki N, Ohashi M, Okazaki T (1992): Centromere protein B assembles human centromeric α-satellite DNA at the 17-bp sequence, CENP-B box. J Cell Biol 116:585–596.

Murphy M, Fitzgerald-Hayes M (1990): Cis- and trans-acting factors involved in centromere function in *Saccharomyces cerevisiae*. Mol Microbiol 4:329–336.

Narayanswami S, Doggett NA, Clark LM, Hildebrand CE, Weier H-U, Hamkalo BA (1992): Cytological and molecular characterization of centromeres in Mus domesticus and *Mus spretus*. Mammal Genome 1:186–194.

Nishimoto T, Uzawa S, Schlegel R (1992): Mitotic checkpoints. Curr Opin Cell Biol 4:174–179.

Niwa O, Matsumoto T, Chikashige Y, Yanagida M (1989): Characterization of Schizosaccharomyces pombe minichromosome deletion derivatives and a functional allocation of their centromere. EMBO J 8:3045–3052.

Ohtsubo M, Okazaki H, Nishimoto T (1989): The RCC1 protein, a regulator for the onset of chromosome condensation locates in the nucleus and binds to DNA. J Cell Biol 109:1389–1397.

Palmer DK, Margolis RL (1987): A 17-kD centromere protein (CENP-A) copurifies with nucleosome core particles and with histones. J Cell Biol 104:805–815.

Palmer DK, O'Day K, Le Trong H, Charbonneau H, Margolis RL (1991): Purification of the centromeric protein CENP-A and demonstration that it is a centromere specific histone. Proc Natl Acad Sci USA 88:3734–3738.

Pankov R, Lemieux M, Hancock R (1990): An antigen located in the kinetochore region in metaphase and on polar microtubule ends in the midbody region in anaphase, characterized using a monoclonal antibody. Chromosoma 99:95–101.

Peterson JB, Ris H (1976): Electron microscopic study of the spindle and chromosome movement in the yeast *Saccharomyces cerevisiae*. J Cell Sci 22:219–242.

Pfarr CM, Coue M, Grissom PM, Hays TS, Porter ME, McIntosh JR (1990): Cytoplasmic dynein is localized to kinetochores during mitosis. Nature. 345:263–265.

Pietras DF, Bennett KL, Siracusa LD, Woodworth-Gutai M, Chapman VM, Griss KW, Kane-Haas C, Hastie ND (1983): Construction of a small *Mus musculus* repetitive DNA library: Identification of a new satellite sequence in *Mus musculus*. Nucleic Acids Res 11: 6965–6983.

Pluta AF, Saitoh N, Goldberg I, Earnshaw WC (1992): Identification of a subdomain of CENP-B that is necessary and sufficient for targeting to the human centromere. J Cell Biol 116:1081–1093.

Polizzi C, Clarke L (1991): The chromatin structure of centromeres from fission yeast: Differentiation of the central core that correlates with function. J Cell Biol 112:191–201.

Rattner JB (1986): Organization within the mammalian kinetochore. Chromosoma 93:515–520.

Rattner JB, Bazett-Jones DP (1989): Kinetochore structure: electron spectroscopic imaging of the kinetochore. J Cell Biol 108:1209–1219.

Rieder CL (1982): The formation, structure and composition of the mammalian kinetochore and kinetochore fiber. Int Rev Cytol 79:1–58.

Rieder CL (1990): Formation of the astral mitotic spindle: Ultrastructural basis for the centrosome-kinetochore interaction. Electron Microsc Rev 3:269–300.

Rieder CL, Alexander SP (1989): The attachment of chromosomes to the mitotic spindle and the production of aneuploidy in newt lung cells. In Resnick MA, Vig BK (eds): Mechanisms of Chromosome Distribution and Aneuploidy. New York: Alan R. Liss, pp 185–194.

Rieder CL, Alexander SP, Rupp G (1989): Kinetochores are transported poleward along a single astral microtubule during chromosome attachment to the spindle in newt lung cells. J Cell Biol 110:81–95.

Rieger R, Michaelis A, Green MM (1976): Glossary of Genetics and Cytogenetics, Classical and Molecular. New York: Springer-Verlag.

Roos UP (1973): Light and electron microscopy of rat kangaroo cells in mitosis. II. Kinetochore structure and function. Chromosoma 41:195–220.

Saitoh H, Tomkiel JE, Cooke CA, Ratrie HR, Maurer M, Rothfield NF, Earnshaw WC (1992): CENP-C, an autoantigen in scleroderma, is a component of the human inner kinetochore plate. Cell 70:115–125.

Saunders MJ, Yeh E, Grunstein M, Bloom K (1990): Nucleosome depeletion alters the chromatin structure of *Saccharomyces cerevisiae* centromeres. Mol Cell Biol 10:5721–5727.

Sawin KE, Mitchison TJ, Wordeman LG (1992): Evidence for kinesin-related proteins in the mitotic apparatus using peptide antibodies. J Cell Sci 101:303–313.

Schlegel R, Pardee AB (1986): Caffeine-induced uncoupling of mitosis from completion of DNA replication in mammalian cells. Science 232:1264–1266.

Schulman I, Bloom KS (1991): Centromeres: An integrated protein/DNA complex required for chromosome movement. Annu Rev Cell Biol 7:311–336.

Shero J, Hieter P (1991): A suppressor of a centromere DNA mutation encodes a putative protein kinase (MCK1). Genes Dev 5:549–560.

Simerly C, Balczon R, Brinkley BR, Schatten G (1990): Microinjected kinetochore antibodies interfere with chromosome movement in meiotic and mitotic mouse embryos. J Cell Biol 111:1491–1504.

Simpson RT (1990): Nucleosome positioning: Occurrence, mechanisms, and functional consequences. Prog Nucleic Acids Res Mol Biol 40:143–184.

Singer MF (1982): Highly repeated sequences in mammalian genomes. Int Rev Cytol 76:67–112.

Steuer ER, Wordeman L, Schroer TA, Sheetz MP (1990): Localization of cytoplasmic dynein to mitotic spindles and kinetochores. Nature 345:266–268.

Sugimoto K, Muro Y, Himeno M (1992): Anti-helix-loop-helix domain antibodies: Discovery of autoantibodies that inhibit DNA binding activity of human centromere protein B (CENP-B). J Biochem 111: 478–483.

Sullivan KF, Glass CA (1991): CENP-B is a highly conserved mammalian centromere protein with homology to the helix-loop-helix family of proteins. Chromosoma 100:360–370.

Takahashi K, Murakami S, Chikashige Y, Niwa O, Yanagida M (1991): A large number of tRNA genes are symmetrically located in fission yeast centromeres. J Mol Biol 218:13–17.

Vandre DD, Davis FM, Rao PN, Borisy GG (1984): Phosphoproteins are components of mitotic microtubule organizing centers. Proc Natl Acad Sci USA 81: 4439–4443.

Wevrick R, Willard HF (1989): Long-range organization of tandem arrays of alpha satellite at the centromeres

of human chromosomes: High frequency array length polymorphism and meiotic stability. Proc Natl Acad Sci USA 86:9394–9398.
Widom J, Klug A (1985): Structure of the 300Å chromatin filament: X-ray diffraction from oriented samples. Cell 43:207–213.
Willard HF (1990): Centromeres of mammalian chromosomes. TIG 6:410–416.
Willard HF, Waye JS (1987): Hierarchical order in chromosome-specific human alpha satellite DNA. Trends Genet 3:192–198.
Wong AKC, Biddle FG, Rattner JB (1990): The chromosomal distribution of the major and minor satellite is not conserved in the genus *Mus*. Chromosoma 99:190–195.
Wong C, Kazazian HH, Stetten G, Earnshaw WC, Van Keuren ML, Antonarakis SE (1989): Molecular mechanism in the formation of a human ring chromosome 21. Proc Natl Acad Sci USA 86:1914–1918.
Wordeman L, Steurer E, Sheetz M, Mitchison T (1991): Chemical subdomains within the kinetochore domain of isolated CHO mitotic chromosomes. J Cell Biol 114:285–294.
Yang CH, Lambie EJ, Snyder M (1992): NuMA: An unusually long coiled-coil related protein in the mammalian nucleus. J Cell Biol 116:1313–1317.
Yen TJ, Compton DA, Earnshaw WC, Cleveland DW (1991): CENP-E, a human centromere associated protein released from chromosomes at the onset of anaphase. EMBO J 10:1245–1991.
Yen TJ, Li G, Schaar B, Szilak I, Cleveland DW (1992): CENP-E is a putative kinetochore motor that accumulates just prior to mitosis. Nature 359:536–539.
Zinkowski RP, Meyne J, Brinkley BR (1991): The centromere-kinetochore complex: A repeat subunit model. J Cell Biol 113:1091–1110.
Zirkle RE (1970a): Involvement of the prometaphase kinetochore in prevention of precocious anaphase. J Cell Biol 47:235a.
Zirkle RE (1970b): UV-microbeam irradiation of newt-cell cytoplasm: Spindle destruction, false anaphase, and delay of true anaphase. Radiat Res 41:516–537.

ABOUT THE AUTHOR

WILLIAM C. EARNSHAW is Professor of Cell Biology and Anatomy at the Johns Hopkins University School of Medicine in Baltimore, Maryland, where he teaches medical and graduate courses dealing with cell biology, histology, and the scientific method. After receiving a B.A. summa cum laude from Colby College in Maine, he pursued doctoral research in the laboratory of Dr. Jonathan King at the Massachusetts Institute of Technology, where he earned a Ph.D. in 1977 for work on the assembly of bacteriophage heads. This was followed by postdoctoral research at the MRC Molecular Biology Laboratory, Cambridge, England, where he studied bacteriophage tail fiber structure with Drs. Aaron Klug and Tony Crowther, and chromatin assembly with Dr. Ron Laskey. He then pursued further postdoctoral studies at the University of Geneva, where he studied human chromosome structure in the laboratory of Dr. Ulrich Laemmli. Since 1982 Dr. Earnshaw's research has focused on the molecular analysis of the role played by non-histone proteins in mitotic chromosome structure and condensation and in the organization of human chromosomal centromeres. He is on the editorial boards of the *Journal of Cell Science, Experimental Cell Research, BioEssays, Chromosoma,* and *Chromosome Research.*

Microtubules: 413–434

The Roles of Microtubules in Chromosome Movement

J. Richard McIntosh

I. INTRODUCTION

Microtubules (MTs) contribute to many facets of chromosome movement. For example, tubulin's polymerization is a prerequisite for mitotic spindle formation, and MT depolymerization is essential for specific mitotic events, such as the motions of chromosomes to the metaphase plate and their subsequent migration to the spindle poles. Moreover, several MT-associated motor enzymes are essential for spindle function; some of these bind to specific mitotic structures, such as spindle poles or kinetochores, while others are distributed throughout the spindle. Their contributions to mitosis are almost certainly based on their mechanochemical interactions with the spindle MTs themselves. Thus chromosome motion involves multiple MT-dependent subprocesses; these act in concert, either in parallel or in series. I will review the action of MTs on chromosomes, analyzing the ways in which individual mitotic subprocesses may fit together to make chromosome segregation reliable and error free.

Two molecular processes underlie MT action during mitosis: the polymerization of tubulin and the mechanochemical action of motor enzymes. MTs form as a result of noncovalent interactions between tubulin dimers. Polymerization is initiated by organizing centers, called *centrosomes,* and modulated by the binding of MT-associated proteins whose affinity for the polymer is regulated by posttranslational modifications [reviewed by Gelfand and Bershadsky, 1991]. MTs display dynamic instability, as discussed by Kirschner and Mitchison [1986], but the pathways of subunit addition and loss and the mechanisms for "catastrophe" and "rescue," i.e., the transitions from a state of growth to a state of rapid shortening and vice versa, are still debated. For the sake of brevity, the properties of MTs that are useful for understanding mitosis will be presented here without a discussion of the current limits of our knowledge.

The mechanisms by which the MT-associated motor enzymes, kinesin and dynein, develop the forces to move over an MT's surface are also not yet understood. What a motor does is clear: It binds by its "tail" to various cellular objects and binds by its motor domain, or "head," to an MT, moving in association with nucleotide hydrolysis in a direction that is defined by the motor itself and by the structural polarity of the MT [Vale and Goldstein, 1990]. In this discussion the motors that move toward the plus end of an MT are called *plus motors*; their counterparts are called *minus motors*. I will not discuss the mechanisms of motor movement, but simply accept these enzymes as machines with defined directions of movement and specific cargos to carry. Analyses of the mechanism for motor action are presented by Scholey and Vale and by Collins (this volume).

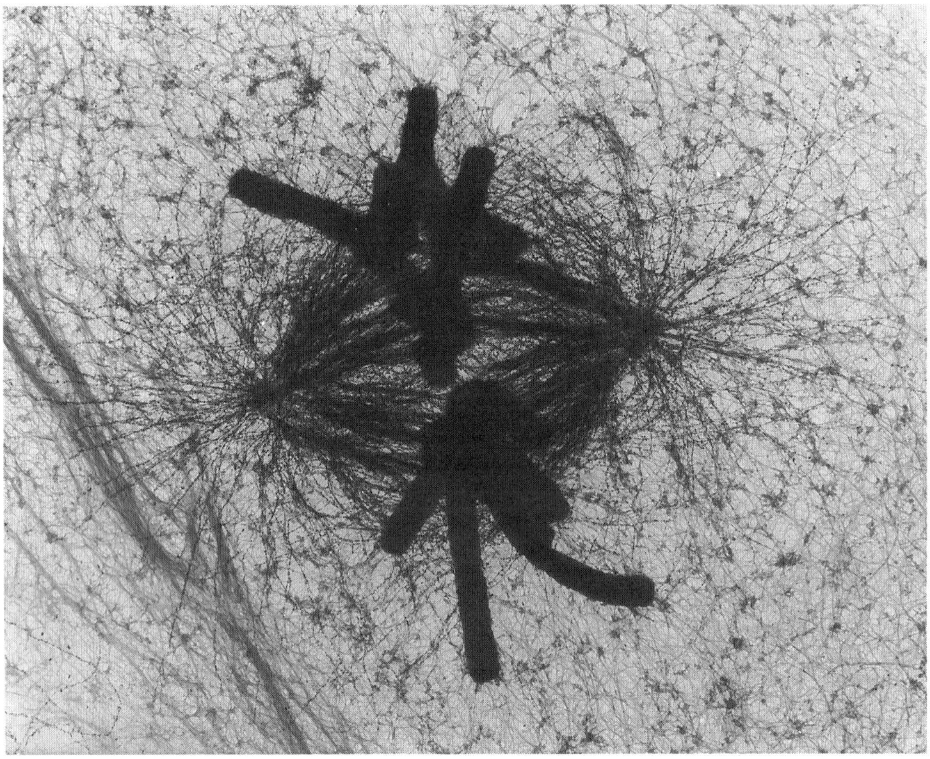

Fig. 1. *A mammalian cell in metaphase. The positions of the poles and chromosomes are characteristic, and the familiar arrangement of spindle MTs is obvious. The length of the astral MTs is greater than is commonly seen in cultured mammalian cells; probably this is a result of the unusually good signal-to-noise ratio and dynamic range of the method used. PtK1 cells were grown on Forvar and carbon-coated gold grids, using standard culture conditions [McIntosh et al., 1975]. The cells were lysed in an MT-stabilizing buffer with Triton X100, then fixed in paraformaldehyde-glutaraldehyde, rinsed, and reacted with antitubulin and a labeled secondary antibody for visualization in a high voltage electron microscope. The MTs have been decorated with 20 nm colloidal gold to make them visible, even at this low magnification.*

A. The Architecture of a Fully Formed Spindle

All eukaryotic cells form a spindle to segregate their chromosomes. The structural diversity of spindles is immense, but there are aspects of spindle design that are universal. All spindles are constructed from MTs, and these polymers are essential for chromosome segregation. All normal spindles have two poles, one at either end; these structures initiate MT assembly and thereby define the position, number, and polarity of the spindle MTs. By metaphase, a spindle is fully formed. It displays a twofold symmetry that underlies its action on the chromosomes (Fig. 1). The MTs that emanate from each pole are oriented with their fast-growing, or plus, ends distal to that pole [Euteneuer and McIntosh, 1981]. Some of these MTs encounter a kinetochore and there become bound to the chromosome. These "kinetochore MTs" (KMTs) bunch together into a fiber that is visible in the light microscope; they are part of the machinery for chromosome motion. Other MTs miss the kinetochores and terminate elsewhere in the body of the spindle, often interdigitating with their counterparts from the opposite pole. Some spindles, like the mammalian metaphase shown in Figure 1, include "astral MTs" that project from the poles into the space outside the spindle. Their functions in mitosis are not certain because many cells lack them, but they probably help to define the plane of cytokinesis in cells that form a cleavage furrow, and they may also function in aspects of spindle elongation [spindle structure is reviewed by Bajer and Mole-Bajer, 1972; Kubai, 1975; Heath, 1980; McDonald, 1989; McIntosh and Hering, 1991].

B. A Note On Notation

Genes, the products of genes, and mutant alleles are referred to with different notations for different organisms. I present data from several species, so it is impossible to be consistent in this discussion with common usage. For clarity, I have adopted the following conventions. All gene names are italicized (e.g., *cut7*), while gene product names are not and are followed by a lower case p (e.g., cut7p). Alleles are indicated by a dash and a number following the gene name (e.g., *BimC-4*), and a mutant strain that contains negligible amounts of functional gene product is denoted by the gene name, followed by a superscript minus sign (e.g., *kar3*$^-$). For *Saccharomyces cerevisiae* and *Drosophila melanogaster* wild-type alleles are designated with upper case letters (e.g., *CIN8*), but for other organisms, only the initial letter is capitalized (e.g., *Eg5*).

II. MICROTUBULE ACTION DURING SPINDLE FORMATION

A. The Early Stages

1. The timing of centrosome duplication and activation. When a cell is "born" at cytokinesis, it inherits a single spindle pole that will usually contain a centrosome that serves as the cell's major MT organizing center during the subsequent interphase. The centrosome duplicates prior to the next mitosis. The details of its duplication depend on centrosome structure, which is quite variable over a broad range of phylogeny [Kalnins, 1991]. Regardless of geometry or timing, however, centrosome duplication precedes the formation of a normal mitotic spindle, because bipolar spindles require one centrosome at each end. The exception to this generalization is the spindles of higher plants, which have unstructured poles; it is not clear what serves to initiate spindle MTs in these cells [De Mey et al., 1982; Smirnova and Bajer, 1992].

After duplication the two centrosomes separate to fulfill their roles as spindle poles. In some cells, such as the blastomeres of echinoderm embryos, the centrosomes separate during prophase before the formation of a spindle [reviewed by Wilson, 1925]. In others, such as cultured newt cells, the centrosomes separate as the spindle forms [Taylor, 1959]. The latter pattern is essentially universal for the small spindles that form in the lower plants and fungi from asymmetric centrosomes [reviewed by Pickett-Heaps, 1975; Heath, 1980]. In cultured mammalian cells, both patterns can be seen during the divisions of sibling cells, and either pathway will lead to a normal metaphase spindle, so the distinction between the two does not seem to be fundamental [McIntosh et al., 1975]. The difference is probably a result of the timing of centrosome separation relative to the increase in centrosome-initiated MTs that occurs when mitosis begins [Snyder and McIntosh, 1975; Kuriyama and Borisy, 1981].

2. The mechanisms of centrosome separation. Centrosome separation in at least some cells depends on the activity of one or more MT-dependent motor enzymes. Several genes that encode kinesin-like proteins (KLPs) have been identified in fungi by temperature-sensitive mutations whose phenotypes include mitotic arrest at the restrictive temperature [reviewed by Roof et al., 1991; Goldstein, 1993; Page and Snyder, 1993]. Analogous processes may contribute to centrosome separation in the cells of higher eukaryotes, because kinesin antigen is localized to the poles of mammalian mitotic spindles [Neighbors et al., 1988; reviewed in Sawin and Scholey, 1991], and the immune precipitation of Eg5p, a KLP, from extracts of *Xenopus* eggs inhibits the ability of this complex protein mixture to promote the formation of bipolar spindles *in vitro* [Sawin et al., 1992].

Some of the KLPs with a role in spindle formation show sufficient similarities in primary structure to have been classified as a specific subset [Goldstein, 1991; Endow, 1991; Sawin and Scholey, 1991], but of these proteins only Eg5p from *Xenopus* has actually been shown to be a motor enzyme *in vitro;* it moves along MTs toward their plus ends [Sawin et al., 1992] and thus may be called a *plus motor.* Eg5p is localized in the vicinity of the centrosomes during spindle formation. BimCp from *Aspergillus* [Enos and Morris, 1990], Cut7p from *Schizosaccharomyces pombe* [Hagan and Yanagida, 1990, 1992], and KAR3p [Meluh and Rose, 1990], KIP1p [Roof et al., 1992], and CIN8p [Hoyt et al., 1992] from *S. cerevisiae* are also localized to the spindle and/or its poles during early mitosis, and their inactivation by mutation leads to a partial or complete failure in spindle pole separation. Their directions of motor action are, however, unknown, so their mechanisms of action remain uncertain.

A mitotic role for minus motors in spindle formation is suggested by both functional and structural studies. Cytoplasmic dynein has been localized to both the centrosomes and the kinetochores of several vertebrate cells [Pfarr et al., 1990; Steuer et al., 1990]. Antibodies to cytoplasmic dynein that inhibit its motile activity *in vitro* have been injected into cultured mammalian cells, and they block centrosome separation with good efficiency, implicating this minus motor in the early stages of spindle formation [Vaisberg et al., 1993]. A KLP from *Drosophila,* NCDp [Endow et al., 1990; McDonald and Goldstein, 1990], has minus motor activity *in vitro* [Walker et al., 1990; McDonald et al., 1990]. The centrosomes separate in meiotic cells of *ncd*$^-$ female fruit flies and in the mitotic spindles of their embryos, but the spindles that form are not normal; they lose chromosomes at elevated rates [Komma et al., 1991; Hatsumi and Endow, 1992], implicating this enzyme in normal spindle organization.

Several hypotheses have been proposed for the mechanisms by which motor enzymes might contribute to the early stages of spindle formation. Figure 2 describes the problem for those fungi in which an intranuclear spindle forms from centrosomes that become part of the nuclear envelope.

1. A cross-bridge or plus motor enzyme could be distributed along the spindle MTs. First it could serve to connect the MTs that splay out from two adjacent centrosomes; the formation of additional cross connections would then provide the energy necessary to reorient the MTs into an antiparallel array, forcing the two spindle poles to face one another across a short span of nucleoplasm; plus motor action would then be able to contribute to centrosome separation and spindle elongation (Fig. 2, path 1).

2. A similar pathway can be constructed from the hypothesis that plus motor enzymes are bound by their tails to the centrosomes, while their heads bind to and push on the MTs. The enzymes that interact with MTs from the opposite pole will again promote centrosome reorientation, the assembly of an interpolar spindle, and the separation of the facing spindle poles (Fig. 2, path 2) [Enos and Morris, 1990; Hagan and Yanagida, 1992].

3. A different understanding of spindle formation is possible with the assumption that a minus motor is involved. If such motors attach by their tails to MTs emanating from one centrosome and with their heads to MTs growing from the other centrosome, they will pull the obliquely oriented polymers into an antiparallel alignment (Fig. 2, path 3) [Vaisberg et al., 1993]. These motors would also pull the centrosomes as close together as possible, given the length of the intervening MTs. This tension might contribute to the stability of the early

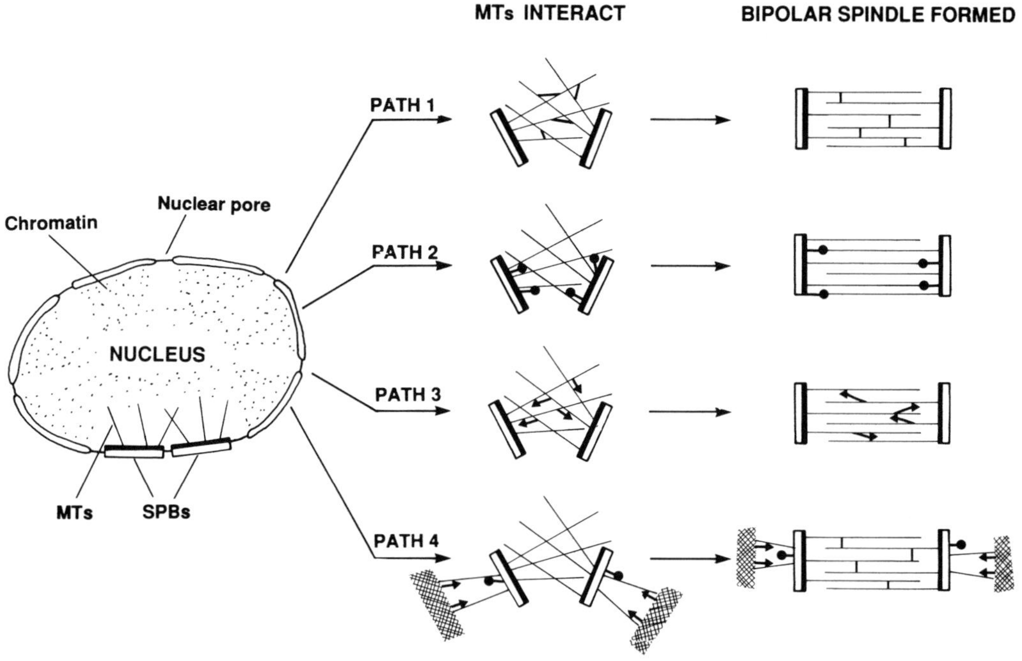

Fig. 2. *Four possibilities are depicted for the pathway that leads from the state in which spindle MTs are initiated by the adjacent spindle pole bodies (the SPBs, as the centrosomes are called in many lower eukaryotes) to the formation of a bipolar spindle. In all paths the MTs emanating from each centrosome are presumed to fan out, allowing the MT from one SPB to encounter at least some of the MTs from the other SPB and perhaps the surface of the other SPB itself. In Path 1 the MTs from sister centrosomes touch, allowing the formation of cross-bridges between the two MT families. Depicted is the situation in which the bonds are formed preferentially between MTs of opposite polarity, even when the MTs are not strictly antiparallel. The transition between the state represented in the middle column, labeled "MTs interact" to the right column, labeled "bipolar spindle formed" would be favored simply by the formation of more cross-bridges, whether the bridges were mechanochemical motors or not. In path 2, the connections are formed by plus end–directed motor enzymes whose inactive end is bound to the centrosomes. Interactions of these motors with the MTs emanating from the same SPB would have no effect on the relative orientation of the centrosomes, but interactions with MTs from the opposite SPB would tend both to push the two centrosomes apart and to reorient the centrosomes so that a maximum number of MTs could link up with motors, as depicted in the right column of this path [Enos and Morris, 1990; Hagen and Yanagida, 1992]. Path 3 posits the existence of minus end–directed motors associated with MTs from both centrosomes and able to interact with MTs from either centrosome. In the diagram, only interactions with MTs from the opposite centrosome are shown, for simplicity. These motors should run along the MTs to which they bind with their active end, pulling the MTs down into an organized bundle [Vaisberg et al., 1993]. While such motor activity would contribute to the formation of a bipolar spindle, an additional motor activity would be required to push the spindle poles apart. Path 4 is based on the assumption of a structure outside the spindle to which centrosome-associated MTs projecting into the cytoplasm can bind (the cross-hatched material in the middle column). If this matrix binds to the cytoplasmic MTs with mechanochemical, minus end–directed links, their motor activity would contribute to both reorientation and separation of the centrosomes [Vaisberg et al., 1993]. If the connections with the matrix are static, then plus end–directed motors at the centrosomes, associated with MT treadmilling, as discussed below, will promote spindle pole separation and reorientation [Enos and Morris, 1990]. Note that these paths are not mutually exclusive, so more than one of these possibilities could pertain simultaneously.*

spindle, but it would clearly be antagonistic to the plus motor action envisioned in Paths 1 and 2 above.

4. Either plus or minus motor enzymes could also contribute to spindle pole reorientation and separation by action outside the spindle (Fig. 2, path 4). If astral MTs can bind at their centrosome-distal ends to some stable structure, like the cell cortex, then minus end motors in the cortex [Aist et al., 1991; Vaisberg et al., 1993] or plus end motors bound to the centrosomes [Enos and Morris, 1990] will contribute to centrosome separation.

The above account is slightly bewildering, not because one cannot see how spindle poles could separate or what enzyme could do the job, but because there are so many legitimate scenarios involving different motors that are all probably part of the early spindle. Indeed, the situation is even more complex than suggested, because genetic studies in *S. cerevisiae* show that some KLPs have overlapping functions [reviewed by Goldstein, 1993]. Cells deficient in *KIP1* are wild type at 26°C, and those deficient in *CIN8* are almost so, but cells lacking both genes are unable to form a spindle at all [Roof et al., 1992; Hoyt et al., 1992]. Moreover, a temperature-sensitive allele of *CIN8* can be shifted from permissive to restrictive conditions during preanaphase spindle pole separation, and the poles then move back together, as if pole separation requires the continued action of this KLP [Saunders and Hoyt, 1992]. However, when this allele is expressed in a *kar3*$^-$ background, then the *cin8*$^-$-dependent loss of pole separation upon switch to restrictive temperature is reduced, suggesting that KAR3p contributes to the decrease in pole separation in the absence of CIN8p. This result can readily be interpreted if CIN8p and KAR3p are assumed to be motor enzymes with opposite polarities of action, but whatever the mechanism, there is clearly an interaction between motor enzymes during spindle formation. We see that a solid understanding of spindle formation will require not just the determination of each motor's mode of action, but also a detailed understanding of the specific role it plays in a given mitotic subprocess.

There are particularly strong amino acid sequence similarities between BimCp, Cut7p, KIP1p, CIN8p, and Eg5p, all of which are involved in spindle formation, encouraging the notion that motor enzymes from different organisms that have similar primary structures may also have similar functions *in vivo*. Motor domain similarities have also been demonstrated between KAR3p of *S. cerevisiae* and KLPAp from *Aspergillus nidulans* [O'Connell et al., 1993]. These two genes are also functionally analogous; a plasmid carrying a wild-type copy of *KLPA* will rescue a *kar3*$^-$ yeast. The methods and tools are now available to make systematic comparisons of motor enzyme function in both homologous and heterologous systems, so a detailed picture of the roles played by each spindle component in each mitotic subprocess is in principle possible. The amount of work involved in such a study is, however, prodigious, and the data will take a while to accumulate.

While the paths for centrosome separation sketched above are described for the lower eukaryotes, equivalent pathways may exist for the early stages of spindle formation in the metazoa. The attention currently given to mitosis in the fungi is certainly well placed, because the detailed work that is possible in such organisms is likely to sort out the multiple processes that contribute to centrosome separation, and what we learn from these studies should help to elucidate the analogous processes in other cells. It is noteworthy, however, that centrosome separation in the cells of many vertebrates does not include the formation of an obvious pole-to-pole spindle, so the action of extra-spindle factors may be particularly important (Fig. 2, path 4). Moreover, spindle formation in higher plants may follow yet another route. Prophase in these organisms is preceded by the formation of a band of cytoplasmic MTs and actin that girdles the cell where cytokinesis will subsequently occur [Traas et al., 1987; Liu and Palevitz, 1992, and the references therein] (see also Lambert and Lloyd, this volume). However, the structures that initiate spindle MTs in such cells are not evident in the electron microscope and are poorly understood [reviewed by Smirnova and Bajer, 1992].

3. Changes in centrosome action that accompany spindle formation. During prophase and prometaphase there are changes in the behavior of both centrosomes and cytoplasmic MTs. The mean MT length shortens and the number of MTs growing from each centrosome increases. Where the process has been studied in some detail, the changes seem to derive from the action of the mitosis-promoting kinase $p34^{cdc2}$ [Buendia et al., 1991]. The centrosome becomes more phosphorylated as its level of MT initiation increases [Vandre et al., 1984], and several mitotic, MT-associated proteins are substrates for spindle-bound kinase activity [Tombes et al., 1991]. Just what the relevant kinase(s) do to the behavior of tubulin or MT-associated proteins has not yet been determined, but the result is an increase in the frequency of catastrophes [Belmont et al., 1990], a decrease in the frequency of rescues [Gliksman et al., 1992], or both [Verde et al., 1992].

By any of several pathways, the MTs then gain access to the chromosomes: through the dispersal of the nuclear envelope, as in all higher plants and animals; through the formation of an intranuclear spindle, as in most lower plants, fungi, and some protozoa; or through the attachment of chromosomes to the inner surface of the nuclear envelope, where the maturation of their kinetochores can promote the attachment of the intranuclear chromosome to a cytoplasmic spindle [reviewed by Kubai, 1975].

B. The Prometaphase Motions of Centromeres

1. Descriptive work. Chromosomes attach to spindle MTs by their kinetochores [reviewed by Mitchison, 1988] (see also Earnshaw, this volume). These specializations bind to the substantial arrays of MTs that emanate from the centrosomes beginning during late prophase or early prometaphase (Fig. 3). The cells of certain vertebrates and higher plants are particularly suitable for a detailed examination of chromosome motion in living cells, because they are thin and their chromosomes are sufficiently large that their kinetochores are visible [reviewed by Bajer and Mole-Bajer, 1972; Rieder et al., 1986]. Three generalizations from this work are pertinent here:

1. During prophase, chromosomes do not interact with the spindle and generally remain at rest, though rapid motions in association with the inner surface of the nuclear envelope have been reported [Rickards, 1975].
2. Prometaphase begins when the centrosome-nitiated MTs start to interact with the kinetochores. The initial movements of kinetochores are generally toward the spindle poles [reviewed by Pickett-Heaps et al., 1982]. These rather rapid motions (0.1–1 μm/sec) appear to occur over the surface of centrosome-initiated MTs and are unlikely to depend on MT depolymerization [Rieder and Alexander, 1990; Merdes and DeMey, 1990]. Their kinematics suggests that they result from the action of minus motor enzymes associated with kinetochores.
3. Following their initial poleward movement, kinetochores often move away from the pole with which they are associated, alternating such motions with additional motions toward the pole. The net effect is a congression of the chromosomes to the equator of the spindle (Fig. 1).

Early descriptions of these bidirectional motions were generally interpreted as the result of sister kinetochores interacting with spindle fibers that emanated from opposite poles. It was suggested that longer kinetochore fibers could produce a stronger pole-directed force, thereby accounting for prometaphase congression to the spindle equator [Ostergren, 1950]. More recent work with other cells or circumstances has demonstrated, however, that a single pole can exert both pulling and pushing forces on a single chromosome, as discussed below.

2. Characterization of the forces acting at kinetochores. The direction of the net force

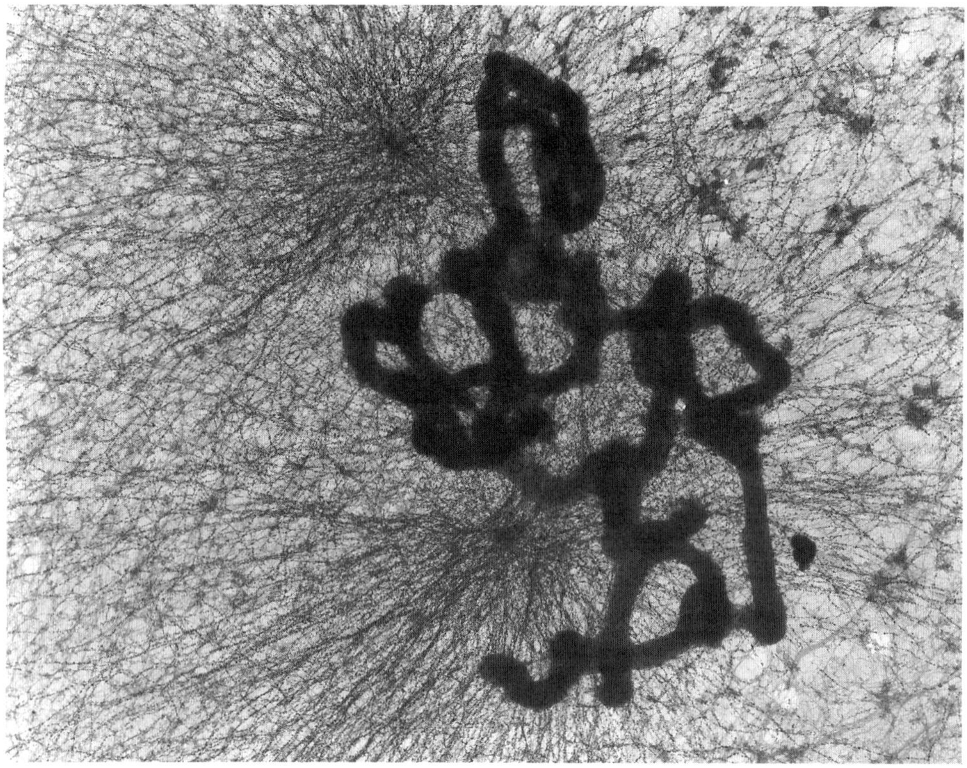

Fig. 3. *The display of MTs characteristic of late prophase or early prometaphase. MTs emanate from both centrosomes and invade much of the rest of the cell, including the space occupied by the chromosomes. Preparation was the same as Figure 1.*

acting at a kinetochore has usually been inferred from the direction of kinetochore motion. Some meiotic chromosomes that lack a homologue with which to pair never congress to the metaphase plate but move poleward in prometaphase, reorient, and then move to the opposite pole, etc., suggesting that there is always a pole-directed force acting on their active kinetochore [Nicklas, 1961]. When a metaphase chromosome is immobile, however, the force acting on each kinetochore is probably not zero, because the ablation of one kinetochore by a pulse of laser light leads to an immediate motion of that chromosome toward the pole connected with the undamaged kinetochore [Rieder et al., 1986]. Apparently, the net force on a kinetochore is usually poleward.

The magnitude of this force has been estimated both by the amount of bend that must be put in a glass needle to retard a chromosome's motions [Nicklas, 1983] and by the maximum velocity with which chromosomes can move through a region of cytoplasm whose viscosity has been estimated by the Brownian motion of nearby particles [Alexander and Rieder, 1991]. The estimates agree that chromosomes experience roughly 10 pN of pole-directed force for each microtubule that is attached to a kinetochore. This force could be generated by three or four motor enzymes per MT, each acting with the mechanical properties of kinesin [Block et al., 1990; Howard et al., 1989; Kuo et al., 1991] or cytoplasmic dynein [Ashkin et al., 1990]. It is also within the range of forces to be expected from a movement driven by MT

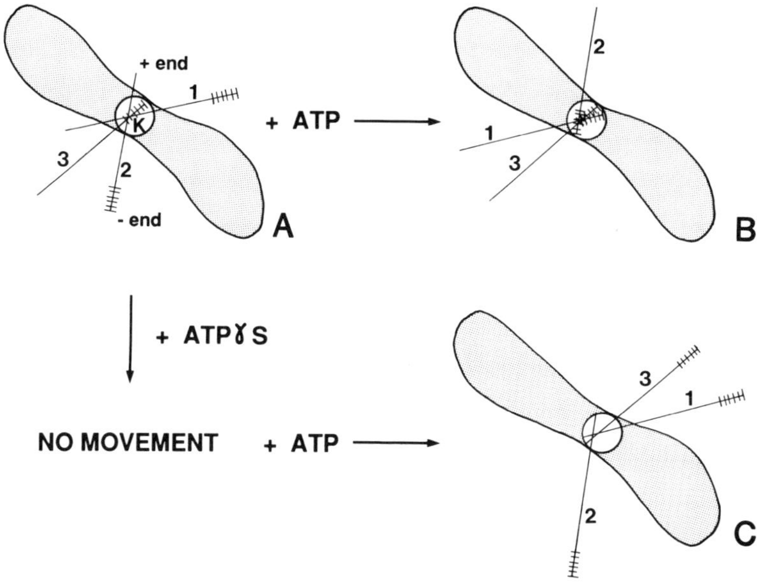

Fig. 4. *Kinetochores have MT motor activity* in vitro. *Isolated chromosomes (irregular stippled area) can be adsorbed to glass, but their kinetochores (K) will sometimes remain active [Hyman and Mitchison, 1991]. MTs (straight lines) will then bind at any place along their length to the kinetochores (**A**). Upon the addition of ATP, e.g., by the liberation of caged ATP through ultraviolet irradiation of the preparation, the predominant motion seen is toward the minus ends of the kinetochore-associated MTs (the cross-hatched MT end; **B**). If, on the other hand, ATPγS is added prior to the addition of ATP, then the motion is predominantly toward the plus ends of the MTs (**C**). Immunofluorescence suggests that the kinetochores of these chromosomes are more highly phosphorylated than those of untreated chromosomes, implicating phosphorylation as a factor in the direction of the predominant kinetochore motor activity.*

disassembly [Inoue, 1981; Hill, 1985; Coue et al., 1991], so the magnitude of the force observed does not exclude any hypotheses about the mechanism for force generation.

3. The site of generation for the forces acting on kinetochores. The forces acting on kinetochores could, in principle, be generated at the junction between the kinetochore and the MTs that bind there, at the junction between kinetochore-associated MTs (KMTs) and the pole, or at any bond between a KMT and a non-KMT or any mechanically sound matrix in the spindle. There is direct evidence for the generation of MT-dependent forces at the kinetochore. When MTs are assembled with a visible marker for their minus ends, they may be used with chromosomes bound to glass to study the motility properties of kinetochores [Hyman and Mitchison, 1991]. This system has revealed that kinetochores possess both minus and plus motor activities. Which of these activities dominates seems to depend upon phosphorylation. The initial addition of ATP promotes a minus end–directed movement, while more prolonged treatment with either ATP or ATPγS induces plus end–directed motion (Fig. 4). Since ATPγS is a poor substrate for most ATPases but a good substrate for most protein kinases, forming thioesters that are in turn poor substrates for protein phosphatases, this ATP analogue often leads to an increase in the level of protein phosphorylation. The phosphorylation of kinetochore components during this treatment has been confirmed by immunocytochemistry, so kinetochore phosphorylation may help to determine which kinetochore motor is most active and thus the direction of chromosome movement.

4. The identity of the force generators at kinetochores. Two polypeptides from cytoplasmic dynein have been localized to the kinetochores: the ~500 Kd heavy chain and a 79 Kd intermediate chain. This motor enzyme may therefore contribute to the pole-directed movements of kinetochores in at least some cells [Pfarr et al., 1990; Steuer et al., 1990; Wordeman et al., 1991]. The KLP CENP-E has also been localized at kinetochores [Yen et al., 1991, 1992]. The predicted amino acid sequence of this protein contains a segment with considerable similarity to the motor domain of plus KLPs, and its tail comprises an unusually long stretch with a high probability of forming an α-helix. While the motile properties of this peptide are not yet known, its primary structure suggests that it may play a role in chromosome motion. Antibodies specific for a different KLP also bind to the kinetochore (Wordeman and Mitchison, personal communication).

A minus motor activity has recently been identified in a protein preparation that binds to an essential segment of centromeric DNA from *S. cerevisiae,* yet the major proteins in this isolate show no sequence similarity to either kinesin or cytoplasmic dynein [Hyman et al., 1992]. There may therefore be other families of spindle MT-dependent motor enzymes yet to be identified. One can imagine that every mitotic motion might have its very own motor [Gelfand and Scholey, 1992], but the redundancy of some spindle KLPs makes this idea unlikely in its strictest form. Instead, the spindle seems to be rather markedly overdetermined [e.g., McIntosh and Hering, 1991; Goldstein, 1993]. Given the importance of doing mitosis right, it is not surprising that cells have resorted to a "belt *and* suspenders" strategy for organizing and segregating chromosomes.

5. Kinetochores permit tubulin polymerization onto attached MTs. Three lines of evidence from living cells support the idea that kinetochores have motor activity and further demonstrate that the attachment between a kinetochore and its KMTs is one that permits addition and loss of tubulin.

1. Labeled tubulin injected into prometaphase cells incorporates into KMTs predominantly at the kinetochore; the length of KMT that is labeled increases with time after tubulin injection. The labeled tubulin added at the kinetochores during prometaphase is then removed as the chromosomes move poleward during anaphase [Mitchison et al., 1986]. Furthermore, during the irregular motions of prometaphase, a chromosome can retain its attachment to the spindle as its fiber lengthens and shortens by addition and loss of subunits, predominantly at its kinetochore [Wise et al., 1991] (Fig. 5). (2) Anaphase chromosomes move closer to a bleach spot made on fluorescent KMTs by a pulse of laser light [Gorbsky et al., 1988]. (3) Anaphase chromosomes also move closer to a cut made in the spindle with a microneedle [Nicklas, 1989]. These data show that a kinetochore is not only a site of motor activity, it is a place where tubulin can add to and come off of MTs while force is still being exerted on the chromosome.

6. Spindle microtubule motions during prometaphase. Observations on chromosome motion relative to MTs are complicated by the fact that many if not all MTs in at least some spindles move poleward throughout mitosis (Fig. 5). Early work with UV microbeams [reviewed by Wilson and Forer, 1989] and with high contrast optics [Hard and Allen, 1977] suggested the existence of a pan-spindle flux of material from equator to pole, but these results were sufficiently indirect that they were difficult to interpret. More recent work with tagged tubulin demonstrates that spindle MTs are in motion, adding subunits at the kinetochores [Mitchison, 1989a; Mitchison and Salmon, 1992] and at the pole-distal ends of non-KMTs [Saxton and McIntosh, 1987; Sawin and Mitchison, 1991]. These motions require that there is also a depolymerization of tubulin from the MTs at or near the poles.

The flux of spindle MTs complicates the interpretation of many observations on spindle physiology because it suggests that each MT exists in a steady-state that involves both a "treadmilling" of the kind first envisioned for tubulin by Margolis and Wilson [1981] and a set of mechanochemical interactions at both the

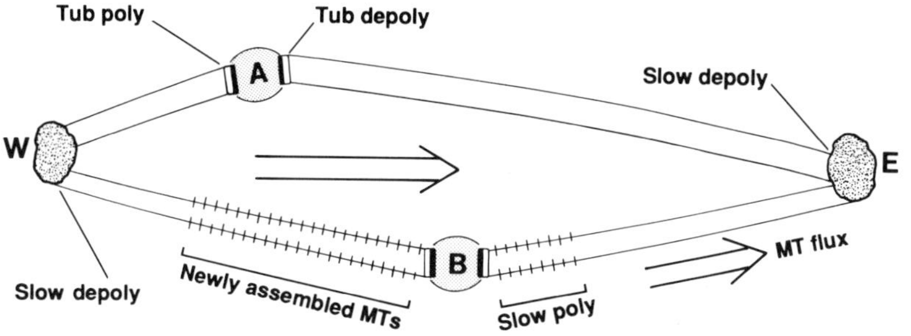

Fig. 5. *A diagram depicting MT behavior during chromosome congression to the metaphase plate. The two spindle poles are labeled W and E, and the centromere regions of two chromosomes are labeled A and B. KMTs are represented by two straight lines, and newly assembled tubulin is indicated by crosshatches on these lines. Centromere A represents a chromosome that has attached to both spindle poles but is not yet properly situated. The large open arrow indicates the motion that is required for congression from A to B. During such motion, new tubulin is added to the elongating KMTs (Tub poly) and removed from the shortening KMTs (Tub depoly) predominantly at the kinetochores. This situation is complicated, however, by the fact that throughout prometaphase and metaphase there is a slow tubulin addition at the kinetochores and a slow tubulin loss at the poles, promoting MT flux from the spindle equator to the poles.*

poles and the kinetochores. These interactions probably include force generation by motor enzymes that are tethered to the material surrounding both the poles and the kinetochores. The spindle MT assembly steady state therefore includes nucleotide flows through the hydrolytic cycles of both motor enzymes and polymerizing tubulin, as well as flows of tubulin and MT-associated protein through the polymers from their plus to their minus ends (Fig. 6) [reviewed by McIntosh and Hering, 1991]. The motor enzymes that act at the poles during metaphase are probably a subset of those implicated in centrosome separation, as discussed above. Since all these actions occur simultaneously, there must be an extensive system of coupling and control to regulate motor action and tubulin polymerization.

7. Microtubule polymerization and chromosome movement. One part of the control on mitotic motors probably derives from the coupling of motor activity to assembly and disassembly of MTs. The idea that tubulin polymerization might regulate the rate of chromosome motion is not new. It has long been recognized that chromosomes move more and more slowly as they form better and better defined connections with the spindle. MTs working as governors on chromosome motion may be one of the factors that keeps mitotic movements slow and enhances the accuracy of chromosome segregation.

It has also been suggested that MT disassembly might contribute energy to the work necessary for chromosome motion [Inoue and Sato, 1967]. More recently it has been demonstrated that MT disassembly can indeed do work on chromosomes that remain bound to these polymers as they shorten [Coue et al., 1991]. Chromosomes and cytoplasmic particles can bind to MTs initiated from a stable seed, then move with the MT ends as the polymers shorten. Such motions continue even in the presence of millimolar vanadate ions and when the concentration of ATP is below 10^{-9} M. The speed of the movements is typical of prometaphase chromosome movements and as much as 20 times that of anaphase. Chromosomes can also be pulled upstream against a rapid flow of buffer, so that at least 1 pN is exerted by the system to overcome the viscous drag on the chromosome. We must therefore add tubulin depolymerization to the list of possible mechanisms for chromosome motion *in vivo*.

It is now clear, however, that chromosome

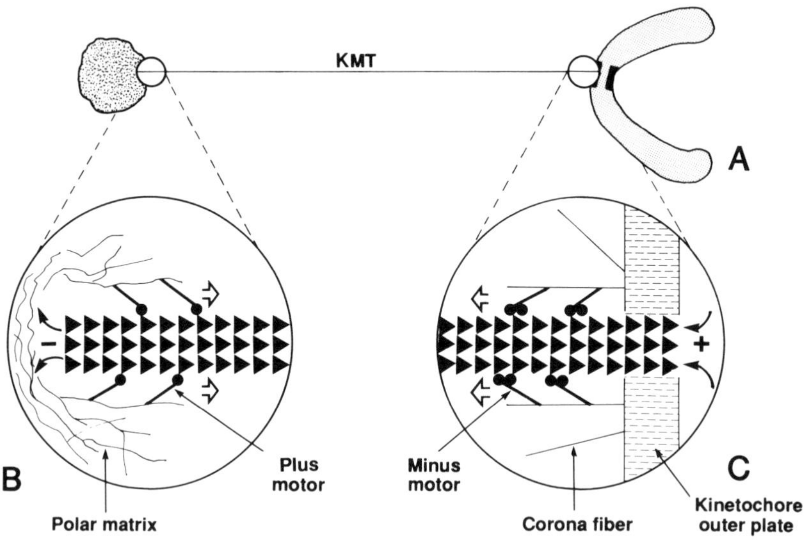

Fig. 6. *A diagram of KMT activities at both a pole and a kinetochore.* **A** *presents an overview of the connection between pole and chromosome. The small circles at each end of the KMT mark the regions diagrammed at higher magnification in* **B** *and* **C**, *in which solid triangles represent the polar arrangement of tubulin subunits, with the MT minus end at the pole (B) and the plus end at the kinetochore (C). KMTs end either at the pole, in the case of small spindles with well-structured centrosomes [e.g., McIntosh et al., 1985; Ding et al., 1993], or in the vicinity of the pole, in the case of mammalian cells with a more dispersed centrosome [McDonald et al., 1992]. In each case electron microscopy suggests that there is structured material to which the KMTs may bind (polar matrix in B). The lack of such a matrix in the spindles of higher plants presents a currently unsolved puzzle [reviewed by Smirnova and Bajer, 1992]. The pole-directed flux of KMTs suggests that the attachment of these polymers to the pole is made by plus motors that can maintain their attachment while the KMTs move poleward and the motor walks toward the kinetochore end of the KMTs (open arrows, B). Near the kinetochore (C) there is probably an analogous action by minus motors that are attached to the kinetochores, perhaps through the "corona" fibers that project from the outer plate of the kinetochore. Thin sections and electron microscopy suggest that metaphase KMTs generally end on the inner surface of the outer plate [Reviewed by Rieder, 1990], so this is probably the place where tubulin is added to permit KMT flux to the pole.*

motion is not due simply to the action of MTs at their kinetochores; there are MT-dependent forces acting on the chromosome as a whole. This action of the spindle was appreciated by Ostergren [1950], based on the positions of metaphase chromosome arms, but it has subsequently been identified as a factor in the position of chromosomes during prometaphase in monopolar spindles [Rieder et al., 1986]. A single centrosome can exert a pulling force on the kinetochores that face it and a pushing force on the arms of the same chromosome, establishing an equilibrium of forces when the chromosome is at some particular distance from the pole (Fig. 7). The force acting away from the pole has been attributed either to a plus motor enzyme loosely associated with the chromosome arms or to the action of MT polymerization during the growth phase of a dynamically unstable polymerization steady state.

Support for the former model comes from the failure in meiotic chromosome organization that occurs in *nod*$^-$ mutants of *Drosophila* [Zhang et al., 1990]. In the absence of this KLP, chromosomes that failed to cross over with their homologue are either lost from the spindle or positioned prematurely at the poles, suggesting that NODp provides a force acting away from the spindle pole [Theurkauf and Hawley, 1992]. In the latter model MT growth

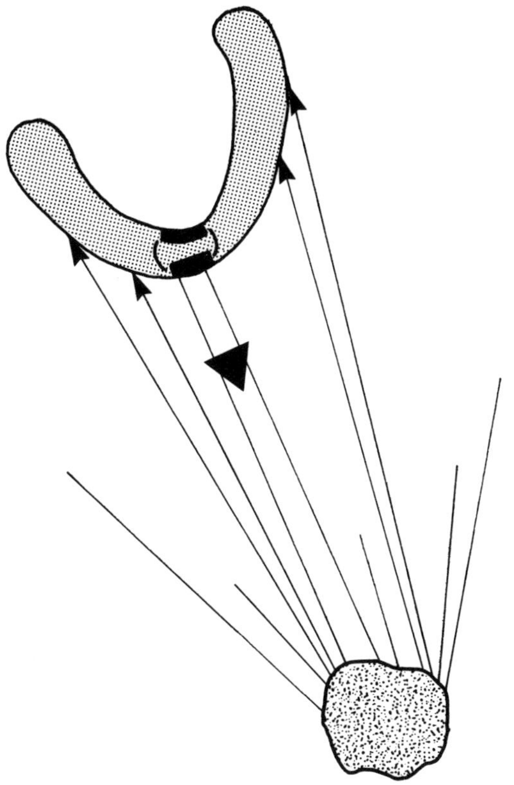

Fig. 7. *A diagram depicting the equilibrium that can exist when a chromosome is attached to a single pole. The kinetochore experiences a pole-directed force (large arrowhead) as a result of factors discussed in Figure 6. Dynamically unstable growth of non-KMTs leads to a situation in which other polymers impinge on the chromatin arms of the same chromosome. As these MTs elongate, they exert a force away from the pole (small filled arrows), but when they shorten, they simply depolymerize, releasing subunits into solution. The net effect is a force acting away from the pole that acts all over the surface of the chromosome. At the right distance from the pole, the number of such general interactions is sufficient to balance the pole-directed force acting at the kinetochore, and the chromosome is in a mechanical equilibrium [Rieder et al., 1986]. Such a force balance could contribute to chromosome congression during prometaphase [Salmon, 1989].*

into regions occupied by chromatin displaces part of that chromatin, while a disassembling MT releases its subunits into solution and shortens out of that region without any impact on chromosome position. The result is a net force acting away from the pole, as observed. MT dynamics are necessary for maintenance of the outward force, supporting the latter model [Ault et al., 1991].

8. Likely mechanisms for prometaphase congression. The "ejection forces" described above have been invoked as the principal players in chromosome congression to the spindle equator [Rieder et al., 1986; Salmon, 1989]. In this model the kinetochore-dependent forces are thought to be independent of position; they change only with the number of MTs bound to that kinetochore. Congression is then interpreted as the result of position-dependent ejection forces that push each chromosome away from each pole with a force that is inversely proportional to the distance from that pole.

The arrangement of chromosomes on monopolar spindles provides important evidence in considering this issue. Both newt cells in culture and echinoderm zygotes will form monopolar spindles under the right conditions, and in each case the chromosomes adopt a position relative to the pole that closely resembles that of the normal metaphase plate [Mazia et al., 1981; Bajer, 1982; Rieder et al., 1986; Leslie, 1992]. These results demonstrate that one pole and its associated material can exert a pushing as well as a pulling force on the chromosomes with which they are associated.

While the behavior of chromosomes on a monopolar spindle is consistent with models for prometaphase based on ejection forces, the results do not prove such models, and there are several arguments against them. The models as stated do not take into account the bidirectional action of kinetochore motors (not surprising, since the models antedate the discovery of this motor action). More important, they are unsatisfactory in accounting for certain cases of prometaphase chromosome position. For example, experimentally generated trivalent chromosomes adopt a position on the metaphase spindle at meiosis I that is nicely consistent with the Ostergren hypothesis [Hays et al., 1982]. The locations and shapes of these chromosomes are poorly consistent with the ejection force hypothesis, because the bulk of the abnormal chromosome lies roughly at the spindle equator. Its two kinetochores that point at one pole lie close to that pole, while the remaining kinetochore lies rather distant from the opposite pole [Hays et al., 1982]. This geometry is not predicted by the postulates of position-dependent ejection forces.

On current evidence, it seems as if different mechanisms for prometaphase congression dominate in different cells. Ejection forces probably play a role in cells whose spindles contain well developed arrays of nonkinetochore MTs, while other factors seem to predominate in most cells. There are two current ideas about what these other factors might be. The first posits a positional control acting on the two polarities of kinetochore motors, i.e., kinetochores are "smart" in the sense that they can know where they are in the spindle [Mitchison, 1989b]. This positional information could reside in the concentration of any factor that is appropriately distributed in the spindle. For example, calmodulin is reversibly bound to spindle MTs in a Ca^{2+}-independent manner [Welsh et al., 1979; Zavortink et al., 1983; Stemple et al., 1988]; it is distributed along KMTs with a gradient in concentration whose maximum lies near the spindle poles [Sweet et al., 1989]. If a calmodulin-dependent kinetochore kinase served either to activate plus motors on kinetochores or to inhibit minus ones, each chromosome should move to the spindle equator. Another protein that is distributed as a gradient from pole to equator is NuMA [Price and Pettijohn, 1986], a protein that has recently been implicated in successful spindle assembly [Yang and Snyder, 1992]. An alternative possibility derives from the fact that MTs near the poles are older than those near the equator, as a result of both dynamic instability and MT flux. Slow-binding spindle components might accumulate where the MTs are older [McIntosh and Hering, 1991].

The second plausible mechanism for congression employs the poleward MT fluxes to allow minus motors bound by their tails along the KMTs to contribute to the force-balance acting on each chromosome. If each motor contributes the same force, on average, then a mechanical equilibrium will be established when the chromosome is at the spindle equator. In this model, the depolymerization of tubulin that is important for congression occurs at the poles, and the prometaphase motions of kinetochores relative to the MTs with which they are bound is seen as an epiphenomenon of unknown function.

Regardless of the mechanism for prometaphase, it is impressive how rarely the spindle makes mistakes in chromosome segregation. The frequency of spindle error has been measured for yeasts at $\sim 10^{-5}$ per chromosome per division [Hartwell and Smith, 1985], while in mammals it appears to be less than 10^{-6} [Waldren, 1983]. The major reason for this level of precision is likely to be the reliability with which sister kinetochores attach to sister centrosomes [reviewed by McIntosh and Hering, 1991], and the principal reason for this reliability seems to be the fact that a chromosome is stably attached to the spindle only when its centromere is under tension [Nicklas, 1988]. A second reason is probably to be found in the arrangement of chromosomes at metaphase. The assembly of the doubled chromosomes into a planar configuration probably reduces the likelihood that one anaphase chromosome will be left out when the nuclear envelope reforms at telophase [Nicklas and Arana, 1992].

III. MICROTUBULE ACTION DURING ANAPHASE

A. The Control of Anaphase Onset

Anaphase begins when the sister chromatids part and segregate toward opposite poles. It is also the beginning of other changes in cell structure and behavior: The KMTs usually shorten while the spindle MTs that interdigitate near the equator elongate [McIntosh et al., 1979, 1985]; the astral MTs elongate [Wilson, 1925]; the centrosomes lose their ability to initiate new MTs [Snyder et al., 1982]; and the special phosphorylation that is characteristic of mitotic centrosomes disappears [Vandre and

Borisy, 1989]. As anaphase progresses, the chromosomes contract and begin to bind the cisternae that will comprise the new nuclear envelopes. Later the cell cortex begins to contract at a position that is dependent on the placement of the two mitotic asters. All these changes are characteristic of a cell's return to interphase. They are linked in some way with diminished activity of the $p34^{cdc2}$ kinase [e.g., Murray and Kirschner, 1989] and the activation of protein phosphatase [e.g., Doonan and Morris, 1989]. Which process (if any) is primary and causal for the others and which processes are parts of a single pathway are not yet known.

It is likely that there is a control on the onset of anaphase, a "checkpoint" that postpones its beginning until the cell is ready to reenter interphase [Hartwell and Weinert, 1989; McIntosh, 1991], but the nature of this control is only now coming under investigation [Li and Murray, 1991; Hoyt et al., 1991]. There is probably a chain of events initiated by the metaphase spindle that leads ultimately to the degradation of cyclin B and the inactivation of $p34^{cdc2}$, followed by the string of events catalogued above. The elucidation of this control pathway will be a major achievement in mitotic research.

Once anaphase begins, there is a significant change in the balance of mitotic forces that is characteristic of metaphase. While the sister chromatids are connected, a spindle can exist in the complex steady state described above, but once the centromeres divide, this balance is broken; there is no force holding the chromosomes at the spindle equator, so anaphase A is a natural result of centromere separation. Spindle elongation too may occur in part because centromere separation leaves nothing to keep the poles together.

The simplicity of "cause and effect" implied in this mechanical model is unlikely, however, to be the whole truth. Several of the events initiated at anaphase onset would potentiate global changes on top of the local force changes. For example, anaphase is accompanied by a slow increase in the concentration of free Ca^{2+} [reviewed by Hepler, 1989]. There is some increase in the concentration of cAMP, accompanied by a decrease in cGMP [Zeilig and Goldberg, 1977]. MT number as well as MT length begins to decrease in the space between the chromosome and the poles. All of these events contribute to the general restructuring of the cell that accompanies chromosome segregation. Together they imply that anaphase is more than simply the segregation of chromosomes, it is the initial phase in the return of the cell to interphase.

B. Chromosome Motion to the Spindle Poles: Anaphase A

A chromosome's approach to the pole requires KMT depolymerization. Some studies suggest that essentially all the requisite subunit loss occurs at the kinetochores [e.g., Gorbsky et al., 1988; Nicklas, 1989], while another study demonstrates some subunit loss (about one-third of the total) at the poles [Mitchison and Salmon, 1992]. For technical reasons the latter report is more likely to be correct. Anaphase A is therefore a multicomponent process that probably involves minus motor activity at the kinetochores, plus motor activity at the poles, and tubulin depolymerization at both kinetochores and poles. An understanding of anaphase is therefore dependent on learning the relative importance of each of these processes.

If ATP is rapidly depleted in a living plant cell, then anaphase A is halted [Helper and Palevitz, 1986]. While ATP dependence might mean that motors are required for chromosome motion, it may also mean that the phosphorylation state of spindle MTs is essential for their controlled disassembly, so the experiment does not discriminate between models. Cultured mammalian cells lysed with digitonin stop anaphase motions, but will resume a form of chromosome-to-pole motion upon the slow solubilization of the spindle MTs by exogenous Ca^{2+} [Spurk and Pickett-Heaps, 1987]. This result could mean that MT disassembly is sufficient for chromosome motion and that motor activity is only a frill, but it is very hard to be sure that all ATP has been washed from a gently lysed cell, so once again the result is hard to interpret. There are currently three principal candidates for the engines that power anaphase A: kinetochore motors, pole motors, and MT disassembly. Sorting out the contribution that each of these makes to anaphase in any particular cell will require mutations or reagents that can shut down that mechanism completely and specifically. Such tools are not yet available, but they can readily be imagined. Their discovery and use may not be too far away.

It would be unwise, however, to assume that an understanding of tubulin assembly and MT-dependent motors will be sufficient to elucidate all aspects of chromosome motion. Already other genes have been identified that are required for anaphase and whose products are located in the spindle; they do not fit into the picture of MTs and their associated proteins. The product of the *Drosophila* gene, *l(1)zw10*, is associated with the kinetochore fibers and is required for mitosis, but it appears to be an independent fibrous structure that contracts onto the kinetochores as the chromosomes separate [Williams et al., 1992]. This and other surprises will probably need clarification for our understanding of chromosome motion to be complete.

C. Chromosome Motion During Spindle Elongation: Anaphase B

At some time during anaphase A, most spindles begin to elongate. This increase in interpolar distance, anaphase B, is accompanied by both MT polymerization and MT movement. The best direct evidence about the mechanism of anaphase B comes from studies on spindles isolated from diatoms [reviewed by Hogan and Cande, 1990]. The interpolar MTs of these spindles are arranged in a well-defined shaft of centrosome-associated polymers that interdigitate near the spindle midplane. During anaphase B, the extent of interdigitation decreases, even though the component MTs elongate [Tippit et al., 1978; McIntosh et al., 1979]. Work with isolated spindles *in vitro* has demonstrated that new tubulin is added to the interdigitating MTs at their pole-distal ends and that the decrease in interdigitation is a result of an ATP-dependent sliding that requires protein phosphorylation [Masuda et al., 1988; Wordeman et al., 1990]. Thus MT assembly and MT-dependent motor activity are once again woven into a single, unified process.

It now seems clear that similar processes occur in cells besides diatoms. Essentially the same observations and experiments have been performed with the fission yeast *Sch. pombe* [Ding et al., 1993; Masuda et al., 1990]. Evidence for MT growth and sliding in the spindle interzone is also available from higher plants [Asada et al., 1991] and animals [Saxton and McIntosh, 1987; Rebhun and Palazzo, 1988; Sheldon and Wadsworth, 1990]. While the interpolar spindle in such organisms is not so well developed, there are clear bundles of MTs that run from one chromosome mass to the other (Fig. 8). Apparently these elongate during anaphase by a combination of plus-end tubulin assembly and MT sliding. A mammalian motor enzyme that cross-links antiparallel MTs and causes them to slide relative to one another with their plus ends trailing has been identified in the anaphase zone of MT interdigitation [Nislow et al., 1990, 1992]. The same region of the spindle contains many additional proteins [reviewed by Rattner, 1992], some of which may also be motor enzymes [Yen et al., 1992], so it seems likely that the zone of MT interdigitation is an active part of the mechanism for anaphase B.

In some cells, however, the force for at least some of spindle elongation is generated outside the spindle. When elongating spindles of certain fungi are treated with pulses of laser light to cut their interpolar spindle fibers, the rate of pole separation increases [Aist and Berns, 1981]. Anaphase irradiation of the astral region in these and other cells slows subsequent spindle elongation, as if the poles were being pulled apart by a front wheel drive [Aist et al., 1991]. Perhaps anaphase should be thought of as a three part process, where anaphase C is the separation of the spindle poles and their associated chromosomes or nuclei by an action from outside the spindle. Certainly there is a time during late anaphase in cultured

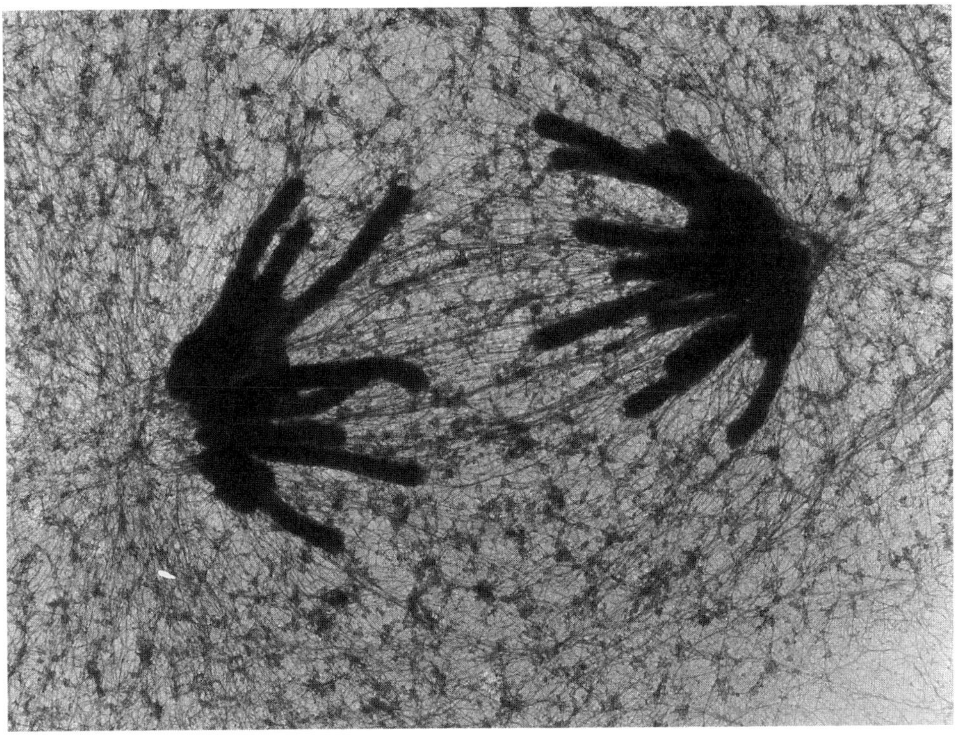

Fig. 8. *A PtK1 cell prepared for electron microscopy as described in Figure 1 after fixation at midanaphase B, except that no antitubulin was applied. Nonetheless, bundles of interzone fibers are readily seen in the interzone between the separating chromosomes. The KMTs are already so short that the chromosomes are essentially at the spindle poles, but the arms of the chromosomes have not yet contracted, which will subsequently facilitate the packaging of the duplicate chromosome sets into sister nuclei.*

mammalian cells when the poles detach from the spindle [Saxton and McIntosh, 1987]. Any subsequent centrosome movement would have to rely on force generation outside the spindle.

The elongation of isolated diatom spindles does not require action from outside the spindle, because there is no structure there to pull on and no astral MTs project from the centrosomes. Likewise, spindle elongation in *S. cerevisiae* does not require astral MTs, because anaphase B is normal in tubulin mutants that lack cytoplasmic MTs [Sullivan and Huffaker, 1992]. Such cells are, however, defective in spindle placement within the cell, so astral MTs may play a positioning role as well as a spindle elongation role. Indeed, MTs interacting with the cell cortex appear to be involved in spindle reorientation during the differentiative divisions of blastomeres in nematode embryos [Hyman, 1989].

The mechanisms of spindle elongation are again matters for research. As with anaphase A, we are faced with an embarrassment of possibilities with which to account for the observed movements. We have tubulin polymerization at the zone of overlap, associated with MT sliding, and we have an extra-spindle mechanism that could work by any or all of the processes depicted in Figure 2, path 4. Once again we must admit that evaluating the relative importance of each process will require the best tools and/or mutants available, and the work will take time. The encouraging aspect of this conundrum, at least for a student of mitosis, is that so much progress has recently been made in identifying realistic possibilities that we are now in an unprecedented position for making progress. The next few years should tell us a lot.

ACKNOWLEDGMENTS

I thank Kent McDonald, Corey Nislow, Eugenie Vaisberg, and Robert West for their careful reading of this paper. The figures were drawn by Dianne Evans, and the micrographs were taken by Mary Morphew. The work from our laboratory described herein was supported by grants GM 33787, GM36663, and RR00592 from the National Institutes of Health.

IV. REFERENCES

Aist JR, Bayles CJ, Tao W, Berns MW (1991): Direct experimental evidence for the existence, structural basis and function of astral forces during anaphase B *in vivo*. J Cell Sci 100:279–288.

Aist JR, Berns MW (1981): Mechanics of chromosome separation during mitosis in *Fusarium (Fungi imperfecti)*: New evidence from ultrastructural and laser microbeam experiments. J Cell Biol 91:446–458.

Alexander SP, Rieder CL (1991): Chromosome motion during attachment to the vertebrate spindle: Initial saltatory-like behavior of chromosomes and quantitative analysis of force production by nascent kinetochore fibers. J Cell Biol 113:805–815.

Asada T, Sonobe S, Shibaoka H (1991): Microtubule translocation in the cytokinetic apparatus of culture tabacco cells. Nature 350:238–241.

Ashkin A, Schutze K, Dziedzic JM, Euteneuer U, Schliwa M (1990): Force generation of organelle transport measured *in vivo* by an infrared laser trap. Nature 348: 346–348.

Ault JG, DeMarco AJ, Salmon ED, Rieder CL (1991): Studies on the ejection properties of asters: Astral microtubule turnover influences the oscillatory behavior and positioning of monooriented chromosomes. J Cell Sci 99:701–710.

Bajer AS (1982): Functional autonomy of monopolar spindle and evidence for oscillatory movement at mitosis. J Cell Biol 93:33–48.

Bajer AS, Mole-Bajer J (1972): Spindle dynamics and chromosome movements. Int Rev Cytol Suppl 3:1–271.

Belmont LD, Hyman AA, Sawin KE, Mitchison TJ (1990): Real-time visualization of cell cycle-dependent changes in microtubule dynamics in cytoplasmic extracts. Cell 62:579–589.

Block SM, Goldstein LS, Schnapp BJ (1990): Bead movement by single kinesin molecules studied with optical tweezers. Nature 348:348–352.

Buendia B, Clarke PR, Felix MA, Karsenti E, Leiss D, Verde F (1991): Regulation of protein kinases associated with cyclin A and cyclin B and their effect on microtubule dynamics and nucleation in *Xenopus* egg extracts. Cold Spring Harbor Symp Quant Biol 56:523–532.

Coue M, Lombillo VA, McIntosh JR (1991): Microtubule depolymerization promotes particle and chromosome movement in vitro. J Cell Biol 112:1165–1176.

De Mey J, Lambert AM, Bajer AS, Moeremans M, De Brabander M (1982): Visualization of microtubules in interphase and mitotic plant cells of Haemanthus endosperm with the immuno-gold staining method. Proc Natl Acad Sci USA 79:1898–1902.

Ding R, McDonald KL, McIntosh JR (1993): Three-dimensional reconstruction and analysis of mitotic spindles from the yeast *Schizosaccharomyces* pombe. J Cell Biol 120:141–152.

Doonan JH, Morris NR (1989): The bimG gene of *Aspergillus nidulans*, required for completion of anaphase, encodes a homolog of mammalian phosphoprotein phosphatase 1. Cell 57:987–996.

Endow S, Henikoff S, Soler-Niedziela L (1990): Mediation of meiotic and early mitotic chromosome segregation in *Drosophila* by a protein related to kinesin. Nature 345:81–83.

Endow SA (1991): The emerging kinesin family of microtubule motor proteins. Trends Biochem Sci 16: 221–225.

Enos AP, Morris NR (1990): Mutation of a gene that encodes a kinesin-like protein blocks nuclear division in *A. nidulans*. Cell 60:1019–1027.

Euteneuer U, McIntosh JR (1981): Structural polarity of kinetochore microtubules in PtK1 cells. J Cell Biol 89:338–345.

Gelfand VI, Bershadsky AD (1991): Microtubule dynamics: Mechanism, regulation, and function. Annu Rev Cell Biol 7:93–116.

Gelfand VI, Scholey JM (1992): Cell biology. Every motion has its motor. Nature 359:480–482.

Gliksman NR, Parsons SF, Salmon ED (1992): Okadeic acid induces interphase to mitotic-like microtubule dynamic instability by inactivating rescue. J Cell Biol 119:1271–1276.

Goldstein L (1991): The kinesin superfamily: Tails of functional redundancy. Trends Cell Biol 1:93–98.

Goldstein LSB (1993): Functional redundancy in mitotic force generation. J Cell Biol 120:1–3.

Gorbsky GJ, Sammak PJ, Borisy GG (1988): Microtubule dynamics and chromosome motion visualized in living anaphase cells. J Cell Biol 106:1185–1192.

Hagan I, Yanagida M (1990): Novel potential mitotic motor protein encoded by the fission yeast cut7$^+$ gene. Nature 347:563–566.

Hagan I, Yanagida M (1992): Kinesin-related cut7 protein associates with mitotic and meiotic spindles in fission yeast. Nature 356:74–76.

Hard R, Allen RD (1977): Behaviour of kinetochore fibres in *Haemanthus katherinae* during anaphase movements of chromosomes. J Cell Sci 27:47–56.

Hartwell LH, Smith D (1985): Altered fidelity of mitotic chromosome transmission in cell cycle mutants of *S. cerevisiae*. Genetics 110:381–395.

Hartwell LH, Weinert TA (1989): Checkpoints: Controls that ensure the order of cell cycle events. Science 246:629–634.

Hatsumi M, Endow SA (1992): Mutants of the microtubule motor protein, nonclaret disjunctional, affect spindle structure and chromosome movement in meiosis and mitosis. J Cell Sci 101:547–559.

Hays TS, Wise D, Salmon ED (1982): Traction force on a kinetochore at metaphase acts as a linear function of kinetochore fiber length. J Cell Biol 93:374–382.

Heath IB (1980): Variant mitoses in lower eukaryotes: Indicators of the evolution of mitosis? Int Rev Cytol 64:1–80.

Hepler PK (1989): Calcium transients during mitosis: Observations. J Cell Biol 109:2567–2573.

Hepler PK, Palevitz BA (1986): Metabolic inhibitors block anaphase A *in vivo*. J Cell Biol 102:1995–2005.

Hill TL (1985): Theoretical problems related to the attachment of microtubules to kinetochores. Proc Natl Acad Sci USA 82:4404–4408.

Hogan CJ, Cande WZ (1990): Antiparallel microtubule interactions: Spindle formation and anaphase B. Cell Motil Cytoskel 16:99–103.

Howard J, Hudspeth AJ, Vale RD (1989): Movement of microtubules by single kinesin molecules. Nature 342:154–158.

Hoyt MA, Totis L, Roberts BT (1991): *S. cerevisiae* genes required for cell cycle arrest in response to loss of microtubule function. Cell 66:507–517.

Hoyt MA, He L, Loo KK, Saunders WS (1992): Two *Saccharomyces cerevisiae* kinesin-related gene products required for mitotic spindle assembly. J Cell Biol 118:109–120.

Hyman AA (1989): Centrosome movement in the early divisions of *Caenorhabditis elegans:* A cortical site determining centrosome position. J Cell Biol 109: 1185–1193.

Hyman AA, Middleton K, Centola M, Mitchison TJ, Carbon J (1992): Microtubule-motor activity of a yeast centromere-binding protein complex. Nature 359: 533–536.

Hyman AA, Mitchison TJ (1991): Two different microtubule-based motor activities with opposite polarities in kinetochores. Nature 351:206–211.

Inoue S (1981): Cell division and the mitotic spindle. J Cell Biol 91:131s–147s.

Inoue S, Sato H (1967): Cell motility by labile association of molecules. The nature of mitotic spindle fibers and their role in chromosome movement. J Gen Physiol (Suppl) 50:259–292.

Kalnins VI (1991): The Centrosome. New York: Academic Press.

Kirschner M, Mitchison T (1986): Beyond self-assembly: From microtubules to morphogenesis. Cell 45:329–342.

Komma DJ, Horne AS, Endow SA (1991): Separation of meiotic and mitotic effects of claret non-disjunctional on chromosome segregation in *Drosophila*. EMBO J 10:419–424.

Kubai D (1975): The evolution of the mitotic spindle. Int Rev Cytol 43:167–227.

Kuo SC, Gelles J, Steuer E, Sheetz MP (1991): A model for kinesin movement from nanometer-level movements of kinesin and cytoplasmic dynein and force measurements. J Cell Sci (Suppl) 14:135–138.

Kuriyama R, Borisy GG (1981): Microtubule nucleating activity of centrosomes in CHO cells is independent of the centriole cycle but coupled to the mitotic cycle. J Cell Biol 91:822–826.

Leslie RJ (1992): Chromosomes attain a metaphase position on half-spindles in the absence of an opposing spindle pole. J Cell Sci 103:125–130.

Li R, Murray AW (1991): Feedback control of mitosis in budding yeast. Cell 66:519–531.

Liu B, Palevitz BA (1992): Organization of cortical microfilaments in dividing root cells. Cell Motil Cytoskel 23:252–264.

Margolis RL, Wilson L (1981): Microtubule treadmills—Possible molecular machinery. Nature 293:705–711.

Masuda H, Hirano T, Yanagida M, Cande WZ (1990): *In vitro* reactivation of spindle elongation in fission yeast nuc2 mutant cells. J Cell Biol 110:417–425.

Masuda H, McDonald KL, Cande WZ (1988): The mechanism of anaphase spindle elongation: Uncoupling of tubulin incorporation and microtubule sliding during in vitro spindle reactivation. J Cell Biol 107:623–633.

Mazia D, Paweletz N, Sluder G, Finze E-M (1981): Cooperation of kinetochores and pole in the establishment of monopolar mitotic apparatus. Proc Natl Acad Sci USA 78:377–381.

McDonald HB, Goldstein LSB (1990): Identification and characterization of a gene encoding a kinesin-like protein in *Drosophila*. Cell 61:991–1000.

McDonald HB, Steward RJ, Goldstein LSB (1990): The kinesin-like NCD protein of *Drosophila* is a minus end-directed microtubule motor. Cell 63:1159–1165.

McDonald KL (1989): Mitotic spindle ultrastructure and design. In Hyams J, Brinkley BR (eds): Mitosis: Molecules and Mechanisms. New York: Academic Press, pp 1–38.

McDonald KL, O'Toole ET, Mastronarde DN, McIntosh JR (1992): Kinetochore microtubules in PTK cells. J Cell Biol 118:369–383.

McIntosh JR (1991): Structural and mechanical control of mitotic progression. Cold Spring Harbor Symp Quant Biol 56:613–619.

McIntosh JR, Cande WZ, Snyder JA (1975): Structure and physiology of the mammalian mitotic spindle. In Inoue S, Stephens RE (eds): Molecules and Cell Movement. New York: Raven Press, pp 31–76.

McIntosh JR, Hering GE (1991): Spindle fiber action and chromosome movement. Annu Rev Cell Biol 7:403–426.

McIntosh JR, McDonald KL, Edwards MK, Ross BM (1979): Three-dimensional structure of the central mitotic spindle in *Diatoma vulgare*. J Cell Biol 83: 428–442.

McIntosh JR, Roos U-P, Neighbors B, McDonald KL (1985): Architecture of the microtubule components of mitotic spindles from *Dictyostelium discoideum*. J Cell Sci 75:93–129.

Meluh PB, Rose MD (1990): KAR3, a kinesin-related gene required for yeast nuclear fusion. Cell 60: 1029–1041.

Merdes A, De Mey J (1990): The mechanism of kinetochore-spindle attachment and polewards movement analyzed in PtK2 cells at the prophase-prometaphase transition. Eur J Cell Biol 53:313–325.

Mitchison TJ (1988): Microtubule dynamics and kinetochore function in mitosis. Annu Rev Cell Biol 4:527–550.

Mitchison TJ (1989a): Polewards microtubule flux in the mitotic spindle: Evidence from photoactivation of fluorescence. J Cell Biol 109:637–652.

Mitchison TJ (1989b): Chromosome alignment at mitotic metaphase: balanced forces or smart kinetochores? In Warner FD, McIntosh JR (eds): Cell Motility, vol 2: Kinesin, Dynein, and Microtubule Dynamics. New York: Alan R. Liss, pp 421–430.

Mitchison TJ, Evans L, Schulze E, Kirschner M (1986): Sites of microtubule assembly and disassembly in the mitotic spindle. Cell 45:515–527.

Mitchison TJ, Salmon ED (1992): Poleward kinetochore fiber movement occurs during both metaphase and anaphase-A in newt lung cell mitosis. J Cell Biol 119:569–582.

Murray AW, Kirschner MW (1989): Dominoes and clocks: The union of two views of the cell cycle. Science 246:614–621.

Neighbors BW, Williams RC, McIntosh JR (1988): Localization of kinesin in cultured cells. J Cell Biol 106:1193–1204.

Nicklas RB (1961): Recurrent pole-to-pole movements of the sex chromosome during pre-metaphase in Melanoplus spermatocytes. Chromosoma 12:97–104.

Nicklas RB (1983): Measurements of force produced by the mitotic spindle in anaphase. J Cell Biol 97:542–548.

Nicklas RB (1988): The forces that move chromosomes in mitosis. Annu Rev Biophys Biophys Chem 17:431–449.

Nicklas RB (1989): The motor for poleward chromosome movement in anaphase is in or near the kinetochore. J Cell Biol 109:2245–2255.

Nicklas RB, Arana P (1992): Evolution and the meaning of metaphase. J Cell Sci 102:681–690.

Nislow C, Lombillo VA, Kuriyama R, McIntosh JR (1992): A plus-end-directed motor enzyme that moves antiparallel microtubules *in vitro* localizes to the interzone of mitotic spindles. Nature 359:543–547.

Nislow C, Sellitto C, Kuriyama R, McIntosh JR (1990): A monoclonal antibody to a mitotic microtubule-associated protein blocks mitotic progression. J Cell Biol 111:511–522.

O'Connell MJ, Meluh PB, Rose MD, Morris NR (1993): Suppression of bimC4 mitotic spindle defect by deletion of klpA, a gene encoding a KAR3-related kinesin-like protein in *Aspergillus nidulans*. J Cell Biol 120: 153–162.

Ostergren G (1950): Considerations on some elementary features of mitosis. Hereditas 36:1–19.

Page BD, Snyder M (1993): Chromosome segregation in yeast. Annu Rev Microbiol (in press).

Pfarr CM, Coue M, Grissom PM, Hays TS, Porter ME, McIntosh JR (1990): Cytoplasmic dynein is localized to kinetochores during mitosis. Comment in Nature 1990 345:206–7. Nature 345:263–265.

Pickett-Heaps JD (1975): The Green Algae. Sunderland MA: Sinauer Assoc. Inc.

Pickett-Heaps JD, Tippit DH, Porter KR (1982): Rethinking mitosis. Cell 29:729–744.

Price CM, Pettijohn DE (1986): Redistribution of the nuclear mitotic apparatus protein (NuMA) during mitosis and nuclear assembly. Properties of purified NuMA protein. Exp Cell Res 166:295–311.

Rattner JB (1992): Mapping the mammalian intercellular bridge. Cell Motil Cytoskel 23:231–235.

Rebhun LI, Palazzo RE (1988): *In vitro* reactivation of anaphase B in isolated spindles of the sea urchin egg. Cell Motil Cytoskel 10:197–209.

Rickards GK (1975): Prophase chromosome movements in living house cricket spermatocytes and their relationship to prometaphase, anaphase and granule movements. Chromosoma 49:407–455.

Rieder CL (1990): Formation of the astral mitotic spindle: Ultrastructural basis for the centrosome–kinetochore interaction. Electron Microsc Rev 3:269–300.

Rieder CL, Alexander SP (1990): Kinetochores are transported poleward along a single astral microtubule during chromosome attachment to the spindle in newt lung cells. J Cell Biol 110:81–95.

Rieder CL, Davison EA, Jensen LCW, Cassimeris L, Salmon ED (1986): Oscillatory movements of monooriented chromosomes and their position relative to the spindle pole result from the ejection properties of the aster and half-spindle. J Cell Biol 103:581–591.

Roof DM, Meluh PB, Rose MD (1991): Multiple kinesin-related proteins in yeast mitosis. Cold Spring Harbor Symp Quant Biol 56:693–703.

Roof DM, Meluh PB, Rose MD (1992): Kinesin-related proteins required for assembly of the mitotic spindle. J Cell Biol 118:95–108.

Salmon ED (1989): Metaphase chromosome congression and anaphase poleward movement. In: Warner FD, McIntosh JR (eds): Cell Movement, vol. 2, Kinesin,

Dynein, and Microtubule Dynamics. New York: Alan R. Liss, pp 431–440.

Saunders WS, Hoyt MA (1992): Kinesin-related proteins required for structural integrity of the mitotic spindle. Cell 70:451–458.

Sawin KE, LeGuellec K, Philippe M, Mitchison TJ (1992): Mitotic spindle organization by a plus-end-directed microtubule motor. Nature 359:540–543.

Sawin KE, Mitchison TJ (1991): Poleward microtubule flux in mitotic spindles assembled *in vitro*. J Cell Biol 112:941–954.

Sawin KE, Scholey JM (1991): Motor proteins in cell division. Trends Cell Biol 1:122–129.

Saxton WM, McIntosh JR (1987): Interzone microtubule behavior in late anaphase and telophase spindles. J Cell Biol 105:875–886.

Shelden E, Wadsworth P (1990): Interzone microtubules are dynamic during spindle elongation. J Cell Sci 97:273–282.

Smirnova EA, Bajer AS (1992): Spindle poles in higher plant mitosis. Cell Motil Cytoskel 23:1–7.

Snyder JA, Hamilton BT, Mullins JM (1982): Loss of mitotic centrosomal microtubule initiation capacity at the metaphase-anaphase transition. Eur J Cell Biol 27:191–199.

Snyder JA, McIntosh JR (1975): Initiation and growth of microtubules in lysed mammalian cells. J Cell Biol 64:744–760.

Spurk TP, Pickett-Heaps JD (1987): On the mechanism of anaphase A: evidence that ATP is needed for microtubule disassembly and not generation of polewards force. J Cell Biol 105:1691–1705.

Stemple DL, Sweet SC, Welsh MJ, McIntosh JR (1988): Dynamics of a fluorescent calmodulin analog in the mammalian mitotic spindle at metaphase. Cell Motil Cytoskel 9:231–242.

Steuer ER, Wordeman L, Schroer TA, Sheetz MP (1990): Localization of cytoplasmic dynein to mitotic spindles and kinetochores. Comment in Nature 1990 345:206–207. Nature 345:266–268.

Sullivan DS, Huffaker TC (1992): Astral microtubules are not required for anaphase in *Saccaromyces cerevisiae*. J Cell Biol 119:379–388.

Sweet SC, Rogers CM, Welsh MJ (1989): Calmodulin is associated with microtubules forming in PTK1 cells upon release from nocodazole treatment. Cell Motil Cytoskel 12:113–122.

Taylor EW (1959): Dynamics of spindle formation and its inhibition by chemicals. J Cell Biol 6:193–222.

Theurkauf WE, Hawley RS (1992): Meiotic spindle assembly in *Drosophila* females: Behavior of nonexchange chromosomes and the effects of mutations in the nod kinesin-like protein. J Cell Biol 116:1167–1180.

Tippit DH, Schulz D, Pickett-Heaps JD (1978): Mitosis in *Fragellaria*. J Cell Biol 79:737–763.

Tombes RM, Peloquin JG, Borisy GG (1991): Specific association of an M-phase kinase with isolated mitotic spindles and identification of two of its substrates as MAP4 and MAP1B. Cell Regul 2:861–874.

Traas JA, Doonan JH, Rawlins DJ, Shaw PJ, Watts J, Lloyd CW (1987): An actin network is present in the cytoplasm throughout the cell cycle of carrot cells and associates with the dividing nucleus. J Cell Biol 105:387–395.

Vaisberg EA, Koonce MP, McIntosh JR (1993): Cytoplasmic dynein plays an essential role in mitotic spindle formation. J Cell Biol (in press).

Vale RD, Goldstein LSB (1990): One motor, many tails: An expanding repertoire of force-generating enzymes. Cell 60:883–885.

Vandre DD, Borisy GG (1989): Anaphase onset and dephosphorylation of mitotic phosphoproteins occur concomitantly. J Cell Sci 94:245–258.

Vandre DV, Davis FM, Rao PN, Borisy GG (1984): Phosphoproteins are components of mitotic microtubule organizing centers. Proc Natl Acad Sci USA 81:4439–4443.

Verde F, Dogterom M, Stelzer E, Karsenti E, Leibler S (1992): Control of microtubule dynamics and length by cyclin A- and cyclin B-dependent kinases in *Xenopus* egg extracts. J Cell Biol 118:1097–1108.

Waldren CA (1983): Mutational analysis in human–hamster hybrid cells. In Hollaender A (ed): Chemical Mutagens: Principles and Methods for Their Detection. New York: Plenum Press, p 256.

Walker RA, Salmon ED, Endow SA (1990): The *Drosophila* claret segregation protein is a minus-end directed motor molecule. Nature 347:780–782.

Welsh MJ, Dedman JR, Brinkley BR, Means AR (1979): Tubulin and calmodulin. Effects of microtubule and microfilament inhibitors on localization in the mitotic apparatus. J Cell Biol 78:624–634.

Williams BC, Karr TL, Montgomery JM, Goldberg ML (1992): The *Drosophila* l(1)zw10 gene product, required for accurate mitotic chromosome segregation, is redistributed at anaphase onset. J Cell Biol 118: 759–773.

Wilson EB (1925): The Cell in Development and Inheritance, 2nd ed. New York: Macmillan.

Wilson P, Forer A (1989): The behaviour of microtubules in chromosomal spindle fibres irradiated singly or doubly with ultraviolet light. J Cell Sci 94:625–634.

Wise D, Cassimeris L, Rieder CL, Wadsworth P, Salmon ED (1991): Chromosome fiber dynamics and congression oscillations in metaphase of PtK2 cells at 23 deg. Cell Motil Cytoskel 18:131–142.

Wordeman L, Davis FM, Rao PN, Cande WZ (1989): Distribution of phosphorylated spindle-associated proteins in the diatom *Stephanopyxis turris*. Cell Motil Cytoskel 12:33–41.

Wordeman LG, Stener E, Sheetz MP, Mitchison TJ (1991): Distinct chemical subdomains within the kinetochore region of isolated CHO chromosomes. J Cell Biol 114:285–294.

Yang CH, Snyder M (1992): The nuclear-mitotic apparatus protein is important in the establishment and maintenance of the bipolar mitotic spindle apparatus. Mol Biol Cell 3:1259–1267.

Yen TJ, Compton DA, Wise D, et al. (1991): CENP-E, a novel human centromere-associated protein required for progression from metaphase to anaphase. EMBO J 10:1245–1254.

Yen TJ, Li G, Schaar BT, Szilak I, Cleveland DW (1992): CENP-E is a putative kinetochore motor that accumulates just before mitosis. Nature 359:536–539.

Zavortink M, Welsh MJ, McIntosh JR (1983): Distribution of calmodulin in living mitotic cells. Exp Cell Res 149:375–385.

Zeilig CE, Goldberg ND (1977): Cell-cycle related changes of the 3′:5′-cyclic GMP levels in Novikoff hepatoma cells. Proc Natl Acad Sci USA 74:1052–1056.

Zhang P, Knowles BA, Goldstein LSB, Hawley RS (1990): A kinesin-like protein required for distributive chromosome segregation in *Drosophila*. Cell 62:1053–1062.

ABOUT THE AUTHOR

J. RICHARD McINTOSH is Professor of Molecular, Cellular, and Developmental Biology at the University of Colorado in Boulder. He teaches one semester of introductory biology and part of a graduate course in cell biology, in addition to seminars related to his own field. He received a B.A. in physics from Harvard College, a Ph.D. in biophysics from Harvard University in 1968, and worked on the biology faculty at that university for three years before moving to Boulder, where has has served ever since. He has worked throughout his career on the mechanisms by which microtubules contribute to cellular morphogenesis and intracellular motility, with particular attention to their role during chromosome movement. His approach to this problem has evolved from a structural one to a more molecular effort, involving biochemistry, molecular biology, and genetics. He has held visiting scientist positions at the MRC Laboratory for Molecular Biology in Cambridge, England, and at Oxford University. He has been the recipient of an Eleanor Roosevelt International Cancer Scholarship, a Guggenheim Fellowship, and grants from the American Cancer Society, the National Institutes of Health, and the National Science Foundation. He has served on the editorial boards of the *Journal of Cell Biology,* the *Journal of Structural Biology,* and *Cell Motility and the Cytoskeleton.* He is currently the Principal Investigator of the Boulder Laboratory for 3-Dimensional Fine Structure and President-Elect of the American Society for Cell Biology.

Index